REVIEWS IN MIN
AND GEOCHEMISTRY

Volume 71 2010

Theoretical and Computational Methods in Mineral Physics: *Geophysical Applications*

EDITORS

Renata Wentzcovitch — *University of Minnesota, Minneapolis, MN, U.S.A.*

Lars Stixrude — *University College London, London, United Kingdom*

ON THE COVER: Crystal structure of the post-perovskite phase of $MgSiO_3$. This phase discovered in 2004 is likely to be the most abundant silicate phase above the core mantle boundary of the Earth.

Series Editor: ***Jodi J. Rosso***

MINERALOGICAL SOCIETY OF AMERICA
GEOCHEMICAL SOCIETY

SHORT COURSE SERIES DEDICATION

Dr. William C. Luth has had a long and distinguished career in research, education and in the government. He was a leader in experimental petrology and in training graduate students at Stanford University. His efforts at Sandia National Laboratory and at the Department of Energy's headquarters resulted in the initiation and long-term support of many of the cutting edge research projects whose results form the foundations of these short courses. Bill's broad interest in understanding fundamental geochemical processes and their applications to national problems is a continuous thread through both his university and government career. He retired in 1996, but his efforts to foster excellent basic research, and to promote the development of advanced analytical capabilities gave a unique focus to the basic research portfolio in Geosciences at the Department of Energy. He has been, and continues to be, a friend and mentor to many of us. It is appropriate to celebrate his career in education and government service with this series of courses.

Reviews in Mineralogy and Geochemistry, Volume 71

Theoretical and Computational Methods in Mineral Physics: Geophysical Applications

ISSN 1529-6466
ISBN 978-0-939950-85-0

THE MINERALOGICAL SOCIETY OF AMERICA
3635 CONCORDE PARKWAY, SUITE 500
CHANTILLY, VIRGINIA, 20151-1125, U.S.A.
WWW.MINSOCAM.ORG

Theoretical and Computational Methods in Mineral Physics:

Geophysical Applications

71 *Reviews in Mineralogy and Geochemistry* **71**

FROM THE SERIES EDITOR

The chapters in this volume represent an extensive review of the material presented by the invited speakers at a short course on *Theoretical and Computational Methods in Mineral Physics* held prior (December 10-12, 2009) to the Annual fall meeting of the American Geophysical Union in San Francisco, California. The meeting was held at the Doubletree Hotel & Executive Meeting Center in Berkeley, California. This meeting was sponsored by the Mineralogical Society of America and the Geochemical Society.

Any supplemental material and errata (if any) can be found at the MSA website *www.minsocam.org*.

Jodi J. Rosso, Series Editor
West Richland, Washington
March 2010

PREFACE

Mineral physics is one of the three pillars of geophysics, the other two being geodynamics and seismology. Geophysics advances by close cooperation between these fields. As such, mineral physicists investigate properties of minerals that are needed to interpret seismic data or that are essential for geodynamic simulations. To be useful, mineral properties must be investigated in a wide range of pressures, temperatures, and chemical compositions. The materials and conditions in the interior of Earth and other terrestrial planets present several challenges. The chemical composition of their mantles is complex with at least five major oxide components and tens of solid phases. Today, these challenges are being addressed by a combination of experimental and computational methods, with experiments offering precise information at lower pressures and temperatures, and computations offering more complete and detailed information at conditions more challenging to experiments. While bulk properties of materials are fundamental to understanding a planet's state, atomistic inspection of these complex materials are fundamental to understanding their properties. A connection is then established between atomic and planetary scale phenomena, which mineral physicists are in a unique position to appreciate.

This book presents a set of review articles offering an overview of contemporary research in computational mineral physics. Fundamental methods are discussed and important

1529-6466/10/0071-0000$05.00 DOI: 10.2138/rmg.2010.71.0

applications are illustrated. The opening chapter by John Perdew and Adrienn Ruzhinszky discusses the motivation, history, and expressions of Kohn-Sham Density Functional Theory (DFT) and approximations for exchange and correlation. This is the established framework for investigation of a condensed matter system's ground state electronic density and energy. It also discusses the recent trend to design higher-level semi-local functionals, with solid state applications in mind. It presents arguments in favor of semi-local approximations for condensed matter and discusses problematic cases where fully non-local approximations are needed. The following article by Yan Zhao and Donald Truhlar, demonstrates current research in search of appropriate exchange and correlation energy functionals. It reviews the performance of families of local, semi-local, and fully non-local exchange and correlation functionals: the so-called "Minnesota" functionals. These new functionals have been designed to give broad accuracy in chemistry and perform very well in difficult cases where popular functionals fail badly. The prospects for their successful applications are encouraging.

Stefano Baroni, Paolo Gianozzi, and Eyvaz Isaev, introduce Density Functional Perturbation Theory, a suitable technique to calculate vibrational properties of extended materials using a combination of density functional theory and linear response techniques. This method gives very accurate phonon frequencies which, in combination with the quasi-harmonic approximation, allow one to study thermal properties of materials. The next chapter by Renata Wentzcovitch, Yonggang Yu, and Zhongqing Wu review the applications of density functional perturbation theory to the investigation thermodynamic properties and phase relations in mantle minerals. The series of studies summarized in this review have explored the accuracy of DFT within its most popular approximations for exchange and correlation energy in combination with the quasiharmonic approximation to offer results with useful accuracy for geophysical studies. The following article by Renata Wentzcovitch, Zhongqing Wu, and Pierre Carrier, summarizes the combination of the quasiharmonic approximation with elasticity theory to investigate thermoelastic properties of minerals at conditions of the Earth interior. Some unfamiliar but essential aspects of the quasiharmonic approximation are discussed. Thermoelastic properties of minerals are essential to interpret seismic observations. Therefore, some examples of interpretation of seismic structures are reviewed.

The article by David Ceperley, returns to the fundamental theme of calculations of ground state energy in condensed matter and introduces Quantum Monte Carlo methods. These methods treat exactly the quantum many-body problem presented by a system of electrons and ions. They treat electrons as particles rather than a scalar charge-density field, as done by DFT. These are computationally intensive methods but the only exact ones. The following article by Lubos Mitas and Jindrich Kolorenc, reviews applications of these methods to transition metals oxides, materials that have some aspects in common with mantle minerals. One of the examined systems, FeO, is a most important component of mineral solid solutions. Matteo Cococcioni continues exploring the same theme. He discusses a modified density functional useful for addressing cases like FeO, which are untreatable by standard DFT. The DFT + Hubbard U method (DFT+U) is a practical approximate method that enables investigations of electronically and structurally complex systems, like minerals. The application of this method to a contemporary and central problem in mineral physics, pressure and temperature induced spin-crossovers in mantle minerals, is reviewed in the next chapter by Han Shu, Koichiro Umemoto, and Renata Wentzcovitch. The geophysical implications of the spin-crossover phenomenon, an electronic transition, are still unclear but some possibilities are suggested.

Michael Ammann, John Brodholt, and David Dobson discuss simulations of bulk ionic diffusion. This property plays an important role in chemical exchange between and within crystalline and melt phases. It plays an important role in the kinetics of phase transitions, compositional zoning, mineral growth, and other important geochemical processes. It can also control rheological properties, especially in the diffusion creep regime, and thus the time scale of mantle convection. This is a very difficult property to investigate at combined pressures and temperature conditions of the mantle, therefore, calculations play a very important role in this area. Phillip Carrez and Patrick Cordier discuss modeling of dislocations and plasticity in deep Earth materials. This article focuses on recent developments in dislocation modeling and applications to our understanding of how the direction of mantle flow is recorded in polycrystalline texture. Next, the article by Stephen Stackhouse and Lars Stixrude, discusses theoretical methods for calculating lattice thermal conductivity in minerals, which controls the cooling of Earth's core. Measurements of thermal conductivity at lower mantle conditions are very challenging to experiments and calculations are a valuable alternative to learning about this property. This article describes the most common methods to calculate this property and presents a review of studies of the lattice thermal conductivity of periclase.

Artem Oganov discusses the prediction of high pressure crystal structures. A genetic algorithm for structural prediction is described and numerous applications predicting new phases with novel properties and phases that can explain experimental data so far not understood is presented. This is a most recent development on the subject of structural predictions, a subject that has been pursued by simulations for several decades now. The possibility of predicting structure and composition by this method is also pointed out. Koichiro Umemoto and Renata Wentzcovitch continue on the same theme of structural prediction by a different approach: combination of phonon calculations and variable cell shape molecular dynamics. The former indicates unstable displacement modes in compressed structures; the latter searches for structures resulting from the superposition of these unstable modes to the compressed lattice. This approach is illustrated with the search of mineral structures at multi-Mbar pressures that are still challenging to static or dynamic compression experiments, but have great interest in view of the discovery of terrestrial exoplanets with several Earth masses. The following chapter by Koichiro Umemoto is on simulations of phase transitions on a different class of planet forming material: H_2O-ice. Ice has a rich phase diagram but many of its phase relations are unknown: large hysteresis precludes their direct measurements in manageable time scales. Therefore, calculations acquire special significance but they are also challenging, the main reasons being the description of hydrogen bond by DFT and hydrogen disorder.

Dario Alfè presents a review of first principles calculations of properties of iron at Earth's core conditions. This chapter includes examples of applications of multiple techniques used in studies of high temperature properties, structure, and melting lines. Results from Quantum Monte Carlo are compared with those from DFT, and results from molecular dynamics simulations are contrasted with predictions of quasiharmonic theory. These comparisons are instructive and illustrate the breadth of research in computational mineral physics. The following chapter by Bijaya Karki turns to DFT based simulations of another type of melt: ionic silicates and oxides. The article discusses the methodology used in these simulations and specially developed methods to analyze the results. The properties of interest are high temperature equations of state, thermodynamics properties, atomic and electronic structure, and self-diffusion and viscosity. Visualization of atomic motion is one of the valuable approaches discussed to gain insight into changes in melt structure with pressure and temperature. These studies are illustrated for 3 melts along the MgO-SiO_2 join.

The following three articles are devoted primarily to the introduction of inter-atomic potentials of broad applicability and relatively high accuracy, and applications to large scale simulations. The first article by Julian Gale and Kate Wright describes the current status of the derivation of force-fields and their applications to static and lattice dynamic calculations in mineral physics. This is done in the context of the General Utility Lattice Program (GULP), which has become quite popular. A selection of applications illustrating the possibilities of this code is then presented. Victor Vinograd and Bjoern Winkler illustrate another important type of application of force-field models: an efficient cluster expansion method to investigate binary mineral solid solutions. The article focuses on a rock-salt system but the technique is general. This type of problem is central to mineral physics and ingenious combinations of first principles methods, force-field models, and purely parameterized free energy expressions, combined with molecular dynamics and Monte Carlo techniques are necessary to address this problem. The predictive treatment of properties of ionic solid solutions is a major challenge in mineral physics. Mark Ghiorso and Frank Spera discuss long duration large scale molecular dynamics simulations using empirical pair-potentials. This article illustrates the concrete requirements on the number of atoms and time scales necessary to obtain information on transport properties such as shear viscosity and lattice thermal conductivity using Green-Kubo theory. These more than 1000-atom and pico-second simulations also improve the statistics in the estimation of equilibrium properties.

Finally, the article by Lars Stixrude and Carolina Lithgow-Bertelloni on the thermodynamics of Earth's mantle, gives an overview of how the elucidation of materials behavior governs planetary processes. It explains how the complexity of the Earth's mantle demands methods that are complementary to first principles calculations and experiments. These methods must allow one to interpolate among and extrapolate from results on minerals with limited compositions to the full chemical richness of the silicate mantle. It then illustrates how the derived properties of multi-phase multi-component systems are used to address mantle heterogeneity on multiple length scales, ranging from that of the subducting slab to the possibility of mantle-wide radial variations in bulk composition.

We hope this volume will stimulate the interest of materials oriented researchers in geophysics and of geophysicists in computational materials physics. The science that has emerged from this mix in the last two decades has been transformative and was unimaginable at the onset. We thank all the authors for contributing to this volume, sometimes leaving aside other important tasks to fulfill deadlines. Their participation in the Short Course is also greatly appreciated. The Short Course was supported by the special NSF/ EAR grant # 0952600, by the Virtual Laboratory for Earth and Planetary Materials (*VLab*) through NSF/ATM grant # 0428774, and by DOE through a grant to the Mineralogical Society of America. We specially thank Jodi Rosso, who managed to put together this volume from articles often provided well outside the guidelines to authors. Her patience and dedication were greatly appreciated. Alex Speer has provided effective guidance throughout the entire organization process, to which we are very grateful. Debbie Schutta has our gratitude for her endurance and help with the organization of the Short Course. Finally, we thank Nancy Ross for her encouragement and support in the initial stages of this project.

Renata Wentzcovitch
Department of Chemical Engineering and Materials Science and
Minnesota Supercomputer Institute
University of Minnesota,
421 Washington Ave, Minneapolis, Minnesota 55455, U.S.A.

Lars Stixrude
Department of Earth Sciences,
University College London,
Gower Street, London WC1E 6BT, United Kingdom

Theoretical and Computational Methods in Mineral Physics: *Geophysical Applications*

71 *Reviews in Mineralogy and Geochemistry* **71**

TABLE OF CONTENTS

1 Density Functional Theory of Electronic Structure: A Short Course for Mineralogists and Geophysicists

John P. Perdew and Adrienn Ruzsinszky

2 The Minnesota Density Functionals and their Applications to Problems in Mineralogy and Geochemistry

Yan Zhao, Donald G. Truhlar

3 Density-Functional Perturbation Theory for Quasi-Harmonic Calculations

Stefano Baroni, Paolo Giannozzi, Eyvaz Isaev

4 Thermodynamic Properties and Phase Relations in Mantle Minerals Investigated by First Principles Quasiharmonic Theory

Renata M. Wentzcovitch, Yonggang G. Yu, Zhongqing Wu

5 First Principles Quasiharmonic Thermoelasticity of Mantle Minerals

Renata M. Wentzcovitch, Zhongqing Wu, Pierre Carrier

6 An Overview of Quantum Monte Carlo Methods

David M. Ceperley

7 Quantum Monte Carlo Studies of Transition Metal Oxides

Lubos Mitas, Jindřich Kolorenč

8 Accurate and Efficient Calculations on Strongly Correlated Minerals with the LDA+U Method: Review and Perspectives

Matteo Cococcioni

9 Spin-State Crossover of Iron in Lower-Mantle Minerals: Results of DFT+*U* Investigations

Han Hsu, Koichiro Umemoto, Zhongqing Wu, Renata M. Wentzcovitch

10 Simulating Diffusion

Michael W. Ammann, John P. Brodholt, David P. Dobson

11 Modeling Dislocations and Plasticity of Deep Earth Materials

Philippe Carrez, Patrick Cordier

12 Theoretical Methods for Calculating the Lattice Thermal Conductivity of Minerals

Stephen Stackhouse, Lars Stixrude

13 Evolutionary Crystal Structure Prediction as a Method for the Discovery of Minerals and Materials

Artem R. Oganov, Yanming Ma, Andriy O. Lyakhov, Mario Valle, Carlo Gatti

14 Multi-Mbar Phase Transitions in Minerals

Koichiro Umemoto, Renata M. Wentzcovitch

15 Computer Simulations on Phase Transitions in Ice

Koichiro Umemoto

16 Iron at Earth's Core Conditions from First Principles Calculations

Dario Alfè

17 First-Principles Molecular Dynamics Simulations of Silicate Melts: Structural and Dynamical Properties

Bijaya B. Karki

18

Lattice Dynamics from Force-Fields as a Technique for Mineral Physics

Julian D. Gale, Kate Wright

19

An Efficient Cluster Expansion Method for Binary Solid Solutions: Application to the Halite-Silvite, NaCl-KCl, System

Victor Vinograd, Björn Winkler

20 Large Scale Simulations

Mark S. Ghiorso, Frank J. Spera

21 Thermodynamics of the Earth's Mantle

Lars Stixrude, Carolina Lithgow-Bertelloni

Reviews in Mineralogy & Geochemistry
Vol. 71 pp. 1-18, 2010

Density Functional Theory of Electronic Structure: A Short Course for Mineralogists and Geophysicists

John P. Perdew and Adrienn Ruzsinszky

Department of Physics and Engineering Physics
Tulane University
New Orleans, Louisiana 70118, U.S.A.

perdew@tulane.edu

INTRODUCTION AND SUMMARY

Mineralogists and geophysicists need to understand and predict the properties of solids and liquids at normal and especially at high pressures and temperatures. For example, they need to know the equilibrium structure, equation of state, phase transitions, and vibrational properties of solids, and the interatomic or intermolecular interaction needed for a molecular dynamics study of liquids (Stixrude et al. 1994; Soederlind and Ross 2000; Karki et al. 2001; Alfè et al. 2002; Steinle-Neumann et al. 2004; Sha and Cohen 2006; Carrier et al. 2007). This information, in sufficient detail, is not always available from experiment. Increasingly, it comes from the simple first-principles Kohn-Sham density functional theory (Kohn and Sham 1965; Parr and Yang 1989; Dreizler and Gross 1990; Perdew and Kurth 2003; Perdew et al. 2009a). Often the ground-state version of this theory suffices, since the electrons can stay close to their ground state even when the nuclear motion is thermally excited; there is however also a temperature-dependent version of the theory (Mermin 1965).

In Kohn-Sham density functional theory, the exact ground-state density and energy of a many-electron system can be found by solving selfconsistent one-electron Schroedinger equations for the orbitals or one-electron wavefunctions. Many different methods of solution are available in many different standard computer codes, which tend to agree when carefully applied. The required computational effort is much less than in approaches that employ a many-electron wavefunction. In practice, one term in the energy as a functional of the electron density, the exchange-correlation energy, must be approximated. Although this term is often a relatively small part of the total energy, it is "nature's glue," responsible for most of the binding of one atom or molecule to another. Simple and computationally undemanding semilocal approximations are accurate enough for the useful description of many atoms, molecules, and solids, and especially for "hard" (strongly-bound) materials under high pressure and temperature. Three levels of semilocal approximations (local spin density approximation, generalized gradient approximation or GGA, and meta-GGA) have been developed nonempirically. These approximations are exact for a uniform electron density, but satisfy (especially at the higher levels) many other known exact constraints on the exchange-correlation energy functional. Fully nonlocal approximations have also been developed.

After a nontechnical explanation of the motivation, history, and expressions of Kohn-Sham density functional theory and its standard approximations for exchange and correlation, and after a discussion of the recent trend to design higher-level semilocal functionals with solid state applications in mind, this article will argue why higher-level semilocal approximations are especially well-suited to dense matter, and will discuss the problematic cases where fully-nonlocal approximations are needed. It will also discuss some special aspects of materials at high pressure and temperature.

1529-6466/10/0071-0001$05.00 DOI: 10.2138/rmg.2010.71.1

The density functional theory we will discuss is nonrelativistic and thus valid when the electrons move at speeds far below that of light. There is however also a relativistic version of density functional theory (Dreizler 2003, and references therein). In particular, the spin-orbit interaction is of relativistic origin and can create noncollinear magnetism. Furthermore, relativistic effects on the densities of the high-density cores in the heavier elements can have an important indirect effect on the valence electrons.

Helpful textbooks are Ashcroft and Mermin (1976) for solid state physics and Martin (2004) for electronic structure theory and methods.

KOHN-SHAM THEORY: THE ORBITALS MAKE IT ACCURATE

Motivation and history of Kohn-Sham theory

Before presenting the full Kohn-Sham density functional theory, we will explain why it is needed and what its historical background is.

The ground-state energy E and electron density $n(\vec{r})$ of an N-electron system can be found from the lowest-energy solution of the Schroedinger equation:

$$H\Psi = E\Psi \tag{1}$$

with

$$n(\vec{r}) = N\int d^3r_2 ... d^3r_N \left|\Psi(\vec{r},\vec{r}_2,...,\vec{r}_N)\right|^2 \tag{2}$$

$n(\vec{r})d^3r$ is the average number of electrons in volume element d^3r at position $\vec{r}$, and also the probability to find one electron in this volume element. However, as a consequence of the electron-electron pair interaction in the Hamiltonian H, Equation (1) is not so easy to solve for $N > 1$, and the required effort can grow rapidly with N. Walter Kohn, who received the 1998 Nobel Prize in Chemistry for the development of density functional theory, explained this as follows (Kohn 1999): The wavefunction has N vector arguments. If the mesh in real space has M values along each direction of three-dimensional space, then the number of wavefunction values to be computed and stored grows as M^{3N}. Clever methods of solution can ameliorate this exponential-growth problem, but unfavorable scaling with N remains.

This problem was recognized very early in the history of quantum mechanics, and two simplifications were developed. The first was the Thomas-Fermi approximation (and various extensions thereof), which replaces the wavefunction as the variational object by the electron density $n(\vec{r})$, and the expectation value of the Hamiltonian H by an approximate density functional. Then the required computation time on a mesh grows only as M^3. This simplification can tell us something about atoms, molecules, and solids, but not much, because there is no adequate density-functional approximation for the kinetic energy functional. In particular, the simple Thomas-Fermi approximation predicts no binding between neutral atoms, so it can at best describe materials under intense pressure. Nevertheless, Hohenberg and Kohn (1964) proved that an exact density functional for the energy exists in principle and its minimization at fixed external potential and electron number yields the ground-state density and energy.

A second early simplification was the Hartree-Fock approximation, which replaces the wavefunction by a single Slater determinant of the best possible orbitals or one-electron wavefunctions, which turn out to be solutions of a one-electron Schroedinger equation with an effective one-electron potential. This approximation, which includes exact exchange and no correlation, is typically much better than the Thomas-Fermi-like approximations, but still misses much of the "glue" that binds atoms together. Slater proposed a simplified Hartree-Fock theory, the X_α approximation, which replaced the nonlocal Hartree-Fock exchange potential by a simple function of the local density (Slater 1974).

All of these ideas came together and were "exactified" by Kohn and Sham (1965). The idea is to find the exact ground-state density from occupied Kohn-Sham orbitals. Since there are roughly $N/2$ such orbitals, the computational effort is formally of order NM^3 in "linear-scaling" implementations (but somewhat worse in most standard implementations). In the 1970's, this theory within the local spin density approximation (LSDA) replaced the Hartree and Hartree-Fock approximations as a foundation for solid state calculations. In the 1980's and 1990's, higher-level approximations (generalized gradient approximations or GGA's, meta-GGA's, and hybrids of these functionals with exact exchange) were found to work better than LSDA for the more inhomogeneous atoms and molecules, and density functional theory became a foundation for quantum chemical calculations as well. The theory was also adopted in mineralogy and geophysics, and today it is increasingly used even in materials engineering. Density functional theory is currently the most widely-used and thus the most-cited branch of physics.

Summary of Kohn-Sham theory

Consider a system of N electrons moving in a spin-dependent ($\sigma = \uparrow$ or $\downarrow$) external scalar multiplicative potential $v_\sigma(\vec{r})$, which can but need not arise from the Coulomb attraction between an electron at position $\vec{r}$ and the nuclei of charge Z_n at R_n:

$$v(\vec{r}) = -\sum\nolimits_n Z_n / \left|\vec{r} - \vec{R}_n\right| \tag{3}$$

We use atomic units which start from the cgs equations and then set the electron charge, the electron mass, and Planck's constant to one; the atomic unit is 1 hartree = 27.21 eV for energy, 1 bohr = 0.5292 Angstrom for distance, and 1 hartree/bohr^3 = 2.94×10^3 GPa for pressure. When the external potential is spin-independent, as in Equation (3), then only the total density ($n(\vec{r}) = n_\uparrow(\vec{r}) + n_\downarrow(\vec{r})$) is needed in principle (von Barth and Hedin 1972), but even then the separate up- and down-spin densities $n_\sigma(\vec{r})$ are needed for an accurate approximate description of open-shell systems (atoms, radical molecules, magnetic solids). When the external potential is spin-dependent (due for example to a magnetic field that couples to electron spin), the spin-density functional theory we shall describe is needed even in principle. But, in most applications to mineralogy and geophysics, the system is spin-unpolarized and the equations presented below are simplified by the substitution $n_\uparrow = n_\downarrow = n/2$

The Hamiltonian is

$$H = -(1/2)\sum\nolimits_i \nabla_i^2 + \sum\nolimits_\sigma \sum\nolimits_i v_\sigma(\vec{r}_i)\delta_{\sigma,\sigma_i} + (1/2)\sum\nolimits_i \sum\nolimits_{j\neq i} 1/\left|\vec{r}_i - \vec{r}_j\right| + C \tag{4}$$

The first term in Equation (4) is the kinetic energy operator, the second is the interaction between the electrons and the external potential, and the third is the electron-electron Coulomb repulsion. C is a term independent of the electron positions, such as the Coulomb repulsion among the nuclei:

$$(1/2)\sum\nolimits_m \sum\nolimits_{n\neq m} Z_m Z_n / \left|\vec{R}_m - \vec{R}_n\right| \tag{5}$$

Kohn and Sham (1965) proved that the ground-state spin densities for this problem can be found from the self-consistent solution of the effective one-electron equations

$$\left[-(1/2)\nabla^2 + v_\sigma(\vec{r}) + (1/2)\int d^3r' n(\vec{r}')/\left|\vec{r}' - \vec{r}\right| + v_{xc,\sigma}([n_\uparrow, n_\downarrow], \vec{r})\right]\psi_{\alpha,\sigma}(\vec{r}) = \varepsilon_{\alpha,\sigma}\psi_{\alpha,\sigma}(\vec{r}) \tag{6}$$

$$n_\sigma(\vec{r}) = \sum\nolimits_\alpha \left|\psi_{\alpha,\sigma}(\vec{r})\right|^2 \theta(\mu - \varepsilon_{\alpha,\sigma}) \tag{7}$$

Here α is a set of non-spin one-electron quantum numbers. The step function θ in Equation (7) provides an occupation number which is 1 for all orbitals with orbital energy $\varepsilon_{\alpha,\sigma}$ below the Fermi level μ, and zero for all above. The ground-state energy is then

$$E = T_s[n_\uparrow, n_\downarrow] + \int d^3r \sum_\sigma n_\sigma(\vec{r}) v_\sigma(\vec{r}) \tag{8}$$
$$+ (1/2)\int d^3r d^3r' n(\vec{r}) n(\vec{r}') / |\vec{r}' - \vec{r}| + E_{xc}[n_\uparrow, n_\downarrow] + C$$

The first term of Equation (8) is the non-interacting kinetic energy

$$T_s[n_\uparrow, n_\downarrow] = \int d^3r \tau(\vec{r}) \qquad \tau(\vec{r}) = \sum_{\alpha,\sigma} (1/2) |\nabla \psi_{\alpha,\sigma}(\vec{r})|^2 \theta(\mu - \varepsilon_{\alpha,\sigma}) \tag{9}$$

The second is the interaction of the spin densities with the external potential. The third is the Hartree electrostatic interaction of the density with itself. The fourth is the exchange-correlation energy, whose functional derivative

$$v_{xc,\sigma}([n_\uparrow, n_\downarrow], \vec{r}) = \delta E_{xc}[n_\uparrow, n_\downarrow] / \delta n_\sigma(\vec{r}) \tag{10}$$

provides the exchange-correlation potential in the Kohn-Sham Equation (6).

The square brackets denote a functional dependence. A spin-density functional is an implicit or explicit rule that produces a number from each input pair of spin-density functions. A simple example is the local spin-density approximation (LSDA) (Kohn and Sham 1965; von Barth and Hedin 1972):

$$E_{xc}^{LSDA}[n_\uparrow, n_\downarrow] = \int d^3r n(\vec{r}) \varepsilon_{xc}^{unif}(n_\uparrow(\vec{r}), n_\downarrow(\vec{r})) \tag{11}$$

where $\varepsilon_{xc}^{unif}(n_\uparrow, n_\downarrow)$ is the exchange-correlation energy per electron for an electron gas of uniform spin densities $n_\uparrow$ and $n_\downarrow$. The functional derivative gives the first-order change of the functional with density:

$$\delta E = \int d^3r \sum_\sigma [\delta E_{xc} / \delta n_\sigma(\vec{r})] \delta n_\sigma(\vec{r}) \tag{12}$$

For example,

$$\delta E_{xc}^{LSDA} / \delta n_\sigma(\vec{r}) = \partial [n \varepsilon_{xc}^{unif}(n_\uparrow, n_\downarrow)] / \partial n_\sigma |_{\vec{r}} \tag{13}$$

In Equation (9), T_s is an explicit functional of the occupied orbitals, but an implicit functional of the spin-densities. That the orbitals $\psi_{\alpha,\sigma}$ are implicit functionals of the spin-density n_σ follows from a theorem of Hohenberg and Kohn (1964), applied to the Kohn-Sham non-interacting system.

The original proofs of Equations (6)-(10) assumed (a) that the ground-state density of the real interacting system is "non-interacting v-representable," i.e., that it is also the ground-state density of a fictitious non-interacting system in a scalar external potential, and (b) that the ground states are nondegenerate. Later work by Levy (1979) showed that nondegeneracy is not required. Non-interacting v-representability is needed if all orbitals below the Fermi level are to be fully occupied, and is not always guaranteed for finite systems (e.g., some transition-metal atoms).

The important advances of Kohn-Sham theory over the earlier Thomas-Fermi-like approximations are two: (1) There is an exact underlying theory, which motivates and guides improved approximations. (2) The non-interacting kinetic energy T_s, a large fraction of the total since typically $E \approx -T_s$, is computed exactly from the orbitals. Only the exchange-correlation energy remains to be approximated, and there are good reasons (to be discussed later) why simple approximations such as Equation (11) work as well as they do.

Uniform electron gas

Consider jellium, an interacting electron gas of uniform spin densities $n_\uparrow$ and $n_\downarrow$, neutralized by a uniform positive background of density $n = n_\uparrow + n_\downarrow$. All properties of this system are functions of $n_\uparrow$ and $n_\downarrow$.

At the Kohn-Sham noninteracting level, the Kohn-Sham orbitals are plane waves of wavevector $\vec{k}$. By a standard counting argument, the occupied electrons of spin σ have wavevectors inside a Fermi sphere of $\vec{k}$-space whose radius is the Fermi wavevector for spin σ

$$k_{F,\sigma} = (6\pi^2 n_\sigma)^{1/3} \tag{14}$$

The Fermi energy for spin σ is

$$\varepsilon_{F,\sigma} = k_{F,\sigma}^2 / 2 \tag{15}$$

and the noninteracting kinetic energy density is

$$n\tau = (3/5)[n_\uparrow \varepsilon_{F,\uparrow} + n_\downarrow \varepsilon_{F,\downarrow}] \tag{16}$$

The exchange energy density is

$$n\varepsilon_x = [-3/(4\pi)][n_\uparrow k_{F,\uparrow} + n_\downarrow k_{F,\downarrow}] \tag{17}$$

Sometimes it is useful to introduce as variables

$$n = 3/(4\pi r_s^3) = k_F^3/(3\pi^2) \qquad \varsigma = (n_\uparrow - n_\downarrow)/n \tag{18}$$

Then $k_F \propto n^{1/3}$, $\varepsilon_F \propto n^{2/3}$, $n\tau \propto n^{5/3}$, $n\varepsilon_x \propto n^{4/3}$, $\varepsilon_x \propto -0.458/r_s$, etc. The correlation energy density is usually written as

$$n\varepsilon_c(r_s . \varsigma) \tag{19}$$

where ε_c is a controlled interpolation (Vosko et al. 1980; Perdew and Zunger 1981; Perdew and Wang 1992) between the known high-density or $r_s \to 0$ ($\propto 0.0311 \ln r_s$) and low-density or $r_s \to \infty$ ($\propto -0.44/r_s$) limits. Note that the ratio $\varepsilon_c/\varepsilon_x$ varies from 0 to about 1 (at $\varsigma = 0$) as r_s increases from 0 (high-density limit) to ∞ (low-density limit).

We should also mention the Fermi wavelength $\lambda_F = 2\pi/k_F$ and the screening length $\lambda_s = 1/k_s$ where $k_s = (4k_F/\pi)^{1/2}$. All of these quantities have local analogs in inhomogeneous systems.

What Kohn-Sham theory promises

It is important to remember that even the exact Kohn-Sham theory can at best deliver the exact ground-state spin densities and energy for electrons in the presence of a static external potential. Properties that cannot be derived from these (such as band gaps from orbital energy differences or band structures: Perdew et al. 1982; Perdew and Levy 1983; Sham and Schlueter 1983; Perdew et al. 2009a) are not expected to be exact, even with the unknown exact functional for the exchange-correlation energy. In particular, energy densities are unmeasurable and non-unique.

Because the nuclei are thousands of times more massive than the electrons, we can often make an adiabatic or Born-Oppenheimer approximation in which the electrons are in their ground state for each set of nuclear positions $\{\vec{R}_n\}$. Then the electronic ground-state energy $E(\vec{R}_1, \vec{R}_2, ...)$ serves as a potential energy or potential energy surface for the nuclei. Minimizing this function yields the equilibrium structure. The second derivatives around the minimum yield elastic constants and harmonic vibrational frequencies. The nuclear motion can be treated either quantum mechanically or even classically, leading to zero-point and thermal vibrational energies for solids, and to molecular dynamics for solids and liquids.

Cohesive or atomization energies, surface and defect formation energies, and ground-state electron removal energies (ionization energies, electron affinities, and work functions) are also accessible to Kohn-Sham theory.

Solving the Kohn-Sham equations

Equations (6), (7), and (10) are to be solved selfconsistently. For example, one might begin with some reasonable first guess for the spin densities, such as $n_\uparrow = n_\downarrow = n/2$ where $n(\vec{r})$ is a superposition of atomic densities. From this, one constructs the effective potential in Equations (6) and (10), and then solves Equation (6) for the occupied (low-energy) orbitals. A new set of spin densities is found by mixing the original set with the new set from Equation (7), and this process is iterated until the spin densities stop changing significantly.

To solve Equation (6), one must specify the number of electrons and the boundary conditions. For example, the density and effective potential might tend to zero far outside the system, as for an atom or molecule, or they might repeat periodically over space, as for a perfect solid. There are many standard codes to solve these equations, assuming either the first boundary condition (as in TURBOMOLE and ADF) or the second (as in BAND, VASP, and WIEN), or a choice of either (as in GAUSSIAN). The first boundary condition is more natural for atoms and molecules, but can also in principle describe a solid as the limit of a sequence of ever-larger clusters of atoms. The second boundary condition is more natural for a perfect solid, but can also describe an atom, molecule, defected solid or surface by employing a large supercell that is repeated periodically. When the supercell is large enough, with enough empty space around a unit, the units no longer interact.

The Kohn-Sham orbitals are canonical symmetry orbitals, e.g., complex Bloch orbitals for periodic boundary conditions. The solution of Equation (6) can be found on a mesh in real space (as in PARSEC) or expanded in a set of basis functions. Localized Gaussian basis functions (GAUSSIAN, TURBOMOLE), localized numerical and Slater functions (ADF, BAND), or delocalized plane waves and augmented plane waves (VASP, WIEN) are all widely used.

Finally, since the deep core electrons do not participate in the binding of one atom to another, the corresponding core orbitals can be found selfconsistently, or frozen as atomic-like orbitals, or removed entirely via electron-ion pseudopotentials. There are many ways to solve the Kohn-Sham equations, but, when applied carefully, they now all tend to give the same answer (e.g., Csonka et al. 2009). One of the authors (JPP) can remember the 1960's and 70's when this was not so.

EXCHANGE-CORRELATION ENERGY: NATURE'S GLUE

Kohn-Sham theory neglecting the exchange-correlation energy (setting $E_{xc}[n_\uparrow,n_\downarrow] = 0$) is the Hartree approximation without self-interaction correction. This approximation binds atoms together, but far too weakly (Kurth and Perdew 2000). Bond lengths and lattice constants are far too large, while elastic constants, vibrational frequencies, and cohesive or atomization energies are far too small. It is primarily the exchange-correlation energy that glues the atoms together. If one also makes the local density or Thomas-Fermi approximation for the kinetic energy, atoms do not bind together at all.

The Hartree approximation makes the total energy too high because it treats the charge on the electrons as a continuous and rigid distribution. In reality, the electrons are point-size objects of quantized charge, and these point charges avoid one another as they move through the average electron density $n(\vec{r})$, for electrons of the same z-component of spin σ as a consequence of the Pauli exclusion principle and for all electrons as a consequence of their mutual Coulomb repulsion. The negative exchange-correlation energy arises primarily from these effects, although it also includes a small positive kinetic energy term arising from Coulomb-correlated motion.

The exact adiabatic connection formula for the exchange-correlation energy (Langreth and Perdew 1975; Gunnarsson and Lundqvist 1976) explains both what it is and why simple approximations to it often work:

$$E_{xc} = (1/2)\int d^3rn(\vec{r})\int d^3r' n_{xc}(\vec{r},\vec{r}')/|\vec{r}'-\vec{r}| \tag{20}$$

Here

$$n_{xc}(\vec{r},\vec{r}') = n_x(\vec{r},\vec{r}') + n_c(\vec{r},\vec{r}') \tag{21}$$

is the density at $\vec{r}'$ of the exact exchange-correlation hole around an electron at $\vec{r}$.

The exact exchange hole density n_x is everywhere nonpositive,

$$n_x(\vec{r},\vec{r}') \leq 0 \tag{22}$$

and integrates to

$$\int d^3r' n_x(\vec{r},\vec{r}') = -1 \tag{23}$$

Apart from its construction from Kohn-Sham and not Hartree-Fock orbitals, the same exchange hole density arises in Hartree-Fock theory. Equation (23) says that, if an electron is located at position $\vec{r}$, it is for that reason missing from the density at all other positions $\vec{r}'$. The correlation hole density n_c satisfies the sum rule

$$\int d^3r' n_c(\vec{r},\vec{r}') = 0 \tag{24}$$

Equation (24) says that Coulomb correlation rearranges the other electrons around a given electron at $\vec{r}$, without changing the number of them. These other electrons are pushed away from the region near the electron at $\vec{r}$, leading to a correlation hole density that is negative at small $|\vec{r}'-\vec{r}|$ (with a Coulomb cusp at $|\vec{r}'-\vec{r}| = 0$) and completely or predominantly positive at large $|\vec{r}'-\vec{r}|$.

In summary, an electron moving through the average electron density is surrounded by a hole in which the density of other electrons is depleted, and this hole reduces the electron-electron Coulomb repulsion energy. Because this effect is greater when the electrons are closer together, it reduces the total energies of molecules and solids more than it reduces the energies of separate atoms, and leads to enhanced atom-atom binding.

The exact exchange-correlation hole densities are related to pair densities that can be found from correlated many-electron wavefunctions. Define

$$n(\vec{r})[n(\vec{r}') + n_{xc}(\vec{r},\vec{r}')] = \int_0^1 d\lambda P_\lambda(\vec{r},\vec{r}') \tag{25}$$

where

$$P_\lambda(\vec{r},\vec{r}') = N(N-1)\int d^3r_3 ... d^3r_N |\Psi_\lambda(\vec{r},\vec{r}',\vec{r}_3,...,r_N)|^2 \tag{26}$$

All the ground-state wavefunctions Ψ_λ have the same ground-state spin densities $n_\uparrow(\vec{r})$ and $n_\downarrow(\vec{r})$, but different electron-electron pair interactions $\lambda/|\vec{r}'-\vec{r}|$ and external potentials $v_\lambda(\vec{r})$. The coupling constant λ is assumed to connect adiabatically the real interacting system ($\lambda = 1$) to the Kohn-Sham non-interacting system ($\lambda = 0$). Given λ, $P_\lambda(\vec{r},\vec{r}')d^3rd^3r'$ is the joint probability to find an electron in d^3r at $\vec{r}$ and another in d^3r' at $\vec{r}'$.

From this and other exact expressions, it is possible to derive many exact constraints (e.g., Levy and Perdew 1985; Levy 1991; Goerling and Levy 1993) on the density functional $E_{xc}[n_\uparrow,n_\downarrow] = 0$, some satisfied by LSDA and some which can be satisfied only by higher-level approximations.

SEMILOCAL AND NONLOCAL APPROXIMATIONS: WHEN CAN WE STAY CLOSE TO HOME?

Local spin density approximation

The LSDA of Equation (11) is the earliest (Kohn and Sham 1965; von Barth and Hedin 1972) and simplest approximation to the spin density functional for the exchange-correlation energy, and the starting point for most subsequent approximations. Its required input, the exchange-correlation energy per electron of an electron gas of uniform spin densities, is accurately known. The standard analytic parametrizations (Vosko et al. 1980; Perdew and Zunger 1981; Perdew and Wang 1992) interpolate between known high- and low-density behaviors, with the interpolation fitted to Quantum Monte Carlo data (Ceperley and Alder 1980).

By construction, LSDA is exact for uniform densities and densities that vary quite slowly over space. This makes LSDA somewhat better suited to solids than to atoms and molecules, whose densities are more inhomogeneous. The early successes of LSDA were for solids, and even today LSDA is widely used in condensed matter physics (but not in quantum chemistry). While LSDA often predicts accurate bond lengths for molecules, it radically overestimates their atomization energies, making LSDA of limited use in chemistry.

The LSDA exchange-correlation hole density around an electron at $\vec{r}$ is clearly just the hole density of a uniform electron gas whose spin-densities are those of the real system at $\vec{r}$. Thus the LSDA hole obeys the exact hole constraints of Equations (22)-(24). As realized early on (Langreth and Perdew 1975; Gunnarsson and Lundqvist 1976), this is an important reason why the LSDA exchange-correlation energy is reasonably good even for densities that are far from uniform. Later, it was found that the LSDA "on-top" exchange hole density $n_x(\vec{r},\vec{r})$ is exact (Ziegler et al. 1977), and the LSDA "on-top" correlation hole density $n_c(\vec{r},\vec{r})$ is often reasonably accurate (Burke et al. 1998). Equation (11) also satisfies some exact constraints on the exchange-correlation energy functional itself, including a density-scaling equality for exchange (Levy and Perdew 1985), a density-scaling inequality for correlation (Levy and Perdew 1985), and the Lieb-Oxford local lower bound (Lieb and Oxford 1981; Perdew 1991).

Gradient expansion

For an electron density that varies slowly over space, the exchange-correlation energy has an expansion in powers of the density gradients (Antoniewicz and Kleinman 1985; Langreth and Vosko 1990; and references therein):

$$E_{xc}[n_\uparrow,n_\downarrow] = E_{xc}^{LSDA}[n_\uparrow,n_\downarrow] + \sum_{\sigma,\sigma'} C_{xc}^{\sigma,\sigma'}(n_\uparrow,n_\downarrow)\nabla n_\sigma \cdot \nabla n_{\sigma'} / [n_\sigma n_{\sigma'}]^{2/3} + \ldots \quad (27)$$

One might expect that this expression, truncated at second order in ∇, would improve upon LSDA, but in fact it is worse than LSDA for typical atoms, molecules, and solids. The reason is that the second-order gradient expansion of the exchange-correlation hole density is not the hole density of any possible system, and strongly violates the exact hole constraints in Equations (22)-(24) (Langreth and Perdew 1980).

Generalized gradient approximation

The generalized gradient approximation (Langreth and Mehl 1983; Perdew and Wang 1986; Becke 1988; Lee et al. 1988; Perdew 1991; Perdew et al. 1996, 1998) is

$$E_{xc}^{GGA}[n_\uparrow,n_\downarrow] = \int d^3r\, n(\vec{r})\varepsilon_{xc}^{GGA}(n_\uparrow,n_\downarrow,\nabla n_\uparrow,\nabla n_\downarrow) \quad (28)$$

The PW86 exchange and PW91 exchange-correlation GGA's were constructed by starting from the second-order gradient expansions for the corresponding hole densities, then sharply cutting off the spurious long-range (large $|\vec{r}'-\vec{r}|$) parts to satisfy the exact hole constraints of Equations (22)-(24), and finally fitting the resulting numerically-defined GGA to a chosen analytic form. The widely-used PBE GGA has two different constructions: one based on the real-space cutoff of the gradient expansion for the hole, and one based on satisfaction of selected exact constraints on the energy. Compared to LSDA, GGA predicts much more accurate atomization energies, so the GGA and its extensions are widely used in chemistry.

GGA's for solids

Standard GGA's including PBE are now widely used in condensed matter physics, but do not always perform satisfactorily for solids: (1) While the LSDA lattice constants of solids are typically 1% too short, the GGA lattice constants are typically 1% too long. Some effects (ferromagnetism, ferroelectricity, bulk moduli and phonon frequencies) are sensitive to lattice constant. (2) While the PBE GGA greatly improves the separate exchange and correlation contributions to the surface energy of jellium over LSDA, there is less error cancellation between exchange and correlation in PBE; the total surface energy is higher and more correct in LSDA than in PBE.

These problems have been addressed by developing new GGA's specifically for solids (Vitos et al. 2000; Armiento and Mattsson 2005; Wu and Cohen 2006; Csonka et al. 2007; Perdew et al. 2008a; Zhao and Truhlar 2008; Csonka et al. 2009), at the cost of worsened accuracy for atoms and molecules. In particular, the PBE construction was found to be biased toward atoms and molecules by the use of sharp cutoffs of the spurious long-range parts of the second-order gradient expansion for the hole density: The exact hole cuts off rather sharply in an atom or small molecule, because the electron density decays exponentially to zero away from the electron, but the hole can be more diffuse in a solid. The PBEsol GGA is a reparameterized PBE based on "restoring the gradient expansion for exchange over a wide range of electron densities," which predicts improved lattice constants and surface energies for solids.

Ladder of density functional approximations: Meta-GGA and fully nonlocal

Most widely-used approximations can be expressed as

$$E_{xc}^{approx}[n_\uparrow, n_\downarrow] = \int d^3r\, n(\vec{r})\,\varepsilon_{xc}^{approx}(n_\uparrow, n_\downarrow, \nabla n_\uparrow, \nabla n_\downarrow, \tau_\uparrow, \tau_\downarrow, \ldots) \tag{29}$$

Here τ of Equation (9) is $\tau_\uparrow + \tau_\downarrow$. Equations (11), (28), and (29) suggest a five-rung ladder of density functional approximations (Perdew and Schmidt 2001). Climbing from one rung to the next involves adding another argument to $\varepsilon_{xc}^{approx}$. Additional arguments or ingredients can be used to satisfy additional exact constraints (as in most of the functionals cited here), or to fit data better (as in Zhao and Truhlar 2006), or both. In Equation (29), $\varepsilon_{xc}^{approx}$ should reduce, for uniform spin densities, to ε_{xc}^{unif}. Functionals that do not, such as the GGA correlation of Lee et al. 1988, can make unacceptable errors for the lattice constants of solids (Kurth et al. 1999).

The first rung or LSDA uses only the local spin densities, but already satisfies many exact constraints on the exchange-correlation hole and energy, which it inherits from the uniform electron gas. The second rung or GGA uses also the gradients of the local spin densities, so it can for slowly-varying densities recover the form or even the proper coefficients of the gradient expansion shown in Equation (27). GGA is also the first rung on which the correlation energy of a finite system (without degeneracy of the Kohn-Sham noninteracting system) can properly scale to a constant (Levy 1991; Goerling and Levy 1993) under uniform density scaling to the high-density limit:

$$n(\vec{r}) \to n_\gamma(\vec{r}) = \gamma^3 n(\gamma\vec{r}) \quad \text{and} \quad \gamma \to \infty \tag{30}$$

These conditions are satisfied by the PBE and PBEsol GGA's. The third rung or meta-GGA (Becke 1996; Tao et al. 2003; Ruzsinszky et al. 2005; Perdew et al. 2009b) adds the positive spin-resolved noninteracting kinetic energy density τ, which can be used to recover the correct fourth-order gradient expansion for the exchange energy and to zero out the correlation energy of any fully-spin-polarized one-electron density. These conditions are satisfied by the TPSS and revTPSS meta-GGA's.

The meta-GGA rung seems to be the first one that can provide an accurate simultaneous description of many atoms, molecules, and solids near equilibrium. The TPSS meta-GGA predicts good atomization energies (much better than the PBE GGA) and good surface energies (also much better than PBE). While the TPSS meta-GGA only slightly improves the lattice constants of typical solids, its revised version, revTPSS, uses the PBEsol idea (restoring the gradient expansion for exchange over a wide range of densities) to produce lattice constants much better than those of the PBE GGA, without degrading the other good results of TPSS.

The first three rungs of the ladder are semilocal functions, in a sense to be defined in the next subsection. For each of these rungs, there is one or more fully nonempirical construction. These rungs are all computationally efficient, and are usefully accurate for solids. The overall best choice seems to be the revTPSS meta-GGA. Selfconsistent LSDA and GGA are available in nearly all standard electronic structure codes. Selfconsistent meta-GGA is available in the GAUSSIAN and TURBOMOLE codes, and should soon be available in the BAND and VASP codes.

The two highest rungs of the ladder are fully nonlocal (and therefore computationally more expensive), and are needed in certain special situations to be described in the next subsection. The fourth or hyper-GGA rung introduces as an ingredient some exact exchange information. For the widely-used *global* hybrid functionals (Becke 1993),

$$E_{xc}^{gh}[n_\uparrow,n_\downarrow] = aE_x^{ex} + (1-a)E_x^{sl} + E_c^{sl} \tag{31}$$

this ingredient is the exact exchange energy E_x^{ex}, which is mixed with a semilocal exchange energy E_x^{sl}. The empirical mixing coefficient a is typically of order 0.2. The *local* hybrid functionals (Perdew et al. 2008b and references therein) employ instead a local mixing parameter $a([n_\uparrow,n_\downarrow],\vec{r})$ constructed in part from the exact exchange energy density, and so can satisfy many more exact constraints. Range-separated hybrids use as an exact-exchange ingredient the exact exchange hole density.

The fifth and highest rung of the ladder is RPA+, which introduces also the unoccupied Kohn-Sham orbitals and their orbital energies. The simplest example is the random phase approximation (RPA), whose performance varies from nonpareil to abysmal (Harl and Kresse 2009; Mori-Sanchez et al. 2009), depending on the system and property. RPA can be 100 times more expensive than GGA, and universally-useful corrections to RPA can be even more expensive.

On the LSDA and GGA rungs, it is easy to find the Kohn-Sham multiplicative effective potential by evaluating the required functional derivative in Equation (12) explicitly. On the meta-GGA and higher rungs, the Kohn-Sham potential can still be found by the optimized effective potential method (Kuemmel et al. 2003, and references therein). However, it is easier and not noticeably less accurate to use an orbital-dependent effective potential instead (e.g., the Hartree-Fock potential for the exact-exchange part of a global hybrid), and this is regularly done (Neumann et al. 1986).

Semilocal approximations: What, why, and when?

The first two rungs of the ladder (LSDA and GGA) are semilocal functionals of the density, in which the integrand of Equation (23) at position $\vec{r}$ is determined by the electron spin densities in an infinitesimal neighborhood of $\vec{r}$. The third rung (meta-GGA) is a fully nonlocal functional of the density, but is still semi-local in this sense: The integrand of Equation (23) at $\vec{r}$ is determined by the density and occupied orbitals in an infinitesimal neighborhood of $\vec{r}$. Since the density and orbitals are available in any Kohn-Sham calculation, the semilocal functionals are computationally efficient, requiring only a single integration over the mesh in real three-dimensional space (with a time requirement proportional to M^3). The fourth (hyper-GGA) and fifth (RPA+) rungs require a double integration over the mesh (with a time requirement proportional to M^6, although clever speed ups are often possible).

The semilocal approximations have been developed nonempirically and tested widely over a long period of time, so it is important to understand why and when they work well, and when the more expensive and less-understood full nonlocality is unavoidable.

Exchange and correlation hole densities are available for nonempirical functionals on all three semilocal rungs of the ladder. For LSDA, these are the known hole densities of the uniform electron gas. For GGA and meta-GGA, the holes are either modeled first to derive the functional, or else "reverse-engineered" (Constantin et al. 2006) to recover the functional while satisfying exact constraints on the hole. In either case, both the approximate exchange and the approximate correlation hole density are compactly localized around or near the electron at $\vec{r}$. This is necessarily so, since the input semilocal information is not enough to determine the electron or hole densities far from the electron This fact suggests that the semilocal functionals (especially the high-level meta-GGA) will work to the extent that the exact holes are also localized.

So, when are the exact holes localized around the electron at $\vec{r}$? Certainly they are localized in an electron gas of uniform or very-slowly-varying density. They are also localized in an atom, because the electron density itself is localized. And these are systems in which a good meta-GGA like revTPSS is typically accurate. Now consider "multi-center" or "multi-atom" systems, such as molecules and solids. In molecules, even at equilibrium geometries, there is some delocalization of the exact exchange hole over two or more centers, but the exact exchange-correlation hole is typically deeper, more short-ranged, and much less delocalized than the exact exchange hole. This is the reason why meta-GGA's give accurate atomization energies for molecules, and why the fraction of exact exchange used in the global hybrid of Equation (31) is small (empirically about 0.1-0.25). There is an error cancellation between semilocal exchange and semilocal correlation, which is lost if we combine full exact exchange with semilocal correlation (Perdew et al. 2008b).

As bond lengths are stretched, the exact exchange and even the exact exchange-correlation holes can become increasingly delocalized over two or more atomic centers, leading to serious failures of the corresponding semilocal functionals. This happens whenever electrons (at the noninteracting or interacting levels, respectively) fluctuate between two or more distant centers (Perdew et al. 1982, 2007). The problem for exchange and correlation together first appears for the transition states of chemical reactions: compound molecules with stretched bonds, for which the semilocal functionals make the hole too localized and thus the energy too low, leading to an underestimation of the reaction barrier heights. (Barriers heights are most difficult for gas-phase reactions, and less so for surface or solid state reactions.) But these barriers can be much improved on the fourth or hyper-GGA rung of the ladder (Perdew et al. 2008b). A similar problem for semilocal functionals presumably appears in the strongly-correlated transition-metal oxides. The problem can be even more severe in the large-bond-length or dissociation limit, where semilocal functionals can place the energy seriously too low for symmetric radicals (Ruzsinszky et al. 2007), and can produce spurious fractional charges on the dissociated atoms of unsymmetric molecules (Ruzsinszky et al. 2006). The Perdew-Zunger 1981 self-interaction correction to LSDA can describe these stretched-bond cases rather well, as a consequence of the particular way in which it makes the functional exact for all one-electron densities, but it is disappointingly inaccurate for the properties of molecules near equilibrium.

The exact exchange-correlation hole is also to a lesser extent delocalized over two or more atomic centers which interact via van der Waals interaction, and here again the semilocal functionals can be wrong. In some equilibrium situations, the semilocal functionals do capture a dominant short-range or overlap contribution to the van der Waals interaction (Tao et al. 2005), but they never capture the long-range part. For that, again, fully nonlocal functionals such as RPA+ are needed.

In some strong-correlation cases, such as the highly-stretched spin-unpolarized hydrogen molecule H_2, the semilocal functionals make the total energy too high. But in these cases there is typically a spin-symmetry breaking in the selfconsistent semilocal solution (e.g., localizing the ↑ electron on one hydrogen atom and the ↓ electron on the other), which corrects the total energy. This works because the on-top exchange-correlation hole density $n_{xc}(\vec{r},\vec{r})$ equals $-n(\vec{r})$ for any fully-spin-polarized system, as it does in the strong-correlation limit (Perdew et al. 1995).

The semilocal functionals PBEsol GGA and revTPSS meta-GGA are especially well-suited for the description of metallic, covalent, or ionic solids near equilibrium or under intense pressure, and thus for most applications in mineralogy and geophysics. Even there, the presence of highly-localized groups of electrons in open shells (e.g., in $3d$ or $4f$ metals) can produce errors that require full nonlocality. For the $3d$ metals, the PBE lattice constants appear to be more correct than the PBEsol or revTPSS values. It remains to be seen to what extent these special-case errors of PBEsol or revTPSS disappear under high pressures of interest.

PRESSURE AND TEMPERATURE IN DENSITY FUNCTIONAL THEORY: PRESSING DOWN AND HEATING UP

Mineralogists would be primarily interested in solids at standard pressure (P = 1 atm = 0.101 GPa) and temperature (T = 293 K). Thus for most mineralogical applications the zero-pressure, ground-state version of density functional theory is very accurate. But geophysicists are interested in solids and liquids at elevated pressures and temperatures up to those present in the earth's core (P = 330 to 360 GPa, T = 5700 K). Under these conditions, all condensed matter is substantially compressed and in many cases (including the iron that is prevalent in the outer core) melted.

Pressure

We will begin by discussing the effects of pressure on a solid at the absolute zero of temperature. Let e and v be the total energy and volume per primitive unit cell (the smallest possible repeated unit of a periodic structure). Let the subscript zero denote the equilibrium value of any property at zero pressure, and define the dimensionless compression ratio or scale factor by

$$x = (v / v_0)^{1/3} \tag{32}$$

Given a crystal structure (such as bcc or body-centered cubic), we can use density functional theory to compute the equation of state $e(x)$, and related quantities such as the pressure

$$P = -de / dv = [-1 / (3v_0 x^2)] de / dx \tag{33}$$

the bulk modulus or elastic stiffness

$$B = -vdP / dv = -(x / 3) dp / dx \tag{34}$$

and the pressure derivative of the bulk modulus

$$B' = dB / dP = -(x / 3) dB / dx \tag{35}$$

Ideally the calculated $e(x)$ would include the zero-point vibrational energies of the ions, but this effect is sometimes neglected. Calculations give a table of values for e as a function of x, which can be fitted to a reasonable analytic form and differentiated analytically. There are many possible analytic forms. One that is based on the simple stabilized jellium model of a metal, and works reasonably well over a wide range of $x < 1$ is the stabilized jellium equation of state (Alchagirov et al. 2001)

$$e^{SJEOS}(x) = a/x^3 + b/x^2 + c/x + d \tag{36}$$

where

$$a = (9/2)B_0 v_0 (B_0' - 3) \tag{37}$$

$$b = (9/2)B_0 v_0 (10 - 3B_0') \tag{38}$$

$$c = (9/2)B_0 v_0 (11 - 3B_0') \tag{39}$$

$$d = e_0 + (9/2)B_0 v_0 (4 - B_0') \tag{40}$$

Atomic units must be used in Equations (36)-(40). For a set of typical solids, B_0 ranges from 2.6 GPa (Cs) to 467 GPa (diamond), and B_0' ranges from 3.2 (Ca) to 6.0 (Sn). Equation (36) diverges to $+\infty$ as $x \to 0$ minimizes at $x = 1$ and tends to d as $x \to \infty$. Of course, no simple analytic form can accurately cover the whole range $0 < x < 1$. The pressure P vanishes at $x = 1$, and increases as x is reduced from 1. The bulk modulus B (and hence also the longitudinal acoustic lattice vibrational frequencies proportional to $B^{1/2}$) also increases as x is reduced from 1.

An interesting feature of Equations (36)-(40) is that the ratio of pressure to equilibrium bulk modulus is determined only by x and B_0':

$$P/B_0 = f(x, B_0') \tag{41}$$

Table 1 shows this relationship, and can be used for a quick estimate of P or x for a given material. For example, the experimental values for bcc Fe are $B_0 = 172$ GPA and $B_0' = 5.0$. At the pressure of the earth's core, $P = 345$ GPa, Table 1 predicts $x = 0.83$, in excellent agreement with the bcc equation of state calculated within GGA by Stixrude, Cohen, and Singh (1994).

Table 1. The ratio P/B_0 of pressure to equilibrium bulk modulus as a function of the compression ratio x and equilibrium pressure derivative of bulk modulus B_0', according to the stabilized jellium equation of state of Equations (36)-(40). Note that for $x < 0.5$ there is likely too much core overlap for these equations to be accurate.

x	B_0'			
	3.	**4.**	**5.**	**6.**
0.50	48.00	120.00	192.00	264.00
0.55	26.82	59.74	92.66	125.58
0.60	15.43	30.86	46.30	61.73
0.65	9.05	16.36	23.67	30.98
0.70	5.35	8.80	12.24	15.68
0.75	3.16	4.74	6.32	7.90
0.80	1.83	2.52	3.20	3.89
0.85	1.01	1.28	1.55	1.82
0.90	0.51	0.59	0.68	0.76
0.95	0.19	0.21	0.22	0.24
1.00	0.00	0.00	0.00	0.00

To determine the equilibrium crystal structure at a given volume per atom v/N_{cell}, we have to compute e from density functional theory for a set of candidate crystal structures, and select the one that minimizes the energy per atom, e/N_{cell}, where N_{cell} is the number of atoms in the primitive cell. If the curves of e/N_{cell} versus v/N_{cell} for two crystal structures cross each other, there can be a structural phase transition. The pressure at which this transition occurs is given by the slope of the common tangent to both curves. At this pressure, the solid can be in a mixed phase where both structures are present.

The evolution under pressure of the band structure (Kohn-Sham orbital energies $\varepsilon_{n\vec{k},\sigma}$ as functions of the band index n and wavevector $\vec{k}$) can also be of interest. Solids tend to be insulators when a band gap separates a filled valence band from an empty conduction band. Under pressure, the bands become increasingly wider in energy, and eventually overlap. Thus all materials are expected to become metals under pressure. The metallic character can also change under pressure. For example, for the simple metals K, Rb, and Cs under pressure, the empty d bands can start to overlap the half-filled s band, turning a simple metal into a transition metal (Soederlind and Ross 2000). Solid Li can show atom pairing at high pressure (Neaton and Ashcroft 1999).

When the pressure is sufficiently high and x is sufficiently small, even the deep core levels overlap and one might expect every solid to start resembling a uniform electron gas, for which the standard semilocal density functional approximations are exact at the nonrelativistic level. Indeed, the high-density limit at the nonrelativistic level seems to be one that is ideally suited for semilocal approximations. In this limit, well-known scaling properties show that the noninteracting kinetic energy T_s dominates over the exchange energy E_x, and E_x dominates over the correlation energy E_c. In this limit also the local Fermi wavelength becomes small compared to the local screening length (semiclassical limit), which means that the effective potential varies slowly on the scale of the local Fermi wavelength, and as a result local density approximations to E_x (and even to T_s) make increasingly smaller relative errors. Of course, when E_c is truly negligible compared to E_x, there can be no error cancellation between semilocal exchange and semilocal correlation, so using exact exchange would be even better than using semilocal exchange. So fully nonlocal functionals, such as the PSTS local hybrid (which formally has 100% of exact exchange) or RPA+, could improve accuracy even under intense pressure. It should be noted however that relativistic effects (Dreizler 2003) must become important at very high electron densities, where they should be included at least in the noninteracting kinetic energy.

Temperature

The equilibrium state of a system minimizes the average energy E at fixed entropy S, and maximizes the entropy at fixed average energy. The balance between these two tendencies is set by the temperature T, so that the equilibrium state minimizes the free energy

$$F = E - TS \tag{42}$$

At the absolute zero of temperature, the equilibrium state is the ground state, which minimizes E. At higher temperatures, entropy is increasingly favored.

Of course E and S should include a nuclear contribution as well as an electronic one. In condensed matter, the characteristic (Debye) temperature of the nuclear motion is typically a few hundred degrees, and thus much smaller than the characteristic (Fermi) temperature of the electronic motion, which can be many tens of thousands of degrees. Thus at temperatures well below the Fermi temperature one can often assume that only the nuclear motions are excited, with the electrons remaining in or near their ground state. Anharmonicity of the potential seen by a vibrating ion can produce zero-point and thermal expansion. As the temperature is increased, a solid can undergo various (first-order) structural phase transitions, and eventually melt into the high-energy, high-entropy liquid state. The phase diagram (Young 1991) of a pure solid, showing the equilibrium structures, can be plotted in the PT plane, while that of a binary alloy requires a third dimension, the concentration c. Nonequilibrium structures can be metastable, but those with imaginary phonon frequencies are unstable (Pollock and Perdew 1998).

At the temperature of the Earth's core, T = 5700 K, the thermal energy k_BT is about 0.5 eV, which is small but not negligible on the scale of the Fermi energy of iron (11.1 eV). The thermal excitation of electrons can be handled by the finite-temperature Kohn-Sham theory of Mermin (1965), in which the occupation numbers become Fermi-Dirac distribution functions. The noninteracting kinetic energy and noninteracting entropy are determined by these occupation numbers in the expected way. At least for the uniform electron gas (and thus for LSDA), the free energy $f_{xc}^{unif}(n_\uparrow,n_\downarrow,T)$ per electron of exchange and correlation is known (Kanhere et al. 1986; Perrot and Dharma-wardana 2000). For low temperatures, the thermal contribution is of order T^2. In the limit of very high temperature, the thermal wavelengths of the electrons become small and the electrons at the nonrelativistic level behave more classically, so that exchange turns off relative to correlation in this limit. Finite-temperature density functionals for exchange and correlation beyond LSDA remain to be developed. A reasonable "guesstimate" might multiply a beyond-LSDA ground-state exchange functional by $f_x^{unif}(n_\uparrow,n_\downarrow,T)/f_x^{unif}(n_\uparrow,n_\downarrow,0)$, and a beyond-LSDA ground-state correlation functional by $f_c^{unif}(n_\uparrow,n_\downarrow,T)/f_c^{unif}(n_\uparrow,n_\downarrow,0)$.

Pressure and temperature together

We can constrain the volume V and temperature T for a fixed number of atoms, so that the free energy F becomes a function of both. Then

$$S=-\partial F/\partial T \qquad P=-\partial F/\partial V \tag{43}$$

The second equation in Equation (43) is the thermal equation of state $P(x,T)$.

ACKNOWLEDGMENTS

This work was supported in part by the National Science Foundation under grant No. DMR-0854769. JPP thanks Renata Wentzcovitch for the invitation to speak at the Mineralogical Society of America meeting, December 10-13, 2009.

REFERENCES

Alchagirov AB, Perdew JP, Boettger JC, Albers RC, Fiolhais C (2001) Energy and pressure versus volume: Equations of state motivated by the stabilized jellium model. Phys Rev B 63:224115

Alfè D, Price GD, Gillan MJ (2002) Iron under Earth's core conditions: Liquid-state thermodynamics and high-pressure melting curve from ab initio calculations. Phys Rev B 65:165118

Antoniewicz PR, Kleinman L (1985) Kohn-Sham exchange potential exact to first order in $\rho(\vec{K})/\rho_0$. Phys Rev B 31:6779-6781

Armiento R, Mattsson AE (2005) Functional designed to include surface effects in selfconsistent density functional theory. Phys Rev B 72:085108

Ashcroft NW, Mermin ND (1976) Solid State Physics. Holt, Rinehart and Winston, New York

Becke AD (1988) Density-functional exchange-energy approximation with exact asymptotic behaviour. Phys Rev A 38:3098-3100

Becke AD (1993) Density-functional thermochemistry. III. The role of exact exchange. J Chem Phys 98:5648

Becke AD (1996) Density-functional thermochemistry. IV. A new dynamical correlation functional and implications for exact-exchange mixing. J Chem Phys 104:1040

Burke K, Perdew JP, Ernzerhof M (1998) Why semi-local density functionals work: Accuracy of the on-top pair density and importance of system averaging. J Chem Phys 109:3760

Carrier P, Wentzcovitch R, Tsuchiya J (2007) First-principles prediction of crystal structures at high temperatures using the quasiharmonic approximation. Phys Rev B 76:064116

Ceperley DM, Alder BJ (1980) Ground state of the electron gas by a stochastic method. Phys Rev Lett 45:566-569

Constantin LA, Perdew JP, Tao J (2006) Meta-generalized gradient approximation for the exchange-correlation hole, with an application to the jellium surface. Phys Rev B 73:205104

Csonka GI, Perdew JP, Ruzsinszky A, Philipsen PHT, Lebegue S, Paier J, Vydrov OA, Angyan JG (2009) Assessing the performance of recent density functionals for solids. Phys Rev B 79:155107

Csonka GI, Vydrov OA, Scuseria GE, Ruzsinszky A, Perdew JP (2007) Diminished gradient dependence of density functionals: Constraint satisfaction and self-interaction correction. J Chem Phys 126:244107

Dreizler RM (2003) Relativistic density functional theory. *In:* A Primer in Density Functional Theory. Fiolhais C, Nogueira F, Marques M (eds) Springer, Berlin, p 123-143

Dreizler RM, Gross EKU (1990) Density Functional Theory. Springer, Berlin

Goerling A, Levy M (1993) Correlation-energy functional and its high-density limit obtained from a coupling-constant perturbation expansion. Phys Rev B 47:13105

Gunnarsson O, Lundqvist BI (1976) Exchange and correlation in atoms, molecules, and solids by the spin-density-functional formalism. Phys Rev B 13:4274-4298

Harl J, Kresse G (2009) Accurate bulk properties from approximate many-body techniques. Phys Rev Lett 103:056401

Hohenberg P, Kohn W (1964) Inhomogeneous electron gas. Phys Rev 136:B864-B871

Kanhere DG, Panat PV, Rajagopal AK, Callaway J (1986) Exchange-correlation potentials for spin-polarized systems at finite temperature. Phys Rev A 33:490-497

Karki BB, Stixrude L, Wentzcovitch R (2001) High-pressure elastic properties of major materials of earth's mantle from first principles. Rev Geophys 39:507-534

Kohn W (1999) Nobel lecture: Electronic structure of matter – wave functions and density functionals. Rev Mod Phys 71:1253-1266

Kohn W, Sham LJ (1965) Selfconsistent equations including exchange and correlation effects. Phys Rev 140:A1133-A1138

Kuemmel S, Perdew JP (2003) Optimized effective potential made simple: Orbital functionals, orbital shifts, and the exact Kohn-Sham exchange potential. Phys Rev B 68:035103

Kurth S, Perdew JP (2000) Role of the exchange-correlation energy: Nature's glue. Int J Quantum Chem 77:814-818

Kurth S, Perdew JP, Blaha P (1999) Molecular and solid state tests of density functional approximations: LSD, GGA's, and Meta-GGA's. Int J Quantum Chem 75:889-909

Langreth DC, Mehl MJ (1983) Beyond the local-density approximation in calculations of ground-state electronic properties. Phys Rev B 28:1809-1834

Langreth DC, Perdew JP (1975) The exchange-correlation energy of a metallic surface. Solid State Commun 17:1425-1429

Langreth DC, Perdew JP (1980) Theory of nonuniform electronic systems: I. Analysis of the gradient approximation and a generalization that works. Phys Rev B 21:5469-5493

Langreth DC, Vosko SH (1990) Response functions and non-local functionals. *In:* Density Functional Theory of Many-Fermion Systems – In celebration of the 25th anniversary of the Hohenberg-Kohn theorem. Trickey SB (ed) Academic, Orlando, p 175-201

Lee C, Yang W, Parr RG (1988) Development of the Colle-Salvetti correlation-energy formula into a functional of the electron density. Phys Rev B 37:785-789

Levy M (1979) Universal variational functionals of electron densities, first-order density matrices, and natural spin-orbitals, and solution of the v-representability problem. Proc Natl Acad Sci USA 76:6062-6065

Levy M (1991) Density functional exchange correlation through coordinate scaling in adiabatic connection and correlation hole. Phys Rev A 43:4637-4646

Levy M, Perdew JP (1985) Hellmann-Feynman, virial, and scaling requisites for the exact universal density functional. Shape of the correlation potential and diamagnetic susceptibility for atoms. Phys Rev A 32:2010-2021

Lieb EH, Oxford S (1981) Improved lower bound on the indirect Coulomb energy. Int J Quantum Chem 19:427-439

Martin RM (2004) Electronic Structure: Basic Theory and Practical Methods. Cambridge University Press

Mermin ND (1965) Thermal properties of the inhomogeneous electron gas. Phys Rev 137:A1441-A1443

Mori-Sanchez P, Cohen AJ, Yang W (2009) Failure of the random phase approximation correlation energy. eprint axXiv:0903.4403

Neaton JB, Ashcroft NW (1999) Pairing in dense lithium. Nature 400:141-144

Neumann R, Nobes RH, Handy NC (1996) Exchange functionals and potentials. Mol Phys 87:1-36

Parr RG, Yang W (1989) Density-Functional Theory of Atoms and Molecules. Oxford University Press, New York

Perdew JP (1991) Unified theory of exchange and correlation beyond the local spin density approximation. *In:* Electronic Structure of Solids. Ziesche P, Eschrig H (eds) Akademie Verlag, Berlin, p 11

Perdew JP, Burke K, Ernzerhof M (1996) Generalized gradient approximation made simple. Phys Rev Lett 77:3865-3868

Perdew JP, Burke K, Wang Y (1998) Generalized gradient approximation for the exchange-correlation hole of a many-electron system. Phys Rev B 54:16533-16539
Perdew JP, Kurth S (2003) Density functionals for non-relativistic Coulomb systems in the new century. *In:* A Primer in Density Functional Theory. Fiolhais C, Noguiera F, Marques M (eds) Springer, Berlin, p 1-55
Perdew JP, Levy M (1983) Physical Content of the exact Kohn-Sham orbital energies: Band gaps and derivative discontinuities. Phys Rev Lett 51:1884-1887
Perdew JP, Parr RG, Levy M, Balduz JL (1982) Density functional theory for fractional particle number: Derivative discontinuities of the energy. Phys Rev Lett 49:1691-1694
Perdew JP, Ruzsinszky A, Constantin LA, Sun J, Csonka GI (2009a) Some fundamental issues in ground-state density functional theory: A guide for the perplexed. J Chem Theory Comput 5:902-908
Perdew JP, Ruzsinszky A, Csonka GI, Constantin LA, Sun J (2009b) Workhorse semilocal density functional for condensed matter physics and quantum chemistry. Phys Rev Lett 103:026403
Perdew JP, Ruzsinszky A, Csonka GI, Vydrov OA, Scuseria GE, Staroverov VN, Tao J (2007) Exchange and correlation in open systems of fluctuating electron number. Phys Rev A 76:040501
Perdew JP, Ruzsinszky A, Tao J, Csonka GI, Constantin LA, Zhou X, Vydrov OA, Scuseria GE, Burke K (2008a) Restoring the gradient expansion for exchange in solids and surfaces. Phys Rev Lett 100:136406
Perdew JP, Savin A, Burke K (1995) Escaping the symmetry dilemma through a pair-density interpretation of spin-density functional theory. Phys Rev A 51:4531-4541
Perdew JP, Schmidt K (2001) Jacob's ladder of density functional approximations for the exchange-correlation energy. *In:* Density Functional Theory and its Applications to Materials. Van Doren VE, Van Alsenoy K, Geerlings P (eds) American Institute of Physics, Melville, New York, p 1-20
Perdew JP, Staroverov VN, Tao J, Scuseria GE (2008b) Density functional with full exact exchange, balanced nonlocality of correlation, and constraint satisfaction. Phys Rev A 78:052513
Perdew JP, Wang Y (1986) Accurate and simple density functional for the electronic exchange energy: Generalized gradient approximation. Phys Rev B 33:8800-8802
Perdew JP, Wang Y (1992) Accurate and simple analytic representation of the electron gas correlation energy. Phys Rev B 45:13244-13249
Perdew JP, Zunger A (1981) Self-interaction correction to density functional approximations for many-electron systems. Phys Rev B 23:5048-5079
Perrot F, Dharma-wardana MWC (2000) Spin-polarized electron liquid at arbitrary temperature: Exchange-correlation energies, electron distribution functions, and the static response functions. Phys Rev B 62:16536-16548
Pollock L, Perdew JP (1998) Exchange-correlation corrections to the lattice dynamics of simple metals and a search for soft modes at normal and expanded volumes. Int J Quantum Chem 69:359
Ruzsinszky A, Perdew JP, Csonka GI (2005) Binding energy curves from nonempirical density functionals I. Shared-electron bonds in closed-shell and radical molecules. J Phys Chem A 109:11015
Ruzsinszky A, Perdew JP, Csonka GI, Vydrov OA, Scuseria GE (2006) Spurious fractional charge on dissociated atoms: Pervasive and resilient self-interaction error of common density functionals. J Chem Phys 125:194112
Ruzsinszky A, Perdew JP, Csonka GI, Vydrov OA, Scuseria GE (2007) Density functionals that are one- and two- are not always many-electron self-interaction-free, as shown for H_2^+, He_2^+, LiH^+, and Ne_2^+. J Chem Phys 126:194112
Sha X, Cohen RE (2006) Lattice dynamics and thermodynamics of bcc iron under pressure: First-principles linear response theory. Phys Rev B 73:104303
Sham LJ, Schlueter M (1983) Density functional theory of the energy gap. Phys Rev Lett 51:1888
Slater JC (1974) The Self-Consistent Field for Molecules and Solids. McGraw-Hill, New York
Soederlind P, Ross M (2000) Rubidium at high pressure and temperature. J Phys Condens Matter 12:921
Steinle-Neumann G, Cohen RE, Stixrude L (2004) Magnetism in iron as a function of pressure. J Phys Condens Matter 16:S1109
Stixrude L, Cohen RE, Singh DJ (1994) Iron at high pressure: Linearized-augmented-plane-wave computations in the generalized gradient approximation. Phys Rev B 50:6442
Tao J, Perdew JP (2005) Test of a nonempirical density functional: Short-range part of the van der Waals interaction in rare-gas clusters. J Chem Phys 122:114102
Tao J, Perdew JP, Staroverov VN, Scuseria GE (2003) Climbing the density functional ladder: Nonempirical meta-generalized gradient approximation designed for molecules and solids. Phys Rev Lett 91:146401 [TPSS]
Vitos L, Johansson B, Kollar J (2000) Exchange energy in the local Airy gas approximation. Phys Rev B 62:10046
von Barth U, Hedin L (1972) A local exchange-correlation potential for the spin-polarized case. (i) J Phys C 5:1629
Vosko SH, Wilk L, Nusair M (1980) Accurate spin-dependent electron liquid correlation energies for local spin-density calculations. Can J Phys 58:1200

Wu ZG, Cohen RE (2006) More accurate generalized gradient approximation for solids. Phys Rev B 73:235116

Young DA (1991) Phase Diagrams of the Elements. University of California Press

Zhao Y, Truhlar DG (2006) The M06 suite of density functionals for main group thermochemistry, thermochemical kinetics, noncovalent interactions, excited states, and transition elements: Two new functionals and systematic tests of four M06-class functionals and 12 other functionals. Theor Chem Acc 120:1432

Zhao Y, Truhlar DG (2008) Construction of a generalized gradient approximation by restoring the density-gradient expansion and enforcing a tight Lieb-Oxford bond. J Chem Phys 128:184109

Ziegler T, Rauk A, Baerends EJ (1977) Calculation of multiplet energies by Hartree-Fock-Slater method. Theor Chim Acta 43:261

Reviews in Mineralogy & Geochemistry
Vol. 71 pp. 19-37, 2010

The Minnesota Density Functionals and their Applications to Problems in Mineralogy and Geochemistry

Yan Zhao

Commercial Print Engine Lab, HP Labs, Hewlett-Packard Co.
1501 Page Mill Road
Palo Alto, California 94304, U.S.A.

yan.zhao3@hp.com

Donald G. Truhlar

Department of Chemistry and Supercomputing Institute
University of Minnesota
Minneapolis, Minnesota 55455, U.S.A.

truhlar@umn.edu

INTRODUCTION

Quantum mechanical electronic structure calculations are playing an increasingly useful role in many areas of mineralogy and geochemistry. This review introduces the density functional method for such calculations, gives an overview of the density functionals developed at the University of Minnesota, and summarizes selected applications using these density functionals that are relevant to mineralogy and geochemistry.

A key reason for the importance of computational methods in mineralogy is the ability to explore problems that cannot easily be studied in the laboratory. For example, it is very difficult to carry out laboratory studies under the real conditions of the Earth's mantle and core because the temperature of the Earth's core ranges up to 6000 K, and the pressure ranges up to 360 GPa. In the past decade, applications of quantum mechanical methods to understanding the properties of minerals and melts in the Earth's interior have become increasingly important (Warren and Ackland 1996; Oganov and Brodholt 2000; Stixrude and Peacor 2002; Brodholt and Vocadlo 2006; Gillan et al. 2006; Tsuchiya et al. 2006). Some specific examples of problems in solid-state geochemistry where electronic structure calculations can be particularly useful are phase equilibria (Tsuchiya et al. 2004; Lay et al. 2005; Schwegler et al. 2008), equations of state, elastic constants, bulk and shear moduli (Li et al. 2006), hydrogen, proton, and water diffusion in minerals (Sakamura et al. 2003; Belonoshko et al. 2004; Pöhlmann et al. 2004), the character and structural properties of hydroxyl groups in minerals and structures of hydrous minerals (Winkler et al. 1994, 1995; Nobes et al. 2000; Brodholt and Refson 2000; Churakov et al. 2003; Walker et al. 2006; Ockwig et al. 2009), hydrolysis and dissolution mechanisms (Strandh et al. 1997, Criscenti et al. 2006; Nangia and Garrison 2008; Morrow et al. 2009), and mineral fractionation (Blanchard et al. 2009).

In the field of geochemistry more broadly, computational quantum mechanics can also be very useful for understanding planetary hydrospheres and atmospheres (Andino et al. 1996; Tossell 2006; Qu et al. 2006; Kuwata et al. 2007; Hegg and Baker 2009; Gao et al. 2009) and oxidation-reduction mechanisms (Neal et al. 2003; Jaque 2007; Kwon et al. 2009) and for

1529-6466/10/0071-0002$05.00 DOI: 10.2138/rmg.2010.71.2

determining the properties of metal nanoparticles and other nanominerals, whose importance in geologic materials is increasingly recognized (Reich et al. 2006; Hochella et al. 2008).

Most of the computational studies in all of these areas have employed Kohn-Sham density functional theory (KSDFT) (Kohn and Sham 1965; Kohn et al. 1996), which is the most popular electronic structure theory in solid-state physics and quantum chemistry. The next section of this review presents an introduction to KSDFT.

DENSITY FUNCTIONALS

KSDFT implicitly begins with the Born-Oppenheimer approximation (Born and Huang 1954; Mead 1988) in which the forces governing atomic motions (that is, internuclear motions, also called nuclear motions) are the negative gradient of a potential energy function that equals the sum of the electronic energy and the nuclear repulsion, and the electronic energy is the ground-state expectation value of the electronic Hamiltonian with the nuclei frozen in position. (Thus the usual potential energy function for nuclear motion is associated with the lowest eigenvalue of the electronic Hamiltonian; higher eigenvalues correspond to electronically excited states and may be calculated for optical spectroscopy or photochemistry.) The Born-Oppenheimer approximation provides a separation of electronic and nuclear motion, and it is expected to be very accurate for most problems in mineralogy and geochemistry, with the main exception being geochemical photochemistry. For the problems where the Born-Oppenheimer separation is valid, the key problem in attaining accurate computations is to make the calculated electronic energy accurate.

To find the electronic energy, KSDFT defines a reference Slater determinant (a Slater determinant is an antisymmetrized product of spin orbitals, where each spin orbital is usually taken as a product of a spin-up or spin-down spin function and a spatial orbital). The spin-up and spin-down electronic densities (which are called the spin densities) associated with this determinant (or, more generally, the spin density matrix associated with it) are the same as the exact spin densities (or space-spin density matrix) (Kohn and Sham 1965; von Barth and Hedin 1972). The electronic energy is given by the sum of the electron-nuclear attraction (and the electron interaction with any other external potential), the classical Coulomb energy of the electron density, and a universal functional of the density. In KSDFT this universal functional is separated into the reference-determinant electronic kinetic energy and an exchange-correlation energy, E_{XC}, which is a functional of the spin densities (or the space-spin density matrix). It is this (unknown) exchange-correlation functional that is usually called the density functional. The spin orbitals are determined by effective potentials computed as functional derivatives of the exchange-correlation energy per electron; this effective potential is discussed further below.

Although KSDFT is an exact many-body theory for the ground-electronic-state properties of systems such as atoms, molecules, surfaces, clusters, nanoparticles, amorphous solids, and crystals, it depends on a universal exchange-correlation functional that is unknown and can only be approximated. The accuracy of a KSDFT calculation depends on the quality of this approximated density functional. In placing KSDFT in context, we note that it differs in approach from the older method of seeking better and better approximations to the many-electron wave function, an approach now known as wave function theory (WFT). Accurate WFT calculations become expensive more rapidly with increasing system size than does KSDFT. Both KSDFT and WFT are in principle exact, but in practice approximations are necessary, and for large and complex systems KSDFT provides a higher accuracy for a given cost. The road to improving its accuracy involves research to design more accurate density functionals.

KSDFT is an example of a self-consistent-field (SCF) method, analogous to the Hartree-Fock method, which is an SCF method in WFT. In an SCF method, each electronic orbital is governed by an effective potential energy function (mentioned above), which is called the field.

For example, in Hartree-Fock theory the field is the average electrostatic field due to the nuclei and to the electronic charge distribution corresponding to the orbitals occupied by the other electrons; this field includes both direct Coulomb interactions and the effect of quantum mechanical exchange of identical electrons. In KSDFT the field is determined from a functional of the density. In either SCF method, the first step is that the orbitals are determined from an approximate field, then more accurate fields are calculated from the resulting orbitals, and new orbitals are found in these new fields; the process is iterated to self-consistency. The Kohn-Sham SCF method is more accurate than the Hartree-Fock one because the latter does not account for dynamical correlation of the electrons, that is, the tendency for two electrons to correlate their motion to minimize their mutual repulsion. In WFT, dynamical correlation is usually determined by post-Hartree-Fock methods that correlate the electrons in non-self-consistent approximations; even though these post-Hartree-Fock steps are not self-consistent, they are very expensive, especially for large and complex systems. In contrast, KSDFT includes dynamical correlation during the SCF step by writing the density functional as the sum of an exchange functional and a correlation functional, where the latter includes dynamical correlation. The field calculated from this density functional incorporates the energetic effect of electron correlation in an approximate way. (It is approximate because we do not know the exact exchange-correlation potential for real systems.) The exchange functional includes electron exchange, which lowers the energy of the electronic subsystem by allowing electrons, which are indistinguishable, to exchange spin orbitals with one another. In contrast, exchange is included exactly in Hartree-Fock theory by obtaining a variationally correct energy for an antisymetrized wave function; this leads to a nonlocal field in Hartree-Fock theory, that is, the field depends not just on local values of the orbitals or density and their local derivatives, but on an integration over all space.

The approximate density functionals in the literature may be classified as local and nonlocal. For local functionals, the exchange-correlation part of the self-consistent field at a given point in space depends on the spin densities and possibly on their derivatives and on the spin-up and spin-down kinetic energy densities at that point in space (some functionals, although none of them discussed in this review, substitute the Laplacian of the spin density for the spin kinetic energy density); for nonlocal functionals, the exchange-correlation part of the self-consistent field at a given point in space involves a nonlocal integral over all space. The only widely studied (to date) method to include nonlocality is to incorporate Hartree-Fock exchange; functionals involving Hartree-Fock exchange are called hybrid. Recently some groups developed nonlocal correlation functionals based on the random-phase approximation (Furche 2008; Scuseria et al. 2008). In order of increasing complexity and accuracy, the three types of local functionals are the local spin density approximation (LSDA) (Kohn and Sham 1965), generalized gradient approximation (GGA) (Langreth and Mehl 1983; Becke 1988; Lee et al. 1988; Perdew et al. 1992, 1996; Zhao and Truhlar 2008b), and meta-GGAs (Becke 1996; Tao et al. 2003; Grüning et al. 2004; Zhao and Truhlar 2006b, 2008c). Nonlocal functionals include hybrid GGAs and hybrid meta functionals. Hybrid GGAs and hybrid meta functionals (both of which include nonlocal Hartree-Fock exchange) have better performance for general-purpose applications in chemistry than local functionals, but tend to be less accurate, all other considerations being equal, for systems with nearly degenerate configurations, such as a transition metal atom with nearly degenerate s^2d^{n-1} and sd^{n-1} configurations. Systems with nearly degenerate configurations are usually called multireference systems (Truhlar 2007).

Note that meta and hybrid terms in a density functional depend on occupied orbitals, not just on the density and its derivatives; the exchange-correlation energy is nevertheless still a functional of the density because formally the orbitals are functionals of the density. The most complete density functionals (such as the random phase ones mentioned above) also include terms that depend on unoccupied orbitals; such functionals are not discussed in detail in the present review.

One hybrid GGA, namely B3LYP (Becke 1988, 1993; Lee et al. 1988; Stephens et al. 1994), has become extraordinary popular (Sousa et al. 2007) in theoretical and computational chemistry. However, B3LYP and other popular functionals have unsatisfactory performance issues such as: 1) underestimation of barrier heights (Zhao et al. 2005a); 2) underestimation of interaction energies for weak noncovalent interactions (Zhao and Truhlar 2008e); and 3) underestimation of bond energies in transition metal compounds (Reiher et al. 2001; Schultz et al. 2005a,b; Harvey 2006). Some recent studies have shown that these shortcomings lead to large systematic errors in the prediction of heats of formation of organic molecules (Woodcock et al. 2002; Check and Gilbert 2005; Izgorodina et al. 2005; Grimme 2006; Schreiner et al. 2006; Wodrich et al. 2006; Zhao and Truhlar 2006a; Wodrich et al. 2007) and incorrect trends in the bond energies of organometallic catalytic systems (Tsipis et al. 2005; Zhao and Truhlar 2007b).

Since 2001, the Minnesota theoretical chemistry group has done extensive work in designing and optimizing more accurate density functionals, and these efforts resulted in several new functionals; eight of those developed in the 2005-2008 time frame are called the Minnesota density functionals (with acronyms beginning with M05, M06, or M08), and the full set of functionals developed at the University of Minnesota may be called University of Minnesota density functionals. In the next section, we review the functionals developed at the University of Minnesota, including some of their validations, and in the rest of the article, we review selected recent applications of University of Minnesota functionals in problems related to mineralogy and geochemistry. For applications of other density functionals and/ or applications in other areas, we recommend some recent reviews (Scuseria and Staroverov 2005; Cramer and Truhlar 2009) as well as a nontechnical introduction to five of the first six Minnesota functionals (Zhao and Truhlar 2008c).

UNIVERSITY OF MINNESOTA FUNCTIONALS

Table 1 lists, in chronological order, the functionals that have been developed by the Minnesota theoretical chemistry group. Although the doubly hybrid functionals (Zhao et al. 2004b) and multicoefficient density functional methods (Zhao et al. 2005b) developed by us can be viewed as generalized density functionals, we did not include them in Table 1 because the energy functional includes unoccupied orbitals, and we have restricted our scope to not discuss such work in detail in the present review.

MPW1K (Lynch et al. 2000) and BB1K (Zhao et al. 2004a) were optimized against a database of barrier heights by using the adiabatic connection method, and both of them were designed for kinetics. In 2004, we also developed MPWB1K and MPW1B95 (Zhao and Truhlar 2004); MPWB1K was designed for kinetics, but it has been shown to have improved performance for noncovalent interactions, and MPW1B95 was designed for main-group thermochemisty. TPSS1KCIS is a byproduct of our work in multicoefficient density functional methods (Zhao et al. 2005b). MPW1KCIS and MPWKCIS1K are byproducts of our nonhydrogen transfer barrier height database work (Zhao et al. 2005a), whereas PBE1KCIS is a byproduct of our noncovalent database work (Zhao and Truhlar 2005a). In 2005, Dahlke and Truhlar (2005) developed three functionals for describing energetics in water clusters, namely PBE1W, PBELYP1W, and TPSSLYP1W.

Also in 2005, PWB6K and PW6B95 (Zhao and Truhlar 2005b) were developed by reoptimizing six parameters in the MPWB1K functional form; PWB6K was shown to have greatly improved performance for noncovalent interactions (Zhao and Truhlar 2005c,d; Zhao et al. 2005d). However, from two benchmark studies in transition metal chemistry (Schultz et al. 2005a,b), we found a dilemma for the performance of density functionals, that is, to obtain more accurate barrier heights, one needed to mix in a high percentage of Hartree-

Table 1. Minnesota functionals.

Functional	Type[a]	Year	Reference
MPW1K	HG	2000	(Lynch et al. 2000)
BB1K	HM	2004	(Zhao et al. 2004a)
MPWB1K	HM	2004	(Zhao and Truhlar 2004)
MPW1B95	HM	2004	(Zhao and Truhlar 2004)
MPW3LYP	HG	2005	(Zhao and Truhlar 2004)
TPSS1KCIS	HM	2005	(Zhao et al. 2005b)
MPW1KCIS	HM	2005	(Zhao et al. 2005a)
MPWKCIS1K	HM	2005	(Zhao et al. 2005a)
PBE1KCIS	HM	2005	(Zhao and Truhlar 2005b)
PW6B95	HM	2005	(Zhao and Truhlar 2005b)
PWB6K	HM	2005	(Zhao and Truhlar 2005b)
MPWLYP1M	HG	2005	(Schultz et al. 2005b)
MOHLYP	G	2005	(Schultz et al. 2005b)
MPWLYP1W	G	2005	(Dahlke and Truhlar 2005)
PBE1W	G	2005	(Dahlke and Truhlar 2005)
PBELYP1W	G	2005	(Dahlke and Truhlar 2005)
TPSSLYP1W	M	2005	(Dahlke and Truhlar 2005)
M05	HM	2005	(Zhao et al. 2005c)
M05-2X	HM	2006	(Zhao et al. 2006)
M06-L	HM	2006	(Zhao and Truhlar 2006b)
M06-HF	HM	2006	(Zhao and Truhlar 2006g)
M06	HM	2008	(Zhao and Truhlar 2008e)
M06-2X	HM	2008	(Zhao and Truhlar 2008e)
SOGGA	G	2008	(Zhao and Truhlar 2008b)
M08-HX	HM	2008	(Zhao and Truhlar 2008d)
M08-SO	HM	2008	(Zhao and Truhlar 2008d)

[a]G: GGA; M: meta-GGA; HG: hybrid GGA; HM: hybrid meta GGA

Fock exchange, whereas transition metal chemistry favors low percentages of Hartree-Fock exchange; MOHLYP and MPWLYP1M are two byproducts of these benchmark studies.

In order to develop a functional that can handle barrier heights, noncovalent interactions, and transition metal chemistry, we developed the M05 functional (Zhao et al. 2005c, 2006) and it has good performance for transition metal chemistry (Zhao and Truhlar 2006e) as well as main-group thermochemistry, barrier heights, and noncovalent interactions. With the same functional form as M05, we developed the M05-2X functional (Zhao et al. 2006) that focuses on main-group chemistry, barrier heights, and noncovalent interactions (Zhao and Truhlar 2006a,c,d,f, 2007b,c).

In 2006, building on all this experience, we have developed four new functionals by combining the functional forms of the M05 (Zhao et al. 2005c, 2006) and VSXC (Voorhis and Scuseria 1998) functionals; they are called the M06-class functionals: (a) M06, a hybrid meta functional, is a functional with good accuracy "across-the-board" for transition metals, main group thermochemistry, medium-range correlation energy, and barrier heights. (b) M06-2X, another hybrid meta functional, is not good for transition metals but has excellent performance for main group chemistry, predicts accurate valence and Rydberg electronic excitation energies, and is an excellent functional for chemical reaction barrier heights and aromatic-aromatic stacking interactions. (c) M06-L is not as accurate as M06 or M06-2X for barrier heights but is the most accurate functional for the energetics of systems containing transition metals and is the only local functional (no Hartree-Fock exchange) with better across-the-board average performance than B3LYP; this is very important because only local functionals are affordable for many demanding applications on very large systems. (d) M06-HF has good performance for valence, Rydberg, and charge transfer excited states with minimal sacrifice of ground-state accuracy. In a recent review (Zhao and Truhlar 2008c), we compared the performance of the M06-class functionals and one M05-class functional to that of some popular functionals for diverse databases and for some difficult cases. For most purposes, the M05 and M06-2X functionals may be considered to simply be earlier versions of the M06 and M06-2X functionals.

In 2008, we explored the limit of accuracy attainable by a global hybrid meta density functional for main-group thermochemistry, kinetics, and noncovalent interactions by using a very flexible functional form called M08 (Zhao and Truhlar 2008d). M08-HX and M08-SO were developed in that exploratory study; they improve on M06-2X, but only a little (Zhao and Truhlar 2008d), and they have excellent performance for barrier heights (Zheng et al. 2009).

In 2008, we also developed a nonempirical GGA, which is exact through second order in a gradient expansion, that is, in terms of an expansion in powers of the gradient of the density, and it is called the second-order generalized gradient approximation (SOGGA). The SOGGA functional differs from other GGAs in that it enforces a tighter Lieb-Oxford bound on the density functional (Zhao and Truhlar 2008b). SOGGA and other functionals have been compared to a diverse set of lattice constants, bond distances, and energetic quantities for solids and molecules.

In chemistry, the quality of a density functional is usually judged on the basis of its predictions for energetic quantities such as thermochemistry and chemical reaction barrier heights, with less emphasis on geometric predictions, which are in some sense easier to predict at least qualitatively correctly. In solid-state chemistry and physics, though, there is often a great emphasis on predicting precise geometrical parameters, such as lattice constants of solids. In many cases, good geometric predictions for molecules (bond lengths, bond angles, etc.) are correlated with good predictions for lattice constants. One somewhat exceptional case is M06-L, which has exceptionally good performance for predicting molecular geometries (Zhao and Truhlar 2008d), but only so-so quality for predicting lattice constants (Zhao and Truhlar 2008b); however the SOGGA functional, which is not as good as M06-L for broad applications, has excellent performance for lattice constants (Zhao and Truhlar 2008b).

In the next section, we review some recent validations and applications of the Minnesota functionals for problems related to mineralogy and geochemistry. We recommend consulting the original references in Table 1 for the general performance of the Minnesota functionals in other areas of chemistry.

The orbital solutions of the SCF equations are usually determined variationally as linear combinations of basis functions. For molecular calculations, the basis functions are usually Gaussians. For calculations on condensed-phase systems (solids, liquids, gas-solid surfaces), the orbitals are sometimes expanded in plane waves. In an ideal world the basis sets would be complete, but in practice they are often truncated, which can contribute some error if the truncation is too severe. Standard truncated sets of Gaussian functions are available, and some examples mentioned in this review are (in order of increasing completeness) 6-31G(d,p), 6-31+G(d,p), TZVP(P), 6-311++(2d,2p), 6-311+G(2df,2p), and aug-cc-pVTZ; we refer the reader to the original publications for further details of these basis sets. When specifying a KSDFT calculation using Gaussian basis functions, a common convention (used in some places in this article) is to label the method as *F/B*, where *F* is the density functional, and *B* is the basis set.

VALIDATIONS AND APPLICATIONS

Water and aqueous chemistry

The chemistry of ground water is one of the most important subjects in geoscience. It is important to benchmark the quality of density functionals for the description of interactions between water molecules, and one of the most definitive ways to do this is to validate them against accurate WFT calculations for small water clusters for which accurate WFT calculations are possible. In 2008, Dahlke et al. (2008b) assessed the performance of 7 density functionals for reaction energies in hydronium, hydroxide, and pure water clusters, and they found that the M06-L functional is very promising for condensed-phase simulations of the transport of hydronium and hydroxide ions in aqueous solution. In another paper, Dahlke et al. (2008a) assessed the accuracy of 11 density functionals for prediction of relative energies and geometries of low-lying isomers of water hexamers, and their calculations show that only three density functionals, M06-L, M05-2X, and M06-2X, are able to correctly predict the relative energy ordering of the hexamers when single-point energy calculations are carried out on geometries obtained with second-order Møller-Plesset perturbation theory (MP2). Of the tested 11 density functionals, the most accurate density functionals for relative energies in water hexamer are PWB6K, MPWB1K, and M05-2X.

More recently, Bryantsev et al. (2009) evaluated the accuracy of the B3LYP, X3LYP, M06-L, M06-2X, and M06 functionals to predict the binding energies of neutral and charged water clusters. They ranked the accuracy of the functionals on the basis of the mean unsigned error (MUE) between calculated benchmark and density functional theory energies, and they found that M06-L (MUE = 0.73 kcal/mol) and M06 (MUE = 0.84) give the most accurate binding energies using very extended basis sets such as aug-cc-pV5Z. For more affordable basis sets, the best methods for predicting the binding energies of water clusters are M06-L/aug-cc-pVTZ (MUE = 1.24 kcal/mol), B3LYP/6-311++G(2d,2p) (MUE = 1.29 kcal/mol), and M06/aug-cc-pVTZ (MUE = 1.33 kcal/mol).

Austin et al. (2009) recently benchmarked several functionals for actinyl complexes and found that the M06 functional is competitive with high-level CCSD(T) methods in the study of the water exchange mechanism of the $[UO_2(OH_2)_5]^{2+}$ ion and of the redox potential of the aqua complexes of $[AnO_2]^{2+}$ (An = U, Np, and Pu).

In order to learn about reactions between oxide minerals and aqueous solutions, Qian et al. (2009) employed the MPWKCIS1K and B3LYP functionals to study the water exchange mechanism of the polyoxocation $GaO_4Al_{12}(OH)_{24}(H_2O)_{12}^{7+}$ in aqueous solution. They found that the reactant, modeled as the polyoxocation already specified plus 15 second-solvation-shell water molecules plus a continuum solvent model for the rest of the aqueous solvent, first loses a water ligand to form an intermediate with a five-coordinated aluminum atom. After that, the incoming water molecule in the second coordination shell attacks the intermediate with a five-coordinated aluminum atom to produce the product. The calculations imply, as should be expected, that both the explicit water molecules in the second hydration sphere and the bulk solvent water have a significant effect on the energy barriers; the need to carry out calculations on a large model is one of the reasons why the relatively inexpensive KSDFT is a preferred method for this kind of calculation.

Tommaso and Leeuw (2009) recently employed the MPW1B95 functional to study the dimerization of calcium carbonate in aqueous solution under natural water conditions. Their calculations suggest that, at T = 298 K and neutral pH conditions, the oligomerization of calcium carbonate is not spontaneous in water. This is an indication that the nucleation of calcium carbonate may not occur through a homogeneous process when calcium-bicarbonate ion pairs are the major source of $CaCO_3$ in the aqueous environment.

Goumans et al. (2009b) employed MPWB1K to study the formation of water on a model dust grain (fosterite) by using an embedded cluster approach. They found that the formation of water on a bare dust grain from hydrogen and oxygen atoms is catalyzed by an olivine surface by stabilizing the reaction intermediates and product.

Many studies of aqueous reactions are not directly relevant to geochemistry but show how complex processes in aqueous solution are becoming more amenable to study. For example, Wu et al. employed BB1K to study the mechanism of aqueous-phase asymmetric transfer hydrogenation of a ketone, in particular acetophenone, with HCOONa catalyzed by ruthenium *N*-(*p*-toluenesulfonyl)-1,2-diphenylethylenediamine (Wu et al. 2008). The reaction has a nonconcerted hydrogen transfer as the rate-limiting step in water, as compared to a rate-limiting step of concerted proton transfer and hydride formation/transfer in isopropanol solvent. The calculations revealed that in aqueous solution water hydrogen bonds to the ketone oxygen at the transition state for hydrogen transfer, lowering the energy barrier by about 4 kcal/mol.

Oxidation-reduction (redox) processes (Winget et al. 2000, 2004; Patterson et al. 2001; Belzile et al. 2004; Lewis et al. 2004, 2007; Anbar et al. 2005; Southam and Saunders 2005; MacQuarrie and Mayer 2005; Charlet et al. 2007; Savage et al. 2008; Helz et al. 2008) are very important in geochemistry. Oxidation and reduction of metal ions in aqueous solution are a particularly important example. Jaque et al. (2007) calculated the standard reduction potential of the $Ru3^{+}|Ru^{2+}$ couple in aqueous solution with 37 density functionals, including BB1K, MPW1B95, MPW1KCIS, MPWKCIS1K, PBE1KCIS, MPWB1K, and TPSS1KCIS, and five basis sets. Either one solvent shell (six water molecules) or two solvent shells (18 water molecules) were treated explicitly, and the rest of the solvent was treated as a dielectric continuum. The calculated difference in solvation free energies of Ru^{3+} and Ru^{2+} varies from −10.56 to −10.99 eV for six waters and from −6.83 to −7.45 eV for 18. The aqueous standard reduction potential is overestimated when only the first solvation shell is treated explicitly, and it is underestimated when the first and second solvation shells are treated explicitly. This study shows the danger of simply choosing a single methodology without examining the dependence of the results on the choice of functional and model.

Atmospheric chemistry

Another important subject in geochemistry is atmospheric chemistry, and the Minnesota functionals have also been applied in this area (Lin et al. 2005; Eillngson et al. 2007; Ellingson and Truhlar 2007).

Isotopic analysis is an important technique for understanding the production, transport, and depletion of various species in the atmosphere (McCarthy et al. 2001, 2003; Keppler et al. 2005), and the fractionation of ^{12}C and ^{13}C in reactions of methane with OH and Cl are important parameters in this analysis. We (Lin et al. 2005; Ellingson et al. 2007) calculated the kinetic isotope effect and its temperature dependence for the reactions of $^{12}CH_4$ and $^{13}CH_4$ with OH using the BB1K, MPW1K, and M06-2X density functionals as well as three high-level multicoefficient methods, including two based in part on density functionals using unoccupied orbitals. Later experimental work led to the conclusion that the tunneling effect is underestimated in these calculations (Sellevåg et al. 2006).

The reaction of OH with H_2S plays an important role in regulating the amount of H_2S in the atmosphere, and therefore its reaction rate is an important input in atmospheric modeling (Perry et al. 1976). This reaction apparently has an unusual temperature dependence, with a negative or very small temperature dependence in at least part of the range of atmospheric temperatures (~200-300 K) and a positive temperature dependence at higher temperatures. A negative temperature dependence is often indicative of a reaction for which the dynamical bottleneck lies in a region where the potential energy of interaction of the reactants is attractive. Thus, in order to model the temperature dependence of this reaction reliably over the whole range, one requires a density functional that is accurate for both reaction barriers and attractive forces. Calculations based on the M06-2X density functional, which is notable for being more accurate than previous density functionals for both of these features, reproduced the unusual temperature dependence found experimentally, whereas calculations based on older density functionals did not (Ellingson and Truhlar 2007).

Recently, Vega-Rodriguez and Alvarez-Idaboy (2009) employed the MPWB1K and M05-2X density functionals to study the mechanism for the reactions of OH with unsaturated aldehydes, which are of importance for atmospheric chemistry. They found that the calculated rate constants with M05-2X are in good agreement with experiments.

Metal oxides

Most metals found in the Earth's crust exist as silicate, oxide, and/or sulfide minerals. Metal oxide materials are particularly interesting for their technological usefulness in areas such as catalysis, photovoltaics, and electronics.

Recently, Sorkin et al. (2008) reported a systematic study of small iron compounds including FeO^-. They found that some of the improvements that were afforded by the semiempirical +U correction (widely employed in condensed-matter physics) could also be accomplished by improving the quality of the exchange-correlation functional.

Benchmark databases for Zn-containing molecules (including ZnO, Zn_2O_2, Zn_3O_3, and Zn_4O_4) have been recently developed by Amin and Truhlar (2008) and Sorkin et al. (2008), including Zn-ligand bond distances, angles, dipole moments, and bond dissociation energies. They tested many density functionals against these benchmark databases, and they found that the best KSDFT method to reproduce dipole moments and dissociation energies of the benchmark Zn compound database is M05-2X.

M05-2X has been employed to investigate the mechanism of the photocatalytic degradation of 1,5-naphthalenedisulfonate (a surfactant) on colloidal TiO_2 (Szabo-Bardos et al. 2008). The degradation is initiated by oxidative attack of an OH radical to make an aromatic-OH adduct. M05-2X calculations allowed a comparative analysis of possible sites of attack by the radical.

The adsorption of CO on the (001) surface of MgO is a challenge for popular density functionals because of its weak interaction character. Valero et al. show that M06-2X and M06-HF are the first two functionals to provide a simultaneously satisfactory description of adsorbate geometry, vibrational frequency shift, and adsorption energy of CO on the (001) surface of MgO (Valero et al. 2008).

Silicates and siliceous minerals

Silicate minerals and silica make up the largest part of the crust of the Earth. The structure, protonation, and reactivity of silicates and compounds containing silicon-oxygen bonds have been widely studied by quantum mechanics (Shambayati et al. 1990; Blake and Jorgensen 1991; Tossell and Sahai 2000; Avramov et al. 2005; Vuilleumier et al. 2009). Zhang et al. (2007) investigated the computational requirements for simulating the structures and proton activity of silicaceous materials. In their study, 14 density functionals in combination with 8 basis sets were tested against high-level wave-function-based methods. They found that the most accurate density functional for both geometries and energetics is M05-2X.

Zwijnenburg et al. (2008) compared the performance of B3LYP and BB1K for the prediction of optical excitations of defects in nanoscale silica clusters. They found that the spatially localized excitations are well described by time-dependent KSDFT, but B3LYP gives excitation energies that are significantly underestimated in the case of the charge-transfer excitations. In contrast, the time-dependent KSDFT calculations with BB1K was found to give generally good excitation energies for the lowest excited states of both localized and charge-transfer excitations.

Goumans et al. (2008) employed MPWB1K to investigate the catalytic effect of a negatively charged silica surface site on addition reactions to an adsorbed unsaturated organic molecule using an embedded cluster approach. Their calculations show that a negatively charged defect on a silica surface (silanolate) effectively catalyzes the addition of H or O atom to the HCCH, $H_2C{=}CH_2$, or CO molecule adsorbed on it. The negative charge polarizes and destabilizes the multiple bonds, which in turn lowers the barriers to addition. These results indicate that complex interstellar molecules could be formed effectively via surface-catalyzed hydrogenation and/or oxidation routes.

The mineral olivine is a magnesium iron silicate with the formula $(Mg,Fe)_2SiO_4$, and it is one of the most common minerals on Earth. Recently MPWB1K has been employed to study the formation of H_2 on an olivine surface (Goumans et al. 2009a), and the results show that the forsterite surface catalyzes H_2 formation by providing chemisorption sites for H atoms. The calculations also indicate that pristine olivine surfaces should be good catalysts for H_2 formation, with low product excitation and high reaction efficiencies.

Morrow et al. (2009) used M05-2X to investigate the dissolution mechanism in aluminosilicate minerals. Their calculations show that Al species from protonated and neutral Al-O-Si sites can leach before Si species. Nangia and Garrison (2009) used the M05-2X density functional to study hydrolysis/precipitation reactions at silica surface sites. The barrier height data successfully explained the experimentally observed dissolution rate over the entire pH range.

Clay minerals such as montmorillonite are of interest as environmentally friendly catalysts supports and also possibly as catalysts on their own (Zhu et al. 2009; Briones-Jurado and Agacino-Valdés 2009).

Zeolites

Zeolites are microporous/nanoporous aluminosilicate minerals, and the zeolitic pores are nanocavities. Zeolites have the capability to catalyze chemical reactions that take place in their internal cages, and they provide a size-selective environment for chemical reactions. Theorefore zeolites are widely used in industry as heterogenenous catalysts. Indeed "every drop of gasoline we burn in our car has seen at least one zeolite catalyst on its way through the refinery" (Sauer 2006). Although industrial zeolites needed for specialized applications are synthesized to obtain the required purity and uniformity and sometimes to obtain unnatural frameworks, many naturally occuring zeolites are known, and zeolite-rich formations have both diagenetic and volcanic origins.

Practical applications of quantum mechanical calculations to reactions in extended systems like zeolites can be carried out by using the ONIOM method or the QM/MM method. In ONIOM calculations (Morokuma 2002), one treats a reaction site (subsystem) by a quantum mechanical method, and the rest of the system (the surrounding atoms and molecules) by a less expensive quantum mechanical method or an analytic potential energy function; when one treats the surroundings by molecular mechanics (that is, an analytical potential energy function—such as the UFF force field (Rappé et al. 1992)—that is parameterized for nonreactive systems or nonreactive subsystems (Hagler and Ewig 1994; Cygan 2001)), this is usually called a combined quantum mechanical and molecular mechanical (QM/MM) calculation (Lin and Truhlar 2007). An ONIOM or QM/MM calculation is usually denoted $A{:}B$, where A denotes the more expensive method used for the reactive subsystem, and B denotes the less expensive method used for the surroundings. In QM/MM calculations, A denotes the QM part, and B denotes the MM part.

BB1K was employed by To et al. (2006b) to investigate the mechanism of the formation of heteroatom active sites in zeolites, and the calculations show that the tetravalent-heteroatom sites preferentially adopt tripodal configurations, whereas the hydrolysis of T-O-Si bridges (T is a heteroatom such as Ti or Sn) helps to relieve lattice strain and to stabilize the structure. Thus hydrolysis and inversion play important roles in the stabilization of heteroatom-based active sites in zeolites. In a subsequent paper, To et al. (2006a) reported QM/MM computations with BB1K for the QM active site in Ti-substituted zeolites (titanium silicalite-1) to study the processes of hydrolysis of Ti-O-Si linkages and inversion of the TiO_4 tetrahedra. The calculated structural features of the tetra- and tripodal Ti moieties are in good agreement with experiments, and suggest that the tripodal species dominate in hydrous conditions, and that this is likely to be the chemically active form. Later, To et al. (2007) employed the BB1K functional to study the interaction of water molecules with Ti sites in titanium silicalite-1, and they found that the hydrations of all Ti centers are exothermic, a result that is in good agreement with experiment and previous theoretical work. In 2008, they (To et al. 2008) employed BB1K to find the active oxidizing species in the H_2O_2/titanium silicalite-1 catalytic system, and their computational results indicate that water is not just a medium for transporting reactants and products; rather it has an active role in stabilizing the peroxo species present on the catalyst.

In 2008, we developed a benchmark database for interactions in zeolite model complexes based on CCSD(T) calculations, and we tested 41 density functionals against the new database (Zhao and Truhlar 2008a). Among the tested density functional methods, M06-L/6-31+G(d,p) gives a mean unsigned error (MUE) without counterpoise correction of 0.87 kcal/mol. With counterpoise corrections, the M06-2X (Zhao and Truhlar 2008e) and M05-2X (Zhao et al. 2006) functionals give the best performance. We also tested 10 functionals against the binding energies of four complexes (two noncovalent and two covalent) of the adsorption of isobutene on a large 16T zeolite model cluster (Fig. 1). The counterpoise corrected binding energies are shown in Table 2. For comparison Table 2 also shows WFT results obtained by treating dynamical correlation by second order perturbation theory; those results are labeled MP2.

Table 2 shows that M05-2X and M06-2X are the best performers for the binding energies in a model 16T zeolite cluster, followed by M06-L and M06, and these four functionals give a smaller MUE than the MP2/TZVP(P) method. The popular B3LYP functional performs poorly with an MUE of 16.6 kcal/mol. These results show that M06-L, M06, M05-2X, and M06-2X are very promising quantum mechanical methods for the QM part of QM/MM simulations of zeolite. This conclusion has been confirmed recently by Maihom et al. (2009). They investigated the mechanisms of ethene methylation with methanol and dimethyl ether in a 128T cluster of ZSM-5 zeolite using the ONIOM(B3LYP/6-31G(d,p):UFF) and ONIOM(M06-2X/6-311+G(2df,2p):UFF) methods with the zeolitic Madelung potential generated by the surface charge representation of the electrostatic embedding potential (SCREEP) method (Vollmer et al. 1999). Their calculations show that the energies for the adsorption of methanol and dimethyl ether on H-ZSM-5 from an ONIOM2(M06-2X/6-311+G(2df,2p):UFF)+SCREEP calculation are in good agreement with the experimental data.

Boekfa et al. (2009) employed four ONIOM methods, namely (MP2:M06-2X), (MP2:B3LYP), (MP2:HF), and (MP2:UFF), to investigate the confinement effect on the adsorption and reaction mechanism of unsaturated aliphatic, aromatic and heterocyclic compounds on H-ZSM-5 zeolite. They found that (MP2:M06-2X) give the best agreement with experiments for the energies of adsorption of ethene, benzene, ethylbenzene, and pyridine on H-ZSM-5.

Kumsapaya et al. (2009) employed the M06-L/6-31G(d,p):UFF ONIOM method to investigate the isomerization of 1,5- to 2,6-dimethylnaphthalene over acidic β zeolite. The results in their study show the excellent performance of a combination of the M06-L functional with the confinement effect represented by the universal force field for investigating the transformations of aromatic species zeolite systems.

Hydroxyl nests are formed in zeolites, especially the all-silicon silacalite-1 (Astorino et al. 1995; Kalipcilar and Culfaz 2000), at a defect involving a missing silicon or aluminum

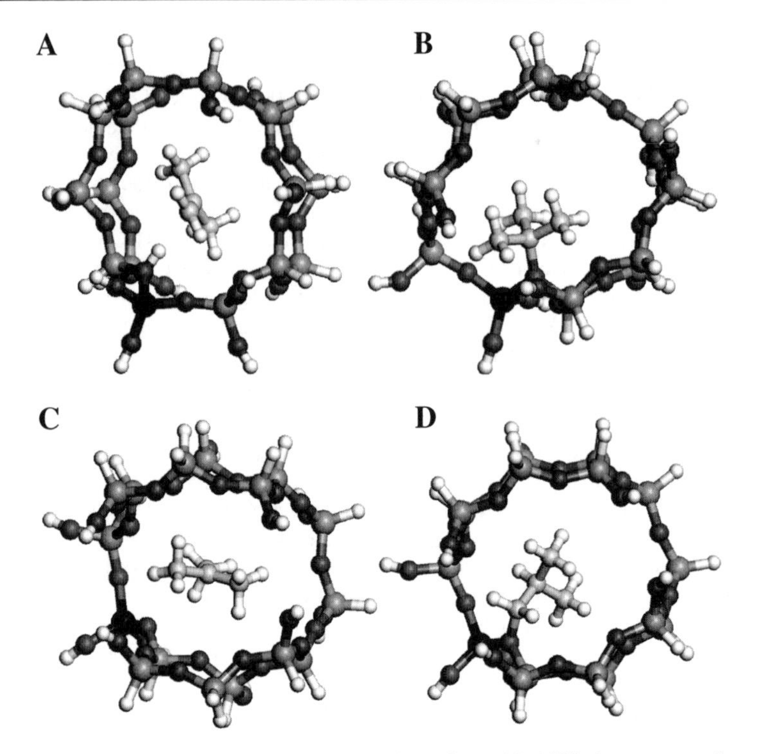

Figure 1. Geometries of the 16T zeolite model complexes. (A) 16T·isobutene π complex, (B) 16T *tert*-butoxide, (C) 16T *tert*-butyl carbenium ion, (D) 16T isobutoxide.

atom; the nest consists of four hydrogen-bonded hydroxyl groups. BB1K has been recently employed to study high-temperature dehydrogenation of defective silicalites (Fickel et al. 2010). The combined experimental and computational study indicates that the product of the dehydroxylation of hydroxyl nests in silicalite-1 is not water but hydrogen.

To et al. (2005) employed BB1K to investigate the ground state and the electronically excited states of the $[AlO_4]^0$ hole in α-quartz and in the siliceous zeolite ZSM-5. Their results show that BB1K is significantly more accurate and reliable than previous popular density functionals, and the BB1K excitation energies show excellent agreement with experiment.

Mineral nanoparticles

Many mineral accumulations are only poorly crystalline or exist as nanoparticles with properties quite different than those of microcrystals or macroscrtystals; transmission electron microscopy can be used to probe their nanomorphology (Penn et al. 2001; Hochella et al. 2008; Isaacson et al. 2009). Computational methods based on electronic structure theory are well suited for studying the structure, dynamics, and phase properties of small particles, such as metal and metal oxide clusters and nanoparticles, as shown for example in recent work on aluminum nanoparticles (Li et al. 2007; Li and Truhlar 2008a,b).

In recent work, the SOGGA, M06, and M06-L density functionals have been found to be uniquely successful for predicting the structures of small anionic gold clusters (Mantina et al. 2009), and in other recent work M06-L was found to be uniquely successful in predicting the structures of small cationic gold clusters (Ferrighi et al. 2009) (M06 and SOGGA were not included in the latter study).

Table 2. Binding energies (kcal/mol) in four complexes involving the adsorption of isobutene on a 16T zeolite cluster model.[a]

Method[b]	π complex	tert-butoxide	tert-butyl carbenium ion	isotutoxide	MUE[c]
Best estimate[d]	15.1	13.9	−9.8	13.9	
M05-2X	11.7	14.9	−8.1	14.0	1.6
M06-2X	12.7	16.6	−9.1	15.6	1.9
M06-L	14.4	15.6	−2.3	13.5	2.6
M06	13.3	16.0	−3.9	14.4	2.6
M06-HF	12.4	18.6	−12.5	18.8	3.7
MP2/TZVP(P)	10.8	10.0	−13.4	9.1	4.1
PBE0	2.9	4.7	−15.6	4.7	9.1
PBE	3.2	2.3	−12.8	2.0	9.6
B97-1	3.9	2.2	−14.3	2.9	9.6
TPSSh	−0.9	1.8	−17.7	1.6	12.1
B3LYP	−2.5	−5.5	−20.7	−4.8	16.6

[a]See Figure 1 for the 16T model cluster and four complexes. All results are taken from a previous paper (Zhao and Truhlar 2008a).

[b]The 6-311+G(2df,2p) basis set is employed for all density functionals in this table; the basis set for the MP2 calculations in indicated after /.

[c]MUE denotes mean unsigned error (same as mean absolute deviation, MAD).

[d]Best estimates were calculated from the two last rows in Table 3 of Tuma and Sauer (2006). The geometries, and MP2/TZVP(P) results are also from Tuma and Sauer (2006). Other results in this table are from the present work.

CONCLUDING REMARKS

In this chapter, we have presented an introduction to the Minnesota density functionals. We reviewed some encouraging validations and applications of the Minnesota functionals in problems important for mineralogy and geochemistry. Since the new-generation Minnesota functionals have been designed to give broad accuracy in chemistry, they perform very well even for many difficult cases where popular functionals fail badly (Zhao and Truhlar 2008c). The prospects for successful further application to a number of problems in mineralogy and geochemistry are encouraging.

ACKNOWLEDGMENTS

The authors are grateful to Shikha Nangia for comments on the manuscript. The work at Minnesota was supported in part by the Air Force Office of Scientific Research (orbital-dependent density functionals) and by the National Science Foundation under grant no. CHE07-04974 (complex systems).

REFERENCES

Amin EA, Truhlar DG (2008) Zn coordination chemistry: Development of benchmark suites for geometries, dipole moments, and bond dissociation energies and their use to test and validate density functionals and molecular orbital theory. J Chem Theory Comput 4:75-85

Anbar AD, Jarzecki AA, Spiro TG (2005) Theoretical investigation of iron isotope fractionation between $Fe(H_2O)_6^{3+}$ and $Fe(H_2O)_6^{2+}$: Implications for iron stable isotope geochemistry. Geochim Cosmochim Acta 69:825-837

Andino JM, Smith JN, Flagan RC, Goodard WA III, Seinfeld JH (1996) Mechanism of atmospherics photooxidation of aromatics. A theory study. J Phys Chem 100:10967-10980

Astorino E, Peri JB, Willey RJ, Busca G (1995) Spectroscopic Characterization of Silicalite-1 and titanium silicalite-1. J Catal 157:482-500

Austin JP, Burton NA, Hillier IH, Sundararajan M, Vincent MA (2009) Which density functional should be used to study actinyl complexes? Phys Chem Chem Phys 11:1143-1145

Avramov PV, Adamovic I, Ho KM, Wang CZ, Lu WC, Gordon MS (2005) Potential energy surfaces of Si_mO_n cluster formation and isomerizations. J Phys Chem A 109:6294-6302

Becke AD (1988) Density-functional exchange-energy approximation with correct asymptotic-behavior. Phys Rev A 38:3098-3100

Becke AD (1993) Density-functional thermochemistry. J Chem Phys 98:5648-5652

Becke AD (1996) Density-functional thermochemistry. IV. A new dynamic correlation functional and implications for exact-exchange mixing. J Chem Phys 104:1040-1046

Belonoshko AB, Rusengren A, Dong Q, Hultquist G, Leygraf C (2004) First-principles study of hydrogen diffusion in α-Al_2O_3 and liquid alumina. Phys Rev B 69:024302

Belzile N, Chen YN, Cal MF, Li Y (2004) A review on pyrrhotite oxidation. J Geochem Explor 84:65-67

Blake JF, Jorgensen SE (1991) Proton affinities and gas-phase basicities of alkyl and silyl ethers. J Org Chem 56:6052-6059

Blanchard M, Poitrasson F, Méheut M, Lazzeri M, Mauri F, Balon E (2009) Iron isotope fractionation between pyrite (FeS_2), hematite (Fe_2O_3) and siderite ($FeCO_3$): A first-principles density functional theory study. Geochim Cosmochim Acta 73:6565-6578

Boekfa B, Choomwattana S, Khongpracha P, Limtrakul J (2009) Effects of the zeolite framework on the adsorptions and hydrogen-exchange reactions of unsaturated aliphatic, aromatic, and heterocyclic compounds in ZSM-5 zeolite: A combination of perturbation theory (MP2) and a newly developed density functional theory (M06-2X) in ONIOM scheme. Langmuir 25:12990-12999

Born M, Huang K (1954) The Dynamic Theory of Crystal Lattices. Clarendon Press, Oxford

Briones-Jurado C, Agacino-Valdés E (2009) Brønsted sites on acid-treated montmorillonite: a theoretical study with probe molecules. J Phys Chem A 113:8994-9001

Brodholt JP, Refson K (2000) An ab initio study of hydrogen in forsterite and a possible mechanism for hydrolytic weakening. J Geophys Res 105:18977-18982

Brodholt JP, Vocadlo L (2006) Applications of density functional theory in the geosciences. MRS Bull 31:675-680

Bryantsev VS, Diallo MS, Duin ACTv, William A. Goddard I (2009) Evaluation of B3LYP, X3LYP, and M06-class density functionals for predicting the binding energies of neutral, protonated, and deprotonated water clusters. J Chem Theory Comput 5:1016-1026

Charlet L, Scheinost AC, Tournassat C, Greneche JM, Géhin A, Fernández-Martínez A, Couert S, Tisserand D, Brendle J (2007) Electron transfer at the mineral/water interface: Selenium reduction by ferrous iron sorbed on clay. Geochim Cosmochim Acta 71:5731-5749

Check CE, Gilbert TM (2005) Progressive systematic underestimation of reaction energies by the B3LYP model as the number of C-C bonds increases. J Org Chem 70:9828-9834

Churakov SV, Khisina NR, Urusov VS, Wirth R (2003) First-principles study of $(MgH_2SiO_4) \cdot n(Mg_2SiO_4)$ hydrous olivine structures. I. Crystal structure modelling of hydrous olivine Hy-2a $(MgH_2SiO_4) \cdot 3(Mg_2SiO_4)$. Phys Chem Minerals 30:1-11

Cramer CJ, Truhlar DG (2009) Density functional theory for transition metals and transition metal chemistry. Phys Chem Chem Phys 46:10757-10816

Criscenti LJ, Kubicki JD, Brantley SL (2006) Silicate glass and mineral dissolution: Calculated reaction paths and activation energies for hydrolysis of Q^3 Si by H_3O^+ using ab initio methods. J Phys Chem A 110:198-206

Cygan RT (2001) Molecular modeling in mineralogy and geochemistry. Rev Mineral Geochem 42:1-35

Dahlke EE, Truhlar DG (2005) Improved density functionals for water. J Phys Chem B 109:15677-15683

Dahlke EE, Olson RM, Leverentz HR, Truhlar DG (2008a) Assessment of the accuracy of density functionals for prediction of relative energies and geometries of low-lying isomers of water hexamers. J Phys Chem A 112:3976-3984

Dahlke EE, Orthmeyer MA, Truhlar DG (2008b) Assessment of multicoefficient correlation methods, second-order Møller-Plesset perturbation theory, and density functional theory for $H_3O^+(H_2O)_n$ (n = 1-5) and $OH^-(H_2O)_n$ (n = 1-4). J Phys Chem B 112:2372-2381

Ellingson BA, Pu J, Lin H, Zha Y, Truhlar DG (2007) Multi-coefficient Gaussian-3 calculation of the rate constant for the OH + CH_4 reaction and its $^{12}C/^{13}C$ kinetic isotope effect with emphasis on the effects of coordinate system and torsional treatment. J Phys Chem A 111:11706-11717

Ellingson BA, Truhlar DG (2007) Explanation of the unusual temperature dependence of the atmospherically important OH + $H_2S \rightarrow H_2O$ + SH reaction and prediction of the rate constant at combustion temperatures. J Am Chem Soc 129:12765-12771

Ferrighi L, Hammer B, Madsen GKH (2009) 2D–3D transition for cationic and anionic gold clusters: a kinetic energy density functional study. J Am Chem Soc 131:10605-10609
Fickel DW, Shough AM, Doren DJ, Lobo RF (2010) High-temperature dehydrogenation of defective silicalites. Microporous Mesoporous Mater 129:156-163, doi: 10.1016/j.micromeso.2009.09.011
Furche F (2008) Developing the random phase approximation into a practical post-Kohn-Sham correlation model. J Chem Phys 129:114105
Gao H, Wang Y, Wan, SQ, Liu JY, Sun CC (2009) Theoretical investigation of the hydrogen abstraction from $CF_3CH_2CF_3$ by OH radicals, F, and Cl atoms: A dual-level direct dynamics study. J Mol Struct THEOCHEM 913:107-116
Gillan MJ, Alfé D, Brodholt J, Vocadlo L, Price GD (2006) First-principles modeling of Earth and planetary materials at high pressures and temperatures. Reps Prog Phys 69:2365-2441
Goumans TPM, Catlow CRA, Brown WA (2008) Catalysis of addition reactions by a negatively charged silica surface site on a dust grain. J Phys Chem C 112:15419-15422
Goumans TPM, Catlow CRA, Brown WA (2009a) Formation of H_2 on an olivine surface: a computational study. Mon Not R Astron Soc 393:1403-1407
Goumans TPM, Catlow CRA, Brown WA, Kastnerz J, Sherwood P (2009b) An embedded cluster study of the formation of water on interstellar dust grains. Phys Chem Chem Phys 11:5431-5436
Grimme S (2006) Seemingly simple stereoelectronic effects in alkane isomers and the implications for Kohn-Sham density functional theory. Angew Chem Int Ed 45:4460-4464
Grüning M, Gritsenko O, Baerends EJ (2004) Improved description of chemical barriers with generalized gradient approximation (GGAs) and meta-GGAs. J Phys Chem A 108:4459-4469
Hagler AT, Ewig CS (1994) On the use of quantum energy surfaces in the derivation of molecular force fields. Comput Phys Commun 84:131-155
Harvey JN (2006) On the accuracy of DFT in transition metal chemistry. Annu Rep Prog Chem Sect C Phys Chem 102:203-226
Hegg DA, Baker MB (2009) Nucleation in the atmosphere. Rep Prog Phys 72:056801
Helz GR, Tossell JA (2008) Thermodynamic model for arsenic speciation in sulfidic waters: A novel use of ab initio computations. Geochim Cosmochim Acta 72:4457-4468
Hochella MF Jr, Lower SK, Maurice PA, Penn RL, Sahai N, Sparks DL, Twining BS (2008) Nanominerals, mineral nanoparticles, and earth systems. Science 319:1631-1635
Isaacson LS, Burton ED, Bush RT, Mitchell DRG, Johnston SG, Macdonald BCT, Sullivan LA, White I (2009) Iron(III) accumulations in inland saline waterways, Hunter Valley, Australia: Mineralogy, micromorphology and pore-water geochemistry. Appl Geochem 24:1825-1834
Izgorodina EI, Coote ML, Radom L (2005) Trends in R-X bond dissociation energies. J Phys Chem A 109:7558-7566
Jaque P, Marenich AV, Cramer CJ, Truhlar DG (2007) Computational electrochemistry: The aqueous $Ru^{3+} | Ru^{2+}$ reduction potential. J Phys Chem C 111:5783-5799
Kalipcilar H, Culfaz A (2000) Synthesis of submicron silicalite-1 crystals from clear solutions. Cryst Res Technol 35:933-942
Keppler F, Harper DBH, Röckmann T, Moore RM, Hamilton JTG (2005) New insight in the atmospheric chloromethane budget gained using stable carbon isotope ratios. Atmos Chem Phys 5:2403-2411
Kohn W, Sham LJ (1965) Self-consistent equations including exchange and correlation effects. Phys Rev 140:A1133-A1138
Kohn W, Becke AD, Parr RG (1996) Density functional theory of electronic structure. J Phys Chem 100:12974-12980
Kumsapaya C, Bobuatong K, Khongpracha P, Tantirungrotechai Y, Limtrakul J (2009) Mechanistic investigation on 1,5- to 2,6-dimethylnaphthalene isomerizationcatalyzed by acidic β zeolite: ONIOM study with an M06-L functional. J Phys Chem C 113:16128-16137
Kuwata RT, Dibble TS, Sliz E, Pedersen EB (2007) Computational studies of intramolecular hydrogen atom transfers in the β-hydroxyethylperoxy and β-hydroxyelhoxy radicals. J Phys Chem A 111:5032-5042
Kwon KD, Rfeson K, Sposito G (2009) On the role of Mn(IV) vacancies in the photoreductive dissolution of hexagonal birnessite. Geochim Cosmochim Acta 73:4142-4150
Langreth DC, Mehl MJ (1983) Beyond the local-density approximation in calculations of ground-state electronic-properties. Phys Rev B 28:1809-1834
Lay T, Heinz D, Ishii, M, Shim, SH, Tsuchiya J, Tsuchiya T, Wentzcovitch R, Yuen D (2005) Multidisciplinary impact of the deep mantle phase transition in perovskite structure. EOS 86:1-5
Lee C, Yang W, Parr RG (1988) Development of the Colle-Salvetti correlation-energy formula into a functional of the electron density. Phys Rev B 37:785-789
Lewis A, Bumpus JA, Truhlar DG, Cramer CJ (2004) Molecular modeling of environmentally important processes: reduction potentials. J Chem Educ 81:596-604
Lewis A, Bumpus JA, Truhlar DG, Cramer CJ (2007) Molecular modeling of environmentally important processes: reduction potentials – CORRECTION. J Chem Educ 84:934

Li L, Weidner DJ, Brodholt J, Alfé D, Price GD, Caracas R, Wentzcovitch R (2006) Elasticity of $CaSiO_3$ perovskite at high pressure and high temperature. Phys Earth Planet Inter 155:249-259

Li ZH, Jasper AW, Truhlar DG (2007) Structures, rugged energetic landscapes, and nanothermodynamics of Al*n* ($2 \leq n \leq 65$) particles. J Am Chem Soc 129:14899-14910

Li ZH, Truhlar DG (2008a) Cluster and nanoparticle condensation and evaporation reactions. Thermal rate constants and equilibrium constants of $Al_m + Al_{n-m} \rightarrow Al_n$ with $n = 2 - 60$ and $m = 1 - 8$. J Phys Chem C 112:11109-11121

Li ZH, Truhlar DG (2008b) Nanosolids, slushes, and nanoliquids. Characterization of nanophases in metal clusters and nanoparticles. J Am Chem Soc 130:12698-12711

Lin H, Truhlar DG (2007) QM/MM: What have we learned, where are we, and where do we go from here? Theor Chem Acc 117:185-199

Lin H, Zhao Y, Ellingson BA, Pu J, Truhlar DG (2005) Temperature dependence of carbon-13 kinetic isotope effects of importance to global climate change. J Am Chem Soc 127:2830-2831

Lynch BJ, Fast PL, Harris M, Truhlar DG (2000) Adiabatic connection for kinetics. J Phys Chem A 104:4811-4815

MacQuarrie KTB, Mayer KU (2005) Reactive transport modeling in fractured rock: A state-of-the-science review. Earth Sci Rev 72:189-227

Mantina M, Valero R, Truhlar DG (2009) Validation study of the ability of density functionals to predict the planar-to-three-dimensional structural transition in anionic gold clusters. J Chem Phys 131:064706

Maihom T, Boekfa B, Sirijaraensre J, Nanok T, Probst M, Limtrakul J (2009) Reaction mechanisms of the methylation of ethene with methanol and dimethyl ether over H-ZSM-5: An ONIOM study. J Phys Chem C 113:6654-6662

McCarthy MC, Connell P, Boering KA (2001) Isotopic fractionation of methane in the stratosphere and its effect on free tropospheric isotopic compositions. Geophys Res Lett 28:3657-3660

McCarthy MC, Boering KA, Rice AL, Tyler SC, Connell P, Atlas EJ (2003) Carbon and hydrogen isotopic compositions of stratospheric methane: 2. Two-dimensional model results and implications for kinetic isotope effects. Geophys Res D 108:4461

Mead CA (1988) The Born-Oppenheimer approximation in molecular quantum mechanics. *In:* Mathematical Frontiers in Computational Chemical Physics. Truhlar DG (ed) Springer, New York, p 1-17

Morokuma K (2002) New challenges in quantum chemistry: Quests for accurate calculations for large molecular systems. Philos Trans R Soc London Ser A 360:1149-1164

Morrow CP, Nangia S, Garrison BJ (2009) Ab initio investigation of dissolution mechanisms in aluminosilicate minerals. J Phys Chem A 113:1343-1352

Nangia S, Garrison BJ (2008) Reaction rates and dissolution mechanisms of quartz as a function of pH. J Phys Chem A 112:2027-2033

Nangia S, Garrison BJ (2009) Ab initio study of dissolution and precipitation reactions from the edge, kink, and terrace sites of quartz as a function of pH. Mol Phys 107:831-843

Neal AL, Rosso KM, Geesey GG, Gorby YA, Little BJ (2003) Surface structure effects on direct reduction of iron oxides by Shewanella oneidensis. Geochim Cosmochim Acta 67:4489-4503

Nobes RH, Akhmatskaya EV, Milman V, White JA, Winkler B, Pickard CJ (2000) An ab initio study of hydrogarnets. Am Mineral 85:1706-1715

Ockwig NW, Greathouse JA, Durkin JS, Cygan RT, Daemen LL, Nenoff TM (2009) Nanoconfined water in magnesium-rich 2:1 phyllosilicates. J Am Chem Soc 131:8155-8162

Oganov AR, Brodholt JP (2000) High-pressure phases in the Al_2SiO_5 system and the problem of aluminous phase in the Earth's lower mantle: Ab initio calculations. Phys Chem Minerals 27:430-439

Oganov AR, Price GD, Scandolo S (2005) Ab initio theory of planetary materials. Z Kristallogr 220:531-548

Patterson EV, Cramer CJ, Truhlar DG (2001) Reductive dechlorination of hexachloroethane in the environment. Mechanistic studies via computational electrochemistry. J Am Chem Soc 123:2025-2031

Penn RL, Zhu C, Xu H, Veblen DR (2001) Iron oxide coatings on sand grains from the Atlantic coastal plain: High-resolution transmission electron microscopy characterization. Geology 29:843-846

Perdew JP, Chevary JA, Vosko SH, Jackson KA, Pederson MR, Singh DJ (1992) Atoms, molecules, solids, and surfaces: Applications of the generalized gradient approximation for exchange and correlation. Phys Rev B 46:6671-6687

Perdew JP, Burke K, Ernzerhof M (1996) Generalized gradient approximation made simple. Phys Rev Lett 77:3865-3868

Perry RA, Atkinson R, Pitts JN Jr (1976) Rate constants for the reactions $OH + H_2S \rightarrow H_2O + SH$ and $OH + NH_3 \rightarrow H_2O + NH_2$ over the temperature range 297-427 K. J Chem Phys 64:3237-3239

Pöhlmann M, Benoit M, Kob W (2004) First-principles molecular-dynamics simulations of a hydrous silica melt: Structural properties and hydrogen diffusion mechanism. Phys Rev B 70:184709

Qian Z, Feng H, Zhang Z, Yang W, Wang M, Wang Y, Bi S (2009) Theoretical exploration of the water exchange mechanism of the polyoxocation $GaO_4Al_{12}(OH)_{24}(H_2O)_{12}^{7+}$ in aqueous solution. Geochim Cosmochim Acta 73:1588-1596

Qu, X, Zhang Q, Wang W (2006) Degradation mechanism of benzene by NO_3 radicals in the atmosphere: A DFT study. Chem Phys Lett 426:13-19

Rappé AK, Casewit CJ, Colwell KS, Goddard WA III, Skiff WM (1992) UFF, a full periodic table force field for molecular mechanics and molecular dynamics simulations. J Am Chem Soc 114:10024-10035

Reich M, Utsunomiya S, Kessler SE, Wang L, Ewing RC, Becker U (2006) Thermal behavior of metal nanoparticles in geologic materials. Geology 34:1033-1036

Reiher M, Salomon O, Hess BA (2001) Reparameterization of hybrid functionals based on energy differences of states of different multiplicity. Theo Chem Acc 107:48-55

Sakamura H, Tsuchiya T, Kawamura K, Otsuki K (2003) Large self-diffusion of water on brucite surface by ab initio potential energy surface and molecular dynamics simulations. Surf Sci 536:396-402

Sauer J (2006) Proton transfer in zeolites. *In:* Hydrogen-Transfer Reactions. Volume 2. Hynes JT, Klinman JP, Limbach HH, Scowen RL (eds) Wiley-VCH, Weinheim, p 685-707

Savage KS, Stefan D, Lehner SW (2008) Impurities and heterogeneity in pyrite: Influences on electrical properties and oxidation products. Appl Geochem 23:103-120

Schreiner PR, Fokin AA, Pascal Jr RA, de Meijere A (2006) Many density functional theory approaches fail to give reliable large hydrocarbon isomer energy differences. Org Lett 8:3635-3638

Schultz N, Zhao Y, Truhlar DG (2005a) Databases for transition element bonding. J Phys Chem A 109:4388-4403

Schultz N, Zhao Y, Truhlar DG (2005b) Density functional for inorganometallic and organometallic chemistry. J Phys Chem A 109:11127-11143

Schwegler E, Sharma M, Gygi F, Galli G (2008) Melting of ice under pressure. Proc Nat Acad Sci USA 105:14779-14783

Scuseria GE, Staroverov VN (2005) Progress in the development of exchange-correlation functionals. *In:* Theory and Application of Computational Chemistry: The First 40 Years. Dykstra CE, Frenking G, Kim KS, Scuseria GE (eds) Elsevier, Amsterdam, p 669-724

Scuseria GE, Henderson TM, Sorensen DC (2008) The ground state correlation energy of the random phase approximation from a ring coupled cluster doubles approach. J Chem Phys 129:1231101

Sellevåg SR, Nyman G, Nielsen CJ (2006) Study of the carbon-13 and deuterium kinetic isotope effects in the Cl and OH reactions of CH_4 and CH_3Cl. J Phys Chem A 110:141-152

Shambayati S, Blake JF, Wierschke SG, Jorgensen WL, Schreiber SL (1990) Structure and basicity of silyl ethers: A crystallographic and ab initio inquiry into the mode of silicon-oxygen interactions. J Am Chem Soc 112:697-703

Sorkin A, Iron MA, Truhlar DG (2008) Density functional theory in transition-metal chemistry: relative energies of low-lying states of iron compounds and the effect of spatial symmetry breaking. J Chem Theory Comput 4:307-315

Sousa SF, Fernandes PA, Ramos MJ (2007) General performance of density functionals J Phys Chem A 111:10439-10452

Southam G, Saunders JA (2005) The geomicrobiology of ore deposits. Econ Geol 100:1067-1084

Stephens PJ, Devlin FJ, Chabalowski CF, Frisch MJ (1994) Ab initio calculation of vibrational absorption and circular dichroism spectra using density functional force fields. J Phys Chem 98:11623-11627

Stixrude L, Peacor DR (2002) First-principles study of illite-smectite and implications for clay mineral systems. Nature 420:165-168

Strandh H, Pettersson LGM, Sjöberg L, Wahlgren U (1997) Quantum chemical studies of the effects on silicate mineral dissolution rates by adsorption of alkali metals. Geochim Cosmochim Acta 61:2577-2587

Szabo-Bardós E, Zsilák Z, Lendvay G, Horváth O, Markovics O, Hoffer A, Töro N (2008) Photocatalytic degradation of 1,5-naphthalenedisulfonate on colloidal titanium dioxide. J Phys Chem B 112:14500-14508

Tao J, Perdew JP, Staroverov VN, Scuseria GE (2003) Climbing the density functional ladder: Nonempirical meta-generalized gradient approximation designed for molecules and solids. Phys Rev Lett 91:146401-146404

To J, Sherwood P, Sokol AA, Bush IJ, Catlow CRA, van Dam HJJ, French SA, Guest MF (2006a) QM/MM modelling of the TS-1 catalyst using HPCx. J Mater Chem 16:1919-1926

To J, Sokol AA, Bush IJ, French SA, Catlow CRA (2007) Formation of active sites in TS-1 by hydrolysis and inversion. J Phys Chem C 111:14720-14731

To J, Sokol AA, Bush IJ, French SA, Catlow CRA (2008) Hybrid QM/MM investigations into the structure and properties of oxygen-donating species in TS-1. J Phys Chem C 112:7173-7185

To J, Sokol AA, French SA, Catlow CRA, Sherwood P, van Dam HJJ (2006b) Formation of heteroatom active sites in zeolites by hydrolysis and inversion. Angew Chem Int Ed 45:1633-1668

To J, Sokol AA, French SA, Kaltsoyannis N, Catlow CRA (2005) Hole localization in $[AlO_4]^0$ defects in silica materials. J Chem Phys 122:144704

Tommaso DD, Leeuw NHd (2009) Theoretical study of the dimerization of calcium carbonate in aqueous solution under natural water conditions. Geochim Cosmochim Acta 73:5394-5404

Tossell JA, Sahai N (2000) Calculating the acidity of silanols and related oxyacids in aqueous solution. Geochim Cosmochim Acta 64:4097-4113
Tossell JA (2006) Boric acid adsorption on humic acids: Ab initio calculations of structures, stabilities, ^{11}B NMR and ^{11}B, ^{10}B isotopic fractionations of surface complexes. Geochim Cosmochim Acta 70:5089-5103
Truhlar DG (2007) Valence bond theory for chemical dynamics. J Comp Chem 28:73-86
Tsipis AC, Orpen AG, Harvey JN (2005) Substituent effects and the mechanism of alkene metathesis catalyzed by ruthenium dichloride catalysts. Dalton Trans 2005:2849-2858
Tsuchiya T, Tsuchiya J, Umemoto K, Wentzcovitch (2004) Phase transition in $MgSiO_3$-perovskite in the Earth's lower mantle. Earth Planet Sci Lett 224:241-248
Tsuchiya T, Wentzcovitch RM, Da Silva CRS, De Gironcoli S (2006) Spin transition in megnesiowüsite in Earth's lower mantle. Phys Rev Lett 96:198501
Tuma C, Sauer J (2006) Treating dispersion effects in extended systems by hybrid MP2:DFT calculations—protonation of isobutene in zeolite ferrierite. Phys Chem Chem Phys 8:3955-3965
Valero R, Gomes JRB, Truhlar DG, Illas F (2008) Good performance of the M06 family of hybrid meta generalized gradient approximation density functionals on a difficult case: CO adsorption on MgO(001). J Chem Phys 129:124710
Vega-Rodriguez A, Alvarez-Idaboy JR (2009) Quantum chemistry and TST study of the mechanisms and branching ratios for the reactions of OH with unsaturated aldehyde. Phys Chem Chem Phys 11:7649-7658
Vuilleumier R, Sator N, Guillot B (2009) Computer modeling of natural silicate melts: What can we learn from ab initio simulations. Geochim Cosmochim Acta 73:6313-6339
Vollmer JM, Stefanovich EV, Truong TN (1999) Molecular modeling of interactions in zeolites: An ab initio embedded cluster study of NH_3 adsorption in chabazite. J Phys Chem B 103:9415-9422
von Barth U, Hedin L (1972) A local exchange-correlation potential for the spin-polarized case: I. J. Phys. C: Solid State Phys 5:1629-1642
Voorhis TV, Scuseria GE (1998) A novel form for the exchange-correlation energy functional. J Chem Phys 109:400-410
Walker AM, Demouchy S, Wright K (2006) Computer modelling of the energies and vibrational properties of hydroxyl groups in α- and β-Mg_2SiO_4. Eur J Mineral 18:529-543
Warren MC, Ackland GJ (1996) Ab initio studies of structural instabilities in magnesium silicate perovskite. Phys Chem Minerals 23:107-118
Winget P, Weber EJ, Cramer CJ, Truhlar DG (2000) Computational electrochemistry: Aqueous one-electron oxidation potentials for substituted anilines. Phys Chem Chem Phys 2:1231-1239 [Amendment – (2000) 2:1871(E)]
Winget P, Cramer CJ, Truhlar DG (2004) Computation of equilibrium oxidation and reduction potentials for reversible and dissociative electron-transfer reactions in solution. Theor Chem Acc 112:217-227
Winkler B, Milman V, Payne MC (1994) Orientation, location, and total energy of hydration of channel H_2O in cordierite investigated by ab-initio total energy calculations. Am Mineral 79:200-204
Winkler B, Milman V, Hennion B, Payne MC, Lee MH, Lin JS (1995) Ab initio total energy study of brucite, diaspore and hypothetical hydrous wadsleyite. Phys Chem Minerals 22:461-467
Wodrich MD, Corminboeuf C, Schleyer PvR (2006) Systematic errors in computed alkane energies using B3LYP and other popular DFT functionals. Org Lett 8:3631-3634
Wodrich MD, Corminboeuf C, Schreiner PR, Fokin AA, Schleyer PvR (2007) How accurate are DFT treatments of organic energies? Org Lett 9:1851-1854
Woodcock HL, Schaefer HF, Schreiner PR (2002) Problematic energy differences between cumulenes and poly-ynes. J Phys Chem A 106:11923-11931
Wu X, Liu J, Tommaso DD, Iggo JA, Catlow CRA, Bacsa J, Xiao J (2008) A multilateral mechanistic study into asymmetric transfer hydrogenation in water. Chem Eur J 14:7699-7715
Zhang Y, Li ZH, Truhlar DG (2007) Computational requirements for simulating the structures and proton activity of silicaceous materials. J Chem Theory Comput 3:593-604
Zhao Y, González-García N, Truhlar DG (2005a) Benchmark database of barrier heights for heavy atom transfer, nucleophilic substitution, association, and unimolecular reactions and their use to test DFT . J Phys Chem A 109:2012-2018
Zhao Y, Lynch BJ, Truhlar DG (2004a) Development and assessment of a new hybrid density functional method for thermochemical kinetics. J Phys Chem A 108(14):2715-2719
Zhao Y, Lynch BJ, Truhlar DG (2004b) Doubly hybrid meta DFT: New multi-coefficient correlation and density functional methods for thermochemistry and thermochemical kinetics. J Phys Chem A 108:4786-4791
Zhao Y, Lynch BJ, Truhlar DG (2005b) Multi-coefficient extrapolated DFT for thermochemistry and thermochemical kinetics. Phys Chem Chem Phys 7:43-52
Zhao Y, Schultz NE, Truhlar DG (2005c) Exchange-correlation functionals with broad accuracy for metallic and nonmetallic compounds, kinetics, and noncovalent Interactions. J Chem Phys 123:161103

Zhao Y, Schultz NE, Truhlar DG (2006) Design of density functionals by combining the method of constraint satisfaction with parametrization for thermochemistry, thermochemical kinetics, and noncovalent interactions. J Chem Theory Comput 2:364-382

Zhao Y, Tishchenko O, Truhlar DG (2005d) How well can density functional methods describe hydrogen bonds to p acceptors? J Phys Chem B 109:19046-19051

Zhao Y, Truhlar DG (2004) Hybrid meta DFT methods for thermochemistry, thermochemical kinetics, and noncovalent interactions. J Phys Chem A 108:6908-6918

Zhao Y, Truhlar DG (2005a) Benchmark databases for nonbonded interactions and their use to test DFT. J Chem Theory Comput 1:415-432

Zhao Y, Truhlar DG (2005b) Design of density functionals that are broadly accurate for thermochemistry, thermochemical kinetics, and nonbonded interaction. J Phys Chem A 109:5656-5667

Zhao Y, Truhlar DG (2005c) How well can new-generation density functional methods describe stacking interactions in biological systems? Phys Chem Chem Phys 7:2701-2705

Zhao Y, Truhlar DG (2005d) Infinite-basis calculations of binding energies for the hydrogen bonded and stacked tetramers of formic acid and formamide and their use for validation of hybrid DFT and ab initio methods. J Phys Chem A 109:6624-6627

Zhao Y, Truhlar DG (2006a) A density functional that accounts for medium-range correlation energies in organic chemistry. Org Lett 8:5753-5755

Zhao Y, Truhlar DG (2006b) A new local density functional for main group thermochemistry, transition metal bonding, thermochemical kinetics, and noncovalent interactions. J Chem Phys 125:194101

Zhao Y, Truhlar DG (2006c) Assessment of density functional theory for π systems: Energy differences between cumulenes and poly-ynes and proton affinities, bond length alternation, and torsional potentials of conjugated polyenes, and proton affinities of conjugated Schiff bases. J Phys Chem A 110:10478-10486

Zhao Y, Truhlar DG (2006d) Assessment of model chemistry methods for noncovalent interactions. J Chem Theory Comput 2:1009-1018

Zhao Y, Truhlar DG (2006e) Comparative assessment of density functional methods for 3d transition-metal chemistry. J Chem Phys 124:224105

Zhao Y, Truhlar DG (2006f) Comparative DFT study of van der Waals complexes. J Phys Chem A 110:5121-5129

Zhao Y, Truhlar DG (2006g) Density functional for spectroscopy: no long-range self-interaction error, good performance for Rydberg and charge-transfer states, and better performance on average than B3LYP for ground states. J Phys Chem A 110:13126

Zhao Y, Truhlar DG (2007a) Attractive noncovalent interactions in the mechanism of Grubbs second-generation Ru catalysts for olefin metathesis. Org Lett 9:1967-1970

Zhao Y, Truhlar DG (2007b) Density functionals for noncovalent interaction energies of biological importance. J Chem Theory Comput 3:289-300

Zhao Y, Truhlar DG (2007c) How well can new-generation density functionals describe protonated epoxides where older functionals fail? J Org Chem 72:295-298

Zhao Y, Truhlar DG (2008a) Benchmark data for interactions in zeolite model complexes and their use for assessment and validation of electronic structure methods. J Phys Chem C 112:6860-6868

Zhao Y, Truhlar DG (2008b) Construction of a generalized gradient approximation by restoring the density-gradient expansion and enforcing a Tight Lieb-Oxford bound. J Chem Phys 128:184109

Zhao Y, Truhlar DG (2008c) Density functionals with broad applicability in chemistry. Acc Chem Res 41:157-167

Zhao Y, Truhlar DG (2008d) Exploring the limit of accuracy of the global hybrid meta density functional for main-group thermochemistry, kinetics, and noncovalent interactions. J Chem Theory Comput 4:1849-1868

Zhao Y, Truhlar DG (2008e) The M06 suite of density functionals for main group thermochemistry, thermochemical kinetics, noncovalent interactions, excited states, and transition elements. Theor Chem Acc 120:215-241

Zheng J, Zhao Y, Truhlar DG (2009) The DBH24/08 database and its use to assess electronic structure model chemistries for chemical reaction barrier heights. J Chem Theory Comput 5:808-821

Zhu L, Letaief S, Liu Y, Gervais F, Detellier C (2009) Clay mineral-supported gold nanoparticles. Appl Clay Sci 43:439-446

Zwijnenburg MA, Sousa C, Sokol AA, Bromley ST (2008) Optical excitations of defects in realistic nanoscale silica clusters: Comparing the performance of density functional theory using hybrid functionals with correlated wavefunction methods. J Chem Phys 129:14706

Reviews in Mineralogy & Geochemistry
Vol. 71 pp. 39-57, 2010

Density-Functional Perturbation Theory for Quasi-Harmonic Calculations

Stefano Baroni

SISSA -- Scuola Internazionale Superiore di Studi Avanzati
via Beirut 2-4, 34151 Trieste Grignano, Italy

CNR-INFM Democritos National Simulation Center
34100 Trieste, Italy

baroni@sissa.it

Paolo Giannozzi

CNR-INFM Democritos National Simulation Center
34100 Trieste, Italy

Dipartimento di Fisica, Università degli Studi di Udine
via delle Scienze 208, 33100 Udine, Italy

paolo.giannozzi@uniud.it

Eyvaz Isaev

Department of Physics, Chemistry and Biology (IFM), Linköping University
581 83 Linköpng, Sweden

Theoretical Physics Department, Moscow State Institute of Steel and Alloys
4 Leninskii prospect, Moscow 119049, Russia

isaev@ifm.liu.se, eyvaz_isaev@yahoo.com

INTRODUCTION

Computer simulations allow for the investigation of many materials properties and processes that are not easily accessible in the laboratory. This is particularly true in the Earth sciences, where the relevant pressures and temperatures may be so extreme that no experimental techniques can operate at those conditions. Computer modeling is often the only source of information on the properties of materials that, combined with indirect evidence (such as seismic data), allows one to discriminate among competing planetary models. Many computer simulations are performed using effective inter-atomic potentials tailored to reproduce some experimentally observed properties of the materials being investigated. The remoteness of the physically interesting conditions from those achievable in the laboratory, as well as the huge variety of different atomic coordination and local chemical state occurring in the Earth interior, make the dependability of semi-empirical potentials questionable. First-principles techniques based on density-functional theory (DFT) (Hohenberg and Kohn 1964; Kohn and Sham 1965) are much more predictive, not being biased by any prior experimental input, and have demonstrated a considerable accuracy in a wide class of materials and variety of external conditions. The importance of thermal effects in the range of phenomena interesting to the Earth sciences makes a proper account of atomic motion essential. Traditionally, this is achieved using molecular dynamics techniques which have been successfully combined with DFT in the first-principles molecular dynamics technique of Car and Parrinello (1985). Well below the melting temperature, the numerical

1529-6466/10/0071-0003$05.00 DOI: 10.2138/rmg.2010.71.3

efficiency of molecular dynamics is limited by the lack of ergodicity, which would require long simulation times, and by the importance of long-wavelength collective motions (phonons), which would require large simulation cells. Both difficulties are successfully dealt with in the quasi-harmonic approximation (QHA) where the thermal properties of solid materials are traced back to those of a system of non-interacting phonons (whose frequencies are however allowed to depend on volume or on other thermodynamic constraints). An additional advantage of the QHA is that it accounts for quantum-mechanical zero-point effects, which would not be accessible to molecular dynamics with classical nuclear motion. The availability of suitable techniques to calculate the vibrational properties of extended materials using a combination of DFT and linear-response techniques (resulting in the so-called density-functional perturbation theory, DFPT; Baroni et al. 1987, 2001) makes it possible to combine the QHA with DFT. The resulting simulation methodology has shown to be remarkably accurate in a wide temperature range, extending up to not very far from the melting line and has been applied to a wide variety of systems, including many which are relevant to the Earth sciences. This paper gives a short overview of the calculation of thermal properties of materials in the framework of the QHA, using DFT. The paper is organized as follows: in the "Thermal Properties and the Quasi-Harmonic Approximation" section, we introduce some of the thermal properties of interest and describe how they can be calculated in the framework of the QHA; in the "*Ab Initio* Phonons" section, we describe the DFPT approach to lattice dynamics; in the "Computer Codes" section, we briefly introduce some of the computer codes that can be used to perform this task; in the "Applications" section, we review some of the application of the first-principles QHA to the study of the thermal properties of materials; finally, the last section contains our conclusions.

THERMAL PROPERTIES AND THE QUASI-HARMONIC APPROXIMATION

The low-temperature specific heat of solids is experimentally found to vanish as the cube of the temperature, with a cubic coefficient that is system-specific (Kittel 1996; Wallace 1998). This is contrary to the predictions of classical statistical mechanics, according to which the heat capacity of a system of harmonic oscillators does not depend on temperature, nor on its spectrum. One of the landmarks of modern solid-state physics, that greatly contributed to the establishment of our present quantum-mechanical picture of matter, is the Debye model for the heat capacity of solids. This model naturally explains the low-temperature specific heat of solids in terms of the (quantum) statistical mechanics of an ensemble of harmonic oscillators, which can in turn be pictorially described as a gas of non-interacting quasi-particles obeying the Bose-Einstein statistics (phonons).

The internal energy of a single harmonic oscillator of angular frequency ω, in thermal equilibrium at temperature T, is:

$$<E>=\frac{\hbar\omega}{2}+\frac{\hbar\omega}{e^{\frac{\hbar\omega}{k_BT}}-1} \tag{1}$$

where k_B is the Boltzmann constant. By differentiating with respect to temperature the sum over all the possible values of the phonon momentum in the Brillouin zone (BZ) of Equation (1), the constant-volume specific heat of a crystal reads:

$$C_V(T)=\frac{1}{V}\sum_{\mathbf{q}\nu}\hbar\omega(\mathbf{q},\nu)n'(\mathbf{q},\nu) \tag{2}$$

where $\omega(\mathbf{q},\nu)$ is the frequency of the ν-th mode (phonon) at point $\mathbf{q}$ in the BZ, $n'(\mathbf{q},\nu) = (\partial/\partial T)[e^{(\hbar\omega(\mathbf{q},\nu)/k_BT)}-1]^{-1}$, and the sum is extended to the first BZ. By assuming that there are three degenerate modes at each point of the BZ, each one with frequency $\omega(\mathbf{q},\nu) = c|\mathbf{q}|$, c being the sound velocity, and converting the sum in Equation (2) into an integral, the resulting expression for the heat capacity, valid in the low-temperature limit, reads:

$$C_V(T) = \frac{1}{\Omega}\frac{12\pi^4}{5}k_B\left(\frac{T}{\Theta_D}\right)^3 \tag{3}$$

where Ω is the volume of the crystal unit cell and $\Theta_D = (2\pi\hbar / k_B)c(3/4\pi\Omega)^{1/3}$ is the so-called Debye temperature.

In the Born-Oppenheimer approximation (Born and Oppenheimer 1927), the vibrational properties of molecules and solids are determined by their electronic structure through the dependence of the ground-state energy on the coordinates of the atomic nuclei (Martin 2004). At low temperature the amplitudes of atomic vibrations are much smaller than interatomic distances, and one can assume that the dependence of the ground-state energy on the deviation from equilibrium of the atomic positions is quadratic. In this, so called *harmonic*, approximation (HA) energy differences can be calculated from electronic-structure theory using static response functions (DeCicco and Johnson 1969; Pick et al. 1970) or perturbation theory (Baroni et al. 1987, 2001) (see the next section).

In the HA, vibrational frequencies do not depend on interatomic distances, so that the vibrational contribution to the crystal internal energy does not depend on volume. As a consequence, constant-pressure and constant-volume specific heats coincide in this approximation, and the equilibrium volume of a crystal does not depend on temperature. Other shortcomings of the HA include its prediction of an infinite thermal conductivity, infinite phonon lifetimes, and the independence of vibrational spectra (as well as related properties: elastic constants, sound velocities etc.) on temperature, to name but a few. A proper account of anharmonic effects on the static and dynamical properties of materials would require the calculation of phonon-phonon interaction coefficients for all modes in the BZ. Although the leading terms of such interactions can be computed even from first principles (Baroni and Debernardi 1994; Debernardi et al. 1995)—and the resulting vibrational linewidths have in fact been evaluated in some cases (Debernardi et al. 1995; Lazzeri et al. 2003; Bonini et al. 2007)—the extensive sampling of the phonon-phonon interactions over the BZ required for free-energy evaluations remains a daunting task. The simplest generalization of the HA, which corrects for most of the above mentioned deficiencies, while not requiring any explicit calculation of anharmonic interaction coefficients, is the QHA.

In the QHA, the crystal free energy is assumed to be determined by the vibrational spectrum via the standard harmonic expression:

$$F(X,T) = U_0(X) + \frac{1}{2}\sum_{\mathbf{q}\nu}\hbar\omega(\mathbf{q},\nu\,|\,X) + k_BT\sum_{\mathbf{q}\nu}\log\left(1 - e^{-\frac{\hbar\omega(\mathbf{q},\nu|X)}{k_BT}}\right) \tag{4}$$

where X indicates any global static constraint upon which vibrational frequencies may depend (most commonly just the volume V, but X may also include anisotropic components of the strain tensor, some externally applied fields, the internal distortions of the crystal unit cell, or other thermodynamic constraints that may be applied to the system), and $U_0(X)$ is the zero-temperature energy of the crystal as a function of X. In the case $X = V$, differentiation of Equation (4) with respect to volume gives the equation of state:

$$\begin{aligned} P &= -\frac{\partial F}{\partial V} \\ &= -\frac{\partial U_0}{\partial V} + \frac{1}{V}\sum_{\mathbf{q}\nu}\hbar\omega(\mathbf{q},\nu)\gamma(\mathbf{q},\nu)\left(\frac{1}{2} + \frac{1}{e^{\frac{\hbar\omega(\mathbf{q},\nu)}{k_BT}} - 1}\right) \end{aligned} \tag{5}$$

where

$$\gamma(\mathbf{q},\nu) = -\frac{V}{\omega(\mathbf{q},\nu)}\frac{\partial\omega(\mathbf{q},\nu)}{\partial V} \tag{6}$$

are the so-called Grüneisen mode parameters. In a perfectly harmonic crystal, phonon frequencies do not depend on the interatomic distances, hence on volume. In such a harmonic crystal Equation (5) implies that the temperature derivative of pressure at fixed volume vanish: $(\partial P/\partial T)_V = 0$. It follows that the thermal expansivity, $\beta = V^{-1}(\partial V/\partial T)_P$, which is given by the thermodynamical relation:

$$\beta = -\frac{1}{V}\frac{(\partial P/\partial T)_V}{(\partial P/\partial V)_T} \tag{7}$$

$$= \frac{1}{B_T}\left(\frac{\partial P}{\partial T}\right)_V \tag{8}$$

$$= \frac{1}{B_T}\sum_{\mathbf{q}\nu}\hbar\omega(\mathbf{q},\nu)\gamma(\mathbf{q},\nu)n'(\mathbf{q},\nu) \tag{9}$$

where $B_T = -V(\partial P/\partial V)_T$ is the crystal bulk modulus, would also vanish for perfectly harmonic crystals. Inspired by Equation (2), let us define $C_V(\mathbf{q},\nu) = \hbar\omega(\mathbf{q},\nu)n'(\mathbf{q},\nu)/V$ as the contribution of the ν-th normal mode at the **q** point of the BZ to the total specific heat, and γ as the weighted average of the various Grüneisen parameters:

$$\gamma = \frac{\sum_{\mathbf{q}\nu}\gamma(\mathbf{q},\nu)C_V(\mathbf{q},\nu)}{\sum_{\mathbf{q}\nu}C_V(\mathbf{q},\nu)} \tag{10}$$

In terms of γ, the thermal expansivity simply reads:

$$\beta = \frac{\gamma C_V}{B_T} \tag{11}$$

The vanishing of the thermal expansivity in the HA would also imply the equality of the constant-pressure and constant-volume specific heats. By imposing that the total differentials of the entropy as a function of pressure and temperature or of volume and temperature coincide, and by using the Maxwell identities, one can in fact show that (Wallace 1998):

$$C_P - C_V = -T\left(\frac{\partial P}{\partial V}\right)_T\left[\left(\frac{\partial V}{\partial T}\right)_P\right]^2 \tag{12}$$

$$= TB_T\beta^2 \tag{13}$$

We conclude this brief introduction to the QHA by noticing that the ansatz given by Equation (4) for the crystal free energy in terms of its (volume-dependent) vibrational frequencies gives immediate access to all the equilibrium thermal properties of the system. Whether this *implicit* account of anharmonic effects through the volume dependence of the vibrational frequency only is sufficient to describe the relevant thermal effects, or else an *explicit* account of the various phonon-phonon interactions is in order, instead, is a question that can only be settled by extensive numeric experience.

AB INITIO PHONONS

Lattice dynamics from electronic-structure theory

Several simplified approaches exist that allow to calculate full (harmonic) phonon dispersions ω(**q**,ν) from semi-empirical force fields or inter-atomic potentials (Brüesch 1982;

Singh 1982). The accuracy of such semi-empirical models is however often limited to the physical conditions (pressure, atomic coordination, crystal structure, etc.) at which the inter-atomic potentials are fitted. Really predictive calculations, not biased by the experimental information used to describe inter-atomic interactions require a proper quantum-mechanical description of the chemical bonds that held matter together. This can be achieved in the framework of electronic-structure theory (Martin 2004), starting from the *adiabatic* or Born and Oppenheimer (BO) approximation, and using modern concepts from DFT (Hohenberg and Kohn 1964; Kohn and Sham 1965) and perturbation theory (Baroni et al. 2001).

Within the BO approximation, the lattice-dynamical properties of a system are determined by the eigenvalues E and eigenfunctions Φ of the Schrödinger equation:

$$\left(-\sum_I \frac{\hbar^2}{2M_I}\frac{\partial^2}{\partial \mathbf{R}_I^2} + E_{BO}(\{\mathbf{R}\})\right)\Phi(\{\mathbf{R}\}) = E\Phi(\{\mathbf{R}\}) \tag{14}$$

where $\mathbf{R}_I$ is the coordinate of the I-th nucleus, M_I its mass, $\{\mathbf{R}\}$ indicates the set of all the nuclear coordinates, and E_{BO} is the ground-state energy of a system of interacting electrons moving in the field of fixed nuclei, whose Hamiltonian—which acts onto the electronic variables and depends parametrically upon $\{\mathbf{R}\}$—reads:

$$H_{BO}(\{\mathbf{R}\}) = -\frac{\hbar^2}{2m}\sum_i \frac{\partial^2}{\partial \mathbf{r}_i^2} + \frac{e^2}{2}\sum_{i\neq j}\frac{1}{|\mathbf{r}_i - \mathbf{r}_j|} + \sum_i V_{\{\mathbf{R}\}}(\mathbf{r}_i) + E_N(\{\mathbf{R}\}) \tag{15}$$

$-e$ being the electron charge, $V_{\{\mathbf{R}\}}(\mathbf{r}) = -\Sigma_I(Z_I e^2/|\mathbf{r}-\mathbf{R}_I|)$ is the electron-nucleus interaction, and $E_N(\{\mathbf{R}\}) = (e^2/2)\Sigma_{I\neq J}(Z_I Z_J/|\mathbf{R}_I-\mathbf{R}_J|)$ the inter-nuclear interaction energy. The equilibrium geometry of the system is determined by the condition that the forces acting on individual nuclei vanish:

$$\mathbf{F}_I \equiv -\frac{\partial E_{BO}(\{\mathbf{R}\})}{\partial \mathbf{R}_I} = 0 \tag{16}$$

whereas the vibrational frequencies, ω, are determined by the eigenvalues of the Hessian of the BO energy, scaled by the nuclear masses:

$$\det\left|\frac{1}{\sqrt{M_I M_J}}\frac{\partial^2 E_{BO}(\{\mathbf{R}\})}{\partial \mathbf{R}_I \partial \mathbf{R}_J} - \omega^2\right| = 0 \tag{17}$$

The calculation of the equilibrium geometry and vibrational properties of a system thus amounts to computing the first and second derivatives of its BO energy surface. The basic tool to accomplish this goal is the Hellmann-Feynman (HF) theorem (Hellmann 1937; Feynman 1939), which leads to the following expression for the forces:

$$\mathbf{F}_I = -\int n_{\{\mathbf{R}\}}(\mathbf{r})\frac{\partial V_{\{\mathbf{R}\}}(\mathbf{r})}{\partial \mathbf{R}_I}d\mathbf{r} - \frac{\partial E_N(\mathbf{R})}{\partial \mathbf{R}_I} \tag{18}$$

where $n_{\{\mathbf{R}\}}(\mathbf{r})$ is the ground-state electron charge density corresponding to the nuclear configuration $\{\mathbf{R}\}$. The Hessian of the BO energy surface appearing in Equation (17) is obtained by differentiating the HF forces with respect to nuclear coordinates:

$$\frac{\partial^2 E_{BO}(\{\mathbf{R}\})}{\partial \mathbf{R}_I \partial \mathbf{R}_J} \equiv -\frac{\partial \mathbf{F}_I}{\partial \mathbf{R}_J} \tag{19}$$

$$= \int \frac{\partial n_{\{\mathbf{R}\}}(\mathbf{r})}{\partial \mathbf{R}_J}\frac{\partial V_{\{\mathbf{R}\}}(\mathbf{r})}{\partial \mathbf{R}_I}d\mathbf{r} + \int n_{\{\mathbf{R}\}}(\mathbf{r})\frac{\partial^2 V_{\{\mathbf{R}\}}(\mathbf{r})}{\partial \mathbf{R}_I \partial \mathbf{R}_J}d\mathbf{r} + \frac{\partial^2 E_N(\{\mathbf{R}\})}{\partial \mathbf{R}_I \partial \mathbf{R}_J} \tag{20}$$

Equation (20) states that the calculation of the Hessian of the BO energy surfaces requires the calculation of the ground-state electron charge density, $n_{\{\mathbf{R}\}}(\mathbf{r})$, as well as of its *linear response* to a distortion of the nuclear geometry, $\partial \mathbf{n}_{\{\mathbf{R}\}}(\mathbf{r})/\partial \mathbf{R}_I$. This fundamental result was first stated in the late sixties by De Cicco and Johnson (1969) and by Pick, Cohen, and Martin (1970). The Hessian matrix is usually called the matrix of the *inter-atomic force constants* (IFC). For a crystal, we can write:

$$C_{ss'}^{\alpha\alpha'}(\mathbf{R}-\mathbf{R}') = \frac{\partial^2 E_{BO}(\{\mathbf{R}\})}{\partial u_s^{\alpha}(\mathbf{R})\partial u_{s'}^{\alpha'}(\mathbf{R}')} \tag{21}$$

where $u_s^{\alpha}(\mathbf{R})$ is the α-th Cartesian components of the displacement of the s-th atom of the crystal unit cell located at lattice site $\mathbf{R}$, and translational invariance shows manifestly in the dependence of the IFC matrix on $\mathbf{R}$ and $\mathbf{R}'$ through their difference only.

Density-functional perturbation theory

We have seen that the electron-density linear response of a system determines the matrix of its IFCs, Equation (20). Let us see now how this response can be obtained from DFT. The procedure described in the following is usually referred to as density-functional perturbation theory (Baroni et al. 1987, 2001).

In order to simplify the notation and make the argument more general, we assume that the external potential acting on the electrons is a differentiable function of a set of parameters, $\lambda \equiv \{\lambda_i\}$ ($\lambda_i \equiv \mathbf{R}_I$ in the case of lattice dynamics). According to the HF theorem, the first and second derivatives of the ground-state energy read:

$$\frac{\partial E}{\partial \lambda_i} = \int \frac{\partial V^{\lambda}(\mathbf{r})}{\partial \lambda_i} n^{\lambda}(\mathbf{r}) d\mathbf{r} \tag{22}$$

$$\frac{\partial^2 E}{\partial \lambda_i \partial \lambda_j} = \int \frac{\partial^2 V^{\lambda}(\mathbf{r})}{\partial \lambda_i \partial \lambda_j} n^{\lambda}(\mathbf{r}) d\mathbf{r} + \int \frac{\partial n^{\lambda}(\mathbf{r})}{\partial \lambda_i} \frac{\partial V^{\lambda}(\mathbf{r})}{\partial \lambda_j} d\mathbf{r} \tag{23}$$

In DFT the electron charge-density distribution, n^{λ}, is given by:

$$n^{\lambda}(\mathbf{r}) = 2\sum_{n=1}^{N/2} |\psi_n^{\lambda}(\mathbf{r})|^2 \tag{24}$$

where N is the number of electrons in the system (double degeneracy with respect to spin degrees of freedom is assumed), the single-particle orbitals, $\psi_n^{\lambda}(\mathbf{r})$, satisfy the Kohn-Sham (KS) equation:

$$\left(-\frac{\hbar^2}{2m}\frac{\partial^2}{\partial \mathbf{r}^2} + V_{SCF}^{\lambda}(\mathbf{r})\right)\psi_n^{\lambda}(\mathbf{r}) = \varepsilon_n^{\lambda}\psi_n^{\lambda}(\mathbf{r}) \tag{25}$$

and the *self-consistent* potential, V_{SCF}^{λ}, is given by:

$$V_{SCF}^{\lambda} = V^{\lambda} + e^2 \int \frac{n^{\lambda}(\mathbf{r}')}{|\mathbf{r}-\mathbf{r}'|} d\mathbf{r}' + \mu_{XC}[n^{\lambda}](\mathbf{r}) \tag{26}$$

where μ_{XC} is the so-called *exchange-correlation* (XC) potential (Kohn and Sham 1965). The electron-density response, $\partial n_{\lambda}(\mathbf{r})/\partial \lambda_i$, appearing in Equation (23) can be evaluated by linearizing Equations (24), (25), and (26) with respect to wave-function, density, and potential variations, respectively. Linearization of Equation (24) leads to:

$$n'(\mathbf{r}) = 4Re\sum_{n=1}^{N/2} \psi_n^{*}(\mathbf{r})\psi'_n(\mathbf{r}) \tag{27}$$

where the prime symbol (as in n') indicates differentiation with respect to one of the λ's. The super-script λ has been omitted in Equation (27), as well as in any subsequent formulas where such an omission does not give rise to ambiguities. Since the external potential (both unperturbed and perturbed) is real, KS eigenfunctions can be chosen to be real, and the sign of complex conjugation, as well as the prescription to keep only the real part, can be dropped in Equation (27).

The variation of the KS orbitals, $\psi'_n(\mathbf{r})$, is obtained by standard first-order perturbation theory (Messiah 1962):

$$(H^{\circ}_{SCF} - \varepsilon^{\circ}_n) | \psi'_n \rangle = -(V'_{SCF} - \varepsilon'_n) | \psi^{\circ}_n \rangle \tag{28}$$

where $H^{\circ}_{SCF} = -(\hbar^2 / 2m)(\partial^2 / \partial \mathbf{r}^2) + V^{\circ}_{SCF}(\mathbf{r})$ is the unperturbed KS Hamiltonian,

$$V'_{SCF}(\mathbf{r}) = V'(\mathbf{r}) + \int \kappa(\mathbf{r},\mathbf{r}') n'(\mathbf{r}') d\mathbf{r}' \tag{29}$$

is the first-order correction to the self-consistent potential, Equation (26), $\kappa(\mathbf{r},\mathbf{r}') = (e^2/|\mathbf{r}-\mathbf{r}'|) + [\delta\mu_{XC}(\mathbf{r})/\delta n(\mathbf{r}')]$ is the *Hartree-plus-XC kernel*, and $\varepsilon'_n = \langle \psi_n^{\circ} | V'_{SCF} | \psi_n^{\circ} \rangle$ is the first order variation of the KS eigenvalue, ε_n. Equations (28–30) form a set of self-consistent equations for the perturbed system completely analogous to the KS equations in the unperturbed case—Equations (24), (25), and (26)—with the KS eigenvalue equation, Equation (25), being replaced by a linear system, Equation (28). The computational cost of the determination of the density response to a single perturbation is of the same order as that needed for the calculation of the unperturbed ground-state density.

The above discussion applies to insulators, where there is a finite gap. In metals a finite density of states occurs at the Fermi energy, and a change in the orbital occupation number may occur upon the application of an infinitesimal perturbation. The modifications of DFPT needed to treat the linear response of metals are discussed in de Gironcoli (1995) and Baroni et al. (2001).

Interatomic force constants and phonon band interpolation

The above discussion indicates that the primary physical ingredient of a lattice-dynamical calculation is the IFC matrix, Equation (20), from which vibrational frequencies can be obtained by solving the secular problem, Equation (17). That phonon frequencies can be classified according to a well defined value of the crystal momentum $\mathbf{q}$ follows from the translational invariance of the IFC matrix, Equation (21). Because of this, the IFC matrix can be Fourier analyzed to yield the so called *dynamical matrix*, prior to diagonalization:

$$\tilde{C}^{\alpha\beta}_{st}(\mathbf{q}) = \sum_{\mathbf{R}} C^{\alpha\beta}_{st}(\mathbf{R}) e^{-i\mathbf{q}\cdot\mathbf{R}} \tag{30}$$

and the squared vibrational frequencies, $\omega(\mathbf{q},\nu)^2$, are the eigenvalues of the $3n \times 3n$ dynamical matrix:

$$D^{\alpha\beta}_{st}(\mathbf{q}) = \frac{1}{\sqrt{M_s M_t}} \tilde{C}^{\alpha\beta}_{st}(\mathbf{q}) \tag{31}$$

n being the number of atoms in the unit cell. The direct computation of the IFCs is unwieldy because it requires the calculation of the crystal electronic linear response to a localized perturbation (the displacement of a single atom or atomic plane), which would in turn break the translational symmetry of the system, thus requiring the use of computationally expensive large unit cells (Martin 2004; Alfè 2010; Parlinski 2010). The IFCs are instead more conveniently calculated in Fourier space, which gives direct access to the relevant $\mathbf{q}$-dependent dynamical matrices (Baroni et al. 2001). Because of translational invariance, the linear response to a *monochromatic* perturbation, i.e., one with a definite wave-vector $\mathbf{q}$, is also monochromatic, and all quantities entering the calculation can be expressed in terms of lattice-periodic quantities

(Baroni et al. 2001). As a consequence, vibrational frequencies can be calculated at any wave-vector in the BZ, without using any supercells, with a numerical effort that is independent of the phonon wave-length and comparable to that of a single ground-state calculation for the unperturbed system.

The accurate calculation of sums (integrals) of lattice-dynamical properties over the BZ (such as those appearing in the QHA formulation of the thermodynamics of crystals in the "Thermal Properties and Quasi-Harmonic Approximation" section) requires sampling the integrand over a fine grid of points. This may be impractical in many cases, and suitable interpolation techniques are therefore called for. The most accurate, and physically motivated, such technique consists in the calculation of real-space IFCs by inverse analyzing a limited number of dynamical matrices calculated on a coarse grid. Dynamical matrices at any arbitrary point in the BZ can then be inexpensively reconstructed by Fourier analysis of the IFC's thus obtained. According to the *sampling theorem* by Shannon (1949), if the IFCs are strictly short-range, a finite number of dynamical matrices, sampled on a correspondingly coarse reciprocal-space grid, is sufficient to calculate them *exactly* by inverse Fourier analysis. The IFCs thus obtained can then be used to calculate exactly the dynamical matrices at any wave-vector not included in the original reciprocal-space grid. In the framework of lattice-dynamical and band-structure calculations this procedure is usually referred to as *Fourier interpolation*. Of course, IFCs are never strictly short-range, and Fourier interpolation is in general a numerical approximation, subject to so-called *aliasing* errors, whose magnitude and importance have to be checked on a case-by-case basis.

Let us specialized to the case of a crystal, in which lattice vectors $\mathbf{R}$ are generated by primitive vectors $\mathbf{a}_1$, $\mathbf{a}_2$, $\mathbf{a}_3$: $\mathbf{R}_{lmn} = l\mathbf{a}_1 + m\mathbf{a}_2 + n\mathbf{a}_3$, with l,m,n integer numbers. The reciprocal lattice vectors $\mathbf{G}$ are generated in an analogous way by vectors $\mathbf{b}_1$, $\mathbf{b}_2$, $\mathbf{b}_3$, such that

$$\mathbf{a}_i \cdot \mathbf{b}_j = 2\pi\delta_{ij} \tag{32}$$

Correspondingly we consider a symmetry-adapted uniform grid of $\mathbf{q}$-vectors:

$$\mathbf{q}_{pqr} = \frac{p}{N_1}\mathbf{b}_1 + \frac{q}{N_2}\mathbf{b}_2 + \frac{r}{N_3}\mathbf{b}_3 \tag{33}$$

where p,q,r are also integers. This grid spans the reciprocal lattice of a supercell of the original lattice, generated by primitive vectors $N_1\mathbf{a}_1$, $N_2\mathbf{a}_2$, $N_3\mathbf{a}_3$. Since wave-vectors differing by a reciprocal-lattice vector are equivalent, all values of p,q,r differing by a multiple of N_1,N_2,N_3 respectively, are equivalent. We can then restrict our grid to $p \in [0, N_1-1]$, $q \in [0, N_2-1]$, and $r \in [0, N_3-1]$. The $\mathbf{q}_{pqr}$ grid thus contains $N_1 \times N_2 \times N_3$ uniformly spaced points and spans the parallelepiped generated by $\mathbf{b}_1$, $\mathbf{b}_2$, $\mathbf{b}_3$. It is often convenient to identify wave-vectors with integer labels spanning the $[-\frac{N}{2}, \frac{N}{2}-1]$ range, rather than $[0, N-1]$. Negative indices can be folded to positive values using the periodicity of discrete Fourier transforms.

Once dynamical matrices have been calculated on the $\mathbf{q}_{hkl}$ grid, IFCs are easily obtained by (discrete) fast-Fourier transform (FFT) techniques:

$$\begin{aligned} C_{st}^{\alpha\beta}(\mathbf{R}_{lmn}) &= \frac{1}{N_1 N_2 N_3} \sum_{pqr} \tilde{C}_{st}^{\alpha\beta}(\mathbf{q}_{pqr}) e^{i\mathbf{q}_{pqr} \cdot \mathbf{R}_{lmn}} \\ &= \frac{1}{N_1 N_2 N_3} \sum_{pqr} \tilde{C}_{st}^{\alpha\beta}(\mathbf{q}_{pqr}) e^{i2\pi\left(\frac{lp}{N_1} + \frac{mq}{N_2} + \frac{nr}{N_3}\right)} \end{aligned} \tag{34}$$

where the bi-orthogonality of the real- and reciprocal-space primitive vectors, Equation (32), is used to get $\mathbf{q}_{pqr} \cdot \mathbf{R}_{lmn} = 2\pi[(lp/N_1) + (mq/N_2) + (nr/N_3)]$. The IFCs thus obtained can be used to calculate dynamical matrices at wave-vectors not originally contained in the reciprocal-space grid. This can be done directly wave-vector by wave-vector, or by FFT techniques, by padding a conveniently large table of IFCs with zeroes beyond the range of those calculated from Fourier analyzing the original coarse reciprocal-space grid.

COMPUTER CODES

In order to implement the QHA from first principles, one needs to compute the complete phonon dispersion of a crystal for different values of the crystal volume. This can be done within DFT by the *direct* or *frozen phonon* method, or by the *linear response* method (Baroni et al. 2001; Martin 2004). The former does not require the use of specialized software beside that needed to perform standard ground-state DFT calculations, but is computationally more demanding. Some software tools that help analyze the output of standard DFT code to produce real-space IFC's and, from these, reciprocal-space dynamical matrices are available (Alfé 2010; Parlinski 2010). As for the linear-response approach, two widely known general-purpose packages exist, QUANTUM ESPRESSO (Giannozzi et al. 2009) and *ABINIT* (Gonze et al. 2002). In the following we briefly describe the former, as well as another code, *QHA*, that can be used as a post-processing tool to perform QHA calculations starting from lattice-dynamical calculations performed with many different methods (semi-empirical as well as first-principles, frozen-phonon, as well as DFPT).

Quantum ESPRESSO

QUANTUM ESPRESSO *(opEn Source Package for Research in Electronic Structure, Simulation, and Optimization*) is an integrated suite of computer codes for electronic-structure calculations and materials modeling, based on DFT, plane waves, pseudopotentials (norm-conserving and ultrasoft) and all-electron Projector-Augmented-Wave potentials (Giannozzi et al. 2009). It is freely available under the terms of the GNU General Public License. QUANTUM ESPRESSO is organized into packages. For the purposes of lattice-dynamical calculations and QHA applications, the two most relevant ones are PWscf and PHonon. The former produces the self-consistent electronic structure and all related computations (forces, stresses, structural optimization, molecular dynamics). The latter solves the DFPT equations and calculates dynamical matrices for a single wave-vector or for a uniform grid of wave-vectors; Fourier interpolation can be applied to the results to produce IFCs up to a pre-determined range in real space. The effects of macroscopic electric field are separately dealt with using the known exact results valid in the long-wavelength limit (Born and Huang 1954). Both the electronic contribution to the dielectric tensor, ε_∞, and the effective charges Z^* are calculated by PHonon and taken into account in the calculation of interatomic force constants. Once these have been calculated, phonon modes at any wave-vector can be recalculated in a quick and economical way. Anharmonic force constants can be explicitly calculated using the D3 code contained in the PHonon package. The volume dependence of the IFCs needed within the QHA is simply obtained numerically by performing several phonon (harmonic) calculations at different volumes of the unit cell.

The QHA code

Once the IFC matrix (or, equivalently, the dynamical matrix over a uniform grid in reciprocal space) has been calculated, thermodynamical properties can be easily calculated using the *QHA* code (Isaev 2010). *QHA* requires in input just a few data: basic information about the system (such as atomic masses, lattice type) and a file containing IFCs, stored in an appropriate format. *QHA* then calculates and several quantities such as the total phonon density of states (DOS), atom-projected DOS, the isochoric heat capacity, the Debye temperature, zero-point vibration energy, internal energy, entropy, mean square displacements for atoms, etc. The DOS is obtained via the tetrahedron method (Lehmann and Taut 1972), while integrals over the frequency are calculated using the Simpson's "3/8 rule."

APPLICATIONS

The first investigations of the thermal properties of materials using *ab initio* phonons and the QHA date back to the early days of DFPT theory, when the thermal expansivity of tetrahedrally coordinated semiconductors and insulators was first addressed (Fleszar and Gonze 1990; Pavone 1991; Pavone et al. 1993). Many other applications have appeared ever since to metals, hydrides, intermetallic compounds, surfaces, and to systems and properties of mineralogical and geophysical interest. Brief reviews of these applications can be found in Baroni et al. (2001) and Rickman and LeSar (2002); this section contains a more up-to-date review, with a special attention paid to those applications that are relevant to the Earth Sciences.

Semiconductors and insulators

One of the most unusual features of tetrahedrally coordinated elemental and binary semiconductors is that they display a negative thermal expansion coefficient (TEC) at very low temperature. This finding prompted the first applications of the QHA to semiconductors, using first a semi-empirical approach (Biernacki and Scheffler 1989), and first-principles techniques in the following (Fleszar and Gonze 1990; Pavone 1991; Pavone et al. 1993; Debernardi and Cardona 1996; Gaal-Nagy et al. 1999; Rignanese et al. 1996; Xie et al. 1999a; Eckman et al. 2000; Mounet and Marzari 2005; Hamdi et al. 2006; Zimmermann et al. 2008). The detailed insight provided by the latter allowed one to trace back this behavior to the negative Grüneisen parameter in the lowest acoustic phonon branch and to its flatness that enhances its weight in the vibrational density of states at low frequency. This behavior is not observed in diamond at ambient conditions—which in fact does not display any negative TEC (Pavone et al. 1993; Xie et al. 1999a)—whereas at pressures larger than ~700 GPa the softening of the acoustic Grüneisen parameters determines a negative TEC. The TEC of diamond calculated by Pavone et al. (1993) starts deviating from experimental points at $T = 600$ K which was explained in terms of enhanced anharmonic effects at higher temperature. However, a recent calculation done with a different XC energy functional (GGA, rather than LDA) (Mounet and Marzari 2005) displayed a fairly good agreement with experiments up to $T = 1200$ K, and with results of Monte-Carlo simulations (Herrero and Ramírez 2000) up to $T = 3000$ K. Graphite shows negative in-plane TEC over a broad temperature range, up to 600K, and the calculated TEC for graphene is negative up to 2000 K (Mounet and Marzari 2005). This is due to a negative Grüneisen parameter of the out-of-plane lattice vibrations along the ΓM and ΓK directions (the so called ZA modes, which plays an important role in the thermal properties of layered materials, due to the high phonon DOS displayed at low frequency because of a vanishing sound velocity (Lifshitz 1952; Zabel 2001). Such an unusual thermal contraction for carbon fullerenes and nanotubes was confirmed by molecular dynamics simulations in Kwon et al. (2004). The heat capacity of carbon nanotubes was calculated in Zimmermann et al. (2008). The out-of-plane TEC calculated for graphite (Mounet and Marzari 2005) is in poor agreement with experiment. This is not unexpected because inter-layer binding is mostly due to dispersion forces which are poorly described by the (semi-) local XC functionals currently used in DFT calculations.

One of the early achievements of DFT that greatly contributed to its establishment in the condensed-matter and materials-science communities was the prediction of the relative stability of different crystal structures as a function of the applied pressure (Gaal-Nagy et al. 1999; Eckman et al. 2000; Correa et al. 2006; Liu et al. 1999; Isaev et al. 2007; Mikhaylushkin et al. 2007; Dubrovinsky et al. 2007). Thanks to the QHA, vibrational effects can be easily included in the evaluation of the crystal free energy, thus allowing for the exploration of the phase diagram of crystalline solids at finite temperature. In Gaal-Nagy et al. (1999) and Eckman et al. (2000), for instance, the *P-T* phase diagram for Si and Ge was studied in correspondence to the diamond $\rightarrow$ β-Sn transition. Noticeable changes in the EOS of ZnSe at finite temperature were shown in Hamdi et al. (2006). The phase boundary between cubic and hexagonal BN has been studied in Kern et al. (1999) using the QHA with an empirical correction to account for the leading (explicit) anharmonic effects. Other applications of the QHA in this area include the low-temperature portion of the *P-T* phase diagram for the diamond → BC8 phase transition (Correa et al. 2006) and the sequence of rhombohedral (223 K) → orthorhombic (378 K) → tetragonal (778 K) → cubic phase transitions in $BaTiO_3$ (Zhang et al. 2006) at ambient pressure.

Simple metals

The QHA has been widely used to investigate the thermal properties of BCC (Quong and Liu 1997; Liu et al. 1999; Debernardi et al. 2001), FCC (Debernardi et al. 2001; Li and Tse 2000; Grabowski et al. 2009; Xie et al. 2000; Narasimhan and de Gironcoli 2002; Xie et al. 1999b; Tsuchiya 2003; Sun et al. 2008), and HCP (Ismail et al. 2001; Althoff et al. 1993) metals. These works generally report a good agreement with experiments as concerns the calculated lattice volume, bulk modulus, TEC, Grüneisen parameter, and high-pressure/high-temperature phase diagram. Some discrepancies in the temperature dependence of C_P and TEC might be connected to the neglect of explicit anharmonic effects at high temperatures, as well as due to overestimated cell volumes when using GGA XC functionals. In Grabowski et al. (2009) it was stressed that implicit quasi-harmonic effects dominate the thermal properties, being almost two orders of magnitude larger than explicit anharmonic ones, irrespective of the XC functional adopted.

The QHA has also been an important ingredient in the calculation of the melting curve of some metals, such as Al (Vocadlo and Alfé 2002), Si (Alfé and Gillan 2002), and Ta (Gülseren and Cohen 2002; Taioli et al. 2007), performed via thermodynamic integration. The vibrational contribution to the low-temperature free energy of the crystal phase was shown to be important for lighter elements (such as Al), whereas it is negligible for heavier ones, such as Ta. The *P-T* phase diagram for HCP-BCC Mg has been obtained in Althoff et al. (1993), where it was shown that a proper account of lattice vibrations improves the prediction of the transition pressure at room temperature. Interestingly, in Xie et al. (1999b) it was noticed that in the QHA equation of state (EOS) of Ag there exists a critical temperature beyond which no volume would correspond to a vanishing pressure—thus signaling a thermodynamic instability of the system—and that this temperature is actually rather close to the experimental melting temperature of Ag. Narasimhan and de Gironcoli (2002) studied the influence of different (LDA and GGA) functionals on the thermal properties of Cu. The contribution of lattice vibrations to the phase stability of Li and Sn has been studied in Liu et al. (1999), Pavone et al. (1998), and Pavone (2001): a proper account of vibrational effects considerably improves the predictions of the low-temperature structural properties of a light element such as Li, which is strongly affected by zero-point vibrations (Liu et al. 1999). The large vibrational entropy associated with low-frequency modes stabilizes the BCC structure of Li (Liu et al. 1999) and β-Sn (Pavone et al. 1998; Pavone 2001) just above room temperature.

Hydrides

One of the best illustrations of the ability of the QHA to account for the effects of lattice vibrations on the relative stability of different crystalline phases is provided by iron and palladium hydrides, FeH and PdH. FeH was synthesized by different experimental groups (Antonov et al. 1980; Badding et al. 1992; Hirao et al. 2004) and its crystalline structure was found to be a double hexagonal structure (DHCP), contrary to the results of ab initio calculations (Elsasser et al. 1998) that, neglecting vibrational effects, would rather predict a simple HCP structure. The puzzle remained unsolved until free-energy calculations for FCC, HCP, and DHCP FeH (Isaev et al. 2007) showed that the hydrogen vibrational contribution to the free energy actually favors the DHCP structure. This is a consequence of the linear ordering of H atoms in HCP FeH, which shifts to higher frequencies the mostly H-like optical band of the system, with respect to the FCC and DHCP phases. The corresponding increase in the zero-point energy makes the DHCP structure—which is the next most favored, neglecting lattice vibrations—the stablest structure at low pressure. The quantum nature of hydrogen vibrations and its influence on the phase stability of hydrides was also clearly demonstrated in Caputo and Alavi (2003 and Hu et al. (2007). First-principles pseudopotential calculations for PdH have shown that tetrahedrally coordinated H (B3-type PdH) is energetically favored with respect to octahedrally coordinated H (B1-type PdH), at variance with experimental findings (Rowe et al. 1972, Nelin 1971). The quantum-mechanical behavior of hydrogen vibrations dramatically affects on the stability of PdH phases at low temperature, favoring the octahedral coordination of hydrogen atoms in PdH (Caputo and Alavi 2003). As another example, the QHA does not predict any $\alpha \rightarrow \beta$ (monoclinic to orthorhombic) phase transition in Na_2BeH_4 (Hu et al. 2007), contrary to the conclusions that were reached from static total-energy calculations. Overall, the structural parameters of most alkaline hydrides calculated using the QHA turned out to be substantially improved by a proper account of zero-point vibrations, both using LDA and GGA XC functionals (more so in the latter case) (Roma et al. 1996; Barrera et al. 2005; Lebègue et al. 2003; Zhang et al. 2007).

Intermetallics

The QHA has been also successfully applied to the thermal properties of intermetallics and alloys. For example, the Grüneisen parameters, isothermal bulk modulus, TEC, and constant-pressure specific heat for Al_3Li have been calculated in Li and Tse (2000). The TEC temperature dependence of the technologically important superalloys B2 NiAl and $L1_2$ Ni_3Al, as well as $L1_2$ Ir_3Nb, have been studied in (Wang et al. 2004; Arroyave et al. 2005; Lozovoi and Mishin 2003; Gornostyrev et al. 2007). This is a very significant achievement of QHA, as it makes it possible very accurate temperature-dependent calculations of the misfit between lattice parameters of low-temperature FCC/BCC alloy and high-temperature $L1_2$/B2 phases, which plays a considerable role in the shape formation of precipitates. It has been found that zero-point vibrations do not affect the type of structural defects in B2 NiAl, nor do they change qualitatively the statistics of thermal defects in B2 NiAl (Lozovoi and Mishin 2003). Ozolins et al. (1998) and Persson et al. (1999) have studied the influence of vibrational energies on the phase stability in Cu-Au and Re-W alloys, using a combination of the QHA and of the cluster-variation method. It turns out that lattice vibrations considerably enhance to the stability of CuAu intermetallic compounds and Cu-Au alloys with respect to phase separation (Ozolins et al. 1998), as well as to the relative stability of the ordered vs. disordered phases at high temperature (Persson et al. 1999).

Surfaces

Ab initio calculations for surfaces coupled with the QHA have been done for the past 10 years. For example, an anomalous surface thermal expansion, the so called surface premelting, has been studied for a few metallic surfaces, such as Al(001) (Hansen et al. 1999), Al(111) (Narasimhan and Scheffler 1997), Ag(111) (Xie et al. 1999c; Narasimhan and Scheffler 1997; Al-Rawi et al. 2001), Rh(001), Rh(110) (Xie and Scheffler 1998), Mg($10\bar{1}0$) (Ismail et al. 2001), Be($10\bar{1}0$) (Lazzeri and de Gironcoli 2002) and Be(0001) (Pohl et al. 1998). Hansen et al. (1999) noticed that the QHA is fairly accurate up to the Debye temperature, above which explicit anharmonic effects, not accounted for in this approximation, become important. While no peculiar effects for the surface inter-layer spacing were found for Al(111) (Narasimhan and Scheffler 1997), for Ag and Rh surfaces it was found that the outermost interlayer distance, d_{12}, is reduced at room temperature, with respect to its bulk value, whereas it is expanded at high temperatures (Narasimhan and Scheffler 1997; Xie and Scheffler 1998; Xie et al.1999c; Al-Rawi et al. 2001). The expansion of d_{12} in the Ag and Rh surfaces, as well as in Be (0001) (Pohl et al. 1998), is related to the softening of some in-plane vibrational modes with a corresponding enhancement of their contribution to the surface free energy. Free energy calculations for Be($10\bar{1}0$) (Lazzeri and de Gironcoli 2002) and Mg($10\bar{1}0$) (Ismail et al. 2001) successfully account for the experimentally observed oscillatory behavior of the interatomic distances. The large contraction of d_{12} in Be($10\bar{1}0$) was explained in terms of a strong anharmonicity in the second layer in comparison with the surface layer (see also Marzari et al. 1999). For Be(0001) no oscillatory behavior in inter-layer spacings was observed in Pohl et al. (1998), but an anomalously large surface thermal expansion does occur.

Earth materials

The extreme temperature and pressure conditions occurring in the Earth interior make many geophysically relevant materials properties and processes difficult, if not impossible, to observe in the laboratory. Because of this, computer simulation is often a premier, if not unique, source of information in the Earth sciences. By increasing the pressure, the melting temperature also increases, so that the temperature range over which a material behaves as a harmonic solid is correspondingly expanded, thus making the QHA a very useful tool to investigate materials properties at Earth-science conditions.

Iron, the fourth most abundant element on Earth and the main constituent of the Earth core, plays an outstanding role in human life and civilization. In Körmann et al. (2008), Sha and Cohen (2006a,b) the thermodynamics and thermoelastic properties of BCC Fe have been treated by means of the QHA and finite-temperature DFT. The temperature dependence of the calculated constant-pressure heat capacity deviates from experiment at room temperature, but a proper inclusion of magnetic effects dramatically improves the agreement up to the Curie temperature (Körmann et al. 2008). The calculated Debye temperature and low-temperature isochoric heat capacity C_V are in good agreement with available experimental data. The magnitude and temperature dependence of the calculated C_{12}, C_{44} elastic constants (Sha and Cohen 2006a) are consistent with experiment (Leese and Lord 1968; Dever 1972; Isaak and Masuda 1995) in the temperature range from 0 K to 1200 K at ambient pressure, while C_{11} is overestimated (Sha and Cohen 2006a), likely because of an underestimated equilibrium volume. The ambient-pressure shear and compressional sound velocities are consistent with available ultrasonic measurements. The *c*/*a* ratio of ε-Fe has been studied in (Sha and Cohen 2006c) up to temperatures of 6000 K and pressures of 400 GPa by using the QHA, resulting in good agreement with previous calculations (Gannarelli et al. 2003) and X-ray diffraction experiment (Ma et al. 2004). A combination of experiments and calculations performed within the QHA was used to show that the FCC and HCP phases of nonmagnetic Fe (Mikhaylushkin et al. 2007) can co-exist at very high temperatures and pressures (–6600 K and 400 GPa), due to quite small free-energy differences.

B1-type MgO and CaO, $MgSiO_3$ perovskite, the aragonite and calcite phases of $CaCO_3$, the various polymorphs of aluminum silicate, Al_2SiO_5, silica, SiO_2 and alumina, Al_2O_3 are very important constituents of the Earth's crust and lower mantle. Besides, it is believed that the Earth's D'' layer is mostly composed of post-perovskite $MgSiO_3$, while γ-spinel Mg_2SiO_4 is the dominant mineral for the lower part of Earth's transition zone. Note that Mg-based minerals do contain some amount of Fe substituting Mg. The high-pressure crystalline structure and stability of these phases are discussed in Oganov (2004 and Oganov et al. (2005). Lattice dynamics and related thermal and elastic properties of B1 MgO have been studied by Strachan et al. (1999), Drummond and Ackland (2002), Oganov et al. (2003), Oganov and Dorogokupets (2003), Karki et al. (1999, 2000), Wu et al. (2008), and Wu and Wentzcovitch (2009). Wentzcovich and co-workers have introduced a *semi-empirical ansatz* that allows for an account of explicit anharmonic contributions to the QHA estimate of various quantities, such as the TEC and C_P (Wu et al. 2008; Wu and Wentzcovitch 2009), resulting in a much improved agreement with experiments. The temperature and pressure dependence of elastic constants of B1 MgO (Isaak et al. 1990; Karki et al. 1999, 2000) calculated within QHA show very good agreement with experimental data (Isaak et al. 1989). Besides, pressure dependence of *ab initio* compressional and shear sound velocities is in consistent with seismic observations for the Earth's lower mantle (Karki et al. 1999). In contrast with these successes, the calculated thermal properties of the B1 and B2 phases of CaO (Karki and Wentzcovitch 2003) are inconsistent with experimental data, and this is most likely due to the too small lattice parameter predicted by the LDA, as later investigations based on a GGA XC functional seem to indicate (Zhang and Kuo 2009).

The thermal properties of $MgSiO_3$ and Mg_2SiO_4 and the phase transition boundary in these minerals (perovskite $\rightarrow$ post-perovskite $MgSiO_3$ and spinel $\rightarrow$ post-spinel Mg_2SiO_4) have been extensively studied (Oganov and Ono 2004; Oganov and Price 2005; Ono and Oganov 2005; Wentzcovitch et al. 2006; Yu et al. 2007, 2008; Wu et al. 2008; Wu and Wentzcovitch 2009) due to their great importance for the Earth's D'' layer and lower mantle, respectively. Improved EOS of B1 MgO (Wu et al. 2008), obtained by means of *renormalized* phonons and QHA, has been used as a new pressure calibration to re-evaluate the high pressure – high temperature phase boundary in $MgSiO_3$ and Mg_2SiO_4 minerals using experimental data from (Fei et al. 2004; Hirose et al. 2006; Speziale et al. 2001).

Alumina, Al_2O_3, plays an important role in high-pressure experiments: for example, it serves as a window material for shock-wave experiments. Cr-doped alumina, ruby, is used as a pressure calibration material in diamond-anvil-cell experiments. Besides, it is a component of solid solutions with $MgSiO_3$ polymorphs that have significantly different thermal properties from pure $MgSiO_3$ minerals. Corundum (α-Al_2O_3) is the most stable phase of alumina at ambient conditions, preceded by the θ phase at lower temperature. The energy difference between the θ and α phases of alumina is rather small, and this raised a question as to whether α-Al_2O_3 is stabilized by phonons. Zero-point vibrations stabilize the corundum phase at low temperatures (Lodziana and Parlinski 2003), whereas free-energy calculations show that the α phase cannot be stabilized by phonons only at room temperature. QHA calculations revealed that at high pressures alumina transforms to $CaIrO_3$- (Oganov and Ono 2005) and U_2S_3-type (Umemoto and Wentzcovitch 2008) polymorphs.

The *P-T* phase diagram for Al_2SiO_5 polymorphs (andalusite, sillimanite, and kyanite) (Winkler et al. 1991) and the thermal properties of $CaCO_3$ polymorphs (calcite and aragonite) (Catti et al. 1993; Pavese et al. 1996) have been studied within the QHA using model inter-atomic potentials. The effect of zero-point vibrations on the equilibrium volume in the calcite phase was found to be quite important and actually larger than the thermal expansion at relatively high temperature (Catti et al. 1993). These calculations (Pavese et al. 1996) were not able to account for the experimentally observed (Rao et al. 1968) negative in-plane TEC in calcite. The heat capacity and entropy calculated for the aragonite phase substantially deviate from experiment. All these problems can be possibly traced back to the poor transferability of model inter-atomic potentials.

The thermal properties of the α-quartz and stishovite phases of SiO_2 have been studied in Lee and Gonze (1995). The heat capacities of both phases were found to be in good agreement with experimental data (Lord and Morrow 1957; Holm et al. 1967), with the stishovite phase having a lower capacity below 480 K. Interestingly, zero-point vibration energy of the stishovite phase affects on thermodynamical properties stronger than in the α-quartz phase (Lee and Gonze 1995). The *P-T* phase diagram of SiO_2 has been examined in Oganov et al. (2005) and Oganov and Price (2005), with emphasis on the stishovite $\rightarrow$ $CaCl_2$ $\rightarrow$ α-PbO_2 $\rightarrow$ pyrite structural changes, resulting in a sequence of transitions that do not correspond to any observed seismic discontinuities within the Earth. Further investigations at ultrahigh temperature and pressure show that SiO_2 exhibits a pyrite $\rightarrow$ cotunnite phase transition at conditions that are appropriate for the core of gas giants and terrestrial exoplanets (Umemoto et al. 2006).

CONCLUSIONS

The QHA is a powerful conceptual and practical tool that complements molecular dynamics in the prediction of the thermal properties of materials not too close to the melting line. In the specific case of the Earth Sciences, the QHA can provide information on the behavior of geophysically relevant materials at those geophysically relevant pressure and temperature conditions that are not (easily) achieved in the laboratory. Large-scale calculations using the QHA for geophysical research will require the deployment of a large number of repeated structure and lattice-dynamical calculations, as well as the analysis of the massive data generated. We believe that this will require the use of dedicated infrastructures that combine some of the features of massively parallel machines with those of a distributed network of computing nodes, in the spirit of the grid computing paradigm. The QUANTUM ESPRESSO distribution of computer codes is geared for exploitation on massively parallel machines up to several thousands of closely coupled processors and is being equipped with specific tools to distribute lattice-dynamical calculation over the grid (di Meo et al. 2009).

ACKNOWLEDGMENTS

The authors wish to thank Renata M. Wentzcovitch for inspiring some of their research in this field, as well as for a critical reading of the manuscript. E.I. thanks the Swedish Research Council VR, the Swedish Foundation for Strategic Research SSF, the MS2E Strategic Research Center and the Göran Gustafsson Foundation for Research in Natural Sciences and Medicine, as well as the Russian Foundation for Basic Researches (grant #07-02-01266) for financial support.

REFERENCES

Al-Rawi AN, Kara A, Staikov P, Ghosh C, Rahman TS (2001) Validity of the quasiharmonic analysis for surface thermal expansion of Ag(111). Phys Rev Lett 86:2074-2077

Alfé D (2010) The PHON code. *http://chianti.geol.ucl.ac.uk/~dario/phon*

Alfé D, Gillan MJ (2002) Electron correlation and the phase diagram of Si. arXiv:cond-mat/0207531v1 [cond-mat.mtrl-sci]

Althoff JD, Allen PB, Wentzcovich RM, Moriarty JA (1993) Phase diagram and thermodynamic properties of solid magnesium in the quasiharmonic approximation. Phys Rev B 48:13253-13260

Antonov VE, Belash IT, Degtyareva VF, Ponyatovsky EG, Shiryayev VI (1980) Obtaining iron hydride under high hydrogen pressure. SovPhys-Dokl 25:490-492

Arroyave R, Shin D, Liu Z-K (2005) Ab initio thermodynamic properties of stoichiometric phases in the Ni-Al system. Acta Materialia, 53:1809-1819

Badding JV, Mao HK, Hemley RJ (1992) High-pressure crystal structure and equation of state of iron hydride: Implications for the Earth's Core. *In:* High Pressure Research in Mineral Physics: Application to Earth and Planetary Sciences. Syono Y, Manghnani MH (eds) Terra Scientific Co., Washington DC, p 363-371

Baroni S, de Gironcoli S, Dal Corso A, Giannozzi P (2001) Phonons and related crystal properties from density-functional perturbation theory. Rev Mod Phys 73(2):515-562

Baroni S, Debernardi A (1994) 3rd-order density-functional perturbation theory - a practical implementation with applications to anharmonic couplings in Si. Solid State Commun 91(10):813-816

Baroni S, Giannozzi P, Testa A (1987) Green's-function approach to linear response in solids. Phys Rev Lett 58(18):1861-1864

Barrera GD, Colognesi D, Mitchell PCH, Ramirez-Cuesta AJ (2005) LDA or GGA? A combined experimental inelastic neutron scattering and ab initio lattice dynamics study of alkali metal hydrides. Chem Phys 317:119-129

Biernacki S, Scheffler M (1989) Negative thermal expansion of diamond and zinc-blende semiconductors. Phys Rev Lett 63(3):290-293

Bonini N, Lazzeri M, Marzari N, Mauri F (2007) Phonon anharmonicities in graphite and graphene. Phys Rev Lett 99(17):176802

Born M, Huang K (1954) Dynamical Theory of Crystal Lattices. Clarendon Press, Oxford, UK

Born M, Oppenheimer JR (1927) Zur Quantentheorie der Moleküle. Ann Physik 84:457-484

Brüesch P (1982) Phonons: Theory and Experiments I; Lattice dynamics and Models of interatomic forces, Springer Series in Solid State Sciences, vol. 343. Springer-Verlag, Berlin, Heidelberg, New York

Caputo R, Alavi A (2003) Where do the H atoms reside in PdH_x systems? Mol Phys 101:1781-1787

Car R, Parrinello M (1985) Unified approach for molecular dynamics and density-functional theory. Phys Rev Lett 55(22):2471-2474

Catti M, Pavese A, Price GD (1993) Thermodynamic properties of $CaCO_3$ calcite and aragonite: a quasi-harmonic calculation. Phys Chem Minerals 19:472-479

Correa AA, Bonev SA, Galli G (2006) Carbon under extreme conditions: Phase boundaries and electronic properties from first-principles theory. Proc Nat Acad Sci USA 103:1204-1208

Debernardi A, Alouani M, Dreyssé H (2001) Ab initio thermodynamics of metals: Al and W. Phys Rev B 63:084305

Debernardi A, Cardona M (1996) Isotopic effects on the lattice constant in compound semiconductors by perturbation theory: An ab initio calculation. Phys Rev B 54:11305-11310

Debernardi A, Baroni S, Molinari E (1995) anharmonic phonon lifetimes in semiconductors from density-functional perturbation theory. Phys Rev Lett 75(9):1819-1822

de Gironcoli S (1995) Lattice dynamics of metals from density-functional perturbation theory. Phys Rev B 51(10):6773-6776 DeCicco PD, Johnson FA (1969) The quantum theory of lattice dynamics. IV. Proc R Soc London A 310(1500):111-119

Dever DJ (1972) Temperature dependence of the elastic constants in -iron single crystals: relationship to spin order and diffusion anomalies. J Appl Phys 43:3293-3301

di Meo R, Dal Corso A, Cozzini S (2009) Calculation of phonon dispersions on the GRID using Quantum ESPRESSO. *In:* Proceedings of the COST School, Trieste. ICTP Lecture Notes Series 24, *http://users.ictp.it/~pub_off/lectures/lns024/10-giannozzi/10-giannozzi.pdf*

Drummond ND, Ackland GJ (2002) Ab initio quasiharmonic equations of state for dynamically stabilized soft-mode materials. Phys Rev B 65:184104

Dubrovinsky L, Dubrovinskaia N, Crichton WA, Mikhaylushkin AS, Simak SI, Abrikosov IA, de Almeida JS, Ahuja R, Luo W, Johansson B (2007) Noblest of all metals is structurally unstable at high pressure. Phys Rev Lett 98:045505

Eckman M, Persson K, Grimvall G (2000) Lattice dynamics and thermodynamic properties of the -Sn phase of Si. Phys Rev B 62:14784-14789

Elsasser C, Zhu J, Louie SG, Meyer B, Fahnle M, Chang CT (1998) Ab initio study of iron and iron hydride: II. Structural and magnetic properties of close-packed Fe and FeH. J Phys Condens Matter 10:5113-5129

Fei Y, Orman JV, Li J, Westrenen WV, Sanloup C, Minarik W, Hirose K, Komabayashi T, Walter M, Funakoshi K (2004) Experimentally determined postspinel transformation boundary in Mg_2SiO_4 using MgO as an internal pressure standard and its geophysical iplications. J Geophys Res 109:B02305

Feynman RP (1939) Forces in molecules. Phys Rev 56(4):340-343

Fleszar A, Gonze X (1990) First-prhinciples thermodynamical properties of semiconductors. Phys Rev Lett 64:2961

Gaal-Nagy K, Bauer A, Schmitt M, Karch K, Pavone P, Strauch D (1999) Temperature and dynamical effects on the high-pressure cubic-diamond-tin phase transition in Si and Ge. Phys Status Solid B 211:275-280

Gannarelli CMS, Alfe D, Gillan MJ (2003) The particle-in-cell model for ab initio thermodynamics: implications for the elastic anisotropy of the Earth's inner core. Phys Earth Planet Inter 139:243-253

Giannozzi P, Baroni S, Bonini N, Calandra M, Car R, Cavazzoni C, Ceresoli D, Chiarotti GL, Cococcioni M, Dabo I, Dal Corso A, de Gironcoli S, Fabris S, Fratesi G, Gebauer R, Gerstmann U, Gougoussis C, Kokalj A, Lazzeri M, Martin-Samos L, Marzari N, Mauri F, Mazzarello R, Paolini S, Pasquarello A, Paulatto L, Sbraccia C, Scandolo S, Sclauzero G, Seitsonen AP, Smogunov A, Umari P, Wentzcovitch RM (2009) QUANTUM ESPRESSO: a modular and open-source software project for quantum simulations of materials. J Phys Condens Matter 21(39):395502 (19 pp)

Gonze X, Beuken J-M, Caracas R, Detraux F, Fuchs M, Rignanese G-M, Sindic L, Verstraete M, Zerah G, Jollet F, Torrent M, Roy A, Mikami M, Ghosez Ph, Raty J-Y, Allan DC (2008) First-principle computation of material properties: the ABINIT software project. Comput Mater Sci 25:478-492

Gornostyrev YuN, Kontsevoi OYu, Khromov KYu, Katsnelson MI, Freeman AJ (2007) The role of thermal expansion and composition changes in the temperature dependence of the lattice misfit in two-phase γ/γ' superalloys. Scr Mater 56:81-84

Grabowski B, Ismer L, Hickel T, Neugebauer J (2009) Ab initio up to the melting point: Anharmonicity and vacancies in aluminum. Phys Rev B 79:134106

Gülseren O, Cohen (2002) RE High-pressure thermoelasticity of body-centered-cubic tantalum. Phys Rev B 65:064103

Hamdi I, Aouissi M, Qteish A, Meskini N (2006) Pressure dependence of vibrational, thermal, and elastic properties of ZnSe: An ab initio study. Phys Rev B 73:174114

Hansen U, Vogl P, Fiorentini V (1999) Quasiharmonic versus exact surface free energies of Al: A systematic study employing a classical interatomic potential. Phys Rev B 60:5055-5064

Hellmann H (1937) Einführung in die Quantenchemie. Deuticke, Leipzig

Herrero CP, Ramírez R (2000) Structural and thermodynamic properties of diamond: A path-integral Monte Carlo study. Phys Rev B 63:024103

Hirao N, Kondo T, Ohtani E, Takemura K, Kikegawa T (2004) Compression of iron hydride to 80 GPa and hydrogen in the Earth's inner core. Geophys Res Lett 31:L06616

Hirose K, Sinmyo R, Sata N, Ohishi Y (2006) Determination of post-perovskite phase transition boundary in $MgSiO_3$ using Au and MgO internal pressure standards. Geophys Res Lett 33:L01310

Hohenberg P, Kohn W (1964) Inhomogeneous Electron Gas. Phys Rev 136(3B):B864-B871

Holm JL, Kleppa OJ, Westrum EF Jr (1967) Thermodynamics of polymorphic transformations in silica. Thermal properties from 5 to 1070 K and pressure-temperature stability fields for coesite and stishovite. Geochim Cosmochim Acta 31:2289-2307

Hu CH, Wang YM, Chen CM, Xu DS, Yang K (2007) First-principles calculations of structural, electronic, and thermodynamic properties of Na_2BeH_4. Phys Rev B 76:144104

Isaak DG, Cohen RE, Mehl MJ (1990) Calculated elastic and thermal properties of MgO at high pressures and temperatures. J Geophys Res 95(B5):7055-7067

Isaak DG, Anderson OL, Goto T (1989) Measured elestic modulus of single-crystal MgO up to 1800 K. Phys Chem Miner 16:704-713

Isaak DG, Masuda K (1995) Elastic and viscoelastic properties of iron at high temperatures. J Geophys Res Solid Earth 100:17689-17698

Isaev E (2010) QHA code to calculate thermodynamics properties. *http://qha.qe-forge.org*

Isaev EI, Skorodumova NV, Ahuja R, Vekilov YK, Johansson B (2007) Dynamical stability of Fe-H in the Earth's mantle and core regions. Proc Nat Acad Sci USA 104:9168-9171

Ismail, Plummer EW, Lazzeri M, de Gironcoli S (2001) Surface oscillatory thermal expansion: Mg. Phys Rev B 63:233401

Karki BB, Wentzcovitch RM, de Gironcoli S, Baroni S (2000) High-pressure lattice dynamics and thermoelasticity of MgO. Phys Rev B 61:8793-8800

Karki BB, Wentzcovitch RM, de Gironcoli S, Baroni S (1999) First-principles determination of elastic anisotropy and wave velocities of MgO at lower mantle conditions. Science 286:1705-1707

Karki BB, Wentzcovitch RM (2003) Vibrational and quasiharmonic thermal properties of CaO under pressure. Phys Rev B 68:224304

Kern G, Kresse G, Hafner J (1999) Ab initio calculation of the lattice dynamics and phase diagram of boron nitride. Phys Rev B 59:8551-8559

Kittel C (1996) Introduction to Solid State Physics, 7th edition. John Wiles & Sons, New York

Kohn W, Sham LJ (1965) Self-consistent equations including exchange and correlation effects. Phys Rev 140(4A):A1133-A1138

Körmann F, Dick A, Grabowski B, Hallstedt B, Hickel T, Neugebauer J (2008) Free energy of bcc iron: Integrated ab initio derivation of vibrational, electronic, and magnetic contributions. Phys Rev B 78:033102

Kwon YK, Berber S, Tománek D (2004) Thermal contraction of carbon fullerenes and nanotubes. Phys Rev Lett 92:015901

Lazzeri M, Calandra M, Mauri F (2003) Anharmonic phonon frequency shift in MgB_2. Phys Rev B 68(22):220509

Lazzeri M, de Gironcoli S (2002) First-principles study of the thermal expansion of Be(100). Phys Rev B 65:245402

Lebègue S, Alouani M, Arnaud B, Pickett WE (2003) Pressure-induced simultaneous metal-insultor and structural-phase transitions in LiH: A quasiparticle study. Europhys Lett 63:562-568

Lee C, Gonze X (1995) Ab initio calculation of the thermodynamic properties and atomic temperature factors of SiO_2 - quartz and stishovite. Phys Rev B 51:8610-8613

Leese J, Lord AE Jr (1968) Elastic stiffness coefficients of single-crystal iron from room temperature to 500 C. J Appl Phys 39:3986-3988

Lehmann G, Taut M (1972) On the numerical calculation of the density of states and related properties. Phys Status Solidi B 54:469-477

Li Z, Tse JS (2000) Ab initio studies on the vibrational and thermal properties of Al_3Li. Phys Rev B 61:14531-14536

Lifshitz IM (1952) Thermal properties of chain and layered structures at low temperatures. Zh Eksp Teor Fiz 22:475-486

Liu AY, Quong AA, Freericks JK, Nicol EJ, Jones EC (1999) Structural phase stability and electron-phonon coupling in lithium. Phys Rev B 59:4028-4035

Lodziana Z, Parlinski K (2003) Dynamical stability of the and phases of alumina. Phys Rev B 67:174106

Lord RC, Morrow JC (1957) Calculation of the heat capacity of - quartz and vitreous silica from spectroscopic data. J Chem Phys 26:230-232

Lozovoi AY, Mishin Y (2003) Point defects in NiAl: The effect of lattice vibrations. Phys Rev B 68:184113

Ma YZ, Somayazulu M, Shen GY, Mao HK, Shu JF, Hemley RJ (2004) In situ X-ray diffraction studies of iron to Earth-core conditions. Phys Earth Planet Inter 143-144:455-467

Martin RM (2004) Electronic Structure: Basic Theory and Practical Methods. Cambridge University Press, Cambridge, UK

Marzari M,Vanderbilt D, De Vita A, Payne MC (1999) Thermal contraction and disordering of the Al(110) surface. Phys Rev Lett 82:3296-3299

Messiah A (1962) Quantum Mechanics. North Holland, Amsterdam
Mikhaylushkin AS, Simak SI, Dubrovinsky L, Dubrovinskaia N, Johansson B, Abrikosov IA (2007) Pure iron compressed and heated to extreme conditions. Phys Rev Lett 99:165505
Mounet N, Marzari N (2005) First-principles determination of the structural, vibrational and thermodynamic properties of diamond, graphite, and derivatives. Phys Rev B 71:205214
Narasimhan S, Scheffler M (1997) A model for the thermal expansion of Ag(111) and other metal surfaces. Z Phys Chem 202:253-262
Narasimhan S, de Gironcoli S (2002) Ab initio calculation of the thermal properties of Cu: Performance of the LDA and GGA. Phys Rev B 65:064302
Nelin G (1971) A neutron diffraction study of palladium hydride. Phys Status Solidi B 45:527-536
Oganov AR (2004) Theory of minerals at high and ultrahigh pressures: structure, properties, dynamics, and phase transitions. *In:* High-Pressure Crystallography in NATO Science Series: II: Mathematics, Physics, and Chemistry. Katrusiak A, McMillan PF (eds) Kluwer Academic Publishers, Dordrecht, p 199-215
Oganov AR, Dorogokupets PI (2003) All-electron and pseudopotential study of MgO: Equation of state, anharmonicity, and stability. Phys Rev B 67:224110
Oganov AR, Gillan MJ, Price GD (2003) Ab initio lattice dynamics and structural stability of MgO. J Chem Phys 118:10174
Oganov AR, Gillan MJ, Price GD (2005) Structural stability of silica at high pressures and temperatures. Phys Rev B 71:064104
Oganov AR, Ono S (2005) The high-pressure phase of alumina and implications for Earth's D″ layer. Proc Nat Acad Sci USA 102:10828-10831
Oganov AR, Ono S (2004) Theoretical and experimental evidence for a post-perovskite phase of $MgSiO_3$ in Earth's D″ layer. Nature 430:445-448
Oganov AR, Price GD (2005) Ab initio thermodynamics of $MgSiO_3$ perovskite at high pressures and temperatures. J Chem Phys 122:124501
Oganov AR, Price GD, Scandolo S (2005) Ab initio theory of planetary materials. Z Kristallogr 220:531-548
Ono S, Oganov AR (2005) In situ observations of phase transition between perovskite and $CaIrO_3$-type phase in $MgSiO_3$ and pyrolitic mantle composition. Earth Planet Sci Lett 236:914-922
Ozolins V, Wolverton C, Zunger A (1998) First-principles theory of vibrational effects on the phase stability of Cu-Au compounds and alloys. Phys Rev B 58:R5897-R5900
Parlinski K (2010) The PHONON software. *http://wolf.ifj.edu.pl/phonon/*
Pavese A, Catti M, Parker SC, Wall A (1996) Modelling of the thermal dependence of structural and elastic properties of calcite, $CaCO_3$. Phys Chem Minerals 23:89-93
Pavone P, Karch K, Schütt O, Windl W, Strauch D, Giannozzi P, Baroni S (1993) Ab initio lattice dynamics of diamond. Phys Rev B 48:3156-3163
Pavone P (1991) Lattice Dynamics of Semiconductors from Density-Functional Perturbation Theory. PhD thesis, SISSA/ISAS, Trieste, Italy
Pavone P (2001) Old and new aspects in lattice-dynamical theory. J Phys Condens Matter 13:7593-7610
Pavone P, Baroni S, de Gironcoli S (1998) $\alpha \leftrightarrow \beta$ phase transition in tin: A theoretical study based on density-functional perturbation theory. Phys Rev B 57:10421-10423
Persson K, Ekman M, Grimvall G (1999) Dynamical and thermodynamical instabilities in the disordered Re_xW_{1-x} system. Phys Rev B 60:9999-10007
Pick RM, Cohen MH, Martin RM (1970) microscopic theory of force constants in the adiabatic approximation. Phys Rev B 1(2):910-920
Pohl K, Cho J-H, Terakura K, Scheffler M, Plummer EW (1998) Anomalously large thermal expansion at the (0001) surface of beryllium without observable interlayer anharmonicity. Phys Rev Lett 80:2853-2856
Quong AA, Liu AY (1997) First-principles calculations of the thermal expansion of metals. Phys Rev B 56(13):7767-7770
Rao KVK, Naidu SVN, Murthy KS (1968) Precision lattice parameters and thermal expansion of calcite. J Phys Chem Solids 29:245-248
Rickman JM, LeSar R (2002) Free -- energy calculations in materials research. Ann Rev Mater Res 32:195-217
Rignanese G-M, Michenaud J-P, Gonze X (1996) Ab initio study of the volume dependence of dynamical and thermodynamical properties of silicon. Phys Rev B 53:4488-4497
Roma G, Bertoni CM, Baroni S (1996) The phonon spectra of LiH and LiD from density-functional perturbation theory. Solid State Commun 98:203-207
Rowe JM, Rush JJ, de Graaf LA, Ferguson GA (1972) Neutron quasielastic scattering study of hydrogen diffusion in a single crystal of palladium. Phys Rev Lett 29:1250-1253
Sha X, Cohen RE (2006a) First-principles thermoelasticity of bcc iron under pressure. Phys Rev B 74:214111
Sha X, Cohen RE (2006b) Lattice dynamics and thermodynamics of bcc iron under pressure: First-principles linear response study. Phys Rev B 73:104303

Sha X, Cohen RE (2006c) Thermal effects on lattice strain in -Fe under pressure. Phys Rev B 74:064103
Shannon CE (1949) Communication in the presence of noise. Proc Institute of Radio Engineers 37(1):10-21
Singh RK (1982) Many body interactions in binary ionic solids. Phys Rep 85:261-401
Speziale S, Zha C, Duffy TS, Hemley RJ, Mao HK (2001) Quasi-hydrostatic compression of magnesium oxide to 52 GPa: implications for the pressure-volume-temperature equation of state. J Geophys Res 106:515-528
Strachan A, Cagin T, Goddard WA III (1999) Phase diagram of MgO from density-functional theory and molecular-dynamics simulations. Phys Rev B 60:15084-15093
Sun T, Umemoto K, Wu Z, Zheng J-C, Wentzcovitch RM (2008) Lattice dynamics and thermal equation of state of platinum. Phys Rev B 78:024304
Taioli S, Cazorla C, Gillan MJ, Alfe D (2007) Melting curve of tantalum from first principles. Phys Rev B 75:214103
Tsuchiya T (2003) First-principles prediction of the P-V-T equation of state of gold and the 660-km discontinuity in Earth's mantle. J Geophys Res 108(B10):2462
Umemoto K, Wentzcovitch RM, Allen PB (2006) Dissociation of MgSiO in the Cores of Gas Giants and Terrestrial Exoplanets. Science, 311:983-986
Umemoto K, Wentzcovitch RM (2008) Prediction of an U_2S_3-type polymorph of Al_2O_3 at 3.7 Mbar. Proc Nat Acad Sci USA 105:6526-6530
Vocadlo L, Alfé D (2002) Ab initio melting curve of the fcc phase of aluminum. Phys Rev B 65:214105
Wallace DC (1998) Thermodynamics of crystals. Dover, New York
Wang Y, Liu Z-K, Chen L-Q (2004) Thermodynamic properties of Al, Ni, NiAl, and Ni_3Al from first-principles calculations. Acta Mater 52:2665-2671
Wentzcovitch RM, Tsuchiya T, Tsuchiya J (2006) MgSiO postperovskite at D'' consitions. Proc Nat Acad Sci USA 103:543-546
Winkler B, Dove MT, Leslie M (1991) Static lattice energy minimization and lattice dynamics calculations on aluminosilicate minerals. Am Mineral 76:313-331
Wu Z, Wentzcovitch RM (2009) Effective semiempirical ansatz for computing anharmonic free energies. Phys Rev B 79:104304
Wu Z, Wentzcovitch RM, Umemoto K, Li B, Hirose K, Zheng J-C (2008) Pressure-volume-temperature relations in MgO: An ultrahigh pressure-temperature scale for planetary sciences applications. J Geophys Res 113:B06204
Xie J, Chen SP, Brand HV, Rabie RL (2000) High-pressure thermodynamic, electronic and magnetic properties of Ni. J Phys Condens Matter 12:8953-8962
Xie J, Chen SP, Tse JS, de Gironcoli S, Baroni S (1999a) High-pressure thermal expansion, bulk modulus, and phonon structure of diamond. Phys Rev B 60:9444-9449
Xie J, de Gironcoli S, Baroni S, Scheffler M (1999b) First-principles calculation of the thermal properties of silver. Phys Rev B 59:965-969
Xie J, de Gironcoli S, Baroni S, Scheffler M (1999c) Temperature-dependent surface relaxations of Ag(111). Phys Rev B 59:970-974
Xie J, Scheffler M (1998) Structure and dynamics of Rh surfaces. Phys Rev B 57:4768-4775
Yu YG, Wentzcovitch RM, Tsuchiya T (2007) First principles investigation of the postspinel transition in Mg_2SiO_4. Geophys Res Lett 34:L10306
Yu YG, Wu Z, Wentzcovitch RM (2008) α-β-γ transformations in Mg_2SiO_4 in Earth's transition zone. Earth Planet Sci Lett 273:115-122
Zabel H (2001) Phonons in layered compounds. J Phys Condens Matter 13:7679-7690
Zhang JY, Zhang LJ, Cui T, Niu YL, Ma YM, He Z, Zou GT (2007) A first-principles study of electron-electron coupling in electron doped LiH. J.Phys Condens Matter 19:425218
Zhang J, Kuo J-L (2009) Phonon and elastic instabilities in rocksalt calcium oxide under pressure: a first-principles study. J Phys Condens Matter 21:015402
Zhang Q, Cagin T, Goddard WA III (2006) The ferroelectric and cubic phases in BaTiO ferroelectric are also antiferroelectric. Proc Nat Acad Sci USA 103:14695-14700
Zimmermann J, Pavone P, Cuniberti G (2008) Vibrational modes and low-temperature thermal properties of graphene and carbon nanotubes: Minimal force-constant model. Phys Rev B 78:045410

Reviews in Mineralogy & Geochemistry
Vol. 71 pp. 59-98, 2010

Thermodynamic Properties and Phase Relations in Mantle Minerals Investigated by First Principles Quasiharmonic Theory

Renata M. Wentzcovitch

Department of Chemical Engineering and Materials Sciences and Minnesota Supercomputer Institute, University of Minnesota, Minneapolis, Minnesota 55455, U.S.A.

wentzcov@cems.umn.edu

Yonggang G. Yu

*Department of Geosciences
Virginia Polytechnic Institute and State University, Blacksburg, Virginia 24060, U.S.A.*

Zhongqing Wu

*Collaboratory for Advanced Computing and Simulations
University of South California
Los Angeles, California 90007, U.S.A.*

INTRODUCTION

This research has been motivated by geophysics and materials physics. The objective of this research has been to advance materials theory and computations for high pressure and high temperature applications to the point that it can make a difference in our understanding of the Earth. Understanding of the mineralogy, composition, and thermal structure of the Earth evolves by close interaction of three fields: seismology, geodynamics, and mineral physics. Earth's structure is imaged by seismology, which obtains body wave velocities and density throughout the Earth's interior (Fig. 1). Interpretation of this data relies on knowledge of aggregate properties of Earth forming materials either measured in laboratory or calculated, in many cases by both. However, the conditions of the Earth's interior, especially at the core, may be challenging for important experiments, and materials computations have emerged to contribute decisively to this field. Earth's evolution is simulated by geodynamics, but these simulations need as input information about rheological and thermodynamic properties of minerals, including phase transformation properties such as Clapeyron slopes. Our research has concentrated on the phases of the Earth's mantle, particularly the deep mantle whose conditions are more challenging for experiments. The mantle accounts for ~83% of the Earth's volume. In this article we will review the essential first-principles approach used to investigate solids at high temperatures and pressures, summarize their performance for mantle minerals (Fig. 2), and point to critical results that have stirred us in the current research path. The success is remarkable, but some limitations point the way to future developments in this field.

The methods used in this research are parameter free, based on density functional theory (DFT) (Hohenberg and Kohn 1964; Kohn and Sham 1965) and the quasiharmonic approximation (Born and Huang 1956). The established framework used to describe the total energy of a system of electrons and ions, DFT, performs very well for most silicates and oxides of the

1529-6466/10/0071-0004$05.00 DOI: 10.2138/rmg.2010.71.4

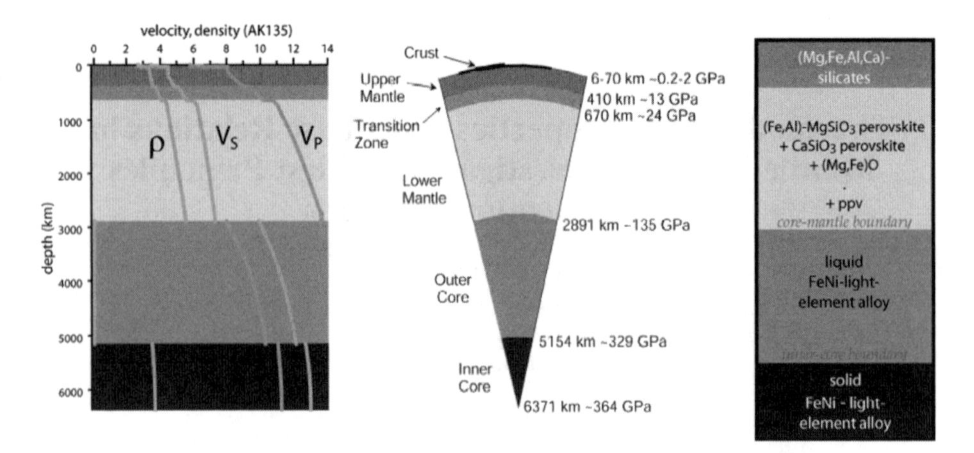

Figure 1. (a) Velocities and density profiles (Kennett et al. 1995), (b) Earth's cross section, and (c) composition and mineralogy of Earth's layers.

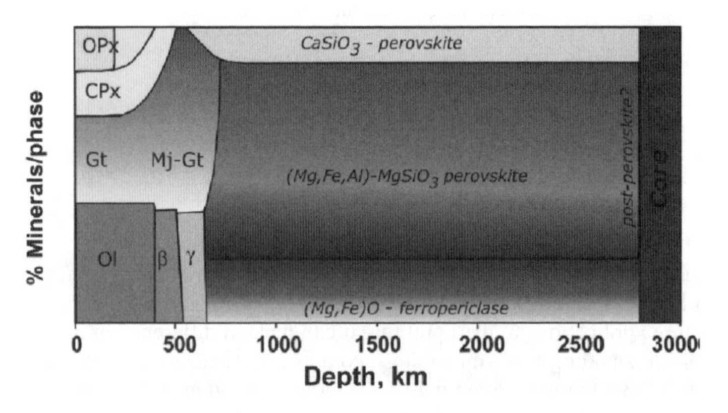

Figure 2. Phases proportions versus depth and or pressure. **Ol** = $(Mg,Fe)_2SiO_4$ olivine, also known as α-phase; β = $(Mg,Fe)_2SiO_4$ wadsleyite, also known as β-phase; γ = $(Mg,Fe)_2SiO_4$ ringwoodite, also known as γ-phase; **Gt** = garnet; **Mj-Gt** = majorite garnet, solid solution between pyrope $Mg_3Al_2(SiO_4)_3$ and majorite $MgSiO_3$ joint, **CPx** = clinopyroxene; **OPx** = orthopyroxene; **Ca-Pv** = $CaSiO_3$-perovskite; **(Mg,Fe,Al)-perovskite** = $(Mg,Al,Fe)(Si,Al)O_3$; **post-perovskite** = $(Mg,Al,Fe)(Si,Al)O_3$ **ferropericlase** = (Mg,Fe)O.

mantle. The theoretical foundation of DFT is reviewed by Perdew and Ruzsinski (2010) in this issue. Several approximations to the exchange-correlation functional are reviewed and tested in molecular clusters by Zhao and Truhlar (2010) in this issue also. For minerals containing iron it is necessary in general, to go beyond DFT. Cococcioni (2010) reviews in this issue the first-principles DFT plus Hubbard U (DFT+U) method (Anisimov et al. 1991). The DFT+U functional is an *ansatz*, but U is not a free parameter. It is determined consistently using linear response theory (Cococcioni and de Gironcoli 2005). Applications of this method are discussed also in the chapter on spin crossovers in iron-bearing minerals in this volume.

A word must be said about pseudopotentials. They are numerically engineered entities that replace the potential generated by the atomic nuclei and core electrons. They can be obtained by several methods. One might contend that they are not parameter free. They use cut off radii in the numerical manipulation of the valence electron wavefunctions in the process of generating the pseudopotential. However, these parameters are not really free. They have optimum values that maximize the pseudopotential transferability to environments other than the atom while simultaneously maximizing their "smoothness." The latter is important to speed up plane-wave calculations which use their Fourier transforms. These two criteria create a tug-of-war in the choice of cut-off radii: smoothness improves by increasing cut-off-radii while transferability improves by decreasing them. The optimum set of radii depends on the degree one is willing to compromise on smoothness or transferability. For this reason several methods to generate pseudopotentials have been proposed (Hamann et al. 1979; Vanderbilt 1990; Troullier and Martins 1991). They have evolved with the aim of decreasing the conflict imposed by these two requirements. Experience matters in generating pseudopotentials. Results are sensitive to their quality. However, a good pseudopotential should represent well an atom core in any environment, in any material. Broadly speaking, this is the ultimate guiding principle for pseudopotential generation.

In this article we review the systematic performance of two popular exchange-correlations functionals, the local density approximation (LDA) (Ceperley and Alder 1980; Perdew and Zunger 1981) and the generalized gradient approximation of Perdew, Burke, and Ernzerhof (PBE-GGA or GGA for short) (Perdew et al. 1996) combined with quasiharmonic theory and first principles phonons obtained using density functional perturbation theory (DFPT) (Baroni et al. 2001). Density functional theory as applied to phonons is reviewed by Baroni et al. (2010) in this volume. The performance of these functionals is examined on mantle silicates and oxides. Probably the most important message here is the critical role played by zero point motion on the structural properties (lattice parameters, compressibility's, etc). Hence, inferences regarding the performance of various exchange correlation functionals should take this effect into account.

Here we need to discuss the quasiharmonic approximation (QHA) and its performance in some details. Its application to minerals at mantle conditions has been viewed with some skepticism in the past (or even today). Obviously, its application must be limited to a certain temperature range before anharmonic effects become non-negligible. Therefore, the question becomes: how small are anharmonic effects for minerals at relevant conditions of the mantle? We have been guided by the relationship between the measured melting temperature, T_M, of mantle phases and the expected mantle temperatures (see Fig. 3). More will be said about this later, but the latter is ~2/3 of the former in most of the mantle, except in the ~300 km above the core-mantle boundary, where the temperature rises to ~4,000 K. Therefore, the QHA is expected to be a good approximation for most of the mantle, but maybe questionable near the core mantle boundary. Comparisons of QHA results with experiments are very helpful in addressing this question. Ideally one would want to compare QHA predictions with results of molecular dynamics (MD) simulations. However, MD results should be converged with respect to the number of atoms, and have small statistical errors to allow for such comparisons. QHA results correspond to dynamics on more than 10^4 atoms.

We have also investigated the magnitude of anharmonic effects at mantle conditions, where experimental results are not usually available. We have developed an approximate functional form for the anharmonic free energy parameterized to *some* experimental data on a material and used it to reproduce *all* available experimental thermodynamic data on the same material (Wu and Wentzcovitch 2009). This anharmonic free energy is then extended to mantle conditions to investigate anharmonic effects on thermodynamic properties and phase boundaries. This was a pedagogical exercise that revealed multiple manifestations of anharmonicity. This approximation assumes that all phonons are stable in the pressure and temperature range investigated. Therefore, it should not be applicable to unstable structures stabilized by anharmonic fluctuations, such as $CaSiO_3$-perovskite (Stixrude et al. 2007).

A disclaimer is in order. We will not discuss extensively first-principles results obtained by others on the phases analyzed here. Conclusions regarding the performance of first-principles

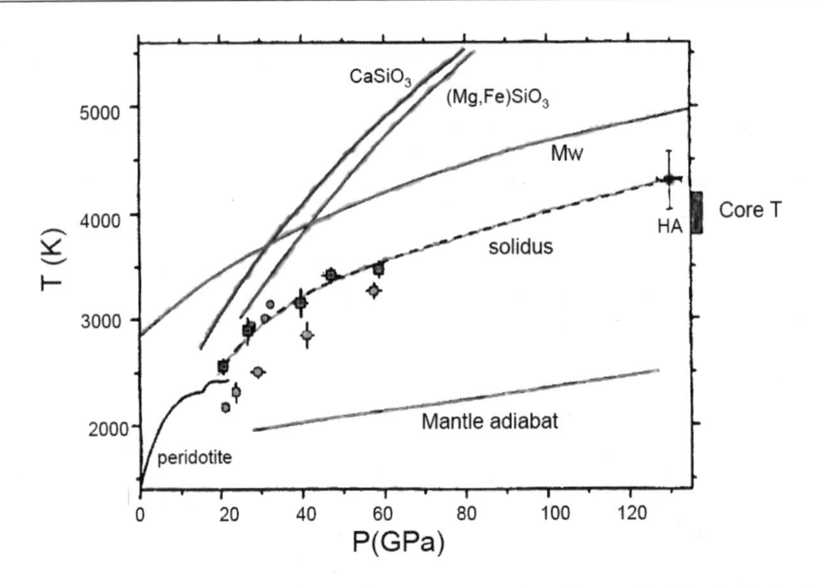

Figure 3. Illustration of melting temperature of mantle minerals (from Zerr et al. 1998). HA is the shock melting point of olivine (Holland and Ahrens 1997). For a more complete overview of experimental melting temperatures of mantle minerals, see Shen and Heinz (1998). Mantle temperature is according to Brown and Shankland (1981). The considerable difference between melting temperatures and expected mantle temperatures, suggested the use of the quasiharmonic approximation (QHA) to investigate thermodynamics properties of minerals at mantle conditions. [Used by permission of Science, from Zerr et al. 1998].

methods can be drawn more safely by comparing results generated systematically, using the same pseudopotentials, and with a homogeneous quality control standard.

THE QUASIHARMONIC APPROXIMATION (QHA)

The QHA is a simple yet effective approximation to thermodynamics properties of crystals. Its effectiveness has become more evident with the advent of first-principles phonons (Gianozzi et al. 1991). This approximation treats phonons as if they did not interact, as a phonon "gas." The system becomes equivalent to a collection of independent harmonic oscillators. If the phonon frequencies are known, the energy levels of the system are well defined, and the partition function and Helmholtz free energy can be calculated. From the latter, all thermodynamic properties can be defined. The partition function, Z, of one oscillator with frequency ω is the summation of the Boltzmann factors of all of its energy levels:

$$Z_i = \sum_{j=0}^{\infty} e^{-\frac{\varepsilon_j(\omega_i)}{k_B T}} \tag{1}$$

where $\varepsilon_j(\omega_i) = (j + \frac{1}{2})\hbar\omega_i$ (i.e., $\frac{1}{2}\hbar\omega_i$, $\frac{3}{2}\hbar\omega_i$, $\frac{5}{2}\hbar\omega_i$, ...) are the energy levels of the oscillator. This summation can be contracted as

$$Z_i = \frac{e^{-\frac{\hbar\omega_i}{2k_B T}}}{1 - e^{-\frac{\hbar\omega_i}{k_B T}}} \tag{2}$$

and the Helmholtz free energy of this oscillator, $F_i^{\text{th}} = -k_B T \ln Z_i$, becomes

$$F_i = \frac{1}{2}\hbar\omega_i + k_B T \ln(1 - e^{-\frac{\hbar\omega_i}{k_B T}}) \tag{3}$$

For a lattice with vibrational modes with frequencies ω_i, the total free energy is

$$F = E + \sum_i F_i^{th} = U + \sum_i \frac{1}{2}\hbar\omega_i + k_B T \sum_i \ln(1 - e^{-\frac{\hbar\omega_i}{k_B T}}) \tag{4}$$

where U is the internal energy, in our case the static DFT energy. As the solid compresses and deforms elastically, E, ω_i's, and F change. With e_{ij}'s as Lagrangian strains defining elastic deformation of a pre-compressed state, the free energy $F(V,T,e_{ij})$ can be used to derive all thermodynamics and thermoelastic properties of the solid. Thermoelasticity is addressed in the article by Wentzcovitch et al. (2010) in this volume, so, here e_{ij}'s are zero. Pressure and entropy are:

$$S = -\left(\frac{\partial F}{\partial T}\right)_V \qquad P = -\left(\frac{\partial F}{\partial V}\right)_T \tag{5}$$

and the thermodynamic potentials can be evaluate in straightforward manner after $V \leftrightarrow P$ variable interchange, if necessary.

$$\begin{aligned} E &= F + TS \\ H &= E + PV \\ G &= F + PV \end{aligned} \tag{6}$$

This is the standard quasiharmonic thermodynamic formalism which we refer to as "statically constrained" QHA (Carrier et al. 2007). In this approach, by construction, phonon frequencies and crystal structures are uniquely related to the volume, irrespective of temperature. This relationship is established by static calculations that optimize crystal structures for a set of chosen pressures (Wentzcovitch et al. 1993). Phonon frequencies are then calculated for these structures, and neither of them changes with temperature after this point. Only pressure changes with temperature at these volumes. This fact has significant implications, particularly for thermoelasticity (Carrier et al. 2007, 2008), but this discussion will be deferred to the next paper on thermoelasticity of minerals. As is, the statically constrained QHA is a very effective approach to study the thermodynamics of solids at low temperatures but not at high temperatures where phonon-phonon interactions, or anharmonic effects, are important. At high T, such as near-melting temperatures, the dynamics of ions is more appropriately treated by classical MD (Oganov et al. 2001). The Debye temperature, Θ_D, is the conventional upper limit of applicability of the QHA, but we have suggested a more direct criterion based on an *a posteriori* inspection of the thermal expansivity (Wentzcovitch et al. 2004a; Carrier et al. 2007).

In this paper we also review phase relations in the most important mantle phases. Direct comparison of the Gibbs free energy of phases establishes their thermodynamic phase boundary. Phase transitions in mantle minerals produce important mantle velocity and density discontinuities. Uncertainties in the experimentally determined phase boundaries are still significant and calculations have a role to play if they can be predictive. Here we summarize the performance of the most popular exchange-correlation functionals, LDA and PBE-GGA (GGA hereafter) in the prediction of phase boundaries between mantle silicates.

THERMODYNAMIC PROPERTIES OF MANTLE PHASES

In first approximation, Earth is spherically homogeneous and has a layered structure. Crust, mantle, and core are the three layers with distinct chemical compositions (Fig. 1). Boundary discontinuities within these layers indicate sharp phase changes induced by pressure and temperature. Lateral velocity variations are more subtle and tell much about the Earth's dynamics. They are primarily related with lateral variations of temperature, subtle chemical composition, and mineralogy. Together with knowledge of the thermodynamic properties of minerals, lateral velocity variations provide essential information for geodynamic modeling. The latter aims to describe the Earth's dynamic evolution.

The main theme of our work is the silicate mantle with its own internal layers and the minerals that comprise this region. At mantle boundary layers, phase transitions between predominant minerals are taking place. These boundaries are ~10 km wide, the width being related with phase loops intrinsic to multi-phase multi-component aggregates of the mantle in thermodynamic equilibrium (see chapter by Stixrude and Lithgow-Bertelloni 2010 in this volume). They also have topography related with lateral variations of temperature, chemical composition or mineral phases (see Fig. 2). Knowledge of these mineral phase boundaries helps to understand conditions and processes taking place at these mantle boundary layers. In contrast with the Earth's crust, which is composed of a tremendous variety of mineral phases, the Earth's mantle appears to be relatively simple and consists of few mineral phases (see Figs. 1 and 2).

Despite the small number of phases, they are solid solutions co-existing in thermodynamic equilibrium. Description of such complex systems poses challenges to first-principles calculations and must be addressed in installments. Semi-empirical approaches to thermodynamic equilibrium, such as that reviewed by Stixrude and Lithgow-Bertelloni (2010) in this volume, are currently the most practical. Mantle solid solutions are primarily magnesium-iron silicates and oxides, some containing aluminum or calcium also. In addition, the iron bearing phases of the lower mantle undergo spin crossovers. Properties of some of these systems are reviewed by Hsu et al. (2010) in this volume. Here we discuss primarily the magnesium end member phases. Our calculations are consistent and systematic using the same set of pseudopotentials with the same convergence criterion (see appendix for computational details including pseudopotential information). These systematic results allow one to assess the performance of the LDA and GGA for these systems and to infer trends reliably.

MgO

MgO in the rocksalt structure is the primary end-member of ferropericlase, $(Mg_{(1-x)}Fe_x)O$ (x~0.18), the second most abundant mineral phase in the Earth's lower mantle after Al-(Mg,Fe) SiO_3 (*Pbnm*) perovskite. As a simple monoxide, it was a good test-case for the first-application of QHA along with the LDA and PBE-GGA. The calculated phonon dispersions of MgO (Karki et al. 1999, 2000a) using DFPT (Baroni et al. 2001) are in excellent agreement with neutron scattering experiments by Sangster et al. (1970) (see Fig. 4). This demonstrates the success of LDA and the linear response theory applied to ionic crystals. The calculated phonon dispersion and density of states are then used along with Equation (4) to compute the Helmholtz free energy from which volume vs. pressure relations at various temperatures are obtained. The result is shown in Figure 5. The dashed line corresponds to static free energy results (0 K DFT results without zero point motion energy). The black solid line corresponds to results at 300 K. The difference between the two curves is caused primarily by zero point motion (2nd term on the r.h.s. of Eqn. 4) which is positive, and to a lesser extent by the thermal excitation energy (3rd term on the r.h.s. of Eqn. 4) which is negative. The free energy shift from static values to 300 K values is more acute at smaller volumes owing to increased values of ω's at high pressures, as shown in Fig. 4. The dependence of individual phonon frequency on volume is quantified by the mode Grüneisen parameter (γ_i), defined as $\gamma_i = (d\ln\omega_i/d\ln V)$ with i labeling modes. With increasing temperature, free energy decreases quickly owing to the 3rd term in Equation (4). Starting from $F(V,T)$, we calculate other thermodynamic potentials and all thermodynamic properties as functions of P and T. These include volume, V, thermal expansivity, α, isothermal bulk modulus and its derivatives, K_T and K_T', isobaric heat capacity, C_p, and thermal Grüneisen parameter, γ_{th} (see Table 1).

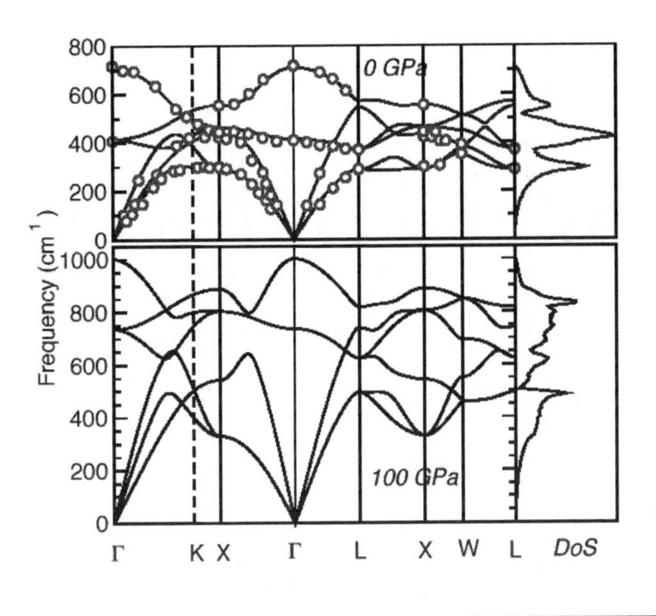

Figure 4. Phonon dispersion of MgO at 0 and 100 GPa (from Karki et al. 2000a). Inelastic neutron scattering data by Sangster et al. (1970) in circles (0 GPa). [Used by permission of APS, from Karki et al. 2000a].

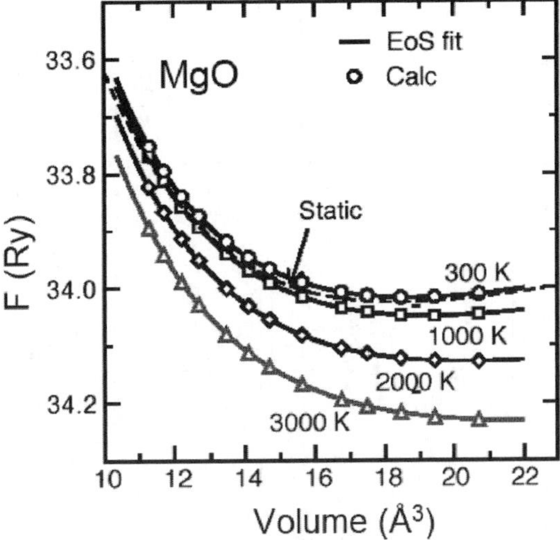

Figure 5. Helmholtz free energy vs. volume of MgO at various temperatures (Karki et al. 1999). Static means 0 K without lattice zero point motion effects.

The LDA results shown here are in excellent agreement with experimental measurements at ambient conditions. The LDA values for V, K_T and K_T', α, C_p, and γ_{th} are all within at most 0.5% from experimental values, while the GGA results deviate more. Specifically, the LDA equilibrium volume at 300 K is only slightly 0.2 % larger than experiments, but GGA overestimates volume by 2.0%. The LDA bulk modulus lies within experimental uncertainties, but the GGA underestimates it by 2%. The predicted LDA thermal expansivity agrees with experiments while the GGA overestimates it by 3%. Similar situation occurs for C_p and γ_{th}. In general, LDA results for thermodynamic, structural, and elastic properties of magnesium silicates and oxides within the QHA are in much better agreement with experiments than the GGA. The opposite is observed for static calculations. This demonstrates the importance of considering vibrational effects before comparing the performance of various exchange-

Table 1. Comparison between measured thermodynamic properties of various magnesium silicates and quasiharmonic LDA and GGA (PBE) results at 300 K and room pressure (3rd order Birch-Murnaghan equation of state was used in this calculation).

	V (A^3/uc)	K_T (GPa)	K_T'	α (10^{-5} K^{-1})	C_p (J/mol/K)	γ
			MgO			
exp	74.8	160(2)	4.15	3.12(16)	37.67	1.54(8)
LDA	75.2	161.7	4.1	3.15	37.62	1.61
GGA	77.5	155.7	3.8	3.22	38.6	1.54
			Forsterite			
exp	289.2 — 291.9	125 — 127.7	4.0	2.48 — 2.83	117.9	1.28
LDA	290.3	127.4	4.3	2.66	119.5	1.25
GGA	302.0	112.2	4.3	3.19	122.8	1.31
			Wadesleyite			
exp	535.3 — 539.3	160(3) — 172(3)	4.3 — 4.8	2.06	114.14	1.26
LDA	541.34	165.7	4.4	2.21	118.1	1.28
GGA	559.2	156.5	3.8	2.47	121.1	1.35
			Ringwoodite			
exp	526.7(3)	183(3)	4.2(3)	1.8 — 2.5	113.0	1.25
LDA	527.5	184.7	4.3	1.97	116.9	1.22
GGA	544.4	169.3	4.2	2.24	120.2	1.33
			Perovskite			
exp	162.5	246 — 272	3.9 — 4.0	1.7 — 2.2	80.6	1.3 — 1.96
LDA*	164.2	243.1	4.1	2.19	82.81	1.61
GGA*	169.7	223.7	3.9	2.46	85.9	1.72
			Post-perovskite			
exp$	81.15 — 81.9	222 — 259.5	3.69 — 4.4	1.70#	—	—
LDA	81.8	223.1	4.2	2.35	82.2	1.65
GGA	85.0	196.2	4.2	2.76	85.5	1.62

			Stishovite			
exp	46.51	309.9	4.59	1.46	42.48	—
LDA	47.0	294.4	4.6	1.18	42.1	1.2
GGA	48.6	267.3	4.3	1.40	43.8	1.25
			LP-Clinoenstatite			
exp	415.4(5) 414.6(1)	108.5(6.3) 111.1(3.3)	4.5(1.3)	—	—	—
LDA	415.2	121.5	5.5	2.5	81.67	1.16
GGA	433.7	102.0	5.5	3.2	83.8	1.27
			HP-Clinoenstatite			
exp	405.1(1.7)	106.4(17.4) 106.9(25.9)	5.4(2.7) 5.3(3.0)	2.64	81.95	—
LDA	397.1	126	5.5	2.46	81.09	1.16
GGA	423.3	94.4	6.4	3.4	83.9	1.12
			Ilmenite			
exp	43.76	212	7.5(1.0), 5.6(1.0)	1.67	78.0	—
LDA	44.2	201.7	4.4	1.92	80.1	1.3
GGA	45.3	207.5	3.7	1.95	81.8	1.35
			Majorite			
exp	759.3	159.8(4.4)	5.8	—	—	—
LDA	759.6	160.7	4.3	2.4	82.8	1.3
GGA	792.2	137.8	4.3	2.9	85.2	1.41

Experimental data are separated by semicolons in each phase from other calculation sources, for most of which the detailed calculated values are not shown here because of inequivalence posed by using different pseudopotentials and other convergence parameters. **MgO**: Fei (1999), Isaak et al. (1989), Touloukian et al. (1977); Karki et al. (1999). **Forsterite**: Guyot et al. (1996), Downs et al. (1996), Gillet et al. (1991), Bouhifd et al. (1996); Li et al. (2007b), Yu et al. (2008), Price et al. (1987), da Silva et al. (1997), Wentzcovitch and Stixrude (1997). **Wadsleyite**: Hazen et al. (1990), Hazen et al. (2000), Horiuchi and Sawamoto (2000), Suzuki et al. (1980), Ashida et al. (1987); Wu and Wentzcovitch (2007). **Ringwoodite**: Meng et al. (1994), Chopelas et al. (1994), Li (2003), Jackson et al. (2000), Weidner et al. (1984), Katsura et al. (2004b) Suzuki et al. (1979), Chopelas (2000); Yu and Wentzcovitch (2006), Kiefer et al. (1997), Piekarz et al. (2002) **Perovskite**: Knittle et al. (1986), Mao et al. (1991), Ross and Hazen (1989), Wang et al. (1994), Chopelas (1999); *Tsuchiya et al. (2005), Karki et al. (2000b), Wentzcovitch et al. (1995) $^{\$}$**Post-perovskite**: both experiments and calculations are incorporated. Murakami et al. (2004), Komabayashi et al. (2008)$^{\#}$, $^{\#}$ thermal expansion value of pv was assumed to be that of ppv at ambient condition; Tsuchiya et al. (2004), Oganov and Ono (2004). **Stishovite**: Akaogi et al. (1995), Andrault et al. (2003); Oganov et al. (2005), Umemoto et al. (2006). **LP-Clinoenstatite & HP- Clinoenstatite**: Angel and Hugh-Jones (1994), Shinmei et al. (1999); Yu et al. (2009), Wentzcovitch (1995), Duan et al. (2001). **Ilmenite**: Horiuchi et al. (1982), Reynard et al. (1996), Chopelas (1999); Karki and Wentzcovitch (2002). **Majorite**: Angel et al. (1989), Gerald Pacalo and Weidner (1997); Downs and Bukowinski (1997), Yu et al. to be published (2010) correlation functionals for structural properties. Complete analysis of these thermodynamic properties at high pressures and temperatures can be found in Karki et al. (2000a).

$MgSiO_3$-perovskite

$MgSiO_3$-perovskite (see Fig. 6a) is the primary end member of the most abundant mineral phase of the lower mantle, $(Mg_{(1-x)},Fe_x)SiO_3$ (x~0.12). Accurate knowledge of high *P-T* thermodynamic properties of this phase is a central ingredient for understanding and modeling the lower mantle. Full LDA phonon dispersions of $MgSiO_3$ perovskite are shown in Figure 7 (Karki et al. 2000b). Predicted Raman and infrared frequencies at 0 GPa compare very well with experimental values (Karki et al. 2000b; Wentzcovitch et al. 2004b). Figure 8 shows the LDA thermal expansivity of perovskite predicted by the QHA and compared with some direct experimental measurements (symbols), some indirect values derived from experimental information (thick inclined dashed lines), and MD calculations using the GGA (horizontal thin dashed lines) (see caption). Figure 8 conveys essential information about the QHA: at low temperatures, its predictions are in excellent agreement with experimental measurements. At high temperatures its prediction departs quickly from the experimentally observed linear temperature dependence. This deviation is suppressed with increasing pressure, and the behavior becomes more linear and agrees better with experimental data up to higher temperatures. As will be seen in the last section, thermal expansivity is most sensitive to anharmonicity. This deviation is caused by the inadequacy of the QHA at high temperatures where phonons interact more strongly. This behavior of the QHA based thermal expansivity naturally suggests a criterion to establish its upper temperature limit of validity. This is determined by the position of inflection point in $\alpha(T,P = cte)$, i.e., by $(\partial^2\alpha/\partial T^2)|_P = 0$. This limit is shown as a function of pressure by the black dots connected by a broken line in Figure 8.

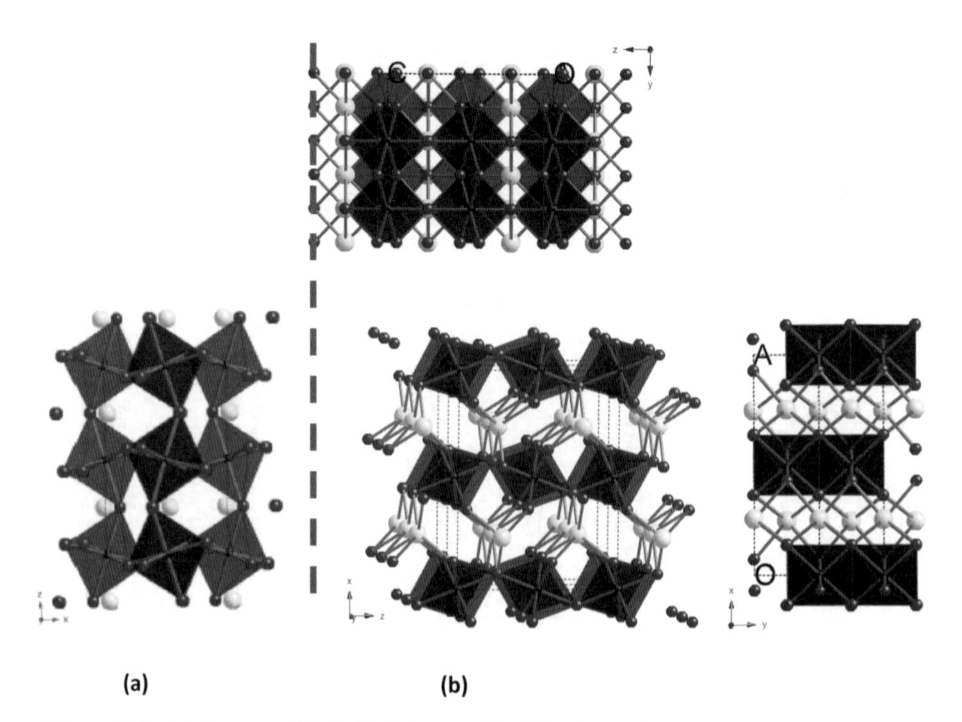

Figure 6. Crystal structures of (left) $MgSiO_3$-perovskite (*Pbnm*) and (right) post-perovskite (*Cmcm*). In post-perovskite, top, bottom left and bottom right views are along [010], [100], and [001] respectively.

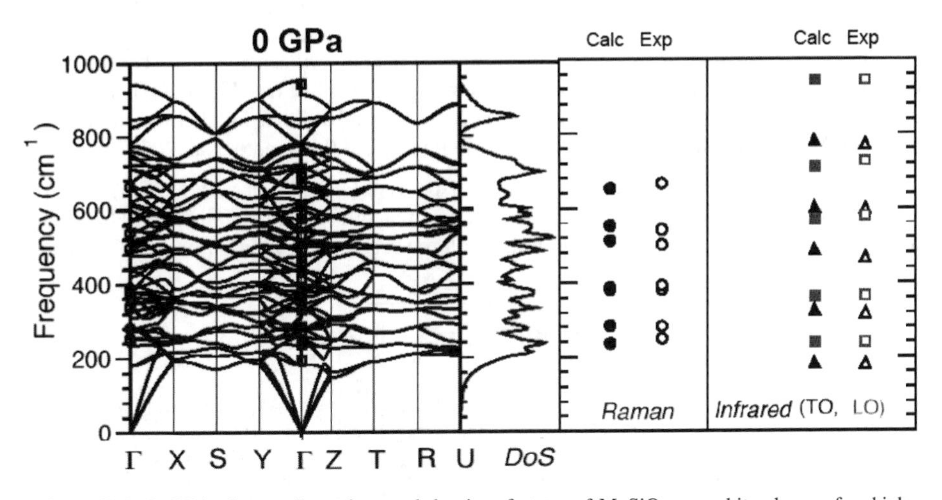

Figure 7. Left: LDA phonon dispersions and density of states of $MgSiO_3$ perovskite along a few high symmetry lines in the first Brillouin zone (Karki et al. 2000b). Right: comparison with experimental Raman and infrared spectroscopy at 0 GPa. Experimental data from Williams et al. (1987), Durben and Wolf (1992), Lu et al. (1994), Chopelas (1996), Lu and Hofmeister (1994). [Used by permission of APS, from Karki et al. 2000b].

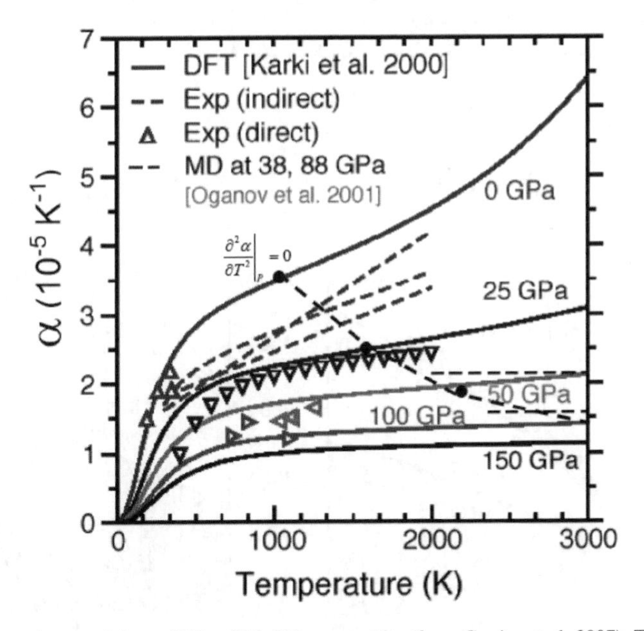

Figure 8. Thermal expansivity, $\alpha(P,T)$, of $MgSiO_3$-perovskite (from Carrier et al. 2007). The black line, labeled "QHA boundary," is defined by the position of the inflection points of $\alpha(P,T)$, as described in the text. [Used by permission of APS, from Carrier et al. 2007].

This criterion places lower mantle temperatures well within the regime of validity of the QHA (see Fig. 9), except perhaps at the entrance of the lower mantle, around 23 GPa and lower pressures, including perhaps ambient conditions. Indeed, $MgSiO_3$ perovskite is a metastable phase below 23 GPa. For other phases such as MgO and a very different material, metallic aluminum, investigation of this criterion shows that it is a very sensible one. The inflection point versus pressure is usually located well below the melting line (see Fig. 10). Therefore, the position of this inflexion point can be used as the lowest bound estimate for melting temperatures at tens of Mbar when measurements are virtually impossible (Umemoto et al. 2006).

The predicted thermodynamic propertied of perovskite (Karki et al. 2000b; Tsuchiya et al. 2005) are in quite good agreement with experimental values, some of which still have large uncertainties. The predicted LDA equilibrium volume at ambient conditions is ~1% larger than

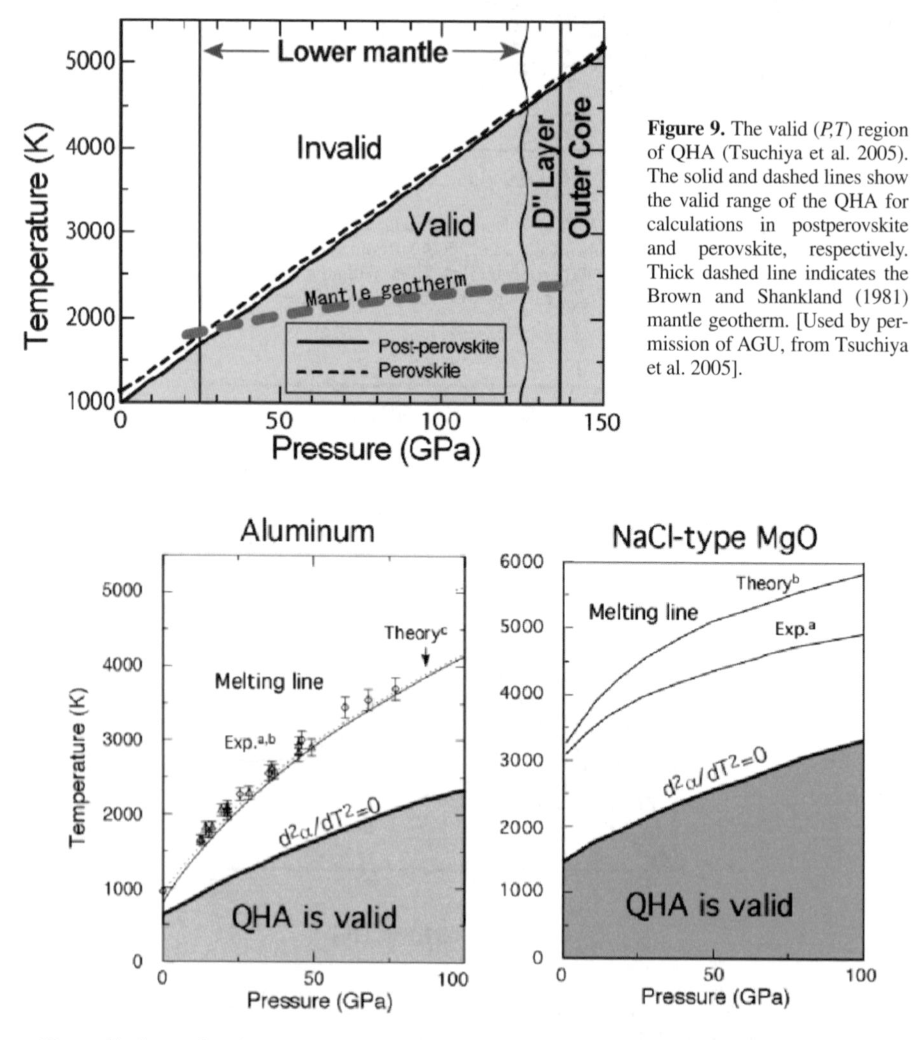

Figure 9. The valid (P,T) region of QHA (Tsuchiya et al. 2005). The solid and dashed lines show the valid range of the QHA for calculations in postperovskite and perovskite, respectively. Thick dashed line indicates the Brown and Shankland (1981) mantle geotherm. [Used by permission of AGU, from Tsuchiya et al. 2005].

Figure 10. Comparison between the limit of QHA validity with melting temperatures for aluminum and periclase. The QHA validity limit was determined according to the criterion based on the inflection point in the thermal expansivity. Aluminum: (a) Boehler and Ross (1997); (b) Shaner et al. (1984); (c) Vočadlo and Alfè (2002). MgO: (a) Zerr and Boehler (1994); Strachan et al. (1999).

the experimental volume. The bulk modulus is expected to be a little smaller, falling in the lower end of the experimental range of values. The thermal expansivity and thermal Grüneisen parameter fall well in the experimental range. The specific heat is ~3% larger. Among the systems investigated this is the worst performance of LDA/QHA found (see Table 1), and we suspect of the non-obvious inadequacy of the QHA at 0 GPa. The GGA predicts considerably worse results. This point is discussed further in this volume in a paper by Wentzcovitch et al. (2010) on thermoelastic properties of minerals. Complete discussion of the thermodynamic properties of perovskite at high pressures and temperatures can be found in Karki et al. (2000b), Wentzcovitch et al. (2004a), and Tsuchiya et al. (2005).

$MgSiO_3$ post-perovskite

$MgSiO_3$ post-perovskite with the $CaIrO_3$ type structure (*Cmcm*), shown in Figure 6b, is the latest found mineral of the lower mantle (Murakami et al. 2004; Tsuchiya et al. 2004; Oganov and Ono 2004). It is presumably the most abundant mineral in the D″ region of the mantle, i.e., ~250 km above the core mantle boundary (CMB). Pressures and temperatures in this region are expected to be above 125 GPa and 2500 K. Its discovery rejuvenated geophysical interest on this region and on the ultra-low velocity zone, the 5-10 km layer adjacent to the CMB. $MgSiO_3$ post-perovskite appears to be very anisotropic at first glance since it has a layered structure. The LDA calculated equation of state parameters are in excellent agreement with those obtained from in situ volume measurement by Komabayashi et al. (2008), while the GGA equation of state is considerably worse. This mineral is nearly unstable at ambient conditions (Tsuchiya et al. 2005). The seeming large discrepancy between the measurement and the LDA prediction of its thermal expansivity, ~30% is misleading, because experimentally post-perovskite is unattainable at ambient conditions and its ambient thermal expansivity was assumed to be the same as that of the perovskite (1.67 K^{-1}) by Komabayashi et al. (2008) in their experimental fittings. At lower mantle conditions, the QHA is expected to predict very well its thermodynamic properties (see Fig. 9). At shallow mantle conditions, where post-perovskite is (almost) unstable, and at the core-mantle boundary, where the temperature can reach 4,000 K, QHA's performance may be questionable. Detailed comparison between the thermodynamic properties of $MgSiO_3$ perovskite and post-perovskite at high pressures and temperatures can be found in Tsuchiya et al. (2005). Briefly, all the thermodynamic properties of these two phases become more similar with increasing pressure. At ~125 GPa, the pressure at the D″ discontinuity, and beyond they are almost indistinguishable (Tsuchiya et al. 2005).

SiO_2 stishovite

Stishovite is one of the high pressure phases of silica (SiO_2). It has the rutile structure (Fig. 11), the natural form of TiO_2 with *$P4_2/mnm$* symmetry. A salient feature in stishovite is that the coordination number of silicon by oxygen is 6 rather than 4, the latter is the case for quartz and coesite, low pressure form of silica. Other silicates that contain SiO_6 octahedra include majorite (partially tetrahedral and octahedral), ilmenite, perovskite, and post-perovskite. Overall good agreement is found between the predicted LDA equilibrium volume and thermodynamic properties and experimental values (Akaogi et al. 1995; Andrault et al. 2003). DFT calculations on stishovite can also be found in Oganov et al. (2005) and Umemoto et al. (2006).

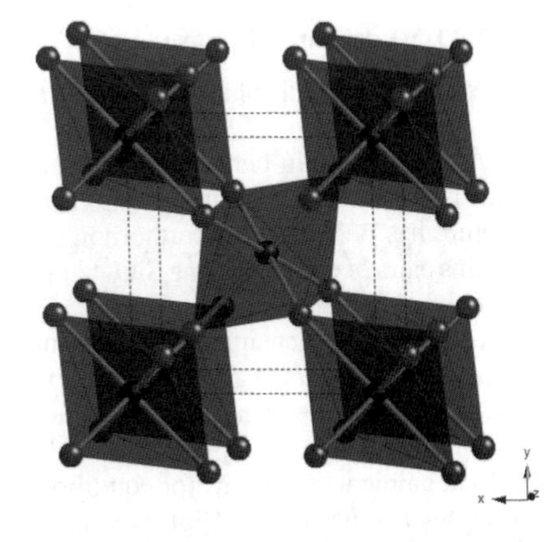

Figure 11. Crystal structures of SiO_2 stishovite (*$P4_2/mnm$*). Silicon sits in the center of oxygen octahedra.

Mg_2SiO_4 forsterite (α-phase)

Forsterite (the α-phase Mg_2SiO_4), is the end member of the dominant phase in the upper mantle, olivine ($(Mg_{(1-x)}Fe_x)_2SiO_4$, with x~0.012). The orthorhombic crystal structure (*Pbnm*) is shown in Figure 12. Oxygens form a distorted hexagonal close packed (HCP) sublattice; SiO_4 tetrahedra form isolated islands; divalent cations, such as magnesium, occupy octahedral sites. Owing to its abundance, Mg_2SiO_4 forsterite has been thoroughly studied experimentally, e.g., its vibrational spectroscopy and thermal elasticity have been measured (Chopelas 1990), and its equation of state parameters determined from experiments (e.g., Guyot et al. 1996; Downs et al. 1996; Gillet et al. 1991; Bouhifd et al. 1996) and calculated by various methods (Price et al. 1987; Li et al. 2006c). The thermodynamic properties of forsterite calculated by LDA and GGA are compared with experiments and shown in Table 1. The ambient conditions LDA results are in excellent agreement with experiments despite the detection of anharmonic effects in some zone center modes (e.g., Gillet et al. 1991). GGA overestimates the volume by 4%, underestimates the thermal expansivity by ~15%, and overestimates the heat capacity by ~4.5% compared with experiments and LDA results. Due to the larger volume predicted by the GGA, the GGA bulk modulus is ~10% smaller than the experimental counterpart. Extensive analysis of the high pressure and high temperature thermodynamic properties of forsterite can be found in Li et al. (2007).

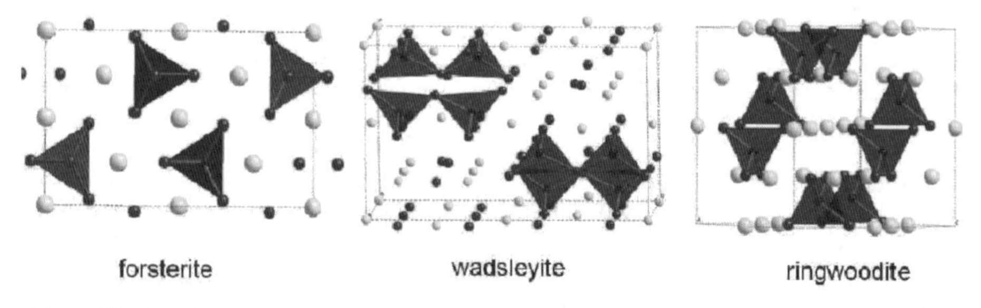

Figure 12. Crystal structures of forsterite (*Pbnm*), α-, wadsleyite (*Imma*), β-, and ringwoodite ($Fd\bar{3}m$), γ-Mg_2SiO_4. Oxygen and magnesium are shown in small and large spheres while silicon atoms sit in the centers of SiO_4 tetrahedra.

Mg_2SiO_4 wadsleyite (β-phase)

Wadsleyite, the β-phase Mg_2SiO_4 has a modified spinel structure. It was discovered by Ringwood and Major (1966) during the process of compressing high-magnesium olivine. It has a stability field in-between that of forsterite and ringwoodite, the true spinel phase. It is a major phase in Earth transition zone (from 410 km to 660 km depth). The wadsleyite crystal structure has body centered orthorhombic symmetry with space group *Imma*. This structure contains pairs of corner sharing SiO_4 tetrahedra forming a Si_2O_7 unit, while leaving one oxygen non-bonded by silicon in the structure (see Fig. 12). This non-silicate oxygen appears to be a good site for hydrogen attachment. For this reason wadsleyite has received great attention as a promising water storage candidate in Earth transition zone (e.g., Smyth 1987). As seen in Table 1, the agreement between LDA predictions (Wu and Wentzcovitch 2007) and measurements (Hazen et al. 1990; 2000; Horiuchi and Sawamoto 2000; Suzuki et al. 1980; Ashida et al. 1987) at ambient conditions for equilibrium volume, bulk modulus and other thermodynamic properties is outstanding. High pressure and high temperature thermodynamic properties of wadsleyite can be found in Wu and Wentzcovitch (2007).

Mg_2SiO_4 ringwoodite (γ-phase)

Mg_2SiO_4 ringwoodite has the spinel structure with $Fd\bar{3}m$ symmetry, which consists of isolated SiO_4 tetrahedra (see Fig. 12). The oxygen sublattice forms an FCC-like structure with magnesium ions occupying interstitial sites between SiO_4 groups. LDA predictions for all thermodynamic properties at ambient condition are in outstanding agreement with experiments (see Table 1). High pressure and high temperature thermodynamic properties of ringwoodite have been analyzed in detailed by Yu and Wentzcovitch (2008). High temperature equations of state and thermodynamics properties are in good agreement with experimental measurements of Meng et al. (1994), Chopelas et al. (1994), Li (2003), Jackson et al. (2000), Weidner et al. (1984), Katsura et al. (2004b), Suzuki et al. (1979), Chopelas (2000). Results are also consistent with the thermodynamics properties deduced from limited Raman and IR spectroscopy by Chopelas et al. (1994). The thermoelastic properties of this phase have also been investigated using MD (Li et al. 2006c).

Low- and high-pressure $MgSiO_3$ clinoenstatite

$MgSiO_3$ low-pressure clinoenstatite (*P2₁/c*, denoted by LP-En) and high-pressure clinoenstatite (*C2/c*, denoted by HP-En) are also important constitutes of Earth upper mantle (CPx in Fig. 2). LP-En exists at ambient conditions while HP-En had eluded unambiguous experimental detection until 1990s because it is unquenchable and converts into LP-En under decompression. Both LP-En and HP-En exist in monoclinic structures with obtuse monoclinic angles (see Fig. 13). In LP-En there are two symmetrically distinct SiO_4 tetrahedral chains (S, and O chain) with two different kinked chain angles, while in HP-En only one exists (O chain, see Fig. 13). A lattice dynamics study of LP-En (Yu and Wentzcovitch 2009) showed that its lowest *Ag* mode softens slightly under pressure. It was also shown that the *Ag* displacement mode can convert the S-chain in LP-En into the O-chain in HP-En. For these two complex and subtle silicate chain structures, we still observe excellent agreement between the predicted LDA thermodynamic properties and experimental measurements (see Table 1) by Angel and Hugh-Jones (1994), Shinmei et al. (1999). Thanks to DFT we also identified one viable transition path related to the *Ag* mode connecting the two enstatite chain structures (Yu and Wentzcovitch 2008). High temperature and high pressure thermodynamic properties of this mineral can be found are extensively discussed in Yu et al. (2010b).

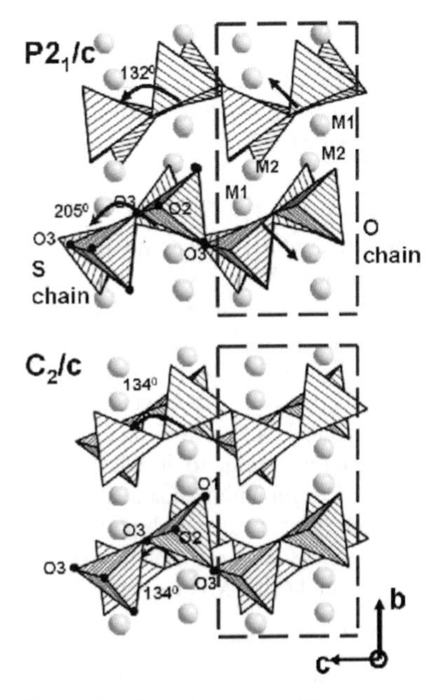

Figure 13. Crystal structure of low pressure ($P2_1/c$) and high pressure ($C2/c$) $MgSiO_3$-clinoenstatite. The spheres represent magnesium in between the SiO_4 tetrahedral chains. [Used by permission of AMS, from Yu and Wentzcovitch 2008].

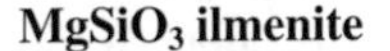

$MgSiO_3$ ilmenite

$MgSiO_3$ ilmenite has a trigonal structure with space group $R\bar{3}$. This structure derives from corundum structure (Al_2O_3) except that the octahedral layers contain alternating cations (see Fig. 14). Oxygens form a distorted HCP sublattice, while both magnesium and silicon are in 6-fold octahedral sites (Fig. 14). Ilmenite is a high pressure phase for $MgSiO_3$ but only at relatively

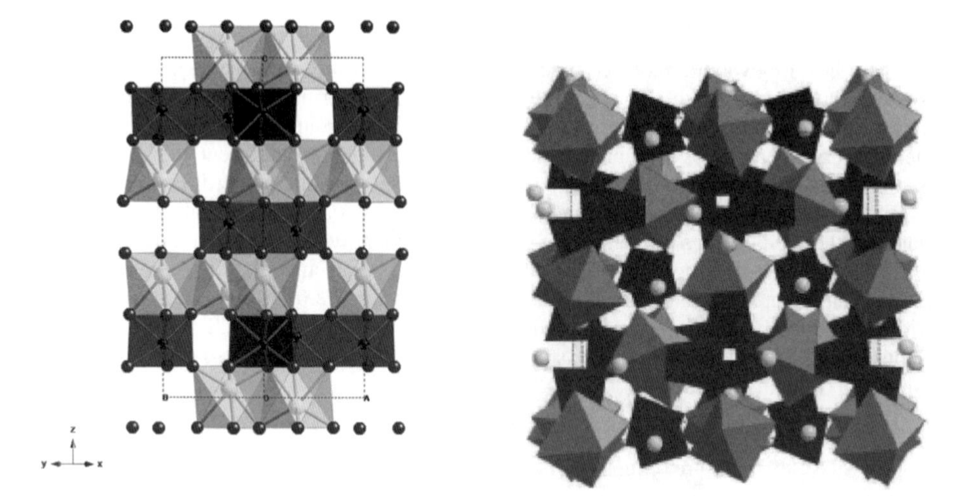

Figure 14. (left) Crystal structure of $MgSiO_3$-ilmenite ($R\overline{3}$) and (right) tetragonal $MgSiO_3$-majorite ($I4_1/a$). Light and dark octahedral in ilmenite represent SiO_6 and MgO_6 units with spheres denoting oxygens. In tetragonal majorite, the 8-fold coordinated magnesium is shown as spheres. 6-fold coordinated ones are at the center of light octahedra. Silicon ions partition between SiO_4 (dark) tetrahedra and SiO_6 (light) octahedra.

low temperatures compared with the mantle geotherm. But it should exist in cold slabs. $MgSiO_3$ ilmenite is stable approximately between 21 to 25 GPa at 1373 K (Ito and Yamada 1981). With increasing temperature and pressure, the ilmenite phase transforms to the *Pbnm* perovskite or to the garnet phase depending on temperature, pressure, and possibly solid solution composition (Ito and Yamada 1981; Liu 1977). Our LDA results are in reasonably good agreement with experiments. Note that depending on different pressure scales, the experimentally determined K' can vary from 7.5 to 5.6 (Reynard et al. 1996). Our calculations are performed in a larger pressure range (e.g., to 50 GPa) than the experiment (to 28 GPa) by Reynard et al. (1996) and the resulting equation of state parameters are expected to be more reliable. Detailed analysis of the high temperature and high pressure thermodynamic properties of this mineral can be found in Karki and Wentzcovitch (2002). The thermoelastic properties of this phase have also been investigated using MD (Li et al. 2009).

$MgSiO_3$ majorite

The magnesium end-member garnet, $Mg_3(MgSi)Si_3O_{12}$, is called majorite. At ~18 GPa and room temperature the stable phase is accepted to be a body-centered tetragonal (bct) phase ($I4_1/a$) with c/a ~ 0.98 (Angel et al. 1989; Heinemann et al. 1997). The single crystal experiment by Angel et al. (1989) revealed twin structures with slightly disordered (~20%) magnesium and silicon octahedral sites. At high temperatures (e.g., 2000 °C), it has been shown that this tetragonal phase transforms to cubic ($Ia\overline{3}d$) with magnesium and silicon completely disordered within the octahedral sites. Also, an intermediate structure with space group $I4_1/acd$ has been proposed in the literature (Nakatsuka et al. 1999; Hofmeister et al. 2004). Figure 14 shows the tetragonal majorite structure in which one quarter of the silicon ions are in (dark) octahedral sites and the other three quarters are in tetrahedral sites. Magnesium ions are also grouped into two types: one type at center of (light) octahedra and the other type at the center of dodecaheddra (spheres). The LDA equation-of-state parameters are in excellent agreement with the single crystal X-ray study of Angel et al. (1989). Thermodynamic properties at high pressures and temperatures will be reported in the near future (Yu et al. 2010a). The thermoelastic properties of this phase have also been investigated using MD (Li et al. 2007a).

CaO

CaO is not a mantle mineral *per se* but it is a component in mantle minerals. Despite the similarity between rocksalt CaO and MgO, CaO appears to be an anharmonic crystal (Karki and Wentzcovitch 2003). It transforms into a CsCl-type structure at ~50-56 GPa (Jeanloz et al. 1979). Phonon calculations indicate that both phases have stable phonons in the 50-60 GPa range (Karki and Wentzcovitch 2003) but phonons soften in NaCl- and become unstable in CsCl-type CaO in the vicinity of the transformation. Therefore the applicability of the QHA in the vicinity of the transition pressure is questionable. In overall, experimental equation of state parameters and thermodynamic properties at 300 K (Oda et al. 1992) are not well reproduced by QHA calculations. V_0 is ~5% smaller, K_T ~10% larger, α ~15% larger, γ_{th} ~25% larger, C_P ~ 35% larger. The more polarizable nature of calcium may induce anharmonicity in the vibrations.

$CaSiO_3$ perovskite

$CaSiO_3$ perovskite is an important lower mantle phase but it is unquenchable to ambient pressure (Wang and Weidner 1994). It is a strongly anharmonic crystal (Stixrude et al. 1996). At lower mantle pressures and temperatures it has cubic symmetry ($Pm\bar{3}m$) but the symmetry is reduced to tetragonal (*I4/mcm*) at high pressures and low temperatures (Shim et al. 2002; Caracas et al. 2005; Caracas and Wentzcovitch 2006; Stixrude et al. 2007) (Fig. 15). At lower mantle conditions the cubic phase is stabilized by anharmonic fluctuations, a picture confirmed by MD simulations (Li et al. 2006a). This behavior renders the QHA inadequate. High pressure and high temperature equation of state parameters and some thermoelastic properties have been obtained by MD (Li et al. 2006b) and by using a Landau potential with parameters determined by first-principles (Stixrude et al. 2007). This system still deserves to be further investigated by fully anharmonic techniques.

PHASE RELATIONS IN SILICATES AND OXIDES

The sharper discontinuities in the mantle are caused primarily by structural phase transformations in minerals. We have investigated systematically the ability of LDA and PBE-GGA exchange-correlation functionals combined with quasiharmonic free energy calculations to predict phase boundaries for polymorphic phase transitions and for one dissociation transition. The possible accuracy of such predictions was unclear. At phase boundary conditions bonds break and reconstruct, atoms may diffuse, phonons may become unstable, and other anharmonic effects may operate. Thermodynamic equilibrium between two phases is established by equating their Gibbs free energy, but these must be computed accurately. In general, large experimental hysteresis indicates the existence of a large energy barrier along structural transformation paths. For these transformations, phonons are still quite stable and harmonic at phase boundary conditions. Therefore, the quasiharmonic free energy is expected to correctly describe phase boundary properties. But the QHA might not always be adequate. Even when it is, there are still uncertainties associated with the exchange-correlation functional that need to be assessed. Only a systematic approach can shed light on the limitations and uncertainties of each of these approximations. Although these approximations may generate large uncertainties for some properties, the uncertainties may be very small for others. Here we point out that broad trends can be extracted regarding the performance of these approximations for mantle silicates and oxides. We will also discuss below results of an anharmonic calculation of the forsterite to wadsleyite phase boundary (Wu and Wentzcovitch 2009).

Phase transitions in Mg_2SiO_4

Earth transition zone is characterized by the 410-km and 660-km seismic density and velocity discontinuities (see Fig. 1). The former can be accounted by the olivine to wadsleyite transition (α-to-β) and the latter is believed to be caused by the ringwoodite dissociation into

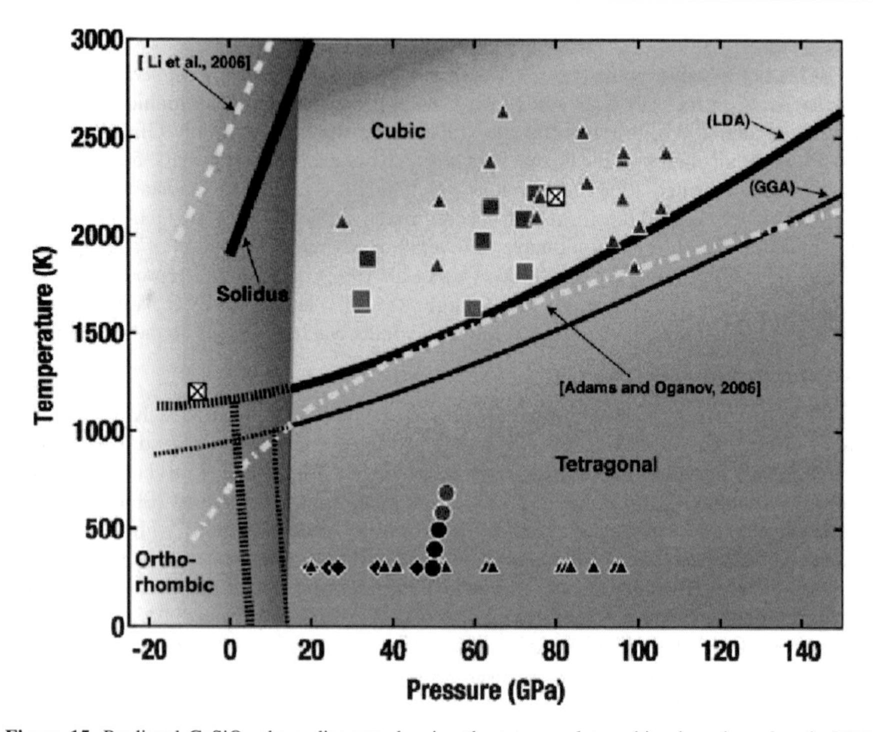

Figure 15. Predicted $CaSiO_3$ phase diagram showing the tetragonal to cubic phase boundary in LDA (thick line) and GGA (thin). Phase boundaries are dashed within the thermodynamic stability field of non-perovskite structures (Akaogi et al. 2004). The orthorhombic to tetragonal transition is indicated by the nearly vertical dashed lines. Location of the transition to the cubic phase from molecular dynamics simulations are shown in yellow: dot-dashed (Adams and Oganov 2006) and dashed (Li et al. 2006b). Experimental data for the tetragonal and cubic phases are in darker and medium symbols (diamonds (Shim et al. 2002), triangles (Ono et al. 2004), circles (Kurashina et al. 2004)). Squares are experimental results for a sample with 5 wt% Al_2O_3 (Kurashina et al. 2004): lighter: orthorhombic, medium: cubic. The solidus curve is the melting line (Zerr et al. 1997). [Used by permission of APS, from Stixrude et al. 2007].

ferripericlase and perovskite (γ-dissociation) (see Fig. 2). This transformation is also called the post-spinel transformation. Another less sharp discontinuity around 520-km depth may be caused by the wadsleyite to ringwoodite transition (β-to-γ). However, other interpretations for the 520 discontinuity also exist (Ita and Stixrude 1992; Weidner and Wang 2000; Saikia et al. 2008), such as $CaSiO_3$-perovskite ex-solution from the marjorite garnet solid solution. Over the past 40 years, starting from early experimental work by Ringwood and Major (1966), many efforts have been devoted to elucidating phase transitions in Mg_2SiO_4, including pinning down the exact phase boundary and their Clapeyron slopes. They have a significant impact on mantle convection models and on our understanding of this region. However, high *P-T in situ* experiments of phase boundary determination are challenging and there have been uncertainties associated with measurements of pressure (pressure scales) and temperature at high pressures (pressure effect on thermocouple). It has been difficult to reach experimental consensus on Clapeyron slopes associated with the transitions expected at 410- and 660-km depths (see Fig. 16). Quasiharmonic first-principles calculations approach the problem from a different angle and avoid uncertainties intrinsic to measurements, such as reaction kinetics.

Systematic first-principles studies of the α-to-β-to-γ-to-pv+pc transitions in the Mg_2SiO_4-$MgSiO_3$-MgO system have been performed recently by Yu and Wentzcovitch (2006) and Yu et

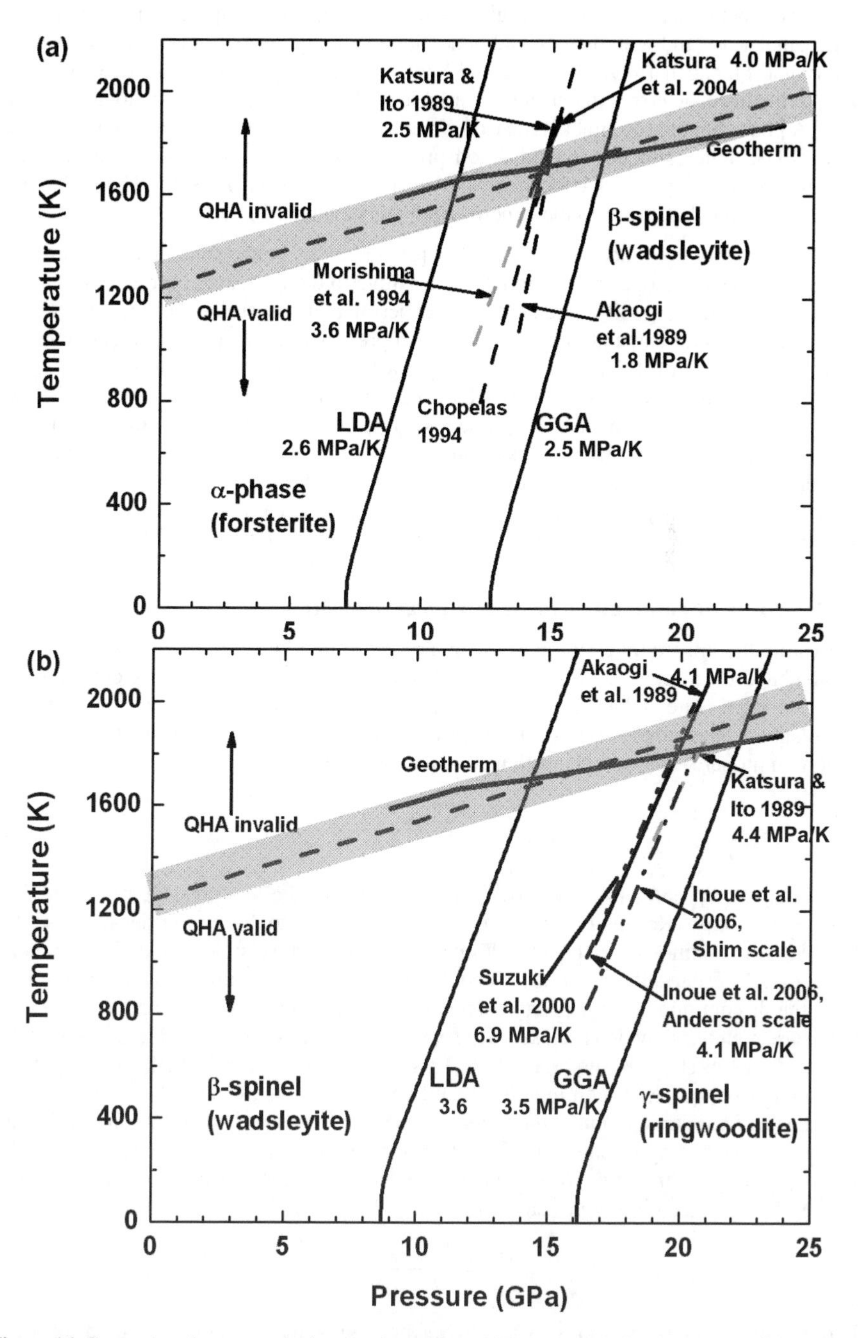

Figure 16. Comparison between calculated and experimental phase boundaries for the α- to β- (top) and β- to γ-phase transitions (bottom) in Mg_2SiO_4 (from Yu et al. 2008). [Used by permission of EPSL, from Yu et al. 2008].

al. (2008). These calculations used the same set of parameters, including pseudopotentials and plane-wave kinetic energy cut-off. Results are documented and compared with high pressure experimental work and with density and bulk sound velocity discontinuity data in the mantle (Shearer and Flanagan 1999; Lawrence and Shearer 2006; Dziewonski and Anderson 1981). Owing to the intrinsic errors in currently available exchange-correlation functionals, the LDA and GGA phase boundaries bracket the experimentally determined ones with a common trend that the LDA underestimates the transition pressure by about 5 to 10 GPa, and the GGA overestimates it by up to ~3 GPa, irrespective of the transition pressure. The GGA phase boundary is somewhat closer to the experimental ones than the LDA.

Despite errors (or uncertainties) in the prediction of static transition pressures, the calculated Clapeyron slopes are essentially insensitive to the exchange-correlation functional used and are in quite good agreement with experimental measurements (see Table 2 and Figs. 16 and 17). Our calculations also agree with previous quasiharmonic calculations—2.7 MPa/K (α-β transition) (Price et al. 1987, using inter-atomic potentials; Chopelas 1991, using measured zone center phonons) and 3.6 MPa/K (β-γ transition) (Price et al. 1987, 4.5 MPa/K). For the post-spinel transition the GGA phase boundary, including the Clapeyron slope, is closer to experimental results by Ito and Takahashi (1989) than to more recent *in situ* synchrotron X-ray data (Fei et al. 2004; Litasov et al. 2005). Nevertheless, calculations and experiments agree better at high temperature (above 1600 °C, Fig. 17) than at low temperature, suggesting that kinetics might be an issue in the experimental determinations of this dissociation phase boundary.

Mantle density discontinuities caused by phase transitions in Mg_2SiO_4

Despite the uncertainties in phase boundary determination, further analysis reveals that the calculated discontinuities in density, bulk modulus, and bulk sound velocity are quite insensitive to pressure and are useful to address the origin of different seismic discontinuities in the mantle, a much more complicated system with other co-existing phases. A discontinuity of a certain property across a phase transition can be defined as:

$$\frac{\Delta x}{x} = 2\left(\frac{x_B - x_A}{x_B + x_A}\right) \tag{7}$$

with x being ρ, K_S, or V_Φ ($V_\Phi = (K_S/\rho)^{1/2}$). Here A denotes the low-pressure phase and B the high-pressure phase. Results for ρ, K_S, and V_Φ, as well as these discontinuities ($\Delta x/x$) at 410-, 520-, 660-km conditions are shown in Table 3 together with seismic data and other mineral physics data. These results are shown also in Figure 18. The lower (upper) bound of the shaded bars is the discontinuity in a property calculated with the LDA (GGA) functional "along the GGA phase boundary," which is closer to the experimental ones. The solid lines in the middle of the shaded regions show the average values. This comparison is made for sake of completeness only. Obviously, mantle discontinuities should be smaller since these transformations occur in aggregates rather than in pure end member minerals. However, the density discontinuity in the mantle produced by these single phase transformations can be estimated (Yu et al. 2006, 2010b). If phase A transforms to A′ in an aggregate of minerals consisting of phases A, B, C, etc., the density discontinuity produced is:

$$\frac{\Delta\rho}{\rho} = \frac{2y}{2/(1-\theta) - y} \tag{8}$$

where y is the volume fraction of phase A in the aggregate, e.g., ~60% of α, β, or γ in an aggregate with pyrolite composition (Ringwood 1979), and θ is the volume ratio of A′ to A, which depends on the pressure and temperature at the transitions at 410-km, etc. For the α-β transition, θ = 0.9504, 0.9506, and 0.9509 (0.9444, 0.9446, and 0.9448) at 1300, 1500, and

Table 2. Comparison of LDA and GGA Clapeyron slopes from this study with those determined by experiments and other calculations.

Series	Transition	GGA	LDA	Exp. and others calc.						
				1	2	3	4	5	6	7
A	α to β	2.7	2.5	3.5	4.1±0.7	2.5	1.8	2.7	3.6	4.0
B	β to γ	3.6	3.5	4.4	4.2	6.9	4.1	6.3	4.5	
C	γ to pc+pv	−2.8	−2.7	−2.8	−2.5	−2.75	−2 - −0.4	−1.3	−0.5	
D	LP-En to HP-En	3.0	2.9	2.0	1.6					
E	pv to ppv	—	—	7.5±0.3	9.6-9.9	5.0-11.5	6	9.7	7.0	

Experimental data and other calculation sources:
α to β — LDA and GGA: Yu et al. 2008, [A1]: Suito 1977, [A2] Watanabe 1982, [A3]: Katsura and Ito 1989, [A4]: Akaogi et al. 1989, [A5]: Price et al. 1987, Chopelas 1991, [A6]: Morishima et al. 1994, Stixrude and Lithgow-Bertelloni 2005, [A7]: Katsura et al. 2004a
β to γ — LDA and GGA: Yu et al. 2008, [B1]: [A3], [B2]: [A4], [B3]: Suzuki et al. 2000, [B4]: Inoue et al. 2006, [B5]: Sawamoto 1987, [B6]: Price et al. 1987
γ to pc+pv — LDA and GGA: Yu et al. 2008, [C1]: Ito and Takahashi 1989, [C2]: Irifune et al. 1998, [C3]: Shim et al. 2001, [C4]: Katsura et al. 2003, [C5]: Fei et al. 2004, [C6]: Litasov et al. 2005
LP-En to HP-En — Yu et al. 2008, [D1]: Gasparik 1990, [D2]: Angel and Hugh-Jones 1994
pv to ppv — [E1]: Tsuchiya et al. 2004, [E2]: Oganov and Ono 2004, [E3]: Hirose 2006 and references therein, [E4]: Sidorin et al. 1999, [E5] Wu et al. 2008, [E6] Ono and Oganov 2005

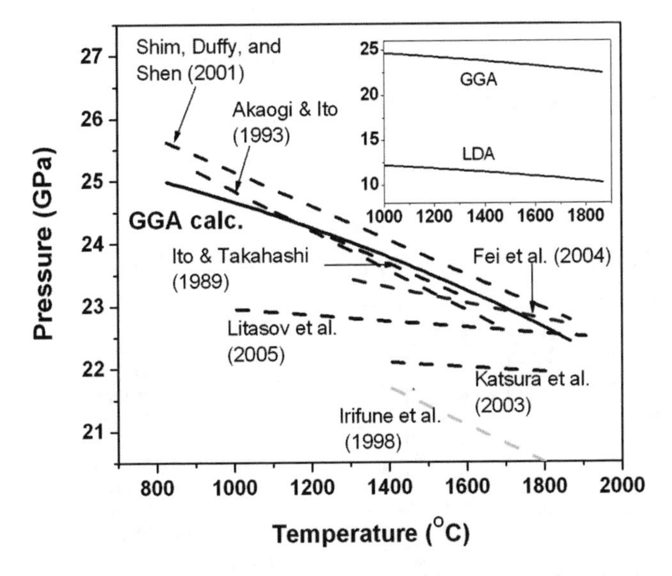

Figure 17. Phase boundary of the post-spinel phase transition (dissociation of ringwoodite into $MgSiO_3$-perovskite and MgO) calculated by LDA and GGA compared to various experimental determinations (from Yu et al. 2007). [Used by permission of AGU, from Yu et al. 2007].

1700 K along the GGA (LDA) boundary; for the β-γ transition, θ = 0.9822, 0.9823, and 0.9824 (0.9796, 0.9797, and 0.9798) at 1500, 1700, and 1900 K along the GGA (LDA) boundary.

Using Equation (8) and the calculated θ along the LDA and GGA phase boundaries, the density discontinuity versus the Mg_2SiO_4 volume fraction at 410-, 550-, and 660-km depth was computed using the LDA functional (Yu et al. 2006, 2008). The results are shown in Figure 19c. According to the seismic impedance data of Shearer and Flanagan (1999), the density discontinuity at 660 km depth is ~5.2%. Such density jump could be produced if the volume fraction of

Table 3. Calculated (LDA) properties of Mg_2SiO_4 along the (GGA) phase boundary for α-β transition (~1500 K, 16.3 GPa), (Yu et al. 2008), β-γ transition (~1700 K, 20.2 GPa) (Yu et al. 2008), and γ-dissociation transition (1900 K and 23.2 GPa) (Yu et al. 2007) compared with seismic data on the 410-, 520-, and 660-km discontinuities and mineral physics experiments.

	ρ			K_S			V_Φ		
transition	**calc.**	**PREM**	**oth.**	**calc.**	**PREM**	**oth.**	**calc.**	**PREM**	**oth.**
				410-km					
Upper 410 (α)	3.47	3.54	—	178.3	173.5	—	7.17	7.00	—
Lower 410 (β)	3.65	3.72	—	218.3	189.9	—	7.73	7.14	—
%Δ 410 (α→β)	5.1%	5.0%	~0.2-4%[a]	20.2%	9.0%	—	7.6%	1.9%	1.5-6.6%[a]
				520-km					
Upper 520 (β)	3.70	3.85[§]	—	231.9	218.1[§]	—	7.91	7.54[§]	—
Lower 520 (γ)	3.77	—	—	249.5	—	—	8.13	—	—
%Δ 520 (β→γ)	1.8%	—	1.3-2.9%[b], 2.5-3.0%[c], 2.5%[d]	7.3%	—	—	2.7%	—	1.3-2.1%[b]
				660-km					
Upper 660 (γ)	3.81	3.99	—	261.1	255.8	—	8.274	8.007	—
Lower 660 (pv+pc)	4.13	4.38	—	282.1 ± 2.8*	299.4	—	8.266 ±0.080*	8.268	—
%Δ 660 (γ→pv+pc)	7.9%	9.3%	5.2%[a]	7.7 ±1.0%*	15.9%	5.5%[a]	−0.10 ±0.48%*	3.2%	0.16%[a]

* uncertainties from VRH average of pv+pc aggregate, [§]PREM at 500-km
[a] Shearer and Flanagan(1999), [b] Lawrence and Shearer (2006), [c] Rigden et al. (1991),
[d] Sinogeikin et al. (2003)

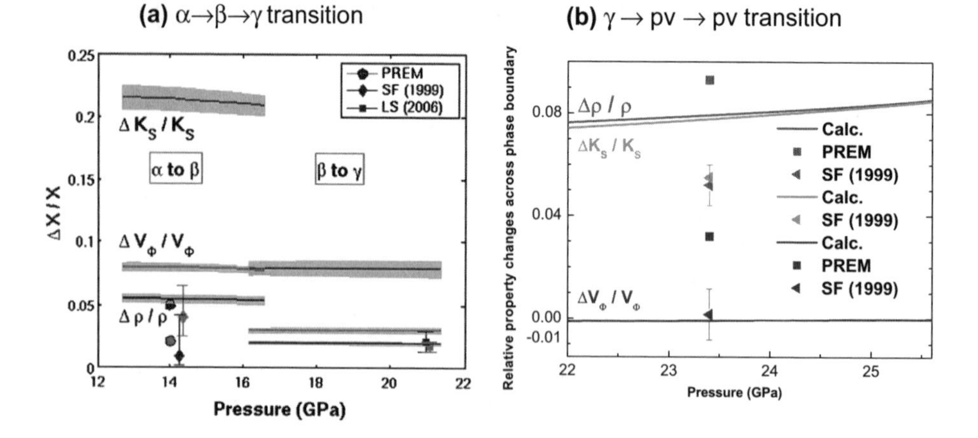

Figure 18. Calculated changes in density, ρ, adiabatic bulk modulus, K_S, and bulk velocity, V_Φ, for phase transitions in Mg_2SiO_4 in the transition zone: the (a) α-β-γ transitions (from Yu et al. 2008) and the (b) γ dissociation (from Yu et al. 2007). See Yu et al. (2008) for an extensive discussion. [Used by permission of AGU and EPSL, from Yu et al. 2007 and Yu et al. 2008, respectively].

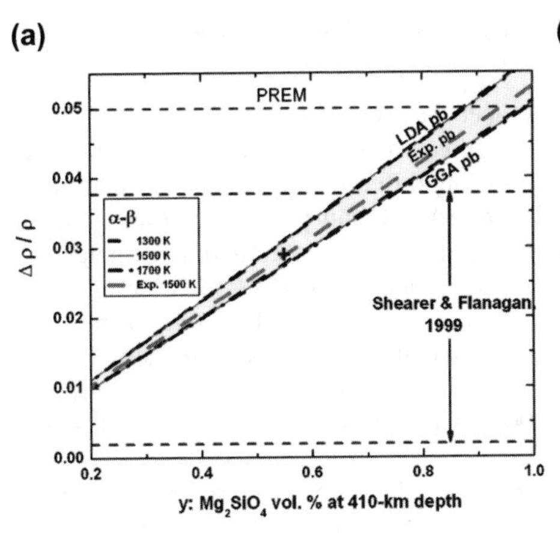

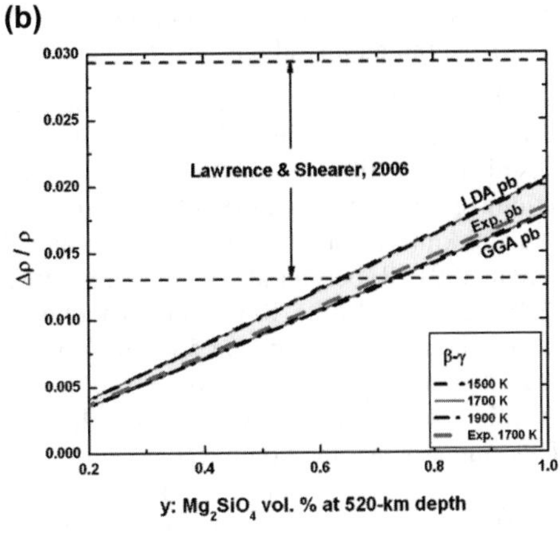

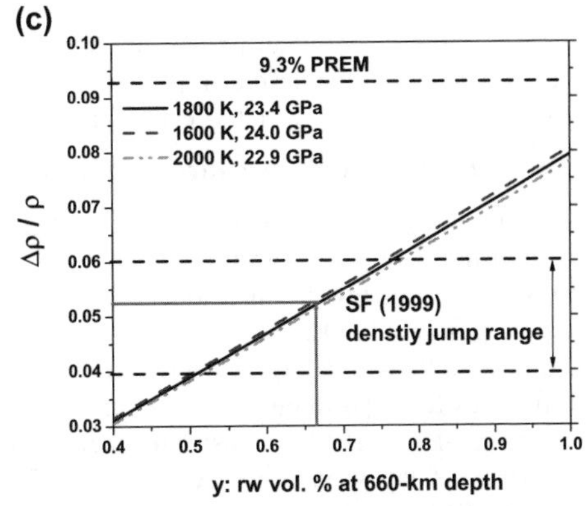

Figure 19. Density discontinuities in a mineral aggregate produced by the (a) α-to-β transition at 410-km versus volume fraction of α, compared to the best estimate from seismic data, 0.9% (Shearer and Flanagan 1999); (b) same for the β-to-γ transition at 520-km, compared with the estimated value of 2.1±0.8% (Lawrence and Shearer 2006), and (c) for the γ dissociation at 660-km, compared to the estimated value of 5.2% (Shearer and Flanagan 1999). Comparison is also made with PREM (Dziewonski and Anderson 1981). [Used by permission of AGU and EPSL, from Yu et al. 2007 and Yu et al. 2008, respectively].

γ in the aggregate were ~67%. This is close but slightly larger than the volume fraction of γ in an aggregate with pyrolite composition which is supposed to have ~60 vol% of γ. The larger density jump across the 660 discontinuity could have contributions from the garnet to perovskite transition which coexists with ringwoodite dissociation, but is known to be broader owing to Al effects (e.g., Ringwood 1975). The production of perovskite by the γ dissociation may alter the thermodynamic equilibrium of the garnet-perovskite system in favor of perovskite, which would also contribute to the density jump presumably (Weidner and Wang 2000).

For the 410-km, the large uncertainties in the seismic impedance data (~0.2-4%, Shearer and Flanagan 1999) prevent an unbiased conclusion about the volume fraction of forsterite or olivine (α) present at 410-km (Fig. 19a). Within this uncertainty it is safe to suggest that at most 80% of Mg_2SiO_4 is needed to produce the 410 discontinuity. The volume fraction in the pyrolite model, 60 vol% would produce a density jump well within the range estimated from the impedance data. However, there could be other factors contributing to the 410 discontinuity. At upper mantle conditions pyroxenes dissolve extensively into garnet over a large pressure range (~7-15 GPa at ~1200 °C). A drastic change in the proportion of pyroxene to garnet occurs near the end of the transformation (Ringwood 1975; Akaogi and Akimoto 1979; Irifune 1987). This can produce a high velocity gradient between 300 and 460-km depth as pointed out by Ringwood (1975). The possibility that the pyroxene-to-garnet transformation coincides with the α-β transition in the mantle and contributes to the 410-km discontinuity is not excluded. In particular, Al, Ca, and Fe, affect phase equilibrium and the sharpness of the transitions in the pyroxene/garnet/Ca-pv system (solid-solid solution) (Stixrude 1997), but it still unclear how (e.g., Weidner and Wang 2000) and more mineral physics studies are needed.

The 520-km discontinuity is considerably weaker than the 410- or 660-km ones, and is also broader (~10-40 km). This discontinuity has been very debated because of controversial seismic observations coming from different schools using different types of seismic data, such as long or short period data, or reflection data at 520-km. Some found global or regional evidence of a 520-km discontinuity (e.g., Wiggins and Helmberger 1973; Shearer 1990; Revenaugh and Jordan 1991; Ryberg et al. 1997), while others did not (Jones et al. 1992; Gossler and Kind 1996; Gu et al. 1998). Deuss and Woodhouse (2001) proposed a 520-km splitting into two discontinuities at 500-km and 560-km depth. Figure 19b compares the predicted density jump caused by the β-γ transition versus volume fraction of Mg_2SiO_4 in the aggregate, with the seismic observation by Lawrence and Shearer (2006). The best seismic estimate is 2.1±0.8%. This suggests that the volume fraction of Mg_2SiO_4 at 520-km is at least 70% (see Fig. 19b), which is excessive by any compositional model. Probably the pyroxene/garnet/Ca-pv system contributes to produce this discontinuity. There is evidence that Ca-pv ex-solves from garnet at ~1500 °C and 17-18 GPa (Gasparik 1989; Canil 1994) and can contribute to the discontinuity at 520-km depth (Weidner and Wang 2000; Saikia et al. 2008). These results verify that it is indispensable to include the impact of the pyroxene/garnet/Ca-pv system to understand the 520-km discontinuity, which presumably changes the width of this discontinuity as well.

Low-pressure to high-pressure $MgSiO_3$ clinoenstatite transition

Unlike reconstructive phase transitions such as the post-spinel transition, the low-pressure to high-pressure clinoenstatite transition in $MgSiO_3$ is a prototype of a displacive phase transition. As mentioned previously, Hp-En results when the lowest Ag displacement mode is superposed to the LP-En structure which is then relaxed. This transition is non-quenchable and reversible, with a large hysteresis in pressure (Angel and Hugh-Jones 1994), which poses difficulties for experimental measurements. The calculated LDA and GGA phase boundaries for the LP-En-to-HP-En transition bracket the experimental hysteresis range and produce a 2.9 MPa/K Clapeyron slope (see Fig. 20). This provides useful information to constrain the LP-En, HP-En, and orthoenstatite triple point. These results suggest that the experimentally constrained triple point (Ulmer and Stalder 2001) should be shifted up in pressure by 1 GPa (Yu et al. 2010b).

Post-perovskite transition in $MgSiO_3$

The post-perovskite transition in $MgSiO_3$ (Murakami et al. 2004; Tsuchiya et al. 2004; Oganov and Ono 2004) is one of the most important findings in mineral physics since the discovery of the perovskite phase in 1974 (Liu 1974). Its geophysical implications are profound. Important new insights have been provided on the nature of the D″ region with large lateral velocity variations and a clear discontinuity with wide topography varying from 0-300 km above the core mantle boundary. Previous experience with first-principles quasiharmonic phase boundary calculations was limited (Karki and Wentzcovitch 2003; Wentzcovitch et al. 2004b) and the uncertainty of the predicted phase boundary for the post-perovskite transition was unclear (Tsuchiya et al. 2004). The LDA and GGA phase boundaries are shown in Figure 21. They differ by ~10 GPa. The Clapeyron slope obtained was 7.5 ± 0.3 MPa/K. At 125 GPa the transition temperature (2500 K) is very close to that observed by Murakami et al. (2004). However, the experimental situation was very uncertain. An extensive review of experimental data on the post-perovskite transformation and on the properties of the post-perovskite phase was written by Hirose (2006). A detailed discussion of the experimental challenges associated with the determination of this phase boundary was presented. The main issue is the uncertainty in the pressure measurement owing to the discrepancies between various pressure markers, whether Pt (Holmes et al. 1989), Au (Jamieson et al. 1982; Tsuchiya 2003), or MgO (Speziale et al. 2001). Using these markers the experimental Clapeyron slope varies from 5 to 11.5

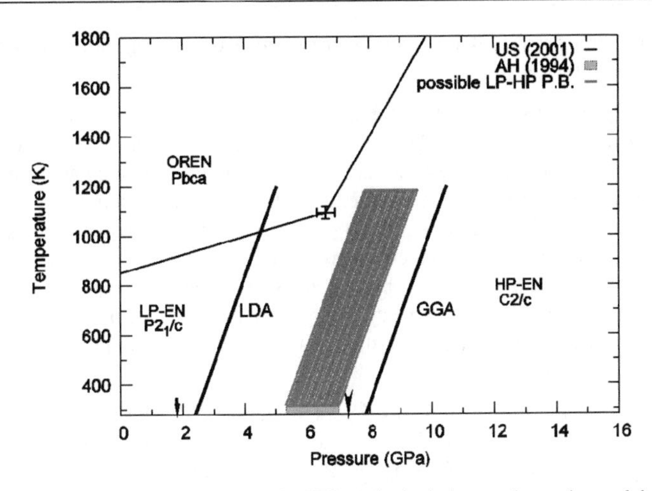

Figure 20. Illustration of how a combination of DFT based calculations and experimental data can be used to constrain the triple point in the LP-En, HP-En, and orthoenstatite system (from Yu et al. 2009). [Used by permission of AGU, from Yu et al. 2009].

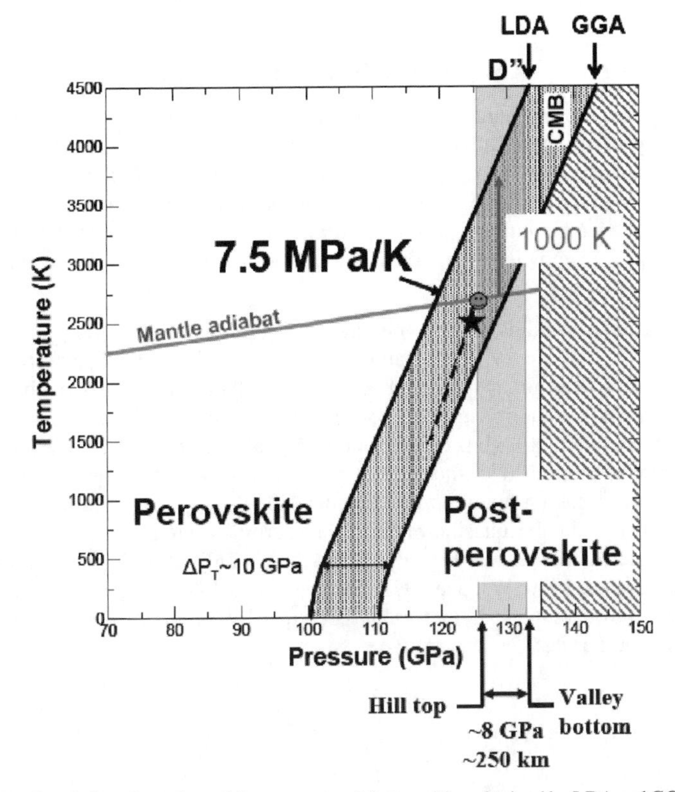

Figure 21. Predicted phase boundary of the post-perovskite transition obtained by LDA and GGA (Tsuchiya et al. 2004). Star is from Murakami et al. (2004); dashed line corresponds to the Clapeyron slope predicted by Sidorin et al. (1999). For an extensive review of results see Hirose (2006). [Used by permission of EPSL, from Tsuchiya et al. 2004].

MPa/K. Another first-principles calculation proposed a Clapeyron slope of 9.6-9.9 MPa/K (Oganov and Ono 2004).

This situation provided motivation for a systematic series of theoretical investigations of other phase transformations (Yu and Wentzcovitch 2006, 2008) that could be contrasted more clearly with experimental data. These studies helped to assert the calculated phase boundary for the post-perovskite transition (Tsuchiya et al. 2004) and provided a concrete sense for the uncertainties involved. More recently, a highly engineered thermal equation of state for MgO combining anharmonic free energy calculations with experimental data (Wu et al. 2008) was used to re-interpret experimental data. It produced a Clapeyron slope equal to 9.7 MPa/K. This latest value reduces the current discrepancy between theoretical calculations and experimental data. Anharmonic effects can still change the predicted Clapeyron slope. This phenomenon is discussed in the next section.

In the review on thermoelastic propeties of minerals (Wentzcovitch et al. 2010) in this volume, an extensive discussion of this transformation and of the discontinuities in density and velocity it produces is given. Thermoelastic properties of this phase have been calculated using first-principles molecular dynamics (Stackhouse and Brodholt 2007). The influence of chemistry on this phase transformation has been investigated by first-principles (Caracas and Cohen 2007). The stability field of this phase, the search for post-post-perovskite polymorphs, and its dissociation into cotunnite-sype SiO_2 and CsCl-type MgO at 11 Mbar has also been reviewed (Wentzcovitch et al. 2007; Umemoto and Wentzcovitch 2010).

ANHARMONIC FREE ENERGY

The QHA does not account for anharmonic effects. In this approximation temperature effects on the crystal structure and on phonon frequencies are accounted by extrinsic volumetric effects only (quasiharmonic thermal expansivity). Because intrinsic temperature effects are more evident at high temperatures, the QHA becomes inadequate beyond a certain T located between the Debye temperature, θ_D, and the melting temperature, T_M (Matsui et al. 1994; Inbar and Cohen 1995; Wentzcovitch et al. 2004a; Carrier et al. 2007). This effect is more visible at lower pressures. (see Fig. 8).

Several parameterizations for the anharmonic free energy have been proposed (Landau and Lifshitz 1980; Wallace 1972; Hui and Allen 1975; Gillet et al. 1999; Holzapfel et al. 2001; Oganov and Dorogokupets 2004) All these methods introduce inevitably volume-sensitive parameters or require explicit knowledge of the renormalized frequencies, none of which are easy to obtain at high pressures and temperatures. Recently, we proposed a semi-empirical parameterization (Wu and Wentzcovitch 2009) that modifies the quasiharmonic phonon frequencies making them temperature dependent, $\Omega^F_{q,j}(V,T)$. These new frequencies can be used directly in the quasiharmonic free energy formula (Eqn. 9). $\Omega^F_{q,j}(V,T)$'s are not the renormalized frequencies $\Omega^S_{q,j}(V,T)$, which can only be used in the quasiharmonic entropy formula (Falk 1970; Wallace 1972; Hui and Allen 1975). Our parameterized phonon frequencies depend on a single parameter that is obtained by comparing predicted and measured thermodynamic properties:

$$\Omega^F_{q,j}(V,T) = \omega(V') \tag{9}$$

with V':

$$V' = V\left\{1 - c\frac{[V - V_0]}{V_0}\right\} \tag{10}$$

V and V_0 are the predicted quasiharmonic volumes at high T and 0 K respectively, both at the same quasiharmonic pressure, $P(V,T)$. c is a constant to be determined empirically by using experimental thermodynamic data. This parameterization embodies the expected anharmonic behavior, which increases with temperature at constant pressure and decreases with pressure at constant temperature: V' tends to V at low temperatures and constant pressure or at high pressures and constant temperatures. Then, the anharmonic free energy is:

$$\begin{aligned} F_A(V,T) &= U(V) + \frac{1}{2}\sum_{q,j}\hbar\omega_{q,j}(V') + k_B T\sum_{q,j}\ln\{1-\exp\left[\frac{-\hbar\omega_{q,j}(V')}{k_B T}\right] \\ &= U(V) + F_H(V',T) - U(V') \end{aligned} \quad (11)$$

The temperature dependence of Ω_i^F ($i \equiv q,j$) can be expressed as (Wu and Wentzcovitch 2009):

$$a_i^F = \left(\frac{\partial \ln \Omega_i^F}{\partial T}\right)_V = \gamma_i(V')\alpha(V,T)c\beta \quad (12)$$

where $\beta = (V^2 K_T)/(V' V_0 K_0)$, $K_0 K_T$ are isothermal bulk moduli at 0 K and finite T, respectively, $\gamma_i(V')$ is the mode Grüneisen parameter, $\alpha(V,T)$ is the quasiharmonic thermal expansivity, and c is the empirically determined constant. At low temperatures, $\beta \sim 1$, a_i^F is proportional to $\alpha(V,T)$. At high temperature, the product $\beta\alpha(V,T)$ increases slightly and linearly with temperature. The temperature dependence of the Ω_i^F and the true renormalized frequencies, Ω_i^S is related approximately by (Wu and Wentzcovitch 2009):

$$a_i^S = \frac{\partial \ln \Omega_i^S}{\partial T} \approx (1+\eta_i)a_i^F \quad (13)$$

where $\eta_i = (2k_B T)/(h\Omega_i^F) - \exp[-h\Omega_i^F/k_B T][(2k_B T/h\Omega_i^F) + 1]$. η_i is positive and increases quickly with temperature if $k_B T < h\Omega_i^F$ and then tends to 1 when $k_B T > h\Omega_i^F$. Both η_i and the product $\beta\alpha$ from Equation (12) increase rapidly with temperature at low temperatures and become weakly temperature dependent at high temperatures. Therefore we expect a noticeable nonlinearity in a_i^S at low temperatures and a nearly linear behavior at high temperatures. Such temperature behavior of the renormalized frequencies has been observed in diamond by Raman scattering (Zouboulis and Grimsditch 1991; Anastassakis et al. 1971). Equations (12) and (13) also show that the vibrational modes with smaller γ_i's have smaller a_i's in absolute values. This behavior of a_i's was indeed observed by Gillet et al. (1991). They found that the absolute value of a_i's for modes involving tetrahedral Si-O bond stretching with smaller γ_i's (Li et al. 2007b) are smaller than for other lattice modes.

As expected, at low temperatures anharmonic effects are negligible irrespective of c (see Fig. 22). Anharmonic effects also become less and less significant with increasing pressure and can be almost entirely overlooked up to 3,000 K at 100 GPa. This is consistent with conclusions by Inbar and Cohen (1995) obtained using the VIB method. The results show great improvement w.r.t. experimental thermodynamics data at high temperatures with increasing c. For $c = 0.1$, the predicted thermal expansivity differs from measured values by less than 1% up to the highest measured temperature, $T = 1700$ K at 0 GPa. γ_{th}, C_p, and K_S also show excellent agreement with experimental values (Isaak et al. 1989). Therefore, anharmonic effects described by Equations (9)-(11) are very close to the expected one. This excellent agreement ensures that the pressure calibration based on P-V-T relations in MgO (Wu et al. 2008) derived using this anharmonic free energy have very precise thermal pressure. The property most affected by anharmonicity is the thermal expansivity, α, followed by γ_{th} and C_P, with K_S being the least sensitive (see Table 4). At conditions of the uppermost lower mantle anharmonicity changes the quasiharmonic K_S of MgO by 0.8%. Anharmonicity decreases with increasing pressure and at core-mantle-boundary conditions the anharmonic contribution to

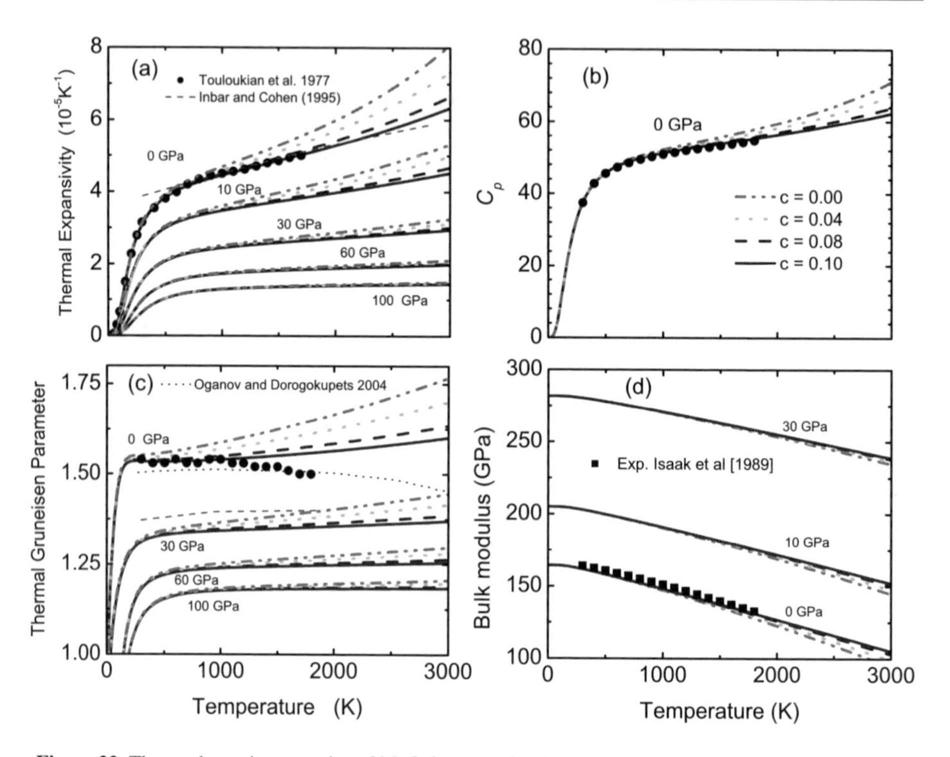

Figure 22. Thermodynamic properties of MgO for several values of c (see Eqn. 10). $c = 0$ corresponds to quasiharmonic results. (a) Thermal expansivity α, (b) thermal Grüneisen parameter, γ_{th}, (c) isobaric specific heat C_p, and (d) adiabatic bulk modulus (from Wu and Wentzcovitch 2009). [Used by permission of APS, from Wu and Wentzcovitch 2009].

the K_S of MgO is only –0.09%. At the transition zone conditions anharmonicity changes the quasiharmonic K_S of forsterite and wadsleyite by less than 0.5%.

One naturally suspects that anharmonicity might be significant at phase boundaries. Their precise determination is necessary to constrain the thermal structure and mineralogy of the mantle. The forsterite to wadsleyite transition, which is responsible for seismic velocity discontinuities around 410 km depth, is a good test case for a couple of reasons: i) there are discrepancies between predicted and measured Clapeyron slopes (CS). All calculations used the QHA and predict similar values (see Table 5); ii) anharmonic effects have long been recognized in forsterite since its heat capacity at constant volume, C_v, can exceed the Dulong-Petit limit at high temperature (Cynn et al. 1996; Anderson et al. 1991; Gillet et al. 1991; Bouhifd et al. 1996); iii) it has been recognized that the conditions at the 410 km discontinuity is at the borderline limit of validity of the QHA (Li et al. 2007b; Yu et al. 2008). Therefore, anharmonicity could be the source of the discrepancy between quasiharmonic predictions and measured or semi-empirical CS's.

We find that anharmonic effects are opposite in forsterite and wadsleyite and this enhances anharmonic effects on their phase boundary (Wu and Wentzcovitch 2009). The parameter c for forsterite is ~ –0.06. This was determined by matching predicted and experimental C_p's since this is most consistent experimental thermodynamics data (see Fig. 23-top). The anharmonic correction considerably improves the predicted heat capacity (see Fig. 23-top for C_p) and thermal expansivity (Matsui and Manghnani 1985; Suzuki et al. 1984; Wu and Wentzcovitch

Table 4. Thermodynamic properties of the investigated phases at relevant mantle conditions, before and after inclusion of anharmonic effects (from Wu and Wentzcovitch 2009).

		MgO at 2000 K and 23 GPa		
c	α (10^{-5}/K)	γ	C_p (J mol^{-1} K^{-1})	K_S (GPa)
0	3.25	1.437	54.07	224.917
0.1	3.04	1.38	52.79	226.822
Change (%)	−6.5	−4.0	−2.4	0.8
		MgO at 4000 K and 135 GPa		
c	α (10^{-5}/K)	γ	C_p (J/mol K)	K_S (GPa)
0	1.31	1.202	52.769	598.19
0.1	1.268	1.178	52.025	597.64
Change (%)	−3.2	−2.2	−1.4	−0.09
		Forsterite at 1700 K and 13.5 GPa		
c	α (10^{-5}/K)	γ	C_p (J/mol K)	K_S (GPa)
0	2.862	1.078	180.549	164.267
−0.06	2.96	1.102	182.008	163.763
Change (%)	3.5	2.2	0.8	−0.3
		Wadsleyite at 1700 K and 13.5 GPa		
c	α (10^{-5}/K)	γ	C_p (J/mol K)	K_S (GPa)
0	2.674	1.173	180.77	202.105
0.08	2.564	1.141	178.85	203.01693
Change (%)	−4.1	−2.7	−1.0	0.45

Table 5. Measured and calculated Clapeyron slopes (MPa/K) of the thermodynamic boundary between forsterite (α) and wadsleyite (β) (from Wu and Wentzcovitch 2009).

High pressure experiments	2.5	Katsura and Ito 1989
	3.5	Suito 1977
	3.6	Morishma et al. 1994
	4.0	Katsura et al. 2003
QHA based calculations	1.8	Akaogi et al. 1989
	2.7	Price et al. 1987; Chopelas 1991; Yu et al. 2008
This study	3.6	using c = −0.06 for α-phase and c = 0.08 for β-phase

2009). Anharmonic corrections with c= −0.06 raise C_v beyond the Dulong-Petit limit at ~2000 K. The experimental crossing temperatures are 1300 ~ 1500 K (Cynn et al. 1996; Anderson et al. 1991; Gillet et al. 1991; Bouhifd et al. 1996). It appears that differences in the crossing temperatures are mainly caused by thermal expansion data used to calculate $C_v(C_p/[1+\alpha\gamma_{th}T])$. The thermal expansivity data adopted in these experiments are much smaller than found by Matsui and Manghnani (1985) and Suzuki et al. (1984). The parameter c for wadsleyite is ~0.08. This was determined by direct comparison of predicted and experimental thermal expansivity (Suzuki et al. 1984) (see Fig. 23-bottom).

The anharmonic contribution to the free energy is negative for forsterite and positive for wadeleyite. This leads to anharmonic stabilization of forsterite at high temperatures. As shown in Figure 24, c_α = −0.06 and c_β = 0.08 increase the transformation pressure by ~1 GPa at 1700 K

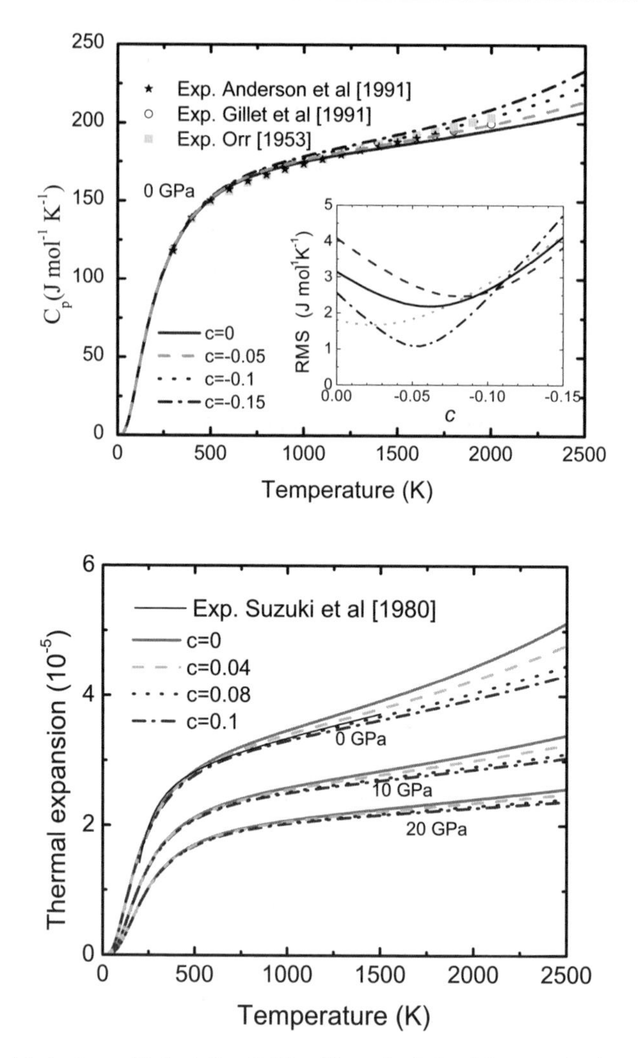

Figure 23. (top) Isobaric specific heat C_p at 0 GPa of forsterite (-phase) for several values of c. The inset shows the root-mean-square (RMS) of the difference between calculated and experimental C_p from Orr (1953) (dashed line), Anderson et al. (1991) (dotted line), Gillet et al. (1991) (dashed dotted line), and all data above (solid line). (bottom) Thermal expansivity, α, of wadsleyite for various values of c (from Wu and Wentzcovitch 2009). [Used by permission of APS, from Wu and Wentzcovitch 2009].

and the CS increases from 2.7 MPa/K ($c_\alpha = c_\beta = 0$) to 3.6 MPa/K (Wu and Wentzcovitch 2009). We also found that, at 1700 K, the CS varies quite linearly with c_α and c_β and can be expressed as: $CS = CS_0 + 7.82(c_\beta - c_\alpha)$, where CS_0 is the slope given by a quasiharmonic calculation ($c_\alpha = c_\beta = 0$). Therefore, anharmonicity may affect the CS particularly strongly when one phase has positive c and the other phase has negative c. In the present case, the discrepancy between direct measurements and calculations of the CS is reconciled after inclusion of anharmonicity in the calculations (see Table 5).

Finally, we found that this parameterization reproduces the correct high and low temperature behaviors of the anharmonic free energy (Wu and Wentzcovitch 2009). Anharmonic free

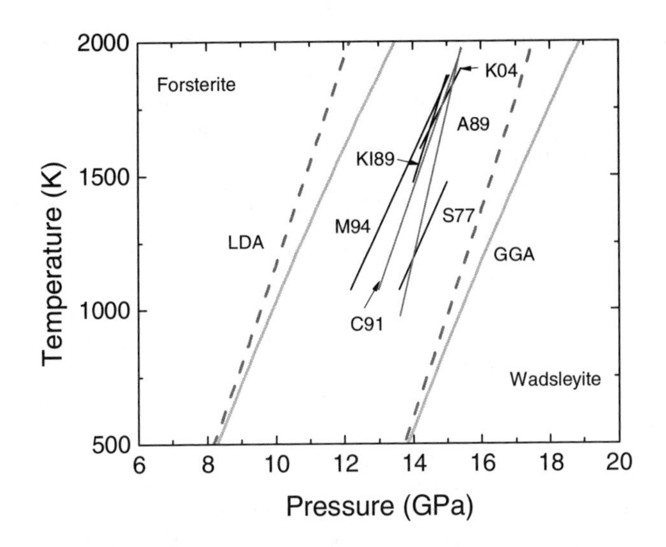

Figure 24. Phase boundary of the forsterite-to-wadsleyite (α-to-β) transformation predicted by LDA (left) and GGA (right) quasiharmonic calculations (dashed lines) and calculations including anharmonic contributions to the free energy according to Equations (9)-(11) using $c_\alpha = -0.06$ and $c_\beta = 0.08$ (thick solid line). Other boundaries are from Katsura et al. (2003), Katsura and Ito (1989), Morishima et al. (1994) (M94), Suito (1977) (S77), Chopelas (1991) (C91) and Akaogi et al. (1989) (A89) (from Wu and Wentzcovitch 2009). [Used by permission of APS, from Wu and Wentzcovitch 2009].

energy can be well fitted by T^2 at high temperature and exhibit the typical T^4 behavior at low temperature for MgO, forsterite and wadsleyite.

SUMMARY

Quasiharmonic theory combined with first-principles phonon density of states gives accurate thermodynamics properties of minerals at the high pressures and temperatures of the Earth interior. Care must be exercised in using this method within its regime of validity. A simple criterion to establish the approximate upper temperature limit of validity of the QHA versus pressure was introduced based on *a posteriori* inspection of the thermal expansivity. This criterion shows that the QHA is a good approximation for minerals at mantle conditions, except for truly anharmonic crystals like $CaSiO_3$-perovskite, and perhaps for minerals at conditions of the core-mantle boundary. In general, the temperature range of applicability of the QHA increases with pressure.

The systems analyzed here, were all investigated consistently and systematically, using the same pseudopotentials, convergence criteria, plane-wave cut-offs, and k- and q-point samplings in LDA and PBE-GGA calculations. The trends extracted from these calculations are therefore reliable. The importance of zero-point-motion effects on structural properties cannot be overemphasized. The LDA exchange-correlation functional gives considerably superior results for 0 K properties after the zero-point-motion energy is included in the calculation. GGA results consistently overestimate volume and underestimate the bulk modulus by several percent. LDA thermodynamic properties at ambient conditions are in excellent agreement with experimental values.

The performance of these exchange-correlation functionals was also investigated for predictions of thermodynamic phase boundaries. In general, the LDA phase boundaries are "shifted" by 5-10 GPa to lower pressures compared to the experimental ones, while GGA boundary are closer to experimental ones but still "shifted" by 2-5 GPa to higher pressures. LDA and GGA Clapeyron slopes are very similar and in fairly good agreement with experimental slopes. Phase boundaries, however, may be more affected by anharmonicity since bond-breaking and bond-reconstruction, phonon softening, diffusion, etc may take place at these pressures and temperatures.

Discontinuities in density and bulk sound velocity for important phase transformations in the mantle transition zone were systematically investigated. Only the magnesium end-members, Mg_2SiO_4, were studied. Predicted discontinuities in density, bulk modulus, and bulk sound velocity are sharp and have useful accuracy for analysis of mantle discontinuities, despite uncertainties in the predicted phase boundary. With the rough premise that the olivine system does not interact with other minerals such as pyroxene/garnet/Ca-perovskite, we were able estimate density discontinuities at 410-km, 520-km, and 660-km depth and compare them with those inferred from seismic data. We conclude that seismic density discontinuities observed at 410-km (0.2-4%) and 660-km (~5.2%) depth can be produced by phase transition in the olivine system alone in a mantle with pyrolite composition (~60 vol% olivine). However, the 520-km discontinuity (2.1±0.8%) cannot. It requires contributions from other mantle transformations, e.g., the Ca-perovskite ex-solution from majorite garnet. We also discuss the post-perovskite phase boundary. Our predicted Clapeyron slope, ~7.5±0.3 MPa/K, differs somewhat from the preferred experimental slopes, > 9.7 MPa/K. This suggests that this phase boundary might be sensitive to anharmonic effects.

We have introduced a semi-empirical *ansatz* to compute anharmonic contributions to the free energy. This method utilizes experimental data at low pressures and high temperatures on *one* property, preferably thermal expansivity, to compute *all* thermodynamic properties at high pressures and high temperatures. It offered excellent results for MgO, and the *P-V-T* relation in this mineral was offered for pressure calibration in diamond-anvil-cells experiments. It was also applied to forsterite (α), wadsleyite (β), and to the α-to-β transformation. The thermodynamic properties of the α- and β-phases improve and are further reconciled with experimental measurements beyond the QHA validity regime after correction for anharmonic effects. This study indicated that anharmonicity manifests differently in different systems, depending whether the "average" phonon frequency increases (β and MgO) or decreases (α) with temperature at constant volume. This difference in behavior affects the Clapeyron slope of the α-to-β transformation, raising it from 2.5-2.6 to 3.6 MPa/K and reconciling it with the latest experimental determinations. Anharmonic effects are most evident in the thermal expansivity, followed by the thermal Grüneisen parameter, constant pressure specific heat, and least evident in the bulk modulus.

APPENDIX

Pseudopotentials used here are the same as those used in Tsuchiya et al. 2004. Pseudopotentials for oxygen and silicon were generated by Troullier and Martin's (1993) method and that of magnesium was generated by the method of von Barth and Car (unpublished). For O, the reference configuration is $2s^22p^4$, r_s and r_p are both 1.45 a.u. p-local; for Si, $3s^23p^23d^0$, $r_s = r_p = r_d = 1.47$ a.u. d-local. Mg pseudopotential was generated from five configurations, $3s^23p^0$, $3s^13p^1$, $3s^13p^{0.5}3d^{0.5}$, $3s^13p^{0.5}$, $3s^13d^1$ with decreasing weights 1.5, 0.6, 0.3, 0.3, and 0.2, respectively, $r_s=r_p=r_d=2.5$ a.u. d-local, nonlinear core correction (Louie et al. 1982) was used.

We have used the same plane-wave energy cut off, 70 Ry, and the same criterion for *k*-point sampling convergence, 5×10^{-5} Ry/atom. The Monkhost-Pack *k*-point mesh adopted in these calculations were 4×4×4, 4×2×4, 4×4×4, 4×4×4, 6×6×4, 4×4×8, 4×4×4, 4×4×4, 4×4×4, and

2×2×2 with a 1/2 shift from the origin for MgO, foresterite wadsleyite, wadsleyite, ringwoodite, post-perovskite, stishovite, LP-enstatite, HP-enstatite, ilmenite, and majorite, respectively. Structure optimizations were performed using variable cell shape MD (Wentzcovtich 1991). Phonons were directly calculated (Baroni et al. 2001) on 2×2×2 q-point and then interpolated for denser meshes to calculate phonon dispersions and density of states. Results on perovskite are from Tsuchiya et al. (2005) where equivalent parameters were used. All calculations were performed in primitive cells (not unit cell in general) using the Quantum ESPRESSO distribution (Gianozzi et al. 2009).

ACKNOWLEDGMENT

This research was supported by NSF grants EAR 0810272, EAR 0635990, ATM 0428774 (VLab), EAR 0757903.

REFERENCES

Adams DJ, Oganov AR (2006) Ab initio molecular dynamics study of $CaSiO_3$ perovskite at P-T conditions of Earth's lower mantle. Phys Rev B 73:184106

Akaogi M, Akimoto S (1979) High-pressure phase equilibria in a garnet lherzolite, with special reference to Mg^{2+}-Fe^{2+} partitioning among constituent minerals. Phys Earth Planet Inter 19:31-51

Akaogi M, Ito E, Navrotsky A (1989) Olivine-modified spinel-spinel transitions in the system Mg_2SiO_4-Fe_2SiO_4: Calorimetric measurements, thermochemical calculation, and geophysical application. J Geophys Res 94:15,671-15,685

Akaogi M, Yano M, Tejima Y, Iijima M, Kojitani H (2004) High-pressure transitions of diopside and wollastonite: phase equilibria and thermochemistry of $CaMgSi_2O_6$, $CaSiO_3$ and $CaSi_2O_5$-$CaTiSiO_5$ system. Phys Earth Planet Inter 143-144:145-156

Akaogi M, Yusa H, Shiraishi K, Suzuki T (1995) Thermodynamic properties of α-quartz, coesite, and stishovite and equilibrium phase relations at high pressures and high temperatures. J Geophys Res 100:22337-22348

Anastassakis E, Hwang HC, Perry CH (1971) Temperature dependence of the long-wavelength optical phonons in diamond. Phys Rev B 4:2493-2497

Anderson OL, Isaak DL, Oda H (1991) Thermoelastic parameters for six minerals at high temperature. J Geophys Res 96:18037-18046

Andrault D, Angel RJ, Mosenfelder JL, Le Bihan T (2003) Equation of state of stishovite to lower mantle pressures. Am Mineral 88:301-307

Angel RJ, Finger LW, Hazen RM, Kanzaki M, Weidner DJ, Leibermann RC, Veblen DR (1989) Structure and twinning of single-crystal $MgSiO_3$ garnet synthesized at 17 GPa and 1800 °C. Am Mineral 74:509-512

Angel RJ, Hugh-Jones DA (1994) Equations of state and thermodynamic properties of enstatite pyroxenes. J Geophys Res 99:19777-19783

Anisimov VI, Zaanen J, Andersen OK (1991) Band theory and Mott insulators: Hubbard U instead of stoner I. Phys Rev B 44:943-954

Ashida T, Kume S, Ito E (1987) Thermodynamic aspects of phase boundary among α-, β-, and γ- Mg_2SiO_4. *In*: High Pressure Research in Mineral Physics. Manghnani M, Syono Y (eds) American Geophysical Union. Geophysical Monograph 39:269-274

Baroni S, Giannozzi P, Isaev E (2010) Density-functional perturbation theory for quasi-harmonic calculations. Rev Mineral Geochem 71:39-57

Baroni S, de Gironcoli S, Corso AD, Giannozzi P (2001) Phonons and related crystal properties from density-functional perturbation theory. Rev Mod Phys 73:515-562

Boehler R, Ross M (1997) Melting curve of aluminum in a diamond cell to 0.8 Mbar: implications for iron. Earth Planet Sci Lett 153:223-227

Born M, Huang K (1956) Dynamical Theory of Crystal Lattices. International Series of Monographs on Physics. Oxford at the Clarendon Press, Hong Kong

Bouhifd MA, Andrault D, Fiquet G, Richet P (1996) Thermal expansion of forsterite up to the melting point. Geophys Res Lett 23:1143-1146

Brown JM, Shankland TJ (1981) Thermodynamic parameters in the Earth as determined from seismic profiles. Geophys J Int 66:579-596

Canil D (1994) Stability of clinopyroxene at pressure-temperature conditions of the transition region. Phys Earth Planet Inter 86:25-34

Caracas R, Cohen R (2007) The effect of chemistry on the physical properties of perovskite and post-perovskite. *In:* Post-perovskite. The Last Mantle Phase Transition. Hirose K, Brodholt J, Lay T, Yuen D (ed) AGU Monograph Series 174:115-128

Caracas R, Wentzcovitch RM (2006) Theoretical determination of the structures of $CaSiO_3$ perovskites. Acta Crystallogr B 62:1025-1030

Caracas R, Wentzcovitch RM, Price GD, Brodholt J (2005) $CaSiO_3$ perovskite at lower mantle pressures. Geohys Res Lett 144:L06306

Carrier P, Justo JF, Wentzcovitch RM (2008) Quasiharmonic elastic constants corrected for deviatoric thermal stresses. Phys Rev B 78:144302

Carrier P, Wentzcovitch R, Tsuchiya J (2007) First-principles prediction of crystal structures at high temperatures using the quasiharmonic approximation. Phys Rev B 76:064116

Ceperley DM, Alder BJ (1980) Ground state of the electron gas by a stochastic method. Phys Rev Lett 45:566-569

Chopelas A (1990) Thermal properties of forsterite at mantle pressure derived from vibrational spectroscopy. Phys Chem Mineral 17:149-156

Chopelas A (1991) Thermal properties of β-Mg_2SiO_4 at mantle pressures derived from vibrational spectroscopy: Implications for the mantle at 400 km depth. J Geophys Res 96:11817-11829

Chopelas A (1996) Thermal expansivity of lower mantle phases MgO and $MgSiO_3$ perovskite at high pressure derived from vibrational spectroscopy. Phys Earth Planet Inter 98:3-15

Chopelas A (1999) Estimates of mantle relevant Clapeyron slopes in the $MgSiO_3$ system from high pressure spectroscopic data. Am Mineral 84:233-244

Chopelas A (2000) Thermal expansivity of mantle relevant magnesium silicates derived from vibrational spectroscopy at high pressure. Am Mineral 85:270-278

Chopelas A, Boehler R, Ko T (1994) Thermodynamics and behavior of Mg_2SiO_4 at high-pressure: Implications for Mg_2SiO_4 phase equilibrium. Phys Chem Mineral 21:351-359

Cococcioni M (2010) Accurate and efficient calculations on strongly correlated minerals with the LDA+U method: review and perspectives. Rev Mineral Geochem 71:147-167

Cococcioni M, de Gironcoli S (2005) Linear response approach to the calculation of the effective interaction parameters in the LDA+U method. Phys Rev B 71:035105

Cynn H, Carnes JD, Anderson OL (1996) Thermal properties of forsterite, including C_v, calculated from αKT through the entropy. J Phys Chem Solids 57:1593-1599

da Silva C, Stixrude L, Wentzcovitch RM (1997) Elastic constants and anisotropy of forsterite at high pressure. Geophys Res Lett 24:1963-1966

Deuss A, Woodhouse J (2001) Seismic observations of splitting of the mid-transition zone discontinuity in Earth's mantle. Science 294:354-357

Downs JW, Bukowinski MST (1997) Variationally induced breathing equations of state of pyrope, grossular, and majorite garnets. Geophys Res Lett 24:1959-1962

Downs RT, Zha CS, Duffy TS, Finger LW (1996) The equation of state of forsterite to 17.2 GPa and effects of pressure media. Am Mineral 81:51-55

Duan W, Karki BB, Wentzcovitch RM, Gu BL (2001) Crystal chemistry of $MgSiO_3$ low-clinoenstatite. Am Mineral 86:762-766

Durben DJ, Wolf GH (1992) High-temperature behavior of metastable $MgSiO_3$ perovskite: a Raman spectroscopic study. Am Mineral 77:890-893

Dziewonski AM, Anderson DL (1981) Preliminary reference Earth model. Phys Earth Planet Inter 25:297-356

Falk H (1970) Inequalities of J. W. Gibbs. J Phys 38:858-869

Fei Y (1999) Effects of temperature and composition on the bulk modulus of (Mg,Fe)O. Am Mineral 84:272-276

Fei Y, Van Orman J, Li J, van Westrenen W, Sanloup C, Minarik W, Hirose K, Komabayashi T, Walter M, Funakoshi K (2004) Experimentally determined postspinel transformation boundary in Mg_2SiO_4 using MgO as an internal pressure standard and its geophysical implications. J Geophys Res 109:2305

Gasparik T (1989) Transformation of enstatite-diopside-jadeite pyroxenes to garnet. Contrib Mineral Petrol 102:389-405

Gasparik T (1990) Phase relations in the transition zone. J Geophys Res 95:15751-15769

Gerald Pacalo RE, Weidner DJ (1997) Elasticity of majorite, $MgSiO_3$ tetragonal garnet. Phys Earth Planet Inter 99:145-154

Giannozzi P, Baroni S, Bonini N, Calandra M, Car R, Cavazzoni C, Ceresoli D, Chiarotti GL, Cococcioni M, Dabo I, Dal Corso A, de Gironcoli S, Fabris S, Fratesi G, Gebauer R, Gerstmann U, Gougoussis C, Kokalj A, Lazzeri M, Martin-Samos L, Marzari N, Mauri F, Mazzarello R, Paolini S, Pasquarello A, Paulatto L, Sbraccia C, Scandolo S, Sclauzero G, Seitsonen AP, Smogunov A, Umari P, Wentzcovitch RM (2009) QUANTUM ESPRESSO: a modular and open-source software project for quantum simulations of materials. J Phys Condens Matter 21:395502

Gillet P, Matas J, Guyot F, Ricard Y (1999) Thermodynamic properties of minerals at high pressures and temperatures from vibrational spectroscopic data. *In:* Microscopic Properties and Processes in Minerals. Wright K, Catlow R (eds) NATO Advanced Studies Institute, Series C: mathematical and physical sciences (Kluwer, Dordrecht) 543:71-92

Gillet P, Richet P, Guyot F, Fiquet G (1991) High-temperature thermodynamic properties of forsterite. J Geophys Res 96:11805-11816

Gossler J, Kind R (1996) Seismic evidence for very deep roots of continents. Earth Planet Sci Lett 138:1-13

Gu Y, Dziewonski A, Agee C (1998) Global de-correlation of the topography of transition zone discontinuities. Earth Planet Sci Lett 157:57-67

Guyot F, Yanbin W, Gillet P, Ricard Y (1996) Quasi-harmonic computations of thermodynamic parameters of olivines at high-pressure and high-temperature. A comparison with experiment data. Phys Earth Planet Inter 98:17-29

Hamann D, Schluter M, Chiang C (1979) Norm-conserving pseudopotentials. Phys Rev Lett 43:1494-1497

Hazen RM, Zhang J, Ko J (1990) Effects of Fe/Mg on the compressibility of synthetic wadsleyite: β-$(Mg_{1-x}Fe_x)_2SiO_4$ ($x \leq 0.25$). Phys Chem Miner 17:416-419

Hazen RM,Weinberger MB, Yang H, Prewitt CT (2000) Comparative high-pressure crystal chemistry of wadsleyite, β-$(Mg_{1-x}Fe_x)_2SiO_4$ with x= 0 and 0.25. Am Mineral 85:770-777

Heinemann S, Sharp TG, Seifert F, Rubie DC (1997) The cubic-tetragonal phase transition in the system majorite ($Mg_4Si_4O_{12}$) pyrope ($Mg_3Al_2Si_3O_{12}$), and garnet symmetry in the Earth's transition zone. Phys Chem Miner 24:206-221

Hirose K (2006) Postperovskite phase transition and its geophysical implications. Rev Geophys 44:3001

Hohenberg P, Kohn W (1964) Inhomogeneous electron gas. Phys Rev 136:864-871

Holland KG, Ahrens TJ (1997) Melting of $(Mg,Fe)SiO_4$ at the core mantle boundary of the Earth. Science 275:1623-1625

Holmes NC, Moriarty JA, Gathers GR, Nellis WJ (1989) The equation of state of platinum to 660 GPa (6.6 MBar). J Appl Phys 66:2962-2967

Holzapfel WB, Hartwig M, Sievers W (2001) Equations of state for Cu, Ag, and Au for wide ranges in temperature and pressure up to 500 GPa and Above. J Phys Chem Ref Data 30:515-528

Horiuchi H, Hirono M, Ito E, Matsui Y (1982) $MgSiO_3$ (ilmenite-type): single crystal X-ray diffraction study. Am Mineral 67:788-793

Horiuchi H, Sawamoto H (2000) β-Mg_2SiO_4: Single-crystal X-ray diffraction study. Am Mineral 66:568-575

Hofmeister AM, Giesting PA, Wopenka B, Gwanmesia GD, Jolliff BL (2004) Vibrational spectroscopy of pyrope-majorite garnets: Structural implications. Am Mineral 89:132-146

Hsu H, Umemoto K, Wu A, Wentzcovitch RM (2010) Spin-state crossover of iron in lower-mantle minerals: results of DFT+U investigations. Rev Mineral Geochem 71:169-199

Hui JCK, Allen PB (1975) Thermodynamics of anharmonic crystals with application to Nb. J Phys C: Solid State Phys 8:2923-2935

Inbar I, Cohen RE (1995) High pressure effects on thermal properties of MgO. Geophys Rev Lett 22:1533-1536

Inoue T, Irifune T, Higo Y, Sanehira T, Sueda Y, Yamada A, Shinmei T, Yamazaki D, Ando J, Funakoshi K, UtsumiW (2006) The phase boundary between wadsleyite and ringwoodite in Mg_2SiO_4 determined by in situ X-ray diffraction. Phys Chem Miner 33:106-114

Irifune T (1987) An experimental investigation of the pyroxene-garnet transformation in a pyrolite composition and its bearing on the constitution of the mantle. Phys Earth Planet Inter 45:324-336

Irifune T, Nishiyama N, Kuroda K, Inoue T, Isshiki M, Utsumi W, Funakoshi K, Urakawa S, Uchida T, Katsura T, Ohtaka O (1998) The postspinel phase boundary in Mg_2SiO_4 determined by in situ X-ray diffraction. Science 279:1698-1700

Isaak DG, Anderson OL, Goto T (1989) Measured elastic moduli of single-crystal MgO up to 1800 K. Phys Chem Miner 16:704-713

Ita J, Stixrude L (1992) Petrology, elasticity, and composition of the mantle transition zone. J Geophys Res 97:6849-6866

Ito E, Takahashi E (1989) Postspinel transformations in the system Mg_2SiO_4-Fe_2SiO_4 and some geophysical implications. J Geophys Res 94:10637-10646

Ito E, Yamada H (1981) Stability relations of silicate spinels, ilmenites and perovskites. *In:* High Pressure Research in Mineral Physics. Akamoto S, Manghnani MH (eds) Center for Academic Publishing of Japan, p 405-419

Jackson J, Sinogeikin SV, Bass J (2000) Sound velocities and elastic properties of γ- Mg_2SiO_4 to 873 K by Brillouin spectroscopy. Am Mineral 85:296-303

Jamieson JC, Fritz JN, Manghnani MH (1982) Pressure measurement at high temperature in x-ray diffraction studies: Gold as a primary standard. *In:* High-Pressure Research in Geophysics. Akimoto S, Manghnani MH (eds) Adv Earth Planet Sci 12:27-48

Jeanloz R, Ahrens T, Mao HK, Bell PM (1979) B1-B2 transition in calcium oxide from shock-wave and diamond-cell experiments. Science 206:829-831

Jones LE, Mori J, Helmberger DV (1992) Short-period constraints on the proposed transition zone discontinuity. J Geophys Res 97:8765-8774

Karki BB, Wentzcovitch RM (2002) First-principles lattice dynamics and thermoelasticity of $MgSiO_3$ ilmenite at high pressure. J Geophys Res 107:2267

Karki BB, Wentzcovitch RM (2003) Vibrational and quasiharmonic thermal properties of CaO under pressure. Phys Rev B 68:224304

Karki BB, Wentzcovitch RM, de Gironcoli S, Baroni S (1999) First-principles determination of elastic anisotropy and wave velocities of MgO at lower mantle conditions. Science 286:1705-1707

Karki BB, Wentzcovitch RM, de Gironcoli S, Baroni S (2000a) High-pressure lattice dynamics and thermoelasticity of MgO. Phys Rev B 61:8793-8800

Karki BB, Wentzcovitch RM, de Gironcoli S, Baroni S (2000b) Ab initio lattice dynamics of $MgSiO_3$ perovskite at high pressure. Phys Rev B 62:14750-14756

Katsura T, Ito E (1989) The system Mg_2SiO_4-Fe_2SiO_4 at high pressures and temperatures: Precise determination of stabilities of olivine, modified spinel, and spinel. J Geophys Res 94:15663-15670

Katsura T, Yamada H, Nishikawa O, Song M, Kubo A, Shinmei T, Yokoshi S, Aizawa Y, Yoshino T, Walter MJ, Ito E, Funakoshi Ki (2004a) Olivine-wadsleyite transition in the system $(Mg,Fe)_2SiO_4$. J Geophys Res 109:2209

Katsura T, Yamada H, Shinmei T, Kubo A, Ono S, Kanzaki M, Yoneda A, Walter MJ, Ito E, Urakawa S, Funakoshi K, Utsumi W (2003) Post-spinel transition in Mg_2SiO_4 determined by high P-T in situ X-ray diffractometry. Phys Earth Planet Inter 136:11-24

Katsura T, Yokoshi S, Song M, Kawabe K, Tsujimura T, Kubo A, Ito E, Tange Y, Tomioka N, Saito K, Nozawa A, Funakoshi Ki (2004b) Thermal expansion of ringwoodite at high pressures. J Geophys Res 109:B12209

Kennett BLN, Engdahl ER, Buland R (1995) Constraints on seismic velocities in the Earth from travel-times. Geophys J Int 122:108-124

Kiefer B, Stixrude L, Wentzcovitch RM (1997) Elastic constants and anisotropy of Mg_2SiO_4 spinal at high pressure. Geophys Res Lett 24:2841-2844

Knittle E, Jeanloz R, Smith GL (1986) Thermal expansion of silicate perovskite and stratification of the earth's mantle. Nature 319:214-216

Kohn W, Sham LJ (1965) Self-consistent equations including exchange and correlation effects. Phys Rev 140:1133-1138

Komabayashi T, Hirose K, Sugimura E, Sata N, Ohishi Y, Dubrovinsky LS (2008) Simultaneous volume measurements of post-perovskite and perovskite in $MgSiO_3$ and their thermal equations of state. Earth Planet Sci Lett 265:515-524

Kurashina T, Hirose K, Ono S, Sata N, Ohishi Y (2004) Phase transition in Al-bearing $CaSiO_3$ perovskite: implications for seismic discontinuities in the lower mantle. Phys Earth Planet Inter 145:67-74

Landau LD, Lifshitz EM (1980) Statistical Physics, Third Edition, Part I. Course of Theoretical Physics, Volume 5. Butterworth & Heinemann: Oxford

Lawrence JF, Shearer PM (2006) Constraining seismic velocity and density for the mantle transition zone with reflected and transmitted waveforms. Geochem Geophys Geosyst 7:Q10012

Li B (2003) Compressional and shear wave velocities of ringwoodite-Mg_2SiO_4 to 12 GPa. Am Mineral 88:1312-1317

Li L, Weidner D, Brodholt J, Alfe D, Price GD, Caracas R, Wentzcovitch R (2006a) Elasticity of $CaSiO_3$ perovskite at high pressure and high temperature. Phys Earth Planet Int 155:249-259

Li L, Weidner D, Brodholt J, Alfe D, Price GD, Caracas R, Wentzcovitch R (2006b) Phase stability of $CaSiO_3$ perovskite at high pressure and high temperature: insights from ab initio molecular dynamics. Phys Earth Planet Int 155:260-268

Li L, Weidner DJ, Brodholt J, Alfè D, Price GD (2006c) Elasticity of Mg_2SiO_4 ringwoodite at mantle conditions. Phys Earth Planet Int 157:181-187

Li L, Weidner DJ, Brodholt J, Alfè D, Price GD (2009) Ab initio molecular dynamics study of elasticity of akimotoite MgSiO3 at mantle conditions. Phys Earth Planet Int 173:115-120

Li L, Weidner DJ, Brodholt J, Price GD (2007a) The effect of cation-ordering on the elastic properties of majorite: An ab initio study. Earth Planet Sc Lett 256:28-35

Li L, Wentzcovitch RM, Weidner DJ, Da Silva CRS (2007b) Vibrational and thermodynamic properties of forsterite at mantle conditions. J Geophys Res 112:B05206

Litasov K, Ohtani E, Sano A, Suzuki A, Funakoshi K (2005) In situ X-ray diffraction study of postspinel transformation in a peridotite mantle: Implication for the 660-km discontinuity. Earth Planet Sci Lett 238:311-328

Liu L-G (1974) Silicate perovskite from phase transformations of pyrope-garnet at high pressure and temperature. Geophys Res Lett 1:277-280

Liu L-G (1977) First occurrence of the garnet-ilmenite transition in silicates. Science 195:990-991

Louie SG, Froyen S, Cohen ML (1982) Nonlinear ionic pseudopotentials in spin-density-functional calculations. Phys Rev B 26:1738-1742

Lu R, Hofmeister AM (1994) Infrared-spectroscopy of $CaGeO_3$ perovskite to 24-GPa and thermodynamic implications. Phys Chem Miner 21:78-84

Lu R, Hofmeister AM, Wang Y (1994) Thermodynamic properties of ferromagnesium silicate perovskites from vibrational spectroscopy. J Geophys Res 99:11795-11804

Mao HK, Hemley RJ, Fei Y, Shu JF, Chen LC, Jephcoat AP, Wu Y, Bassett WA (1991) Effect of pressure, temperature, and composition on lattice parameters and density of $(Fe,Mg)SiO_3$-perovskites to 30 GPa. J Geophys Res 96:8069-8079

Matsui M, Price GD, Patel A (1994) Comparison between the lattice dynamics and molecular dynamics methods: Calculation results for $MgSiO_3$ perovskite. Geophys Rev Lett 21:1659-1662

Matsui T, Manghnani MH (1985) Thermal expansion of single-crystal forsterite to 1023 K by Fizeau interferometry. Phys Chem Minerals 12:201-210

Meng Y, Fei Y, Weidner DJ, Gwanmesia GD, Hu J (1994) Hydrostatic compression of γ-Mg_2SiO_4 to mantle pressures and 700 K: Thermal equation of state and related thermoelastic properties. Phys Chem Miner 21:407-412

Morishima H, Kato T, Suto M, Ohtani E, Urakawa S, Utsumi W, Shimomura O, Kikegawa T (1994) The Phase Boundary Between α- and β-Mg_2SiO_4 determined by in situ x-ray observation. Science 265:1202-1203

Murakami M, Hirose K, Kawamura K, Sata N, Ohishi Y (2004) Post-perovskite phase transition in $MgSiO_3$. Science 304:855-858

Nakatsuka A, Yoshiasa A, Yamanaka T, Ohtaka O, Katsura T, Ito E (1999) Symmetry change of majorite solid-solution in the system $Mg_3Al_2Si_3O_{12}$-$MgSiO_3$. Am Mineral 84:1135-1143

Oda H, Anderson OL, Isaak DG, Suzuki I (1992) Measurements of elastic properties of single-crystal CaO up to 1200 K. Phys Chen Miner 19:96-105

Oganov AR, Brodholt JP, Price GD (2001) The elastic constant of $MgSiO_3$ perovskite at pressure and temperatures of the Earth's mantle. Nature 411:934-937

Oganov AR, Dorogokupets PI (2004) Intrinsic anharmonicity in equations of state and thermodynamics of solids. J Phys Condens Matter 16:1351-1360

Oganov AR, Gillan MJ, Price GD (2005) Structural stability of silica at high pressures and temperatures. Phys Rev B 71:064104

Oganov AR, Ono S (2004) Theoretical and experimental evidence for a post-perovskite phase of $MgSiO_3$ in Earth's D″ layer. Nature 430:445-448

Ono S, Oganov AR (2005) In situ observations of phase transition between perovskite and $CaIrO_3$-type phase in $MgSiO_3$ and pyrolitic mantle composition. Earth Planet Sci Lett 236:914- 932

Ono S, Ohishi Y, Mibe K, (2004) Phase transition of Ca-perovskite and stability of Al-bearing Mg-perovskite in the lower mantle. Am Mineral 89:1480-1485

Orr RL (1953) High temperature heat contents of magnesium orthosilicate and ferrous orthosilicate. J Am Chem Soc 75:528-529

Perdew JP, Burke K, Ernzerhof M (1996) Generalized gradient approximation made simple. Phys Rev Lett 77:3865-3868

Perdew JP, Ruzsinszky A (2010) Density functional theory of electronic structure: a short course for mineralogists and geophysicists. Rev Mineral Geochem 71:1-18

Perdew JP, Zunger A (1981) Self-interaction correction to density-functional approximations for many electron systems. Phys Rev B 23:5048-5079

Piekarz P, Jochym PT, Parlinski K, Lazewski J (2002) High pressure and thermal properties of γ-Mg_2SiO_4 from first-principles calculations. J Chem Phys 117:3340-3344

Price GD, Parker SC, Leslie M (1987) The lattice dynamics and thermodynamics of the Mg_2SiO_4 polymorphs. Phys Chem Miner 15:181-190

Revenaugh J, Jordan TH (1991) Mantle layering from ScS reverberations. 2. The transition zone. J Geophys Res 96:19763-19780.

Reynard B, Fiquet G, Itie JP, Rubie DC (1996) High-pressure X-ray diffraction study and equation of state of $MgSiO_3$ ilmenite. Am Mineral 81:45-50

Rigden SM, Fitz Gerald JD, Jackson I, Gwanmesia GD, Liebermann RC (1991) Spinel elasticity and seismic structure of the transition zone of the mantle. Nature 354:143-145

Ringwood AE (1975) Composition and Petrology of the Earth's Mantle. McGraw-Hill, New York

Ringwood AE (1979) Composition and origin of the Earth. *In*: The Earth, its Origin, Structure and Evolution. McElhinny MW (ed) Academic Press, New York, p 1-58

Ringwood AE, Major A (1966) Synthesis of Mg_2SiO_4-Fe_2SiO_4 spinel solid solutions. Earth Planet Sci Lett 1:241-245

Ross NL, Hazen RM (1989) Single crystal X-ray diffraction study of $MgSiO_3$ perovskite from 77 to 400 K. Phys Chem Miner 16:415-420

Ryberg T, Wenzel F, Egorkin AV, Solodilov L (1997) Short-period observation of the 520 km discontinuity in northern Eurasia. J Geophys Res 102:5413-5422

Saikia A, Frost DJ, Rubie DC (2008) Splitting of the 520-kilometer seismic discontinuity and chemical heterogeneity in the mantle. Science 319:1515-1518

Sangster MJL, Peckham G, Saunderson DH (1970) Lattice dynamics of magnesium oxide. J Phys C: Solid State Phys 3:1026-1036

Sawamoto H (1987) Phase diagram of $MgSiO_3$ at pressures up to 24 GPa and temperatures up to 2200 °C: Phase stability and properties of tetragonal garnet. *In*: High Pressure Research in Mineral Physics. Manghnani M, Syono Y (eds) American Geophysical Union. Geophysical Monograph 39:209-219

Shaner JW, Brown JM, McQueen RG (1984) Melting of metals above 100 GPa. *In*: High Pressure in Science and Technology, Part I: Collective Phenomena and Transport Properties (Volume 22). Homan C, MacCrone RK, Whalley E, (eds) Elsevier Science Ltd, p 137-141

Shearer PM (1990) Seismic imaging of upper-mantle structure with new evidence for a 520-km discontinuity. Nature 344:121-126

Shearer PM, Flanagan MP (1999) Seismic velocity and density jumps across the 410- and 660-kilometer discontinuities. Science 285:1545-1548

Shen G, Heinz D (1998) High pressure melting of deep mantle and core materials. Rev Mineral Geochem 27:369-396

Shim SN, Duffy TS, Shen G (2001) The post-spinel transformation in Mg_2SiO_4 and its relation to the 660-km seismic discontinuity. Nature 411:571-574

Shim SN, Jeanloz R, Duffy TS (2002) The tragonal structure of $CaSiO_3$ perovskite above 20 GPa. Geophys Res Lett 29:2166-2169

Shinmei T, Tomioka N, Fujino K, Kuroda K, Irifune T (1999) In situ X-ray diffraction study of enstatite up to 12 GPa and 1473 K and equations of state. Am Mineral 84:1588-1594

Sidorin I, Gurnis M, Helmberger DV (1999) Evidence for a ubiquitous seismic discontinuity at the base of the mantle. Science 286:1326-1331

Sinogeikin SV, Bass JD, Katsura T (2003) Single-crystal elasticity of ringwoodite to high pressures and high temperatures: implications for 520 km seismic discontinuity. Phys Earth Planet Inter 136:41-66

Smyth JR (1987) β-Mg_2SiO_4: A potential host for water in the mantle? Am Mineral 72:1051-1055

Speziale S, Zha C, Duffy TS, Hemley RJ, Mao HK (2001) Quasi-hydrostatic compression of magnesium oxide to 52 GPa: Implications for the pressure-volume-temperature equation of state. J Geophys Res 106:515-528

Stackhouse S Brodholt J (2007) High temperature elasticity of $MgSiO_3$ post-perovskite. *In:* Post-Perovskite, The Last Mantle Phase Transition. Hirose K, Brodholt J, Lay T, Yuen D (eds) AGU Monograph Series 174:99-114

Stixrude L (1997) Structure and sharpness of phase transitions and mantle discontinuities. J Geophys Res 102:14835-14852

Stixrude L, Cohen RE, Yu R, Krakauer H (1996) Prediction of phase transition in $CaSiO_3$ perovskite and implications for lower mantle structure. Am Mineral 81:1293-1296

Stixrude L, Lithgow-Bertelloni C (2005) Mineralogy and elasticity of the oceanic upper mantle: Origin of the low-velocity zone. J Geophys Res 110:B3204

Stixrude L, Lithgow-Bertelloni C (2010) Thermodynamics of the Earth's mantle. Rev Mineral Geochem 71:465-484

Stixrude L, Lithgow-Bertelloni C, Kiefer B, Fumagalli P (2007) Phase stability and shear softening in $CaSiO_3$ perovskite at high pressure. Phys Rev B 75:024108

Strachan A, Cagin T, Goddard WA (1999) Phase diagram of MgO from density-functional theory and molecular-dynamics simulations. Phys Rev B 60:15084-15093

Suito K (1977) Phase relations of pure Mg_2SiO_4 up to 200 kilobars. *In*: High Prssure Research—Applications to Geophysics. Manghnani MH, Akimoto S (eds) Academic Press, New York, p 255-266

Suzuki A, Ohtani E, Morishima H, Kubo T, Kanbe Y, Kondo T, Okada T, Terasaki H, Kato T, Kikegawa T (2000) In situ determination of the phase boundary between wadsleyite and ringwoodite in Mg_2SiO_4. Geophys Res Lett 27:803-806

Suzuki I, Ohtani E, Kumazawa M (1979) Thermal expansion of γ-Mg_2SiO_4. J Phys Earth 27:53-61

Suzuki I, Ohtani E, Kumazawa M (1980) Thermal expansion of modified spinel, β-Mg_2SiO_4. J Phys Earth 28:273-280

Suzuki I, Takei H, Anderson OL (1984) Thermal expansion of single-crystal forsterite Mg_2SiO_4. *In:* Thermal Expansion 8 by International Thermal Expansion Symposium (8th 1981 Gaithersburg, Md.). Hahn TA (ed) Plenum Press, NY. p 79-88

Touloukian YS, Kirby RK, Taylor RE, Lee TYR (1977) Thermal Expansion -- Nonmetallic Solids (Thermophysical Properties of Matter, Vol. 13). Plenum, New York

Troullier N, Martins JL (1991) Efficient pseudopotentials for plane-wave calculations. Phys Rev B 43:1993-2006

Tsuchiya J, Tsuchiya T, Wentzcovitch RM (2005) Vibrational and thermodynamic properties of $MgSiO_3$ postperovskite. J Geophys Res 110:2204

Tsuchiya T (2003) First-principles prediction of the P-V-T equation of state of gold and the 660-km discontinuity in Earth's mantle. J Geophys Res 108:2462

Tsuchiya T, Tsuchiya J, Umemoto K, Wentzcovitch RM (2004) Phase transition in $MgSiO_3$ perovskite in the earth's lower mantle. Earth Planet Sci Lett 224:241-248

Umemoto K, Wentzcovitch RM, Allen PB (2006) Dissociation of $MgSiO_3$ in the cores of giants and terrestrial exoplanets, Science 311:983-986

Umemoto K, Wentzcovitch RM (2010) Multi-Mbar phase transitions in minerals. Rev Mineral Geochem 71:299-314

Ulmer P, Stalder R (2001) The $Mg(Fe)SiO_3$ orthoenstatite-clinoenstatite transitions at high pressures and temperatures determined by Raman-spectroscopy on quenched samples. Am Mineral 86:1267-1274

Vanderbilt D (1990) Soft self-consistent pseudopotentials in a generalized eigenvalue formalism. Phys Rev B 41:7892-7895

Vočadlo L, Alfè D (2002) Ab initio melting curve of the fcc phase of aluminum. Phys Rev B 65:214105

Wallace DC (1972) Thermodynamics of Crystal. Wiley, New York

Wang Y, Weidner DJ, Liebermann RC, Zhao Y (1994) P-V-T equation of state of $(Mg,Fe)SiO_3$ perovskite: constraints on composition of the lower mantle. Phys Earth Planet Inter 83:13-40

Wang YB, Weidner DJ (1994) Thermoelasticity of $CaSiO_3$ perovskite and implications for the lower mantle. Geophys Res Lett 21:895-898

Watanabe H (1982) Thermochemical properties of synthetic high pressure compounds relevant to the earth's mantle. *In*: High Pressure Research in Geophysics. Akimoto S, Manghnani M (eds) Center for Academic Publishing of Japan, p 441-464

Weidner D, Sawamoto H, Sasaki S, Kumazawa M (1984) Single-crystal elastic properties of the spinel phase of Mg_2SiO_4. J Geophys Res 89:7852-7860

Weidner DJ, Wang Y (2000) Phase transformations: Implications for mantle structure. *In:* Earth's Deep Interior: Mineral Physics and Tomography From the Atomic to the Global Scale. Geophys Monogr 117:215-235

Wentzcovitch RM (1991) Invariant molecular-dynamics approach to structural phase transitions. Phys Rev B 44:2358-2361

Wentzcovitch RM, Hugh-Jones DA, Angel RJ, Price GD (1995) Ab initio study of $MgSiO_3$ C2/c enstatite. Phys Chem Mineral 22:453-460

Wentzcovitch RM, Karki BB, Cococcioni M, de Gironcoli S (2004a) Thermoelastic properties of $MgSiO_3$-perovskite: insights on the nature of the earth's lower mantle. Phys Rev Lett 92:018501

Wentzcovitch RM, Martins JL, Price GD (1993) Ab initio molecular dynamics with variable cell shape: application to $MgSiO_3$. Phys Rev Lett 70:3947

Wentzcovitch RM, Stixrude L (1997) Crystal chemistry of forsterite: a first-principles study. Am Mineral 82:663-671

Wentzcovitch RM, Stixrude L, Karki B, Kiefer B (2004b) Akimotoite to perovskite. Geohys Res Lett 31:L10611

Wentzcovitch RM, Umemoto K, Tsuchiya T, Tsuchiya J (2007) Thermodynamic properties and stability field of $MgSiO_3$ post-perovskite. In: Post-Perovskite. The Last Mantle Phase transition. Hirose K, Brodholt J, Lay T, Yuen D (eds) AGU Monograph Series 174:79-98

Wentzcovitch RM, Wu Z, Carrier P (2010) First principles quasiharmonic thermoelasticity of mantle minerals. Rev Mineral Geochem 71:98-128

Wiggins RA, Helmberger DV (1973) Upper mantle structure of the western United States. J Geophys Res 78:1870-1880

Williams Q, Jeanloz R, McMillan P (1987) Vibrational spectrum of $MgSiO_3$ perovskite: zero pressure Raman and mid-infrared spectra to 27 GPa. J Geophys Res 92:8116-8128

Wu Z, Wentzcovitch RM (2007) Vibrational and thermodynamic properties of wadsleyite: A density functional study. J Geophys Res 112:B12202

Wu Z, Wentzcovitch RM (2009) Effective semiempirical ansatz for computing anharmonic free energies. Phys Rev B 79:104304
Wu Z, Wentzcovitch RM, Umemoto K, Li B, Hirose K, Zheng J-C (2008) Pressure-volume-temperature relations in MgO: An ultrahigh pressure-temperature scale for planetary sciences applications. J Geophys Res 113:B06204
Yu YG, Angel RJ, Wentzcovitch RM (2010a) Thermodynamics properties of $MgSiO_3$ majorite. in preparation
Yu YG, Wentzcovitch RM (2006) Density functional study of vibrational and thermodynamic properties of ringwoodite. J Geophys Res 111:B12202
Yu YG, Wentzcovitch RM (2008) Low pressure clino- to high pressure clinoenstatite phase transition: a phonon related mechanism. Am Mineral 94:461-466
Yu YG, Wentzcovitch RM, Angel RJ (2010b) Thermodynamics properties of and phase boundary between low pressure and high pressure clinoenstatite. J Geophys Res 115:B02201
Yu YG, Wentzcovitch RM, Tsuchiya T, Umemoto K, Weidner DJ (2007) First principles investigation of the postspinel transition in Mg_2SiO_4. Geophys Res Lett 34:L10306
Yu YG, Wu Z, Wentzcovitch RM (2008) α-β-γ transformations in Mg_2SiO_4 in Earth transition zone. Earth Planet Sci Lett 273:115–122
Zerr A, Boehler R (1994) Constraints on the melting temperature of the lower mantle from high pressure experiments on MgO and magnesiowüstite. Nature 371:506-508
Zerr A, Diegeler A, Boehler R (1998) Solidus of Earth's deep mantle. Science 281:243-246
Zerr A, Serghiou G, Boehler R (1997) Melting of $CaSiO_3$ perovskite to 430 kbar and first in situ measurements of lower mantle eutectic temperatures. Geophys Res Lett 24:909-912
Zhao Y, Truhlar DG (2010) The Minnesota density functionals and their applications to problems in mineralogy and geochemistry. Rev Mineral Geochem 71:19-37
Zouboulis ES, Grimsditch M (1991) Raman scattering in diamond up to 1900 K. Phys Rev B 43:12490-12493

Reviews in Mineralogy & Geochemistry
Vol. 71 pp. 99-128, 2010

First Principles Quasiharmonic Thermoelasticity of Mantle Minerals

Renata M. Wentzcovitch

Department of Chemical Engineering and Materials Sciences and
Minnesota Supercomputer Institute
University of Minnesota
Minneapolis, Minnesota, 55455, U.S.A.

wentzcov@cems.umn.edu

Zhongqing Wu

Department of Physics & Astronomy
University of Southern California
Los Angeles, California, 90089-0484, U.S.A.

Pierre Carrier

Department of Computer Science
University of Minnesota
Minneapolis, Minnesota, 55455, U.S.A.

INTRODUCTION

Thermodynamic (Anderson 2005) and elastic properties (Musgrave 1970) of minerals provide the fundamental information needed to analyze seismic observations and to model Earth's dynamic state. The connection between pressure, temperature, chemical composition, and mineralogy that produce seismic velocity gradients, heterogeneities, and discontinuities, can be established with knowledge of thermoelastic and thermodynamic equilibrium properties of single phases and their aggregates.

There is a broad-based need in solid Earth geophysics for these thermoelastic properties to model the Earth. However, the materials and conditions of the Earth's interior present several challenges. The chemical composition of the Earth's mantle is complex with at least five major oxide components and tens of solid phases. Today, this type of challenge is more effectively addressed by a combination of experimental and computational methods. Experiments offer accurate information at lower pressures and temperatures, while computations offer more complete and detailed information at higher pressures and temperatures, where experimental uncertainties are large and conditions difficult to control in the laboratory.

The overwhelming success of density functional theory (Hohenberg and Kohn 1964; Kohn and Sham 1965) combined with the quasiharmonic approximation (QHA) (Born and Huang 1956; Wallace 1972) for computations of thermodynamic properties of major mantle minerals has been reviewed in this book (Wentzcovitch et al. 2010). Although these properties were cited and compared with experiments only at ambient conditions, in general, quasiharmonic calculations perform even better at higher pressures (see references in Wentzcovitch et al. 2010). In this paper we review the extension of these calculations to high pressure and high temperature elasticity. There are important exceptions, such as $CaSiO_3$-perovskite, whose high temperature structure is stabilized by anharmonic fluctuations (Stixrude et al. 1996). Such

1529-6466/10/0071-0005$05.00 DOI: 10.2138/rmg.2010.71.5

cases cannot be addressed using the QHA and one must resort of molecular dynamics (MD) in some form. Great progress has been made in the last decade in this area of first principles computations of elastic constants, whether by MD (Oganov et al. 2001; Li et al. 2005, 2006b,c, 2009; Stackhouse et al. 2005) or QHA calculations (Karki et al. 1999; Wentzcovitch et al. 2004, 2006). This progress resulted from the maturing of theoretical and computational methods (Car and Parrinello 1985; Giannozzi et al. 1991; Wentzcovitch and Martins 1991; Payne et al. 1992; Baroni et al. 2001) and from the availability of stable and well-tested public domain software (Payne et al. 1992; Kresse and Furthmüller 1996; Gonze et al. 2002; Giannozzi et al. 2009) for materials computations.

The purpose of our effort has been to provide the most significant and accurate information needed to understand the fine- and large-scale structure of the mantle. A wide variety of geophysical issues of fundamental importance remain to be resolved, e.g., the geodynamic implications of upper mantle anisotropy, the causes and significance of an upper mantle low velocity zone, the significance of seismic discontinuities and high velocity gradients in the transition zone for possible mantle stratification, and whether heterogeneity in the lower mantle is primarily of chemical/mineralogical versus thermal in origin. Experimental elasticity and thermodynamic data for upper mantle and transition zone minerals are more numerous. The pressures and temperatures in the upper mantle (including the transition zone) can be more closely approached experimentally and with relatively high accuracy. While it is possible to produce extremely high pressures and temperatures in the laser heated diamond anvil cell (DAC), elasticity of relevant phases at these conditions are less well established experimentally for various reasons, including uncertainties in pressure scales, thermal gradients across laser-heated samples, and other technical difficulties in characterizing the most extreme pressures and temperatures. Therefore, the emphasis of our effort has been on lower mantle phases at extreme conditions. This is the pressures and temperatures regime where theory can make the greatest impact in augmenting experimental data and producing essential information.

THEORETICAL BACKGROUND

Several fundamental formal papers on elasticity theory have been written throughout the years (Barron and Klein 1965; Thomsen 1972; Davies 1974). There are also excellent textbooks on more fundamental aspects and one usually starts research on elasticity by studying them (Musgrave 1970; Nye 1985). Here we present a distilled version of the basic formalism for calculations of thermoelastic properties of materials. Because our calculations are performed numerically and *access directly the free energy* computed by first principles, the need for analytical treatment and approximations is minimal. The tricks are in the numerical implementation, not in the analytical development. The simplicity of the basic formalism contrasts with elaborate formal developments addressing analytically the consequences of the free energy expansion (Thomsen 1972; Davies 1974), by necessity truncated, in a power series of strains. Formal developments originated in the practical *impossibility of measurements to access directly the free energy*, but only its derivative with respect to volume, i.e., pressure, or the elastic constants via velocity measurements. Reviews of static (Karki et al. 2001a) and semi-empirical (Stixrude and Lithgow-Bertelloni 2005) approaches to elasticity of minerals discuss several useful details of elasticity calculations.

Here we focus on details of high-pressure and high-temperature QHA calculations that have not been explicitly presented yet. We treat a crystal as a homogeneous, anisotropic medium and assume that stress and strain are anisotropic. The propagation of acoustic waves in an aeolotropic medium is described by Cristoffel's equation (Musgrave 1970):

$$[C_{ijkl}n_j n_l - \rho v^2 \delta_{ik}]p_k = 0 \qquad (1)$$

C_{ijkl} are the components of the (4th rank) elastic constant tensor in Cartesian coordinates, with i, j, k, and l as one of the Cartesian indexes x, y, or z, n_j's are the direction cosines of the propagation direction, and p_k's are the components of the polarization vector. This equation is true at any temperature or pressure and the eigen-solutions are three waves, one quasi-longitudinal and two quasi-transversal. Purely longitudinal and transversal polarizations occur only in isotropic materials. We will return to this point at the end of this section.

These calculations of C_{ijkl} require the definition of a finite strain induced by a large compression and of an additional superposed infinitesimal strain. The large finite strain of choice is the frame invariant Eulerian strain E_{ij} :

$$E_{ij} = \frac{1}{2}\left(f_{ij} + f_{ji} - f_{ik}f_{jk}\right) \tag{2}$$

with

$$X_i - a_i = f_{ij}X_j \tag{3}$$

where $\vec{a}$ is the position vector of a point in a material in the unstrained state and $\vec{X}$ is the position vector of the same point after the material is compressed. E_{ij} is invariably assumed to be isotropic in finite strain expansions of the free energy used to derive equations of state. This is assumed here as well.

$$E_{ij} = -f\delta_{ij} \tag{4a}$$

$$f = \frac{1}{2}\left[\left(\frac{V_0}{V}\right)^{2/3} - 1\right] \tag{4b}$$

where V_0 is a reference volume and $f > 0$ (< 0) corresponds to compression (extension). The Helmholtz free energy of a compressed state is then expanded in a Taylor series in these strains.

$$F(f) = a + bf + cf^2 + df^3 + \ldots \tag{5}$$

This expansion includes a linear term, meaning, the reference state with $V = V_0$ is not the zero pressure state. The temperature dependence of this free energy is temporarily omitted. The second infinitesimal strain, u_{ij}, is a Lagrangian strain associated with the deformation of the pre-compressed state:

$$x_i - X_i = u_{ij}X_j \tag{6}$$

The choice of Lagrangian versus Eulerian strain is indifferent. Lagrangian strains seem more natural in practical calculations. Its frame invariant form is:

$$e_{ij} = \frac{1}{2}\left(u_{ij} + u_{ji} - u_{ik}u_{jk}\right) \tag{7a}$$

or simply

$$e_{ij} = \frac{1}{2}\left(u_{ij} + u_{ji}\right) \tag{7b}$$

Superposition of a deformation on pre-compressed states changes Equation (5) into:

$$F(f, e_{ij}) = a(e_{ij}) + b(e_{ij})f + c(e_{ij})f^2 + d(e_{ij})f^3 + \ldots \tag{8}$$

It is implicit in this formula that volume is expressed also in terms of two independent strains, f and e_{ij}. Therefore, V_0 in Equation (4b) is $V_0(0, e_{ij})$.

The isothermal stress-strain coefficient tensor at a certain pressure and temperature $C^T_{ijkl}(P,T)$ is:

$$C^T_{ijkl}(P,T) = \frac{1}{V}\left(\frac{\partial^2 G}{\partial e_{ij}\partial e_{kl}}\right)_{P,T} \tag{9a}$$

with (in contracted notation)

$$P(V,T)\big|_T = -\frac{\partial F(V,T)}{\partial f}\frac{\partial f}{\partial V}\bigg|_{e_{ij},T} - \frac{\partial F(V,T)}{\partial e_{ij}}\frac{\partial e_{ij}}{\partial V}\bigg|_{f,T} \tag{9b}$$

For volume conserving combinations of e_{ij}, the second term on the right-hand side of Equation (9b) essentially vanishes. The first term is essentially independent of e_{ij} for any infinitesimal strain. Then, Equation (9a) becomes:

$$C^T_{ijkl}(P,T) = \frac{1}{V}\left(\frac{\partial^2 F}{\partial e_{ij}\partial e_{kl}}\right)_{f,T} + P\delta^{ij}_{kl} \tag{10a}$$

with

$$\delta^{ij}_{kl} = \frac{1}{V}\left(\frac{\partial^2 V}{\partial e_{ij}\partial e_{kl}}\right)_{f,T} = \frac{1}{2}\left(2\delta_{ij}\delta_{kl} - \delta_{ik}\delta_{jl} - \delta_{il}\delta_{jk}\right) \tag{10b}$$

For these strains, one can also write changes in the free energy in the following convenient form (Wallace 1972):

$$\frac{\delta F}{\mathrm{V}}\bigg|_{f,T} = \sigma_{ij}e_{ij} + \frac{1}{2}C^T_{ijkl}e_{ij}e_{kl} \tag{11a}$$

or, because of Equations (9a,b):

$$\frac{\delta G}{\mathrm{V}}\bigg|_{P,T} = \frac{1}{2}C^T_{ijkl}e_{ij}e_{kl} \tag{11b}$$

Although the natural variables of the Gibbs free energy are temperature, T, and stresses, σ_{ij} (see Stixrude and Lithgow-Bertelloni 2005), the use of strains, e_{ij}, is possible as long as one can calculate this free energy. This is possible because pressure, i.e., the first term in the right-hand side of Equation (9b) is basically independent of e_{ij}.

Our implementation of quasiharmonic elasticity calculations proceeds as follows: a) equilibrate the crystal structure at several pressures (several f's) (Wentzcovitch et al. 1993); b) apply a series of combinations of strains, e_{ij}, positive and negative, as many as necessary, for each equilibrium structure and relax the atomic coordinates with fixed cell shape; c) compute the vibrational density of states for equilibrium and strained configurations (Baroni et al. 2001); d) the quasiharmonic free energy is then computed for all configurations generated:

$$F(f,e_{i,j},T) = E(f,e_{i,j}) + \frac{1}{2}\sum_q \hbar\omega_q(f,e_{i,j}) + k_BT\sum_q \ln\left(1-\exp\left(-\frac{\hbar\omega_q(f,e_{i,j})}{k_BT}\right)\right) \tag{12}$$

where $E(f,e_{ij})$ is the static total energy from first principles calculations; e) fit Equation 8 for each strain e_{ij} and temperature T. The temperature grid is as dense as desired. The pressure grid (read compressive Eulerian strain grid) is sparse. For example, to cover the entire pressure range of the mantle, up to ~135 GPa, the calculation in $MgSiO_3$- perovskite required 10-15 pressures -20 GPa $< P <$ 200 GPa. This enlarged pressure interval warrants smoothness of free energy derivatives at the end of the desired pressure interval and provides information on expanded volumes necessary to describe the state of the system at low pressures and high temperatures. Once these free energies are available one can proceed with the finite difference calculation of the second derivatives in Equation (9a) or Equation (10). Thermodynamic properties are obtained simultaneously from $F(V(f),T)$. This procedure for calculating high-P, T elasticity is analogous to that of static elasticity calculations (Wentzcovich et al. 1995; Karki et al. 1997, 2001a), although in static calculations one obtains the first principles stress tensor (Nielsen and Martin 1985) directly.

High temperature isothermal stresses can also be calculated (Davies 1974):

$$\sigma^T_{ij}(P,T) = \frac{1}{V}\frac{\partial F}{\partial e_{ij}}\bigg|_{f,T} \tag{13}$$

The isothermal constants given by Equation (9a) are relevant for comparisons with measurements in static compression experiments only. The time scale of deformation in seismic events, or in laboratory measurements of acoustic wave velocities using Brillouin scattering or resonant ultra-sonic spectroscopy, is much shorter than that of thermal diffusion for relevant length scales. Therefore, one needs to compute instead the adiabatic elastic constants. Standard algebraic manipulations involving changes of variables give:

$$C^S_{ijkl}(P,T) = C^T_{ijkl}(P,T) + \frac{VT\lambda_{ij}\lambda_{kl}}{C_V} \tag{14a}$$

with

$$\lambda_{ij}(P,T) = \left[\frac{\partial S(P,T)}{\partial e_{ij}}\right]_{f,T} \tag{14b}$$

where $S(P,T)$ is the entropy. Similarly, adiabatic stresses become:

$$\sigma^S_{ij}(f,T) = \frac{1}{V}\frac{\partial U}{\partial \varepsilon_{ij}}\bigg|_{S,f} = \sigma^T_{ij}(f,T) + \frac{ST\lambda_{ij}}{C_V} \tag{15}$$

In Cartesian notation, there are 81 independent elastic constants, but this number is reduced to 21 by the requirement that C_{ijkl} are symmetric with respect to interchanges (i,j), (k,l), and (ij,kl). This allows the replacement of a pair of Cartesian indices ij by a single index α, the Voigt index, according to: $11 \rightarrow 1$, $22 \rightarrow 2$, $33 \rightarrow 3$, 23 or $32 \rightarrow 4$, 13 or $31 \rightarrow 5$, 12 or $21 \rightarrow 6$. Simultaneously, one should also replace Cartesian strains by Voigt strains, ε_α, such that: $e_{ii} = \varepsilon_\alpha$, $e_{ij} = e_{ji} = \varepsilon_\alpha/2$, and $\sigma_{ij} = \sigma_{ji} = \sigma_\alpha$, with the relationship between $ij \rightarrow \alpha$ shown above. With this replacement, one can re-write Equations (11a,b) using also Voigt indices varying from 1 to 6:

$$\frac{\delta F}{V}\bigg|_{f,T} = \sigma_\alpha\varepsilon_\alpha + \frac{1}{2}C^T_{\alpha\beta}\varepsilon_\alpha\varepsilon_\beta \tag{16a}$$

and

$$\frac{\delta G}{V}\bigg|_{P,T} = \frac{1}{2}C^T_{\alpha\beta}\varepsilon_\alpha\varepsilon_\beta \tag{16b}$$

We are interested also in computing elastic properties of polycrystalline aggregates. This is an extensive subject thoroughly discussed in textbooks dedicated to elasticity (Musgrave 1970). Besides, we follow a treatment of this topic that is standard and extensively used in mineral physics. Therefore, we simply mention it here.

A purely isotropic material is viewed as a polycrystalline aggregate without preferred orientation of grains. The average elastic constants of this material are defined by the relationship between macroscopic stresses and strains. There are basically two ways to think about and calculate this relationship: i) assuming the stress is uniform across grains and strain is not or ii) assuming strain is uniform and stress is not. The uniform stress case corresponds to a state with internal relaxation at grain boundaries and gives lower bounds for the elastic constants (Musgrave 1970). This is the so-called Reuss average. The second case is the Voigt average and gives upper bounds. The average of the two averages is the Hill or Voigt-Reuss-Hill (VRH) average (Watt et al. 1976). The Voigt and the Reuss averages consist in computing the orientational averages of the elastic constant tensor, C, and compliance tensor, $S = C^{-1}$, respectively. The average elastic constant tensor so computed, contains only two independent elastic constants with their respective bounds: $K_{VRH} \pm \delta K_{VRH}$ and $G_{VRH} \pm \delta G_{VRH}$, with $\delta M_{VRH} = ½(M_V - M_R)$. Elastic constant tensors with preferred lattice orientations can also be calculated this way. For a detailed presentation of this topic, see Musgrave (1970). Isotropic elastic constants with tighter bounds can be obtained using the Hashin-Shtrickman variational principle (Hashin and Strikman 1962). Results for several crystalline systems have been summarized by Watt (1980).

Finally, in a isotropic medium there are two distinct wave velocities obtained from Crystoffel's equation: the longitudinal or primary velocity, V_P, and the twofold degenerate shear velocity, V_S:

$$V_P = \sqrt{\frac{K + \frac{4}{3}G}{\rho}} \quad \text{and} \quad V_S = \sqrt{\frac{G}{\rho}} \tag{17}$$

from which the bulk sound velocity can be defined:

$$V_\varphi = \sqrt{\frac{K}{\rho}} = \sqrt{V_P^2 - \frac{4}{3}V_S^2} \tag{18}$$

These are the basic ingredients of our calculations. We now summarize results on the high temperature elastic properties of the major lower mantle phases, MgO, $MgSiO_3$-perovskite, and $MgSiO_3$-post-perovskite, and some geophysical consequences derived from these results. Although there is by now a considerable number of papers on high temperature elasticity of mantle phases computed using MD (Oganov et al. 2001; Li et al. 2005, 2006b,c, 2009; Stackhouse et al. 2005), here we restrict ourselves to review quasiharmonic calculations only. These calculations offer elastic properties in a continuum of pressure and temperature, which is important when computing gradients of these properties. This is quite common in geophysics, particularly when trying to distinguish thermal versus compositional effects, as will be pointed out below.

ELASTICITY OF LOWER MANTLE PHASES

Earth's internal structure, temperature, and mantle phases have been summarized in the previous paper (see Figs. 1, 2, and 3 in Wentzcovitch et al. 2010). We have been primarily concerned with the elastic properties of lower mantle phases: (Mg,Fe,Al)(Si,Al)O_3-perovskite, (Mg,Fe,Al)(Si,Al)O_3-post-perovskite, (Mg,Fe)O, and $CaSiO_3$-perovskite. Here we review the high temperature quasiharmonic elasticity of $MgSiO_3$-perovskite, $MgSiO_3$-post-perovskite, and MgO. Some properties of the (Mg,Fe) phases have been reviewed by Hsu et al. (2010) in this volume. Static elastic properties of the perovskite and post-perovskite containing aluminum and iron have been reported by Caracas and Cohen (2007). High temperature properties of $CaSiO_3$-perovskite cannot be investigated using the QHA. It is a very anharmonic phase and its high temperature structure is stabilized by anharmonic fluctuations (Stixrude et al. 1996; Caracas et al. 2005). Its elastic properties have been investigated by molecular dynamics (Li et al. 2005) and by using a free energy expansion in terms of its structural parameters (Stixrude et al. 2007). In first approximation Earth is spherical and homogeneous but there are important deviations of both. Seismic tomography (Woodhouse and Dziewonski 1984; Grand 1994; van der Hilst et al. 1997; Masters et al. 2000) reveals the 3D velocity structure of the mantle, local and global. The starting point for understanding these observations is knowledge of individual phases' elastic properties at *in situ* conditions.

These calculations have used the local density approximation (LDA) (Ceperley and Alder 1980; Perdew and Zunger 1981) combined with quasiharmonic theory (Born and Huang 1956; Wallace 1972) and density functional perturbation theory (DFPT) (Baroni et al. 2001) for phonon calculations. Variable cell shape MD is used for structural equilibration at desired pressures (Wentzcovitch 1991). The same pseudopotentials of Troullier and Martins (1991), Vanderbilt (1990) and Von Barth and Car (unpublished), cited in the appendix of Wentzcovitch et al. (2010), were used here. Computations were performed using codes of the Quantum ESPRESSO distribution (Giannozzi et al. 2009).

MgO

Periclase (MgO) was the first system to have its high temperature elastic properties investigated by this fully first principles approach (Karki et al. 1999). The original results, including some small discrepancies from experiments, have been reproduced since then. MgO is the most anisotropic phase of the lower mantle (Wentzcovitch et al. 2006) and its anisotropy is strongly pressure dependent. Its thermodynamic properties (Karki et al. 2000b), including anharmonic effects (Wu et al. 2008; Wu and Wentzcovitch 2009) were summarized in Wentzcovitch et al. (2010). Although temperature effects on its elasticity at ambient conditions are known to be substantial and to counteract the effect of pressure (Isaak et al. 1989; Chen et al. 1998), elasticity at deep lower mantle conditions have not been measured so far. *A posteriori* inspection of the quasiharmonic thermal expansivity (see Wentzcovitch et al. 2010) indicates that this is a good approximation for MgO at 0 GPa up to ~1,100 K and even better at geophysically relevant conditions (Wu and Wentzcovitch 2009).

The elastic constants of this materials with rocksalt structure (see Wentzcovitch et al. 2010), C_{11}, C_{12}, and C_{44}, were obtained up to 150 GPa and 3,000 K (Karki et al. 1999) by calculating the free energies for strained lattices. For cubic systems, volume conserving strains (tetragonal and trigonal) allow the determination of C_{44} and of $C_s = (C_{11} - C_{12})/2$, for which isothermal and adiabatic values are identical. Isothermal or adiabatic values of the bulk modulus, $K = (C_{11} + 2\ C_{12})/3$, are then used together to obtain the corresponding values of C_{11} and C_{12}. The predicted ambient values and their initial pressure and temperature gradients agree quite well with measurements (Isaak et al. 1989; Sinogeikin and Bass 1999) (see Table 1, Figs. 1 and 2). As seen in Table 1, except for C_{11}, the predicted C_{ij} at ambient conditions are smaller than the experimental values. This is typical of DFT calculations that overestimate the equilibrium volume (see Table 1), in this case ~0.5%. So far, LDA has proved to be the best functional for calculations of structural and elastic properties of minerals at finite temperatures (Wentzcovitch et al. 2010) but the small volume overestimation causes the elastic constants to be underestimated. C_{44} in particular is the most underestimated. The reason is unclear but this result is reproduced still today. However, the temperature dependence of C_{44} is best reproduced (see Fig. 1). The temperature gradients of C_{ij} at 0 GPa are well predicted for all C_{ij}, but above ~1,250 K the QHA gradients, especially of C_{11} and C_{12} start deviating from experimental values. As discussed in Wentzcovitch et al. (2010) this is caused by unharmonic effects. The predicted cross *P-T* derivatives of C_{ij} are smaller (with opposite sign) than those obtained by experiments on MgO to 8 GPa and 1600 K (Chen et al. 1998). However, QHA results are consistent with earlier data to 0.8 GPa and 800 K (Spetzler 1970) (Table 1) and potential-induced breathing (PIB) model calculations (Isaak et al. 1990).

Table 1. Adiabatic elastic moduli (M) of MgO and their pressure and temperature derivatives at ambient conditions. The modulus c_{110} corresponds to elastic modulus for the longitudinal wave along [110] direction. Numbers in parentheses are experimental uncertainties. From Karki et al. (1999).

	c_{11}	c_{12}	c_{44}	c_{110}	K_S	G
LDA+QHA (V_0 = 18.81Å^3/cell)						
M (GPa)	298	94	147	344	162	128
$\partial M/\partial P$	9.56	1.45	1.03	6.39	4.15	2.44
$\partial M/\partial T$ (GPa/K)	−0.0598	0.0089	−0.0088	−0.0343	−0.0140	−0.0216
$\partial M^2/\partial P\partial T$ (10^{-3}/K)	0.56	−0.06	0.20	0.45	0.14	0.44
Experiments (V_{exp} = 18.69Å^3/cell)						
M (GPa) *	297.9(15)	95.8(10)	154.4(20)	351.3(22)	163.2(10)	130.2(10)
$\partial M/\partial P$ *	9.05(20)	1.34(15)	0.84(20)	6.04(37)	4.0(1)	2.4(1)
$\partial M/\partial$T (GPa/K) +	−0.0585	0.0075	−0.0126	−0.0381	−0.0145	−0.024
$\partial M^2/\partial P\partial T$ # (10^{-3}/K) $	−1.3(4)	5.1(24)	−0.2(3)	1.7(7)	3.0(15)	−1.8(10)
	0.1(4)	0.1(3)	0.1(1)	0.0(1)	0.1(3)	0.1(2)

* Sinogeikin et al. (1998), + Isaak et al. (1989), # Chen et al. (1998), $ Spetzler (1970)

The pressure-induced change of sign in the single-crystal anisotropy, as expressed by the anisotropy factor:

$$A = \frac{(2C_{44} - C_{11} + C_{12})}{C_{11}} \tag{19}$$

is characteristic of MgO (Sinogeikin and Bass 1999; Duffy et al. 1995). This quantity expresses the breakdown of the Cauchy relation that holds for isotropic materials. Our results show that the strong pressure dependence of the anisotropy in MgO is preserved at high temperatures (Fig. 3). Temperature effects counteract pressure effects and are monotonically suppressed with increasing pressure. MgO is the most anisotropic phase of the lower mantle at core-mantle boundary (CMB) conditions, as will be seen below. Therefore, lattice-preferred orientation (LPO) in MgO is a possible cause of anisotropy observed in this region.

The isotropic longitudinal (V_P) and shear (V_S) wave velocities of MgO along several isotherms are shown in Figure 4. The pressure dependence of V_S agrees very well with 300 K experiments up to ~100 GPa (Murakami et al. 2009). In the lower mantle, the properties of MgO are modified by the presence of iron (see Hsu et al. 2010). The wave velocities of ferropericlase, (Mg,Fe)O, are smaller and the longitudinal velocity is affected by the high spin to low spin crossover of iron at lower mantle pressures (Wentzcovitch et al. 2009; Hsu et al. 2010). In fact all thermodynamic properties are affected by this crossover (Wu et al. 2009).

The seismic parameters:

$$R_{S/P} = \left.\frac{\partial(\ln V_S)}{\partial(\ln V_P)}\right|_P, \quad R_{\varphi/S} = \left.\frac{\partial(\ln V_\phi)}{\partial(\ln V_S)}\right|_P, \quad \text{and} \quad R_{\rho/S} = \left.\frac{\partial(\ln \rho)}{\partial(\ln V_S)}\right|_P \tag{20}$$

express the relative magnitude of lateral (isobaric) variations in V_S, V_P, V_ϕ (Eqn. 17 and 18), and density ρ in the mantle. Seismic tomography (Dziewonski and Woodhouse 1987; Grand 1994; Masters et al. 2000), as summarized by Karato and Karki (2001), indicates that $R_{S/P}$ increases

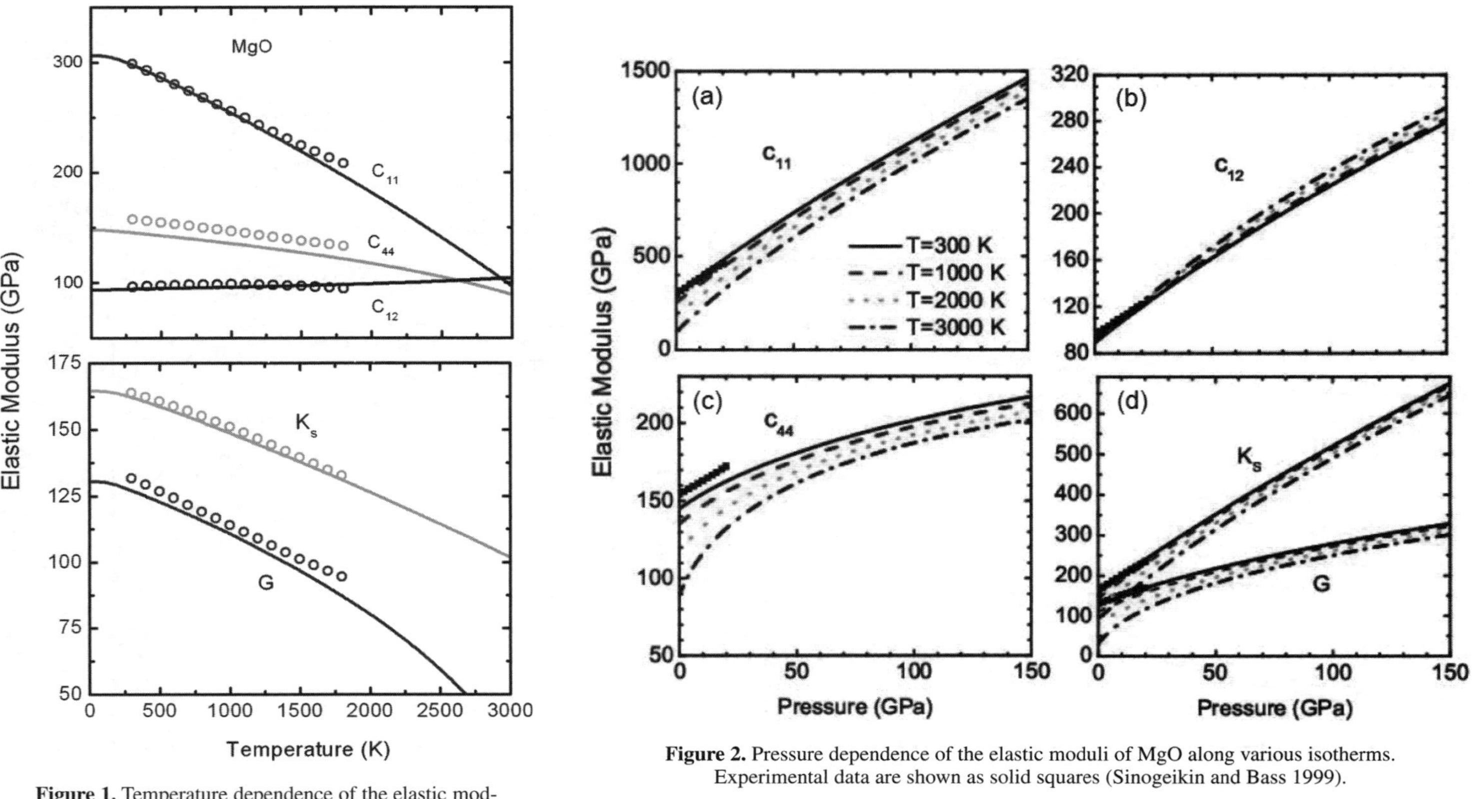

Figure 1. Temperature dependence of the elastic moduli of MgO at 0 GPa. Experimental data (Isaak et al. 1989) are shown as open circles.

Figure 2. Pressure dependence of the elastic moduli of MgO along various isotherms. Experimental data are shown as solid squares (Sinogeikin and Bass 1999).

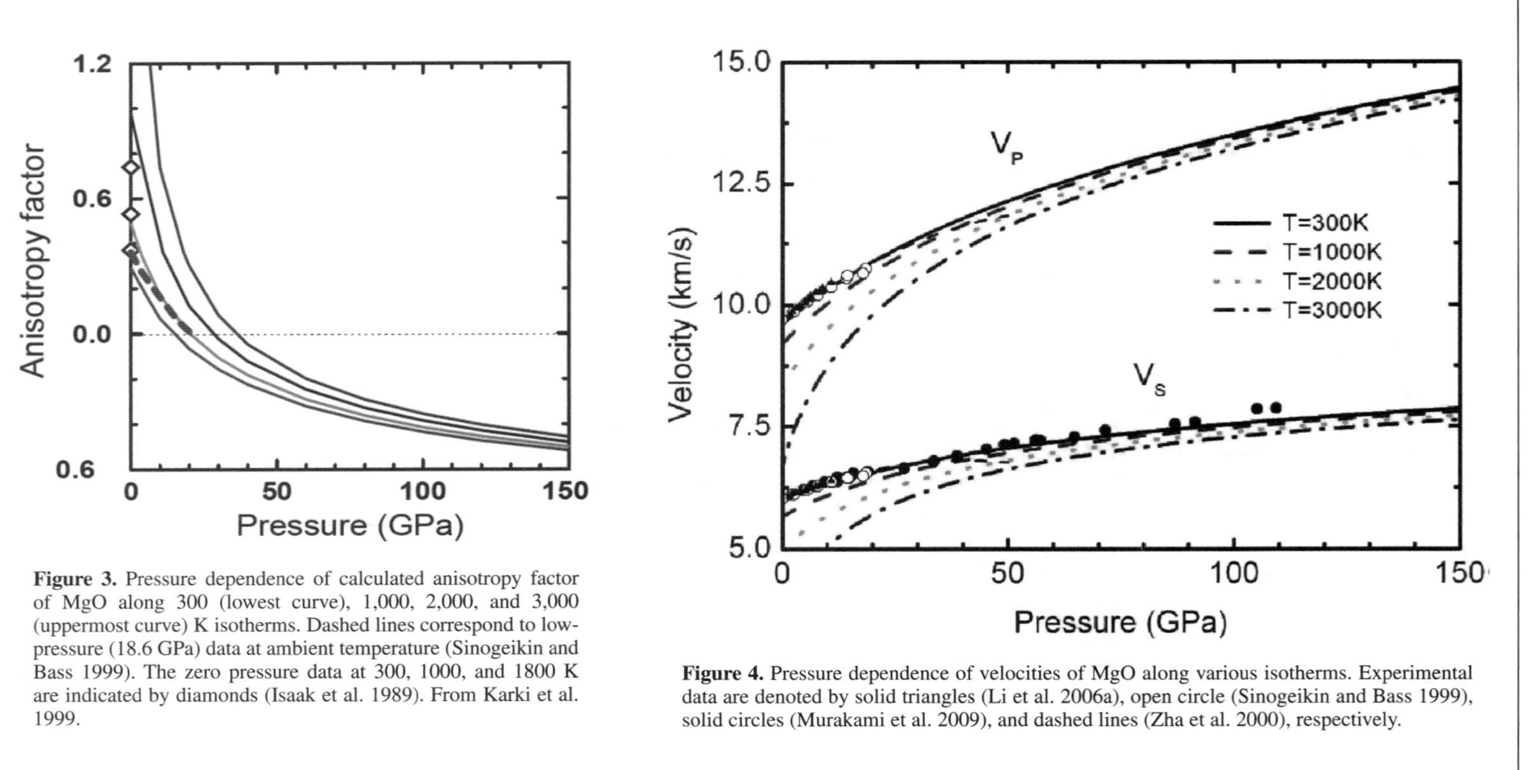

Figure 3. Pressure dependence of calculated anisotropy factor of MgO along 300 (lowest curve), 1,000, 2,000, and 3,000 (uppermost curve) K isotherms. Dashed lines correspond to low-pressure (18.6 GPa) data at ambient temperature (Sinogeikin and Bass 1999). The zero pressure data at 300, 1000, and 1800 K are indicated by diamonds (Isaak et al. 1989). From Karki et al. 1999.

Figure 4. Pressure dependence of velocities of MgO along various isotherms. Experimental data are denoted by solid triangles (Li et al. 2006a), open circle (Sinogeikin and Bass 1999), solid circles (Murakami et al. 2009), and dashed lines (Zha et al. 2000), respectively.

from more than ~2.0 to more than 3.0 from the top to the bottom of the lower mantle. Possible causes of lateral velocity variations include variations of temperature, composition, phase, etc. High temperature experiments at ambient pressure have yielded $R_{S/P}$ ~1.5 for thermally induced v (Isaak et al. 1989). These QHA velocities constrain "thermal" $R_{S/P}$ in MgO to vary from ~1.4 at the top to ~1.9 at the bottom of the lower mantle. Earlier PIB calculations estimated $R_{S/P}$ ~ 2.5 in this region (Isaak et al. 1990). As will be seen below, lateral temperature change in the bottom of the mantle can induce the post-perovskite transition (Tsuchiya et al. 2004; Oganov and Ono 2004; Wentzcovitch et al. 2010). Lateral temperature *and* phase change can increase considerably this parameter (Wentzcovitch et al. 2006). For a more extensive discussion of lateral variations in the mantle see Stixrude and Lithgow-Bertelloni (2007).

$MgSiO_3$-perovskite

Owing to its abundance in the Earth lower mantle (LM) (see Fig. 1 and 2 in Wentzcovitch et al. 2010) the elastic properties of $MgSiO_3$-perovskite (perovskite henceforth), more precisely $(Mg,Fe)SiO_3$, determine to a great extent the properties of this region. *In situ* conditions, with temperatures varying from ~1,900 K to 4,000 K, are still challenging for elasticity measurements, even though elasticity measurements in the entire pressure range of the mantle is now possible (Murakami et al. 2009). Therefore, these first principles calculations have been essential to advance understanding of the state of the lower mantle. Single crystal and aggregate elastic properties of perovskite and of other mantle phases have been used to predict acoustic velocities of hypothetical lower mantle aggregates. Comparison with the spherically averaged seismic velocity profiles of this region (Dziewonski and Anderson 1981) have then made to test the plausibility of homogeneous mineralogical models (Stixrude et al. 1992; da Silva et al. 2000; Karki et al. 2001b; Wentzcovitch et al. 2004; Xu et al. 2008). These comparisons aim to resolve, for instance, the problem of compositional stratification in the mantle, with implications for the style of thermochemical convection that has been operating in the Earth.

Perovskite has orthorhombic (*Pbnm*) symmetry and 9 independent elastic constants (C_{ij} henceforth), 3 diagonal (C_{11}, C_{22}, C_{33}), 3 off-diagonal (C_{12}, C_{13}, C_{23}), and 3 shear (C_{44}, C_{55}, C_{66}) C_{ij}. Its adiabatic C_{ij} at ambient conditions are summarized and compared to experimental data in Table 2. As can be seen, the LDA/QHA (Wentzcovitch et al. 2004) equilibrium volume at ambient conditions is ~1.2% larger than the experimental volume (Karki et al. 2000b). This "expanded" volume causes substantial underestimation of C_{ij}. The thermodynamics properties of perovskite were summarized in Wentzcovitch et al. (2010) and the source of this discrepancy does not appear to be the QHA. The generalized gradient approximation (GGA) (Wang and Perdew 1991; Perdew et al. 1996) overestimates this volume even further (Carrier et al. 2007) (on these two issues, see discussion in Wentzcovitch et al. 2010). However, this discrepancy should decrease with pressure and these predictions should be more reliable, especially predictions of pressure and temperature gradients. C_{ij} at LM conditions are shown in Figure 5 compared with results of MD simulations using the GGA (Oganov et al. 2001). Both sets of results "shifted" pressure to bring ambient condition values into agreement with experimental values. The overall good agreement between these results is more than coincidence. Pressure shift to correct problems with DFT results is not a good practice (Wu et al. 2008) since the problem caused by DFT decreases with pressure (increasing density). A better perspective of the discrepancy between calculations and measurements of aggregated elasticity and velocities is shown in Figure 6, where no shift is applied to QHA/LDA results. In the overall scale of changes in C_{ij} caused by pressure and temperature, as well as experimental uncertainties, the discrepancy is relatively small, and the accuracy of these results is still useful.

Without losing perspective of these errors and uncertainties, it is still possible to extract sensible information about the state of the lower mantle. Figure 7 displays the adiabatic bulk and shear moduli, K_S and G, for isotropic and homogeneous mineralogical models along a standard adiabatic geotherm (Brown and Shankland 1981) that nearly coincides with the isentropes

Table 2. Adiabatic elastic constants, their pressure and temperature derivatives, and wave velocities of $MgSiO_3$-perovskite predicted by LDA+QHA: (1) results at ambient conditions (Wentzcovitch et al. 2004); equilibrium volume is listed in parentheses in first column; (2) range of values obtained in static density functional theory calculations (Wentzcovitch et al. 1998; Kiefer et al. 2002); (3) experimental values obtained by Brillouin (Yeganeh-Haeri 1994), ultrasonic measurements ($^{+}$) (Sinelnikov et al. 1998), (*) (Li and Zhang 2005), and ($^{\#}$) (Sinogeikin et al. 2004) the experimental equilibrium volume is listed at parentheses of first column; (4) at ambient conditions and (4′) at 300 K and 100 GPa; (5) experimental values at ambient conditions obtained by ultrasonic measurements (Li and Zhang 2005). From Wentzcovitch et al. (2004).

	c_{11}	c_{22}	c_{33}	c_{44}	c_{55}	c_{66}	c_{12}	c_{13}	c_{23}	K_S	G	V_P	V_S
(1) M (164.1Å^3)	444	496	428	185	165	138	121	127	142	238	162	10.58	6.32
(2) M	491 ⋮ 477	560 ⋮ 524	474 ⋮ 456	203 ⋮ 198	186 ⋮ 173	153 ⋮ 145	134 ⋮ 128	144 ⋮ 135	156 ⋮ 144	263 ⋮ 257	179 ⋮ 175	10.94 ⋮ 10.89	6.53 ⋮ 6.51
(3) M (162.3Å^3)	482	537	485	204	186	147	144	147	146	264, 253$^{*,\#}$	177, 173*, 175$^{+,\#}$	11.04, 10.88$^{\#}$ 10.86*	6.57, 6.56+, 6.49*,6.53$^{\#}$
(4) $\partial M/\partial P$	7.6	8.7	9.4	2.5	2.0	2.7	4.5	3.4	3.7	5.07	2.1	0.07	0.028
(4′) $\partial M/\partial P$	3.1	5.0	5.0	1.3	0.7	1.4	3.2	2.3	2.3	3.11	1.0	0.020	0.003
(5) $\partial M/\partial P$										4.4	2.0		
(4) $\partial M/\partial T$	−0.058	−0.072	−0.073	−0.025	−0.009	−0.030	−0.018	−0.002	−0.001	−0.027	−0.0231	−0.0003	−0.0002
(4′) $\partial M/\partial T$	−0.015	−0.031	−0.020	−0.014	0.001	−0.016	−0.002	0.002	0.004	−0.006	−0.01	−0.0001	−0.0001
(5) $\partial M/\partial T$										−0.021	−0.028		
(4) $\partial M^2/\partial P\partial T$	0.0012	0.0009	0.0015	0.0003	0.0002	0.0004	0.0004	0.0001	0.00	0.0006	0.0005	0.00002	0.00001

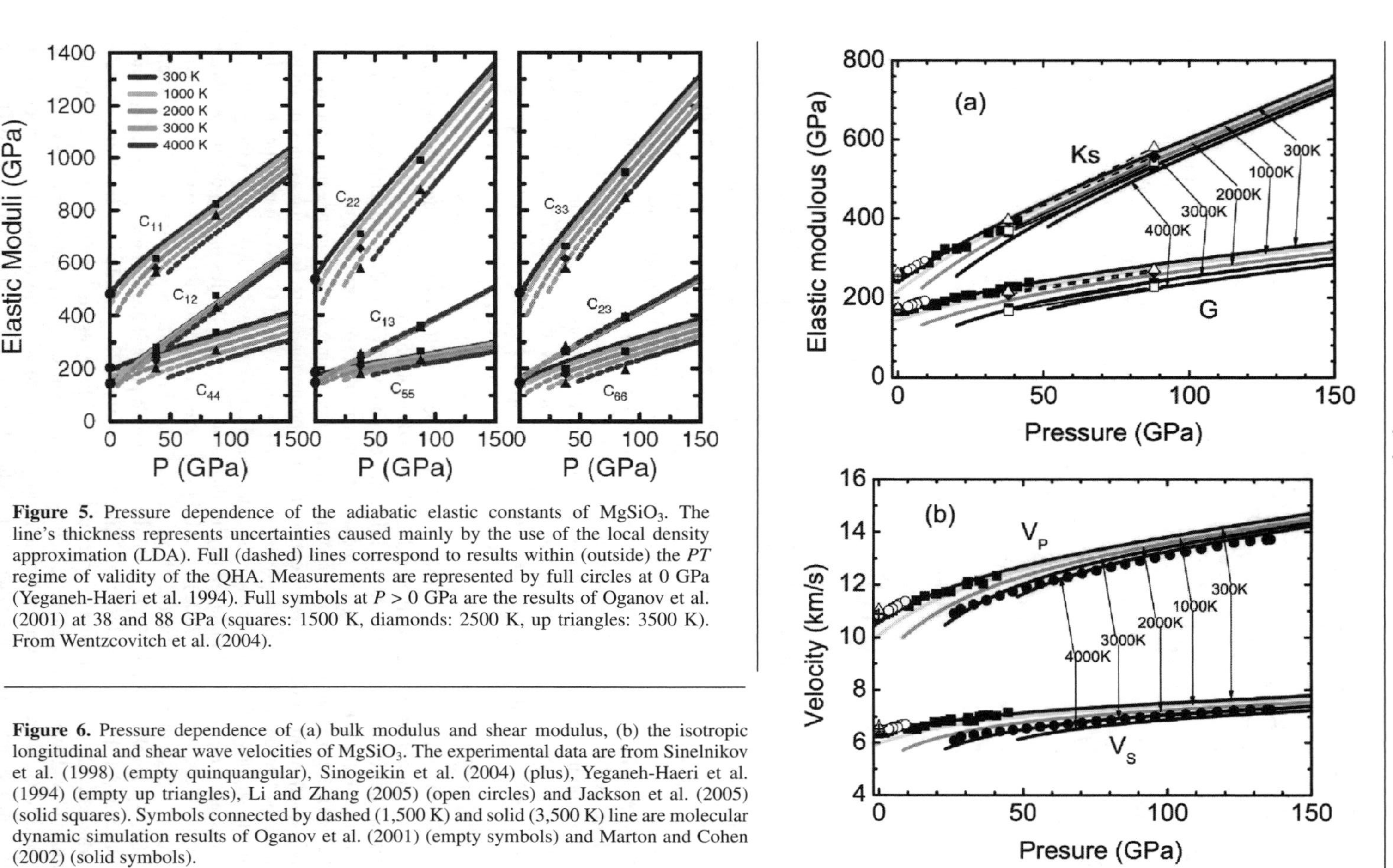

Figure 5. Pressure dependence of the adiabatic elastic constants of $MgSiO_3$. The line's thickness represents uncertainties caused mainly by the use of the local density approximation (LDA). Full (dashed) lines correspond to results within (outside) the *PT* regime of validity of the QHA. Measurements are represented by full circles at 0 GPa (Yeganeh-Haeri et al. 1994). Full symbols at $P > 0$ GPa are the results of Oganov et al. (2001) at 38 and 88 GPa (squares: 1500 K, diamonds: 2500 K, up triangles: 3500 K). From Wentzcovitch et al. (2004).

Figure 6. Pressure dependence of (a) bulk modulus and shear modulus, (b) the isotropic longitudinal and shear wave velocities of $MgSiO_3$. The experimental data are from Sinelnikov et al. (1998) (empty quinquangular), Sinogeikin et al. (2004) (plus), Yeganeh-Haeri et al. (1994) (empty up triangles), Li and Zhang (2005) (open circles) and Jackson et al. (2005) (solid squares). Symbols connected by dashed (1,500 K) and solid (3,500 K) line are molecular dynamic simulation results of Oganov et al. (2001) (empty symbols) and Marton and Cohen (2002) (solid symbols).

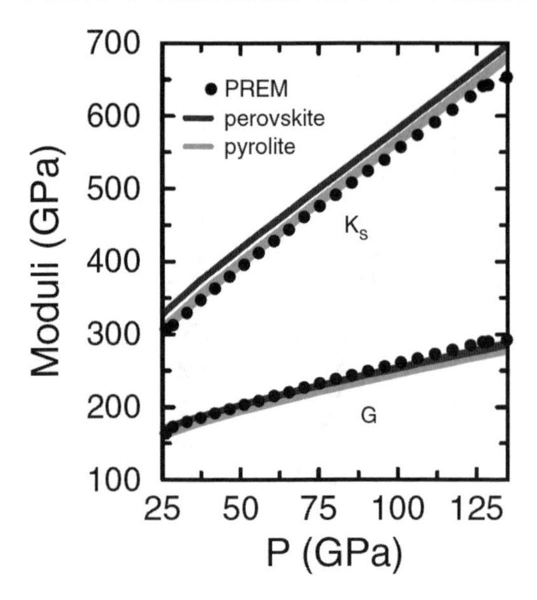

Figure 7. Bulk (K_S) and shear (G) moduli for $(Mg_{(1-x)}Fe_x)SiO_3$-perovskite and pyrolite with $20\% < V_{fp} < 30\%$ (ferropericlase = $Mg_{(1-y)}Fe_yO$) along the Brown and Shankland (1981) geotherm. $0.0 < x < 0.12$ and $1 < y/x < 4$. The effect of iron on the elastic moduli of perovskite and ferropericlase were extracted from Kiefer et al. (2002) and Jackson (1998), respectively. From Wentzcovitch et al. (2004).

of $MgSiO_3$ and MgO (Karki et al. 2000a,b). The mineralogical models consist of pure perovskite, $(Mg_{(1-x)}Fe_x)SiO_3$, with x~0.12 and pyrolite (Jackson 1998), which contains ~20 vol% of ferropericlase, $(Mg_{(1-y)}Fe_y)O$, with y~0.18. Iron partitioning (y/x) between these phases was allowed to vary in the range $1<y/x<4$ for fixed Fe/Mg ratio. The effect of iron on the elasticity of perovskite was taken from static calculations (Kiefer et al. 2002) and was assumed to be temperature independent. The effect of iron on the elasticity of ferropericlase was taken from 300 K experiments (Jackson 1998) (see Fig. 7 caption). The effect of Al_2O_3 in solid solution with perovskite (Tschermak substitution, with an Al-Al pair replacing a pair of Mg-Si in perovskite) on its elastic moduli, was assumed to be small in view of the large difference between the elastic moduli of perovskite and ferropericlase and was disregarded. The elastic moduli of $CaSiO_3$-perovskite in the LM (~5 mol%) was also taken to be similar to that of perovskite in view of the same differences. Voigt-Reuss-Hill aggregate averages for K_S and G result in the profiles shown in Figure 7.

The overall smaller K_S of pyrolite agrees reasonably well with K_{PREM}, the bulk modulus of the preliminary reference Earth model (PREM) (Dziewonski and Anderson 1981), in the upper part of the lower mantle (down to ~1,400 km depth or 55 GPa). Both mineralogical models describe G_{PREM} reasonably well in the upper lower mantle for reasonable ranges of x and y/x. However, regardless of the content of ferropericlase, x, or, y/x, with increasing depth K_{PREM} and G_{PREM} seem to depart consistently and in opposite directions from the values predicted along the geotherm. For depths greater than ~1,400 km, K_S and G develop pressure gradients that are slightly larger and somewhat smaller than PREM values. Modification of the geotherm would alter K_S and G simultaneously in the same direction. Inclusion of possible anelastic effects (Karato and Karki 2001) would make it even more difficult to reconcile G_{PREM} with values of G predicted for reasonable geotherms and compositions in this range. The elasticity $CaSiO_3$-perovskite and of Al_2O_3 in solid solution with perovskite, still need to be incorporated in the model. However, unless their elastic behaviors differ considerably from perovskite's, particularly their pressure and temperature gradients, these results suggested that the deep and the shallow lower mantle differ somehow. In particular, G_{PREM} appears to be larger than the predicted G. There are several possible geophysical reasons for this, besides the number of approximations involved in this calculation. Radial chemical and/or phase heterogeneity are some of the possibilities. Indeed, we know today that there is a new phase present in the deep mantle, i.e., $MgSiO_3$-post-perovskite (to be discussed next). The shear modulus of this phase is larger than that of perovskite by ~20 GPa at D″ conditions, which extends up to ~300 km above the CMB. PREM does not show the discontinuity most likely associated with this phase change. It has wide topography (~300 km wide) and the positive velocity jump in G associated with this phase change is smoothed by PREM. This topic is reviewed next.

$MgSiO_3$-post-perovskite

Post-perovskite is almost unanimously accepted to be the predominant phase of $MgSiO_3$ just above the core mantle boundary (CMB), the D″ region. Its discovery (Murakami et al. 2004; Oganov and Ono 2004; Tsuchiya et al. 2004) has had multidisciplinary impact in geophysics. High temperature elasticity calculations of this phase produced revealing results that, when compared with seismic data, suggest its presence in the D″ layer. Therefore, in this section will make reference to geophysical issues more than in any other. The importance of these calculations to geophysics will be abundantly demonstrated here.

Post-perovskite has orthorhombic structure (*Cmcm*) and 9 independent C_{ij}, like perovskite (Tsuchiya et al. 2004; Wentzcovitch et al. 2006). They are quite different from perovskite's (see Table 3). Post-perovskite has a layered structure (see Fig. 6b in Wentzcovitch et al. 2010), expands anisotropically, and has complex pressure- and temperature-dependent elastic behavior. However, its aggregate moduli do not differ so much from those of perovskite. Thermodynamic properties of both phases, K_S, are quite similar beyond ~80 GPa (Tsuchiya et al. 2005). The shear modulus of post-perovskite, G (Fig. 8a), is larger and has larger pressure and temperature gradients. This result and the fact that post-perovskite is ~1.5% denser than perovskite at deep lower mantle conditions determine the velocity contrasts between these phases. The shear velocity, V_S, of post-perovskite and its gradients (Fig. 8b) are larger than those of perovskite (Fig. 6b). Because of G, V_S of post-perovskite is also larger and has larger gradients. In contrast, the bulk velocity, V_ϕ (Eqn. 18), is smaller than that of perovskite because of their similar K_S and post-perovskite's larger density. Figure 8c shows the velocity jumps across the calculated phase boundary (see Fig. 21 in Wentzcovitch et al. 2010) (see Fig. 8c caption). These results are consistent with the increase in seismic velocities observed ~200–300 km above the core–mantle boundary in certain places (Lay and Helmberger 1983) but most easily detected beneath regions of past subduction, presumably colder places, such as beneath Central America (Lay and Helmberger 1983; Wyssession et al. 1998; Garnero 2004). There, $V_S > V_P$ and ΔV_S ~ 2–3% is observed, but this observation is clearly a regional property of a notably heterogeneous layer.

Another baffling property of D″ revealed by global tomographic models (Woodhouse and Dziewonski 1984; Grand 1994; Masters et al. 2000) is the anti-correlation between lateral (isobaric) heterogeneities in V_ϕ and V_S. The likely causes are usually addressed by comparing the seismic parameters, $R_{S/P}$, $R_{\phi/S}$, and $R_{\rho/S}$ (see Eqn. 20) to theoretical or experimental predictions of these ratios at relevant conditions (Karato and Karki 2001). It is known that in the shallow lower mantle $R_{S/P}$ ~ 2.3 and $R_{\phi/S}$ ~ 0.0, whereas in D″ $R_{S/P}$ ~ 3.4 and $R_{\phi/S}$ ~ −0.2 (Masters et al. 2000). Velocity anomaly ratios produced by isobaric temperature changes in pure post-perovskite and perovskite aggregates are displayed in Figure 9a,b along with the seismic parameters extracted from Karato and Karki (2001). In pure perovskite aggregates, $R_{S/P}$ increases with pressure and temperature but reaches at most ~2.3 at 135 GPa and 4,000 K, whereas $R_{\phi/S}$ is approximately pressure independent and slightly decreases with temperature to reach ~0.16 at similar conditions (see Fig. 2a). In post-perovskite $R_{S/P}$ decreases with pressure, but because of its larger $(\partial G/\partial T)_P$, it increases more rapidly with temperature to reach ~ 2.8 at the same conditions. $R_{\phi/S}$ is smaller than that of perovskite, ~0.1, at these conditions (see Fig. 9a,b). It has been argued that anelasticity, anisotropy, and lateral variations in calcium content in the deepest mantle might be necessary to produce these large $R_{S/P}$ and negative values for $R_{\phi/S}$ (Karato and Karki 2001). Figure 9c compares the seismic parameters with the computed ratios of velocity anomalies

Table 3. Elastic properties in GPa, km/s, and K of $MgSiO_3$ (1) perovskite and post-perovskite ((2), (4), (5), (6)) at 125 GPa and 2500 K and (3) 140 GPa and 4000 K. Diagonal c_{ij}s are typically underestimated by ~2%, off diagonal ones by ~1.5%, and shear ones by less than 1%. B_S and G are Voigt-Reuss-Hill averaged adiabatic bulk and shear moduli. Typically they are underestimated by 2% (random deviations in individual c_{ij} are averaged out). V_P, V_S and V_Φ are compressional, shear, and bulk velocities. [From Wentzcovitch et al. 2006.]

	c_{11}	c_{22}	c_{33}	c_{12}	c_{13}	c_{23}	c_{44}
(1) M	874	1095	1077	539	436	469	311
(2) M	1146	888	1139	454	418	507	311
(3) M	1119	900	1131	498	486	536	343
(4) $\partial M/\partial P$	6.4	4.5	6.3	2.9	2.3	2.5	3.1
(5) $\partial M/\partial T$	–0.083	–0.037	–0.069	0.0011	0.022	–0.0054	–0.011
(6) $\partial^2 M/\partial P\partial T$	0.00061	0.00017	0.00026	-0.000063	0.000089	–0.00013	0.00076

	c_{55}	c_{66}	B_S	G	V_P	V_S	V_Φ
(1) M	255	296	655	276	13.9	7.2	11.1
(2) M	238	352	656	294	14.0	7.4	11.1
(3) M	231	326	685	284	14.0	7.3	11.3
(4) $\partial M/\partial P$	2.2	2.7	3.6	2.2	0.032	0.022	0.022
(5) $\partial M/\partial T$	–0.028	–0.045	–0.017	–0.030	–0.00029	–0.0003	–0.000073
(6) $\partial^2 M/\partial P\partial T$	0.00030	0.00014	0.000063	0.00032	0.0000035	0.0000053	0.0000012

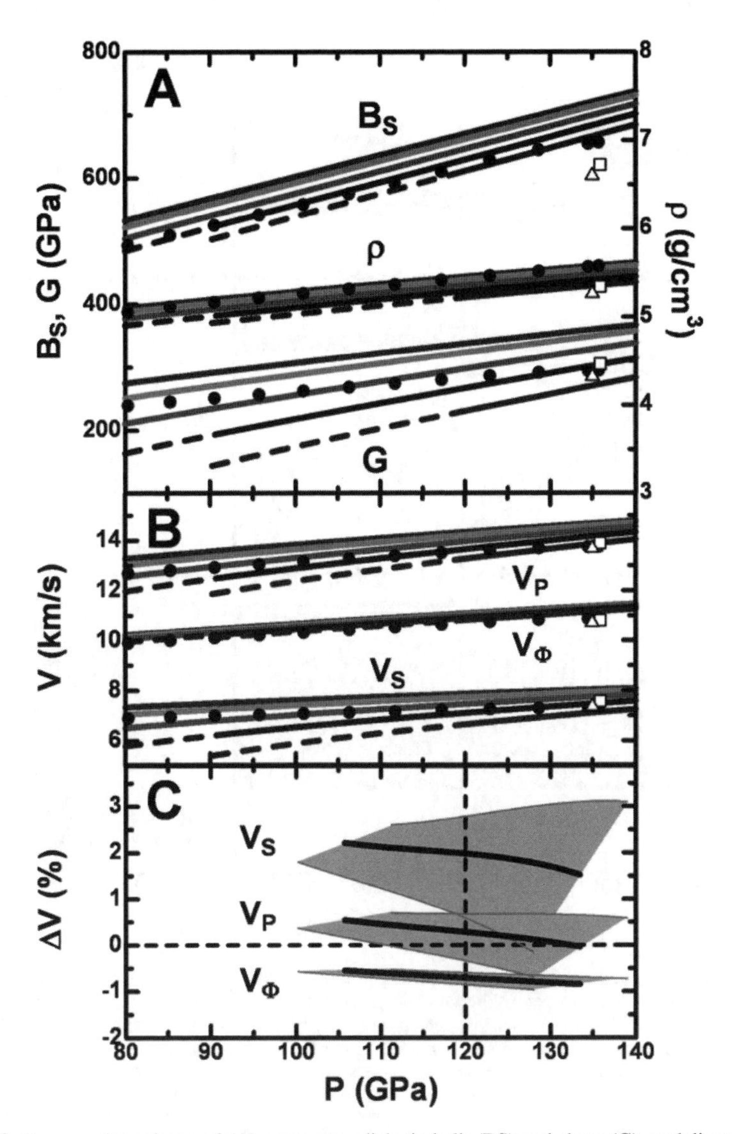

Figure 8. Pressure dependence of (A) aggregate adiabatic bulk (BS) and shear (G) moduli and density (ρ) of post-perovskite along the 300 K (uppermost), 1000 K (second from top), 2000 K (thrid from top), 3000 K (fourth from top) and 4000 K (fifth from top) isotherms; (B) isotropic longitudinal ($V_P = [(B_S + 4G/3)/\rho]^{1/2}$), shear ($V_S = (G/\rho)^{1/2}$) and bulk ($V_S = (B/\rho)^{1/2}$) wave velocities. Filled circles in (A,B) are PREM values (Dziewonski and Anderson 1981) for comparison; (C) velocity jumps across the previously obtained phase boundary (Tsuchiya et al. 2004). Thick black lines represent the jumps at the center of our phase boundary with a DFT related uncertainty of 10 GPa. Shaded areas are possible values throughout the boundary uncertainty domain (Tsuchiya et al. 2004)). Open triangles and squares are respectively results from MD simulations at 135 GPa and 4,000K and 136 GPa and 3,000 K (Stackhouse et al. 2005). The vertical dashed line indicates approximately the topmost location of the D″ discontinuity. From Wentzcovitch et al. (2006).

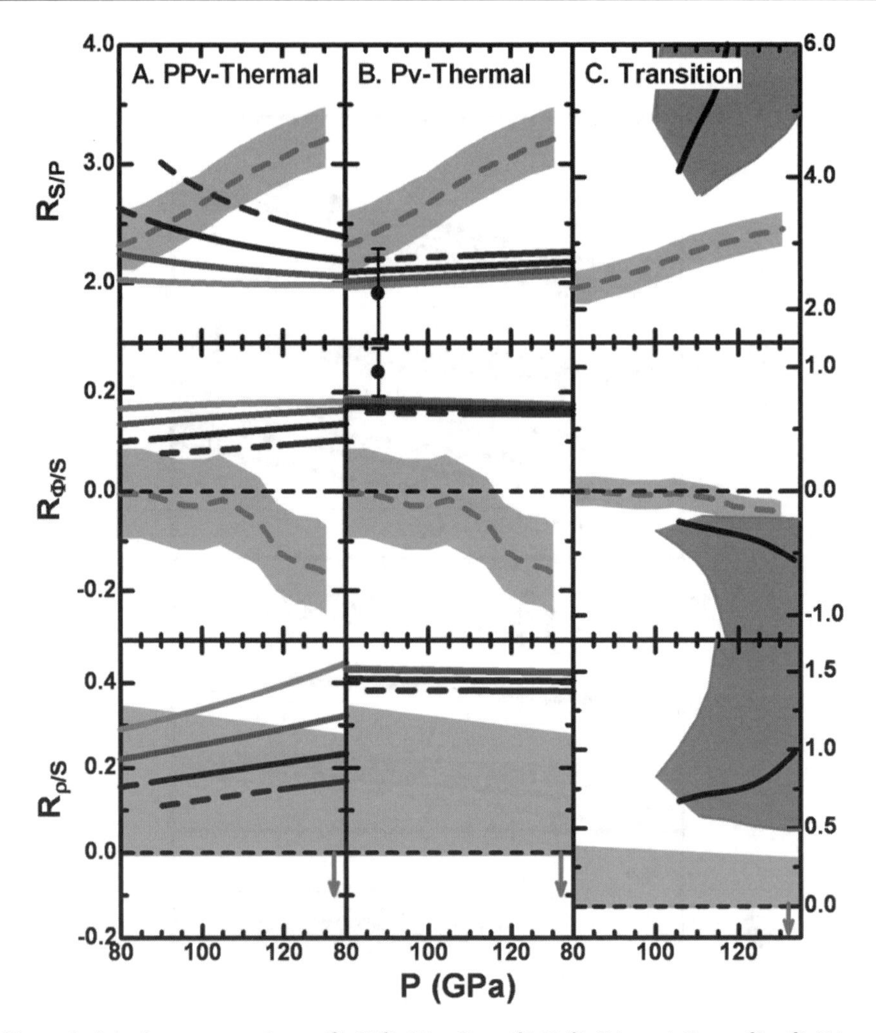

Figure 9. Seismic parameters $R_{S/P} = (\partial \ln V_S/\partial \ln V_P)_P$, $R_{\phi/S} = (\partial \ln V_\phi/\partial \ln V_P)_P$, and $R_{\rho/S} = (\partial \ln \rho/\partial \ln V_S)_P$ at several temperatures for post-perovskite (A) and perovskite (B) (sequence same as in Fig. 8 but now starting from 1000 K). These values are presented for comparison. Filled circles with error bar (± 20%) are from MD simulations (Stackhouse et al. 2005). (C) Contribution of phase transition to these parameters with uncertainties (shaded blue areas) related to phase boundary computation (same as in Fig. 8C). Seismic values extracted from (Master et al. 2000) and uncertainties (gray shaded areas) were summarized in (Karato and Karki 2001). From Wentzcovitch et al. 2006.

caused by the post-perovskite transition along our phase boundary (Fig. 21 in Wentzcovitch et al. 2010). Very large values for $R_{S/P}$ (~6) and negative values for $R_{\phi/S}$ (−0.5) result from this phase change. Therefore, lateral variation in phase abundances enhances the seismic parameters in the correct directions. The presence of secondary phases, such as ferropericlase and $CaSiO_3$, should decrease these ratios. Besides, the real multicomponent system should produce a *PT* domain for co-existence between these phases, decreasing these anomalies. Nevertheless, lateral variation in the abundances of these phases produces robust results in the right direction.

As pointed out from the outset, the topography of D″ is consistent with a solid-solid phase transition with positive Clapeyron slope (4-10 MPa/K) induced by lateral temperature variations

(Sidorin et al. 1999). Interestingly, the correlation between geographic location and nature of V_ϕ and V_S anomalies is also consistent with this assumption if post-perovskite is present in D″. Beneath the Central Pacific V_ϕ is faster and V_S is slower than their spherical mantle averages, whereas beneath the Circum-Pacific the opposite holds (see Masters et al. 2000 and references therein). These regions are generally considered to be respectively hotter and colder than average. The positive Clapeyron slope of this transition implies that hotter regions should contain less post-perovskite and therefore have faster V_ϕ and slower V_S, whereas colder regions should be enriched in the latter and have slower V_ϕ and faster V_S, as observed. Lateral variations in phase abundances alone do not account for the complex structure and properties of D″. However, this perspective should be useful as a reference model from which interpretations of deviations might be sought.

These results indicate that $R_{\rho/S}$, the third parameter in Equation (20), caused by the post-perovskite transition, should be positive and perhaps increase with depth in the lowermost mantle (see bottom row in Fig. 9), unless anelasticity or chemical heterogeneities occur simultaneously. Some 3D density models do not support this prediction (Ishii and Tromp 1999), whereas others do (Romanowicz 2001). Estimates of this parameter obtained by joint inversions of seismic and geodynamic data also tend to offer a positive ratio (Forte et al. 1994). It appears that until a consensus on the 3D structure of ρ is reached, this issue will remain open. It has been pointed out that an average excess density of 0.4% in the lowermost mantle is possible (Masters and Gubbins 2003), in agreement with our expectations.

Another pressing issue is the anisotropy of D″, a case of boundary layer anisotropy. Flow in these regions have a significant horizontal component and can align grains with preferred orientation if some of the phases present are in the dislocation creep regime. Despite multiple anisotropy styles observed (Wysession et al. 1998) and the numerous possible sources of anisotropy (Kendall and Silver 1998), evidences suggest that anisotropy in certain places of D″ could result from lattice preferred orientation in largely strained aggregates produced be mantle circulation. Beneath regions of past and present subduction, such as the Caribbean, where colder than average and largely deformed paleo-plates are expected to reside and horizontal flow is expected, transverse anisotropy with $V_{SH} > V_{SV}$ is generally observed ($V_{SH(SV)}$ is the velocity of horizontally (vertically) polarized shear waves propagating horizontally). It is plausible that in post-perovskite the primary slip system involves (010) (silica "layers" reside in this plane). Lateral material displacement could then align mainly the silica layers parallel to the horizontal plane. However, this simple picture did not provide a satisfactory explanation for the anisotropy in D″ (Tsuchiya et al. 2004).

Figure 10 shows "shear wave splittings" at conditions of the thermodynamic post-perovskite phase boundary (see Figure 21 in Wentzcovitch et al. 2010) for some preferred alignments of the crystalline axes of MgO, perovskite, and post-perovskite in the vertical direction of transversely anisotropic aggregates. Such aggregates have a particular crystalline axis oriented vertically and random orientation of axes in the horizontal plane (see Fig. 10 caption). It is seen that at relevant conditions (*i*) the vertical alignment of [001] in post-perovskite produces the largest positive ($V_{SH} - V_{SV}$) splitting; (*ii*) the shear wave splittings in perovskite and post-perovskite have similar magnitudes regardless of orientation; (*iii*) vertical alignment of perovskite's [100] produces positive ($V_{SH} - V_{SV}$); and (*iv*) horizontal alignment of MgO's {100}, its primary slip plane at high *P*s (Yamazaki and Karato 2002), produces considerably larger ($V_{SH} - V_{SV}$). Despite lack of direct information on the slip systems of these phases at D″ conditions, it appears that unlikely that post-perovskite can be a more significant source of anisotropy in D″ than perovskite. Despite being less abundant (20 vol%) periclase is the most anisotropic phase. It is also weaker than the other phases. It is therefore likely to undergo more extensive deformation (Yamazaki and Karato 2002) and be a more important source of anisotropy in D″.

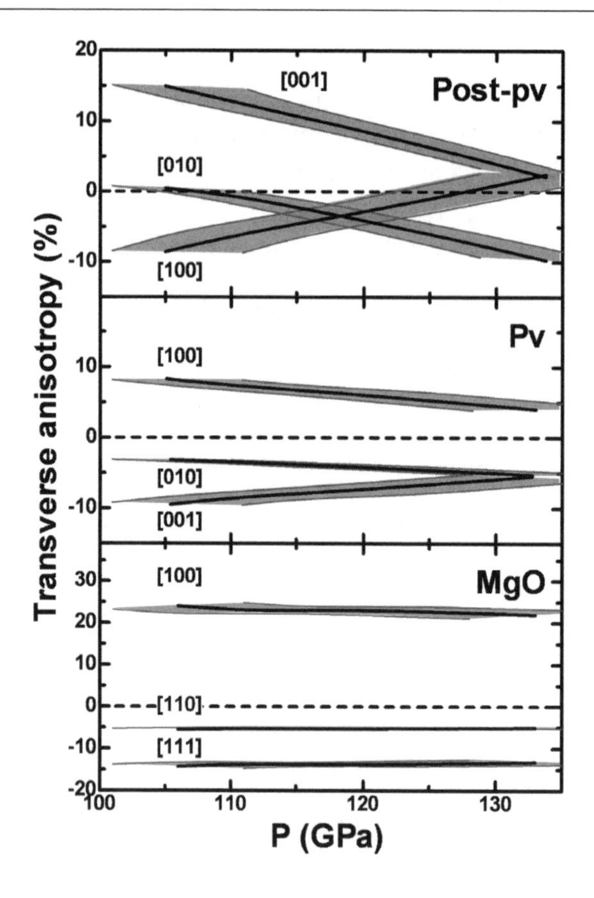

Figure 10. Shear wave splittings, $(V_{SH} - V_{SV})$, in $MgSiO_3$ post-perovskite and perovskite at the post-perovskite phase boundary conditions with respective uncertainty. $V_{SH(SV)}$ is the velocity of a horizontally (vertically) polarized shear wave propagating horizontally in aggregates with transverse anisotropy. Such aggregates have one crystalline axis aligned vertically and randomly oriented of axes in the horizontal plane. For [001] aligned vertically, $V_{SH} = (N/\rho)^{1/2}$ and $V_{SV} = (L/\rho)$ where, $N = (1/8)(C_{11} + C_{22}) - (1/4)C_{12} + (1/2)C_{66}$ and $L = (1/2)(C_{44} + C_{55})$ (Wentzcovitch et al. 1998). Velocities for aggregates with [100] and [010] aligned vertically can be obtained by cyclic permutation of indices. From Wentzcovitch et al. (2006).

SELF-CONSISTENT QHA

The QHA, as discussed here and in the previous article (Wentzcovitch et al. 2010), is a simple and powerful method for evaluating free energies using phonon frequencies obtained by first principles (Baroni et al. 2010 in this volume). There is still another useful consequence of this approximation not often appreciated: it can predict also complex crystal structures at high pressures and temperatures using only results obtained for equilibrium ($e_{ij} = 0$) configurations (Carrier et al. 2007). For clarity sake we repeat here the expression of the QHA free energy:

$$F(V,T) = E(V) + \frac{1}{2}\sum_q \hbar\omega_q(V) + k_BT\sum_q \ln\left(1 - \exp\left(-\frac{\hbar\omega_q(\mathrm{V})}{k_BT}\right)\right) \tag{21}$$

$U(V)$ is the DFT static energy versus volume obtained after full structural relaxation under isotropic pressure and $\omega_q(V)$ is the phonon spectrum for these fully relaxed structures. Crystal structure parameters and phonon frequencies are uniquely related to volume since their values under pressure have been determined by static calculations and are given at $P_{stat}(V)$ only. At $T=T'$, the pressure $P'(V,T')$, contains the following contributions:

$$P' = P_{stat} + P_{ZP} + P_{th} \tag{22}$$

where, P_{stat}, P_{ZP}, and P_{th} are the negative volume derivatives of the three terms in the right-hand side of Equation (21), respectively. No further structural relaxation is performed at T'. The function $V(P',T')$ is obtained by inverting $P'(V,T')$. Therefore, if $V(P',T') = V(P_{stat})$, structural parameters at P',T' are equal to those at P_{stat}. This is the *statically constrained QHA* (sc-QHA). The validity of this approximation can be tested by comparing its predictions with some available high P,T experimental data. If the result is favorable, it asserts the validity of the sc-QHA in that P,T range. In general, this is expected if the temperature is in the domain of validity of the QHA. This can be assessed by *a posteriori* inspection of the predicted QHA thermal expansivity (see Fig. 8 in Wentzcovitch et al. 2010).

$MgSiO_3$-perovskite is one of the materials most studied at high pressures and temperature. There is a wealth of crystallographic data (Fiquet et al. 1998; Funamori et al. 1996; Ross and Hazen 1989, 1990; Utsumi et al. 1995) available up to 60 GPa and ~2,670 K. Its orthorhombic structure has 20 atoms per primitive cell and 10 degrees of freedom: 3 lattice parameters and 7 internal parameters. Figure 11 (see caption) displays experimental lattice parameters (data points) obtained at various conditions plotted against volume, irrespective of pressure and temperature, and compares them with parameters determined by static LDA calculations (full line) (Carrier et al. 2007). Clearly *all* experimental lattice parameters obtained at various

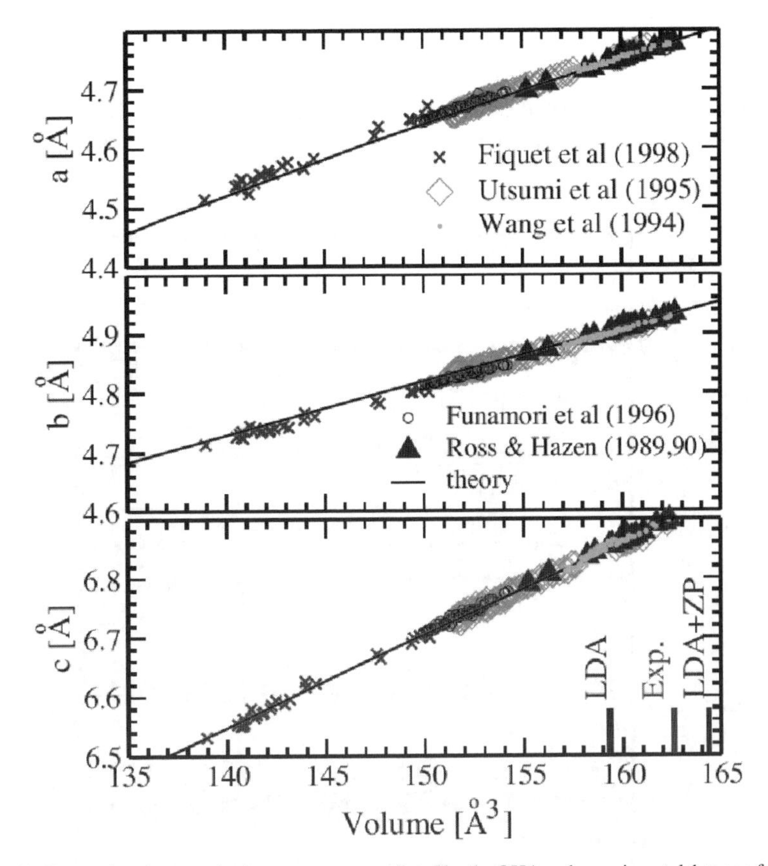

Figure 11. Comparison between lattice parameters predicted by the QHA and experimental data as a function of volume. The LDA and LDA + zero-point-motion-energy equilibrium volumes are also compared to the experimental equilibrium volume at 0 GPa. Experimental data are from Figuet et al. (1998), Funamori et al. (1996), Ross and Hazen (1989, 1990), Utsumi et al. (1995), Wang et al. (1994). Temperatures vary from 295 to 1024 K for Wang et al. (1994) from 293 to 2668 K for Fiquet et al. (1998), from 298 to 1173 K for Utsumi et al. (1995), from 293 to 2000 K for Funamori et al. (1996) and from 77 to 400 K for Ross and Hazen (1989, 1990). Solid lines are from static calculations. From Carrier et al. (2007).

P,T's lie on or very close to the theoretical lines. Although $V_{LDA} < V_{exp}$ at 0 GPa, measured and computed structural parameters *at the same volume* agree very well. The effect of zero point motion energy on the calculated zero pressure volume is indicated as well. It increases volume significantly and should be taken into account in the performance analysis of exchange-correlation energy functionals. Closer examination of Figure 11 at small volumes indicates a small but systematic deviation of experimental data from the theoretical lines. Some of these data points include very high temperature data. For instance, both data of Fiquet et al. (1998) and of Funamori et al. (1996) for the lattice parameter **a** are slightly larger (~0.3% or 0.01 Å) than the QHA prediction. The opposite is observed for **b** (~ −0.1% or −0.005 Å). No clear discrepancies are noticeable in the lattice parameter **c**. Since these points are the highest T data sets, one might suspect the validity of the QHA is questionable, but this is not the case.

The origin of these discrepancies can be traced to deviatoric thermal stresses. P_{stat} in Equation (22) is isotropic by construction, since structures were optimized under hydrostatic conditions (Wentzcovitch et al. 1993). However, P_{ZP} and P_{th} are not necessarily isotropic. Deviatoric thermal stresses (in Voigt notation) are:

$$\delta\sigma_i^T(T,P) = P + \frac{1}{V}\frac{\partial F}{\partial e_i}\bigg|_{f,T} \tag{23}$$

These stresses along [100], [010], and [001] are shown in Figure 12. Negative (positive) values indicate the structure should expand (contract) along the corresponding direction. One can see

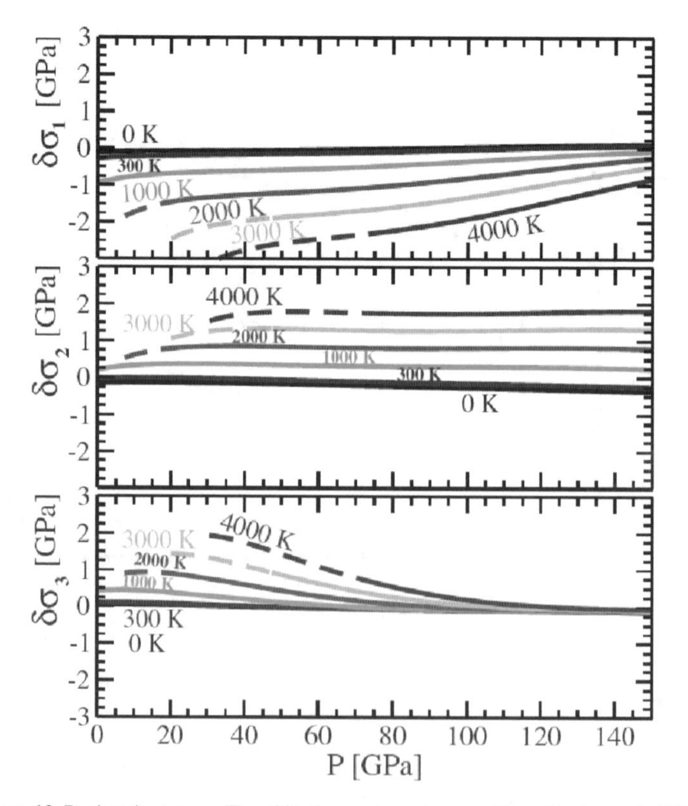

Figure 12. Deviatoric stresses (Eqn. 23) along **a**, **b**, and **c** axes. From Carrier et al. (2007).

that there is a natural relation between the high T deviations seen in Figure 11 and the deviatoric stresses in Figure 12. These deviatoric stresses can be relaxed to first order if one knows the compliance tensor:

$$\kappa_{ij}(T,P) = C_{ij}^{-1}(T,P) \tag{24}$$

Since C_{ij} are available for perovskite (Fig. 5), deviatoric stresses in Equation (23) can be relaxed by application of the following strains:

$$\varepsilon_i(P,T) = \kappa_{ij}(P,T)\delta\sigma_j(P,T) \tag{25}$$

Figure 13 displays the resulting corrections on the lattice parameters as a function of volume at various temperatures, combined with the experimental data of Fiquet et al. (1998) and Funamori et al. (1996). The agreement between predicted and measured experimental data improves considerably despite fluctuations in the experimental data. The relaxed volume, $V^{(1)}(P,T)$, under "hydrostatic" conditions differs from $V_{stat}(P)$. Therefore, phonon frequencies $\omega_q(V^{(1)})$, might be recalculated and also become temperature dependent. This dependence is

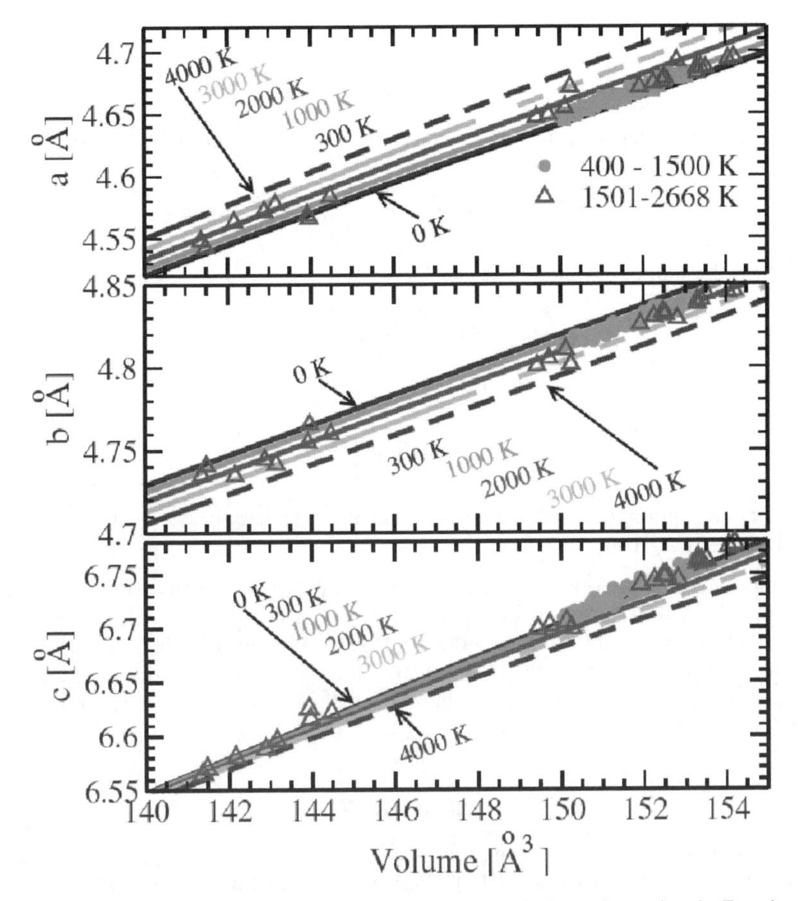

Figure 13. Corrected lattice parameters $[\mathbf{a}-\varepsilon_1\mathbf{a}]$,$[\mathbf{b}-\varepsilon_2\mathbf{b}]$, and $[\mathbf{c}-\varepsilon_3\mathbf{c}]$ where ε_i's are given by Equation (25), compared with the data of Funamori et al. (1996) data between 400 and 1500 K (dots) and Fiquet et al. (1998) data above 1500 K (triangles). Solid lines are from theory. As indicated by the Fiquet et al. (1989) data, experimental errors increase with temperature. From Carrier et al. (2007).

not caused by the usual phonon-phonon interaction but by the non-isotropic nature of the thermal pressure. This procedure should be repeated until deviatoric thermal stresses vanish, at which point all thermally induced forces should also vanish. This is the *self-consistent QHA* (sc-QHA).

Corrections of elastic constants to first order are also possible and have been carried out for perovskite and post-perovskite (Carrier et al. 2008) at D″ conditions ($P \sim 120$ GPa and $T \sim 3{,}000$ K). For these *orthorhombic* materials, the diagonal and off diagonal components of c_{ij} expanded in a Taylor series of strains (in Voigt's notation) are:

$$C_{ij}(P,T,\varepsilon) = C_{ij}(P,T,0) + \sum_{k=1}^{6} \left.\frac{\partial C_{ij}}{\partial \varepsilon_k}\right|_{P,T} \varepsilon_k + \ldots \tag{26}$$

or

$$C_{ij}(P,T,\varepsilon) = C_{ij}(P,T,0) + \sum_{k=1}^{6}\sum_{m=1}^{6} \left.\frac{\partial C_{ij}}{\partial P'}\right|_{P,T} \left.\frac{\partial P'}{\partial \sigma_m}\right|_{P,T} \left.\frac{\partial \sigma_m}{\partial \varepsilon_k}\right|_{P,T} \varepsilon_k \ldots \tag{27}$$

where

$$P' = \frac{1}{3}\sum_{m=1}^{3} \sigma_m \tag{28}$$

Therefore the first-order corrected elastic constants at the strains given by Equation (25) are:

$$C_{ij}(P,T,\varepsilon) = C_{ij}(P,T,0) + \left.\frac{\partial C_{ij}}{\partial P'}\right|_{P,T} \frac{\delta\sigma_\Sigma}{3} + \ldots \tag{29}$$

where $\delta\sigma_\Sigma$ is the sum of deviatoric stresses:

$$\delta\sigma_\Sigma = \sum_{m=1}^{3} \delta\sigma_m \tag{30}$$

each given by:

$$\delta\sigma_m = \sum_{k=1}^{6} C_{mk}\varepsilon_k \tag{31}$$

This correction requires knowledge of dC_{ij}/dP and of the deviatoric stresses, both of which are known from the sc-QHA calculation.

Changes in the isothermal bulk and shear moduli of perovskite and post-perovskite after one cycle of thermal stress relaxation are displayed in Figure 14. In perovskite, these changes are within 0.5 GPa up to 150 GPa and 4,000 K. This is within the margin of uncertainty of the calculations. In post-perovskite the changes are a little larger and negative. The changes are largest at the highest *P,T* but still relatively very small compared with values shown in Figs. 6a and 8a.

SUMMARY

We have presented the formalism used to compute thermoelastic properties of solids using the statically constrained QHA (sc-QHA). Combined with first principles LDA calculations it is a simple and accurate although computationally intensive method to compute thermoelastic properties of crystals from which elasticity of aggregates can be obtained within bounds. Results for MgO, $MgSiO_3$-perovskite, and $MgSiO_3$-post-perovskite and comparisons with high

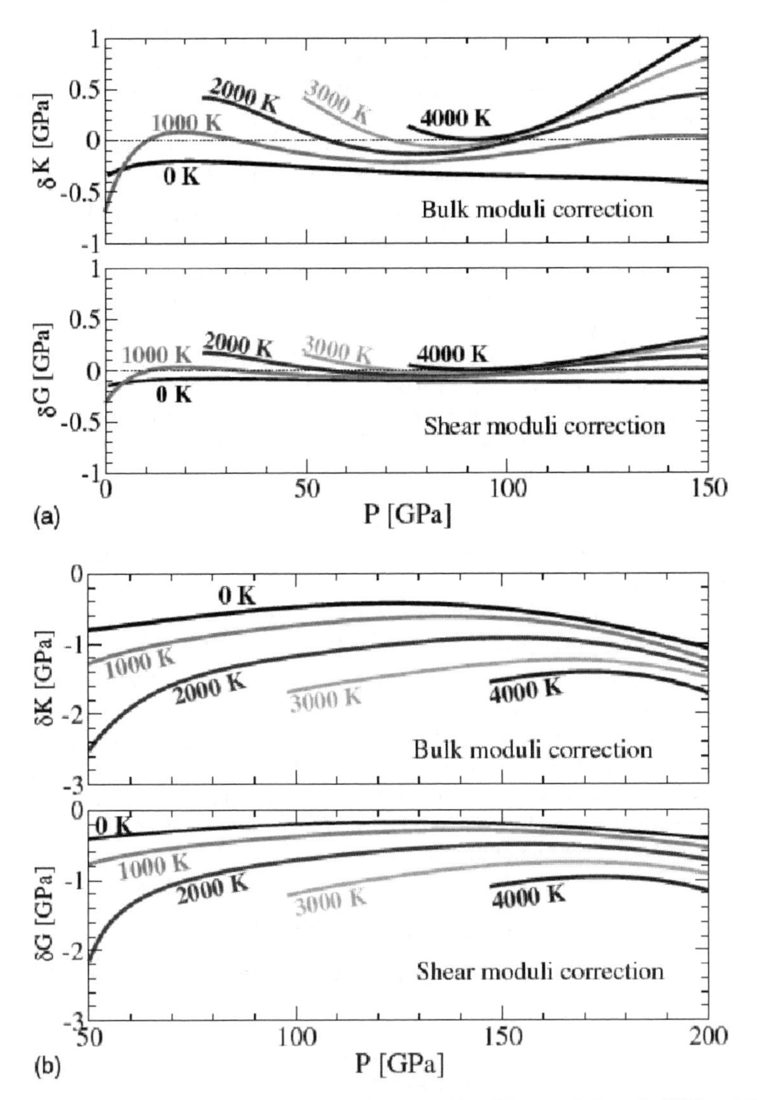

Figure 14. Corrections to bulk and shear moduli of (a) pervskite (Wentzcovitch et al. 2004) and (b) post-perovskite (Wentzcovitch et al. 2006). From Carrier et al. (2008).

pressure and high temperature experimental data have been reviewed to illustrate the accuracy possible to achieve.

An interesting byproduct of QHA calculations is the prediction of high temperature crystal structures. This is possible because, by construction, the QHA relates crystal structure parameters and (non-interacting) phonon frequencies uniquely with volume and such structure-volume relationship is well established by static calculations. In non-cubic solids, sc-QHA calculations develop deviatoric thermal stresses at high temperatures. *Thermal pressure is not isotropic*. Relaxation of these stresses leads to a series of corrections to the structure that may be taken to any desired order, up to self-consistency. This is the self-consistent QHA (sc-QHA). We have shown how to correct elastic constants for deviatoric stresses generated by the sc-QHA. We have illustrated the procedure by correcting to first order the elastic constants of perovskite and post-perovskite, the major silicate phases of the Earth's lower mantle. This correction is very satisfactory for obtaining the aggregate elastic constants and velocities of these minerals at *in situ* conditions of the lower mantle. This procedure can also be used to predict elastic constants in the presence of deviatoric stresses, or to correct elasticity measurements performed under non-hydrostatic conditions, as often happen in diamond-anvil cells, if deviatoric stresses are known.

Several insights of geophysical significance have been obtained from high temperature elasticity calculations: 1) MgO is the most anisotropic phase in the lower mantle. Since it is also the weakest, it is a potentially important source of anisotropy in aggregates with lattice preferred orientation produced by mantle flow; 2) aggregates with typical pyrolite Mg/Si ratio, i.e., ~20 vol% of MgO and ~80 vol% of perovskite, appear to reproduce the elastic properties of the lower mantle better than the pure perovskite aggregate, at least down to ~1,600 km depth. This suggests that shallow to mid lower mantle has the same chemistry of the upper mantle (as far as Mg/Si ratio is concerned); 3) in the deep lower mantle, post-perovskite free aggregates appear to slowly develop elastic properties that deviate from those of the Preliminary Reference Earth Model (PREM). The elastic properties of post-perovskite suggest that this deviation is caused by its presence in the D″ layer, the deepest ~300 km of the mantle; 4) the post-perovskite transition causes velocity jumps similar to those detected in some places in D″ beneath subduction zones. This is consistent with expectations based on the post-perovskite phase boundary with positive Clapeyron slope; 5) velocity changes across the post-perovskite transition suggest that the anti-correlation between lateral bulk and shear velocity changes in D″ could be caused by lateral changes in the perovskite/post-perovskite abundances.

For almost two decades the D″ region has remained an enigma. It is a complex layer at the interface of two chemically distinct regions, mantle and core. Post-perovskite should co-exist with other solid phases, and probably also with melts, in this region. There are still several mysteries, in the details, to be resolved; but, the insights offered by these calculations have advanced considerably knowledge of the puzzling D″ layer. These successes show that first principles calculations of thermoelastic properties of mantle minerals are poised to contribute much more to our understanding of the planet in the years ahead.

ACKNOWLEDGMENTS

This research was supported by NSF grants EAR 0810272, EAT 0635990, ATM 0428774 (VLab), and EAR 0757903.

REFERENCES

Anderson GM (2005) Thermodynamics of Natural Systems. 2nd edition. Cambridge Univ. Press, Cambridge, UK

Baroni S, Giannozzi P, Isaev E (2010) Density-functional perturbation theory for quasi-harmonic calculations. Rev Mineral Geochem 71:39-57

Baroni S, Dal Corso A, Giannozzi P, de Gironcoli S (2001) Phonons and related crystal properties from density-functional perturbation theory. Rev Mod Phys 73:515-562

Barron THK, Klein ML (1965) Second-order elastic constants of a solid under stress. Proc Phys Soc 85:523-532

Born M, Huang K (1956) Dynamical Theory of Crystal Lattices. International Series of Monographs on Physics. Oxford at the Clarendon Press, Hong Kong

Brown JM, Shankland TJ (1981) Thermodynamic parameters in the Earth as determined from seismic profiles. Geophys J R Astron Soc 66:579-596

Car R, Parrinello M (1985) Unified approach for molecular dynamics and density functional theory. Phys Rev Lett 55:2471-2474

Caracas R, Cohen R (2007) The effect of chemistry on the physical properties of perovskite and post-perovskite. *In:* Post-perovskite. The Last Mantle Phase Transition. Hirose K, Brodholt J, Lay T, Yuen D (eds) AGU Monograph Series 174:115-128

Caracas R, Wentzcovitch RM, Price GD, Brodholt J (2005) $CaSiO_3$ perovskite at lower mantle pressures. Geohys Res Lett 144:L06306

Carrier P, Justo JF, Wentzcovitch RM (2008) Quasiharmonic elastic constants corrected for deviatoric thermal stresses. Phys Rev B 78:144302

Carrier P, Wentzcovitch RM, Tsuchiya J (2007) First principles prediction of crystal structures at high temperatures using the quasiharmonic approximation. Phys Rev B 76:064116

Ceperley DM, Alder BJ (1980) Ground state of the electron gas by a stochastic method. Phys Rev Lett 45:566-569

Chen G, Liebermann RC, Weidner DJ (1998) Elasticity of single crystal MgO to 8 gigapascals and 1600 Kelvin. Science 280:1913-1916

da Silva CRS, Wentzcovitch RM, Patel A, Price GD, Karato S-I (2000) The composition and geotherm of the lower mantle: constraints from the calculated elasticity of silicate perovskite. Phys Earth Planet Inter 118:103-109

Davies GF (1974) Effective elastic-moduli under hydrostatic stress. 1. Quasiharmonic theory. J Phys Chem Solids 35:1513-1520

Duffy TS, Hemley RJ, Mao HK (1995) Equation of state and shear strength at multi-Mbar pressures: magnesium-oxide to 227 GPa. Phys Rev Lett 70:1371-1374

Dziewonski AM, Anderson DL (1981) Preliminary reference Earth model. Phys Earth Planet Inter 25: 297-356

Dziewonski AM, Woodhouse JH (1987) Global images of the Earth interior. Science 236:37-48

Fiquet G, Andrault D, Dewaele A, Charpin T, Kunz M, Haüsermann D (1998) P-V-T equation of state of $MgSiO_3$ perovskite. Phys Earth Planet Inter 105:21-31

Forte AM, Woodward RL, Dziewonski AM (1994) Joint inversions of seismic and geodynamic data for models of 3 dimensional mantle heterogeneity. J Geophys Res 99:21857-21878

Funamori N, Yagi T, Utsumi W, Kondo T, Uchida T, Funamori M (1996) Thermoelastic properties of $MgSiO_3$ perovskite determined by in situ x-ray observations up to 30 GPa and 2000 K. J Geophys Res 101:8257-8269

Garnero E (2004) A new paradigm for Earth's core-mantle boundary. Science 304:834-836

Giannozzi P, Baroni S, Bonini N, Calandra M, Car R, Cavazzoni C, Ceresoli D, Chiarotti GL, Cococcioni M, Dabo I, Dal Corso A, de Gironcoli S, Fabris S, Fratesi G, Gebauer R, Gerstmann U, Gougoussis C, Kokalj A, Lazzeri M, Martin-Samos L, Marzari N, Mauri F, Mazzarello R, Paolini S, Pasquarello A, Paulatto L, Sbraccia C, Scandolo S, Sclauzero G, Seitsonen AP, Smogunov A, Umari P, Wentzcovitch RM (2009) QUANTUM ESPRESSO: a modular and open-source software project for quantum simulations of materials. J Phys Condens Matter 21:395502

Giannozzi P, de Gironcoli S, Pavone P, Baroni S (1991) Ab initio phonon dispersions in elemental semiconductors Phys Rev Lett 43:7231-7234

Gonze X, Beuken J-M, Caracas R, Detraux F, Fuchs M, Rignanese G-M, Sindic L, Verstraete M, Zerah G, Jollet F, Torrent M, Roy A, Mikami M, Ghosez Ph, Raty J-Y, Allan DC (2002) First-principles computation of material properties : the ABINIT software project. Comput Mater Sci 25:478-492

Grand SP (1994) Mantle shear structure beneath the America and surrounding oceans. J Geophys Res 99:11591-11621

Hashin Z, Strikman S (1962) A variational approach to the theory of the elastic behavior of polycrystals. J Mech Phys Solids 10:343-352

Hohenberg P, Kohn W (1964) Inhomogeneous electron gas. Phys Rev 136:B864-B871

Hsu H, Umemoto K, Wu A, Wentzcovitch RM (2010) Spin-state crossover of iron in lower-mantle minerals: results of DFT+U investigations. Rev Mineral Geochem 71:169-199

Isaak DG, Andersen OL, Goto T (1989) Measured elastic moduli of single-crystal MgO up to 1800 K. Phys Chem Miner 16:704-713

Isaak DG, Cohen RE, Mehl MJ (1990) Calculated elastic and thermal properties of MgO at high pressures and temperatures. J Geophys Res 95:7055-7067

Ishii M, Tromp J (1999) Normal-mode and free-air gravity constraints on lateral variations in velocity and density of Earth's mantle. Science 285:1231-1236

Jackson I (1998) Elasticity, composition and temperature of the Earth's lower mantle: a reappraisal. Geophys J Int 134:291-311

Jackson JM, Zhang J, Shu J, Sinogeikin SV, Bass JD (2005) High-pressure sound velocities and elasticity of aluminous $MgSiO_3$ perovskite to 45 GPa: Implications for lateral heterogeneity in Earth's lower mantle. Geophys Rev Lett 32:L21305

Karato S-I, Karki BB (2001) Origin of lateral variation of seismic wave velocities and density in the deep mantle. J Geophys Res 106:B21771-B21783

Karki BB, Stixrude L, Clark SJ, Warren MC, Ackland GJ, Crain J (1997) Structure and elasticity of MgO at high pressure. Am Mineral 82:51-60

Karki BB, Stixrude L, Wentzcovitch RM (2001a) High-pressure elastic properties of major materials of earth's mantle from first principles. Rev Geophys 39: 507-534

Karki BB, Wentzcovitch RM, de Gironcoli S, Baroni S (1999) First-principles determination of elastic anisotropy and wave velocities of MgO at lower mantle conditions. Science 286:1705-1707

Karki BB, Wentzcovitch RM, de Gironcoli S, Baroni S (2000a) Ab initio lattice dynamics of $MgSiO_3$-perovskite. Phys Rev B 62:14750-14756

Karki BB, Wentzcovitch RM, de Gironcoli S, Baroni S (2000b) High-pressure lattice dynamics and thermoelasticity of MgO. Phys Rev B 61:8793-8800

Karki BB, Wentzcovitch RM, de Gironcoli S, Baroni S (2001b) First principles thermoelasticity of $MgSiO_3$-perovskite: consequences for the inferred properties of the lower mantle. Geophys Res Lett 28:2699-2702

Kendall J-M, Silver PG (1998) Investigating causes of D″ anisotropy. *In:* Core-Mantle Boundary Region. Geodynamics Series. Volume 28. Gurnis M, Wysession ME, Knittle E, Buffet B (eds) Am Geophys Union, Washington, DC, p 97-118

Kiefer B, Stixrude L, Wentzcovitch RM (2002) Elasticity of $(Mg,Fe)SiO_3$-Perovskite at high pressures. Geophys Res Lett 29:14683

Kohn W, Sham LJ (1965) Self-consistent equations including exchange and correlation effects. Phys Rev 140:1133-1138

Kresse G, Furthmuller J (1996) Efficient iterative schemes for ab initio total-energy calculations using a plane-wave basis set. Phys Rev B 54:11169-11186

Lay T, Helmberger DV (1983) A lower mantle S-wave triplication and the shear velocity structure of D″. Geophys J R Astron Soc 75:799-838

Li B, Zhang J (2005) Pressure and temperature dependence of elastic wave velocity of $MgSiO_3$ perovskite and the composition of lower mantle. Phys Earth Planet Inter 151:143-154

Li BS, Woody K, Kung J (2006a) Elasticity of MgO to 11 GPa with an independent absolute pressure scale: Implication for pressure calibration. J Geophys Res 111:B11206

Li L, Brodholt J, Stackhouse S, Weidner DJ, Alfredsson M, Price GD (2005) Elasticity of $(Mg, Fe)(Si, Al)O_3$-perovskite at high pressure. Earth Planet Sci Lett 240:529-536

Li L, Weidner DJ, Brodholt J, Alfè D, Price GD (2006b) Elasticity of Mg_2SiO_4 ringwoodite at mantle conditions. Phys Earth Planet Inter 157:181-187

Li L, Weidner DJ, Brodholt J, Alfè D, Price GD (2009) Ab initio molecular dynamics study of elasticity of akimotoite $MgSiO_3$ at mantle conditions. Phys Earth Planet Inter 173:115-120

Li L, Weidner DJ, Brodholt J, Alfè D, Price GD, Caracas R, Wentzcovitch R (2006c) Elasticity of $CaSiO_3$ perovskite at high pressure and high temperature. Phys Earth Planet Inter 155:249-259

Marton FC, Cohen RE (2002) Constraints on lower mantle composition from molecular dynamics simulations of $MgSiO_3$ perovskite. Phys Earth Planet Inter 134:239-252

Masters G, Gubbins D (2003) On the resolution of density within the Earth. Phys Earth Planet Inter 140:159-167

Masters G, Laske G, Bolton H, Dziewonski A (2000) The relative behavior of shear velocity, bulk sound speed, and compressional velocity in the mantle: implications for chemical and thermal structure. *In:* Earth's Deep Interior: From Mineral Physics and Tomography from Atomic to the Global Scale. Karato S-I, Forte AM, Liebermann RC, Masters G, Stixrude L (eds) American Geophysical Union, Washington DC. Geophysical Monograph Series 117:63-87

Murakami M, Hirose K, Kawamura K, Sata N, Ohishi Y (2004) Post-perovskite phase transition in $MgSiO_3$. Science 304:855-858

Murakami M, Ohishi Y, Hirao N, Hirose K (2009) Elasticity of MgO to 130 GPa: Implications for lower mantle mineralogy. Earth Planet Sci Lett 277:123-129

Musgrave MJP (1970) Crystal Acoustics. Holden-Day, Boca Raton, FL

Nielsen OH, Martin RM (1985) Quantum mechanical theory of stress and force. Phys Rev B 32:5780-3891

Nye JF (1985) Physical Properties of Crystals: Their Representation by Tensors and Matrices, 2nd ed. Oxford University Press, Oxford

Oganov AR, Brodholt JP, Price GD (2001) The elastic constant of $MgSiO_3$ perovskite at pressure and temperatures of the Earth's mantle. Nature 411:934-937

Oganov AR, Ono S (2004) Theoretical and experimental evidence for a post-perovskite phase of $MgSiO_3$ in Earth's D″ layer. Nature 430: 445-448

Payne MC, Teter MP, Allan DC, Arias TA, Joannopoulos JD (1992) Iterative minimization techniques for ab initio total-energy calculations: molecular dynamics and conjugate gradients. Rev Mod Phys 64:1045-1097

Perdew J, Zunger A (1981) Self-interaction correction to density-functional approximations for many electron systems. Phys Rev B 23: 5048–5079

Perdew JP, Burke K, Ernzerhof M (1996) Generalized gradient approximation made simple. Phys Rev Lett 77:3865-3868

Romanowicz B (2001) Can we resolve 3D density heterogeneity in the lower mantle? Geophys Res Lett 28:1107-1110

Ross N, Hazen R (1989) Single crystal X-ray diffraction study of $MgSiO_3$ perovskite from 77 to 400 K. Phys Chem Miner 16:415-420

Ross N, Hazen R (1990) High-pressure crystal chemistry of $MgSiO_3$ perovskite. Phys Chem Miner 17:228-237

Sidorin I, Gurnis M, Helmberger D (1999) Evidence for a ubiquitous seismic discontinuity at the base of the mantle. Science 286:1326-1331

Sinelnikov YD, Chen G, Neuville DR, Vaughan MT, Liebermann RC (1998) Ultrasonic shear wave velocities of $MgSiO_3$ perovskite at 8 GPa and 800 K and lower mantle composition. Science 281:677-679

Sinogeikin SV, Bass JD (1999) Single-crystal elasticity of MgO at high pressure. Phys Rev B 59:R14141-R14144

Sinogeikin SV, Bass JD, Katsura T (1998) Single crystal elasticity of MgO at high pressure. Phys Rev B 59: R14141-R14144

Sinogeikin SV, Zhang J, Bass JD (2004) Elasticity of single crystal and polycrystalline $MgSiO_3$ perovskite by Brillouin spectroscopy. Geophys Rev Lett 31:L06620

Spetzler HA (1970) Equation of state of polycrystalline and single-crystal MgO to 8 kbars and 800 K. J Geophys Res 75:2073-2087

Stackhouse S, Brodholt JP, Wookey J, Kendall JM, Price GD (2005) The effect of temperature on acoustic anisotropy of the perovskite and post-perovskite polymorphs of $MgSiO_3$. Earth Planet Sci Lett 230:1-10

Stixrude L, Cohen RE, Yu RC, Krakauer H (1996) Prediction of phase transition in $CaSiO_3$ perovskite and implications for lower mantle structure. Am Mineral 81:1293-1296

Stixrude L, Hemley RJ, Fei Y, Mao HK (1992) Thermoelasticity of silicate perovskite and magnesiowustite and stratification of the Earth mantle. Science 257:1099-1101

Stixrude L, Lithgow-Bertelloni C (2005) Thermodynamics of mantle minerals. I. Physical properties. Geophys J Int 162:610-632

Stixrude L, Lithgow-Bertelloni C (2007) Influence of phase transformations on lateral heterogeneity and dynamics in Earth's mantle. Earth Planet Sci Lett 263:45-55

Stixrude L, Lithgow-Bertelloni C (2010) Thermodynamics of the Earth's mantle. Rev Mineral Geochem 71:465-484

Stixrude L, Lithgow-Bertelloni C, Kiefer B, Fumagalli P (2007) Phase stability and shear softening in $CaSiO_3$ perovskite at high pressure, Phys Rev B 75:24108

Thomsen L (1972) Fourth-order anharmonic theory—elasticity and stability. J Phys Chem Solids 33:363-378

Troullier N, Martins JL (1991) Efficient pseudopotentials for plane-wave calculations. Phys Rev B 43:1993-2006

Tsuchiya J, Tsuchiya T, Wentzcovitch RM (2005) Vibrational and thermodynamic properties of $MgSiO_3$ post-perovskite. J Geophys Res 110:B02204, doi:10.1029/2004JB003409

Tsuchiya T, Tsuchiya J, Umemoto K,Wentzcovitch RM (2004) Phase transition in $MgSiO_3$ perovskite in the earth's lower mantle. Earth Planet Sci Lett 224:241–248

Utsumi W, Funamori N, Yagi T, Ito E, Kikegawa T, Shimomura O (1995) Thermal expansivity of $MgSiO_3$ perovskite under high pressures up to 20 GPa. Geophys Res Lett 22:1005-1008

van der Hilst RD, Widiyantoro S, Engdahl ER (1997) Evidence for deep mantle circulation from global tomography. Nature 386:578-584

Vanderbilt D (1990) Soft self-consistent pseudopotentials in a generalized eigenvalue formalism. Phys Rev B 41:7892-7895

Wallace DC (1972) Thermodynamics of Crystals. 1st Edition. John Wiley and Sons, New York

Wang Y, Perdew J (1991) Correlation hole of the spin-polarized electron-gas, with exact small-wave-vector and high-density scaling. Phys Rev B 24:13298-13307

Wang Y, Weidner DJ, Liebermann RC, Zhao Y (1994) P-V-T equation of state of $(Mg,Fe)SiO_3$ perovskite: constraints on composition of the lower mantle. Phys Earth Planet Inter 83:13-40

Watt JP (1980) Hashin-Shtrikman bounds on the effective elastic moduli of polycrystals with monoclinic symmetry. J Appl Phys 51:1520-1524

Watt JP, Davies GF, Connell RJO (1976) Elastic properties of composite materials. Rev Geophys Space Phys 14:541-563

Wentzcovitch RM (1991) Invariant molecular dynamics approach to structural phase transitions. Phys Rev B 44:2358-2361

Wentzcovitch RM, Justo JF, Wu Z, da Silva CRS, Yuen D, Kohlstedt D (2009) Anomalous compressibility of ferropericlase throughout the iron spin crossover. Proc Nat Acad Sci USA 106:8447-8452

Wentzcovitch RM, Karki BB, Cococcioni M, de Gironcoli S (2004) Thermoelastic properties of $MgSiO_3$-perovskite: insights on the nature of the Earth's lower mantle. Phys Rev Lett 92:018501

Wentzcovitch RM, Karki BB, Karato SI, da Silva CRS (1998) High pressure elastic anisotropy of $MgSiO_3$-perovskite and geophysical implications. Earth Planet Sci Lett 164:371-378

Wentzcovitch RM, Martins JL (1991) First principles molecular dynamics of Li: Test of a new algorithm. Solid State Commun 78:831-834

Wentzcovitch RM, Martins JL, Price GD (1993) Ab initio molecular dynamics with variable cell shape: Application to $MgSiO_3$ perovskite. Phys Rev Lett 70:3947-3950

Wentzcovitch RM, Ross N, Price GD (1995) Ab initio investigation of $MgSiO_3$ and $CaSiO_3$-perovskites at lower mantle pressures. Phys Earth Planet Inter 90:101-112

Wentzcovitch RM, Tsuchiya T, Tsuchiya J (2006) $MgSiO_3$ post perovskite at D″ conditions. Proc Nat Acad Sci 103:543-546

Wentzcovitch RM, Yu YG, Wu Z (2010) Thermodynamic properties and phase relations in mantle minerals investigated by first principles quasiharmonic theory. Rev Mineral Geochem 71:59-98

Woodhouse JH, Dziewonski AM (1984) Mapping the upper mantle: three-dimensional modeling of earth structure by inversion of seismic waveforms, J Geophys Res 89:5953-5986

Wu Z, Justo JF, da Silva CRS, de Gironcoli S, Wentzcovitch RM (2009) Anomalous thermodynamic properties in ferropericlase throughout its spin crossover transition. Phys Rev B 80:014409

Wu Z, Wentzcovitch RM (2009) Effective semiempirical ansatz for computing anharmonic free energies. Phys Rev B 79:104304

Wu Z, Wentzcovitch RM, Umemoto K, Li B, Hirose K (2008) P-V-T relations in MgO: an ultrahigh P-T scale for planetary sciences applications. J Geophys Res 113:B06204

Wysession ME, Lay T, Revenaugh J, Williams Q, Garnero EJ, Jeanloz R, Kellogg LH (1998) The D″ discontinuity and its implications. *In:* Core-Mantle Boundary Region. Geodynamics Series. Volume 28. Gurnis M, Wysession ME, Knittle E, Buffet B (eds) Am Geophys Union, Washington, DC, p 273-297

Xu WB, Lithgow-Bertelloni C, Stixrude L, Ritsema J (2008) The effect of bulk composition and temperature on mantle seismic structure. Earth Planet Sci Lett 275:70-79

Yamazaki D, Karato S (2002) Fabric development in (Mg,Fe)O during large strain, shear deformation: implications for seismic anisotropy in Earth's lower mantle. Earth Planet Sci Lett 131:251-267

Yeganeh-Haeri A (1994) Synthesis and re-investigation of the elastic properties of single-crystal magnesium silicate perovskite. Phys Earth Planet Inter 87:111-121

Zha CS, Mao HK, Hemley R (2000) Elasticity of MgO and a primary pressure scale to 55 GPa. Proc Nat Acad Sci USA 97:13494-13499

Reviews in Mineralogy & Geochemistry
Vol. 71 pp. 129-135, 2010

An Overview of Quantum Monte Carlo Methods

David M. Ceperley

Department of Physics and National Center for Supercomputing Applications
University of Illinois Urbana-Champaign
Urbana, Illinois, 61801, U.S.A.

ceperley@ncsa.uiuc.edu

In this brief article, various types of quantum Monte Carlo (QMC) methods are introduced, in particular, those that are applicable to systems in extreme regimes of temperature and pressure. References to longer articles have been given where detailed discussion of applications and algorithms appear.

MOTIVATION

One does QMC for the same reason as one does classical simulations; there is no other method able to treat exactly the quantum many-body problem aside from the direct simulation method where electrons and ions are directly represented as particles, instead of as a "fluid" as is done in mean-field based methods. However, quantum systems are more difficult than classical systems because one does not know the distribution to be sampled, it must be solved for. In fact, it is not known today which quantum problems can be "solved" with simulation on a classical computer in a reasonable amount of computer time. One knows that certain systems, such as most quantum many-body systems in 1D and most bosonic systems are amenable to solution with Monte Carlo methods, however, the "sign problem" prevents making the same statement for systems with electrons in 3D. Some limitations or approximations are needed in practice. On the other hand, in contrast to simulation of classical systems, one does know the Hamiltonian exactly: namely charged particles interacting with a Coulomb potential. Given sufficient computer resources, the results can be of quite high quality and for systems where there is little reliable experimental data. For this reason, QMC methods, though more expensive, are useful to benchmark and validate results from other methods.

The two main applications discussed in this review are the "electronic structure problem"; computing the energy of the interacting systems of electrons and fixed ions, and the problem of quantum effects of the nuclei.

RANDOM WALK AND MARKOV CHAINS

Most QMC algorithms are based on random walks, known mathematically as Markov chains. This is a general class of algorithms, introduced in the famous work by Metropolis, the Rosenbluths and the Tellers (1953) for sampling any probability distribution. Let us denote the desired distribution as $\Pi(S)$, where the variable "S" represents the state of the walk, for example the coordinates of all of the particles. In a random walk, one invents some moving rules (transition probabilities) for changing the state of the system so that the probability of being in state S_{n+1} in the $(n+1)$ step, given that the random walk was S_n, is given by $T(S_n; S_{n+1})$. By enforcing the detailed balance condition, and the ergodicity of random walk, one can guarantee that the asymptotic distribution will be $\Pi(S)$ (Kalos and Whitlock 1986).

1529-6466/10/0071-0006$05.00 DOI: 10.2138/rmg.2010.71.6

A typical Monte Carlo simulation starts out the state of the system at some reasonable configuration; for example an appropriate number of electrons near each ion. The random walk algorithm is then applied many times. Runs of 10^6 to 10^{12} steps are typical. During the random walk, system averages are blocked together, for example the energy of system during a thousand steps. These block averages are saved for later analysis. During the analysis one examines the various properties to see, whether after an initial "warm-up," the properties stabilize about some mean value, and fluctuate about it, hopefully in a normal fashion. According to the central limit theorem, the mean value of the simulation will converge to the exact value: i.e. $<E> = \int dS\, \Pi(S)\, E(S)$ within an "error bar" which scales as $M^{-1/2}$ where M is the number of steps of the random walk. The estimated error is determined by the fluctuations in the block averages from the overall mean, and their autocorrelation. All we need to do to get another decimal place of accuracy is to run the system one hundred times longer, or to borrow another 99 processors from your colleagues. The steady progress of computer technology and algorithms has made calculations presented here feasible and even routine.

VARIATIONAL MONTE CARLO

Variational Monte Carlo (VMC) is a simple generalization of classical Monte Carlo, introduced by McMillan (1965) for liquid helium. Ceperley (1977, 1978) has a detailed description of VMC for fermions and for the pure electron gas. In classical MC for an interaction $V(R)$, one samples the coordinates, $R = (\mathbf{r}_1, \mathbf{r}_2, \ldots, \mathbf{r}_N)$ from the Boltzmann distribution: $\exp(-\beta V(R))/Z$ where $\beta = 1/(k_B T)$. In VMC, we sample the electron coordinates from the squared modulus of a trial function; $|\phi_T(R)|^2$. Since electrons are fermions, one needs to use an antisymmetric function for $\phi_T(R)$; without correlation one would take a single Slater determinant. Typically there are $N/2$ spatial orbitals in the spin up determinant and the same number in the spin down determinant. If the wavefunction were a simple determinant, then we would have a Hartee-Fock result, and there would be no need to use a Monte Carlo procedure to determine the energy. The advantage of MC simulation is that one can introduce correlation directly by multiplying by a correlation factor, for example the well-known "Jastrow" form: $\phi_T(R) = \exp(-U(R))\ \mathrm{Det}(R)$. The Jastrow function $U(R) = \Sigma u(r_{ij})$ is usually a sum of two-body and three-body terms, it has a form similar to the potential energy function. The "cusp condition" gives the exact value of the wavefunction derivative as two charged particles approach each other. The case of two spin-unlike electrons is especially important, since the Slater determinant does not correlate electrons in different spin states. Typically, the function: $u(r) = a/(b + r)$ is used to correlate electrons, where (a,b) are variational parameters, or can be set by analytic conditions. But if we set $du(r_i - r_j)/dr = -1/2$ we restore the correct value near the co-incident point $r_i \sim r_j$ and thereby recover most of the correlation energy missing in the independent electron method. In this equation, and in what follows, we use atomic units.

This is called variational Monte Carlo because one can vary parameters in the trial function to minimize the variational energy, defined as the expectation of the Hamiltonian, H, over the trial function. By the usual theorem, this is always an upper bound to the exact energy and gives the exact energy when the trial function is an exact eigenstate. The variational energy is computed as $E_V = <E_L(R)>$: we average the "local energy" over the VMC random walk and $E_L(R) = \phi_T(R)^{-1}\ H\ \phi_T(R)$ is the "local energy" of the trial function. One important aspect of QMC is the zero variance principle (Ceperley 1977): as the trial function tends to an exact eigenstate of H, the fluctuations in the local energy tend to zero, and hence the number of MC steps needed to achieve a given error bar decreases to zero. As a result of this principle, QMC calculations can achieve much higher accuracy than classical MC calculations. Of course in electronic structure problems, one needs much higher accuracy.

The key problem in VMC is to come up with increasingly accurate trial wavefunctions and to efficiently optimize the parameters in them. There has been recent progress on both of these issues. Concerning better trial functions one can incorporate pair (geminal) (Casula and Sorella 2003) and backflow transformation of orbitals (Holzmann et al. 2003) Optimization methods can now treat thousands of free parameters (Umrigar et al. 2007) within the context of a VMC random walk.

Another issue is how to treat pseudopotentials, particularly non-local ones, because QMC methods scale quite badly with the atomic number. We need to get rid of the core states. In VMC one uses an auxiliary integration in the core; this can be generalized for the other quantum method, e.g., as in Mitas et al. (1991). However, the use of pseudopotentials is a key approximation in Quantum Monte Carlo because typical pseudopotentials have not been constructed for or tested for correlated calculations.

The VMC methods are quite efficient, and scale with the number of electrons as N^3 for bulk properties and N^4 for energy gaps. Methods have been introduced to reduce this scaling, using correlated sampling for example, or for localized systems such as insulators with large band gaps, sparse matrix algorithms (Williamson et al. 2001). The difficulty with VMC is that it favors simple, well-understood phases over less well-characterized phases, and this can bias computation of something like a pressure-induced phase transition. However, it is a very valuable method, because the trial function that comes out of VMC is necessary for the algorithm that we discuss next.

DIFFUSION MONTE CARLO

The primary zero temperature QMC method is projector MC, where a function of the Hamiltonian takes the trial function and projects out the ground state. The most common implementation of this method is diffusion MC where one computes a modified trial function defined as $\phi(R,t) = \exp(-Ht)\ \phi_T(R)$ where the trial function $\phi_T(R)$ is also the initial state. For large projection time "t" $\phi(R,t)$ will converge to the lowest state that has an overlap with $\phi_T(R)$. The first question is how to implement the projection numerically. There are two different methods commonly used.

With importance sampling, one can show that one can interpret the projection as a diffusing, branching, random walk and is known as Diffusion MC (DMC). This method, with "importance sampling" was first used in the work of Ceperley and Alder (1980) and is described in detail for molecular systems in Reynolds et al. (1982) or for solid state systems by Foulkes (2001).

A more recent implementation stores the whole evolution as a path in phase space. Computer time is then used to wiggle this path. This is known under several names, such as Variational Path Integrals (Ceperley 1995), or Reptation Quantum Monte Carlo (Baroni and Moroni 1999). The path integral approach has the advantage that all static properties can be computed without bias, while in the DMC approach, only the energy is unbiased by the trial function. Also, it is a more convenient approach to estimating energy differences (Pierleoni and Ceperley 2006).

In either case, the algorithm can be interpreted as a stochastic process for fermions only with an approximation, because the underlying integrand must be made positive to avoid the "fermion sign problem." The most common approach is called the fixed-node approximation. Here again, the trial functions enters. One searches for the lowest energy state ϕ with the restriction that $\phi(R)\phi_T(R) \geq 0$ for all possible coordinate values R. To implement this, one simply requires that the random walk does not change the sign of $\phi_T(R)$. The resulting upper bound is much lower than the starting variational energy. In systems such as the electron gas, or for a simple metal, one can obtain very accurate results, because the wavefunction nodes are highly constrained by symmetry. Estimates of the fixed-node error are obtained either from expensive "nodal-release" calculations (Ceperley and Alder 1980) or from comparison to experiment or results from other theoretical approaches. Systems for which the fixed-node approach is less successful are molecular systems with a ground state having contributions from several Slater determinants, such as the molecule C_2. Of course if the trial function includes the needed determinants, then the fixed-node answer can be arbitrarily accurate.

In general, one would like to use complex wavefunctions, both because this allows one to work with the full symmetry of the problem and because this allows a faster extrapolation to the infinite system, with the so-called Twist Average Boundary Condition (Lin et al. 2001). In this method, aimed at metallic systems, one integrates over the Brillouin zone of the simulation cell (or supercell), thereby reducing finite-size effects. The complex wavefunctions can be used either in VMC and DMC, using the so-called fixed-phase method, (Ortiz et al. 1993). This method reduces to the fixed-node method for real trial functions, but for arbitrary phases gives an upper bound which is exact if the phase of the trial function is exact. Further progress on the reducing finite size effects from the Coulomb interaction are obtained by consideration of properties of the static structure factor for long wavelengths (Chiesa et al. 2006).

PATH INTEGRAL MC

The two previous methods discussed above were methods for zero temperature, single many-body eigenstates. We now discuss a method, useful for quantum calculations at non-zero temperatures. Path Integral MC (PIMC) is based on Feynman's (1953) imaginary time path integral description of liquid ^{4}He. In this method, the quantum statistical mechanics of distinguishable particle or bosonic systems maps into a classical statistical problem of interacting polymers. Hence the probability distribution to sample is:

$$\prod(S) = \mathrm{C}\exp\left\{-\sum\frac{(R_t - R_{t-1})^2}{4\lambda\tau} - \tau V(R_t)\right\}$$

where R_t is the path and it is cyclic in the time variable $0 \le t \le \beta = 1/(k_B T)$ so that $R_0 = R_\beta$. The PIMC procedure consists of sampling the path with the random walk algorithm using this action. Here C is a constant and $\lambda = \hbar^2/2m$. For details of this method see Ceperley (1995).

For bosons, since they are indistinguishable, one has to consider paths which are cyclic up to a permutation of path labels, so that $R_0 = \mathrm{P}R_\beta$ where P is a permutation operator acting on particle labels. Figure 1 on the left, illustrates paths coming from distinguishable particle

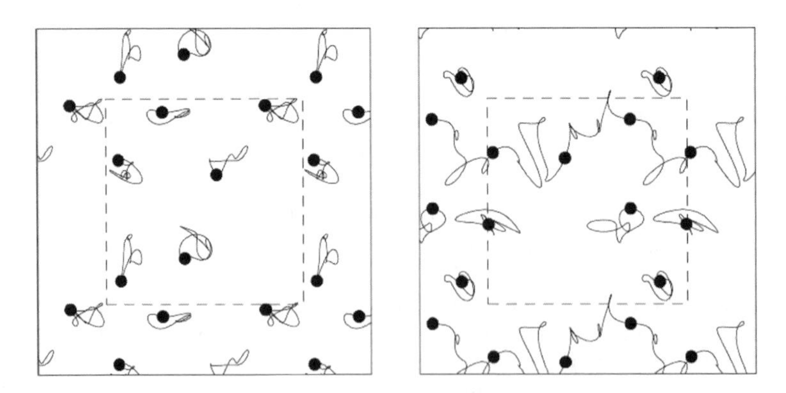

Figure 1. An example of a typical "imaginary-time path" for 6 hard spheres. The black dots represent the beginning of the path. *Left*: the identity permutation appropriate for distinguishable particles. *Right*: a permutation of 3 atoms winding across the periodic box, appropriate to systems involving bose condensation, represents superfluid flow in PIMC.

statistics (i.e., P = 1) and, on the right, bosonic statistics with exchange. Both the path R_t and the permutation P are sampled in a random walk.

For fermions one has to include a minus sign for odd permutations so that those paths contribute negatively to any averages. At low enough temperature, all permutations are equally likely, so the sign becomes random, uncorrelated with the configuration, so all averages become zero over zero; in practice the signal to noise ratio decreases as $\exp(-2\, N\, E_F/k_B T)$ where E_F is the Fermi energy (Ceperley 1996). The fixed-node approximation can also be used with path integrals allowing one to treat electrons using path integrals (Ceperley 1992). Such calculations have been carried out for hot dense hydrogen, as for example occurs in the interior of the giant planets. See for example Magro et al. (1996) and Militzer and Ceperley (2001). All quantum effects are included in such a calculation, including those of the proton quantum motion. However, we have not been successful in going to low temperatures with this approach; i.e. temperatures less than about 20% of the Fermi energy, because of the difficulty of sampling the paths.

Another very useful application for PIMC, besides its primary application for all types of bosonic systems, is to treat the effects of nuclear degrees of freedom, e.g., the effects of zero point motion in lithium (Filippi et al. 1998). The same type of approach is also employed for calculations of water that include zero point motion: a molecular dynamics calculation is done for both the electronic degrees of freedom and the imaginary time path of the protons.

COUPLED ELECTRON ION MC

In the field of Molecular Dynamics simulations of atomic and molecular systems, a great leap-forward was taken when Car and Parrinello (1985), introduced the *ab initio* MD method, replacing an *ad hoc* intermolecular potential with a potential computed using density functional theory (DFT) "on the fly." We now describe a recent attempt to do something similar using potentials coming from QMC, in particular from reptation QMC. The Coupled Electron-Ion MC (CEIMC) assumes the Born-Oppenheimer potential and makes either a classical or a quantum (Path Integral) description of the ions (Pierleoni et al. 2006). Assuming one is doing a classical MC of the ions, after a move of the ionic coordinates from S to S', in Metropolis MC one accepts the move with probability $\min[1,\exp(-(V(S') - V(S))/k_B T)]$. With CEIMC one computes the electronic energy difference $\Delta V = V(S') - V(S)$ using the Projector MC technique described above. A possible difficulty is that this energy difference will be noisy, since it is determined by a MC procedure, and unless the noise is much smaller than $k_B T$, the results will be biased, since the acceptance formula is non-linear. We have shown (Ceperley and Dewing 1999) that by adding a "penalty" of $\sigma^2/(2T^2)$ in the above exponent, that this bias can be avoided. Here σ is the statistical error in ΔV. The basic idea of CEIMC is to compute, using correlated sampling QMC methods, accurate energy differences.

CEIMC should be especially useful for systems where DFT functionals are less reliable, such as close to a metal-insulator transition. Most of our work on this has focused on warm dense hydrogen. It is possible to construct quite accurate many-body trial functions for hydrogen, avoiding the use of pseudopotentials. Using this method we have recently examined the equation of state of liquid hydrogen (Morales et al. 2009) the atomic-molecular and melting transition in hydrogen and the demixing transition in hydrogen-helium mixtures. One is able to achieve simulations at quite low temperatures (say 500K) and use high accuracy trial wavefunctions.

Some of the advantages of CEIMC, compared with AIMD, are that the MC procedure is more robust than M, the energy functional includes fully correlated electrons without problems such as self-interaction arising in DFT, pseudopotentials are not needed for light atoms and finally, one can average over other degrees of freedom without slowing down the computation. Two examples of this last point are in the average over twisted boundary conditions and in doing quantum versus classical protons. Using this method on massively parallel computers, with hundreds of thousands weakly coupled processors, one can perform a simulation of higher accuracy than AIMD, to verify its predictions. The development of QMC to calculate forces, so that one can do a dynamical simulation, should allow prediction of dynamical properties as well.

AUXILLARY FIELD MC

The Auxiliary-Field Monte Carlo (AFQMC) method (Zhang et al. 2003) is a significantly different alternative to DMC. Its random walks have a state space of (single particle) Slater determinants that are subject to a fluctuating external potential. The sign problem is eliminated by requiring that the phase of the determinant remain close to the phase of the trial determinant. Applications show that this often reduces the reliance of the solution on the quality of the trial wave function. Because one represents the random walkers in a single-particle basis, similar to DFT and other standard electronic structure methods in chemistry and physics, much of the technology developed from decades of effort in standard one-particle methods can be directly used for correlated calculations. In complex systems such as transition metal oxides, the use of an effective basis in which to carry out the random walks could lead to significant gains in efficiency, though in practice the calculations are more time consuming than DMC calculations. Further, the AF method allows us to do QMC calculations on both "simple" and realistic model Hamiltonians in which higher and irrelevant energy scales can be systematically removed. To date, the new method has been applied to close to 100 systems, several recent calculations have treated solids (Kwee et al. 2008). The method demonstrated excellent accuracy, better than 0.1 eV in most cases, and was consistently able to correct errors in the mean-field trial wave function.

PROSPECTIVES

QMC methods are already very valuable for calculations of high pressure bulk properties such as the equation of state of materials. Path integral Monte Carlo is useful for high temperature electronic structure calculations and calculation of quantum properties involving light ions. Coupled electron Ion Monte Carlo is good for low temperature properties, where one needs to go beyond methods based on Density Functional Theory particularly for disordered systems. There have been many important algorithmic developments in the last decade which allow calculations on realistic material systems. That, combined with access to computer systems with millions of hours of computation, has resulted in a growing number of QMC calculations relevant to geo-physics and astrophysics. As methods and computers develop, the QMC algorithms can take advantage of massive parallelism to deliver results of high quality, and to benchmark less expensive methods, e.g., to decide which DFT functional is appropriate for a given problem.

REFERENCES

Baroni S, Moroni S (1999) Reptation quantum Monte Carlo: a method for unbiased ground-state averages imaginary-time correlations. Phys Rev Lett 82:4745-4748

Car R, Parrinello M (1985) Unified Approach for molecular dynamics and density-functional theory. Phys Rev Lett 55:2471-2474

Casula M, Sorella S (2003) Geminal wave functions with Jastrow correlation: a first application to atoms. J Chem Phys 119: 6500-6511

Ceperley D, Chester GV, Kalos MH (1977) Monte Carlo simulation of a many-Fermion system. Phys Rev B 16:3081-3099

Ceperley D (1978) Ground state of the Fermion one-component plasma: a Monte Carlo study in two three dimensions. Phys Rev B 18:3126-3138

Ceperley DM, Alder BJ (1980) Ground state of the electron gas by a stochastic method. Phys Rev Lett 45:566-569

Ceperley DM (1992) Path integral calculations of normal liquid ^{3}He. Phys Rev Lett 69:331-334

Ceperley DM (1995) Path integrals in the theory of condensed helium. Rev Mod Phys 67:279-355

Ceperley D (1996) Path integral Monte Carlo methods for Fermions in Monte Carlo molecular dynamics of condensed matter systems. *In:* Monte Carlo and Molecular Dynamics of Condensed Matter Systems. Binder K, Ciccotti G (eds) Editrice Compositori, Bologna, Italy, p 1-23, *http://people.physics.illinois.edu/Ceperley/papers/103.pdf*

Ceperley DM Dewing M (1999) The penalty method for random walks with uncertain energies. J Chem Phys 110:9812-9820

Chiesa S, Ceperley DM, Martin RM, Holzmann M (2006) Finite size error in many-body simulations with long-ranged interactions. Phys Rev Lett 97 076404:1-4

Feynman RP (1953) Atomic theory of the λ transition in helium. Phys Rev 91:1291-1301

Filippi C, Ceperley DM (1998) Path integral Monte Carlo calculation of the kinetic energy of condensed lithium. Phys Rev B 57:252-257

Foulkes WMC, Mitas L, Needs RJ, Rajagopal G (2001) Quantum Monte Carlo simulations of solids. Rev Mod Phys 73:33-84

Holzmann M, Ceperley DM, Pierleoni C, Esler K (2003) Backflow correlations for the electron gas metallic Hydrogen. Phys Rev E 68 046707:1-15

Kalos MH, Whitlock PA (1986) Monte Carlo Methods. Wiley

Kwee H, Zhang S, Krakauer H (2008) Finite-size correction in many-body electronic structure calculations. Phys Rev Lett 100:126404

Lin C, Zong FH, Ceperley DM (2001) Twist-averaged boundary conditions in continuum Quantum Monte Carlo algorithms. Phys Rev E 64:016702

Magro WR, Ceperley DM, Pierleoni C, Bernu B (1996) Molecular dissociation in hot, dense hydrogen. Phys Rev Lett 76:1240-1243

McMillan WL (1965) Ground state of liquid ^{4}He. Phys Rev 138:A442-A451

Metropolis N, Rosenbluth A, Rosenbluth M, Teller A, Teller E (1953) Equation of state calculations by fast computing machines. J Chem Phys 21:1087-1091

Militzer B, Ceperley DM (2001) Path integral Monte Carlo simulation of the low-density hydrogen plasma. Phys Rev E63:66404

Morales MA, Pierleoni C, Ceperley DM, (2009) Equation of state of metallic hydrogen from coupled electron-ion Monte Carlo simulations. submitted to Phys Rev E,; arXiv:0906.1594

Mitas L, Shirley EL, Ceperley DM (1991) Nonlocal pseudopotentials in diffusion Monte Carlo. J Chem Phys 95:3467

Ortiz G, Ceperley DM, Martin RM (1993) New stochastic method for systems with broken time-reversal symmetry; 2-D Fermions in a magnetic field. Phys Rev Lett 71:2777-2780

Pierleoni C, Bernu B, Ceperley DM, Magro WR (1994) Equation of state of the hydrogen plasma by path integral Monte Carlo simulation. Phys Rev Lett 73:2145-2149

Pierleoni C, Ceperley DM (2006) The coupled electron-ion Monte Carlo Method. *In:* Computer Simulations in Condensed Matter Systems: From Materials to Chemical Biology. Ferrario M, Ciccotti G, Binder K (eds) Springer Berlin Heidelberg. Lecture Notes in Physics 703:641-683

Reynolds PJ, Ceperley DM, Alder BJ, Lester Jr WA (1982) Fixed-node quantum Monte Carlo for molecules. J Chem Phys 77:5593-5603

Umrigar CJ, Toulouse J, Filippi C, Sorella S, Hennig RG (2007) Alleviation of the Fermion-sign problem by optimization of many-body wave functions. Phys Rev Lett 98:110201

Williamson AJ, Hood RQ, Grossman JC (2001) Linear-Scaling Quantum Monte Carlo Calculations. Phys Rev Lett 87:246406

Zhang S, Krakauer H (2003) Quantum Monte Carlo method using phase-free walks with Slater determinants. Phys Rev Lett 90:136401

Reviews in Mineralogy & Geochemistry
Vol. 71 pp. 137-145, 2010

Quantum Monte Carlo Studies of Transition Metal Oxides

Lubos Mitas

Center for High Performance Simulation and Department of Physics
North Carolina State University
Raleigh, North Carolina, 27695-8202 U.S.A.

lmitas@ncsu.edu

Jindřich Kolorenč

University of Hamburg, Jungiusstrasse 9
20355 Hamburg, Germany

Institute of Physics, ASCR, Na Slovance 2
CZ-18221 Prague, Czech Republic

ABSTRACT

Electronic structure of the manganese and iron oxide solids is studied by the quantum Monte Carlo (QMC) methods. The trial wavefunctions are built using orbitals from unrestricted Hartree-Fock and Density Functional Theory, and the electron-electron correlation is recovered by the fixed-node QMC. The correlation effects are significant and QMC estimations of the gap and cohesion show a very good agreement with experiment. Comparison with hybrid functional results points out the importance of the exact exchange for improvement of the Density Functional description of transition metal oxide systems. The QMC methodology have enabled us to calculate the FeO equation of state as well as the transition pressure between the low- and high-pressure FeO structures.

INTRODUCTION

Transition metal compounds and transition metal oxides (TMOs) in particular belong to the most complex and important types of solid materials. TMOs exhibit a multitude of collective effects such as ferro-, ferri- and anti-ferromagnetism, ferroelectricity, superconductivity, in addition to a host of structural transitions resulting from temperature and pressure changes or doping (Brandow 1977, 1992). In addition, several of these systems, such as FeO, can be identified as parent compounds of materials which are ecnountered in the Earth interior and therefore are of central interest for geophysics.

Electronic structure of TMOs poses a well-known challenge both for theory and experiment and this challenge has remained on a forefront of condensed matter research for decades (Svane and Gunnarsson 1980; Oguchi et al. 1983; Terakura et al. 1984; Zaanen et al. 1985; Norman 1991; Anisimov et al. 1991; Szotek et al. 1993; Hugel and Kamal 1996; Pask et al. 2001; Kasinathan et al. 2007). In particular, systems such as MnO, FeO, CoO, NiO and some other similar oxides have become paradigmatic examples of inuslators with strong electron-electron correlation effects and antiferromagnetic ordering. Let us for simplicty consider just the MnO solid for a moment. Since the transition elements in these systems have open

1529-6466/10/0071-0007$05.00 DOI: 10.2138/rmg.2010.71.7

d-shells, such solids should be nominally metals. However, for a long time it has been known and experiments have shown that these systems are *large gap insulators*. In addition, they exhibit antiferromagnetic ordering with the Neel temperatures of the order of a few hundred Kelvins. Interestingly, the gap remains present even at high temperatures when there is no long-range magnetic order. Using qualitative classification Zaanen and coworkers (1985) suggested that there are basically two relevant mechanisms: Mott-Hubbard and charge transfer. In the Mott-Hubbard picture, the gap opens because of large Coulomb repulsion associated with the double occupancy of the strongly localized d-states. The spin minority bands are pushed up in the energy, leading to gap opening. In the charge transfer mechanism, $4s$ electrons of the transition metal atom fill the unoccupied p-states of oxygen with resulting gap opening as well. In the real materials both of these mechanisms are present to a certain extent and therefore the systems exhibit insulating behavior. The antiferromagnetism happens on the top of this and is caused by weak super-exchange interactions between the moments of neighboring transition metal atoms mediated by bridging oxygens. The localized d subshells feature unpaired spins which at low temperatures order into an antiferromagnetic AF II insulator with alternating spin (111) planes in cubic rocksalt structure. At low temperatures, magnetostriction distortions favor lower symmetry structures such as the rhombohedral structure in the cases of MnO and FeO.

The electronic structure of TMOs have been studied by a number of theoretical approaches, most notably by the spin-polarized Density Functional Theory (DFT) with several types of functionals and more recently, by methods which treat electronic correlations using several new ideas. Augmented spherical wave local density approximation calculations (Oguchi et al. 1983) and subsequent work (Terakura et al. 1984) explained the stability of the AF II ordering in NiO and MnO. The rhombohedral distortion of MnO was successfully described by accurate DFT calculations (Pask et al. 2001). However, commonly used DFT functionals are less reliable for predicting many other properties of strongly correlated systems. For example, DFT band gaps can be significantly underestimated or even absent, leading to false metallic states. A number of beyond-the-DFT approaches have been proposed to fix this deficiences such as self-interaction correction (Svane and Gunnarsson 1980; Szotek et al. 1993), orbital-polarization corrections (Norman 1991), on-site Coulomb interaction (LDA+U approach) (Anisimov et al. 1991; Hugel and Kamal 1996) or dynamical mean-field theory (Kotliar et al. 2006). These approaches have led to significant improvements in reproducing the experiments and provided better insights into both electronic and structural properties of these systems.

It is also known that spin-unrestricted Hartree-Fock, despite its simplicity and deficiences, provides a useful insight into the nature of the electronic structure in these systems. Towler et al. (1994) studied MnO and NiO with unrestricted Hartree-Fock theory (UHF), seemingly a rather poor choice since it neglects the electron correlation completely. Nevertheless, for transition elements the exchange, which is treated explicitly in the HF theories, is at least as important as correlation, especially for metallic ions with an effective d-subshells occupation close to half-filling. The UHF results confirmed the crucial role of exchange in TMOs and provided a complementary picture to DFT with overestimated gaps and underestimated cohesion, but also with the correct AF order, magnetic moments within 10% from experiments and reasonably accurate lattice constants. Moreover, unlike DFT approaches which predict insulator only for the AF II ground state, UHF keeps the gap open also for the ferromagnetic or any spin-disordered phases. This agrees with experiment which shows that MnO is an insulator well above the Néel temperature $T_N - 118$ K since the spin ordering Mn-O-Mn superexchange mechanism is very weak and enables localized moment flips without disturbing the overall gapped phase.

In this short review we present some results from our earlier quantum Monte Carlo (QMC) calculations of MnO which were done some time ago (Lee et al. 2004), as well as

QMC calculations of FeO ambient and high-pressure phases equations of state, which were carried out very recently.

QMC approaches are introduced in another contribution to this volume (Ceperley 2010) and more details can be found, for example, in Foulkes et al. (2001) so we just briefly mention the key points. QMC methodology has proved to be a powerful technique for studies of quantum many-body systems and also real materials. In essence, QMC has a number of advantages when compared with other approaches:

- direct and explicit many-body wave function framework for solving the stationary Schrödinger equation;
- favorable scaling with the systems size;
- wide range of applicability;
- the effectiveness of the fixed-node approximation which enables to obtain 90-95% of the correlation effects;
- scalability on parallel machines;
- new insights into the many-body phenomena.

The remaining fixed-node errors account for small, but still important fraction of correlations. The disadvantage of QMC is approximately three orders of magnitude slowdown in comparison with traditional methods based on orbitals or density functionals. Also, in its simplest form, QMC total energies enable the estimation of gaps or other energy differences, however, the spectral information beyond this is very limited or very inefficient to obtain.

Our aim in using QMC for TMO has been to understand and quantify the impact of explicit treatment of both exact exchange and correlation on the key properties such as cohesion, band gap and several other properties. Our calculations further expand on earlier attempts to apply QMC to NiO system using variational approach (Tanaka 1993, 1995) or large-core pseudopotentials (Needs and Towler 2002).

We employ both variational Monte Carlo and the diffusion Monte Carlo (DMC) methods. The trial/variational wavefunction is expressed as a product of Slater determinants of single-particle spin-up and spin-down orbitals ($\{\phi_\alpha\},\{\phi_\beta\}$) multiplied by a Jastrow correlation factor (Umrigar et al. 1988; Moskowitz and Schmidt 1992)

$$\Psi_T = \mathrm{Det}\{\phi_\alpha\}\mathrm{Det}\{\phi_\beta\}\exp\left[\sum_{I,i<j} u(r_{iI}, r_{jI}, r_{ij})\right] \tag{1}$$

where I corresponds to the ions, i and j to the electrons and r_{iI}, r_{jI}, r_{ij} to the distances. Similarly to the previous work (Tanaka 1993; Mitas and Martin 1994; Foulkes et al. 2001) the correlation function consists of a linear combination of electron-electron and electron-ion terms and involves nine variational parameters. The DMC method is used to remove the major part of variational bias which is inherent to the variational methods. DMC is based on the property that the projection $\lim_{\tau\to\infty}\exp(-\tau H)\Psi_T$ is proportional to the ground state for an arbitrary Ψ_T with the same symmetry and non-zero overlap (Foulkes et al. 2001). The fermion sign problem caused by the antisymmetry of electrons is avoided by the commonly used fixed-node approximation.

In our calculations the core electrons are eliminated by pseudopotentials (Dolg et al. 1987; Y. Lee, private communication). For transition elements the choice of accurate pseudopotentials is far from trivial and we use small-core pseudopotentials for the transition elements with $3s$ and $3p$ states in the valence space.

MnO CALCULATIONS

For the MnO solid we first carried out calculations with the spin-unrestricted Hartree-Fock and DFT (B3LYP and PW86) methods using the CRYSTAL98/03 packages (Dovesi et al. 1998/2003). The orbitals were expanded in Gaussian basis sets with (12*s*,12*p*,7*d*) Gaussians contracted to [3*s*,3*p*,2*d*] and (8*s*,8*p*,1*d*) contracted to [4*s*,4*p*,1*d*] for Mn and O atoms, respectively. Figure 1 shows the band structure of MnO solid which is obtained from UHF (a), B3LYP (b), PW86 (c) methods. Note that B3LYP hybrid functional contains 20% of the Hartree-Fock exchange so that it "interpolates" between the exact HF exchange and the effective local DFT exchange limits (Becke 1993) and it often provides an improved picture of excitations both in molecules and solids.

In QMC the MnO solid is represented by a supercell with periodic boundary conditions. This way of simulating an infinite solid involves finite size errors which scale as $1/N$ where N is the number of atoms in the supercell (Ceperley 1978; Kent et al. 1999). The finite size errors affect mainly the estimation of cohesive energy where one needs to calculate the energy per primitive cell vs. isolated atoms. In order to filter out the finite size bias we have carried out VMC calculations of supercells with 8, 12, 16, 20 and 24 atom/supercell. Only Γ-point for each supercell and Ewald energies extrapolations were used to estimate the total energy in the thermodynamic limit. The most accurate and extensive fixed-node DMC calculations were carried out with B3LYP orbitals for 16 and 20 atoms in the supercell. The cohesive energy obtained by the DMC method shows an excellent agreement with experiment (Table 1). To evaluate the impact of the correlation on the gap we have estimated the energy of the $\Gamma \rightarrow B$ excitation by an exciton calculation (Mitas and Martin 1994). The QMC result for excitation energy is less perfect with the difference from experiment being a fraction of an eV. This clearly shows that the wavefunction and corresponding fixed-node error is larger for an excited state. Nevertheless, the differential energy gain for excited vs. ground state from correlation of ≈ 8 eV is substantial and demonstrates the importance of this effect both in qualitative and quantitative sense. Due to large computational demands, similar but statistically less precise DMC calculations were carried out also with the UHF orbitals. While for the ground state the difference between the two sets of orbitals was marginal, the excited state with UHF orbitals appeared higher in energy approximately by $\approx 1.5(0.5)$ eV indicating thus, not surprisingly, even larger fixed-node bias for the excited state in the UHF approach.

It is quite encouraging that the fixed-node DMC with the simplest possible single-determinant wavefunction leads to a consistent and parameter-free description of the basic properties of this strongly correlated system. An obvious question is whether one-determinant is sufficiently accurate for an antiferromagnet since the wavefunction with different spin-up and spin-down orbitals is manifestly not an eigenfunction of the square of the total spin operator. In order to eliminate the spin contamination one would need to explore wavefunction forms beyond the single-determinant Slater-Jastrow, for example, generalized valence bond wavefunctions. However, since the actual mechanism is the Mn-O-Mn superexchange, one can expect the resulting effect to be small and very difficult to detect within our error bars.

The analysis of orbitals indicates that the nature of the top valence bands is rather similar in all approaches with both *p* and *d* states having significant weights in these states across the Brillouin zone. This is supported also by the Mullikan population analysis which shows effective magnetic moments on Mn atoms in UHF, B3LYP and PW86/PW91/PBE methods to be 4.92, 4.84 and 4.78 μ_B, respectively; these values are quite close to each other and border the range of experimental estimates of 4.58-4.78 μ_B.

The bottom of the conduction band is free-electron-like Γ state with significant amplitudes from atomic O(3s) and Mn(4s) orbitals and it is this state which is responsible for the DFT gap closing in ferromagnetic or spin-disordered phases. For the ferromagnetic phase B3LYP

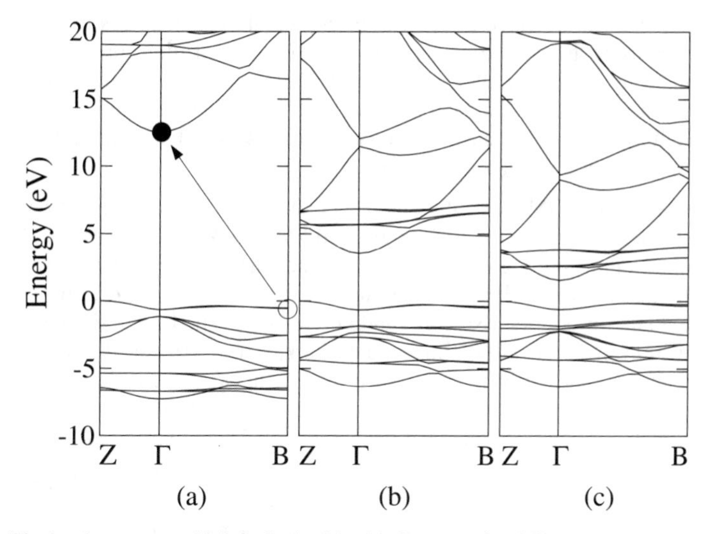

Figure 1. The band structures of MnO obtained by (a) the unrestricted Hartree-Fock method, (b) DFT/B3LYP functional, and (c) DFT/PW86 functional. The PW91 and PBE functionals provide essentially the same picture as the PW86 functional. The calculated excitation in QMC is indicated by an arrow on the UHF plot and the corresponding one-particle states are denoted by open and filled circles.

Table 1. The MnO solid cohesive energy and $B \rightarrow \Gamma$ excitation energy calculated by UHF, DFT and DMC methods compared with experiment. The determinant part of the DMC wavefunction used the B3LYP one-particle orbitals.

	UHF	PW86	B3LYP	DMC	Experiment
E_{coh}	6.03	11.00	9.21	9.40 (5)	9.50
$B \rightarrow \Gamma$	13.5	1.2	4.0	4.8 (2)	≈ 4.1

exhibits a gap of $\approx$ 2.4 eV and it is straightforward to check that by decreasing the weight of exact exchange the gap decreases. For example, with 10% of exact exchange in B3LYP the gap lowers to $\approx$ 1.2 eV. The functionals without the exact exchange, such as PW86/PW91/PBE, lead to ferromagnetic metals due to the overlaps with the uppermost valence bands and subsequent rehybridization of states around the Fermi level.

FeO CALCULATIONS

If one applies mainstream DFT approaches to FeO, the results are highly unsatisfactory. First, the equilibrium atomic structure which comes out is not correct. Instead of the rocksalt antiferromagnet another structure, a tetrahedral distortion dubbed iB8, appears to have the lowest energy at equilibrium conditions. Interestingly, iB8 is actually a high-pressure phase of FeO. This discrepancy is surprising since DFT methods typically lead to correct equilibrium structures and very reasonable geometric parameters such as lattice constants. Another problem appears in the electronic structure since for the correct structure and geometry DFT leads to a metallic state. More sophisticated methods beyond DFT have been applied to this system in order to reconcile some of the results with experiments, nevertheless, a number of questions remain unanswered. For example, at high pressures, FeO undergoes a structural transition into the iB8 phase; however, the value of the transition pressure which agrees with experiment is difficult to obtain using the mainstream approaches. Similar transition appears in MnO, where recent benchmarking of several DFT and post-DFT approaches provided transition pressure estimates between 65 and 220 GPa, i.e., more than 300% spread in the predictions (Kasinathan et al. 2007). Significant discrepancies between experiments and DFT results exist also for CoO and more complicated transition metal compounds.

We have carried out QMC calculations of FeO using supercells with periodic boundary conditions to model the infinite solid (Kolorenč and Mitas 2008). Several supercell sizes were calculated and k-point sampling of the Brillouin zone was carried out by the so-called twist averaging with the purpose of eliminating finite size effects. The core electrons were replaced by pseudopotentials for both Fe (Ne-core) and O (He-core). The largest simulated supercells had more than 300 valence electrons and the total energies were sizeable due to the presence of "semicore" $3s$ and $3p$ states of Fe in the valence space, what have made the calculations rather demanding. The wave function had the Slater-Jastrow form and the orbitals were obtained from unrestricted (spin-polarized) calculations within DFT with hybrid functionals and HF.

Table 2 shows the QMC calculated equilibrium parameters. Note that QMC identifies the correct equilibrium structure and provides very accurate value of the cohesive energy. Cohesive energies are very difficult to calculate since the DFT methods show typical bias of 15-30%. At present, there is basically no other method besides QMC which can get this level of accuracy. Note also good agreement with experiments for the other quantities including the band gap. In QMC the band gap is calculated as a difference of two total energies—ground state and excited state, where the excited state is formed by promoting an electron from the top valence band into the conduction band.

In addition, the equations of state have been calculated for both the equilibrium structure and also for the high-pressure phase, see Figure 2. The estimated transition pressure is 65(5) GPa, at the lower end of the experimental range 70-100 GPa (Fang et al. 1999; Zhang 2000). Clearly, the agreement with experiment is not perfect and reflects several idealizations used in our calculations. We checked that the rhombohedral distortion did not change the results within our error bars so it appears unlikely that this was the dominant contribution. The fact that the experimental results also vary significantly suggests that there other reasons need to be considered. For example, it is well known that FeO is basically always slightly nonstoichiometric what could affect the experimental pressures significantly. Another reason might be the presence of defects which could also push the experimental observations towards higher pressures. Clearly, this question will have to be resolved with further effort on both theoretical and experimental fronts.

The calculations clearly illustrate the capabilities of QMC methods and considering that only the simplest trial wave function of Slater-Jastrow type was employed, the results are remarkable and very encouraging. Note that the calculations do not have any free non-variational parameters. It is simply the best possible solution within the trial function nodes and the given Hamiltonian.

The results presented here therefore provide new insights into electronic structure of TMOs. In the Mott-Hubbard picture the origin of the band gap is the large on-site repulsion between the d electrons which basically relates the gap to the two-site $d^n d^n \rightarrow d^{n-1} d^{n+1}$ type of excitation. On the other hand, the charge transfer favors the ionic picture of $Mn^{++}O^{--}$ with the oxygen p states at the top of the valence band the gap given by the $p \rightarrow d$ excitation energy. Our QMC results quantify that UHF produces qualitatively correct wavefunctions although biased towards the charge transfer limit, especially for excited states. The hybrid B3LYP and PBE0 functionals offer more balanced zero-order theory and appear to provide much more balanced and less biased electronic structure than the DFT non-hybrid functionals. This view

Table 2. Comparison of the calculated structural properties of FeO solid in DFT and in the fixed-node DMC with experimental data. The energy difference E_{iB8}-E_{B1} is evaluated at the experimental lattice constant 4.334 Å. DFT results are from Fang et al. (1998, 1999). Experimental data is from the following: CRC Handbook (2007), McCammon and Liu (1984), Zhang (2000) and, Bowen et al. (1975).

Method/quantity	DFT/PBE	FN-DMC	Experiment
E_{iB8}-E_{B1} [eV]	–0.2	0.5(1)	> 0
cohesion energy [eV]	~ 11	9.66(4)	9.7
lattice constant [Å]	4.28	4.324(6)	4.334
bulk modulus [GPa]	180	170(10)	~ 180
band gap [eV]	~ 0	2.8(3)	2.4

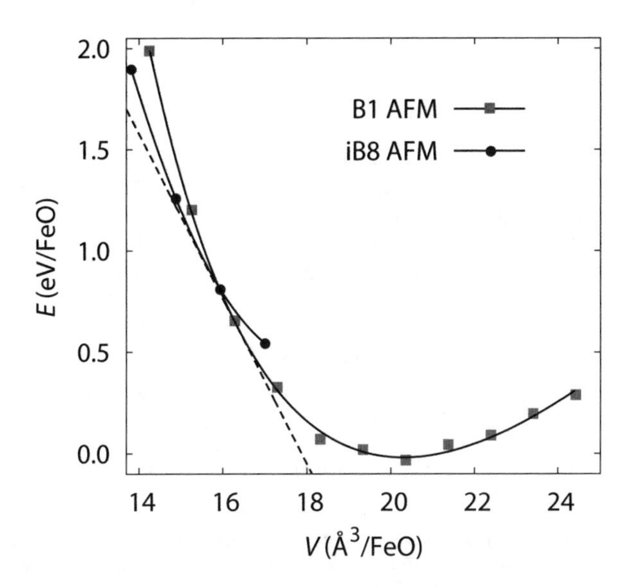

Figure 2. FN-DMC Energy as a function of volume for FeO for B1 (squares) and iB8 (circles) phases. Lines are fits with Murnagham equation of state.

is supported also by the orbital analysis proposed by Brandow (1997, 1992), and, in effect, also by introduction of on-site terms such as Hubbard U which restore some of the Hartree-Fock character. The hybrid functional alleviates the DFT biases by i) eliminating part of the self-interaction and by ii) introducing the exchange "pull-down" attraction resulting in lower energies of localized states. Using such orbitals in QMC correlated framework enables us to obtain results which are close to experiment without any additional parameters.

It is tempting to consider the usefulness of a hybrid functional which would be, however, derived from a fundamental theory instead of a fit to a testing set of molecules underlying B3LYP (Becke 1993). Although such approach would not fix all the deficiencies of the approximate DFT functionals it could serve, for example, as a cost effective method for providing more appropriate sets of one-particle orbitals for building accurate wavefunctions.

In fact, for molecular systems such as TiO and MnO we were able to directly optimize the weight of the exact exchange within a QMC framework. By varying the weight of the exact exchange in the functional and by iterating fixed-node DMC calculations we found the best set of one-particle orbitals which provided the lowest fixed-node energy in DMC (Wagner and Mitas 2003); we have used similar idea also in calculations of FeO and for solids more thorough analysis in this direction has been carried out very recently (Kolorenĉ and Mitas, unpubl. work).

CONCLUSION

In conclusion, we have performed calculations of the MnO and FeO solids in the variational and diffusion Monte Carlo methods. We have evaluated the cohesive and excitation energies which show excellent agreement with experiment. The results clearly show a crucial role of both exchange and correlation and their accurate description not only for MnO and FeO but obviously for other transition metal oxide systems as well. We have calculated the equation of state for the equilibrium and high-pressure structures of FeO by the QMC solution of the many-body Schrödinger equation without any non-variational parameters.

ACKNOWLEDGMENTS

Support by NSF EAR-05301110, DMR-0804549 and OCI-0904794 grants and by DOE DE-FG05-08OR23336 Endstation grant is gratefully acknowledged. We acknowledge also allocations at ORNL through INCITE and CNMS initiatives as well as allocations at NSF NCSA and TACC centers. We would like to thank J.-W. Lee and L. K. Wagner for their contributions to the mentioned projects and Yoonseok Lee for providing the TM pseudopotentials.

REFERENCES

Anisimov VI, Zaanen J, Andersen OK (1991) Band theory and Mott insulators: Hubbard U instead of Stoner I. Phys Rev B 44:943-954

Becke AD (1993) Density-functional thermochemistry. III. The role of exact exchange. J Chem Phys 98:5468

Bowen HK, Adler D, Auker BH (1975) Electrical and optical properties of FeO. J Solid State Chem 12:355-359

Brandow BH (1977) Electronic structure of Mott insulators. Adv Phys 26:651-808

Brandow B (1992) The physics of Mott electron localization. J Alloys Compd 181:377-396

Ceperley DM (1978) Ground state of the fermion one-component plasma: A Monte Carlo study in two and three dimensions. Phys Rev B 18:3126-3138

Ceperley DM (2010) An overview of quantum Monte Carlo methods. Rev Mineral Geochem 71:129-135

CRC Handbook of Chemistry and Physics (2007), CRC Press, Boca Raton

Dolg M, Wedig U, Stoll H, Preuss H (1987) Energy-adjusted ab initio pseudopotentials for the first row transition elements. J Chem Phys 86:866

Dovesi R, Saunders VR, Roetti C, Causa M, Harrison NM, Oralndo R, Zicovich-Wilson CM (1998/2003), CRYSTAL98/03 User's Manual, University of Torino, Torino. *http://www.crystal.unito.it/*

Fang Z, Terakura K, Sawada H, Miyazaki T, Solovyev I (1998) Inverse versus normal NiAs structures as high-pressure phases of FeO and MnO. Phys Rev Lett 81:1027-1030

Fang Z, Solovyev IV, Sawada H, Terakura K (1999) First-principles study on electronic structures and phase stability of MnO and FeO under high pressure. Phys Rev B 59:762-774

Foulkes WMC, Mitas L, Needs RJ, Rajagopal G (2001) Quantum Monte Carlo simulations of solids. Rev Mod Phys 73:33

Hugel J, Kamal M (1996) Electronic ground state of MnO, FeO, CoO and NiO within the LSDA + U approximation. Solid State Commun 100:457-461

Kasinathan D, Kunes J, Koepernik K, Diaconu CV, Martin RL, Prodan ID, Scuseria GE, Spaldin N, Petit L, Schulthess TC, Pickett WE (2007) Mott transition of MnO under pressure: A comparison of correlated band theories. Phys Rev B 74:195110

Kent PRC, Hood RQ, Williamson AJ, Needs RJ, Foulkes WMC (1999) Finite-size errors in quantum many-body simulations of extended systems. Phys Rev B 59:1917-1929

Kolorenč J, Mitas L (2008) Quantum Monte Carlo Calculations of Structural Properties of FeO Under Pressure. Phys Rev Lett 101:185502

Kotliar G, Savrasov SY, Haule K, Oudovenko VS, Parcollet O (2006) Electronic structure calculations with dynamical mean-field theory. Rev Mod Phys 78:865

Lee J-W, Mitas, Wagner LK (2004) Quantum Monte Carlo study of MnO solid. arXiv:cond-mat/0411247v2

McCammon CA L, Liu L-G (1984) The effects of pressure and temperature on nonstoichiometric wüstite, Fe_xO: The iron-rich phase boundary. Phys Chem Minerals 10:106-113

Mitas L, Martin RM (1994) Quantum Monte Carlo of nitrogen: atom, dimer, atomic, and molecular solids. Phys Rev Lett 72:2438-2441

Needs RJ, Towler MD (2002) The diffusion quantum Monte Carlo method: designing trial wave functions for NiO. *In:* Proceedings of the 11th International Conference on Recent Progress in Many-Body Theories. Bishop RF, Brandes T, Gernoth KA, Walet NR, Xian Y (eds) World Scientific, Singapore, p. 434-444

Norman MR (1991) Crystal-field polarization and the insulating gap in FeO, CoO, NiO, and La_2CuO_4. Phys Rev B 44:1364-1367

Oguchi T, Terakura K, Williams AR (1983) Band theory of the magnetic interaction in MnO, MnS, and NiO. Phys Rev B 28:6443

Pask JE, Singh DJ, Mazin II, Hellberg CS, Kortus J (2001) Structural, electronic, and magnetic properties of MnO. Phys Rev B 64:024403

Moskowitz JW, Schmidt KE (1992) Correlated Monte Carlo wave functions for some cations and anions of the first row atoms. J Chem Phys 97:3382; doi:10.1063/1.463938

Svane A, Gunnarsson O (1980) Transition-metal oxides in the self-interaction–corrected density-functional formalism. Phys Rev Lett 65:1148-1151

Szotek Z, Temmerman WM, Winter H (1993) Application of the self-interaction correction to transition-metal oxides. Phys Rev B 47:4029-4032

Tanaka S (1993) Cohesive energy of NiO: a quantum Monte Carlo approach. J Phys Soc Jpn 62:2112-2119

Tanaka S (1995) Variational quantum Monte Carlo approach to the electronic structure of NiO. J Phys Soc Jpn 64:4270-4277

Terakura K, Oguchi T, Williams AR, Kübler J (1984) Band theory of insulating transition-metal monoxides: Band-structure calculations. Phys Rev B 30:4734-4747

Towler MD, Allan NL, Harrison NM, Saunders VR, Mackrodt W, Aprà E (1994) Ab initio study of MnO and NiO. Phys Rev B 50:5041-5054

Umrigar C, Wilson KG, Wilkins JW (1988) Optimized trial wave functions for quantum Monte Carlo calculations. Phys Rev Lett 60:1719-1721

Wagner LK, Mitas L (2003) A quantum Monte Carlo study of electron correlation in transition metal oxygen molecules. Chem Phys Lett 370:412-417

Zaanen J, Sawatzky G, Allen J (1985) Band gaps and electronic structure of transition-metal compounds. Phys Rev Lett 55:418-421

Zhang J (2000) Effect of defects on the elastic properties of wüstite. Phys Rev Lett 84:507-510

Reviews in Mineralogy & Geochemistry
Vol. 71 pp. 147-167, 2010

8

Accurate and Efficient Calculations on Strongly Correlated Minerals with the LDA+U Method: Review and Perspectives

Matteo Cococcioni

Department of Chemical Engineering and Materials Science
University of Minnesota
Minneapolis, Minnesota 55455, U.S.A.

matteo@umn.edu

ABSTRACT

The LDA+U has become the method of choice to perform predictive and affordable calculations of the properties of minerals in the Earth's interior. In fact, the ubiquitous presence of transition metals (especially Fe) in several compounds of the mantle and of the core of our planet imposes an accurate description of the effects of electronic correlation. At the same time the investigation of the thermodynamic and mechanical properties of these materials at various and extreme conditions of temperature and pressure requires a method to evaluate energies and energy derivatives with a low computational cost. LDA+U is one of the few approach (if not the only one) to satisfy both these requirements. In this paper I will review the general formulation of this approach, will present some successful paradigmatic applications, and finally will discuss possible developments to improve its accuracy.

INTRODUCTION

Strongly correlated systems represent a significant challenge for computational techniques based on Density Functional Theory (DFT) (Hohenberg and Kohn 1964; Kohn and Sham 1965). Transition-metal-bearing minerals, ubiquitous in the interior of the Earth, fall within this category and sometimes represent particularly difficult cases. In fact, some of their valence electrons are located on very localized (almost atomic) *d* states; thus they assume a "correlated" ground state and many-body terms of the electronic interactions become dominant. Although DFT is exact in principles, the explicit expression of the exchange-correlation energy functional (E_{xc}), that is supposed to contain all the necessary corrections to the electronic energy beyond the Hartree term (the classical Coulomb interaction between charge densities) is not known. In most commonly used approximations to the exact functional, the Local Density Approximation (LDA) (Ceperley and Alder 1980; Ortiz and Ballone 1994; Perdew and Zunger 1981) and the Generalized Gradient Approximation (GGA) (Becke 1988; Perdew 1991; Perdew et al. 1996), the potential acting on one electron is a functional of the electronic charge density (and, possibly, of its derivatives). While this drastic simplification is quite accurate for systems where electrons are delocalized on covalent bonds (as in band semiconductor, like Si and GaAs) or on extended states (like in simple metals), when localization is more pronounced many-body terms of the electronic interaction, become dominant and the accuracy of the calculation rapidly deteriorates. The consequences on the description of structural properties can be quite severe; thus, the calculation of the thermoelastic properties of correlated minerals, that are crucial to translate seismic data into information about the composition, mineralogy and temperature of our planet interior, require extra care.

1529-6466/10/0071-0008$05.00 DOI: 10.2138/rmg.2010.71.8

Among the corrective schemes, introduced over the last decades to overcome these difficulties, the LDA+U (Anisimov and Gunnarson 1991; Anisimov et al. 1997), prototype of Hubbard-modeled corrections, is one of the simplest, yet most effective ones, and has rapidly become the method of choice to perform predictive calculations on models of transition-metal-bearing minerals (TMM) of realistic complexity. What makes this approach particularly appealing compared to other corrective schemes (e.g., DFT+Dynamical Mean Field Theory (Georges and Kotliar 1992; Georges et al. 1996)) or even non DFT-based techniques, as Monte Carlo, are the limited computational cost (approximately the same as standard DFT calculations) and the simplicity of its expression that makes the total energies and its derivatives (e.g., forces, stresses, force constants, etc) relatively easy and inexpensive to obtain. In fact, these quantities are crucial to evaluate, for example, the relative stability of different phases, to determine the equilibrium structure, to describe structural, magnetic and phase transitions, to compute excitations spectra and, thus, to effectively introduce finite temperature effects.

The paper is organized as follows: in the "The LDA+U Functional" section I will review the general formulation of the LDA+U approach detailing the differences between the two flavors most commonly used in literature; in the "Energy Derivatives" section I will present the implementation (in a plane-wave pseudo-potential DFT code) of LDA+U energy derivatives as forces and stress; in the "Calculation of *U*" section I will describe our linear response approach to calculate the Hubbard *U* and present some paradigmatic applications of it to molecules and crystals; the "Transition-Metal-Bearing Minerals" section is dedicated to transition-metal bearing minerals and illustrates the improvement obtained with LDA+U compared to standard DFT approximations for this class of materials; in the "Extended LDA+U+V Approach" section I will briefly describe a recent extension to the LDA+U functional that contains on-site and inter-site electronic interactions; finally I will discuss possible future developments to improve the accuracy and versatility of this corrective approach.

THE LDA+U FUNCTIONAL

The LDA+U approach (all that follows applies to other approximate DFT energy functionals as well, as the generalized gradient approximation or GGA) was formulated and developed during the 1990's (Anisimov and Gunnarson 1991; Anisimov et al. 1997) to improve the accuracy of LDA in describing systems characterized by localized, strongly correlated valence electrons. LDA+U consists of a simple, additive correction to the DFT energy functional that is shaped on the Hubbard Hamiltonian. In its simplest, single-band formulation, the Hubbard Hamiltonian can be written as follows:

$$H_{Hub} = t \sum_{<i,j>,\sigma} (c_{i,\sigma}^{\dagger} c_{j,\sigma} + h.c.) + U \sum_{i} n_{i,\uparrow} n_{i,\downarrow} \tag{1}$$

where $\langle i,j \rangle$ denotes nearest-neighbor sites, $c_{i,\sigma}^{\dagger}$, $c_{j,\sigma}$, and $n_{i,\sigma}$ are electronic creation, annihilation and number operators for the electron of spin σ on site i. This model is normally used to study systems in the limit of strong localization (on atomic orbitals). In fact, in Equation (1) the motion of electrons is described by the hopping between neighbor sites (whose amplitude t is to the Fourier transform of the bandwidth of the system and represents single-particle terms of the electronic energy, i.e., the kinetic energy and the interaction with the nuclei) while the Coulomb repulsion, whose strength is represented by the parameter U (named the "Hubbard U"), is only significant between electrons on the same atom. Despite its simplicity, this model is able to capture the fundamental mechanism of Mott localization: depending on whether one-body (t) or many-body (U) terms of the electronic Hamiltonian are dominant, the electrons in the system minimize their energy by spreading over extended states or localizing on atomic orbitals. The general idea of LDA+U is to incorporate this physics into DFT by a corrective functional that explicitly accounts for short-range (i.e., on-site) electron-electron interactions. In fact, while one-body terms of the energy are quite well represented in commonly used approximations to (the exact) DFT functionals, the interaction between localized electrons, being difficult to capture through a functional of the total charge density (and, possibly, of its derivatives), is not equally well accounted for. A correction to DFT functionals to alleviate this difficulty must act on the interaction part of the Hamiltonian and selectively operate on localized states. The LDA+U total energy functional of a system is generally formulated as follows:

$$E_{LDA+U} = E_{LDA} + E_U = E_{LDA} + E_{Hub} - E_{dc} \tag{2}$$

In this equation E_{Hub} contains electron-electron interactions as modeled in the Hubbard Hamiltonian. E_{dc} is a mean-field approximation to E_{Hub} and models the electronic correlation already contained in E_{DFT}. This term is subtracted from the total functional to avoid double-counting of the interactions contained in E_{Hub}. In its original formulation the corrective functional E_U was not invariant with respect to rotations of the atomic orbitals used to define the "on-site" occupations (corresponding to the number operators in Equation (1)). A fully basis-set invariant formulation of LDA+U was first introduced in Liechtenstein et al. (1995) where the Hubbard and double-counting terms appearing in Equation (2) are respectively defined as follows:

$$E_{Hub}[\{n^I_{mm'}\}] = \frac{1}{2}\sum_{\{m\},\sigma,I}[< m,m'' | V_{ee} | m',m''' > n^{I\sigma}_{mm'} n^{I-\sigma}_{m''m'''} + \tag{3}$$

$$(< m,m'' | V_{ee} | m',m''' > - < m,m'' | V_{ee} | m''',m' >) n^{I\sigma}_{mm'} n^{I\sigma}_{m''m'''}]$$

$$E_{dc}[\{n^I\}] = \sum_I \{\frac{U}{2} n^I (n^I - 1) - \frac{J}{2}[n^{I\uparrow}(n^{I\uparrow} - 1) + n^{I\downarrow}(n^{I\downarrow} - 1)]\} \tag{4}$$

In these equations $\mathbf{n}^{I\sigma}$ is a matrix representing the occupation of atomic states of a definite angular momentum l (generally 2 or 3) on the atomic site I with spin σ. These quantities correspond to the number operators appearing in Equation (1); in practice, however, occupation matrices are generally defined as projections of Kohn-Sham (KS) states onto atomic orbitals:

$$n^{I\sigma}_{m,m'} = \sum_{k,v} f_{kv} < \psi^\sigma_{kv} | \varphi^I_{m'} >< \varphi^I_m | \psi^\sigma_{kv} > \tag{5}$$

In Equation (5) m is the magnetic quantum number associated with l ($-l \le m \le l$), while f^σ_{kv} are the occupations of the Kohn-Sham orbitals ψ^σ_{kv} based on the distribution of their energies around the Fermi level. The one given in Equation (5) is not the only possible definition for atomic occupations but is a particularly convenient one for plane-wave pseudo-potential implementations of DFT (as the one contained in the PWscf code of the Quantum-Espresso package (Giannozzi et al. 2009) to which I will refer throughout this article) where localized orbitals are not explicitly involved in the construction of the energy functionals. In Equation (4) U and J are screened Coulomb and exchange (average) interactions. The V_{ee} parameters in Equation (3) represent the Coulomb interaction among electrons occupying localized orbitals of the same atom. Their formal expression (and the ratios between different terms) is usually borrowed from the expansion of the $e^2/|r - r'|$ Coulomb potential in terms of spherical harmonics (see Anisimov et al. 1997 and references quoted therein):

$$< m,m'' | V_{ee} | m',m''' > = \sum_k a_k(m,m',m'',m''') F^k \tag{6}$$

where $0 \le k \le 2l$ (l is the angular moment of the Hubbard electrons) and

$$a_k(m,m',m'',m''') = \frac{4\pi}{2k+1}\sum_{q=-k}^{k} < lm | Y_{kq} | lm' >< lm'' | Y^*_{kq} | lm''' > \tag{7}$$

The F^k coefficients, that in HF theory are the radial Slater integrals representing the *bare* electron-electron interaction, in the formulation of the LDA+U functional are treated as parameters that represent *screened* electronic couplings.

Through evaluating E_{dc} as the mean-field approximation to E_{Hub} (which corresponds to assuming all the occupation matrices equal to the mean value of their trace), a precise relationship can be obtained between U, J and the V_{ee} integrals appearing in Equation (3) or the Slater integrals F^k. For d electrons only F^0, F^2, and F^4 are necessary (odd k and higher order term being all 0):

$$U = \frac{1}{(2l+1)^2} \sum_{m,m'} < m,m' | Vee | m,m' > = F^0$$
$$J = \frac{1}{2l(2l+1)} \sum_{m \neq m',m'} < m,m' | Vee | m',m > = \frac{F^2 + F^4}{14} \tag{8}$$

For f electrons F^6 is also non-zero and enters the expression of J. These equations are often used in literature to reconstruct the needed F^k (and, thus, the V_{ee} integrals) from the average interactions U and J that are either fitted or computed. This procedure is usually based on the assumption that the ratios between F^k preserve their atomic values.

A commonly used simplified expression to the fully invariant LDA+U energy functional discussed so far was introduced by Dudarev et al. (1998). Within this approximation, that can be obtained by simply neglecting the anisotropy of the electronic interactions ($U = F^0$, $J = F^2 = F^4 = 0$), the total corrective functional results:

$$E_U = E_{Hub} - E_{dc} = \sum_{I,\sigma} \frac{U_{eff}}{2} Tr\left[\mathbf{n}^{I\sigma}\left(1 - \mathbf{n}^{I\sigma}\right)\right] \tag{9}$$

The effective electronic interaction U_{eff} normally assumed to be corrected by the exchange interaction: $U_{eff} = U - J$; in fact, only spin-diagonal terms appear in Equation (9). This simplified expression of the corrective functional is indeed very useful to understand how it acts on the electronic wavefunctions. To this purpose it is convenient to compute the Hubbard correction to the Kohn-Sham potential that can be easily obtained as the functional derivative of the corrective term to the energy with respect to $(\psi_{kv}^\sigma)^*$:

$$V_U | \psi_{kv}^\sigma > = \frac{\delta E_{Hub}}{\delta(\psi_{kv}^\sigma)^*} = \sum_{I,m,m'} \frac{U_{eff}}{2}\left(\delta_{mm'} - 2n_{m'm}^{I\sigma}\right) | \varphi_{m'}^I >< \varphi_m^I | \psi_{kv}^\sigma >$$
$$= \sum_{I,m} \frac{U_{eff}}{2}\left(1 - 2\lambda_m^{I\sigma}\right) | \tilde{\varphi}_m^I >< \tilde{\varphi}_{m'}^I | \psi_{kv}^\sigma > \tag{10}$$

where the last equality exploits the rotational invariance of the functional (preserved in the simpler formulation of Eqn. 9) to express this contribution in the diagonal basis. From Equation (10) it is evident that the corrective potential is attractive for occupations ($\lambda_m^{I\sigma}$) exceeding 1/2, while it is repulsive for occupations lower than this threshold value. Fully occupied atomic orbitals are thus stabilized at the expense of the population of the ones that are less than half-occupied. The difference between the potential acting on empty and full states and, nominally, the energy gap of the system, is approximately equal to U_{eff} (although this is only true if one neglects the relaxation of the Kohn-Sham orbitals due to the Hubbard potential). It is important to notice that the full occupation of some atomic orbitals results in the suppression of their hybridization with neighbor atoms; in fact, hybridized states would lead to fractional atomic occupations (see the definition in Eqn. 5) that are energetically unfavorable.

The effect of the Hubbard correction on the electronic structure of a system depends on the precise formulation of the functional and, specifically, on the approximation used for the double-counting term. The one introduced in Equation (4), and used throughout this paper, corresponds to the so-called "Fully-Localized-Limit" (FLL) and is particularly well suited to describe systems in which strong electronic correlation results in the localization of electrons on atomic orbitals. On the opposite limit, the "Around-Mean-Field" (AMF) approach is based on a double-counting term constructed in the hypothesis that all the states of the same atom are equally occupied. The total Hubbard correction to the DFT energy functional contributes, in this case, a negative term that encourages fluctuations around the average occupation:

$$E_U^{AMF} = -\sum_{I,\sigma} \frac{U_{eff}}{2} Tr\left(\delta\mathbf{n}^{I\sigma} \cdot \delta\mathbf{n}^{I\sigma} \right) \tag{11}$$

where $\delta n_{mm'}^{I\sigma} = n_{mm'}^{I\sigma} - \delta_{mm'} Tr(n^{I\sigma}) / (2l+1)$. In Lechermann et al. (2004) and Petukhov et al. (2003) the two approaches, FLL and AMF, are discussed and compared and the AMF double counting is used to study the interplay between magnetic and structural properties in FeAl. In Petukhov et al. (2003) a modified LDA+U is proposed that is a linear combination between the two limits. A very detailed and comprehensive discussion of possible LDA+U functionals (and, specifically, double-counting terms) is presented in Ylvisaker et al. (2009). In this paper authors focus, in particular, on the role of anisotropy and of the exchange coupling J arguing that proper account for the orbital dependence of the various interaction parameters (through the use of a fully rotationally invariant formulation given in Eqns. 3 and 4) is necessary in some materials to obtain an accurate description of their magnetic properties and possible transitions in between different states as the insulator-insulator transition in MnO due to a magnetic moment collapse accompanied by a volume contraction. In this case the simple approximation corresponding to the functional in Equation (9) is expected to be significantly less accurate.

In this paper I will refer to the simplified FLL functional (Eqn. 9) that was used to obtain all the results presented in the following sections.

ENERGY DERIVATIVES

One of the major advantages of LDA+U consists in the simple formulation of the corrective functional it is based on, which allows for the easy implementation of energy derivatives, as forces and stresses, that are essential quantities to identify and characterize the equilibrium structure of materials in different conditions. In this section, I will review the implementation of forces and stresses that are contained in the plane-wave pseudopotential total energy and Car-Parrinello molecular dynamic codes (PWscf and CP) of the QUANTUM-ESPRESSO package (Giannozzi et al. 2009).

The Hubbard forces

The Hubbard forces are defined as the derivative of the Hubbard energy with respect to ionic displacements. The force acting on the atom α in the direction i is defined as:

$$F_{\alpha,i}^U = -\frac{\partial E_U}{\partial \tau_{\alpha i}} = -\sum_{I,m,m',\sigma} \frac{\partial E_U}{\partial n_{m,m'}^{I\sigma}} \frac{\partial n_{m,m'}^{I\sigma}}{\partial \tau_{\alpha i}} = -\frac{U}{2} \sum_{I,m,m',\sigma} (\delta_{mm'} - 2n_{m'm}^{I\sigma}) \frac{\partial n_{m,m'}^{I\sigma}}{\partial \tau_{\alpha i}} \tag{12}$$

where $\delta\tau_{\alpha i}$ is the atomic displacement, E_U and $n_{m,m'}^{I\sigma}$ are the Hubbard energy and the atomic occupation matrix element defined in the previous section. Based on the definition of the atomic occupations (Eqn. 5), the derivatives in Equation (12) can be further decomposed as:

$$\frac{\partial n_{m,m'}^{I\sigma}}{\partial \tau_{\alpha i}} = \sum_{k,v} f_{kv}^{\sigma} [\frac{\partial}{\partial \tau_{\alpha i}} \left(< \phi_{mk}^I | \psi_{kv}^{\sigma} > \right) < \psi_{kv}^{\sigma} | \phi_{m'k}^I > + < \phi_{mk}^I | \psi_{kv}^{\sigma} > \frac{\partial}{\partial \tau_{\alpha i}} < \psi_{kv}^{\sigma} | \phi_{m'k}^I >] \tag{13}$$

so that the problem is reduced to determine the quantities

$$\frac{\partial}{\partial \tau_{\alpha i}} < \phi_m^I | \psi_{kv}^{\sigma} > \tag{14}$$

for each I, m, m', σ, k and v. Since the Hellmann-Feynman theorem applies, no response of the electronic wave function has to be considered while varying the ionic positions. The quantities in Equation (14) can thus be calculated considering only the derivative of the atomic wave functions with respect to the atomic displacement:

$$\frac{\partial}{\partial \tau_{\alpha i}} < \phi^{I}_{mk} | \psi^{\sigma}_{v} > = < \frac{\partial \phi^{I}_{m}}{\partial \tau_{\alpha i}} | \psi^{\sigma}_{kv} > \tag{15}$$

Although the atomic occupations are defined on localized atomic orbitals, the product with Kohn-Sham wavefunctions at a given **k**-vector (Eqn. 5) selects the Fourier component of the atomic wavefunction at the same k-point. This component can be constructed as a Bloch sum of localized atomic orbitals:

$$\begin{aligned} \phi^{at}_{i,\mathbf{k},I}(\mathbf{r}) &= \frac{1}{\sqrt{N}} \sum_{\mathbf{R}} e^{-i\mathbf{k}\cdot\mathbf{R}} \phi^{at}_{i,I}(\mathbf{r}-\mathbf{R}-\tau_I) \\ &= e^{-i\mathbf{k}\cdot\mathbf{r}} \frac{1}{\sqrt{N}} \sum_{\mathbf{R}} e^{i\mathbf{k}\cdot(\mathbf{r}-\mathbf{R})} \phi^{at}_{i,I}(\mathbf{r}-\mathbf{R}-\tau_I) \end{aligned} \tag{16}$$

In this equation i is the cumulative index for all the quantum numbers $\{n,l,m\}$, defining the atomic state, N is total number of direct lattice vectors **R**. The second factor in the right hand side of Equation (16) is a function with the periodicity of the lattice. Its Fourier spectrum thus contains only reciprocal lattice vectors:

$$\phi^{at}_{i,\mathbf{k},I}(\mathbf{r}) = \frac{1}{\sqrt{V}} \sum_{\mathbf{G}} e^{-i(\mathbf{k}+\mathbf{G})\cdot\mathbf{r}} c_{i,I}(\mathbf{k}+\mathbf{G}) \tag{17}$$

(**G** are the reciprocal lattice vector: $\mathbf{G}\cdot\mathbf{R} = 0$, V is the volume of the system: $V = N\Omega$) the $\mathbf{k} + \mathbf{G}$ Fourier component of the atomic Bloch sum results

$$\begin{aligned} c_{i,I}(\mathbf{k}+\mathbf{G}) &= \frac{1}{\sqrt{V}} \int_V d\mathbf{r} e^{i(\mathbf{k}+\mathbf{G})\cdot\mathbf{r}} \phi^{at}_{i,\mathbf{k},I}(\mathbf{r}) \\ &= \frac{1}{N\sqrt{\Omega}} \sum_{\mathbf{R}} \int_V d\mathbf{r} e^{i(\mathbf{k}+\mathbf{G})\cdot(\mathbf{r}-\mathbf{R})} \phi^{at}_{i,I}(\mathbf{r}-\mathbf{R}-\tau_I) \\ &= \frac{1}{N\sqrt{\Omega}} e^{i(\mathbf{k}+\mathbf{G})\cdot\tau_I} \sum_{\mathbf{R}} \int_V d\mathbf{r} e^{i(\mathbf{k}+\mathbf{G})\cdot(\mathbf{r})} \phi^{at}_{i,I}(\mathbf{r}) \end{aligned} \tag{18}$$

This expression of the Fourier component is particularly useful because the dependence on the atomic position is explicit. The response to the ionic displacement can thus be calculated straightforwardly and we obtain:

$$\frac{\partial \phi^{at}_{i,\mathbf{k},I}}{\partial \tau_{\alpha j}} = \delta_{I,\alpha} \frac{i}{\sqrt{V}} \sum_{\mathbf{G}} e^{-i(\mathbf{k}+\mathbf{G})\cdot\mathbf{r}} c_{i,\alpha}(\mathbf{k}+\mathbf{G})(\mathbf{k}+\mathbf{G})_j \tag{19}$$

where $(\mathbf{k} + \mathbf{G})_j$ is the component of the vector along direction j and i is the imaginary unit. Due to the presence of the Kronecker δ in front of its expression, the derivative of the atomic wave function is different from zero only in the case it is centered on the ion which is being displaced. Thus, the derivative in Equation (15) only contributes to forces acting on "Hubbard" atoms. Finite off-site terms in the expression of the forces arise when using ultrasoft pseudopotentials as projections between orbitals normally involve an augmentation part. However this case is not explicitly treated in the present work.

The Hubbard stresses

Starting from the expression for the Hubbard energy functional, given in Equation (9), we can compute the contribution to the stress tensor. This quantity is defined as

$$\sigma^U_{\alpha\beta} = -\frac{1}{\Omega}\frac{\partial E_U}{\partial \varepsilon_{\alpha\beta}} \tag{20}$$

where Ω is the unit cell volume (the energy is also given per unit cell), $\varepsilon_{\alpha,\beta}$ is the strain applied to the crystal:

$$\mathbf{r}_\alpha \to \mathbf{r}'_\alpha = \sum_\beta (\delta_{\alpha\beta} + \varepsilon_{\alpha\beta})\mathbf{r}_\beta \tag{21}$$

where $\mathbf{r}$ is a coordinate internal to the unit cell. The procedure, already developed for the forces (see Eqns. 12 and 13), can be applied to the case of stresses as well. The problem thus reduces to evaluating the derivative

$$\frac{\partial}{\partial \varepsilon_{\alpha\beta}} < \phi^I_{mk} \mid \psi^\sigma_{kv} > \tag{22}$$

In order to determine the functional dependence of atomic and KS wavefunctions on the strain we deform the lattice accordingly to Equation (21) and study how these functions are modified by this distortion. Very small distortions will be considered so that expansions to first order around the undeformed values will be sufficiently accurate. As a consequence, the distortion of the reciprocal lattice is opposite to the real space one:

$$\mathbf{k}_\alpha \to \mathbf{k}'_\alpha = \sum_\beta (\delta_{\alpha\beta} - \varepsilon_{\alpha\beta})\mathbf{k}_\beta \tag{23}$$

Thus, the products $(\mathrm{k} + \mathrm{G})\cdot\mathrm{r}$ appearing in the plane wave (PW) expansion of the wave functions (see for example Eqn. 17) remain unchanged.

Let's first study the modification of the atomic wavefunctions taking in consideration the expression given in Equation (17). The volume appearing in the normalization is transformed as follows:

$$V \to V' = |1+\varepsilon| V \tag{24}$$

where $|1 + \varepsilon|$ is the determinant of the matrix $\delta_{\alpha\beta} + \varepsilon_{\alpha\beta}$ describing the deformation of the crystal. Applying the strain defined in Equation (21), to the expression of the $\mathbf{k} + \mathbf{G}$ Fourier component of the atomic wave function we obtain:

$$\begin{aligned} c_{i,I}(\mathbf{k}'+\mathbf{G}') &= \frac{1}{N\sqrt{|1+\varepsilon|\Omega}} e^{i(\mathbf{k}'+\mathbf{G}')\cdot\tau'_I} \int_{V'} d\mathbf{r}' e^{i(\mathbf{k}'+\mathbf{G}')\cdot\mathbf{r}'} \phi^{at}_{i,I}(\mathbf{r}') \\ &= \frac{1}{\sqrt{|1+\varepsilon|}} \frac{1}{N\sqrt{\Omega}} e^{i(\mathbf{k}+\mathbf{G})\cdot\tau_I} \int_{V'} d\mathbf{r} e^{i(\mathbf{k}'+\mathbf{G}')\cdot\mathbf{r}} \phi^{at}_{i,I}(\mathbf{r}) \end{aligned} \tag{25}$$

As the integral appearing in this expression does not change upon distorting the integration volume if we define:

$$\tilde{c}_{i,I}(\mathbf{k}+\mathbf{G}) = c_{i,I}(\mathbf{k}+\mathbf{G}) e^{-i(\mathbf{k}+\mathbf{G})\cdot\tau_I} \tag{26}$$

we obtain:

$$\tilde{c}'_{i,I}(\mathbf{k}'+\mathbf{G}') = \frac{1}{\sqrt{|1+\varepsilon|}} \tilde{c}_{i,I}(\mathbf{k}'+\mathbf{G}') = \frac{1}{\sqrt{|1+\varepsilon|}} \tilde{c}_{i,I}((1-\varepsilon)(\mathbf{k}+\mathbf{G})) \tag{27}$$

From this expression we easily obtain:

$$\begin{aligned}\phi^{at}_{i,\mathbf{k},I}(\mathbf{r}) &= \frac{1}{\sqrt{V}}\sum_{\mathbf{G}} e^{-i(\mathbf{k}+\mathbf{G})\cdot\mathbf{r}} e^{i(\mathbf{k}+\mathbf{G})\cdot\tau_I}\tilde{c}_{i,I}(\mathbf{k}+\mathbf{G}) \rightarrow \frac{1}{\sqrt{V'}}\sum_{\mathbf{G}'} e^{-i(\mathbf{k}'+\mathbf{G}')\cdot\mathbf{r}'} e^{i(\mathbf{k}'+\mathbf{G}')\cdot\tau'_I}\tilde{c}'_{i,I}(\mathbf{k}'+\mathbf{G}') \\ &= \frac{1}{|1+\varepsilon|}\frac{1}{\sqrt{V}}\sum_{\mathbf{G}} e^{-i(\mathbf{k}+\mathbf{G})\cdot\mathbf{r}} e^{i(\mathbf{k}+\mathbf{G})\cdot\tau_I}\tilde{c}_{i,I}((1-\varepsilon)(\mathbf{k}+\mathbf{G}))\end{aligned} \tag{28}$$

According to the Bloch theorem Kohn-Sham (KS) wavefunctions can be expressed as follows:

$$\psi^{\sigma}_{kv}(\mathbf{r}) = \frac{1}{\sqrt{V}}\sum_{G} e^{-i(\mathbf{k}+\mathbf{G})\cdot\mathbf{r}} a^{\sigma}_{kv}(\mathbf{G}) \tag{29}$$

where

$$a^{\sigma}_{kv}(\mathbf{G}) = \frac{1}{\sqrt{V}}\int_V d\mathbf{r} e^{i(\mathbf{k}+\mathbf{G})\cdot\mathbf{r}}\psi^{\sigma}_{kv}(\mathbf{r}) \tag{30}$$

Upon distorting the lattice as described in Equation (21) the electronic charge density is expected to rescale accordingly. One can thus imagine the electronic wave function in a point of the strained space to be proportional to the value it had in the corresponding point of the undistorted lattice:

$$\psi^{\sigma}_{kv}(\mathbf{r}) \rightarrow \psi'^{\sigma}_{k'v}(\mathbf{r}') = \alpha\psi^{\sigma}_{kv}((1-\varepsilon)\mathbf{r}') \tag{31}$$

where the proportionality constant α is to be determined by normalizing the wave function in the strained crystal. By a simple change of the integration variable we obtain:

$$1 = \int_{V'} d\mathbf{r}' |\psi'^{\sigma}_{k'v}|^2 = |1+\varepsilon| \int_V d\mathbf{r} |\alpha\psi^{\sigma}_{kv}|^2 = |1+\varepsilon|\alpha^2 \tag{32}$$

from which, choosing α real, we have

$$\alpha = \frac{1}{\sqrt{|1+\varepsilon|}} \tag{33}$$

Using this result we can determine the variation of the $\mathbf{k} + \mathbf{G}$ Fourier component of ψ^{σ}_{kv} (Eqn. 30). We easily obtain:

$$\begin{aligned}a^{\sigma}_{k'v}(\mathbf{G}') &= \frac{1}{\sqrt{V'}}\int_{V'} d\mathbf{r}' e^{i(\mathbf{k}'+\mathbf{G}')\cdot\mathbf{r}'}\psi'^{\sigma}_{k'v}(\mathbf{r}') \\ &= \frac{1}{\sqrt{|1+\varepsilon|}}\frac{1}{\sqrt{V}}\int_V |1+\varepsilon| d\mathbf{r} e^{i(\mathbf{k}+\mathbf{G})\cdot\mathbf{r}}\frac{1}{\sqrt{|1+\varepsilon|}}\psi^{\sigma}_{k'v}(\mathbf{r}) = a^{\sigma}_{kv}(\mathbf{G})\end{aligned} \tag{34}$$

We can now evaluate the first order variation of the scalar products between atomic and Kohn-Sham wavefunctions:

$$<\phi^I_{mk}|\psi^{\sigma}_{kv}>' = \frac{1}{\sqrt{|1+\varepsilon|}}\sum_{G} e^{i(\mathbf{k}+\mathbf{G})\cdot\tau_I}[c^I_{mk}((1-\varepsilon)(\mathbf{k}+\mathbf{G}'))]^* a^{\sigma}_{kv}(G) \tag{35}$$

The expression of the derivative follows immediately (for small strains $|1+\varepsilon| \sim 1+Tr(\varepsilon)$):

$$\frac{\partial}{\partial\varepsilon_{\alpha\beta}} <\phi^I_{mk}|\psi^{\sigma}_{kv}>|_{\varepsilon=0} = -\frac{1}{2}\delta_{\alpha\beta} <\phi^I_{mk}|\psi^{\sigma}_{kv}> - \sum_{G} e^{i(\mathbf{k}+\mathbf{G})\cdot\tau_I} a^{\sigma}_{kv}(G)\partial_{\alpha}[c^I_{mk}(\mathbf{k}+\mathbf{G}')]^*(\mathbf{k}+\mathbf{G})_{\beta} \tag{36}$$

The explicit expression of the derivative of the Fourier components of the atomic wavefunctions won't be detailed here. In fact this quantity depends on the particular definition of the atomic orbitals that can vary in different implementations.

CALCULATION OF *U*

The strength of the "+U" correction described in the previous paragraphs is controlled by the effective on-site electronic interaction — the Hubbard U — whose value is not known *a priori*. While it is common practice, the semiempirical evaluation of this parameter is not completely satisfactory and does not allow to appreciate the variations it could possibly undergo e.g., during chemical reactions, structural transformations or under different physical conditions.

To overcome these difficulties we recently introduced a method to calculate the Hubbard U from linear-response theory (Cococcioni and de Gironcoli 2005). This calculation renders LDA+U completely *ab initio*. Within this scheme U is computed as the second derivative of the DFT energy with respect to the atomic occupations n^I. After subtracting a contribution due to the rehybridization of the electronic orbitals (not related to electron-electron interactions) the final expression of the effective interaction is obtained:

$$U^I = \frac{\partial^2 E}{\partial (n^I)^2} - \frac{\partial^2 E_{KS}}{\partial (n_0^I)^2} \tag{37}$$

In Equation (37), E is the self-consistent total energy while E_{KS} is its Kohn-Sham counterpart — the energy of a non-interacting electron system that has the same ground state charge density of the interacting one — whose second derivative is related to the rehybridization of orbitals. The second derivatives in Equation (37) are calculated through linear response theory evaluating the susceptibility matrices $\chi_{IJ} = \partial n^I/\partial\alpha^J$, and $(\chi_0)_{IJ} = \partial n_0^I/\partial\alpha^J$ $\partial n^I/\partial\alpha^J$ that measure, respectively, the self-consistent and the Kohn-Sham response of the total occupation of "Hubbard" atoms to perturbations in the potential acting on all the atomic states of Hubbard sites ($\Delta V^I = \alpha^I \Sigma_i|\phi_i^I\rangle\langle\phi_i^I|$). It can be demonstrated (Cococcioni and de Gironcoli 2005) that the inverse of these matrices correspond to the second derivatives appearing in Equation (37) and the effective Hubbard U can be obtained as:

$$U^I = (\chi^{-1} - \chi_0^{-1})_{II} \tag{38}$$

In fully consistency with linear-response theory, U results to be the effective electronic coupling when electrons are described through the atomic orbital occupations. This approach introduced for the first time the possibility of consistent, completely *ab initio* evaluations of U from the same DFT ground state that is to be corrected. It thus allows us to take into account the variations of the on-site coupling during chemical reactions, structural and magnetic transitions and to appreciate the differences between atoms in crystallographically distinct positions or in different magnetic configurations. The approach outlined above to evaluate the on-site electronic repulsion U from linear response theory proved to be extremely effective in a large number of cases involving transition-metal compounds (Zhou et al. 2004a,b,c; Cococcioni and de Gironcoli 2005; Kulik et al. 2006; Sit et al. 2006, 2007; Scherlis et al. 2007) that will be outlined in the following paragraphs.

Li-ion batteries cathodes

The calculation of the electrochemical properties of transition-metal compounds that are considered as candidate materials for Li-ion batteries cathodes, are often quite challenging for "standard" DFT approximations as electrons in these materials tend to localize on atomic d states and to assume a correlated motion. The nominal voltage of the battery (with respect to a metal Li anode), the chemical stability of its electrodes through charge and discharge processes, and also the microscopic mechanisms at the basis of Li and electronic conduction are quite poorly described by DFT. In Li_xFePO_4 iron phosphate, one of the most interesting materials for this application, at intermediate Li concentrations ($0 < x < 1$) the ground state is characterized by the (Mott) localization of part of the valence electrons on some of the Fe ions (Fe^{2+}) that leaves the rest of them in their 3+ oxidation state (Zhou et al. 2004c). This localization of electrons corresponds to a phase separation into fully lithiated $LiFePO_4$ and a de-lithiated $FePO_4$ end-members (few stable intermediate compounds were observed only recently (Delacourt et al. 2005; Dodd et al. 2006). Unfortunately standard DFT approximations fail to reproduce the correlation-driven electronic localization and result in an unphysical ground state with electrons equally spread among $Fe^{+\alpha}$ ($2 < \alpha < 3$, depending on Li concentration) ions. Intermediate Li-concentration compounds are thus stabilized with a significant (and systematic) underestimation of the voltage. The use of the LDA+U functional is crucial to obtain the correct ground state with electrons localized on Fe^{2+} ions. Using our consistent linear-response-based evaluation of the Hubbard U we could predict positive formation energies (i.e., unstable compounds) at intermediate Li concentrations, consistently with the observed separation in a Li-rich and a de-lithiated phase, and we could estimate the nominal voltage (of the ideal Li-anode device) in excellent quantitative agreement with experimental measurements (Yamada and Chung 2001; Yamada et al. 2001). Equally accurate results were also obtained for a broad class of Li_xMPO_4 (M = Mn, Fe, Co, Ni) olivine, Li_xMO_4 (M = Co, Ni) layered (Zhou et al. 2004b) and for (Li_xMSiO_4) silicate (Zhou et al. 2004a) cathode materials. As evident from the graphs in Figure 1, computing the on-site Hubbard repulsion was critical to quantitatively reproduce the experimentally measured voltage of devices having the chosen materials as cathodes. This represents a large, systematic improvement over standard DFT results (corresponding to U = 0 in Fig. 1) that stems from the capability of the LDA+U approach to correctly address the localization of strongly correlated electrons. In a more recent work (Zhou et al. 2006) LDA+U was also successfully employed to explain the formation of some stable compounds at specific intermediate concentrations of Li.

Water-solvated transition-metal ions

Electron-transfer processes are central to our understanding of electrochemical reactions. Unfortunately, due to lack of cancellation of the electronic self-interaction, standard DFT energy functionals generally fail in providing a quantitatively accurate description of these processes and result in the formation of fictitious delocalized states. In recent works (Sit et al. 2006, 2007) we used LDA+U together with a constraint functional to improve the description of electronic localization and to study electron-transfer processes in a more quantitative way. To this purpose the LDA+U was implemented in the Car-Parrinello Molecular Dynamic (MD) code of the Quantum-Espresso package (Giannozzi et al. 2009) that was used to sample the diabatic energy surface of the electron-trasfer process between Fe^{2+} and Fe^{3+} ions solvated in

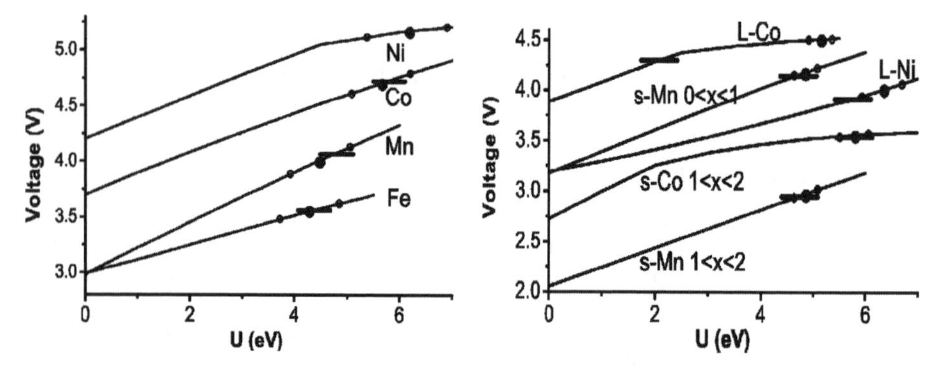

Figure 1. Voltage as a function of U for $LiMPO_4$ materials in the olivine structure (*left graph*) and for $LiMO_2$ layered and spinel structures (*right graph*). The short horizontal lines indicate the experimental voltage of each material. The two small open circles on a curve represent the voltage for U calculated in the de-lithiated and fully-lithiated states. The big solid circle represents the voltage at the average of the two U values. Figure is from Zhou et al. (2004b).

water. "Standard" DFT approximations predict an unphysical sharing of the electron between two $Fe^{2.5+}$ ions in this case. As reported by Sit et al. (2006), thanks to the constraint functional, the *d* electrons of Fe were correctly localized around the nuclei and the calculated diabatic reorganization energy of the aqua-iron system was obtained in excellent agreement with the prediction of Marcus theory (Newton and Sutin 1984; Marcus and Sutin 1985; Warshel and Parson 1991; Marcus 1993).

Chemical reactions on transition-metal complexes

Homogeneous catalytic reactions are particularly challenging to study with "standard" DFT functionals and are generally performed using wavefunction-based quantum-chemistry approaches that are significantly more computationally demanding than DFT. The lack of cancellation of the spurious self-interaction in most approximate DFT energy functionals leads to the over-delocalization of the electronic wavefunctions that precludes or seriously compromises the correct representation of bond breaking/formation phenomena, electron-transfer steps, etc. LDA+U can significantly improve the description of these systems encouraging the localization of electrons on atomic orbitals and thus partially eliminating the effects of self-interactions. The use of the corrective approach for these systems, however, needs extra care as in many cases the DFT (LDA or GGA) ground state is qualitatively different from the correlated (LDA+U) one (stabilizing, e.g., covalent bonds instead of localized atomic states) and the response of its electronic structure, from which the Hubbard *U* is obtained, can vary as well. A refinement in the approach to compute the Hubbard *U* was thus formulated to allow for the evaluation of the effective electronic coupling self-consistently from a "correlated" (LDA+U) ground state (Kulik et al. 2006). In the new scheme the on-site electronic coupling is updated until the computed value is consistent with the one used to enforce the correlated ground state of the system. This approach, particularly useful for molecular systems, for which DFT approximate functional usually overestimates electronic delocalization, was applied to the study of Fe_2 and Fe_2^- dimers and the H addition-elimination reaction on a FeO^+ dimer. For the Fe_2 and Fe_2^- dimers our "self-consistent" LDA+U approach shows a striking and systematic agreement with more sophisticated (and computationally expensive) quantum-chemistry (MRCI – Hubner and Sauer 2002; CCSD(T) – Kulik et al. 2006) calculations correctly identifying the lowest anion state as $^8\Sigma_u^-$ and the first excited state, $^8\Delta_g$, 0.38 eV above; the lowest singly-ionized neutral states as $^9\Sigma_g^-$ and $^7\Sigma_g^-$ with a splitting of 0.6 eV in between them comparing very well with the theoretical (MRCI and CCSD(T)) and experimental results (0.53 eV); equilibrium distances and vibrational frequencies in agreement with experiments. In contrast with observations, GGA predicts, instead, Δ ground states for both systems. In the case of the gas-phase molecular hydrogen addition-elimination reaction on FeO^+ the challenge mainly consisted in describing the steepness of the spin-energy surface close to reactants and products (along the reaction path) explaining the low yield of the reaction. GGA results in a quite poor representation of the process stabilizing FeO^+ in a $^4\Delta$ ground state and failing to predict a spin-crossover before the delivery of products (top graph of Fig. 2). Our GGA+U approach, instead, favors a $^4\phi$ state for FeO^+ (in agreement with CCSD(T)) and correctly reproduces the spin transition at the intermediate state 3. Relative energies of different intermediate and transition states along the reaction pathway (bottom graph of Fig. 2), structural and vibrational properties of reactant molecules were also obtained in excellent agreement with CCSD(T) (Kulik et al. 2006).

In a more recent study the "self-consistent" LDA+U was also used to describe the ground and low lying states of Fe heme complexes for the unligated configuration and in presence of a diatomic molecule coordinating the iron center (Scherlis et al. 2007). Besides their central role in organometallic chemistry, these compounds represent a clear case where LDA, GGA, and common hybrid functionals fail to reproduce the experimental magnetic splittings. The "self-consistent" GGA+U demonstrated higher accuracy than GGA, Hartree-Fock and some

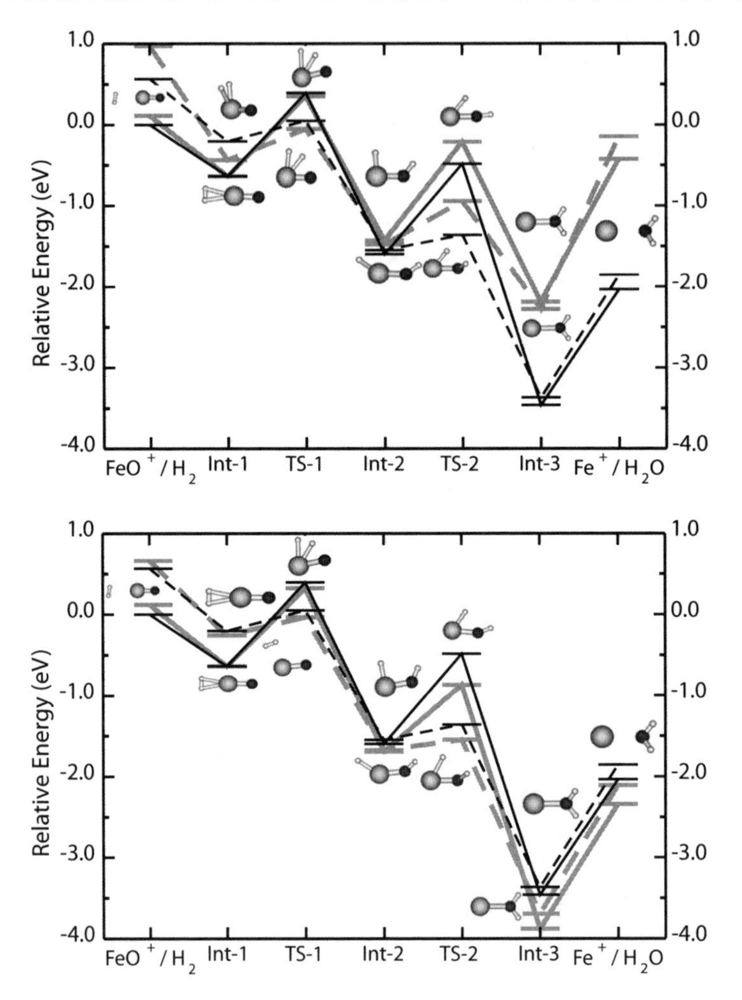

Figure 2. Potential energy surface and geometries for the $FeO^+ + H_2$ reaction using GGA (*top*) and GGA+U (*bottom*) (thick gray lines) as compared against CCSD(T) reference calculations (thin black lines). Solid lines indicate sextet while dashed lines indicate quartet spin states. Figure is from Kulik et al. (2006).

commonly used hybrid functionals, (e.g., B3LYP), and predicted spin-transition energies and molecular geometries in quantitative agreement with experiments for both the considered configurations (Scherlis et al. 2007). These results strongly point to the "self-consistent" LDA+U as an excellent tool to describe organometallic complexes. In fact, because of the predictive power demonstrated in the cases described above and the reduced computational costs (compared to quantum chemistry approaches), the "self-consistent" LDA+U method and, in particular, its application to iron molecular complexes have fixed a standard to which other functionals, being developed for DFT calculations, are compared (Sorkin et al. 2008).

TRANSITION-METAL-CONTAINING MINERALS

Studying Fe-bearing minerals is crucial to understand the physical behavior of the Earth's interior. Unfortunately DFT is not able to capture the physics of these systems due to the presence of d valence electrons for which correlation plays a crucial role. While the use of LDA+U for this class of materials is not new (Fang et al. 1998, 1999; Gramsh et al. 2003), the self-consistent evaluation of the interaction parameter (the Hubbard U) proved to be very important to characterize the variation of their properties with pressure and to predict structural and magnetic transitions in agreement with experimental results (Cococcioni and de Gironcoli 2005; Tsuchiya et al. 2006; Hsu et al. 2009).

For Fe_2SiO_4 fayalite, end-member of the Mg-Fe olivine mineral $Mg_xFe_{1-x}SiO_4$, particularly abundant in the Earth's upper mantle, standard approximations to DFT, while providing a description of the magnetic and structural properties in reasonable agreement with available experimental data (Cococcioni and de Gironcoli 2005; Cococcioni et al. 2003), fail to reproduce the observed insulating state and predicts metallic properties. Favoring the complete localization of Fe d electrons, the LDA+U corrective scheme is able to fix this problem and to reproduce the Mott-insulating electronic ground state. The precise evaluation of the Hubbard U as a function of atomic positions and cell parameters was crucial to obtain quantitatively accurate results and to predict the electronic energy band gap, the crystal structure and the ground state magnetic order in excellent agreement with available experiments (Cococcioni and de Gironcoli 2005).

In the case of FeO, the Fe-rich end-member of the $Mg_xFe_{1-x}O$ iron-magnesium wüstite, one of the most abundant minerals of the lower mantle of the Earth, the application of the LDA+U corrective scheme was even more critical to obtain a description of this system consistent with experimental results. In fact, standard DFT approximations, not only predict a metallic ground state (in contrast with the observed insulating character), but also overestimates the rhombohedral distortion this material undergoes under the effect of pressure. A better account for electronic correlation and the precise evaluation of its strength (the Hubbard U) as a function of volume were essential to fix the description of both these aspects. Mott localization of Fe d electrons, induced by the Hubbard correction to the Hamiltonian, resulted in a band gap in quantitative agreement with the one measured in experiments (Balberg and Pinch 1978). An orbital-ordered electronic ground state (Fig. 3, top and middle-right panels) was also found that, stabilized by correlation (and thus not accessible within standard DFT), is responsible for the structural deformation of this material (Cococcioni and de Gironcoli 2005). In fact, the localization of the minority-spin d electrons of Fe ions on the triangular (111) plane suppresses their overlap with O p states and determines the elongation of the structure along the [111] direction. Within this ground state the rhombohedral distortion of FeO under pressure was predicted in excellent quantitative agreement with available experiments (Willis and Rooksby 1953; Yagi et al. 1985) (Fig. 3, graph at the bottom).

In a more recent work Tsuchiya et al. (2008) used LDA+U to study the pressure-driven spin transition in magnesiowüstite $(Mg_{1-x}Fe_x)O$. Authors showed that the High-Spin (HS) to Low-Spin (LS) transition in this material takes place smoothly as pressure increases with a continuous variation of properties as observed in experiments. This smooth transition to the LS, lower-volume phase is realized through the conversion of a larger and larger fraction of Fe ions from a HS to a LS configuration. The LDA+U approach was used in this study with the Hubbard U consistently evaluated as a function of pressure for both spin configurations. The use of this scheme was crucial to obtain a reliable energetics of the two spin phases and to predict the structural reorganization (e.g. the distortion of the O octahedron cage around each Fe) accompanying the spin transition. Furthermore, appreciating the variation of this parameter with pressure allowed to obtain an accurate estimate of the spin state of the material (i.e., the number of HS ions vs. LS ones) across a broad range of pressure and temperature conditions characteristic of the lower mantle of the Earth.

A further confirmation of the importance of the consistent evaluation of the Hubbard U comes from the work by Hsu et al. (2009) where the low-spin ground state of $LaCoO_3$

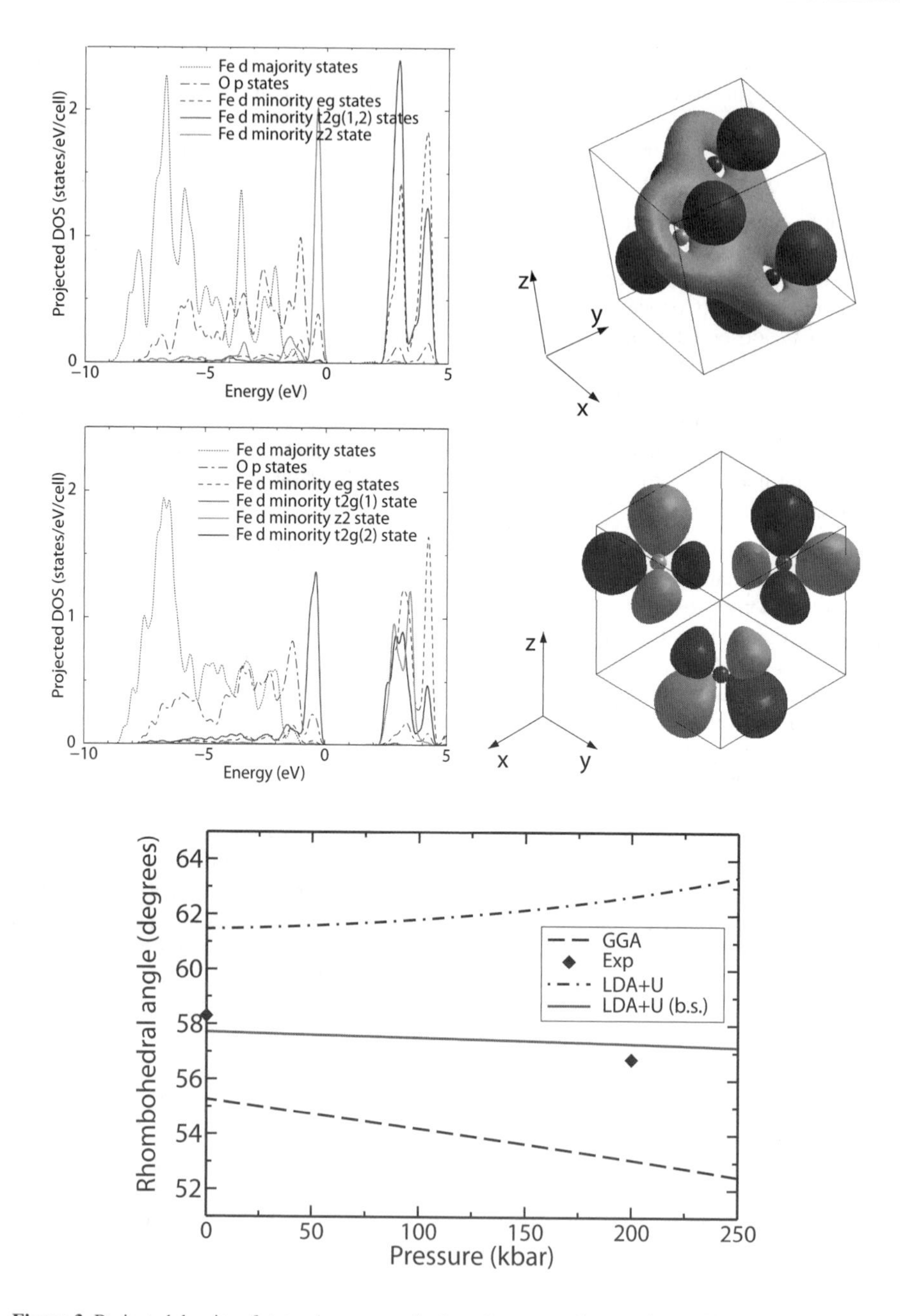

Figure 3. Projected density of states (*upper panels*; for colors see online version or Fig. 12 in Cococcioni and de Gironcoli 2005) and highest energy occupied orbital of FeO (*upper panels*) in the unbroken symmetry (*top-row panels*) and broken symmetry states (*middle-row panels*). The graph at the bottom plots the rhombohedral angle as a function of pressure. Solid black line describes LDA+U results in the broken-symmetry phase (from Cococcioni and de Gironcoli 2005). Diamonds represent the result of the extrapolation to 0 GPa of the experimental data from Willis and Rooksby (1953) and Yagi et al. (1985). Figure is from Cococcioni and de Gironcoli (2005).

was considered. This paper investigates, in particular, the structural reorganization of this material under pressure comparing the results with available experimental data. The Hubbard interaction parameter, in this case, is computed self-consistently with the structure of the system through an iterative procedure that alternates a calculation of the effective interaction parameter and an LDA+U structural relaxation (using the last obtained value of U) that is cycled until both structural and electronic parameters reach convergence. In general, the more accurate description of electronic localization through the LDA+U is very important to refine the bonding properties of the material (especially around the Co site) and produces structural parameters (e.g., the bonding angles) in closer agreement with experiment than GGA. The accurate evaluation of the variation of U with the volume of the system is also used to refine the calculation of pressure (the derivative of the total energy $E(V)$) and this contribution proves to be very significant to obtain the electronic and the structural properties as functions of pressure and the equation of state of the material in excellent agreement with experiments.

EXTENDED LDA+U+V APPROACH

The standard LDA+U corrective functional contains only on-site (intra-atomic) electron-electron interactions. This simplification of the original Hubbard model stems from the idea that electronic correlation is generally associated with Mott localization of electrons on atomic orbitals (Austin and Mott 1970). In fact, the corrective "+U" energy functional described in Equation (9) forces the occupations of atomic states to be either 1 or 0 and suppresses the hybridization between orbitals belonging to different atomic sites. However, this could represent a quite drastic approximation for some materials whose ground state is characterized by the formation of covalent bonds or, in general, by more itinerant electronic states. In a recent work (Tomczak and Biermann 2007), it has been argued, for example, that the physics of the low-temperature phase of VO_2 is a combination of Mott (correlated) and Peierls (band) scenarios. In NiO the hybridization between the d states of transition metal ions and the p states of oxygen is largely responsible for the charge-transfer character of this material and controls other features of its electronic structure as well (e.g., the position of satellite peaks of the photoemission spectrum) (Miura and Fujiwara 2008). Although the subject is still matter of debate (Anderson 1987; Anderson et al. 1987), some numerical studies conducted with model Hamiltonians have shown that in high T_c superconductors the electronic interactions between electrons on different atomic sites play an important role in determining the pairing mechanism at the origin of superconductivity (Hirsch et al. 1988; Imada 1991; Jursa et al. 1996; Szabó and Gulácsi 1996; de Mello 1999; Thakur and Das 2007); thus, a numerical approach based on the on-site Hubbard model is probably unable to describe effectively the physics of these systems.

To eliminate the over-suppression of hybridization and define a computational scheme that, while able to address Mott localization of electrons on atomic orbitals, can also accurately describe more delocalized ground states, we recently reformulated the corrective Hamiltonian using the extended Hubbard model (Hubbard 1965, 1967) with on-site and inter-site electronic interactions. The novel corrective functional reads:

$$E_{U+V} = E_{Hub} - E_{dc} = \sum_I \frac{U^I}{2} Tr\left[\mathbf{n}^I\left(1-\mathbf{n}^I\right)\right] - \sum_{IJ} \frac{V^{IJ}}{2} Tr\left(\mathbf{n}^{IJ}\mathbf{n}^{JI}\right) \tag{39}$$

where $n_{m,m'}^{IJ\sigma} = \Sigma_{k,v} f_{kv}^{\sigma} \langle\phi_m^I|\psi_{kv}^{\sigma}\rangle\langle\psi_{kv}^{\sigma}|\phi_{m'}^J\rangle$ and I and J denote distinct atomic sites. The inter-site electronic interaction V^{IJ} favors the occupation of states with finite components on neighbor atoms (hybridized). This tendency competes with the occupation of atomic orbitals stabilized by the on-site coupling U^I. Thus, more general localization regimes (e.g. on molecular orbitals) can set in depending on the detailed balance between the two electronic parameters.

Fortunately, the inter-site coupling V^{IJ} can be computed simultaneously to the on-site U^I and *at no additional cost* (in fact, it is the off-diagonal term of the interaction matrix given in Eqn. 37). The new functional has been used to calculate the electronic and the structural properties of quite diverse systems as NiO (strongly-correlated, charge-transfer insulator), Si and GaAs (hybridized, band semiconductors). The electronic structure of NiO has been thoroughly investigated in a large number of papers (Bengone et al. 2000; Dudarev et al. 2000; Moreira et al. 2002; Feng and Harrison 2004; Rohrbach et al. 2004; Ren et al. 2006; Kunes et al. 2007; Miura and Fujiwara 2008; Gougoussis et al. 2009). In these works it was shown that LDA+U or higher order approximations are needed to obtain an accurate description of the correlated ground state and, in particular, of its charge-transfer insulating character and its spectroscopic properties. For Si and GaAs, approximate DFT functionals provide structural (Filippi et al. 1994; Juan et al. 1995; Nielsen and Martin 1985) and vibrational properties (Giannozzi et al. 1991) in excellent agreement with experiments; however, the accurate representation of the band structure requires higher order approximations (Rohlfing et al. 1993; Städele et al. 1999; Aulbur et al. 2000a,b; Heyd et al. 2005; Nguyen and de Gironcoli 2009). For all these materials GGA+U+V results in significant improvements over GGA and on-site GGA+U. The extended GGA+U+V approach results in a band gap (1.36 eV and 0.9 eV for Si and GaAs respectively) in much better agreement with the experimental value (1.17 eV and 1.42 eV) than GGA (0.64 eV and 0.19 eV). On-site-only GGA+U produces, instead, a band gaps of 0.39 and 0.0 eV for Si and GaAs respectively. Similar improvement in the agreement with experiments is also obtained for the band gap of NiO (with respect to GGA) and for the structural properties (equilibrium lattice parameter and bulk modulus) of all systems as shown in Table 1. The encouraging results shown above, demonstrate that systems as diverse as band and Mott/charge-transfer insulators can be accurately described by the same computational approach. LDA+U+V thus represents a general, unifying theoretical framework able to effectively capture very different electronic localization regimes and, in particular, many intermediate situations that are usually the most challenging ones. The presence of inter-site interactions also represents a first step towards a covariant formulation and avoids the *ad-hoc* construction of a localized basis set sometimes used to address the localization of electrons on, e.g., molecular orbitals (Kresse et al. 2003; Köhler and Kresse 2004; Dabo et al. 2007). Many systems where correlated orbitals are significantly hybridized with extended ones as, for example, in semiconductors doped with magnetic impurities, or phenomena where transitions between different localization regimes take place (as, e.g., in the formation or collapse of chemical bonds) will be described with higher accuracy than possible with DFT or LDA+U.

Table 1. Data from Campo and Cococcioni (2010). The equilibrium lattice parameter, (a, in Å), the bulk modulus (B, in GPa), and the band gap (E_g, in eV) of NiO, Si and GaAs. Results obtained from different numerical approaches (GGA, GGA+U, GGA+U+V) are compared with experimental data [^{a} = Lide (1998); ^{b} = Huang et al. (1994); ^{c} = Fujimori and Minami (1984) and Sawatzky and Allen (1984)]. Experimental data for Si and GaAs were obtained from Levinstein et al. (1996, 1999). Data from calculations is adapted from Campo and Cococcioni (2010).

	NiO			Si			GaAs		
	a	B	E_g	a	B	E_g	a	B	E_g
GGA	4.20	188	0.6	5.47	83	0.64	5.774	58.4	0.19
GGA+U	4.27	181	3.2	5.363	93.9	0.39	5.736	52.6	0.00
GGA+U+V	4.23	197	3.2	5.370	102.5	1.36	5.654	67.7	0.90
Exp	4.18^a	166-208^b	3.1-4.3^c	5.431	98.0	1.17	5.653	75.3	1.42

CONCLUSIONS AND OUTLOOK

LDA+U is nowadays one of the most widespread computational schemes to perform *ab initio*, DFT-based calculations on transition-metal bearing minerals and correlated systems in general. The popularity of this approach is certainly due to the modest computational cost (roughly the same of standard DFT approximations) and to the simplicity of the corrective functional it is based on that guarantees the possibility to easily implement and compute energy derivatives, like atomic forces and stresses, that are key quantities to predict the structural properties and the possible phase transitions taking place in the interior of the Earth. Despite the LDA+U approach has been evolved to a quite flexible and versatile computational approach and has often shown a remarkable improvement over standard DFT approximations in predicting the properties of strongly correlated materials, there is still a large space for improvement in the formulation of the corrective functional. In the following I will describe the extensions to the LDA+U energy functional that would be particularly important for the study of minerals in the Earth's interior.

The specific focus this kind of investigation poses on structural phase transitions requires very reliable energetics (e.g., to assess the relative stability of different phases). This relies on two factors: the precise evaluation of the interaction parameters and the definition of the double-counting term of the corrective functional. While the linear-response approach to calculate the electronic couplings is a well consolidated and reliable technique, the definition of double-counting term represents a more delicate part of the theory and no formulation to date has been able to conform the very different degrees of electronic localization. To reach a higher level of flexibility requires a more precise mapping of the DFT energy functional and a more sensitive evaluation of the size of the correlation hole around each electron on "Hubbard" atoms. A more consistent evaluation of the Hubbard U and V would also be beneficial to obtain reliable energetics; in fact, the electronic interaction parameters should be evaluated in close consistency with the equilibrium electronic and crystal structures. This last condition requires the interaction parameter to be calculated from a correlated (LDA+U) ground state. While a method for the self-consistent evaluation of U has been developed by Kulik et al. (2006) the procedure needs to be optimized and significantly simplified to treat systems with different Hubbard species.

The possible (and quite common) anisotropy of crystals is another reason for further improvement as the present functional only includes one electronic interaction parameter per Hubbard site or between each pair of sites (in the case of LDA+U+V). A more refined formulation of the functional, able to resolve, the difference in the interaction between electrons on orbitals of different symmetry (e.g., t_{2g} and e_g states of a transition metal in octahedral coordination with oxygen) is needed to give a more accurate description of anisotropic systems and will be based on the exploitation of symmetry to evaluate the relative weight of different terms of the interaction potential. The fully orbital-dependent formulation should be constructed, in particular, in a strictly covariant fashion to eliminate any dependence of results on the particular choice of the basis set of localized orbitals used to define atomic occupations. This condition will require the efficient evaluation of all the (symmetry-independent) interaction parameters. A method to efficiently compute the necessary electronic couplings at runtime would be, in this context, highly desirable as it would significantly speed up this kind of simulations and increase their reliability.

Finally some algorithms and functionals already developed for standard DFT functionals should be extended to the LDA+U Hamiltonian. Two of them would be particularly useful for the study of transition-metal minerals: the development of a non-collinear magnetic formulation and the extension of Density-Functional-Perturbation-Theory to the LDA+U Hamiltonian to compute vibrational frequencies and modes from the correlated ground state of these systems. Both lattice vibrations and magnetic fluctuations (magnons) are very important low-energy excitations and their calculation is a fundamental ingredient to predict the behavior of these systems at finite (and high) temperature and pressure. The implementation of non-collinear spin formalism would allow, in particular, for the proper treatment of effects related to the interplay between structural and magnetic degrees of freedom that can be important to understand the origin of specific transition in the planet interior.

In summary LDA+U is a method of enormous potential to perform efficient and reliable calculations on correlated systems; the extensions proposed above are crucial to further improve its accuracy, especially for TM-containing minerals, and their development will be the object of an intense future research activity.

ACKNOWLEDGMENTS

The author is grateful to the Minnesota Supercomputing Institute for providing computational resources to develop the extended LDA+U+V functional. Partial suport from NSF grant EAR-0810272 is also gratefully acknowledged.

REFERENCES

Anderson PW (1987) The resonating valence bond state in La_2CuO_4 and superconductivity. Science 235:1196-1198

Anderson PW, Baskaran G, Zou Z, Hsu T (1987) Resonating-valence-bond theory of phase transitions and superconductivity in La_2CuO_4-based compounds. Phys Rev Lett 58:2790-2793

Anisimov VI, Gunnarson O (1991) Density-functional calculation of effective Coulomb interactions in metals. Phys Rev B 43:7570-7574

Anisimov VI, Aryasetiawan F, Liechtenstein AI (1997) First-principles calculations of the electronic structure and spectra of strongly correlated systems: the LDA+U method. J Phys Condens Matt 9:767-808

Aulbur WG, Jönsson L, Wilkins JW (2000a) Quasiparticle calculations in solids. Solid State Physics 54:1-218

Aulbur WG, Städele M, Görling A (2000b) Exact-exchange-based quasiparticle calculations. Phys Rev B 62:7121-7132

Austin IG, Mott NF (1970) Metallic and nonmetallic behavior in transition metal oxides. Science 168:71

Balberg I, Pinch HL (1978) The optical absorption of iron oxide. J Magn Magn Mater 7:12-15

Becke AD (1988) Density-functional exchange-energy approximation with correct asymptotic behavior. Phys Rev A 38:3098-3100

Bengone O, Alouani M, Blöchl P, Hugel J (2000) Implementation of the projector augmented-wave LDA+U method: Application to the electronic structure of NiO. Phys Rev B 62:16392-16401

Campo VL, Cococcioni M (2010) Extended DFT+U+V method with on-site and inter-site electronic interactions. J Phys Condens Matter 22:055602

Ceperley DM, Alder BJ (1980) Ground state of the electron gas by a stochastic method. Phys Rev Lett 45:566-569

Cococcioni M, de Gironcoli S (2005) Linear response approach to the calculation of the effective interaction parameters in the LDA+U method. Phys Rev B 71:35105-35120

Cococcioni M, Corso AD, de Gironcoli S (2003) Structural, electronic, and magnetic properties of Fe_2SiO_4 fayalite: Comparison of LDA and GGA results. Phys Rev B 67:94102-94109

Dabo I, Wieckowski A, Marzari N (2007) Vibrational recognition of adsorption sites for CO on platinum-and platinum-ruthenium surfaces. J Am Chem Soc 129:11045-11052

de Mello EVL (1999) The extended Hubbard model applied to phase diagram and the pressure effects in $Bi_2Sr_2CaCu_2O_{8+y}$. Brazilian J Phys 29:551-556

Delacourt C, Poizot P, Tarascon J-M, Masquellier C (2005) The existence of a temperature-driven solid solution in Li_xFePO_4 for $0 \le x \le 1$. Nat Mater 4:254-260

Dodd JL, Yazami R, Fultz B (2006) Phase diagram of Li_xFePO_4. Electrochem Solid State Lett 9:A151-A1155

Dudarev SL, Botton GA, Savrasov SY, Humphreys CJ, Sutton AP (1998) Electron-energy loss spectra and the structural stability of nickel oxide: An LSDA+U study. Phys Rev B 57:1505-1509

Dudarev SL, Peng L-M, Savrasov SY, Zuo J-M (2000) Correlation effects in the ground-state charge density of Mott insulating NiO: A comparison of ab initio calculations and high-energy electron diffraction measurements. Phys Rev B 61:2506-2512

Fang Z, Terakura K, Sawada H, Miyazaki T, Solovyev I (1998) Inverse versus normal NiAs structures as high-pressure phases of FeO and MnO. Phys Rev Lett 81:1027-1030

Fang Z,. Solovyev IV, Sawada H, Terakura K (1999) First-principles study on electronic structures and phase stability of MnO and FeO under high pressure. Phys Rev B 59:762-774

Feng X-B, Harrison NM (2004) Metal-insulator and magnetic transition of NiO at high pressures. Phys Rev B 69:035114-035117

Filippi C, Singh DJ, Umrigar CJ (1994) All-electron local-density and generalized-gradient calculations of the structural properties of semiconductors. Phys Rev B 50:14947-14951

Fujimori A, Minami F (1984) Valence-band photoemission and optical absorption in nickel compounds. Phys Rev B 30:957-971

Georges A, Kotliar G (1992) Hubbard model in infinite dimensions. Phys Rev B 45:6479-6483

Georges A, Kotliar G, Krauth W, Rozenberg M (1996) Dynamical mean-field theory of strongly correlated fermion systems and the limit of infinite dimensions. Rev Mod Phys 68:13-125

Giannozzi P, de Gironcoli S, Pavone P, Baroni S (1991) Ab initio calculation of phonon dispersions in semiconductors. Phys Rev B 43:7231-7242

Giannozzi P, Baroni S, Bonini N, Calandra M, Car R, Cavazzoni C, Ceresoli D, Chiarotti GL, Cococcioni M, Dabo I, Dal Corso A, de Gironcoli S, Fabris S, Fratesi G, Gebauer R, Gerstmann U, Gougoussis C, Kokalj A, Lazzeri M, Martin-Samos L, Marzari N, Mauri F, Mazzarello R, Paolini S, Pasquarello A, Paulatto L, Sbraccia C, Scandolo S, Sclauzero G, Seitsonen AP, Smogunov A, Umari P, Wentzcovitch RM (2009) Quantum espresso: a modular and open-source software project for quantum simulations of materials. J Phys Cond Matt 21(39):395502-395520; *http://www.quantum-espresso.org*

Gougoussis C, Calandra M, Seitsonen A, Brouder C, Shukla A, Mauri F (2009) Intrinsic charge transfer gap in NiO from Ni /it K-edge x-ray absorption spectroscopy. Phys Rev B 79:045118-045122

Gramsh SA, Cohen RE, Savrasov SY (2003) Structure, metal-insulator transitions, and magnetic properties of FeO at high pressures. Am Mineral 88:257-261

Heyd J, Peralta JE, Scuseria GE, Martin RL (2005) Energy band gaps and lattice parameters evaluated with the Heyd-Scuseria-Ernzerhof screened hybrid functional over a large benchmark set of solids. J Chem Phys 123:174101

Hirsch JE, Loch E, Scalapino DJ, Tang S (1988) Antiferromagnetism and superconductivity: Can a Hubbard U do it all by itself? Physica C 153-155:549-554

Hohenberg P, Kohn W (1964) Inhomogeneous electron gas. Phys Rev 36:B864-B871

Hsu H, Umemoto K, Cococcioni M, Wentzcovitch RM (2009) First-principles study for low-spin $LaCoO_3$ with a structurally consistent Hubbard U. Phys Rev B 79:125124-125132

Huang E, Jy K, Yu S-G (1994) Bulk modulus of NiO. J Geophys Soc China 37:7-17

Hubbard J (1965) Electron correlations in narrow energy bands. IV. The atomic representation. Proc R Soc London A 285:542-560

Hubbard J (1967) Electron correlations in narrow energy bands. V. A perturbation expansion about the atomic limit. Proc R Soc London A 296:82-99

Hubner O, Sauer J (2002) Confirmation of ${}^9\Sigma_g^-$ and ${}^8\Sigma_u^-$ ground states of Fe_2 and Fe_2^- by CASSCF/MRCI. Chem Phys Lett 358:442-448

Imada M (1991) Superconducting correlation of two-dimensional Hubbard model near half-filling. J Phys Soc Jpn 60:2740-2747

Juan Y-M, Kaxiras E, Gordon RC (1995) Use of the generalized gradient approximation in pseudopotential calculations of solids. Phys Rev B 51:9521-9525

Jursa R, Wermbter S, Czycholl G (1996) Superconductivity within the Extended Hubbard Model. *In:* Proceedings of the 21st International Conference on Low Temperature Physics, page 613

Köhler L, Kresse G (2004) Density functional study of CO on Rh(111). Phys Rev B 70:165405

Kohn W, Sham KJ (1965) Self-consistent equations including exchange and correlation effects. Phys Rev A 140:A1133-A1138

Kresse G, Gil A, Sautet P (2003) Significance of single-electron energies for the description of CO on Pt(111). Phys Rev B 68:073401-073404

Kulik HJ, Cococcioni M, Scherlis DA, Marzari N (2006) Density functional theory in transition-metal chemistry: a self-consistent Hubbard U approach. Phys Rev Lett 97:103001-103004

Kunes J, Anisimov VI, Skornyakov SL, Lukoyanov AV, Vollhardt D (2007) NiO: correlated band structure of a charge-transfer insulator. Phys Rev Lett 99:156404-156407

Lechermann F, Fähnle M, Meyer B, Elsässer C (2004) Electronic correlations, magnetism, and structure of Fe-Al subsystems: An LDA+U study. Phys Rev B 69:165116-165122

Levinstein M, Rumyantsev S, Shur M (eds) (1996, 1999) Handbook Series on Semiconductor Parameters. Volumes 1 and 2. World Scientific, London. *http://www.ioffe.ru/SVA/NSM/Semicond/*

Lide DR (ed) (1998) CRC Handbook of Chemistry and Physics. CRC, Boca Raton

Liechtenstein AI, Anisimov VI, Zaanen J (1995) Density-functional theory and strong interactions: Orbital ordering in Mott-Hubbard insulators. Phys Rev B 52:R5467-R5470

Marcus RA (1993) Electron transfer reactions in chemistry. Theory and experiment. Rev Mod Phys 65:599-610

Marcus RA, Sutin N (1985) Electron transfers in chemistry and biology. Biochim Biophys Acta 881:265-322

Miura O, Fujiwara T (2008) Electronic structure and effects of dynamical electron correlation in ferromagnetic bcc Fe, fcc Ni, and antiferromagnetic NiO. Phys Rev B 77:195124-195135

Moreira IPR, Illas F, Martin R (2002) Effect of Fock exchange on the electronic structure and magnetic coupling in NiO. Phys Rev B 65:155102-155115

Newton MD, Sutin N (1984) Electron transfer reactions in condensed phases. Annu Rev Phys Chem 35:437-480

Nguyen H-V, de Gironcoli S (2009) Efficient calculation of exact exchange and RPA correlation energies in the adiabatic-connection fluctuation-dissipation theory. Phys Rev B 79:205114-205125

Nielsen OH, Martin RM (1985) Stresses in semiconductors: Ab initio calculations on Si, Ge, and GaAs. Phys Rev B 32:3792-3805

Ortiz G, Ballone P (1994) Correlation energy, structure factor, radial distribution function, and momentum distribution of the spin-polarized uniform electron gas. Phys Rev B 50:1391-1405

Perdew JP (1991) Unified theory of exchange and correlation beyond the local density approximation. *In:* Electronic Structure of Solids '91. Proceedings of the 75th WE-Heraeus-Seminar. Ziesche P, Eschrig H (eds) Akademie Verlag, Berlin, p 11-20

Perdew JP, Zunger A (1981) Self-interaction correction to density-functional approximations for many-electron systems. Phys Rev B 23:5048-5079

Perdew JP, Burke K, Ernzheroff M (1996) Generalized gradient approximation made simple. Phys Rev Lett 77:3865-3868

Petukhov AG, Mazin II, Chioncel L, Liechtenstein AI (2003) Correlated metals and the LDA+U method. Phys Rev B 67:153106-153109

Ren X, Leonov I, Keller G, Kollar M, Nekrasov I, Vollhardt D (2006) LDA+DMFT computation of the electronic spectrum of NiO. Phys Rev B 74:195114-195121

Rohlfing M, Krüger P, Pollmann J (1993) Quasiparticle band-structure calculations for C, Si, Ge, GaAs, and SiC using Gaussian-orbital basis sets. Phys Rev B 48:17791-17805

Rohrbach A, Hafner J, Kresse G (2004) Molecular adsorption on the surface of strongly correlated transition-metal oxides: A case study for CO/NiO(100). Phys Rev B 69:075413-075425

Sawatzky GA, Allen JW (1984) Magnitude and origin of the band gap in NiO. Phys Rev Lett 53:2339-2342

Scherlis DA, Cococcioni M, Sit PH-L, Marzari N (2007) Simulation of Heme using DFT+U: a step toward accurate spin-state energetics. J Phys Chem B 111:7384-7391

Sit PH-L, Cococcioni M, Marzari N (2006) Realistic, quantitative descriptions of electron-transfer reactions: diabatic surfaces from first-principles molecular dynamics. Phys Rev Lett 97:028303-028306

Sit PH-L, Cococcioni M, Marzari N (2007) Car-Parrinello molecular dynamics in the DFT+U formalism: Structure and energetics of solvated ferrous and ferric ions. J Electroanalytical Chem 607:107-112

Sorkin A, Iron MA, Truhlar DG (2008) Density functional theory in transition-metal chemistry: relative energies of low-lying states of iron compounds and the effect of spatial symmetry breaking. J Chem Theory Comp 4:307–315

Städele M, Moukara M, Majewski JA, Vogl P, Görling A (1999) Exact exchange Kohn-Sham formalism applied to semiconductors. Phys Rev B 59:10031-10043

Szabó Z, Gulácsi Z (1996) Superconducting phases of the extended Hubbard model for doped systems. *In:* Proceedings of the 21st International Conference on Low Temperature Physics. Czech J Phys Suppl 46(2):609-610

Thakur JS, Das MP (2007) Superconducting order parameters in the extended Hubbard model: A simple MEAN-FIELD study. Int J Mod Phys B 21:2371-2383

Tomczak JM, Biermann S (2007) Effective band structure of correlated materials - the case of VO_2. J Phys.: Condens Matter 19:365206

Tsuchiya T, Wentzcovitch RM, da Silva CRS, de Gironcoli S (2006) Spin transition in Magnesiowüstite in earth's lower mantle. Phys Rev Lett 96:198501-198504

Warshel A, Parson WW (1991) Computer simulations of electron-transfer reaction in solution and in photosynthetic reaction centers. Annu Rev Phys Chem 42:279-309

Willis TM, Rooksby HP (1953) Change of structure of ferrous oxide at low temperature. Acta Crystallogr 6:827-831

Yagi T, Suzuki T, Akimoto S (1985) Static compression of wüstite ($Fe_{0.98}O$) to 128 GPa. J Geophys Res 90:8784-8788

Yamada A, Chung S-C (2001) Crystal chemistry of the olivine-type $Li(Mn_yFe_{1-y})PO_4$ and $(Mn_yFe_{1-y})PO_4$ as possible 4V cathode materials for lithium batteries. J Electrochem Soc 148:A960-A967

Yamada A, Chung S-C, Hinokuma K (2001) Optimized $LiFePO_4$ for lithium battery cathodes. J Electrochem Soc 148:A224-A229

Ylvisaker ER, Pickett WE, Koepernik K (2009) Anisotropy and magnetism in the LSDA+U method. Phys Rev B 79:35103-35114

Zhou F, Cococcioni M, Kang K, Ceder G (2004a) The Li intercalation potential of $LiMPO_4$ and $LiMSiO_4$ olivines with M = Fe, Mn, Co, Ni. Electrochem Commun 6:1144-1148

Zhou F, Cococcioni M, Marianetti AC, Morgan D, Ceder G (2004b) First-principles prediction of redox potentials in transition-metal compounds with LDA+U. Phys Rev B 70:235121-235128

Zhou F, Marianetti AC, Cococcioni M, Morgan D, Ceder G (2004c) Phase separation in Li_xFePO_4 induced by correlation effects. Phys Rev B 69:201101-201104

Zhou F, Maxisch T, Ceder G (2006) Configurational electronic entropy and the phase diagram of mixed-valence oxides: the case of Li_xFePO_4. Phys Rev Lett 97:155704-155707

Reviews in Mineralogy & Geochemistry
Vol. 71 pp. 169-199, 2010

Spin-State Crossover of Iron in Lower-Mantle Minerals: Results of DFT+*U* Investigations

Han Hsu

Department of Chemical Engineering and Materials Science
University of Minnesota
Minneapolis, Minnesota 55455-0132, U.S.A.

hsuhan@cems.umn.edu

Koichiro Umemoto

Department of Geology and Geophysics
University of Minnesota,
Minneapolis, Minnesota 55455-0132, U.S.A.

umemoto@cems.umn.edu

Zhongqing Wu

Department of Physics
University of South California
Los Angeles, California 90007, U.S.A.

zhongwu@usc.edu

Renata M. Wentzcovitch

Department of Chemical Engineering and Materials Science
University of Minnesota
Minneapolis, Minnesota 55455-0132, U.S.A.

wentzcov@cems.umn.edu

INTRODUCTION

The lower mantle is the largest layer of the Earth. Pressures and temperatures there vary from 23 GPa to 135 GPa and ~1,900 K to 4,000 K. Aluminous perovskite, Al-$Mg_{1-x}Fe_xSiO_3$, and ferropericlase, $Mg_{1-x}Fe_xO$, are the most important phases of the Earth's lower mantle, comprising ~62 vol% and ~35 vol% of this region, respectively. The remaining consists of $CaSiO_3$-perovskite, according to the pyrolitic compositional model (Ringwood 1982). Silicate perovskite transforms into another polymorph, post-perovskite, at conditions expected to occur near the D″ discontinuity in the deep slower mantle, i.e., 2,500 K and 125 GPa (Murakami et al. 2004; Oganov and Ono 2004; Tsuchiya et al. 2004; Wentzcovitch et al. 2006). The spin-state crossover (also referred to as "spin pairing transition" or "spin(-state) transition") of iron in ferropericlase under pressure was observed in 2003 by X-ray emission spectroscopy (XES) (Badro et al. 2003). In the following year a similar phenomenon was also identified in iron-bearing perovskite ($Mg_{1-x}Fe_xSiO_3$) by the same technique (Badro et al. 2004). This phenomenon had been predicted for decades (Fyfe 1960; Gaffney and Anderson 1973; Ohnishi 1978; Sherman 1988, 1991; Sherman and Jansen 1995; Cohen et al. 1997) but the pressure conditions were challenging for this type of experiment and it took several decades to observe it. The implications of this phenomenon for the properties of this region, or of the entire planet, are yet to be understood. Today, theoretical studies are

1529-6466/10/0071-0009$05.00 DOI: 10.2138/rmg.2010.71.09

making decisive contributions to this problem. This article reviews the main theoretical results on the spin-state crossover in the lower mantle phases.

Spin changes in strongly correlated oxides and silicates under pressure is the type of problem that has challenged electronic structure theory for decades. Ferrous (Fe^{2+}) and ferric (Fe^{3+}) irons are strongly correlated ions. The established first principles framework used to describe electronic states in solids, density functional theory (DFT) (Hohenberg and Kohn 1964; Kohn and Sham 1965), has not so far been able to explain some of the most fundamental properties of materials containing these ions: whether they are insulators or metals. For decades this challenge has inspired the development of new methods for electronic structure computations that may be reaching maturity just in time to address this problem. In this paper, we offer a review of theoretical/computational studies that have been performed to date on these two materials and analyze the performance of different approaches by contrasting results with each other and with experimental measurements. We also emphasize the geophysical implications of these results when possible. The results summarized here were obtained using standard DFT within the local density approximation (LDA) (Ceperley and Alder 1980; Perdew and Zunger 1981), the generalized gradient approximation (GGA) (Perdew et al. 1996), or the internally consistent DFT+U approach (Cococcioni and de Gironcoli 2005; Cococcioni et al. 2010). DFT, with its various approximations to exchange-correlation energy, is reviewed in this volume (Perdew and Ruzsinszky 2010; Zhao and Truhlar 2010). The DFT+U method (see Cococcioni 2010) is sufficiently developed to the point that can address structurally complex strongly correlated minerals. For a general review of the spin crossover in ferropericlase and perovskite, including experiments up to 2007-08, we refer to a recent paper, Lin and Tsuchiya (2008). Experimental information that has become available since the publication of this review and that is relevant for comparison with calculations is reviewed here.

THE SPIN-PAIRING PHENOMENON

The 5-fold degeneracy of the $3d$ atomic states of iron splits in a crystalline field. The nature and ordering of the new energy levels depend primarily on the atomic arrangement of the neighboring ions and their distances. A comprehensive study of the electronic states of transition metal ions in crystalline fields can be found in the book by Sugano et al. (1970). The electronic states of iron in Earth's forming silicates and oxides at ambient conditions are relatively localized. In stoichiometric compounds such as FeO, Fe_2O_3, etc., the magnetic moments of iron are usually ordered anti-ferromagnetically. The solid solutions of the lower mantle are not magnetically ordered. In ferropericlase and iron-bearing perovskite, the energetic order, symmetry, and occupation of the $3d$ orbitals of iron can be understood on the basis of crystal field symmetry and Hunds rule. Ferrous iron has six $3d$ electrons. In the high-spin (HS) state, five of them have spin +1/2 (up) and one has spin −1/2 (down). Their angular momentum is essentially quenched, spin orbit interaction is practically non-existent, and the total-spin quantum number, S, is equal to 2. The intermediate-spin (IS) state has four $3d$ electrons with spin up and two with spin down, for a total spin $S = 1$. The low-spin (LS) state, has three electrons up, three down, and $S = 0$. Ferric iron has five $3d$ electrons; the HS state has five spin up electrons, for $S = 5/2$; the IS and LS states have $S = 3/2$ and 1/2 respectively. The charge density distribution of the partially occupied $3d$ electrons is non-spherical and is related to the ordering of the energy levels established by the crystalline field and the number of $3d$ electrons. This non-spherical distribution distorts the first coordination shell of ligands, the so-called Jahn-Teller (J-T) effect.

In the Earth forming minerals we will review, iron is in the HS state at zero pressure. Under pressure, the Fe-O bond-length decreases, the crystal field increases, and the energy splitting between the d-orbital derived states increases. This leads to a change in the occupation of these orbitals, i.e., the spin pairing phenomenon characterized by a change in the occupation of the lowest lying *d*-orbitals and a change in the total spin S. How these occupations change under pressure is not obvious. In principle, spins could pair gradually, with a change of S from 2 to 1 to 0 in ferrous iron and from 5/2 to 3/2 to 1/2 in ferric iron. In fact, it depends on unpredictable factors (e.g., chemistry, stoichiometry, etc.) and it is often difficult to anticipate or explain the nature of these spin changes. For instance, spin changes in (Mg,Fe)SiO_3-perovskite is a current topic of controversy.

One should have in mind that in the Hubbard model, usually invoked to describe the electronic structure of these solids, it is the ratio U/Δ, where U is the Coulomb correlation energy and Δ is the *d*-bandwidth, that characterizes the correlation strength, with $U/\Delta \gg 1$ ($U/\Delta \ll 1$) corresponding to the strong (weak) correlation limit. Under pressure the band width Δ increases and the strength of the correlation decreases.

THEORETICAL APPROACH

The spin crossover systems we will review are paramagnetic solid solutions, insulators with localized moments on the iron site. Besides, the iron concentration, X_{Fe}, is relatively low ($X_{Fe} < 0.2$). There are a few challenges in addressing properties of these systems: description of their electronic structure, treatment of the solid solution problem, and calculation of their vibrational properties, without which no thermodynamics treatment is possible. The methods chosen to address these issues are outlined below.

Spin- and volume-dependent Hubbard *U*

The strongly correlated behavior of iron in the oxides and silicates of the mantle has deterred the use of standard DFT for predictive studies. DFT results are particularly poor for ferropericlase, which is predicted to be metallic with DFT, rather than insulator (Tsuchiya et al. 2006). The corrections for the electron-electron interaction on the iron sites can be included by using the so-called DFT+U method. Here we used a rotationally invariant version of the DFT+U approach implemented in the plane wave pseudopotential method, where U is calculated in an internally consistent way (Cococcioni and de Gironcoli 2005). Details of this method are also discussed by Cococcioni (2010) in this volume. In short, the total energy functional in the DFT+U is:

$$E_{DFT+U} = E_{DFT} + \frac{U}{2}\sum_{I,\sigma} Tr\left[\mathbf{n}^{I\sigma}\left(\mathbf{1}-\mathbf{n}^{I\sigma}\right)\right] \tag{1}$$

where E_{DFT} is the DFT ground-state energy of the structure, and $\mathbf{n}^{I\sigma}$ is the occupation matrix of the atomic site I with spin σ. The atomic-like orbitals used in defining $\mathbf{n}^{I\sigma}$ are arbitrary, but the calculations of U and E are consistent with this definition (Cococcioni and de Gironcoli 2005; Fabris et al. 2005). The Hubbard U is computed using linear response theory. In practice this procedure starts by computing the DFT ground state of a given structure. The occupation matrix $\mathbf{n}^{I\sigma}$ at site I is determined from the DFT ground state. The next step is to apply perturbations to the potential localized at the Hubbard site I. The perturbed states lead to different occupation matrices. The linear response of the occupation matrix $\mathbf{n}^{I\sigma}$ to the local potential shift is used to determine the Hubbard U. In this scheme, the value of U depends on both the spin state and the unit-cell volume (Cococcioni and de Gironcoli 2005; Tsuchiya et al. 2006; Hsu et al. 2009a).

The calculation of pressure in the DFT+U method presents an issue that is pervasive in this method: the calculation of gradients. Since U depends on volume, the pressure P determined by the negative derivative of the total energy E_{DFT+U} (see Eqn. 1) with respect to volume V has the following form:

$$P = -\frac{\partial E_{DFT}}{\partial V} - \frac{U}{2}\sum_{I,\sigma}\frac{\partial}{\partial V} Tr\left[\mathbf{n}^{I\sigma}\left(\mathbf{1}-\mathbf{n}^{I\sigma}\right)\right] - \frac{\partial U}{2\partial V}\sum_{I,\sigma} Tr\left[\mathbf{n}^{I\sigma}\left(\mathbf{1}-\mathbf{n}^{I\sigma}\right)\right] \tag{2}$$

The first two terms are obtained from a modified form of the first principles stress (Nielsen and Martin 1983) in a plane wave basis set. However, pressure depends also on the volume dependence of U (last term in Eqn. 2), which is not obtained *a priori*. The same problem arises in the calculation of other structural gradients. U should be structurally consistent and currently its gradient with respect to positions, strains, etc, is not calculated. Nevertheless, the first two terms in Equation (2) are calculated and they can be used to move atoms, in the case of forces, or change volume, in the case of stress/pressure, after which U can be recomputed until the structural degrees of freedom and U converge. Pressure and other structural gradients in this method are currently calculated numerically by finite differences from E_{DFT+U}, with structurally consistent U (Hsu et al. 2009a).

It should be mentioned that the Hubbard U can be determined more rigorously by perturbing the ground state of a series of DFT+U ground state with several chosen input Us. In this approach, the consistency between the response and the DFT+U ground states should be achieved. The Hubbard U determined this way is called self-consistent U (Kulik et al. 2006). This method, however, was not used in calculations reviewed here.

Thermodynamic treatment of the mixed spin state

At finite temperatures, ferropericlase or perovskite with iron concentration X_{Fe} can have irons in multiple spin states, i.e., in a mixed-spin (MS) state. Knowledge of the equilibrium population of each spin state at any given temperature and pressure is crucial to study the consequences of the spin crossover on their properties. Since X_{Fe} of these minerals in actual lower mantle are relatively small (~0.18 and ~0.12 for ferropericlase and perovskite respectively), we treat the ferropericlase and perovskite in the MS state as the ideal solid solution of the mixture of each pure spin state. (NOTE: this is not the same as a solid solution of MgO and FeO). The Gibbs free energy of the MS state is thus written as

$$G(n_\sigma,P,T)=\sum_\sigma n_\sigma(P,T)G_\sigma(P,T)+G^{mix}(P,T) \tag{3}$$

where n_σ and G_σ are the fraction and Gibbs free energy of each spin state, respectively. The subscript σ denotes spin state, i.e., LS, IS, HS, etc. The term G^{mix} results from the entropy of mixing solution, and it is given by

$$G^{mix}=k_BTX_{Fe}\sum_\sigma n_\sigma \log n_\sigma \tag{4}$$

The Gibbs free energy of each spin state G_σ has three contributions:

$$G_\sigma=G_\sigma^{stat+vib}+G_\sigma^{mag}+G_\sigma^{site} \tag{5}$$

where $G_\sigma^{stat+vib}$ is the Gibbs free energy containing static and vibrational contributions. The calculation of the vibrational free energy will be addressed in the next subsection. If this contribution is disregarded, $G_\sigma^{stat+vib}$ reduces to the static enthalpy of spin state σ,

$$G_\sigma^{stat+vib}=H_\sigma \tag{6}$$

The second term in Equation (5) derives from the magnetic entropy,

$$G_\sigma^{mag}=-k_BTX_{Fe}\log\left[m_\sigma(2S_\sigma+1)\right] \tag{7}$$

where m_σ and S_σ are the orbital degeneracy and the spin quantum number of each spin state, respectively. The third term comes from the site entropy

$$G_\sigma^{site}=-k_BTX_{Fe}\log N_\sigma^{site} \tag{8}$$

where N_σ^{site} is the number of equivalent equilibrium sites for irons in each spin state.

By minimizing the total Gibbs free energy (Eqn. 3) with respect to n_σ under the constraint

$$\sum_\sigma n_\sigma = 1 \tag{9}$$

the following expressions for each spin state are then obtained:

$$n_\sigma(P,T) = n_{HS} \times \frac{N_\sigma^{site} m_\sigma (2S_\sigma + 1)}{N_{HS}^{site} m_{HS} (2S_{HS} + 1)} \exp\left(-\frac{\Delta G_\sigma^{stat+vib}}{k_B T X_{Fe}}\right) \quad \text{for } \sigma \neq \text{HS}$$

$$n_{HS}(P,T) = \left[1 + \frac{N_{LS}^{site} m_{LS} (2S_{LS} + 1)}{N_{HS}^{site} m_{HS} (2S_{HS} + 1)} \exp\left(-\frac{\Delta G_{LS}^{stat+vib}}{k_B T X_{Fe}}\right) + \frac{N_{IS}^{site} m_{IS} (2S_{IS} + 1)}{N_{HS}^{site} m_{HS} (2S_{HS} + 1)} \exp\left(-\frac{\Delta G_{IS}^{stat+vib}}{k_B T X_{Fe}}\right)\right]^{-1} \tag{10}$$

where $\Delta G_\sigma^{stat+vib} \equiv G_\sigma^{stat+vib} - G_{HS}^{stat+vib}$. Again, when the lattice vibrational contribution is disregarded, we have $\Delta G_\sigma^{stat+vib} = \Delta H_\sigma \equiv H_\sigma - H_{HS}$.

The vibrational free energy: the Vibrational Virtual Crystal Model (VVCM)

The quasiharmonic approximation (QHA) (Wallace 1972) is a good approximation for most mantle minerals at relevant conditions. Its performance is reviewed in this volume (Wentzcovitch et al. 2010). Within the QHA, the free energy is given by:

$$F(V,T) = \left[U(V) + \sum_{\mathbf{q},j} \frac{\hbar\omega_{\mathbf{q},j}(V)}{2}\right] + k_B T \sum_{\mathbf{q},j} \log\left[1 - \exp\left(-\frac{\hbar\omega_{\mathbf{q},j}(V)}{k_B T}\right)\right] \tag{11}$$

where $U(V)$ and $\omega_{\mathbf{q},j}(V)$ are the volume-dependent static internal energy and the phonon spectrum. Once $F(V,T)$ is known, other thermodynamic quantities, such as pressure and Gibbs free energy, can be derived. For the systems that can be treated using standard DFT, such as MgO, the interatomic force constants $D_{\mu\nu}^{ij}$, which is defined as

$$D_{\mu\nu}^{ij} = \frac{\partial^2 E}{\partial R_i^\mu \partial R_j^\nu} \tag{12}$$

can be calculated by first principles using density-functional perturbation theory (DFPT) (Baroni et al. 2001). In Equation (12), R_i^μ and R_j^ν denote the μth and ν^{th} Cartesian components of the positions of atom i and j, respectively. With the computed interatomic force constants, the phonon dispersion relation can be determined by the formula

$$\det\left|\frac{\tilde{D}_{\mu\nu}^{ij}(\mathbf{q})}{\sqrt{M_i M_j}} - \omega_{\mathbf{q}}^2\right| = 0 \tag{13}$$

where the dynamical matrix $\tilde{D}_{\mu\nu}^{ij}(\mathbf{q})$ is defined as

$$\tilde{D}_{\mu\nu}^{ij}(\mathbf{q}) \equiv \sum_{i,j} \exp\left[-i\mathbf{q}\cdot(\mathbf{R}_\mathrm{i} - \mathbf{R}_\mathrm{j})\right] D_{\mu\nu}^{ij} \tag{14}$$

Unfortunately, DFPT is not currently implemented in combination with DFT+U. Moreover, calculations of vibrational density of states (VDoS) of solid solutions are not straightforward. Therefore, another reasonably accurate approach to calculate the vibrational VDoS of these systems is needed. Wu et al. (2009) developed a vibrational virtual-crystal model (VVCM) to calculate the phonon spectrum and thermodynamic properties of ferropericlase at temperatures up to 4000 K and pressures up to 150 GPa.

In the VVCM, the cations forming the solid solution, i.e., magnesium and iron, are replaced by an "average" cation that reproduces the same vibrational properties of the solid solution,

$$M_{VC}^{cation} = (1 - X_{Fe})M_{Mg} + X_{Fe}M_{Fe} \quad (15)$$

where M_{Mg} and M_{Fe} are the mass of magnesium and iron, respectively. Therefore, the VVCM has only two atoms per unit cell of the rocksalt structure. The next question is how to determine the interatomic force constants between the atoms in the virtual crystal. Within DFT+U, the elastic constants of the actual solid solution, $C_{\sigma\tau\alpha\beta}$, can be calculated using stress-strain relation. The acoustic wave velocities, v, along several directions can then be determined using Crystoffel's equation,

$$\det\left|C_{\sigma\tau\alpha\beta}n_\tau n_\beta - \rho v^2\delta_{\sigma\alpha}\right| = 0 \quad (16)$$

where $\mathbf{n}$, ρ, and δ are the wave propagation direction, mass density, and Kroenecker delta, respectively. Starting from the elastic constants of MgO, its largest nearest neighbor interatomic force constants of the virtual crystal are varied. The dynamical matrix, phonon dispersions, and the acoustic wave velocities

$$v[\mathbf{n}] = \frac{d\omega_{\mathbf{q}}}{d\mathbf{q}}[\mathbf{n}]\Big|_{\mathbf{q}=0} \quad (17)$$

vary accordingly. The appropriate interatomic force constants of the virtual crystal are those that give the wave velocities (Eqn. 17) matching the wave velocities of the actual solid solution (Eqn. 16). With this set of appropriately chosen interatomic force constants, the phonon spectrum and the vibrational density of states can be determined accordingly. The thermodynamic properties of ferropericlase calculated using this method will be shown in the next section.

Computational details

In ferropericlase (Tsuchiya et al. 2006; Wentzcovitch et al. 2009; Wu et al. 2009), the computations were performed using the LDA (Ceperley and Alder 1980; Perdew and Zunger 1981). The oxygen pseudopotential was generated by the method of Troullier-Martins (Troullier and Martins 1991) with core radii $r(2s) = r(2p) = 1.45$ a.u in the configuration $2s^2 2p^4$ with p locality. The magnesium pseudopotential was generated by the method of von Barth-Car. Five configurations, $3s^2 3p^0$, $3s^1 3p^1$, $3s^1 3p^{0.5} 3d^{0.5}$, $3s^1 3p^{0.5}$, and $3s^1 3d^1$ with decreasing weights 1.5, 0.6, 0.3, 0.3, and 0.2, respectively, were used. Core radii were $r(3s) = r(3p) = r(3d) = 2.5$ a.u. with d locality. The ultrasoft pseudopotential for iron was generated using the modified Rappe-Rabe-Kaxieas-Joannopoulos (RRKJ) scheme (Rappe et al. 1990). Core radii were $r(4s) = (2.0, 2.2)$, $r(4p) = (2.2, 2.3)$, and $r(3d) = (1.6, 2.2)$ a.u. in the configuration $3d^7 4s^1$, where the first and second numbers in each parenthesis represent the norm-conserving core radius and ultrasoft radius, respectively. All-electron potential pseudized at $r_c < 1.7$ a.u. was taken as a local potential. The plane wave energy cutoff was 70 Ry. Brillouin zone sampling for electronic states was carried out on 8 k-points for the cubic supercell containing 64 atoms. Equivalent k-point mesh were used in 128 and 216 atom calculations for ferropericlase with $X_{Fe} = 0.0625$ and 0.03125.

For $(Mg,Fe)SiO_3$ perovskite (Umemoto et al. 2008, 2009; Hsu et al. 2009b), the computations used both LDA (Ceperley and Alder 1980; Perdew and Zunger 1981) and (PBE) GGA (Perdew et al. 1996). The pseudopotentials for Fe, Si, and O were generated by Vanderbilt's method (Vanderbilt 1990). The valence electronic configurations used were $3s^2 3p^6 3d^{6.5} 4s^1 4p^0$, $3s^2 3p^1$, and $2s^2 2p^4$ for Fe, Si, and O, respectively. The pseudopotential for Mg is the same as the one used for ferropericlase. The plane wave cutoff energy was 40 Ry. In each supercell being used, the k-point mesh is fine enough to achieve convergence within 1 mRy/Fe in the total energy. Variable-cell-shape molecular dynamics (Wentzcovitch et al.

1993) implemented in QUANTUM-ESPRESSO package (Giannozzi et al. 2009) is used for structural optimizations. In Hsu et al. (2009b), the electric field gradient (EFG) tensors are calculated using the augmented plane wave + local orbitals (APW+lo) method (Madsen et al. 2001) implemented in the WIEN2k code (Blaha et al. 2001), and they are converted to the QS values using Q = 0.16 barn (Petrilli et al. 1998).

SPIN-STATE CROSSOVER IN FERROPERICLASE

The pressure-induced spin-state crossover (HS-to-LS) of iron in ferropericlase was already observed in many types of experiments, including experiments that probe the electronic states of ferropericlase, such as X-ray emission spectroscopy (XES) (Badro et al. 2003; Lin et al. 2005, 2006a, 2007a,c; Vanko and de Groot 2007) and X-ray absorption near-edge spectroscopy (XANES) (Kantor et al. 2009; Narygina et al. 2009), and experiments that probe the asymmetry of electron charge distribution around the iron nucleus, such as Mössbauer spectroscopy (Gavriliuk et al. 2006; Kantor et al. 2006, 2007, 2009; Lin et al. 2006a; Speziale et al. 2005). Several physical quantities of ferropericlase are affected by the spin-state crossover, including the enhancement of density (reduction of volume) observed in X-ray diffraction (Fei et al. 2007; Lin et al. 2005; Speziale et al. 2005, 2007), variation of sound velocity observed in nuclear resonant inelastic X-ray scattering (Lin et al. 2006b), enhanced absorption in the mid- and near-infrared range (Goncharov et al. 2006; Keppler et al. 2007), change of elastic properties observed using impulsive stimulated scattering (Crowhurst et al. 2008), and decrease of electrical conductivity (Lin et al. 2007b). Using Mössbauer spectroscopy, the effect of temperature and pressure on the quadrupole splitting of the iron nucleus was studied as well (Kantor et al. 2009; Lin et al. 2009). The transition pressures observed in these experiments are not exactly the same, depending on both the iron concentration and the experimental techniques. The room-temperature XES spectra of samples with different iron concentration show that the transition pressure increases with iron concentration (Lin et al. 2005, 2006a, 2007a; Vanko and de Groot 2007). For the various experiments (XES, Mossbauer, and optical absorption) with iron concentration close to that in the lower mantle (17%), the transition pressure at room temperature mainly ranges from 50 to 70 GPa (Goncharov et al. 2006; Kantor et al. 2006; Lin et al. 2005, 2006a; Speziale et al. 2005; Vanko and de Groot 2007). A lower transition pressure of 30 GPa, however, is also observed in the X-ray diffraction experiment by Fei et al. (2007).

The dependence of the (static) transition pressure on iron concentration was studied theoretically using GGA+U (Persson et al. 2006) and LDA+U methods (Tsuchiya et al. 2006). In Persson et al. (2006), high iron concentrations ($X_{Fe} > 0.25$) were considered. The Hubbard U was chosen to be 3.0 and 5.0 eV. Irrespective of U, the transition pressure increased with X_{Fe}, while the transition pressures for $U = 3.0$ eV were smaller than those given by $U = 5.0$ eV, irrespective of X_{Fe}. In Tsuchiya et al. (2006), the Hubbard U was determined by first principles (Cococcioni and de Gironcoli 2005). The transition pressure was not sensitive to X_{Fe} for $X_{Fe} < 0.1875$ for uniformly distributed iron configurations. In these configurations irons interacted weakly, which suggested the use of an ideal solid solution model would be a good approximation to investigate the thermodynamics of the spin crossover (Tsuchiya et al. 2006). The sensitivity of the transition pressure to X_{Fe} for $X_{Fe} > 0.25$ might be cause by iron-iron interaction (Kantor et al. 2009) but might also be understood, at least in part, as the effect of chemical pressure caused by magnesium. The latter is smaller than ferrous iron in the HS state. The lattice parameter of the (Mg,Fe)O solid solution increases with X_{Fe}. Therefore, solid solutions with large X_{Fe} need to be compressed further to induce the HS-to-LS crossover triggered by a critical Fe-O distance. The effect of iron-iron interaction in ferropericlase still deserves further investigation to clarify this point. A comparison between the DFT+U calculations (Persson et al. 2006; Tsuchiya et al. 2006) and experiments using XES (Lin et al. 2005, 2006a) and Mössbauer spectroscopy (Badro et al. 1999; Pasternak et al. 1997;

Speziale et al. 2005) is shown in Figure 1 (Persson et al. 2006). Thermodynamic properties of ferropericlase at various temperatures and pressures were computed by first principles in Wu et al. (2009) and Wentzcovitch et al. (2009), using the Hubbard U obtained by Tsuchiya et al. (2006). These results are summarized in the next sub-sections.

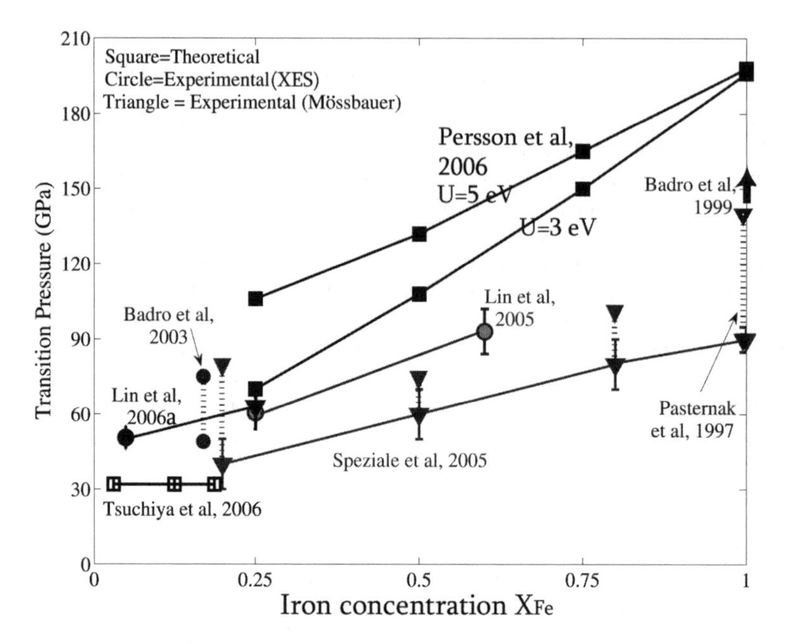

Figure 1. Dependence of the HS-to-LS transition pressure of ferropericlase on the iron concentration. The computational results by Persson et al. (2006) and Tsuchiya et al. (2006) are both shown along with various experiments. The transition pressured from Speziale et al. 2005 were measured at 6-300 K, while others were all room temperature measurements. [Used by permission of AGU, from Persson et al. (2006)].

Static LDA+U calculation

In Tsuchiya et al. (2006), the LDA+U method was used to calculate the atomic and electronic structure of ferropericlase with various iron concentration and different unit-cell volumes. The Hubbard U was determined from the linear response of the occupation matrix to a potential shift at the iron site. As shown in Figure 2, the computed Hubbard U is spin and volume dependent. The effect of iron concentration, X_{Fe}, is less significant and the Hubbard Us can be fitted to a single function of unit-cell volume, V, to first order. The distinguishing factor is the spin state.

When the on-site Coulomb interaction (Hubbard U) is included, desirable atomic structure and orbital occupancies are obtained, as shown in Figure 3. In the HS case, the five majority-spin d-electrons occupy the five d-orbitals to give spherical majority charge distribution, while the minority-spin electron occupies the d_{xy} orbital and induces greater Fe-O distance on the xy plane. This Jahn-Teller distortion does not happen in LDA. In LDA, all Fe-O distances are the same, and all three t_{2g} orbitals are degenerate and equally occupied by the minority-spin electron. Due to the partially-filled t_{2g} band, the HS ferropericlase is half-metallic in LDA. Such difference in orbital occupancy between LDA and LDA+U can be understood on the basis of the DFT+U energy functional of Equation (1). When U is included, each d-orbital tends to be completely filled or empty to minimize E. Thus, the minority-spin electron in HS iron tends to completely occupy one of the t_{2g} orbitals and induce J-T distortion, instead of partially

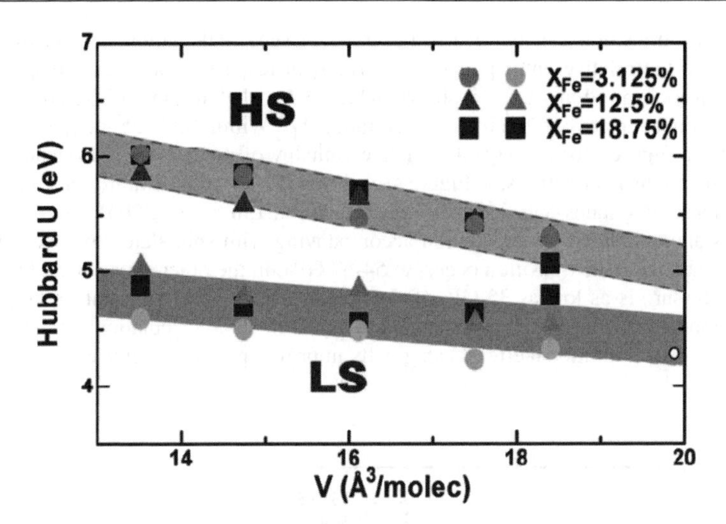

Figure 2. Computed Hubbard U of iron in ferropericlase as a function of unit-cell volume. The Hubbard U is not very sensitive to the iron concentration. Open circle is the U of HS iron in FeO. [Used by permission of APS, from Tsuchiya et al. (2006)].

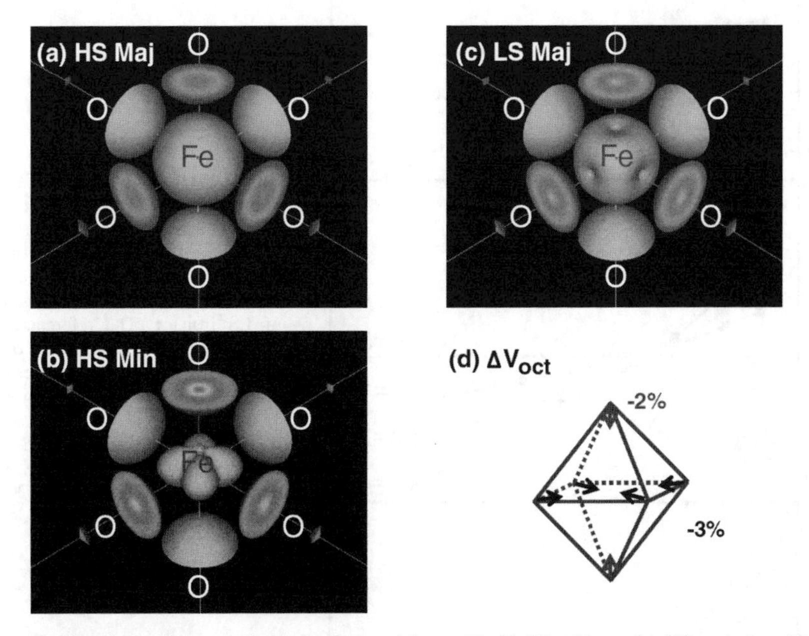

Figure 3. Charge density around iron in ferropericlase with 12.5% of iron. (a) HS majority; (b) HS minority; (c) LS majority; and (d) depiction of the polyhedral volume collapse across the spin transition. The isosurface value is 0.3 e/Å^3. [Used by permission of APS, from Tsuchiya et al. (2006)].

occupying three degenerate t_{2g} orbitals as in LDA. The J-T distortion splits the t_{2g} derived level into a singlet, the d_{xy} "band", and a doublet, the d_{yz} and d_{xz} "bands". With a completely filled d_{xy} band, HS ferropericlase is an insulator in LDA+U. The charge distribution shown in Figures 3(a-c) also implies a sudden decrease of volume when the HS-to-LS crossover occurs. The octahedral volume reduction is ~8% (see Fig. 3d).

Spin-state transition can be clearly observed in static calculation, as shown in Figure 4. By plotting the relative enthalpies of pure HS with respect to pure LS ferropericlase for different iron concentrations, X_{Fe}, it is shown in Figure 4(a) that the HS-to-LS static "transition" pressure P_t is not sensitive to the iron concentration X_{Fe}. Within $0.03125 < X_{Fe} < 0.1875$, P_t is about 32 GPa. Figure 4(b) displays the relative enthalpy of mixed-spin (MS) states for $X_{Fe} = 0.1875$ with various LS fractions, n. Figure 4(c) shows the zero temperature static compression curves of these MS states and the experimental data in Lin et al. (2005). The experimental data shows anomalous volume reduction accompanying with spin-state crossover. In this set of experimental data, the transition occurs at 54-67 GPa. In the other experiment, however, the transition pressure is as low as 35 GPa (Fei et al. 2007). These experiments were performed at room temperature while these results were obtained in static calculations. As shall be seen below, inclusion of vibrational effects can greatly improve agreement between calculations and experiments.

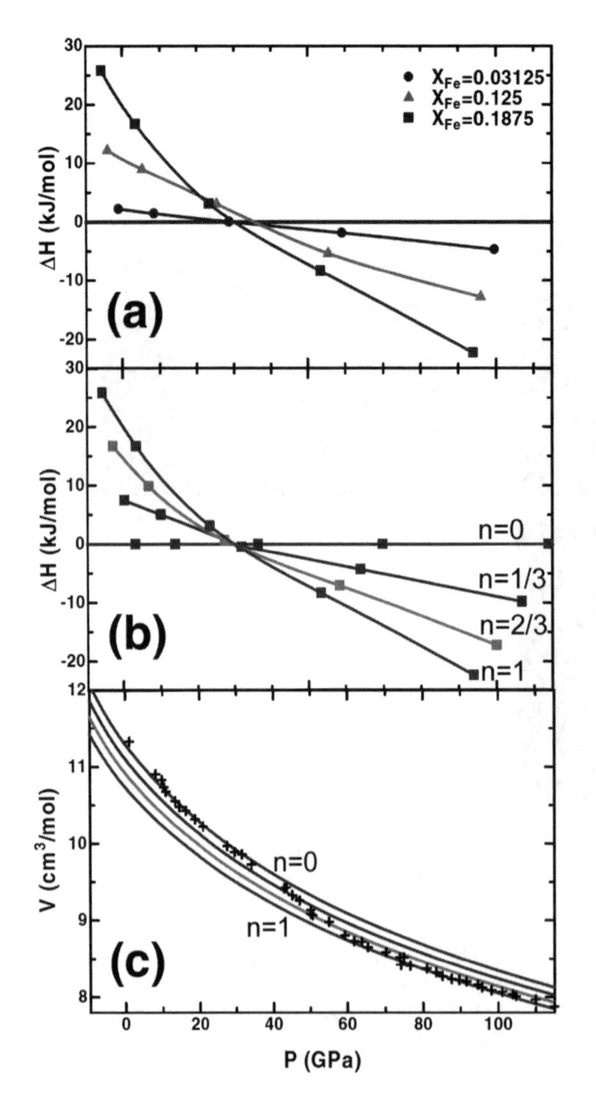

Figure 4. (a) Static enthalpy difference between LS and HS phases at three different iron concentrations. (b) Enthalpy difference in ferropericlase with iron concentration of 18.75% between states with various fractions of LS iron, n. The reference line corresponds to the enthalpy of all irons in HS state ($n = 0$). (c) Pressure-volume curves of ferropericlase with 18.75% of iron at different fraction of LS iron. The plus signs are experimental results at 300 K with 17% of iron by Lin et al. (2005). [Used by permission of APS, from Tsuchiya et al. (2006)].

VVCM for ferropericlase

A vibrational virtual crystal model was engineered to give the average vibrational properties of and acoustic velocities of the ferropericlase solid solution of in pure HS and LS states. Only the VDoSs of the pure spin states are necessary to proceed with the thermodynamics calculation outlined in the previous section. Both ferropericlase and MgO are cubic systems and have only three elastic constants, C_{11}, C_{12}, and C_{44} (in Voigt notation). The longitudinal (v_L) and transverse (v_T) wave velocities along the [100] direction are

$$\begin{aligned} v_L &= \omega_L / q = \sqrt{C_{11}/\rho} \\ v_T &= \omega_T / q = \sqrt{C_{44}/\rho} \end{aligned} \quad (18)$$

and along the [100] direction are

$$\begin{aligned} v_L &= \sqrt{(C_{11}+C_{12}+C_{44})/2\rho} \\ v_{T1} &= \sqrt{C_{44}/\rho} \\ v_{T2} &= \sqrt{(C_{11}-C_{12})/2\rho} \end{aligned} \quad (19)$$

For ferropericlase, all three static elastic constants can be calculated with LDA+U. The next step is to choose the force constant in the virtual crystal to reproduce these wave velocities. The most relevant force constants, $D^{ij}_{\mu\nu}$, in the virtual crystal are shown in Figure 5. They are: (a) the Mg-O nearest-neighbor longitudinal constant, D^{12}_{xx}, (b) the Mg-O nearest-neighbor transverse constant, D^{23}_{xx}, and (c) the Mg-Mg nearest neighbor magnesium interaction constant, D^{24}_{xy}. Their values are listed in Table 1. All other force constants have minor effects in the acoustic dispersion. The three force constants, D^{13}_{xx}, D^{16}_{xx}, and D^{13}_{zz}, are between the oxygen

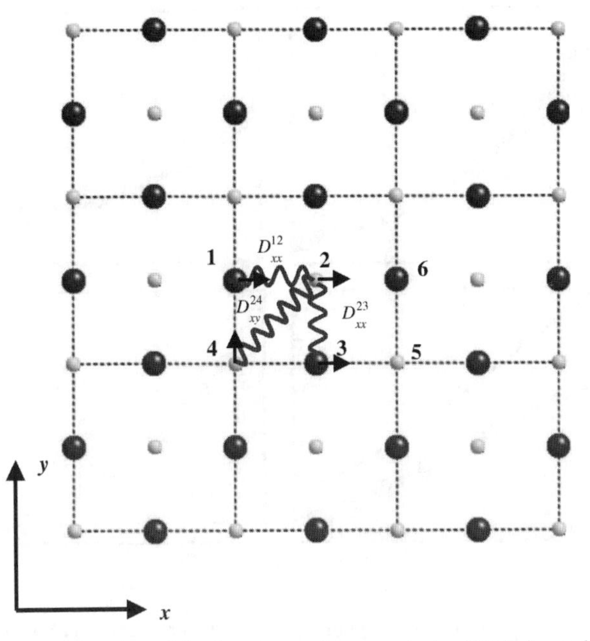

Figure 5. Schematic representation of the three largest force constants of MgO in xy plane. See text for the definition of these constants. The oxygen and magnesium atoms are represented by the large and small spheres, receptively. [Used by permission of APS, from Wu et al. (2009)].

Table 1. The eight largest force constants of MgO and their modified values for ferropericlase in HS and LS states. These values were obtained for the lattice constant of 4.07 Å (Wu et al. 2009).

	D^{12}_{xx}	D^{24}_{xy}	D^{23}_{xx}	D^{13}_{xx}	D^{45}_{xx}	D^{24}_{xx}	D^{16}_{xx}	D^{13}_{zz}
MgO	−0.1517	−0.0153	0.0117	−0.0078	−0.0054	−0.0047	0.0046	−0.0034
HS	−0.1570	−0.0167	0.0173	−0.0078	−0.0054	−0.0047	0.0046	−0.0034
LS	−0.1571	−0.0118	0.0162	−0.0078	−0.0054	−0.0047	0.0046	−0.0034

atoms, and the replacement of Mg by Fe in the virtual crystal affects these forces very little. D^{45}_{xx} and D^{24}_{xx} are between Mg and O, but they are much smaller than the three major force constants, D^{12}_{xx}, D^{24}_{xy}, and D^{23}_{xx}. Other force constants that are not presented are even smaller than the ones shown in Table 1. Therefore, only the three major force constants are modified to describe ferropericlase in pure spin states.

The elastic constants and the bulk modulus obtained by first-principles in LDA+*U* and the ones obtained from the reproduced phonon velocities in the virtual crystals are shown in Figure 6. They agree with each other very well. Therefore the VVCM's acoustic phonon dispersions are precisely the same as those of HS and LS ferropericlase. The VDoS obtained using VVCM for the case with lattice constant 4.21 Å is shown in Figure 7. The agreement between the result obtained using VVCM method and LDA+*U* (Fig. 6) ensures that the thermodynamic properties are calculated using the correct VDoS at low frequencies and with a reasonably

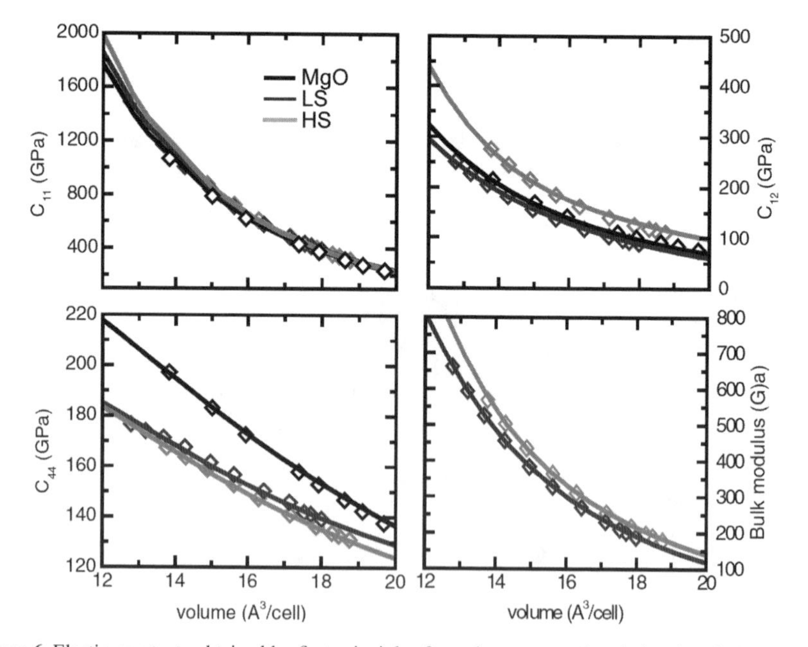

Figure 6. Elastic constants obtained by first principles from the stress-strain relation (symbols) and from phonon velocities (lines).The acoustic phonon velocities of MgO are computed directly using DFPT. The phonon velocities of ferropericlase in HS and LS were obtained by modifying the force constants of MgO. [Used by permission of APS, from Wu et al. (2009)].

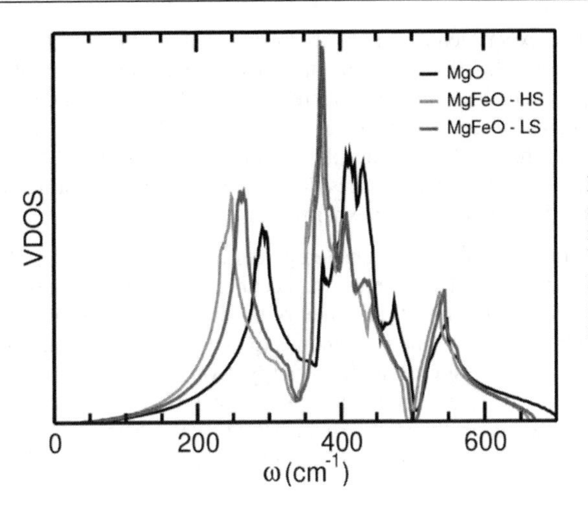

Figure 7. Vibrational density of states of MgO and ferropericlase in HS and LS states. These VDoS were computed with lattice parameter a = 4.21 Å. [Used by permission of APS, from Wu et al. (2009)].

good, i.e., representative, VDoS at high frequencies. The VVCM in conjunction with the QHA should provide more suitable vibrational and thermodynamic properties than a Mie-Debye-Gruneisen model.

Before presenting the thermodynamics properties of ferropericlase, the pressure and temperature range of the validity of QHA should be addressed. It is determined by *a posteriori* inspection of the thermal expansion coefficient, $\alpha \equiv (1/V)(\partial V/\partial T)|_P$, of ferropericlase in MS state. This method is discussed in the paper on quasiharmonic thermodynamics properties of minerals in this book. For ferropericlase in MS state, the volume as a function of the LS iron fraction $n \equiv n_{LS}(P,T)$ is determined by the derivative of Gibbs free energy with respect to pressure at a fixed temperature,

$$V(n) = \left.\frac{\partial G(n)}{\partial P}\right|_T = \left.\frac{\partial G(n)}{\partial P}\right|_{T,n} + \left.\frac{\partial G(n)}{\partial n}\right|_{T,P} \left.\frac{\partial n}{\partial P}\right|_T \tag{20}$$

At equilibrium, the term $\partial G/\partial n|_{T,P}$ vanishes, so the volume is

$$V(n) = nV_{LS} + (1-n)V_{HS} \tag{21}$$

where $V_{LS/HS} \equiv \partial G_{LS/HS}/\partial P|_T$ are the volumes of the pure LS/HS states, and the HS fraction $n_{HS} = 1 - n$. The thermal expansion coefficient of the MS ferropericlase is then

$$\alpha V(n) = nV_{LS}\alpha_{LS} + (1-n)V_{HS}\alpha_{HS} + (V_{LS} - V_{HS})\left.\frac{\partial n}{\partial T}\right|_P \tag{22}$$

The upper temperature limit of the validity of QHA at a certain pressure can be indicated by the inflection point of α with respect to temperature, namely, $\partial^2\alpha/\partial T^2|_P = 0$ (Carrier et al. 2008; Wentzcovitch et al. 2004). Beyond this inflection point, the thermal expansivity deviates from the usually linear behavior of experimental measurements. The temperature limits for ferropericlase in the pure LS and HS states are first established. The thermal expansion coefficient of MS state is determined from Equation (22). The maximum temperature limit for the predictions in the MS state is chosen as the minimum of those established for the HS and LS states.

Thermodynamic properties of ferropericlase

All methods discussed in previous sections provide a fine description of ferropericlase in the MS state at finite temperatures and pressures, as can be confirmed by the agreement between the computed (Wu et al. 2009) and experimental (Lin et al. 2005; Fei et al. 2007) isothermal compression curves shown in Figure 8. At 300 K, the calculated isotherm displays anomaly in $V(P,T,n)$ consistent with the experimental results. The theoretical equilibrium volume at 300 K, 11.46 cm^3/mol, is slightly larger than the experimental one, 11.35 cm^3/mol. This volume difference is consistent with the difference of iron concentration in the experiments (0.17) and calculation (0.1875). The predicted volume reduction resulting from the spin-state transition is 4.2% for $X_{Fe} = 0.1875$, compared to about 3-4% in experiments.

The fraction of LS irons as a function of pressure and temperature, $n(P,T)$, in ferropericlase with $X_{Fe} = 0.1875$ is shown in Figure 9. At 0 K, the transition occurs at ~36 GPa, and it is very sharp. As the temperature increases, the pressure at the center of the crossover range increases, and the crossover broadens. The black (white) line corresponds to $n(P,T) = 0.5$ in calculations including (without) the vibrational contribution to the free energy. Even at 0 K the transition pressure increases by 2.5 GPa because of zero-point motion energy. At 300 K, the center of the crossover occurs at 39.5 GPa, which agrees with data by Fei et al. (2007), but differs from many other experiments showing a transition pressure around 50-55 GPa (Goncharov et al. 2006;

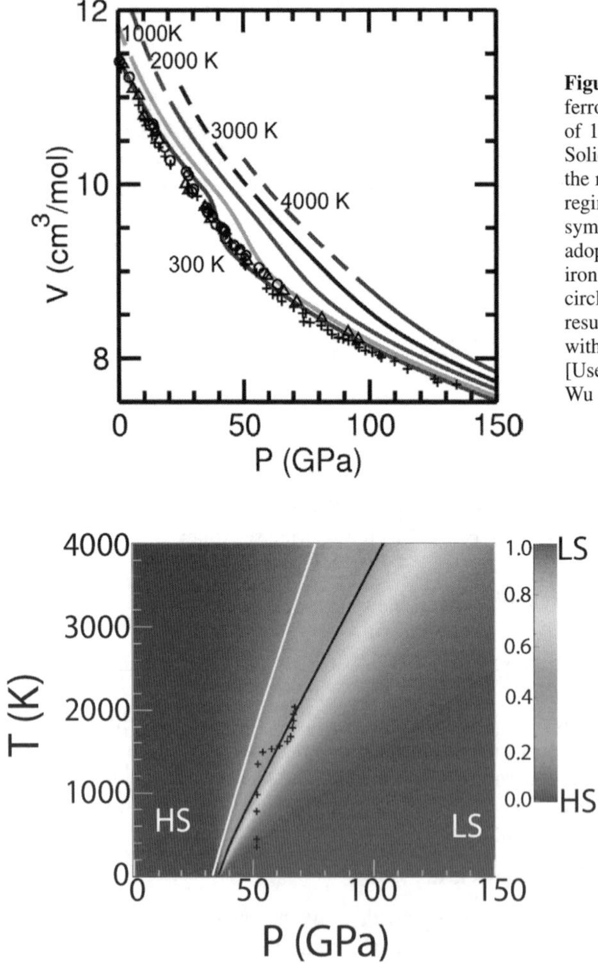

Figure 8. Compression curves of ferropericlase with iron concentration of 18.75% along several isotherms. Solid (dashed) lines correspond to the results within (outside) the (P,T) regime of the validity of QHA. Plus symbols are experimental results adopted from Lin et al. (2005), with iron concentration of 17%. The circle and triangle symbols are the results presented in Fei et al. (2007), with iron concentration of 20%. [Used by permission of APS, from Wu et al. (2009)].

Figure 9. The LS iron fraction as a function of pressure and temperature, $n(P,T)$, in ferropericlase with iron concentration of 18.75%. The black (white) line corresponds to the middle point of the crossover, namely, $n(P,T) = 0.5$, with (without) the inclusion the vibrational contribution in the free energy. In other words, along the black (white) line, the relation $\Delta G_{LS} = 0$ ($\Delta H_{LS} = 0$) holds. The plus symbols are the experimental data corresponding to $n(P,T) = 0.5$ presented in Lin et al. (2007c). [Used by permission of APS, from Wu et al. (2009)].

Kantor et al. 2006; Lin et al. 2005, 2006a; Speziale et al. 2005; Vanko and de Groot 2007). The computed transition pressure may still improve after a more complete treatment of the solid solution and after a more accurate Hubbard *U* is calculated self-consistently, as mentioned earlier.

Figure 10 shows the thermal expansion coefficient α, constant-pressure heat capacity C_P, and thermal Grüneisen parameter γ, of ferropericlase along several isobars. At very low or very high pressure, α looks "normal" since in these two extremes ferropericlase is in essentially pure HS and LS states, respectively. The normal α of HS ferropericlase is the same as that of MgO (Touloukian et al. 1977). In the temperature range when QHA is valid ($T < 1500$ K), the computed α agrees with that of ferropericlase up to $X_{Fe} = 0.36$ (van Westrenen et al. 2005). Comparing Figures 9 and 10, we can see that anomalies occur in the crossover region, namely, in the MS state. Similar anomalous behavior can be observed in heat capacity and thermal Grüneisen parameter in the same pressure and temperature range, only that the anomaly in C_P is not as dramatic as in α and γ.

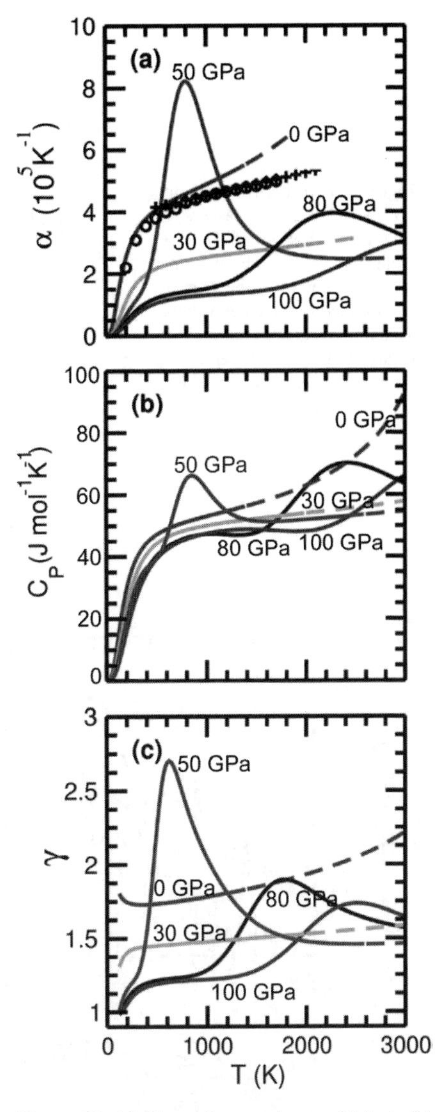

Figure 10. (a) Thermal expansion coefficient; (b) constant-pressure heat capacity; and (c) thermal Grüneisen parameter of ferropericlase with iron concentration of 18.75% along several isobars. Solid (dashed) lines correspond to the results within (outside) the (*P*, *T*) regime of validity of the QHA. Circles and crosses are the experimental values at 0 GPa for ferropericlase with iron concentration of 0% (Touloukian et al. 1977) and 36% (van Westrenen et al. 2005), respectively. [Used by permission of APS, from Wu et al. (2009)].

The fraction of LS irons along several isotherms is shown in Figure 11. Along with Equations (20)-(22), it provides useful information to understand the anomalous behavior in several physical quantities shown in Figure 12, including the softening of the adiabatic bulk modulus K_s, the sudden decrease in bulk wave velocity, v_φ, and the enhancement of density ρ. The transition is sharper at lower temperature (Fig. 11), and greater anomalies can be observed (Fig. 12) at those temperatures. Below 35 GPa, the calculated ρ at 300 K agrees very well with experiments using samples with $X_{Fe} = 0.17$ (Lin et al. 2005). The remaining difference is consistent with the difference in iron concentrations. The reduction of K_s and v_φ is consistent with experimental data on samples containing $X_{Fe} = 0.06$ (Crowhurst et al. 2008). The larger anomalies in the calculation are consistent experimental data on samples with X_{Fe} three times smaller.

The effect of spin-state crossover in ferropericlase along a typical geotherm (Boehler 2000) is shown in Figure 13(a-b). The anomalies in K_s (25 ± 6%) and v_φ (15 ± 7%) predicted

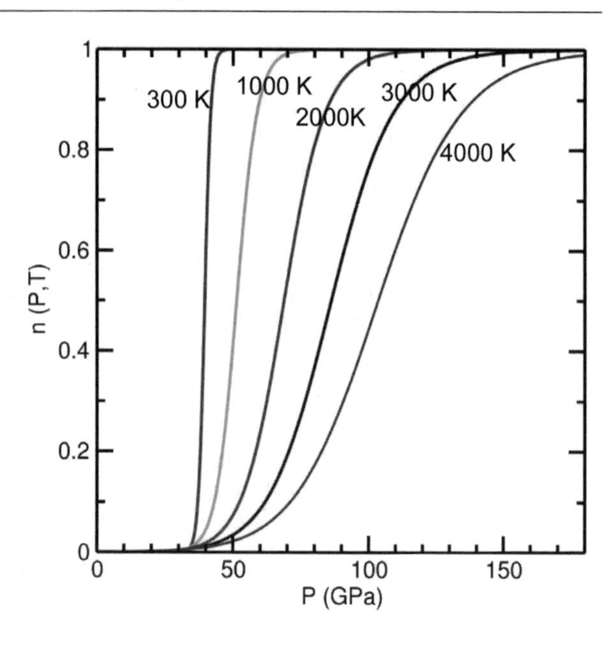

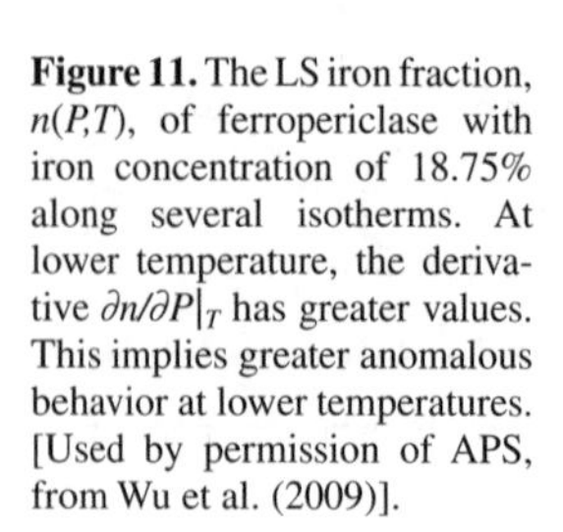

Figure 11. The LS iron fraction, $n(P,T)$, of ferropericlase with iron concentration of 18.75% along several isotherms. At lower temperature, the derivative $\partial n/\partial P|_T$ has greater values. This implies greater anomalous behavior at lower temperatures. [Used by permission of APS, from Wu et al. (2009)].

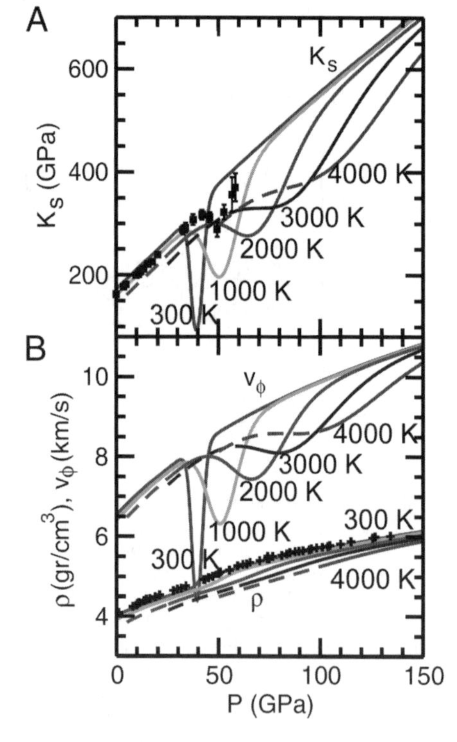

Figure 12. Pressure dependence of the calculated adiabatic bulk modulus K_S (A), bulk wave velocity v_φ, and density ρ (B) of ferropericlase with 18.75% of iron along several isotherms. Solid (dashed) lines correspond to the results within (outside) the (P,T) regime of validity of the QHA. Also presented in (A) are the experimental data of ferropericlase with 6% of iron (Crowhurst et al. 2008). The calculated anomaly is about three times greater than the experiment, agree with the difference of iron concentration. Crosses shown in (B) are the data of ferropericlase with 17% of iron at room temperature (Lin et al. 2005). [Used by permission of National Academy of Sciences, from Wentzcovitch et al. (2009)].

by a purely elastic model start at ≈ 40 GPa (≈ 1000 km depth) and are most pronounced at ≈ 70 ± 20 GPa (≈ 1600 ± 400 km depth), but the crossover continues down to the core-mantle boundary (CMB) pressure with a possible reentrance into the HS state because of the thermal boundary layer above the CMB (Boehler 2000). However, one should have in

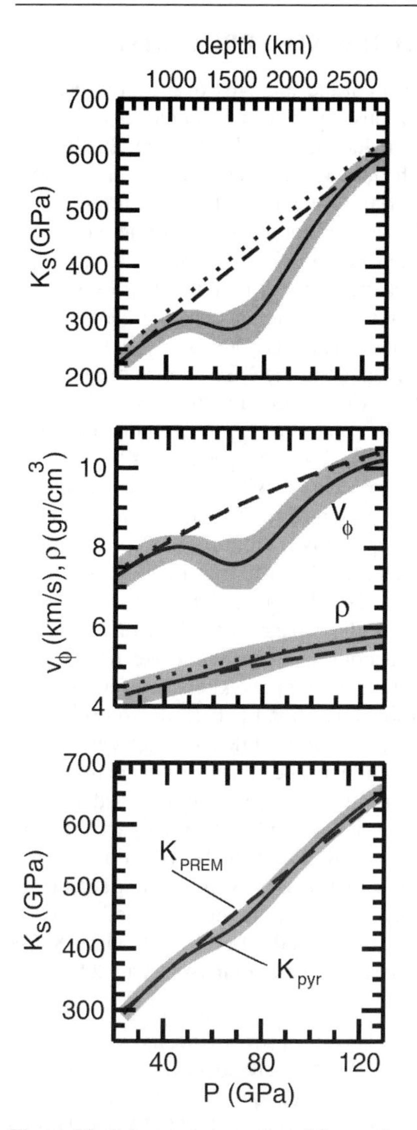

Figure 13. Calculated properties of ferropericlase with 18.75% of iron along a lower mantle geotherm (Boehler 2000). (A) adiabatic bulk modulus K_S; (B) bulk wave velocity v_φ, and density ρ; and (C) bulk modulus of an aggregate with pyrolite composition (McDonough and Sun 1995), K_{pyr}, and PREM's bulk modulus, K_{PREM} (Dziewonski and Anderson 1981). In (A) and (B), solid, dashed, and dotted lines correspond to the properties computed in the MS, HS, and LS states, respectively. Shaded regions represent the uncertainties caused mainly by the uncertainty in the computed enthalpies of HS and LS states and transition pressure at 0 K. [Used by permission of National Academy of Sciences, from Wentzcovitch et al. (2009)].

mind that the bulk modulus softens. In contrast, density increases smoothly throughout the entire pressure range of the lower mantle. The shaded areas correspond to uncertainties caused by uncertainties in the calculated static transition pressure and the narrower crossover pressure range (Wentzcovitch et al. 2009).

The net effect of the spin crossover in ferropericlase on the bulk modulus of a uniform aggregate with pyrolite composition (McDonough and Sun 1995), K_{pyr}, along a typical geotherm (Boehler 2000) is shown in Figure 13(c). This comparison is made to elucidate and highlight an effect that may be quite subtle. The bulk modulus of perovskite calculated by first-principles is adopted (Wentzcovitch et al. 2004) in the calculation of K_{pyr}. Compared with the bulk modulus of the Preliminary Reference Earth Model (PREM) (Dziewonski and Anderson 1981), K_{PREM}, a subtle reduction of about 4% is observed in K_{pyr}, which appears to be smoothed or cut through by K_{PREM}. The uncertainly of K_{pyr} is quite large and permits the signature of the spin-state crossover in ferropericlase to fall within the uncertainty of global seismic constraints. In this sense K_{PREM} does not appear to be inconsistent with the calculated K_{pyr} along the geotherm.

An intriguing possibility of a viscosity anomaly caused by this crossover in the mantle has been raised (Wentzcovitch et al. 2009). In a well mixed mantle, phase separation between ferropericlase and perovskite is expected to occur owing to the contrast in their rheological properties (Barnhoorn et al. 2005; Holtzman et al. 2003; Yamazaki and Karato 2001). In this situation, the softer phase, ferropericlase, is expected to dominate the rheology of the region. The softening of the bulk modulus in ferropericlase is most enhanced at ~1,500 km depth (Fig. 13). Viscosity profiles for the mantle with viscosity minima around 1,500 have been proposed (Forte and Mitrovica 2001; Mitrovica and Forte 2004) and have been difficult to explain. This spin crossover in ferropericlase may offer an interpretation for this proposed viscosity minimum in the mantle, but in-depth studies of this phenomenon are still needed.

SPIN-STATE CROSSOVER IN FERROSILICATE PEROVSKITE

In contrast to ferropericlase, the detailed mechanism of the spin crossover in iron-bearing magnesium silicate perovskite is still controversial. Various experimental techniques have been used to study this phenomenon in iron in perovskite, including XES (Badro et al. 2004; Li et al. 2004), XANES (Narygina et al. 2009), and Mössbauer spectroscopy (Jackson et al. 2005; Lin et al. 2008; McCammon et al. 2008). In contrast to the conclusive HS-to-LS transition in ferropericlase observed in experiments, there is still a lack of consensus regarding the nature of the spin state change in perovskite. In Badro et al. (2004), the XES spectra of $(Mg_{0.9}Fe_{0.1})SiO_3$ suggested a two-step transition. At 70 GPa, iron changes from pure HS to a MS state, and at 120 GPa, all irons fall to the LS state. The sample in this experiment contained ferrous and ferric iron. In Li et al. (2004), both $(Mg_{0.92}Fe_{0.09})SiO_3$ and $(Mg_{0.87}Fe_{0.09})(Si_{0.94}Al_{0.10})O_3$ samples show a broad crossover, and the non-vanishing satellite peak in the XES spectra measured at 100 GPa was attributed to the existence of IS iron. In XANES spectra (Narygina et al. 2009), a crossover in the 30-87 GPa range was observed, but the spin state of iron was not unambiguously determined. The Mössbauer studies by Jackson et al. (2005) suggested that ferrous iron remains in HS state at all pressures, and the spin crossover occurs only in ferric iron. In contrast, the Mössbauer spectra of the $(Mg_{0.88}Fe_{0.12})SiO_3$ sample presented in McCammon et al. (2008) showed the existence of a "new" species of iron with a quadrupole splitting (QS) of 3.5 mm/s. It was suggested that this high-QS iron is in the IS state, the "HS-to-IS transition" occurs at 30 GPa, and the iron remains in the IS state up to 110 GPa. The IS iron at high pressure was also inferred from a Mössbauer study (Lin et al. 2008). These discrepancies result mainly from the more complex nature of perovskite compared with ferropericlase. In ferropericlase, ferrous iron replaces magnesium and occupies clearly the octahedral site. In perovskite, however, iron can be in ferrous or ferric states, without an accurately known partitioning. Ferrous iron is widely believed to occupy the 8-12 coordinated dodecahedral site (A site), while ferric iron can occupy the dodecahedral or the octahedral site (B site). The fraction and occupancy of ferric irons also depend on the presence of aluminum. To say the least, the existence of an intermediate-spin ferrous iron is uncertain. While some experiments were interpreted as providing evidence in support of IS iron, (Lin et al. 2008; McCammon et al. 2008), theoretical calculations clearly indicate the opposite. First-principles calculations show that IS iron is the least energetically favorable state, and the HS-to-IS crossover should not occur in the pressure range of the lower mantle (Bengtson et al. 2009; Hsu et al. 2009b, Umemoto et al. 2009b). Also, the computed QS of IS ferrous iron is much smaller than that observed in experiments (Bengtson et al. 2009; Hsu et al. 2009b), as shall be discussed later.

In an attempt to better understand the spin-state crossover in perovskite and interpret the experimental results, several calculations were performed, but they are not always in agreement with each other (Li et al. 2005; Hofmeister 2006; Zhang and Oganov 2006; Stackhouse et al. 2007a; Bengtson et al. 2008, 2009; Umemoto et al. 2008, 2009; Hsu et al. 2009b). The HS-to-LS crossover pressure in ferrous iron in $(Mg,Fe)SiO_3$ determined by static calculations strongly depends on the choice of exchange-correlation functional among other things. In general, GGA gives higher transition pressure than LDA by ~50 GPa (Hofmeister 2006; Zhang and Oganov 2006; Stackhouse et al. 2007; Bengtson et al. 2008; Umemoto et al. 2008; Hsu et al. 2009b), as has been seen in other non-strongly-correlated systems as well (Tsuchiya et al. 2004; Yu et al. 2007, 2008, Wentzcovitch et al. 2010). It also depends on other factors, such as the magnetic and atomic order (Umemoto et al. 2008), the site degeneracy of irons (Umemoto et al. 2009), and the inclusion of on-site Coulomb interaction, Hubbard U (Hsu et al. 2009b). In all these calculations, the HS-to-IS transition was not observed. The spin-state transition of ferric iron is also discussed in some theoretical works (Li et al. 2005; Zhang and Oganov 2006; Stackhouse et al. 2007). The results of these calculations are summarized (with selected iron concentrations x) in Table 2. Since ferrous iron has greater population in perovskite and it occupies the A site, our work (Umemoto et al. 2008, 2009; Hsu et al. 2009b) have focused on Al-free iron-bearing

Table 2. Computed HS-to-LS transition pressures of Fe^{3+} and Fe^{2+}. There are two transition pressures adopted from Umemoto et al. (2008) for each method. The higher transition pressure corresponds to the configuration with uniformly distributed iron cations, and the iron-(110) configuration (see text) gives the lower transition pressure shown in table.

Composition	x	Method	Iron	HS-to-LS	Reference
$(Mg,Fe)(Si,Al)O_3$	0.0625	GGA	Fe^{3+}	105 GPa	Li et al. 2005
$(Mg,Fe)(Si,Al)O_3$	0.03125	GGA	Fe^{3+}	134 GPa	Zhang & Oganov 2006
$(Mg,Fe)(Si,Fe)O_3$	0.0625	GGA	Fe^{3+} (A site)	76 Gpa	Zhang & Oganov 2006
$(Mg,Fe)(Si,Fe)O_3$	0.125	GGA	Fe^{3+} (A site)	60 GPa	Stackhouse et al. 2007
$(Mg,Fe)SiO_3$	0.125	GGA	Fe^{2+}	130 GPa	Stackhouse et al. 2007
$(Mg,Fe)SiO_3$	0.125	GGA	Fe^{2+}	202 GPa	Bengtson et al. 2008
$(Mg,Fe)SiO_3$	0.25	LDA	Fe^{2+}	96 GPa	Bengtson et al. 2008
$(Mg,Fe)SiO_3$	0.125	LDA	Fe^{2+}	56-97 GPa	Umemoto et al. 2008
$(Mg,Fe)SiO_3$	0.125	GGA	Fe^{2+}	117-160 GPa	Umemoto et al. 2008
$(Mg,Fe)SiO_3$	0.125	LDA+U	Fe^{2+}	~ 300 GPa	Hsu et al. 2009b
$(Mg,Fe)SiO_3$	0.125	GGA+U	Fe^{2+}	> 300 GPa	Hsu et al. 2009b

perovskite, namely, $(Mg_{1-x}Fe_x)SiO_3$. The results of these calculations are discussed below. In the lower mantle $x \approx 0.1$, and we set $x = 0.125$ in our calculations.

Quadrupole splitting of ferrous iron in perovskite

The stability of IS iron and its QS is the first issue we address. The QS is directly proportional to the product of the nuclear quadrupole moment and the electric field gradient (EFG) at the center of iron nucleus. The EFG depends on the asymmetry of the charge density around the nucleus and is not determined solely by the spin state. In $(Mg,Fe)SiO_3$ perovskite, ferrous iron substitutes for magnesium in the large 8-12 coordinated cage. This is a large cage and iron might find metastable sites, in addition to the lowest energy site. The ligand field, the *d*-orbital occupancy, and the electron charge distribution might differ among these sites. Such difference, even if small, can change the EFG and thus lead to different QSs, irrespective of spin state. Therefore, interpretation of the spin state on the basis of QSs observed by Mössbauer spectroscopy is not unambiguous.

Bengtson et al. (2009) conducted a random search for possible equilibrium HS iron sites by displacing them from their "previously known" equilibrium position and subsequently relaxing them. This GGA investigation found two metastable sites for HS iron in $(Mg_{0.75}Fe_{0.25})SiO_3$. Their QSs were 3.2 and 2.3 mm/s. At 0 GPa, the site with small QS was 8 meV/Fe more stable than the HS QS site. The atomic configurations of these two sites around iron at 0 GPa are shown in Figure 14 (Bengtson et al. 2009). They produce different ligand fields. No other metastable site was found for HS iron. While the energy difference between these HS states was noticed to decrease with pressure, the enthalpy crossing between them was not demonstrated (Bengtson et al. 2009). They also indicated that the QS of (the manually constrained) IS iron was 0.7 mm/s, far smaller than the 3.5-4.0 mm/s observed by McCammon et al. (2008). The QS for the LS state calculated by Bengtson et al. (2009) was 0.8 mm/s.

A systematic search for metastable equilibrium sites of iron in $MgSiO_3$ perovskite in HS, LS, and IS states was conducted by Hsu et al. (2009b). These calculations started searching for unstable phonons (GGA) in a cubic perovskite structure with 20 atoms supercell. The

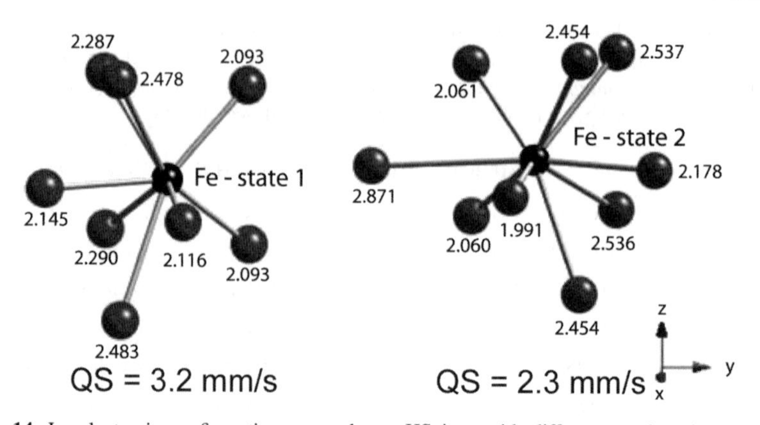

Figure 14. Local atomic configurations around two HS iron with different quadrupole splittings in $(Mg_{0.75}Fe_{0.25})SiO_3$ perovskite at zero pressure. The state with QS = 2.3 mm/s has lower energy than the state with QS = 3.3 mm/s by 0.8 meV/Fe at zero pressure. [Used by permission of AGU, from Bengtson et al. (2009)].

displacement modes of the unstable phonons were subsequently added to the atomic positions followed by full structural relaxations. Phonon frequencies were recalculated and this process was repeated until no more unstable phonons were present. Three HS and three LS sites were found in $(Mg_{0.75}Fe_{0.25})SiO_3$. It turned out that only the HS states with QS of 2.3 and 3.3 mm/s and the LS state with QS 0.8 mm/s are in the relevant energy range, with the two relevant HS states are competing energetically with each other. The local relevant atomic configurations of HS iron found in this work are similar to those shown in Figure 14 (Bengtson et al. 2009). At 0 GPa, the low-QS (2.3 mm/s) HS structure has slightly larger volume, and its energy is lower than that of the high-QS (3.3 mm/s) state by 24.3 meV/Fe. The high-QS HS structure is the same HS structure obtained by Umemoto et al. (2008, 2009). A stable IS state with QS of 1.4 mm/s is found in Hsu et al. (2009b). This IS state is the same as that obtained in Umemoto et al. (2009). In the 30-150 GPa pressure range, the magnetic moment of this state converged spontaneously to $2\mu_B$ ($S = 1$). This QS value, although higher than that obtained in Bengtson et al. (2009), is still considerably lower than 3.5-4.0 mm/s measured by MaCammon et al. (2008).

The reason why the QSs of the two competing HS states differ was discussed in Hsu et al. (2009b). The QS value depends on the EFG at the center of the iron nucleus, and the EFG is produced by the electron density. For both HS states, the spin-up electrons fill all five d-orbitals. Their contribution to the EFG is negligible since their charge density is essentially spherical. The EFG thus results primarily from the spin-down electron density. The major component of the EFG tensor, V_{zz}, depends on the orbital occupancy according to:

$$V_{zz} \propto \frac{e}{r^3}\left(2n_{xy} - n_{yz} - n_{xz} - 2n_{z^2} + 2n_{x^2-y^2}\right)$$

where n_{xy}, n_{yz},… correspond to the occupancy of the d_{xy}, d_{yz}, …, orbitals respectively (Chen and Yang 2007). Electrons occupying d_{xy}, $d_{x^2-y^2}$, or d_{z^2} orbitals contribute twice as much to the EFG as those in d_{xz} or d_{yz} orbitals. In the high-QS HS state, the spin-down occupancies are ~0.47 and ~0.62 for the $d_{x^2-y^2}$ and d_{xy} orbitals, respectively, and essentially 0.0 for the other orbitals. In the low-QS HS structure, only the d_{yz} orbital has significant occupancy, ~0.97. This difference in d-orbital occupancies of the HS states is the origin of their different QSs.

The effect of X_{Fe}, exchange-correlation functional (GGA, LDA), and Hubbard U on the low-to-high QS and spin-state crossover was also investigated by Hsu et al. (2009). The

Hubbard U was determined by the linear response of the corresponding DFT ground state to the local perturbation at the iron site (Cococcioni and de Gironcoli 2005). The pressure at which the enthalpies of the high- and low-QS HS states cross depends on all these factors. However, it always occurred below 24 GPa. The QSs increase with Hubbard U to 2.5 and 3.5 mm/s for the determined Us. The HS-to-IS enthalpy crossing was never observed between 0 and 150 GPa. The high enthalpy of IS was also demonstrated by Umemoto et al. (2009), as shall be discussed in the next sub-section.

Dependence of the spin crossover pressure on the site and orbital degeneracies

The spin-state crossover in $(Mg_{0.875}Fe_{0.125})SiO_3$ at finite temperatures is discussed in this sub-section (Umemoto et al. 2009). The effect of vibrations is disregarded, as in Tsuchiya et al. (2006), and only the effect of orbital and site degeneracies (Eqns. 7 and 8) are considered. The Gibbs free energy of each spin state (Eqn. 5) is then given by $G_\sigma = H_\sigma + G_\sigma^{mag} + G_\sigma^{site}$. The states involved are: the high-QS HS state, and the LS and IS relevant states. Only the high-QS HS state was relevant in this study (Umemoto et al. 2009) because this was an LDA calculation and the high- and low-QS HS states merged into a single one with high-QS at ~10 GPa (Hsu et al. 2009b), well before the HS-to-LS crossover occurs, ~100 GPa.

HS iron in perovskite is on a xy mirror plane, but the IS and LS are not. They are displaced from the xy mirror plane along the $\pm z$-axis, as shown in Figure 15. Therefore, there are two possible sites for IS and LS iron if irons are not close to each other. This means $N_{HS}^{site} = 1$, and $N_{IS}^{site} = N_{LS}^{site} = 2$ in Equation (8). The LDA enthalpy of IS iron is much higher than those of HS and LS iron, as shown in Figure 16 (Umemoto et al. 2009), and this makes the IS iron fraction n_{IS} negligible ($< 3\%$) in 0-3000 K temperature range (Umemoto et al. 2009). The effect of site degeneracy on the LS iron fraction n_{LS} is demonstrated in Figure 17. To make this point clear the magnetic entropy was temporarily disregarded. Site entropy does not affect n_{LS} at 0 K, as expected. At finite temperatures, it helps to stabilize the LS state with respect to the HS state. Without site entropy, the HS-to-LS crossover pressure (at which $n_{LS} = 0.5$) is 97 GPa at all temperatures. With site entropy, the crossover pressure changes to 95, 88, 80, and 72 GPa at the temperature of 300, 100, 2000, and 3000 K, respectively. Therefore, at lower-mantel temperatures, site entropy should decrease noticeably the onset of the spin-state crossover. It should be noted that this calculation is based on the assumption of ideal solid solution with uniformly distributed irons. For higher iron concentrations or other atomic configurations where iron-iron interactions are not negligible (Umemoto et al. 2008), this effect is probably smaller.

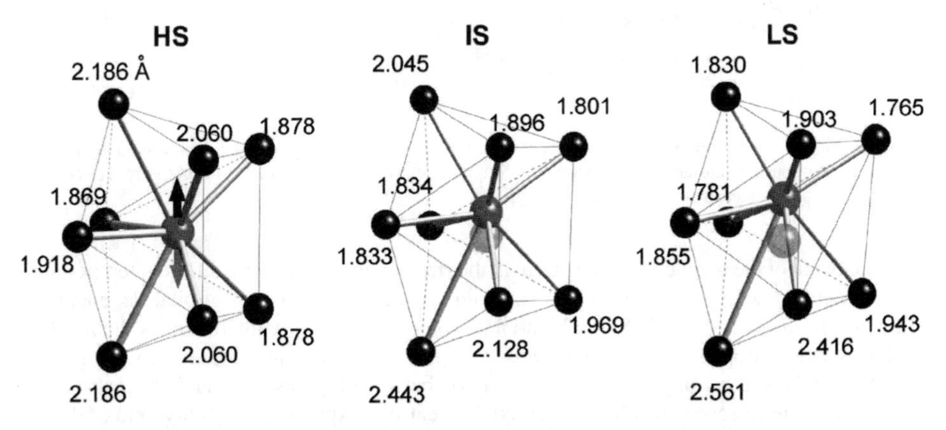

Figure 15. Local atomic configurations around HS, IS, and LS iron at 120 GPa optimized for $(Mg_{0.875}Fe_{0.125})SiO_3$ with LDA. Numbers next to oxygens are Fe-O distances (in Å). Pale spheres represent symmetrically equivalent sites for IS and LS states. [Used by permission of Elsevier, from Umemoto et al. (2009)].

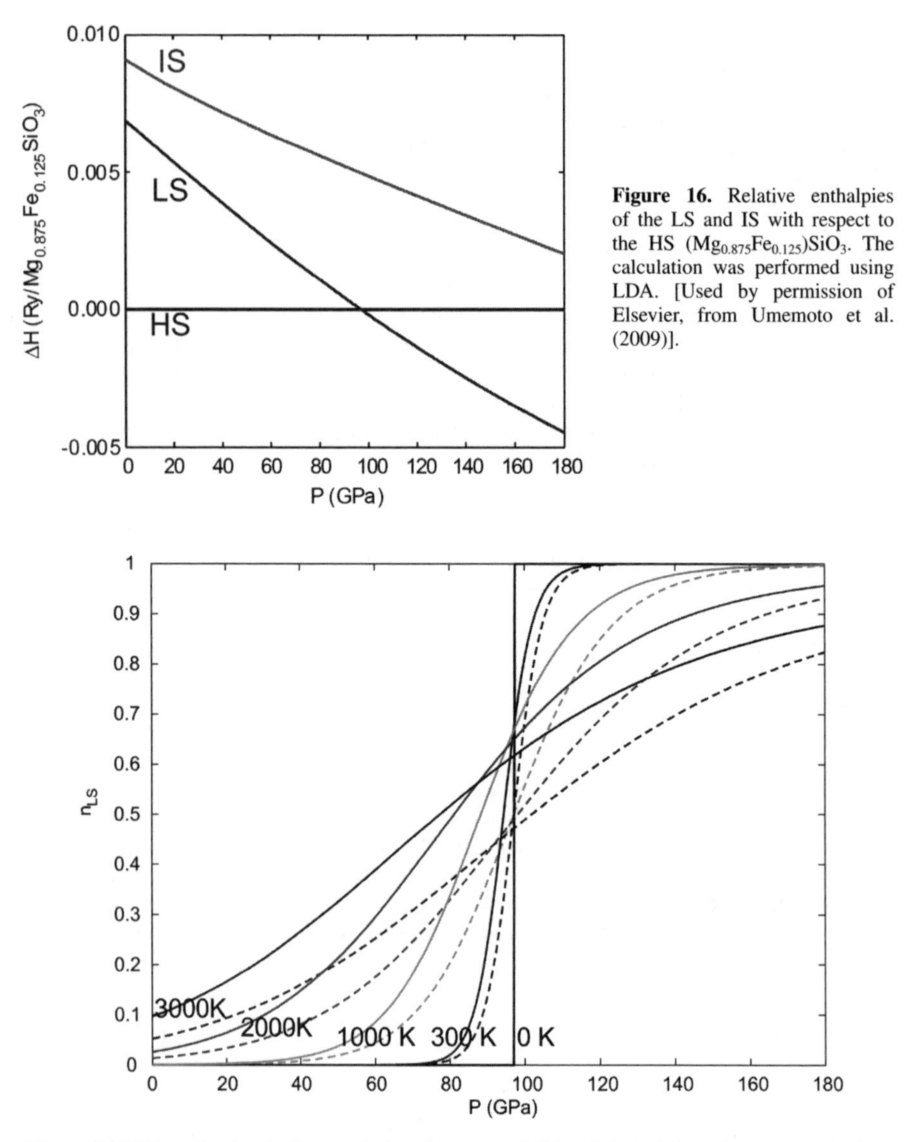

Figure 16. Relative enthalpies of the LS and IS with respect to the HS $(Mg_{0.875}Fe_{0.125})SiO_3$. The calculation was performed using LDA. [Used by permission of Elsevier, from Umemoto et al. (2009)].

Figure 17. LS iron fraction (n_{LS}) at several temperatures. Solid and dashed lines denote n_{LS} calculated with and without LS site entropy, respectively. Magnetic entropy is disregarded in this figure. [Used by permission of Elsevier, from Umemoto et al. (2009)].

Inclusion of the magnetic entropy term in the free energy introduces a different temperature dependence in the crossover pressure. The spin quantum numbers and orbital degeneracies (Eqn. 7) are: $S_{LS} = 0$, $S_{IS} = 1$, $S_{HS} = 2$, and $m_{LS} = m_{IS} = m_{HS} = 1$. Figure 18 shows the $n_{LS}(P,T)$ with magnetic entropies. Since the term $m_{HS}\,(2S_{HS}+1) > N_{LS}^{site}$, the magnetic entropy has greater effects than the site entropy. However, comparison between Figures 18(a) and 18(b) show that the effect of site entropy can still be observed. Site entropy expands the stability field of the LS state, partially compensating the effect of magnetic entropy, which favors the HS state.

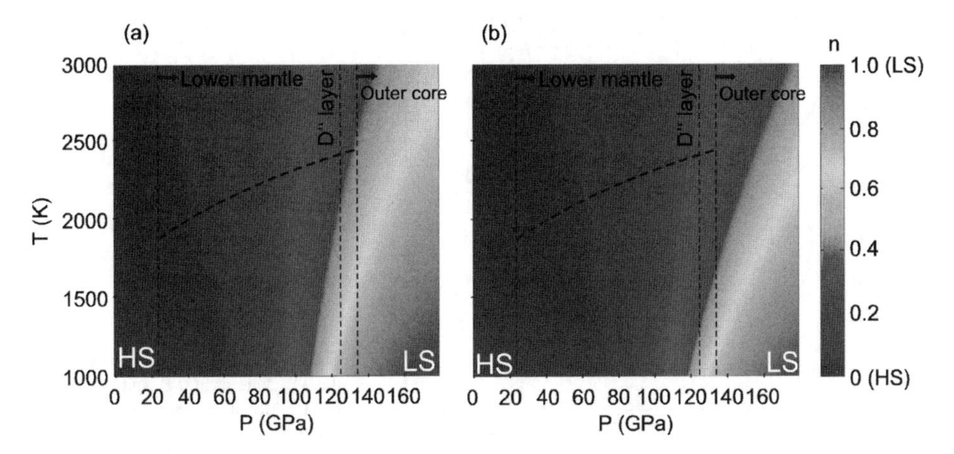

Figure 18. LS iron fraction (n_{LS}) with magnetic entropy. In (a) the LS site entropy is included, while in (b) it is not. Dashed lines denote a mantle geotherm derived by Brown and Shankland (1981). [Used by permission of Elsevier, from Umemoto et al. (2009)].

The compression curves and the bulk modulus of $(Mg_{0.875}Fe_{0.125})SiO_3$ in pure and mixed-spin states are shown in Figures 19 and 20, respectively. While the calculated volumes at 300 K (271.55 a.u.^{3}/f.u), are underestimated in LDA calculations (experimental value 275.50 a.u.^{3}/f.u.) owing to the absence of vibrational zero point motion effects in this calculation, the bulk modulus reproduces the experimental value 257 GPa (Lundin et al. 2008) quite well. The volume difference between the LS and HS states is quite small (about 0.3%). Therefore, even at 300 K, the volume reduction accompanying the spin-state crossover is difficult to observe, let alone the case at 2000 K, as can be seen in Figure 19. This is very different from ferropericlase discussed earlier. In perovskite, the softening in bulk modulus "*for this particular atomic configuration*" is pronounced at 300 K but barely noticeable at 2000 K, as shown in Figure 20. Therefore, at lower-mantle conditions, the elastic properties of perovskite should be in practice unaffected by the spin-state crossover.

Dependence of transition pressure on the concentration and distribution of iron

Our discussions so far have been focused on the calculations with low concentration (X_{Fe} = 0.125) of iron distributed uniformly in the perovskite. Iron-iron interaction is weak in this case and the ideal solid solution approximation is fine for sake of understanding different effects. However, iron-iron interaction is non-negligible at higher iron concentrations or for low iron concentration in clustered iron configurations. A thorough study with a variety of spatial distributions of irons at various concentrations ($0.0625 \leq X_{Fe} \leq 1.0$) was conducted by Umemoto et al. (2008). Two extreme configurations at X_{Fe} = 0.125 and 0.5 are shown in Figure 21: one configuration with uniformly distributed irons and the other with irons ordered in the (110) plane. The latter is referred to as iron-(110) hereafter. There can be ferromagnetic (FM) and anti-ferromagnetic (AFM) order in both configurations. The effect of magnetic order is not important. The HS-to-LS crossover pressure only differs by a few GPa (Umemoto et al. 2008). Therefore, we only mention results for the FM case.

Iron-iron interaction reduces the transition pressure. This interaction has a substantial elastic component. In either configuration, the transition pressure decreases with iron concentration, by about 100 GPa when X_{Fe} increases from 0.0625 to 1.0, as shown in Figure 22. In the range of $0.0625 \leq X_{Fe} \leq 0.125$, where iron-iron interactions are weak, the transition pressure is not sensitive to iron concentration for uniformly distributed configurations. For low

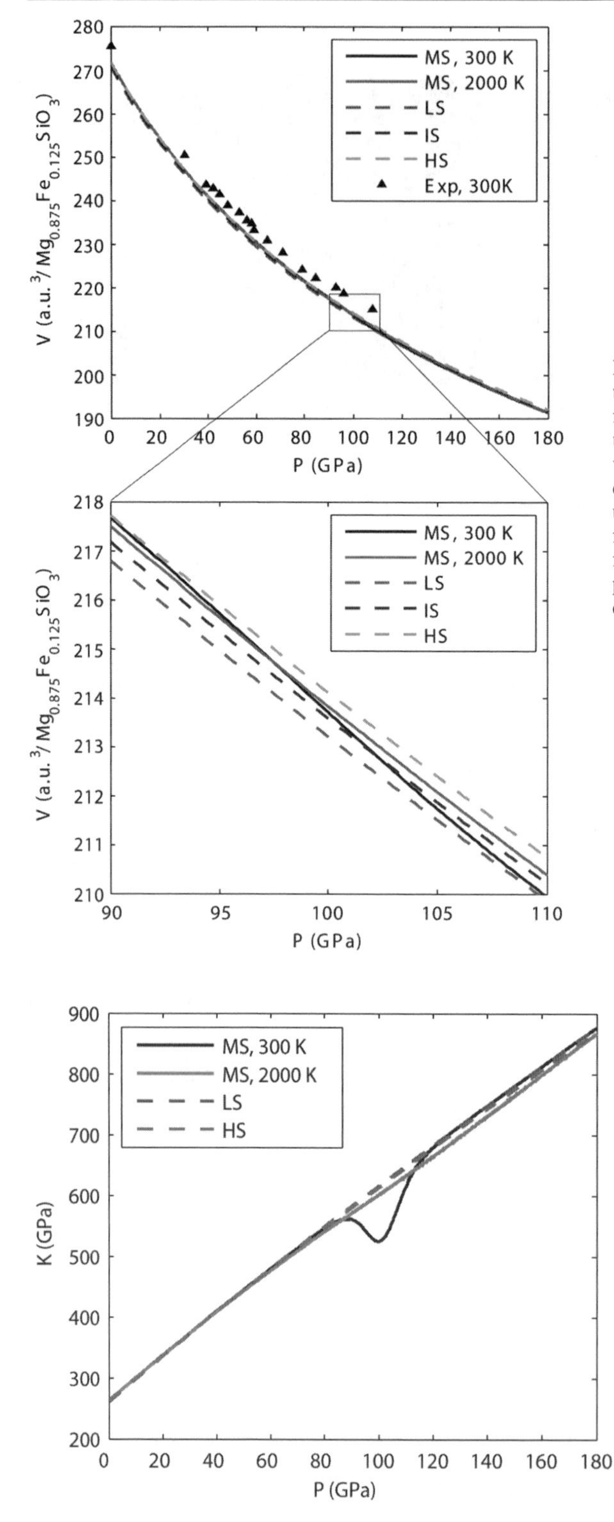

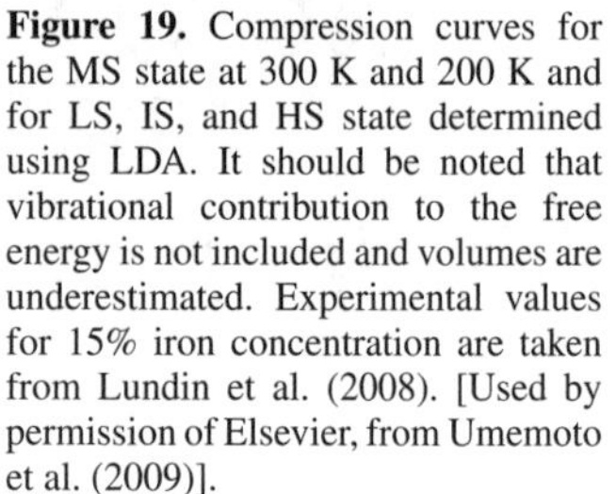

Figure 19. Compression curves for the MS state at 300 K and 200 K and for LS, IS, and HS state determined using LDA. It should be noted that vibrational contribution to the free energy is not included and volumes are underestimated. Experimental values for 15% iron concentration are taken from Lundin et al. (2008). [Used by permission of Elsevier, from Umemoto et al. (2009)].

Figure 20. Pressure dependence of the calculated bulk modulus (LDA) of the MS state at 300 K and 2000 K and for the LS and HS states. [Used by permission of Elsevier, from Umemoto et al. (2009)].

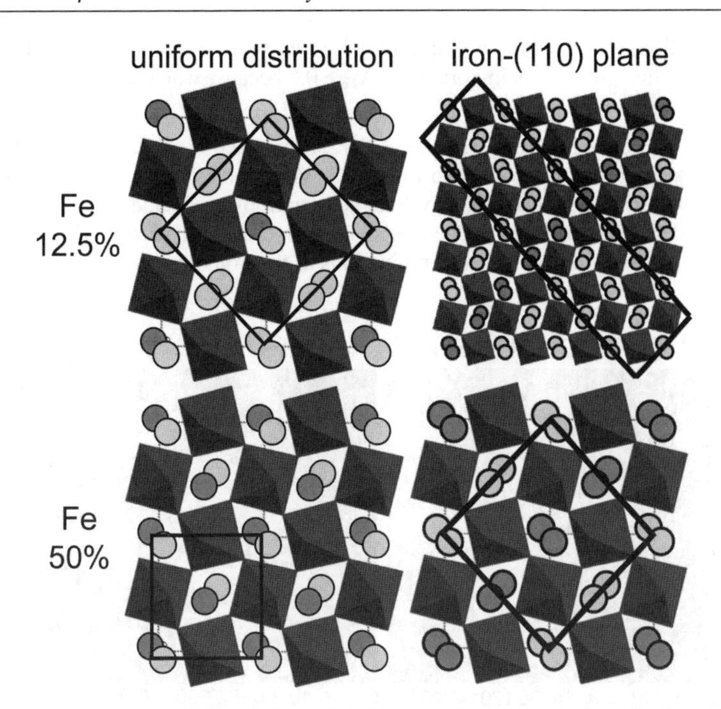

Figure 21. $(Mg_{0.5}Fe_{0.5})SiO_3$ and $(Mg_{0.875}Fe_{0.125})SiO_3$ in the configurations with iron distributed uniformly and iron placed on the (110) plane. [Used by permission of Elsevier, from Umemoto et al. (2008)].

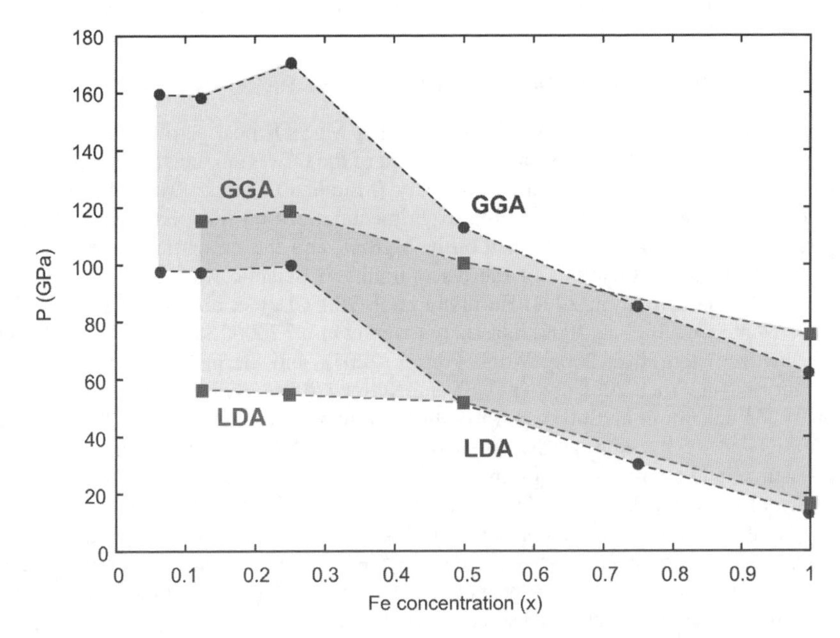

Figure 22. Calculated HS-to-LS transition pressure of $(Mg_{0.875}Fe_{0.125})SiO_3$ as a function of iron concentration in the case of iron distributed uniformly (circle) and the case of iron placed on the (110) plane (square). Both GGA and LDA are presented. Dashed lines and shaded bands are guides to the eye. [Used by permission of Elsevier, from Umemoto et al. (2008)].

iron concentration, the transition pressure in these two types of configurations differs by 50 GPa (Fig. 22). This effect is not important for $X_{Fe} \geq 0.5$ because irons are close and interacting to start with.

For both iron configurations with $X_{Fe} = 0.125$, the fraction of LS iron as a function of pressure and temperature are plotted in Figure 23. Vibrational effects are not present in this calculation. Only the orbital degeneracy (magnetic entropy) is included. It can be expected that the site degeneracies have greater effect on the transition pressure in the iron-(110) configuration because there are more combinations of the iron displacement.

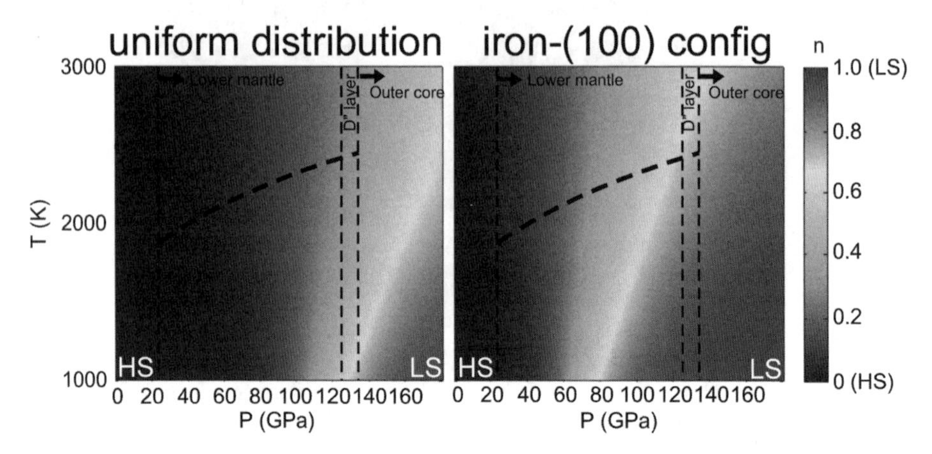

Figure 23. LS iron fraction (n_{LS}) of $(Mg_{0.875}Fe_{0.125})SiO_3$ in the case of iron distributed uniformly and the case of iron placed on the (110) plane. [Used by permission of Elsevier, from Umemoto et al. (2008)].

SPIN-STATE CROSSOVER IN POST-PEROVSKITE

The spin state and its crossover in iron-bearing $MgSiO_3$ post-perovskite are still very unclear. This is expected to be the major constituent of the D″ region, i.e, the region extending up to ~250 km above the core-mantle boundary (Murakami et al. 2004; Oganov and Ono 2004; Tsuchiya et al. 2004). Most of the experimental work on this phase was focused on the perovskite-to-post-perovskite transition in $MgSiO_3$ and the properties of $MgSiO_3$-post-perovskite, including estimations of the discontinuity in seismic wave speed produced by this transition (Garnero et al. 2004; Hernlund et al. 2005; Lay et al. 2006; van der Hilst et al. 2007; Wentzcovitch et al. 2006), elasticity (Merkel et al. 2006; Stackhouse and Brodholt 2007b; Wentzcovitch et al. 2006; Wookey et al. 2005), and thermal conductivity (Buffett 2007; Matyska and Yuen 2005a,b; Naliboff and Kellogg 2006). Synthesis (Mao et al. 2004), chemistry (Murakami et al. 2005), and Fe-Mg partitioning (Kobayashi et al. 2005) between $(Mg,Fe)SiO_3$-post-perovskite, perovskite, and ferropericlase have also been investigated. As to the valence- and spin-state of iron in $(Mg,Fe)SiO_3$-post-perovskite, there are very limited experimental works (Jackson et al. 2009; Lin et al. 2008) and they are not in agreement with each other. The experimental data presented in Lin et al. (2008) showed no ferric iron in the sample and ferrous iron appeared to be in the IS state at 134 GPa in the temperature range of 300-3200 K. However, ferric iron appears to be the dominant species in other experiments at 120 GPa and room temperature (Jackson et al. 2009).

First-principles GGA calculations agree that ferrous iron should remain in the HS state at least up to 150 GPa (Stackhouse et al. 2006; Zhang and Oganov 2006; Caracas and Cohen 2008), and IS should not be observed. As to ferric iron, it should also remain in the HS state

(Zhang and Oganov 2006). In summary, no spin-state crossover is observed in calculations, consistent with the experimental results by Jackson et al. (2009). However, the transition pressure should depend strongly on the exchange correlation functional and on the inclusion of on-site Coulomb interaction (Hubbard *U*). Clearly, much more needs to be investigated in this system to better understand this complex system.

SUMMARY

Results of several first-principles calculations on the spin-state crossover of iron in lower-mantle minerals have been reviewed. The LDA+*U* method gives desirable atomic and electronic structure in ferropericlase, (Mg,Fe)O. Both low-spin and high-spin ferropericlase are insulating. Jahn-Teller distortion is observed around iron in the high-spin state. A vibrational virtual crystal model (VVCM) permitted calculations of thermodynamic properties of this system at lower mantle conditions. Predictions are overall in good agreement with several experimental data sets. They display anomalies in the bulk modulus and allowed predictions of anomalies in thermodynamic properties and of the elastic signature of this phase in the mantle. An intriguing possibility of a viscosity anomaly caused by this crossover in the mantle has been raised. Improvements in these calculations to go beyond the ideal HS-LS solid solution are still desirable, as well as self-consistent calculations of the Hubbard *U*. These upgrades should improve agreement between predictions and measurements of crossover pressure ranges.

(Mg,Fe)SiO_3 perovskite is a more difficult system to investigate, therefore more controversial. Currently, there is lack of consensus regarding the existence of IS iron in perovskite. All calculations, irrespective of exchange-correlation functional used, agree on one issue: the IS state is not energetically competitive and no HS-to-IS crossover is expected to occur at lower-mantle pressures. The calculated HS-LS crossover pressure in ferrous iron strongly depends on the exchange-correlation functional (LDA, GGA, or DFT+*U*), and on the iron distribution in the supercell. This makes it more difficult to compare results with or interpret experimental data. A non-ideal HS-LS solid solution treatment appears to be essential for this system. On the positive side, the HS-LS crossover does not appear to affect the compressibility of this system to experimentally detectable levels. In the lower mantle, the change in compressibility of this system should be even less detectable. A thorough study of ferric iron using a more appropriate exchange-correlation functional or the DFT+*U* method is still needed for more extensive comparison between experimental data and theoretical results.

(Mg,Fe)SiO_3 post-perovskite is the least understood phase. Existing experimental data appear contradictory, and computational work is limited. The spin-state crossover in (Mg,Fe)SiO_3-post-perovskite is still a wide open question.

ACKNOWLEDGMENT

This research was supported by NSF grants EAR 0810272, EAR 0635990, ATM 0428774, EAR 0757903, and by the MRSEC Program of the National Science Foundation under Awards No. DMR-0212302 and No. DMR-0819885.

REFERENCES

Badro J, Fiquet G, Guyot F, Rueff JP, Struzhkin VV, Vanko G, Monaco G (2003) Iron partitioning in Earth's mantle: toward a deeo lower mantle discontinuity. Science 300:789-791

Badro J, Rueff JP, Vanko G, Monaco G, Fiquet G, Guyot F (2004) Electronic transitions in perovskite: possible nonconvecting layers in the lower mantle. Science 305:383-386

Badro J, Struzhkin VV, Shu J, Hemley RJ, Mao H, Kao C, Rueff J, Shen G (1999) Magnetism in FeO at megabar pressure from X-ray emission spectroscopy. Phys Rev Lett 83:4101-4104

Barnhoorn A, Bystricky M, Kunze K, Burlini L, Burg JP (2005) Strain localisation in bimineralic rocks: Experimental deformation of synthetic calcite-anhydrite aggregates. Earth Planet Sci Lett 240:748-763

Baroni S, de Gironcoli S, Dal Corso A, Giannozzi P (2001) Phonons and related crystal properties from density-functional perturbation theory. Rev Mod Phys 73:515-562

Bengtson A, Li J, Morgan D (2009) Mössbauer modeling to interpret the spin sate of iron in (Mg,Fe)SiO_3. Geophys Res Lett 36:L15301

Bengtson A, Persson K, Morgan D (2008) *Ab initio* study of the composition dependence of the pressure-induced spin crossover in perovskite (Mg_{1-x},Fe_x)SiO_3. Earth Planet Sci Lett 265:535-545

Blaha P, Schwarz K, Madsen GKH, Kvasnicka D, Luitz J (2001) *WIEN2k*, An Augmented Plane Wave + Local Orbitals Program for Calculating Crystal Properties (Karlheinz Schwarz, Techn. Universität Wien, Austria). ISBN 3-9501031-1-2

Boehler R (2000) High-pressure experiments and the phase diagram of lower mantle and core materials. Rev Geophys 38:221-245

Brown JM, Shankland TJ (1981) Thermodynamic parameters in the Earth as determined from seismic profiles. Geophys J R Astr Soc 66:579-596

Buffett BA (2007) A bound on her flow below a double crossing of the perovskite-post-perovskite phase transition. Geophys Res Lett 34:L17302

Caracas R, Cohen RE (2008) Ferrous iron in post-perovskite from first-principles calculations. Earth Planet Sci Lett 168:147-152

Carrier P, Justo JF, Wentzcovitch RM (2008) Quasiharmonic elastic constants corrected for deviatoric thermal stresses. Phys Rev V 78:144302

Ceperley DM, Alder BJ (1980) Ground state of the electron gas by a stochastic method. Phys Rev Lett 45:566-569

Chen YL, Yang DP (2007) Mössbauer Effect in Lattice Dynamics. Wiley-VCH, Weinheim

Cococcioni M (2010) Accurate and efficient calculations on strongly correlated minerals with the LDA+U method: review and perspectives. Rev Mineral Geochem 71:147-167

Cococcioni M, de Gironcoli S (2005) Linear response approach to the calculation of the effective interaction parameters in the LDA+*U* method. Phys Rev B 71:035105

Cohen RE, Mazin II, Isaak DG (1997) Magnetic collapse in transition metal oxides at high pressure: Implications for the Earth. Science 275:654-657

Crowhurst J, Brown JM, Goncharov A, Jacobsen SD (2008) Elasticity of (Mg,Fe)O through the spin transition of iron in the lower mantle. Science 319:451-453

Dziewonski AM, Anderson DL (1981) Preliminary references Earth model (PREM). Phys Earth Planet Interiors 25:297-356

Ekstrom G, Dziewonski AM (1998) The unique anisotropy of the Pacific upper mantle. Nature 394:168-172

Fabris S, de Gironcoli S, Baroni S, Vicario G, Balducci G (2005) Taming multiple valency with density functionals: A case study of defective ceria. Phys Rev B 71:041102

Fei Y, Zhang L, Corgne A, Watson H, Ricolleau A, Meng Y, Prakapenka V (2007) Spin transition and equation of state of (Mg,Fe)O solid solutions. Geophys Res Lett 34:L17307

Forte AM, Mitrovica JK (2001) Deep-mantle high-viscosity flow and thermochemical structure inferred from seismic and geodynamic data. Nature 410:1049-1056

Fyfe WS (1960) The possibility of d-electron uncoupling in olivine at high pressures. Geochim Cosmochim Acta 19:141-143

Gaffney ES, Anderson DL (1973) Effect of low-spin Fe^{2+} on the composition of the lower mantle. J Geophys Res 78:7005-7014

Garnero EJ, Maupin V, Lay T, Fouch M (2004) Variable azimuthal anisotropy in earth's lowermost mantle. Science 306:259-261

Gavriliuk AG, Lin JF, Lyubutin IS, Struzhkin VV (2006) Optimization of the conditions of synchrotron Mössbauer experiment for studying electron transition at high pressures by the example of (Mg,Fe)O magnesiowüstite. J Exp Theor Phys Lett 84:161-166

Giannozzi P, Baroni S, Bonini N, Calandra M, Car R, Cavazzoni C, Ceresoli D, Chiarotti GL, Cococcioni M, Dabo I, Dal Corso A, de Gironcoli S, Fabris S, Fratesi G, Gebauer R, Gerstmann U, Gougoussis C, Kokalj A, Lazzeri M, Martin-Samos L, Marzari N, Mauri F, Mazzarello R, Paolini S, Pasquarello A, Paulatto L, Sbraccia C, Scandolo S, Sclauzero G, Seitsonen AP, Smogunov A, Umari P, Wentzcovitch RM (2009) QUANTUM ESPRESSO: a modular and open-source software project for quantum simulations of materials. J Phys Condens Matter 21:395502

Goncharov AF, Struzhkin VV, Jacobsen SD (2006) Reduced radiative conductivity of low-spin (Mg,Fe)O in the lower mantle. Science 312:1205-1208

Grand SP, van der Hilst RD, Widyantoro S (1997) Global seismic tomography: A snapshot of convection in the earth. Geol Soc Am Today 7:1

Hernlund JW, Thomas C, Tackley PJ (2005) A doubling of the post-perovskite phase boundary and structure of the Earth's lowermost mantle. Nature 434:882-886

Hofmeister AM (2006) Is low-spin Fe^{2+} present in Earth's mantle? Earth Planet Sci Lett 243:44-52

Hohenberg P, Kohn W (1964) Inhomogeneous electron gas. Phys Rev 136:B864-871

Holtzman BK, Kohlstedt DL, Zimmerman ME, Heidelbach F, Hiraga T, Hustoft J (2003) Melt segregation and strain partitioning: Implications for seismic anisotropy and mantle flow. Science 301:1227-1230

Hsu H, Umemoto K, Cococcioni M, Wentzcovitch RM (2009a) First-principles study for low-spin $LaCoO_3$ with structurally consistent Hubbard *U*. Phys Rev B 79:125124

Hsu H, Umemoto K, Blaha P, Wentzcovitch EM (2009b) Spin states and hyperfine interactions of iron in $(Mg,Fe)SiO_3$ perovskite under pressure. Earth Planet Sci Lett (in review)

Jackson JM, Sturhahn W, Shen G, Zhao J, Hu MY, Errandonea D, Bass JD, Fei Y (2005) A synchrotron Mössbauer spectroscopy study of $(Mg,Fe)SiO_3$ perovskite up to 120 GPa. Am Mineral 90:199-205

Jackson JM, Sturhahn W, Tschauner O, Lerche M, Fei Y (2009) Behavior of iron in $(Mg,Fe)SiO_3$ post-perovskite assemblages at Mbat pressures. Geophys Res Lett 36:L10301

Kantor I, Dubrovinsky LS, McCammon CA (2006) Spin crossover in (Mg,Fe)O: A Mössbauer effect study with an alternative interpretation of X-ray emission spectroscopy data. Phys Rev B 73:100101(R)

Kantor I, Dubrovinsky LS, McCammon CA (2007) Reply to "Comment on 'Spin crossover in (Mg,Fe)O: A Mössbauer effect study with an alternative interpretation of X-ray emission spectroscopy data'". Phys Rev B 75:177103

Kantor I, Dubrovinsky LS, McCammon CA, Steinle-Neumann G, Kantor A (2009) Short-range order and Fe clustering in $Mg_{1-x}Fe_xO$ under pressure. Phys Rev B 80:014204

Keppler H, Kantor I, Dubrovinsky LS (2007) Optical absorption spectra of ferropericlase to 84 GPa. Am Mineral 92:433-436

Kobayashi Y, Kondo T, Ohtani E, Hirao N, Miyajima N, Yagi T, Nagase T, Kikegawa T (2005) Fe-Mg partitioning between $(Mg,Fe)SiO_3$ post-perovskite, perovskite, and magnesiowüstite in the Earth's lower mantle. Geophys Res Lett 32:L19301

Kohn W, Sham LJ (1965) Self-consistent equations including exchange and correlation effects. Phys Rev 140:A1133-1138

Kulik H, Cococcioni M, Scherlis DA, Marzari N (2006) Density functional theory in transition metal chemistry: A self-consistent Hubbard *U* approach. Phys Rev Lett 97:103001

Lay T, Hernlund J, Garnero EJ, Thorne M (2006) A post-perovskite lens and D″ hear flux beneth the central pacific. Science 314:1272-1276

Li J, Struzhkin V, Mao H, Shu J, Memeley R, Fei Y, Mysen B, Dera P, Pralapenka V, Shen G (2004) Electronic spin state of iron in lower mantle perovskite. Proc Nat Acad Sci USA 101:14027-14030

Li L, Brodholt JP, Stackhouse S, Weidner DJ, Alfredsson M, Price GD (2005) Electronic spin state of ferric iron in Al-bearing perovskite in the lower mantle. Geophys Res Lett 32:L17307

Lin JF, Gavriliuk AG, Struzhkin VV, Jacobsen SD, Sturhahn W, Hu M, Chow P, Yoo CS (2006a) Pressure-induced electronic spin transition of iron in magnesiowüstite-(Mg, Fe)O. Phys Rev B 73:113107

Lin JF, Gavriliuk AG, Sturhahn W, Jacobsen SD, Zhao J, Lerche M, Hu M (2009) Synchrotron Mössbauer spectroscopy of ferropericlase at high pressure and temperatures. Am Mineral 94:594-599

Lin JF, Jacobsen SD, Sturhahn W, Jackson JM, Zhao J, Yoo CS (2006b) Sound velocities of ferropericlase in Earth's lower mantle. Geophys Res Lett 33:L22304

Lin JF, Struzhkin VV, Gavriliuk AG, Lyubutin I (2007a) Comment on "Spin crossover in (Mg, Fe)O: A Mössbauer effect study with an alternative interpretation of X-ray emission spectroscopy data". Phys Rev B 75:177102

Lin JF, Struzhkin VV, Jaconsen SD, Hu MY, Chow P, Kung J, Liu H, Mao Hk, Hemley R (2005) Spin transition of iron in magnesiowüstite in the Earth's lower mantle. Nature 436:377-380

Lin JF, Tsuchiya T (2008) Spin transition of iron in the earth's lower mantle. Phys Earth Planet Inter 170:248-259

Lin JF, Vanko G, Jacobsen SD, Iota V, Struzhkin VV, Prakapenka VB, Kuznetsov A, Yoo CS (2007c) Spin transition zone in earth's lower mantle. Science 317:1740-1743

Lin JF, Watson H, Vanko G, Alp EE, Prakapenka VB, Dera P, Struzhkin V, Kubo A, Zhao J, McCammon C, Evans WJ (2008) Intermediate-spin ferrous iron in lowermost mantle post-perovskite and perovskite. Nature Geosci 1:688-691

Lin JF, Weir ST, Jackson DD, Evans WJ, Yoo CS (2007b) Electrical conductivity of the low-spin ferropericlase in the Earth's lower mantle. Geophys Res Lett 34:L16305

Lundin S, Catalli K, Santillan J, Shim SH, Prakapenka VB, Kunz M, Meng Y (2008) Effect of Fe on the equation of state of mantle silicate perovskite over 1 Mbar. Phys Earth Planet Inter 168:97-102

Madsen G, Blaha P, Schwarz K, Sjoestedt E, Nordstroem L (2001) Efficient linearization of the augmented plane-wave method. Phys Rev B 64:195134

Mao WL, Shen G, Prakapenka VB, Meng Y, Campbell AJ, Heinz DL, Shu J, Hemley R, Mao H (2004) Ferromagnesian postperovekite silicates in the D″ layer of the Earth. Proc Nat Acad Sci USA 101:15867-15869

Matyska C, Yuen DA (2005a) The importance of radiative heat transfer on superlimes in the lower mantle with the new post-perovskite phase change. Earth Planet Sci Lett 234:71-81

Matyska C, Yuen DA (2005b) Lower mantle dynamics with the post-perovskite phase change, radiative thermal conductivity, temperature- and depth-dependent viscosity. Phys Earth Planet Inter 196-207

McCammon C, Kantor I, Narygina O, Rouquette J, Ponkratz U, Sergueev I, Mezouar M, Prakapenka V, Dubrovinsky L (2008) Stable intermediate-spin ferrous iron in lower-mantle perovskite. Nature Geosci 1:684-687

McDonough WF, Sun S (1995) The composition of Earth. Chem Geol 120:223-253

Merkel S, Kubo A, Miyagi L, Speziale S, Duffy T, Mao H, Wenk H (2006) Plastic deformation of MgGeO3 post-perovskite at lower mantle pressures. Science 311:644-646

Mitrovica JK, Forte AM (2004) A new inference of mantle viscosity based on joint inversion of concecion and glacial isostatic adjustment data. Earth Planet Sci Lett 225:177-189

Murakami M, Hirose K, Kawamura K, Sata N, Ohishi Y (2004) Post-perovskite phase transition in $MgSiO_3$. Science 304:855-858

Murakami M, Hirose K, Sata N, Ohishi Y (2005) Post-perovskite phase transition and mineral chemistry in the pyrolitic lowermost mantle. Geophys Res Lett 32:L03304

Naliboff JB, Kellogg LH (2006) Dynamics effects of a step-wise increase in thermal conductivity and viscosity in the lowermost mantle. Geophys Res Lett 33:L12S09

Narygina O, Mattesini M, Kantor I, Pascarelli S, Wu X, Aquilanti G, McCammon C, Dubrovinsky L (2009) High-pressure experimental and computational XANES studies of $(Mg,Fe)(Si,Al)O_3$ perovskite and (Mg,Fe)O ferropericlase as in the Earth's lower mantle. Phys Rev B 79:174115

Nielsen OH, Martin RM (1983) First-principles calculations of stress. Phys Rev Lett 50:697-700

Oganov AR, Ono S (2004) Theoretical and experimental evidence for a post-perovskite phaseof $MgSiO_3$ in Earth's D″ layer. Nature 430:445-448

Ohnishi S (1978) A theory of pressure-induced high-spin-low-spin transition of transition metal oxides. Phys Earth Planet Inter 17:130-139

Pasternak MP, Taylor RD, Li X, Nguyen JH, McCammon CA (1997) High pressure collapse of magnetism in $Fe_{0.94}O$: Mössbauer spectroscopy beyond 100 GPa. Phys Rev Lett 79:5046-5049

Perdew JP, Burke K, Ernzerhof M (1996) Generalized gradient approximation made simple. Phys Rev Lett 77:3865-3868

Perdew JP, Ruzsinszky A (2010) Density functional theory of electronic structure: a short course for mineralogists and geophysicists. Rev Mineral Geochem 71:1-18

Perdew JP, Zunger A (1981) Self-interaction correction to density-functional approximations for many-electron systems. Phys Rev B 23:5048-5079

Persson K, Bengtson A, Ceder G, Morgan D (2006) *Ab initio* study of the composition dependence of the pressure-induced spin transition in the $(Mg_{1-x},Fe_x)O$ system. Geophys Res Lett 33:L16306

Petrilli HM, Blöchl PE, Blaha P, Schwarz K (1998) Electric-field-gradient calculations using the projector augmented wave method. Phys Rev B 57:14690-14697

Rappe AM, Rabe KM, Kaxrias E, Joannopoulos JD (1990) Optimized pseudopotentials. Phys Rev B 41:1227-1230

Ringwood AE (1982) Phase transformations and differentiation in subducted lithosphere: implications for mantle dynamics basalt petrogenesis and crustal evolution. J Geol 90:611-642

Sherman DM (1988) High-spin to low-spin transition of iron(II) oxides at high pressures: possible effects on the physics and chemistry of the lower mantle. *In*: Structural and Magnetic Phase Transition in Minerals. Ghose S, Coey JMD, Salje E (eds) Springer Verlag, New York, p 113-118

Sherman DM (1991) The high-pressure electronic structure of magnesiowüstite (Mg,Fe)O: applications to the physics and chemistry of the lower mantle. J Geophys Res 96(B9):14299-14312

Sherman DM, Jansen HJF (1995) First-principles predictions of the high-pressure phase transition and electronic structure of FeO: implications for the chemistry of the lower mantle and core. Geophys Res Lett 22:1001-1004

Speziale S, Lee VE, Clark SM, Lin JF, Pasternak MP, Jeanloz R (2007) Effects of Fe spin transition on the elasticity of (Mg,Fe)O magnesiowüstites and implications for the seismological properties of the Earth's lower mantle. J Geophys Res 112:B10212

Speziale S, Milner A, Lee VE, Clark SM, Pasternak M, Jeanloz R (2005) Iron spin transition on Earth's lower mantle. Proc Nat Acad Sci USA 102:17918-17922

Stackhouse S, Brodolt JP (2007b) The high-temperature elasticity of $MgSiO_3$ post-perovskite. *In*: Post-perovskite: The Last Mantle Phase Transition, Geophysical Monograph. Vol. 174. Hirose K, Brodholt J, Lay T, Yuen D (eds), American Geophysical Union, Washington DC, p 99-113

Stackhouse S, Brodolt JP, Dobson DP, Price GD (2006) Electronic spin transitions and the seimic properties of ferrous iron–bearing $MgSiO_3$ post-perovskite. Geophys Res Lett 33:L12S03

Stackhouse S, Brodolt JP, Price GD (2007a) Electronic spin transitions in iron-bearing $MgSiO_3$ perovskite. Earth Planet Sci Lett 253:282-290

Sugano S, Tanabe Y, Kamimura H (1970) Multiplets of Transition-Metal Ions in Crystals. Academic Press, New York and London

Touloukian YS, Kirdby RK, Taylor RE, Lee TYR (1997) Thermophysical Properties of Matter. Volume 13. Plenum Press, New York

Troullier N, Martines JL (1991) Efficient pseudopotentials for plane-wave calculations. Phys Rev B 43:1993-2006

Tsuchiya T, Tsuchiya J, Umemoto K, Wentzcovitch RM (2004) Phase transition in $MgSiO_3$ perovskite in the Earth's lower mantle. Earth Planet Sci Lett 224:241-248

Tsuchiya T, Wentzcovitch RM, da Silva CRS, de Gironcoli S (2006) Spin transition in magnesiowüstite in earth's lower mantle. Phys Rev Lett 96:198501

Umemoto K, Hsu H, Wentzcovitch RM (2009) Effect of site degeneracies on the spin crossover in $(Mg,Fe)SiO_3$ perovskite. Phys Earth Planet Inter (in press). doi: 10.1016/j.pepi.2009.10.014

Umemoto K, Wentzcovitch RM, Yu YG, Requist R (2008) Spin transition in $(Mg,Fe)SiO_3$ perovskite under pressure. Earth Planet Sci Lett 276:198-206

van der Hilst RD, de Hoop MV, Wang P, Shim SH, Ma P, Tenorio L (2007) Seismostratigraphy and thermal structure of earth's core-mantle boundary region. Science 315:1813-1817

van Westrenen W, Li J, Fei Y, Frank MR, Hellwig H, Komabayashi T, Mibe K, Minarik WG, van Orman JA, Watson HC, Funakoshi K, Schmidt W (2005) Thermoelastic properties of $(Mg_{0.64}Fe_{0.36})O$ ferropericlase based on in situ X-ray diffraction to 26.7 GPa and 2173 K. Phys Earth Planet Interiors 151:163-176

Vanderbilt D (1990) Soft self-consistent pseudopotentials in a generalized eigenvalue formalism. Phys Rev B 41:R7892-7895

Vanko G, de Groot FMF (2007) Comment on "Spin crossover in (Mg,Fe)O: A Mössbauer effect study with an alternative interpretation of X-ray emission spectroscopy data". Phys Rev B 75:177101

Wallace D (1972) Thermodynamics of Crystals. Wiley, New York

Wentzcovitch RM, Justo JF, Wu Z, da Silva CRS, Yuen DA, Kohlstedt D (2009) Anomalous compressibility of ferropericlase throughout the iron spin cross-over. Proc Nat Acad Sci USA 106:8447-8452

Wentzcovitch RM, Karki BB, Cococcioni M, de Gironcoli S (2004) Thermoelastic peroperties of MgSiO3-perovskite: Insights on the nature of the earth's lower mantle. Phys Rev Lett 92:018501

Wentzcovitch RM, Martins JL, Price GD (1993) Ab initio molecular dynamics with variable cell shape: application to MgSiO3. Phys Rev Lett 70:3947

Wentzcovitch RM, Tsuchiya T, Tsuchiya J (2006) $MgSiO_3$ postperovskite at D″ conditions. Proc Nat Acad Sci USA 103:543-546

Wentzcovitch RM, Yu YG, Wu Z (2010) Thermodynamic properties and phase relations in mantle minerals investigated by first principles quasiharmonic theory. Rev Mineral Geochem 71:59-98

Wookey J, Stackhouse S, Kendall JM, Brodholt JP, Price GD (2005) Efficacy of post-perovskite as an explanation for lowermost-mantle seismic properties. Nature 438:1003-1007

Wu Z, Justo JF, da Silva CRS, de Gironcoli S, Wentzcovitch RM (2009) Anomalous thermodynamic properties in ferropericlase throughout its spin crossover transition. Phys Rev B 80:014409

Yamazaki D, Karato S-i (2001) Some mineral physics constraints on the rheology and geotherm structure of Earth's lower mantle. Am Mineral 86:381-391

Yu Y, Wentzcovitch RM, Tsuchiya T (2007) First-principles investigation of the postpinel transition in Mg_2SiO_4. Geophys Res Lett 34:L10306

Yu Y, Wu Z, Wentzcovitch RM (2008) α-β-γ transformations in Mg_2SiO_4 in Earth's transition zone. Earth Planet Sci Lett 273:115-122

Zhang F, Oganov AR (2006) Valence state and spin transitions of iron in Earth's mantle silicates. Earth Planet Sci Lett 249:436-443

Zhao Y, Truhlar DG (2010) The Minnesota density functionals and their applications to problems in mineralogy and geochemistry. Rev Mineral Geochem 71:19-37

Reviews in Mineralogy & Geochemistry
Vol. 71 pp. 201-224, 2010

Simulating Diffusion

Michael W. Ammann, John P. Brodholt, and David P. Dobson

Department of Earth Sciences
University College London
London, WC1E 6BT, United Kingdom

m.ammann@ucl.ac.uk

INTRODUCTION

Ionic or atomic diffusion controls chemical exchange between and within different crystalline, melt and fluid phases. It can control the kinetics of phase transitions, the rate at which minerals grow, the degree of compositional zoning in minerals, and other important geochemical processes. Diffusion also plays a central role in the rheological properties of minerals and melts. In diffusion creep stress is accommodated through ions migrating from regions of high stress to regions of low stress. This is achieved either through bulk diffusion or grain boundary diffusion. In dislocation creep the rate-limiting step is often dislocation climb—the process where a dislocation has to migrate out of the dislocation plane to avoid an obstacle—and this requires ionic diffusion.

There are two main reasons for studying diffusion theoretically. The first is to determine the atomistic mechanism of diffusion. This can provide understanding of the underlying mechanisms, and allow one to extrapolate results to other systems. When different sets of experimental results disagree, theory can often help to decide which is correct or explain the differences. But the most valuable reason for using theory is to predict diffusion properties for systems or conditions where no data exist. As will be shown, theoretical calculations can be used to predict absolute diffusion rates very accurately—perhaps as accurately as is obtainable in high-pressure and high-temperature experiments. However, let it be understood that we are not advocating abandoning experimentation for theory.

In this paper we will use our recent *ab initio* calculations on the absolute diffusion rates of periclase and perovskite as an example of what can be done and how to do it.

BASIC METHODS

Ab initio vs. empirical potentials

The basis for the techniques described here are atomistic; the system is described as set of interacting atoms or ions, and it is these interactions that govern its behavior. The interactions can either be described from first principles (i.e., quantum mechanically) or empirically. The first principles method treats the system as a set of interacting nuclei and electrons, and uses (approximate) solutions to Schrödinger's equations to obtain energies and forces. Currently, almost all work involving first principles methods, such as the one that will be discussed in this chapter, make use of density functional theory (DFT, Hohenberg and Kohn 1964; Kohn and Sham 1965). The empirical approach uses a predefined inter-atomic (or inter-molecular) potential, which is fit to some experimental property or to a first-principles result. There are also the so-called semi-empirical methods, which are based on quantum mechanics but contain many approximations and include experimentally derived parameters. The different methods

1529-6466/10/0071-0010$05.00 DOI: 10.2138/rmg.2010.71.10

have their pluses and minuses. The empirical approach allows us to look at far larger systems than the first principles (or *ab initio*) methods. The downside is that the empirically-derived potentials may not work accurately in systems and atomic environments for which they were not originally designed. This is a particular worry for diffusion studies since atoms migrate from their normal configuration through a very different coordination environment than their equilibrium position. Moreover, some empirical potentials have bond-angle dependent parameters and it is not clear what to do with those as atoms migrate. For instance, potential models for Si-O interaction often have a three-body bond angle term in order to maintain the correct coordination with oxygen. For tetrahedrally coordinated silicon, this would be 109.47°. But as the silicon ion moves out of the tetrahedron, it changes coordination, and the bond-angle term may impose inappropriate forces. The *ab initio* techniques do not suffer from this in the same way, although they do have other approximations which add uncertainty to the results.

Predicting diffusion coefficients in fluids and melts

Although this paper will concentrate on modeling diffusion in crystalline phases, it is worth mentioning how diffusion is modeled in fluids and melts. Using atomistic level simulations to predict the diffusion properties of fluids and melts is, in principle, simpler than in solids. Molecular dynamics simulations, for instance, allows one to follow the path of all the atoms as they move as function of time. By keeping track of the mean-squared-displacement of each of atom in the simulation, one obtains a direct measurement of the diffusion coefficient for each atomic species. The only complication for viscous silicate melts is that the simulation must be run for sufficient time to allow adequate sampling of space by the atoms (e.g., Wan et al. 2005; Mookherjee et al. 2008; Nevins et al. 2009). Less viscous fluids, such as aqueous solution, need far less time. This molecular dynamic approach has also been attempted for solids, but diffusion coefficients in solids are many orders of magnitude slower than even viscous melts and jumps are very rare. This means that the simulation times are generally prohibitively long, and can only be used with empirical potentials and for simple jumps. Nevertheless, this technique has been used for MgO (see below) and for diffusion of relatively small impurity ions such as H and He (e.g., Reich et al. 2007).

Predicting diffusion coefficients in crystalline phases

As mentioned in the previous section, the situation for mineral phases is not so simple; diffusion time scales are orders of magnitude slower than in melts and fluids such that atomic jumps in a direct MD simulation are very rare events. With slower diffusing systems or when using *ab initio* methods, the normal approach for calculating diffusion coefficients is to break the diffusion process into parts and calculate the different parts individually. The process is given pictorially in Figure 1.

Basically the process is one of calculating the number of times a vacancy jumps between two sites per unit time. This is known as the jump frequency, Γ. For an individual defect the jump frequency is given by (e.g., Poirier 1985)

$$\Gamma = \nu e^{\frac{\Delta G}{kT}} = \nu e^{\frac{\Delta S}{k}} e^{\frac{-\Delta H}{kT}} \tag{1}$$

where ΔG, ΔS and ΔH are the migration free energy, the entropy and enthalpy respectively. ν is a characteristic attempt frequency of the atom trying to jump over the barrier. T and k have their usual meanings of temperature and the Boltzmann constant respectively. The diffusion coefficient for a species is then given as

$$D = N_V \frac{Z}{6} l^2 \Gamma \tag{2}$$

where N_V is the vacancy concentration in the crystal at the particular temperature and pressure, l is the jump distance, and Z is a geometric factor incorporating the number of possible jumps.

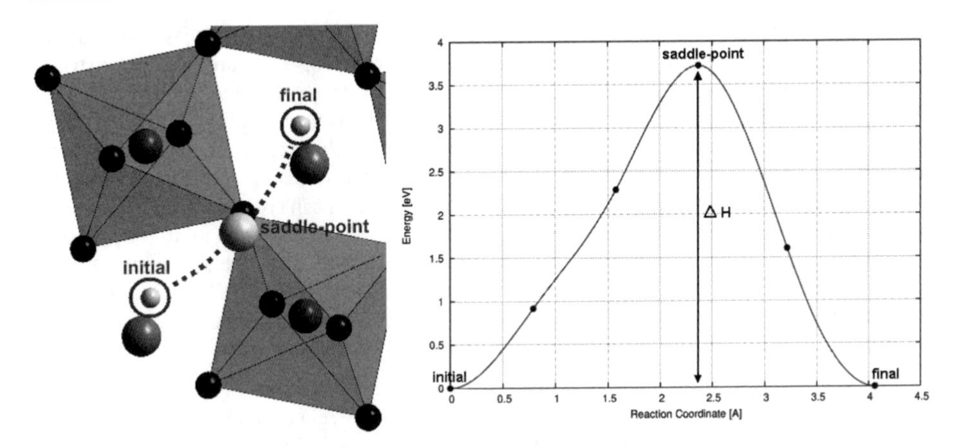

Figure 1. Left hand side: a magnesium ion migrates in $MgSiO_3$ perovskite from its initial to its final position by overcoming a saddle-point. Right hand side: The energy barrier encountered by the migrating ion as calculated with CI-NEB. The saddle point is at the maximum of the energy barrier and the energy difference from the initial state defines the migration enthalpy ΔH.

The enthalpy of migration, ΔH, is obtained from the difference between the energy of the system at the saddle point and its ground state energy before attempting a jump:

$$\Delta H = E_{saddle} - E_{groundstate} \tag{3}$$

where the simulation is performed at constant volume. The issue of how to compare thermodynamic quantities measured in experiments (at constant pressure) with the ones obtained from theoretical calculations (at constant volume) is discussed at length by, e.g., Catlow et al. (1981) or Harding (1989).

For systems with high symmetry (i.e., MgO), finding the saddle point is reasonably straightforward since it often lies on a line of symmetry. For lower symmetry materials, such as perovskite, it is not obvious where the saddle point is. It is possible to use trial and error, or a grid searching method to find the saddle point, but this is quite inefficient. A better approach is to use a method such as the Nudged-Elastic-Band; this is described later.

The attempt frequency and the migration entropy are obtained together using harmonic transition-state theory. This is also called Vineyard Theory (Vineyard 1957). In this theory, many-body effects are incorporated by calculating the full 3N vibrational frequencies at the equilibrium point and the $3N$–1 non-negative vibrational frequencies at the saddle point. It can be shown that ν and ΔS are simultaneously obtained from

$$v^* = ve^{\Delta S/k} = \frac{\prod_{j=1}^{3N} f_j}{\prod_{j=1}^{3N-1} f_j'} \tag{4}$$

where N is the number of atoms in the simulation cell, f_j are the vibrational frequencies at the equilibrium site, and f_j' are the frequencies at the saddle point. The phonon spectrum at the saddle point contains one negative frequency; this is due to the unstable atom sitting exactly on the saddle point. The product of frequencies at the saddle point only contains, therefore, N–1 positive frequencies. The phonon spectra can be calculated using a standard lattice dynamics approach, either using *ab initio* forces or empirical potentials.

Vineyard-theory is not an exact theory. It assumes that the energy surfaces at the saddle-point and at the equilibrium position are perfectly harmonic. This does not have to be strictly true and one has then to correct for the anharmonicity (Sangster and Stoneham 1984). However, in our DFT calculations we find that anharmonicity is negligible (<1% deviation). Another idealization is that Vineyard-theory assumes that each jump is successful—i.e., each ion jumping towards a vacancy will reach the vacancy once it has crossed the saddle-point. Again, for real systems this is not the case as the jumping atom can cross the saddle-point only to immediately return back to its initial state. This can happen because the jump is a complex, dynamical many-body process (dynamical correlation). Luckily, unsuccessful jumps are rare and are unlikely to exceed 10% as has been shown by Flynn and Jacucci (1982). This is because the time after which the system loses its memory of previous jumps (onset of randomization) is generally much shorter than any vibrational period of any particle in the system.

In order to ensure that there is only one negative frequency at the saddle point, it is essential to find the saddle-point exactly—i.e., the crystal must be fully relaxed and all atomic forces are relaxed to within a very small tolerance (typically $<10^{-6}$ eV/Å). This is another reason to use a method such as the nudged-elastic band, rather than trial and error or random searching.

The last unknown required to calculate absolute diffusion rates is the number of vacancies, N_V. Estimating N_V is non-trivial. Vacancy concentrations can vary significantly depending on the experimental conditions (i.e., pressure, temperature, oxygen fugacity, impurity concentration, sample history, etc.). It is possible to estimate the intrinsic vacancy concentration directly and self-consistently by calculating the vacancy energies. For instance, the number of Schottky vacancies n (where an atom is taken from the bulk of the crystal and added to the surface) is given as

$$n = Ne^{-\frac{\Delta G_V}{kT}} \tag{5}$$

where N is the number of atoms in the crystal and ΔG_V is the Gibbs free energy of vacancy formation. The energy can, if necessary, be calculated directly from first principles. However, for most Earth materials, the intrinsic concentration is very small, even at very high temperatures. The vacancy concentration of the crystal is more likely to be set by the number of impurities (i.e., the extrinsic concentration). For instance, Fe^{3+} on a normal 2+ cation site could be charge balanced by cation vacancies. Similarly, oxygen vacancies may be set by the concentration of 1+ ions. For well-characterized experimental samples, these can be estimated reasonably accurately, and allow us to compare our calculated diffusion rates directly with experimental measurements. Estimating vacancy concentrations for minerals in the Earth's mantle, however, is subject to considerable uncertainty. This has to be kept in mind when applying results to the Earth.

Defect calculations: Mott-Littleton, super-cells and embedded clusters

Crystalline defects exert a strong distance dependent perturbation on the crystal. Firstly, defects in crystals are often charged; this results in a Coulomb interaction that decays slowly as $1/r$ (r being the distance). Neutral defects can also have a slowly decaying electrostatic interaction, since they often possess strong dipole and quadrupole moments. Secondly, the crystal lattice is distorted due to relaxation of the ions around the defect (repelled or attracted depending on their charge relative to the defect). It is assumed that this lattice distortion decays as $1/r^3$. When modeling defects, especially highly charged vacancies, care has to be taken to ensure that these long-range interactions are taken into account, and that they do not artificially bias the results.

The classic approach is the Mott-Littleton Method (Mott and Littleton 1938). In this method, the defective crystal is divided into two separately treated spherical regions. The inner region surrounds the defect and is treated accurately by calculating the relaxations and distortions on the atoms from interatomic forces. The outer region is treated less accurately (as atoms only interact with the distortion and the charge of the defect but not with each other), and is used to shield the charge and distortion caused by the defect (polarization). This approach has only been used with (semi-)empirical interatomic potentials. Codes such as GULP (*https://www.ivec.org/gulp/*; Gale 1997; Gale and Rohl 2003) make this a relatively routine procedure, subject to the accuracy of the interatomic-potentials (Cherry et al. 1995a,b; Blanchard et al. 2005; Lowitzer et al. 2008; Ball et al. 2008; Béjina et al. 2009).

If we wish to use *ab initio* methods, in particular DFT, the most straightforward implementations make use of super-cells and periodic boundary conditions (PBC). PBC mean that the system is repeated infinitely in space. This is especially useful for crystalline lattices since it is then sufficient to calculate the properties of a single unit cell. Periodic boundary conditions yield the same result as if the unit cell has been repeated infinitely in all directions, forming a perfect, infinitely sized crystal. This approach, however, has its drawbacks when it comes to defect calculations, since the defect is also repeated infinitely along all directions. This leads to a very high concentration of defects, giving rise to spurious elastic interactions between neighboring cells (mirror images of the simulation). This interaction should scale as $1/L^3$, where L is the cell size. The effect of the elastic interaction can be reduced by using sufficiently large supercells (a large simulation unit built up from several unit cells), such that deformations at the cell-boundaries are negligible. However, the relaxation is almost never completely removed by the edge of the supercell, and the calculations contain a small artificial contribution from this. Nevertheless, this contribution is small and the super-cell method using DFT forces and energies has been used successfully on defect calculations in Earth materials such as olivine (Brodholt 1997; Brodholt and Refson 2000), perovskite (Brodholt 2000; Karki and Khanduja 2006a), post-perovskite (Karki and Khanduja 2007) and periclase (Karki and Khanduja 2006b).

An intermediate approach between using pure DFT with super-cells and Mott-Littleton methods for defect calculations is the so-called embedded cluster method. This again divides up space into regions: a central region, which treats the defect at the quantum mechanical level, a surrounding region that is treated classically via interatomic potentials, and a third outer region which is just a set a of fixed point-charges. This method is implemented in codes such as GUESS (Sushko et al. 1999, 2000a,b) and has been used to study defects in olivine (Braithwaite et al. 2002, 2003; Walker et al. 2006, 2009).

In our diffusion calculations which we discuss in this chapter, we used the supercell approach. First of all we wish to avoid the use of empirical pair-potentials and use *ab initio* energies and forces. We can use the tried and tested pseudopotentials (and closely related PAW parameters (Blöchl 1994; Kresse and Joubert 1999) in codes like VASP (Kresse and Hafner 1993; Kresse and Furthmüller 1996a,b). Secondly, super-cells of up to a few hundred atoms can now be routinely calculated. And thirdly, routines are readily available for the lattice dynamics calculations (Parlinski 2008; Baroni et al. 2001; Alfè 2009) required for the jump frequencies, and nudged-elastic-bands (Henkelman et al. 2000) for finding saddle points of the migrating atoms—something that is very important in complicated migrating pathways.

As mentioned above, PBC in combination with defective systems has its own difficulties. The introduction of charged point defects or defect clusters results in an artificial electrostatic self-interaction between the supercell and all its images. This electrostatic self-interaction can have rather large effects on the calculated defect energies. It is therefore worthwhile to briefly discuss how one can reduce the error made on the defect energetics when using PBC.

Firstly, Leslie and Gillan (1985) proposed a simple correction to the self-interaction of charged point defects. A charged defect within a cell with PBC is equivalent to a periodic array of charged defects. However, a periodically repeating array of charged supercells does not have a well-defined energy. This difficulty can be overcome by introducing a uniformly distributed background charge (so-called jellium) compensating the charge of the supercell. The correction term is hence given by the energy of the charged defect array embedded in the compensating jellium. Assuming that the space between the defects is large enough, i.e., the supercell is large enough; a macroscopic approximation can be made: the defect array and the jellium are immersed in a structureless dielectric, whose dielectric constant ε is equal to that of the perfect crystal. The energy of such an array, and hence the correction term is given by $E_{array} = -\alpha Q^2/(2\varepsilon L)$ where α is the appropriate Madelung constant, Q the charge of the defect and L is the lattice parameter of the periodic array. Clearly, large charges result in larger corrections. This correction requires the correct permittivity ε; this can now also be calculated reasonably straightforwardly using the same *ab initio* methods as in the defect calculation (e.g., Karki et al. 2000; Oganov et al. 2003). The charge-interaction correction can then be applied completely self-consistently.

Secondly, neutral or charged defect clusters can also introduce large dipole-moments into the supercell, giving rise to similar (although less strong) interactions as in the case of charged point defects discussed in the previous paragraph. Makov and Payne (1995), and more thoroughly Kantorovich (1999) and Kantorovich and Tupitsyn (1999), showed that the correcting factor for dipole-dipole interaction is given by $E_{dipole} = 2\pi P^2/(3V_c)$ where P is the total dipole moment of the supercell and V_c is the supercell volume.

With these corrections in hand, defect energies can be readily calculated using the supercell approach. It is important to note that when calculating migration enthalpies, the energy of the two similarly charged systems are subtracted from each other. Since the charge-interaction corrections presented above are the same for each system, they cancel. However, in fact the two systems have very different ionic arrangements (one having two vacancies and a migrating ion at its saddle-point, and the other with a single vacancy) and, therefore, they have different higher order electrostatic moments. A small error in the migration energy is expected from this.

The climbing image nudged elastic band method

As discussed previously, a key aspect to obtaining diffusion rates is finding the saddle point. This is necessary for the migration enthalpy and for ensuring that there is only one negative vibrational frequency for the Vineyard theory. The importance of this is exemplified by recent calculations and experiments on $MgSiO_3$ perovskite.

Empirical potential calculations on silicon diffusion in $MgSiO_3$ perovskite had estimated the migration enthalpy as being about 9 eV (Wright and Price 1993); this was much higher than for oxygen and magnesium, and agreed with the idea that silicon was the slowest diffusing species—and therefore rate limiting the rheology—in mantle perovskites. However, a later experimental study found a significantly lower migration energy of only 3.5 eV (Yamazaki et al. 2000). An identical value was also obtained by Dobson et al. (2008) but using a different procedure. The calculations were repeated by Karki and Khanduja (2007), but in this case using DFT to calculate the necessary energies. Their calculated migration enthalpies for oxygen and magnesium agreed with experimental values, but they also found a very high value for silicon of 8.33 eV (at room pressure). They suggested, therefore, that silicon diffusion must occur via some sort of cooperative mechanism involving oxygen vacancies. However, the reason is more prosaic than that; both theoretical studies chose the wrong migration pathway. Figure 2 is a contour map of computed migration enthalpies in a plane of possible saddle-points for silicon diffusing along the [110] pathway at two different pressures (see Ammann et al. 2009 for details). The two minimum energy points shown are 3.58 eV and 3.52 eV and are the most likely place through which the migrating atom would pass. The pathway chosen by Wright and Price (1993) and Karki and Khanduja (2007) are not shown, but are slightly higher on the Z-axis than the 3.52 eV (and 6.59 eV) point. Note that you do not have to be very far from the minimum energy point for the apparent (and erroneous) migration energy to increase substantially.

The climbing image nudged elastic band (CI-NEB) method (Henkelman et al. 2000) has transformed the search for saddle points. The method is shown graphically in Figure 3. Each

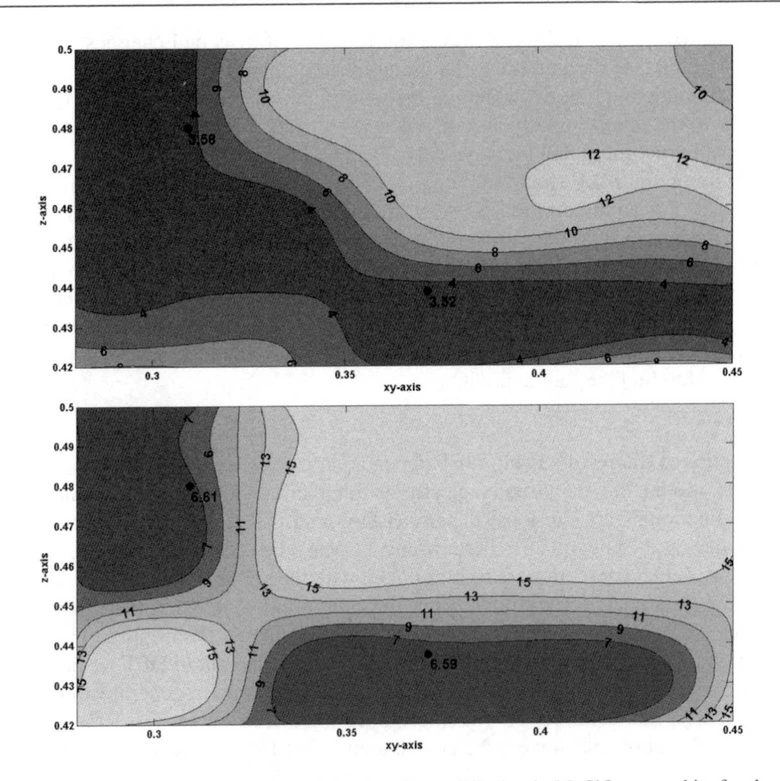

Figure 2. Contour maps of migration enthalpies for silicon diffusion in $MgSiO_3$ perovskite for the direct jump along <110> at 3.3 GPa (upper) and 151.7 GPa (lower). The calculated minima are marked with black dots and labeled with their corresponding value. In calculations, a small deviation from the minimum energy migration pathway (black dots) can result in a substantial increase of the migration enthalpy shown [With kind permission from Springer Science+Business Media: *Physics and Chemistry of Minerals*, "DFT study of migration enthalpies in $MgSiO_3$ perovskite." Vol. 36, 2009, p 151-158, Ammann et al., Figure 3.]

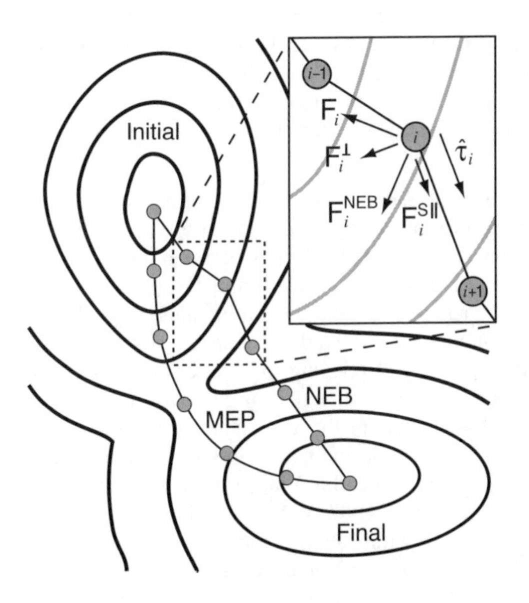

Figure 3. In order to find the minimum energy pathway (MEP) one can use the nudged elastic band (NEB) method. The nudged elastic band force $\mathbf{F}^{NEB}$ is the sum of the spring force $\mathbf{F}^{S\|}$ along the tangent τ plus the perpendicular force from the potential $\mathbf{F}^{\perp}$. The unprojected force due to the potential $\mathbf{F}$ is also shown [Reprinted with permission from Sheppard et al. (2008) *Journal of Chemical Physics*, Vol. 128, Issue 13, 134106. Copyright 2008, American Institute of Physics.]

of the points corresponds to a set of atoms for which the forces and energies are calculated. The two end points are the initial and final equilibrium positions. The points in-between are possible configurations of atoms which the system might take as an atom hops from one side of the barrier to the other. Initially atomic positions may be just linearly interpolated between their positions in the initial and final systems. The idea is to evolve the different systems such that they all lie on the MEP (minimum energy pathway) by minimizing the NEB-force $\mathbf{F}^{NEB}$. The image (set of atomic coordinates) with the highest energy is pushed upwards along the NEB until it reaches exactly the saddle point. The CI-NEB method provides a recipe for doing this. Moreover, there is a toolkit for use with the DFT code VASP available from *http://theory.cm.utexas.edu/vtsttools/*. Since we are only interested in the saddle point and not the whole minimum energy pathway, we use only four images. Nevertheless, for a low-symmetry pathway like silicon in perovskite, this requires on the order of 200 individual electronic minimizations of the super-cell to find the saddle point.

LDA vs. GGA

The quality of the results using DFT depends on the ability of the exchange-correlation functional to model the many-body electronic interactions. The most common exchange correlation functionals are the local density (LDA) and the general gradient approximation (GGA) (Perdew and Zunger 1981; Perdew and Y. Wang 1992; Perdew and Burke 1996). On the other hand, defect energetics are tightly linked with the cell-volume (as is evident from the strong pressure dependence(see below)). It is, thus, important that experimental and theoretical cell-volumes are equal (or at least similar). We have calculated the imposed pressure on the supercell within GGA. In order to estimate the uncertainty inherent to DFT we have compared GGA with LDA calculations performed at the same cell-volume. In general we find that LDA produces slightly smaller values for the migration enthalpy, by about 5 to 10%. Similarly, we find that ν^0 is also about 5% lower when calculated with LDA than with GGA.

RESULTS ON MANTLE PHASES

Results on MgO

Magnesiowüstite (Fe, Mg)O is thought to be the second most abundant mineral in the lower mantle accounting for about 20% of the volume (e.g., Jackson 1998). The investigation of the diffusional properties of its pure endmember periclase MgO is hence of major importance for Earth Sciences. It crystallizes in the rock-salt structure even under lower mantle conditions (Lin 2002; Oganov et al. 2003). Periclase is also an important industrial material with a wide range of applications and often serves as a prototype material for other ionic oxides (Kotomin et al. 1996).

MgO is a simple cubic oxide for which high-pressure and high-temperature experimental diffusion measurements exist. Since it crystallizes in the rock-salt structure, the location of the saddle point is given trivially by symmetry and is located half-way between the initial and final position. This is, however, only strictly true at high pressures. We find at low pressures, in agreement with Vočadlo et al. (1995), that the magnesium saddle-point bifurcates perpendicular to the jump trajectory. But also these saddle-points can be readily found by offsetting from the half-way position (a second-order saddle-point) and relaxing the migrating ion into the bifurcated first order saddle points (the lowering in energy is though negligible). In contrast with Vočadlo et al. (1995), we have not found any bifurcation along the oxygen jump trajectory. Thus, the CI-NEB method is not required for MgO. The ease of finding the saddle point has meant that absolute diffusion rates in MgO have been calculated within Vineyard-theory a number of times (e.g., Sangster and Stoneham 1984; Harding et al. 1987; Vočadlo et al. 1995). The pre-exponential factors from the different studies agree to within about 1 order of magnitude; some of this scatter can be attributed to different potentials and some to different methods.

Even more studies concentrated on the migration enthalpies and all obtained migration enthalpies of about 2 eV for both magnesium and O, which is in agreement with experiments. Some experimental and theoretical values of the migration enthalpies are given in Table 1 (which is not exhaustive) and attempt frequencies are given in Table 2.

Table 3 summarizes calculated migration enthalpies in periclase at high pressures. Ita and Cohen (1997, 1998) as well as Ito and Toriumi (2007) performed molecular dynamics simulations using interatomic potentials to calculate the absolute diffusion rates in MgO under lower mantle conditions. While both studies are in agreement with the available experimental data (up to 35 GPa), they disagree with each other at elevated pressures (above 60 GPa). Ita and Cohen (1997, 1998) observe a continuous increase of the migration enthalpy with increasing pressure, while in contrast Ito and Toriumi (2007) find that the migration enthalpy decreases after reaching a maximum around 50 GPa. The difference between their results could simply be linked to their different interatomic potentials. Our results agree with those of Ita and Cohen (1997, 1998). Nevertheless, it should be noted that all three studies have an activation volume that is in agreement with the available experimental data (Yamazaki and Irifune 2003; Holzapfel et al. 2003; Van Orman et al. 2003) at pressures up to 35 GPa.

Table 1. Migration enthalpies in MgO at 1 bar from experiments and theory (for single vacancy diffusion). DFT: density functional theory, LDA: local density approximation, GGA: general gradient approximation, MD: molecular dynamics, HF: Hartree-Fock, LD: lattice dynamics, ML: Mott-Littleton.

Theory	ΔH_{Mg} [eV]	ΔH_O [eV]	**Method**
This study	1.93	2.05	DFT GGA
	1.82	1.93	DFT LDA
Ito and Toriumi (2007)	2.09	2.23	MD
Gilbert et al. (2007)	2.20	2.31	DFT LDA
Karki and Khanduja (2007)	2.26	2.42	DFT LDA
Kotomin and Popov (1998)	2.43	2.50	HF
Ita and Cohen (1997, 1998)	1.70	1.97	MD
Vočadlo et al. (1995)	1.99	2.00	LD
De Vita et al. (1992)	2.39	2.48	DFT LDA
Harding et al. (1987)	—	2.1	ML
Sangster and Stoneham (1984)	—	2.26	ML
Sangster and Rowell (1981)	2.07	2.11	ML
Mackrodt and Stewart (1979)	2.16	2.38	ML
Experiments			
Yoo et al. (2002)	—	3.24±0.13	
Shirasaki and Hama (1973)	—	2.43±0.21	
Oishi and Kingery (1960)	—	2.71±0.26	
Mackwell et al. (2005)	2.17±0.07	—	Mg-Fe interdiffusion
Holzapfel et al. (2003)	2.64±0.17	—	Mg-Fe interdiffusion (8-23 GPa)
Yamazaki and Irifune (2003)	2.34±0.33	—	Mg-Fe interdiffusion (7-35 GPA)
Yang and Flynn (1994,1996)	2.52	6.91	Ca-diffusion (2.33 this study)
Sempolinsky and Kingery (1980)	2.28±0.21	—	
Duclot and Departes (1980)	2.20	—	
Wuensch et al. (1973)	2.76±0.08	—	
Lindner and Parfitt (1957)	3.44±0.13	—	

Table 2. Attempt frequencies for magnesium (ν_{Mg}), oxygen (ν_O), calcium (ν_{Ca}) and bound divacancies (ν_{MgO}) in MgO for the single jumps at 0 GPa. The same labels for the methods as in Table 1 are used.

	ν_{Mg} [THz]	ν_O [THz]	ν_{Ca} [THz]	ν_{MgO} [THz]	Method
This study	12.12	13.82	25.1	87.9	DFT GGA
Ita and Cohen (1997, 1998)	5.2	5.2	—	—	MD
Vočadlo et al. (1995)	15.95	8.55	—	—	LD
Harding et al. (1987)	23.15	—	—	—	ML
Sangster and Stoneham (1984)	32.9	—	—	—	ML

Table 3. Migration enthalpies of MgO at pressures P of the lower mantle. ΔH_{MgO} is the migration enthalpy for a divacancy (the higher of the two barriers for magnesium and oxygen hops).

	P [GPa]	ΔH_{Mg} [eV]	ΔH_O [eV]	ΔH_{MgO} [eV]
This study	5.3	2.06	2.2	2.81
	30.3	2.61	2.82	2.83
	136.6	3.88	4.27	2.67
Karki and Khanduja (2007)	20	3	2.9	—
	50	3.3	3.15	—
	150	4.3	3.95	—
Ito and Toriumi (2007)	20	2.59	2.66	—
	50	2.98	3.02	—
	140	2.12	2.31	—
Ita and Cohen (1997,1998)	20	2.34	2.55	—
	80	3.22	3.64	—
	140	3.99	4.31	—

Our results for MgO are shown in Figure 4. In order to compare our theoretical results with experiments, an estimate of the vacancy concentration in the experiments must be made. The Schottky-formation energy, ΔH_S, and Frenkel-formation energy, ΔH_F, have been calculated several times over the last 30 years and are given in Table 4 (the list is not exhaustive). For the charged defect correction, we adopted the value of the permittivity calculated by (Oganov et al. 2003). Formation energies continuously increase with pressure (Karki and Khanduja 2006b).

Our calculations and previous results find formation energies of Schottky and Frenkel-pair defects to be between 6.45-7.7 eV and 10.35-15.2 eV respectively. The equilibrium concentration of intrinsic magnesium vacancies is, therefore, small. However, the presence of heterovalent impurities will result in the formation of extrinsic vacancies in order to maintain charge neutrality. In fact, only a small concentration of impurities (much less than a few ppm at 2000 K) is sufficient for the number of extrinsic magnesium-vacancies to greatly exceed the number of intrinsic vacancies, and generally, the number of extrinsic magnesium-vacancies is assumed to dominate the number of defects in MgO by several orders of magnitude. It is possible, therefore, to constrain the experimental magnesium-vacancy concentration from their measured impurity concentrations. As shown in Figure 4, magnesium diffusion in MgO can readily be explained by our results using reasonable vacancy concentrations.

The experiments on oxygen diffusion are more difficult to explain than for magnesium. First of all oxygen diffusion is significantly slower than magnesium, and secondly, the different

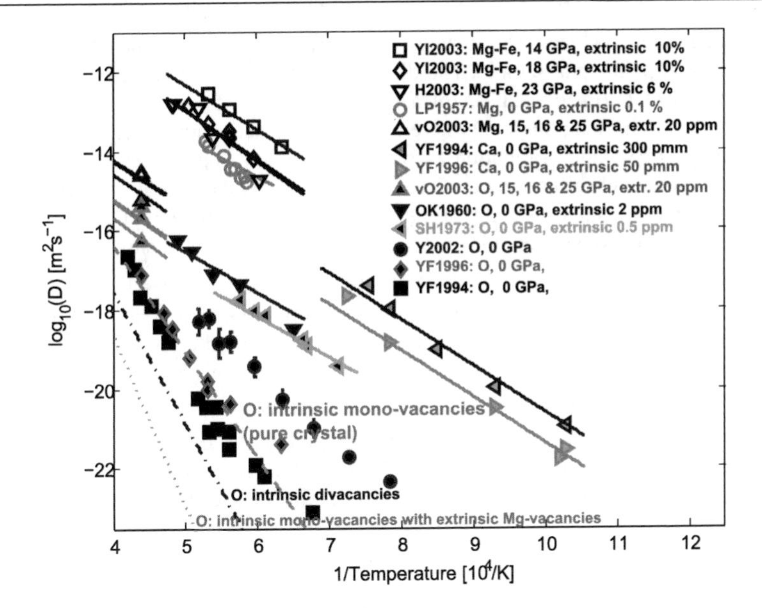

Figure 4. Lines are the results of our calculations using vacancy concentrations appropriate for the experiments. Empty symbols are magnesium (calcium) diffusion, filled symbols are for oxygen. In all experiments, magnesium diffusion can be readily explained assuming extrinsic vacancy hopping. YI2003: Yamazaki and Irifune (2003); H2003: Holzapfel et al. (2003); LP1957: Linder and Parfitt (1957); YF1996, YF1994: Yang and Flynn (1994, 1996); vO2003: van Orman et al. (2003); OK1960: Oishi and Kingery (1960); SH1973: Shirasaki and Hama (1973); Y2002: Yoo et al. (2002). The lowest oxygen diffusion data is not explained by theory: free oxygen vacancies are expected to be suppressed by extrinsic magnesium vacancies (for YF1994 and YF1996, they are constrained via Ca-diffusion data) resulting in the dotted line. Intrinsically formed divacancies are at least a magnitude too low (dash-dotted line). Intrinsically formed oxygen (mono-)vacancies could explain the Yang and Flynn data (dashed line) if no impurities (an extrinsic magnesium vacancy below 10^{-8}) are around for suppression (which is not the case).

Table 4. Theoretical Schottky ΔH_S and Frenkel ΔH_F defect formation energies in MgO at 0 GPa. The comment (bound) is used for the formation of divacancies and the same labels for the methods as in Table 1 are used.

	ΔH_S [eV]	ΔH_F [eV]	Method
This study	6.45 3.88 (bound)	14.21 (O)	DFT GGA
Gilbert et al. (2007)	5.97	10.35 (Mg) 12.17 (O)	DFT LDA
Karki and Khanduja (2006)	6.83	—	GGA LDA
Alfe and Gillan (2005)	7.5±0.53	—	QMC
Ita and Cohen (1997, 1998)	6.48 4.92 (bound)	—	MD
De Vita et al. (1992)	6.88	—	DFT LDA
Jacobs and Vernon (1990)	7.7 5.0	12.4 (Mg) 11.6 (O)	
Mackrodt (1988)	7.66	—	ML
Sangster and Rowell (1981)	7.72	—	MD
Mackrodt and Stewart (1979)	7.5 4.95 (bound)	11.9 (Mg) 15.2 (O)	ML

studies produce different slopes. The slowness is mostly due to the low concentration of oxygen vacancies in the experimental sample. The different slopes suggest that different mechanisms of migration and of intrinsic vacancy formation are at work. Some experiments have been thought to be in the intrinsic regime, where the measured activation energy also contains an activation energy of vacancy formation as well as migration (high slopes), and some are in the extrinsic regime, and so the measured activation energy is the migration energy only (low slopes). However, our results suggest that it cannot be intrinsic diffusion that is responsible for the higher slopes, but another, yet unknown, extrinsic diffusion mechanism.

The experiments of Oishi and Kingery (1960) as well as of Shirasaki and Hama (1973) are easiest to explain. They were performed on relatively impure samples, and the measured diffusion was assumed to be in the extrinsic regime. Plotted on the Figure 4 are our absolute diffusion rates, assuming an extrinsic vacancy concentration of about 2 ppm and 0.5 ppm respectively. These fit the experimental data well.

The experiments of Yang and Flynn (1994, 1996) are not as easy to interpret. The dashed line is our prediction for oxygen diffusion if we assume the oxygen vacancies are being formed intrinsically. The concentration of oxygen and magnesium vacancies is given by

$$C_{V_O} C_{V_{Mg}} = e^{\frac{-\Delta H_S}{kT}} \tag{6}$$

where, for no extrinsic vacancies, $C_{V_O} = N_{V_O}/N$ and $C_{V_{Mg}} = N_{V_O}/N$ are the oxygen and magnesium vacancy concentrations respectively. If we assume the oxygen and magnesium vacancies are charge balancing each other, then the diffusion coefficient is given as

$$D = N_V \frac{Z}{6} l^2 \nu^* e^{\frac{-\Delta H_M}{kT}} = N \frac{Z}{6} l^2 \nu^* e^{\frac{-(\Delta H_s/2 + \Delta H_M)}{kT}} \tag{7}$$

where the defect formation enthalpy, ΔH_S, the migration enthalpy, ΔH_M, and the effective jump frequency, ν^*, are all calculated from first principles. Although this fits the experimental data well, the good fit is, in fact, fortuitous. Yang and Flynn also measured Ca diffusion in the same MgO samples. These are also shown on Figure 4, and are very well described by an extrinsic vacancy diffusion mechanism (i.e., the slope is just the migration enthalpy), with an extrinsic cation vacancy concentration of about 50 to 300 ppm. The problem is that in samples with significant extrinsic concentrations of one of the Schottky pairs, Equation (6) above shows that the concentration of the other vacancy is proportionally reduced. For $C_{V_{Mg}}$ of a few 10s of ppm, C_{V_O} is reduced to a tiny amount, and there are simply not enough oxygen vacancies to produce the diffusion coefficients shown in Figure 4. The oxygen diffusion coefficients for a sample containing 50 ppm magnesium vacancies are shown in Figure 4 (dashed line); it is many orders of magnitude too slow.

This problem with rationalizing the cation and anion diffusion results simultaneously was realized by Yang and Flynn, and they suggested that their diffusion experiments were actually measuring the diffusion of bound MgO vacancy pairs (or divacancies). The concentration of MgO divacancies is given, for the intrinsic case, by

$$C_{V_{MgO}} = e^{\frac{-(\Delta H_S + \Delta H_B)}{kT}} \tag{8}$$

where ΔH_B is the energy of binding the two vacancies together (which is negative since they are opposite charge and, therefore, attractive). To calculate the mobility of the bound pair, we have calculated the migration enthalpies and frequency factors for the bound vacancies individually, and assumed that the slowest species limits the diffusion of the bound pair (see Tables 2, 3 and 4). The binding energy is about −2.6 eV and the effective migration energy for the slowest vacancy is 2.8 eV. Our theoretical absolute diffusion rates of the divacancies are shown in Figure 4 (dashed-dot lines); they too are much slower than measured in experiments. To date, we are unable to constrain the oxygen diffusion mechanism for the experiments of Yang and Flynn (1994, 1996).

The recent measurement of oxygen diffusion by Yoo et al. (2002), revealed yet another activation energy, and much higher rates than the previous experiments. They suggest tentatively that their measured diffusion rates are actually for interstitial oxygen, although they acknowledge that these are unlikely to be in significant concentrations in MgO. Again, we have tested this by calculating the migration enthalpy for the oxygen interstitial and its formation energy. Interestingly, the migration energy for the interstitial is low: 0.7 eV. However, the formation enthalpy is, as expected, very large (~7 eV; half the Frenkel defect formation energy). Therefore, we conclude that there is not a significant concentration of oxygen interstitials in MgO. What the diffusion mechanism for the experiments of Yoo et al. (2002) is, also remains unclear.

Further complicating the assessment of the oxygen vacancy concentration and oxygen diffusion mechanism is the experimental finding that oxygen diffusion along dislocations (pipe-diffusion) has a similar activation energy (Narayan and Washburn 1973, Sakaguchi et al. 1992) as mono-vacancy diffusion. This has likely been observed by Yang and Flynn (1994, 1996) (and maybe also by Yoo et al. 2002) in their low temperature data of oxygen diffusion. At the same time, this might suggest that even the data of Oishi and Kingery (1960) as well as of Shirasaki and Hama (1973) represents pipe-diffusion instead of extrinsic oxygen diffusion as we have assumed here.

Although it is not possible to explain all the experimental results for oxygen diffusion in MgO, we are able to model successfully experiments where the migration mechanism is simple and unambiguous. For instance we very accurately model the direct vacancy hopping mechanism for cations when the vacancies are extrinsically controlled. Similarly, we can accurately model oxygen diffusion when it is in the extrinsic regime. In other words, there is good reason to expect that the migration barriers and frequency factors calculated via DFT are accurate to within a few tenths of an eV.

Results on $MgSiO_3$ perovskite

$MgSiO_3$ perovskite is the dominant phase in the Earth's lower mantle comprising up to 80% of its volume and it is thus of fundamental importance for understanding the thermochemical evolution of the Earth (e.g., Jackson 1998)

Computational studies of defect formation energies using the Mott-Littleton method suggest that the dominant defect in perovskite is the MgO partial Schottky defect (Price et al. 1989; Wright and Price 1993). Frenkel defects (interstitials) are energetically unfavorable. The same result has been found by first principles calculations (Karki and Khanduja 2006a, 2007). The high intrinsic formation energies (7.4 eV—MgO partial Schottky; 20.8 eV—full Schottky); rapidly increasing with increasing pressure) imply that vacancy concentrations in experiments and in the Earth's mantle are controlled extrinsically, i.e., by impurity content.

Magnesium and oxygen diffusion. We find that diffusion of both oxygen and magnesium in perovskite occurs via simple vacancy hopping. The migration enthalpies and jump frequencies are given in Tables 5 and 6. Figure 5 shows our predicted diffusion rates plotted against those found experimentally. The upper and lower bounds are those found from LDA and GGA respectively. In order to make this comparison, we need to estimate the vacancy concentrations in the experiments. For the oxygen diffusion experiments of Dobson (2003) this is relatively straightforward since he doped his sample with 0.6% Na in order to extrinsically control oxygen vacancies. Since each oxygen vacancy is charge balancing two Na^+ ions, this results in 0.1% vacancies (per unit cell). At high temperatures, a change in the conduction mechanism has been

Table 5. Migration enthalpies of all species in $MgSiO_3$ perovskite from experiments and theory. Our values are averages over the migration enthalpies of all different jumps (8 for oxygen, 3 for magnesium). For silicon our value is the maximum energy of the barrier of the six jump cycle.

	P [GPa]	ΔH_{Mg} [eV]	ΔH_{Si} [eV]	ΔH_O [eV]	Method
This study	24	3.81	3.64LDA	1.06	DFT GGA
	140	6.43	5.98LDA	2.23	
Karki and Khanduja (2007)	30	4.78	9.1	1.41	DFT LDA
	120	7.71	10.48	2.57	
Wright and Price (1993)	0	4.57	9.2	0.96	ML
	60	6.13	—	—	
	125	7.43	10.32	—	
Price et al. (1989)	0	4.6	—	0.8	ML
Holzapfel et al. (2005)	24	4.29±0.64	—	—	Mg-Fe interdiff.
Dobson et al. (2008)	25	—	3.6±0.76	—	
Yamazaki et al. (2000)	25	—	3.48±0.38	—	
Dobson (2003)	25	—	—	1.35±0.2	Na-doped
Xu and McCammon (2002)	25	—	—	1.47	Al-bearing

Table 6. Average attempt frequencies for magnesium and oxygen diffusion in $MgSiO_3$ perovskite. The attempt frequencies for the six jump cycle are shown in Figure 4.

P [GPa]	ν_{Mg} [THz]	ν_O [THz]
24	4.95	5.94
140	17.39	14.04

observed (probably from oxygen ionic to intrinsic electronic) and we subtracted this contribution from the experimental data in order to obtain the pure oxygen ionic conduction. Thereby our predicted diffusion rates for oxygen are only slightly lower than those found by Dobson (2003) but with a migration enthalpy almost exactly the same. Earlier studies of the conductivity of perovskite (Katsura et al. 1998; Xu et al. 1998; Xu and McCammon 2002) contained iron and hence the electrical conductivity is dominated by small-polaron electronic conduction. Nevertheless, at the highest temperature the study of Xu et al. (1998) showed a contribution from oxygen ionic conduction. Xu and McCammon (2002) analyzed this oxygen ionic component and found an activation energy in good agreement with our migration enthalpy.

For the Mg-Fe exchange experiments of Holzapfel et al. (2005) the concentration of magnesium vacancies is controlled by the amount of ferric iron. For an Al-free system, we can use the geochemical experiments of Lauterbach et al. (2000) and the expected oxygen fugacity in the multi-anvil experiments. Using the expression of Lauterbach et al. (2000), we estimate that the extrinsic concentration of magnesium vacancies should be about 0.2%. Using this concentration, our predictions for magnesium diffusion are in good agreement with the experiments.

Silicon diffusion. Silicon diffusion in $MgSiO_3$ perovskite, however, is complicated by the fact that we find that silicon diffusion does not occur via a simple vacancy method, where silicon jumps from one site directly into an adjacent vacancy, as with magnesium and oxygen. The lowest migration enthalpy for the direct jump is 5.2 eV, substantially higher than the values of ~3.6 eV obtained experimentally. Rather, we find that it occurs via a so-called

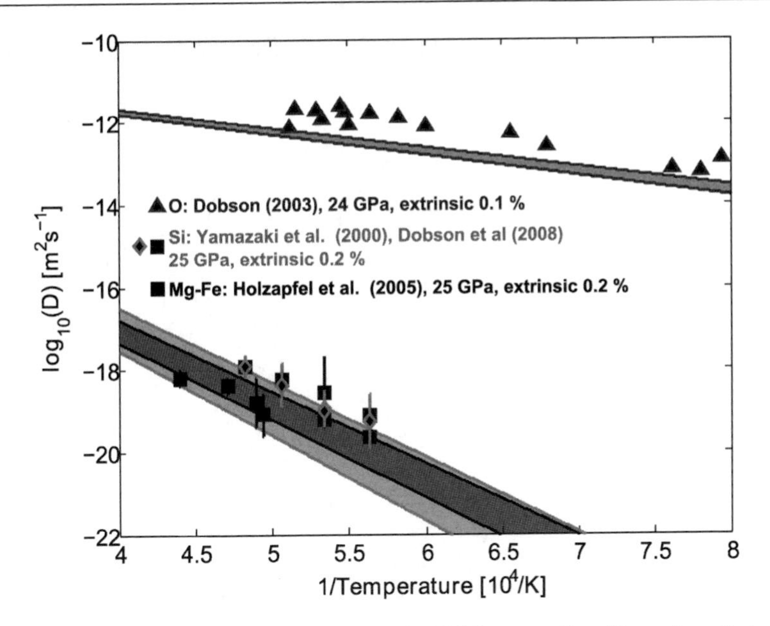

Figure 5. Comparison of experimental diffusion rates in $MgSiO_3$ perovskite with our theoretical calculations. Oxygen diffusion experiments in perovskite (Dobson 2003) had an estimated vacancy concentration of 0.1% (controlled via Na-content). We have subtracted the conductivity contributed by another mechanism in the high-temperature regime observed in the experiments. We calculated the vacancy concentration for the experiments (Holzapfel et al. 2005) of magnesium-iron interdiffusion in perovskite at 24 GPa at reducing conditions from the geochemical experiments of Lauterbach et al. (2000) to be 0.1-0.2%. Silicon vacancy concentrations in the experiments (Yamazaki et al. 2000; Dobson et al. 2008) have been estimated to be approximately 0.2% making reasonable assumptions (see text for details), however, they are not very certain.

six-jump cycle. This is common in some binary alloys (e.g., Elcock and McCombie 1957; Huntington et al. 1961; Debiaggi et al. 1996; Divinski and Herzig 2000; Duan 2006). The six jump cycle is shown in Figure 6. In a normal vacancy hopping mechanism, the migrating ion hops directly to an adjacent vacancy. But in the six-jump cycle, a magnesium ion jumps into the silicon vacancy, making an antisite defect and a magnesium vacancy. The adjacent silicon ion then jumps into the magnesium vacancy, leaving a new silicon vacancy. The situation then repeats itself, with a magnesium jumping into the new silicon vacancy, followed by the first antisite magnesium jumping into the vacated magnesium site. The silicon ion then jumps into the silicon vacancy from the magnesium site, and the cycle finishes with the magnesium on the silicon site jumping into the adjacent magnesium vacancy. The silicon vacancy has jumped to an adjacent site via six different intermediate hops, each with its own activation enthalpy. To analyze this complex diffusion process, we first have to consider what the effective activation energy of the six jumps is, and secondly, we have to consider how many times the cycle is broken by a vacancy hopping to a site that is not part of the cycle.

An analytical solution to the rate of the six jump cycle can be obtained using the approach of Arita et al. (1989). This method provides an effective jump-frequency, Γ, for the complete cycle. Also, by calculating the jump-frequency at different temperatures, it is possible to obtain an apparent activation (or migration) enthalpy for the cycle. It does, however, require the twelve individual jump-frequencies (obtained from Vineyard theory) and their migration energies. These are shown in Figure 6 for the fastest silicon cycle. The effective activation energy lies somewhere between the migration energy of the single largest jump, and the energy difference between the original site and the highest energy saddle point. As shown by Arita

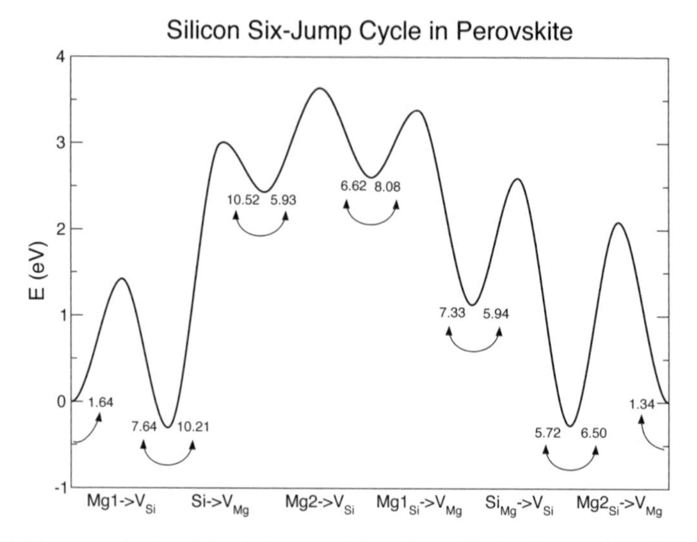

Figure 6. The energy barrier of the six-jump cycle for silicon diffusion in perovskite. The numbers are the attempt frequencies towards neighboring states as indicated by the arrows.

et al. (1989), the former is appropriate at high temperatures, while the latter is appropriate at low temperatures. The details depend on the particular system. We have used this approach for the cycle shown in Figure 6 and find that at all reasonable temperatures, the apparent migration energy is equal to 3.6 eV. This is the same as the maximum saddle point energy. The apparent migration enthalpy does decrease towards the largest single jump (3.3 eV for the silicon jumping into the magnesium vacancy), but only at very high temperatures. This is as expected since the maximum saddle-point energy (3.6 eV) is very similar to the single largest jump. This migration enthalpy agrees very well with the experimentally derived values of 3.61 eV and 3.5 eV obtained by Dobson et al. (2008) and Yamazaki et al. (2000) respectively. The entire cycle for one silicon vacancy to migrate can be described as a single Arrhenius process

$$D_V = \nu_a l^2 e^{\frac{-\Delta H}{kT}} = 3.5\times10^{12} l^2 e^{\frac{-3.6}{kT}} \tag{9}$$

Figure 5 shows the available experimentally obtained diffusion coefficients of silicon compared to our results. In order to make such a comparison, we have used a silicon vacancy concentration of 0.2%. This seems unexpectedly high for silicon vacancies, but it seems unavoidable that the experiments do indeed have very high concentrations of silicon vacancies. Using a nearest neighbor jump distance of 2.4 Å, a representative jump frequency of 10 THz, and the experimentally determined migration enthalpy of 3.6 eV, we can invert each of the experimental diffusion coefficients of Dobson et al. (2008) for the silicon vacancy concentration. These range from 0.35% to 0.075%. A similar range is found from the data of Yamazaki et al. (2000). The range is also in rough agreement with the defect calculations of Hirsch and Shankland (1991). So a vacancy concentration of about 0.2% is perfectly consistent with the fast measured diffusion rates.

Although the experimental diffusion rates can be explained by the six-jump cycle, there is an additional complication to be considered. At each step of the cycle there is a possibility that one of the vacancies takes a hop to a site not in the cycle. This, therefore, breaks the cycles until another vacancy comes along. For instance, the first step in the cycle produces a magnesium vacancy. The activation barrier for it to take the next step in the cycle is about 3.3 eV. However, the energy barrier for it to hop to one of the other neighboring magnesium sites instead of the silicon site, is only between 3.6 and 4.1 eV (LDA and GGA respectively); there is, therefore, a non-negligible probability that it will take this jump. These processes of breaking the cycle must be taken into account when calculating silicon diffusion coefficients.

We can do this using Kinetic Monte Carlo (reviewed by Voter 2005). Given a set of rate constants (diffusion constants in this case), KMC is a way of propagating a dynamic system through a possibly complex set of paths (or phase space). If we image a vacancy at a certain position, it may have a number of possible paths. Normally it will take the one that is most probable (generally the lowest migration barrier), but sometimes it will take another. Once it has moved on, it is faced with another set of possibilities and associated probabilities. KMC provides a way of moving the system through the phase space and determining the overall rate constant (or diffusion constant). We could use this technique as an alternative to the analytic technique for obtaining the effective diffusion coefficient for the full cycle, ignoring possible breaks in the cycle; we do indeed get the same result using both methods. But KMC has to be used when a diffusing species has a choice of paths.

We are interested in how many times the six-jump cycle is broken. In order to use KMC to do this, we not only need the jump-frequencies for the possible first jumps off the cycle, we also need the next jumps in order to ensure that it doesn't jump back into the cycle. This rapidly becomes an impossibly large number of calculations, so we have restricted the calculation to the first two jumps off the cycle, after which we assume the vacancy is gone.

Our results show that the cycle is broken somewhere between about 1% and 20% of the time, depending on whether LDA or GGA jump-rates are used. This is mainly because of the change in the energy-barrier of the single jumps. LDA results in lower energy-barriers than GGA, however, the relative change of the energy barriers (from GGA to LDA) for the cycle-breaking jumps is smaller than the change in the cycle-barriers. The cycle is mostly broken at the second step. The migration energy for the magnesium vacancy to jump into the silicon site is about 3.3 eV; this is only slightly less than the migration energy of 3.6 eV for it to move into an adjacent magnesium site that is not in the cycle. The next jump is even lower in energy and the vacancy migrates away. The reason the cycle breaks can be easily understood: the silicon cycle starts with a silicon vacancy (charge –4), which is then occupied by a magnesium ion forming a magnesium on a silicon site (charge –2) and a magnesium vacancy (charge –2). The antisite and the magnesium vacancy have the same charge sign, and hence repel each other electrostatically. The energy of the reaction

$$Mg_{\mathrm{Mg}}^{x} + V_{\mathrm{Si}}^{''''} = V_{\mathrm{Mg}}^{''} + Mg_{\mathrm{Si}}^{''} \tag{10}$$

is only –0.3 eV when they are adjacent to each other, however, it decreases to about –2.2 eV (depending on the functional and the permittivity chosen for the charged defect correction) when the magnesium antisite and the magnesium are allowed to be infinitely separated. In other words, regardless of whether or not the cycle breaks, there should only ever be a vanishingly small concentration of silicon vacancies to begin with. This is completely at odds with the large number of vacancies required by fast diffusion coefficients observed experimentally.

One way to overcome the electrostatic repulsion of antisite and magnesium vacancy is to neutralize one of them; the most obvious candidates to do this are protons. We have, therefore, calculated the energy of the following reaction,

$$Mg_{\mathrm{Mg}}^{x} + 2H_{\mathrm{Si}}^{''} = V_{\mathrm{Mg}}^{''} + (Mg+2H)_{\mathrm{Si}}^{x} \tag{11}$$

where, as with the previous reaction, the two species on the right-hand side are infinitely apart. We find that this reaction is strongly endothermic, with an energy of 2.44 eV. In other words, protons increase the concentration of silicon vacancies, something that has been suggested for other silicates such as forsterite (Brodholt 1997). Obviously the presence of protons may change the jump frequencies, but jumps into sites inhabited with protons become substantially more complicated calculations and we have not attempted these yet.

At present we are able to explain the high silicon diffusion rates seen experimentally in two independent studies as a six-jump cycle with about 0.2% silicon vacancies stabilized by protons. The activation enthalpy of the direct jump is too high. So far, appreciable amounts of water have not been experimentally verified in perovskite, with some studies showing negligible water solubility (Litasov et al. 2003; Bolfan-Casanova 2005). Regardless of whether it is water, or some other extrinsic mechanism, the experiments are only consistent with high concentrations of silicon vacancies.

Other investigated silicon diffusion mechanisms

Previous studies (Wright and Price 1993; Karki and Khanduja 2007) on silicon migration barriers in perovskite found a decrease in the migration enthalpy in the presence of an oxygen vacancy. However, for our direct pathway (Ammann et al. 2009), the migration energy in the presence of a oxygen vacancy is increased to about 6.2 eV. This finding is not surprising as an oxygen-vacancy has the same charge-sign as the migrating silicon ion.

In many other silicates, silicon diffusion occurs via an interstitial mechanism and migration enthalpy and pre-exponential factor are linearly correlated (Béjina and Jaoul 1997) (compensation law). As observed by Béjina and Jaoul (2003), silicon diffusion in perovskite also satisfies this compensation law and they therefore suggested that silicon in perovskite diffuses via an interstitial mechanism. We find that there is a stable split interstitial configuration in which two silicon interstitials are located on opposite faces of the oxygen octahedron (slightly elevated above the centre of the triangle faces), around a vacancy at the centre of the octahedron. Split interstitials are very common in many materials and for various species—e.g., in forsterite (Walker et al. 2009; Béjina et al. 2009), quartz (Roma et al. 2001), various metals (Schilling 1978) and semi-conductors (Lee et al. 1998). However, our results indicate that also this split-interstitial mechanism cannot explain the experimental findings as the formation energy of this state is about 10 eV at 24 GPa.

Finally, we consider antisite migration. The idea is that the cation sublattice can be partially inverted, i.e., that magnesium ions occupy silicon vacancies and silicon ions occupy magnesium vacancies. The intrinsic generation of such an antisite pair, i.e., magnesium and silicon swap sites at the same time is energetically unlikely (12 eV at 24 GPa). The formation energy of such a bound antisite-pair is, however, only 3.1 eV (at 24 GPa) above the perfect crystal.

Silicon jumps, onto magnesium vacancies and vice versa, are the energetically most favorable cation jumps (among the ones investigated). We have found three mechanisms which make use of these antisites and are energetically feasible: I) the direct jump to second nearest neighbor vacancies in the presence of a magnesium vacancy, II) the antistructure bridge and III) the six-jump cycle.

(1) In order to migrate silicon efficiently through the crystal by the direct mechanism, magnesium vacancies need to be neighboring silicon vacancies. Unsurprisingly, this is not the case as they carry the same charge and thus repel each other.

(2) The antistructure bridge mechanism (also known in binary alloys, e.g., Duan 2006) starts with a silicon-vacancy plus a silicon-antisite atom. The antisite ion jumps into the vacancy effectively changing the type of the vacancy and a nearby silicon ion jumps from its silicon site onto the new magnesium vacancy creating again an antisite plus a silicon vacancy. The analogous process works for magnesium antisites. While we can expect to have quite some inversion for silicon vacancies (occupied with magnesium ions, formation energy ~-2.5 eV), there will be only very few silicon atoms occupying magnesium vacancies (formation energy ~8 eV). While this mechanism might well

contribute to magnesium diffusion, the resulting silicon diffusion rate would be much too slow compared with experiments. One could also envisage a situation in which a magnesium and a silicon swap their sites forming two neighboring antisites allowing a new mechanism. Such a pair would be electrostatically bound and could be moving via silicon and magnesium vacancy sites in a corporate manner. However, such pairs are rare as the formation energy is about 3.1 eV and the migration barrier is about 4.5 eV.

(3) The six jump cycle discussed above is the only mechanism we have found which can explain the experiments.

IMPLICATIONS FOR THE EARTH'S LOWER MANTLE

Viscosity of the lower mantle

We can use our absolute diffusion rates to estimate the viscosity of the lower mantle. At this point we assume that the volumetrically greatest phase, i.e., perovskite, controls its viscosity.

The lower mantle is probably deforming in the diffusion creep regime (Karato and Wenk 1995). Firstly, strain rates are low, temperatures are high, and grain sizes are thought to be small (Solomatov et al. 2002). Secondly, there is almost no evidence of shear-wave splitting in the lower mantle, which would be indicative of lattice preferred orientation developed in the dislocation creep regime. And thirdly, viscosity models from glacial rebound studies indicate a linear viscosity in the lower mantle, consistent with diffusion creep. We can estimate the viscosity of the lower mantle deforming via diffusion creep by using the Nabarro-Herring expression (Poirier 1985):

$$\eta = \frac{G^2 kT}{\alpha D_{eff} \Omega} \tag{12}$$

where G is the characteristic grain size, α is a geometrical factor, Ω is the molecular volume, and D_{eff} is the effective diffusion coefficient. The effective diffusion coefficient is simply the stoichiometrically averaged diffusion coefficients, D_i, of the individual ionic species

$$D_{eff} = \left(\sum_i \frac{n_i}{D_i} \right)^{-1} \tag{13}$$

where n_i is the stoichiometric factor for each species. The diffusion coefficient for each species should also contain a contribution from grain-boundary diffusion. Grain boundary diffusion is, however, small in both periclase and perovskite for large enough grains ($G > 500$ μm in periclase (van Orman et al. 2003) and $G > 10$ μm in perovskite (Yamazaki et al. 2000)). Therefore, for the expected grain size in the lower mantle (Solomatov et al 2002), grain boundary diffusion is negligible.

The viscosity profile of the lower mantle, using our absolute diffusion results, is shown in Figure 7. The upper bound corresponds to the product $G^2N_V = 2\times10^{-5}$ mm^{-1}. In other words for a grain size of 1 mm and a vacancy concentration of 2×10^{-5}, or alternatively, a grain size of 0.1 mm and a vacancy concentration of 2×10^{-7}. The lower bound corresponds to the $G^2N_V = 2\times10^{-3}$ mm^{-1}; that is a two orders of magnitude greater vacancy concentration than the upper bound (for the same grain size). The geotherm is taken from Stacey (1992). Superimposed on our predicted viscosity profile is that obtained by Mitrovica and Forte (2004) from a joint inversion of convection and glacial rebound data. The steep weakening seen at the base of the lower mantle is due to rapidly increasing temperatures in the thermal boundary layer. The agreement with the viscosity profile of Mitrovica and Forte (2004) is striking, especially when realizing that no experimental data have gone into the predicted viscosity.

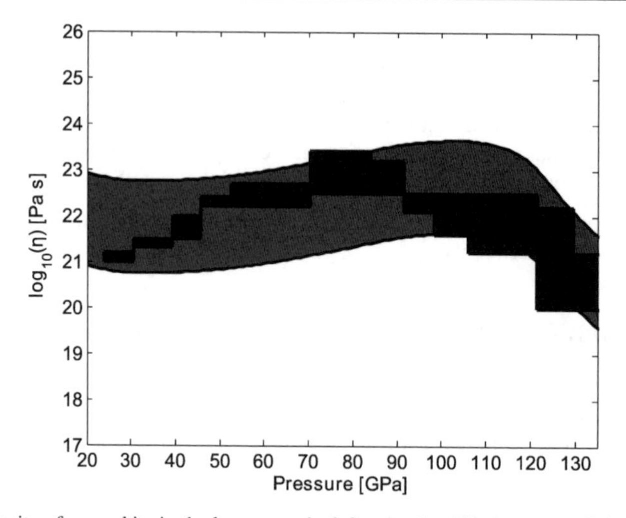

Figure 7. Viscosity of perovskite in the lower mantle deforming by diffusion creep. Calculated viscosity profile of the lower mantle with a composition of 100% $MgSiO_3$ perovskite (curved). Upper bounds correspond to either a grain size of G = 1 mm and a vacancy concentration of $N_v = 2\times10^{-5}$ or G = 0.1 mm and $N_v = 2\times10^{-7}$. Lower bounds correspond to either G = 1 mm and $N_v = 2\times10^{-3}$ or G = 0.1 mm and $N_v = 2\times10^{-5}$. Vacancy concentrations were assumed to be fixed throughout the entire lower mantle. Superimposed are the results of Mitrovica and Forte (2004) from inversion modeling.

It is clear that further work needs to be done in order to investigate the effects of impurities such as iron and alumina on the diffusion and therewith on rheology. However, the presence of impurities will result in an increased number of vacancies which should, therefore, lower the viscosity.

CONCLUSIONS

When the diffusion mechanism is simple, and when the concentration of vacancies in the experimental sample is reasonably clear, we have shown that we are able to predict from first principles the absolute diffusion in mantle minerals very accurately. For instance, we predict cation diffusion (magnesium and calcium) in all MgO experiments very well. We also predict oxygen diffusion in MgO, when the concentrations are extrinsically controlled. In perovskite we accurately reproduce the available experimental data on magnesium and oxygen diffusion as a simple vacancy hopping mechanism. Silicon diffusion in perovskite is more complicated since the direct jump does not seem to be energetically favoured. Rather we explain the data via the so-called six-jump-cycle. However, we are unable to explain why the concentration of vacancies in the experimental sample is so high. Our best explanation is that they are stabilized by water, as in other silicates.

ACKNOWLEDGMENTS

We are grateful to the EU for the Marie Cure "c2c" research-training network (Contract No. MRTN-CT-2006-035957). Computations were performed on HECToR, the UK national HPC facility, and on Legion, the UCL cluster.

REFERENCES

Alfè D (2009) PHON: A program to calculate phonons using the small displacement method. Comput Phys Commun 180:2622-2633, doi:10.1016/j.cpc.2009.03.010

Alfè D, Gillan MJ (2005) The Shottky defect formation energy in MgO calculated by diffusion Monte Carlo. Phys Rev B 71:220101:1-3

Ammann MW, Brodholt JP, Dobson DP (2009) DFT study of migration enthalpies in $MgSiO_3$ perovskite. Phys Chem Miner 36:151-158

Arita M, Koiwa M, Ishioka S (1989) Diffusion mechanisms in ordered alloys - a detailed analysis of six-jump vacancy cycle in the B2 type lattice. Acta Metall 37:1363-1374

Ball AJ, Murphy ST, Grimes RW, Bacorisen D, Smith R, Uberuaga BP, Sickafus KE (2008) Defect processes in $MgAl_2O_4$ spinel. Solid State Sci 10:717-724

Baroni S, de Gironcoli S, Dal Corso A, Giannozzi P (2001) Phonons and related crystal properties from density-functional perturbation theory. Rev Mod Phys 73:515-562

Béjina F, Blanchard M, Wright K, Price GD (2009) A computer simulation study of the effect of pressure on Mg diffusion in forsterite. Phys Earth Planet Inter 172:13-19

Béjina F, Jaoul O (1997) Silicon diffusion in minerals. Earth Planet Sci Lett 153:229-238

Béjina F, Jaoul O, Liebermann RC (2003) Diffusion in minerals at high pressure: a review. Phys Earth Planet Inter 139:3-20

Blanchard M, Wright K, Gale JD (2005) Atomistic simulation of Mg_2SiO_4 and Mg_2GeO_4 spinels: a new model. Phys Chem Miner 32: 332-338

Blöchl PE (1994) Projector augmented-wave method. Phys Rev B 50:17953-17979

Bolfan-Casanova N (2005) Water in the Earth's mantle. Mineral Mag 69:229-257

Braithwaite JS, Sushko PV, Wright K, Catlow RCA (2002) Hydrogen defects in forsterite: A test case for the embedded cluster method. J Chem Phys 116:2628-2635

Braithwaite JS, Wright K, Catlow RCA (2003) A theoretical study of the energetics and IR frequencies of hydroxyl defects in forsterite. J Geophys Res 108:2284

Brodholt J (1997) *Ab initio* calculations on point defects in forsterite (Mg_2SiO_4) and implications for diffusion creep. Am Mineral 82:1049-1053

Brodholt J (2000) Pressure-induced changes in the compression mechanism of aluminous perovskite in the Earth's mantle. Nature 407:20-622

Brodholt J, Refson K (2000) An *ab initio* study of hydrogen in forsterite and a possible mechanism for hydrolytic weakening. J Geophys Res 105:18,977-18,982

Catlow CRA, Corish J, Jacobs PWM, Lidiard AB (1981) The thermodynamics of characteristic defect parameters. J Phys C Solid State Phys 14:L121-L125

Cherry M, Islam MS, Gale JD, Catlow CRA (1995a) Computational studies of protons in perovskite-structured oxides. J Phys Chem 99:14614-14618

Cherry M, Islam MS, Catlow CRA (1995b) Oxygen ion migration in perovskite-type oxides. J Solid State Chem 118:125-132

De Vita A, Gillan MJ, Lin JS, Payne MC, Stich I, Clarke LJ (1992) Defect energetics in MgO treated by first-principles methods. Phys Rev B 46:12964-3322

Debiaggi SB, Decorte PM, Monti AM (1996) Diffusion by vacancy mechanism in Ni, Al, and Ni_3Al: calculation based on many-body potentials. Phys Status Solidi 195:37-54

Divinski S, Herzig C (2000) On the six-jump cycle mechanism of self-diffusion in NiAl. Intermetallics 8:1357-1368

Dobson DP (2003) Oxygen ionic conduction in $MgSiO_3$ perovskite. Phys Earth Planet Inter 139:55-64

Dobson DP, Dohmen R, Wiedenbeck M (2008) Self-diffusion of oxygen and silicon in $MgSiO_3$ perovskite. Earth Planet Sci Lett 270:125-129

Duan J (2006) Atomistic simulations of diffusion mechanisms in stoichiometric Ni_3Al. J Phys Condens Matter 18:1381-1394

Duclot M, Departes C (1980) Effect of impurities on cationic conductivity of magnesium-oxide single-crystals. J Solid State Chem 31:337-385

Elcock EW, McCombie CW (1957) Vacancy diffusion in binary ordered alloys. Phys Rev 109:605-606

Flynn CP, Jacucci G (1982) Dynamical corrections to rate theory. Phys Rev B 25:6225-6234

Gale JD (1997) GULP: A computer program for the symmetry-adapted simulation of solids. J Chem Soc Faraday Trans 93:629-637

Gale JD, Rohl AL (2003) The general utility lattice program (gulp). Mol Simul 29:291-341

Gibson A, Haydock R, LaFemina JP (1994) Stability of vacancy defects in MgO: The role of charge neutrality. Phys Rev B 50:2582-2592

Gilbert CA, Kenny SD, Smith R, Sanville E (2007) *Ab initio* study of point defects in magnesium oxide. Phys Rev B 76:184103:1-10

Harding JH, Sangster MJL, Stoneham AM (1987) Cation diffusion in alkaline-Earth oxides. J Phys C: Solid State Physics 20: 5281-5292
Harding JH (1989) Calculation of the entropy of defect processes in ionic solids. J Chem Soc Faraday Trans 2 85:351-365
Henkelman G, Uberuaga BP, Jónsson H (2000) A climbing image nudged elastic band method for finding saddle points and minimum energy paths. J Chem Phys 113:9901-9904
Hirsch LM, Shankland TJ (1991) Point defects in (Mg,Fe)SiO_3 perovskite. Geophys Res Lett 18:1305-1308
Hohenberg P, Kohn W (1964) Inhomogenous electron gas. Phys Rev 136:B864-B871
Holzapfel C, Rubie DC, Frost DJ, Langenhorst F (2005) Fe-Mg interdiffusion in (Mg,Fe)SiO_3 perovskite and lower mantle reequilibration. Science 309:1707-1710
Holzapfel C, Rubie DC, Mackwell S, Frost DJ (2003) Effect of pressure on Fe-Mg interdiffusion in (Fe_xMg_{1-x}) O, ferropericlase. Phys Earth Planet Inter 139:21-34
Huntington HB, Miller NC, Nerses V (1961) Self-diffusion in 50-50 gold-cadmium. Acta Metall 9:749-754
Ita J, Cohen RE (1997) Effects of pressure on diffusion and vacancy formation in MgO from nonempirical free-energy integrations. Phys Rev Lett 79:3198-3201
Ita J, Cohen RE (1998) Diffusion in MgO at high pressure: implications for lower mantle rheology. Geophys Res Lett 25:1-4
Ito Y, Toriumi M (2007) Pressure effect of self-diffusion in periclase (MgO) by molecular dynamics. J Geophys Res 112:B04206, doi:10.1029/2005JB003685
Jackson I (1998) Elasticity, composition and temperature of the Earth's lower mantle: a reappraisal. Geophys J Int 134:291-311
Jacobs PWM, Vernon ML (1990) Defect energies for magnesium oxide and lithium. J Chem Soc Faraday Trans 86:1233-1238
Kantorovich LN (1999) Elimination of the long-range interaction in calculations with periodic boundary conditions. Phys Rev B 60:15475-15479
Kantorovich LN, Tupitsyn II (1999) Coulomb potential inside a large finite crystal. J Phys Condens Matter 11: 6159-6168
Karato S, Wenk HR (1995) Superplasticity in Earth's lower mantle: evidence from seismic anisotropy and rock physics. Science 270:458-461
Karki BB, Khanduja G (2006a) Computer simulation and visualization of vacancy defects in $MgSiO_3$ perovskite. Phys Rev B 61:8793-8800
Karki BB, Khanduja G (2006b) Vacancy defects in MgO at high pressure. Am Mineral 91: 511-516
Karki BB, Khanduja G (2007) A Computational study of ionic vacancies and diffusion in $MgSiO_3$ perovskite and post-perovskite. Earth Planet Sci Lett 260:201-211
Karki BB, Wentzcovitch RM, de Gironcoli S, Baroni S (2000) *Ab initio* lattice dynamics of $MgSiO_3$ perovskite at high pressure. Phys Rev B 62:14750-14756
Katsura T, Sato K, Ito E (1998) Electrical conductivity of silicate perovskite at lower-mantle conditions. Nature 395:493-495
Kohn W, Sham LJ (1965) Self-consistent equations including exchange and correlation effects. Phys Rev 140:A1133-A1138
Kotomin EA, Kuklja MM, Eglitis RI, Popov AI (1996) Quantum chemical simulations of the optical properties and diffusion of electron centres in MgO. Mater Sci Eng B 37:212-214
Kotomin EA, Popov AI (1998) Radiation-induced point-defects in simple oxides. Nucl Instrum Methods Phys Res B 141:1-15
Kresse G, Furthmüller J (1996a) Efficient iterative schemes for *ab initio* total-energy calculations using a plane-wave basis set. Phys Rev B 54:11169-11186
Kresse G, Furthmüller J (1996b) Efficiency of ab-initio total energy calculations for metals and semiconductors using a plane-wave basis set. Comput Mater Sci 6:15-50
Kresse G, Hafner J (1993) *Ab initio* molecular dynamics for liquid metals. Phys Rev B 47:558-561
Kresse G, Joubert J (1999) From ultrasoft pseudopotentials to the projector augmented wave method. Phys Rev B 59:1758-1775
Lauterbach S, McCAmmon CA, van Aken P, Langenhorst F, Seifert F (2000) Mössbauer and ELNES spectroscopy of (Mg,Fe)(Si,Al)O_3 perovskite: A highly oxidised component of the lower mantle. Contrib Mineral Petrol 138:17-26
Lee W-C, Lee S-G, Chang KJ (1998) First-principles study of the self-interstitial diffusion mechanism in silicon. J Phys Condens Matter 10:995-1002
Leslie M, Gillan MJ (1985) The energy and elastic dipole tensor of defects in ionic crystal calculated by the supercell method. J Phys C Solid State Phys 18:973-982
Lin J-F, Heinz DL, Mao H, Hemley RJ, Devine JM, Li J, Shen G (2002) Stability of magnesiowüstite in Earth's lower mantle. Proc Nat Acad Sci USA 100:4405-4408

Lindner R, Parfitt GD (1957) Diffusion of radioactive magnesium in magnesium oxide crystals. J Chem Phys 26:182-185
Litasov K, Ohtani E, Langenhorst F, Yurimoto H, Kubo T, Kondo T (2003) Water solubility in Mg-perovskites and water storage capacity in the lower mantle. Earth Planet Sci Lett 211:189-203
Lowitzer S, Wilson DJ, Winkler B, Milman V, Gale JD (2008) Defect properties of albite. Phys Chem Miner 35:129-135
Mackrodt WC (1988) Temperature dependence of lattice and defect properties of MgO and Li_2O. J Mol Liq 39:121-136
Mackrodt WC, Stewart RF (1979) Defect properties of ionic solids. III. The calculation of the point-defect structure of the alkaline-earth oxides and CdO. J Phys C Solid State Phys 12:5015-5036
Mackwell S, Bystricky M, Sproni C (2005) Fe-Mg interdiffusion in (Mg,Fe)O. Phys Chem Miner 32:418-425
Makov G, Payne MC (1995) Periodic boundary conditions in *ab initio* calculations. Phys Rev B 51:4014-4022
Mitrovica JX, Forte AM (2004) A new inference of mantle viscosity based upon joint inversion of convection and glacial isostatic adjustment data. Earth Planet Sci Lett 225:177-189
Mookherjee M, Stixrude L, Karki B (2008) Hydrous silicate melt at high pressure. Nature 452:983-986
Narayan J, Washburn J (1973) Self-diffusion in magnesium oxide. Acta Metall 21:533-538
Nevins D, Spera FJ, Ghiorso MS (2009) Shear viscosity and diffusion in liquid $MgSiO_3$: Transport properties and implications for terrestrial planet magma oceans. Am Mineral 94:975-980
Oganov AR, Gillan MJ, Price GD (2003) *Ab initio* lattice dynamics and structural stability of MgO. J Chem Phys 118:10174-10182
Oishi Y, Kingery WD (1960) Oxygen diffusion in periclase crystals. J Chem Phys 33:905-906
Parlinski K (2008) Software PHONON. *http://wolf.ifj.edu.pl/phonon/*
Perdew JP, Burke K (1996) Comparison shopping for a gradient-corrected density functional. Int J Quantum Chem 57:309-319
Perdew JP, Zunger A (1981) Self-interaction correction to density-functional approximations for many-electron systems. Phys Rev B 23:5048-5079
Perdew JP, Wang Y (1992) Accurate and simple analytic representation of the electron-gas correlation energy. Phys Rev B 45:13244-13249
Poirier JP (1985) Creep of Crystals. Cambridge University Press
Price GD, Wall A, Parker SC (1989) The properties and behaviour of mantle minerals: a computer-simulation approach. Philos Trans R Soc London A 328:391-407
Reich M, Ewing RC, Ehlers TA, Becker U (2007) Low-temperature anisotropic diffusion of helium in zircon: Implications for zircon (U-Th)/He thermochronometry, Geochim Cosmochim Acta 71:3119-3130
Roma G, Limoge Y, Baroni S (2001) Oxygen self-diffusion in α-quartz. Phys Rev Lett 86:4564-4567
Sakaguchi I, Yurimoto H, Sueno S (1992) Self-diffusion along dislocations in single-crystal MgO. Solid State Commun 84:889-893
Sangster MJL, Rowell DK (1981) Calculation of defect energies and volumes in some oxides. Philos Mag A 44:613-624
Sangster MJL, Stoneham AM (1984) Calculation of absolute diffusion rates in oxides. J Phys C Solid State Phys 17:6093-6104
Sheppard D, Terrell R, Henkelman G (2008) Optimization methods for finding minimum energy paths. J Chem Phys 128:34106:1-10
Schilling W (1978) Self-interstitial atoms in metals. J Nucl Mater 69-70:465-489
Sempolinsky DR, Kingery WD (1980) Ionic conductivity and magnesium vacancy mobility in magnesium oxide. J Am Ceram Soc 63:664-669
Shirasaki S, Hama M (1973) Oxygen-diffusion characteristics of loosely sintered polycrystalline MgO. Chem Phys Lett 20:361-365
Solomatov VS, El-Khozondar R, Tikare V (2002) Grain size in the lower mantle: constraints from numerical modeling of grain growth in two-phase systems. Phys Earth Planet Inter 129:265-282
Stacey FD (1992) Physics of the Earth. Brooksfield Press
Sushko PV, Shluger AL, Catlow CRA, Baetzold RC (1999) Embedded cluster approach: application to complex defects. Radiat Eff Defects Solids 151:215-221
Sushko PV, Shluger AL, Baetzold RC, Catlow CRA (2000a) Embedded cluster calculations of metal complex impurity defects: properties of the iron cyanide in NaCl. J Phys Condens Matter 12:8257-8266
Sushko PV, Shluger AL, Catlow CRA (2000b) Relative energies of surface and defect states: *ab initio* calculations for the MgO (001) surface. Surf Sci 450:153-170
Tsuchiya T, Tsuchiya J, Umemoto K, Wentzcovitch RM (2004) Phase transition in $MgSiO_3$ perovskite in the Earth's lower mantle. Earth Planet Sci Lett 224:241-248
Van Orman JA, Fei Y, Hauri EH, Wang J (2003) Diffusion in MgO at high pressures: constraints on deformation mechanisms and chemical transport at the core-mantle-boundary. Geophys Res Lett 30:28:1-4

Vineyard GH (1957) Frequency factors and isotope effects in solid state rate processes. J Phys Chem Solids 3:121-127

Vočadlo L, Wall A, Parker SC, Price GD (1995) Absolute ionic diffusion in MgO - computer calculations via lattice dynamics. Phys Earth Planet Inter 88:193-210

Voter AF (2005) Introduction To The Kinetic Monte Carlo Method. Springer, NATO Publishing Unit

Walker AM, Demouchy S, Wright K (2006) Computer modeling of the energies and vibrational properties of hydroxyl groups in α- and β-Mg_2SiO_4. Eur J Mineral 18:529-543

Walker AM, Woodley SM, Slater B, Wright K (2009) A computational study of magnesium point defects and diffusion in forsterite. Phys Earth Planet Inter 172:20-27

Wan JTK, Duffy TS, Scandolo S, Car R (2007) First principles study of density, viscosity and diffusion coefficients of liquid $MgSiO_3$ at conditions of the Earth's deep mantle. J Geophys Res 112:B03208, doi:10.1029/2005JB004135

Wright K, Price GD (1993) Computer simulation of defects and diffusion in perovskite. J Geophys Res 98:22.245-22.253

Wuensch BJ, Steele WC, Vasilos T (1973) Cation self-diffusion in single-crystal MgO. J Chem Phys 58:5258-5266

Xu Y, McCammon C (2002) Evidence for ionic conductivity in lower mantle (Mg,Fe)(Si,Al)O_3 perovskite. J Geophys Res 107: doi:10.1029/2001JB000677

Xu Y, McCammon C, Poe BT (1998) The effect of alumina on the electrical conductivity of silicate perovskite. Science 282:922-924

Yamazaki D, Irifune T (2003) Fe-Mg interdiffusion in magnesiowüstite up to 35 GPa. Earth Planet Sci Lett 216:301-311

Yamazaki D, Kato T, Yurimoto H, Ohtani E, Toriumi M (2000) Silicon self-diffusion in $MgSiO_3$ perovskite at 25 GPa. Phys Earth Planet Inter 119:299-309

Yoo H-I, Wuensch BJ, Petuskey WT (2002) Oxygen self-diffusion in single-crystal MgO: Secondary-ion mass spectrometric analysis with comparison of results from gas-solid and solid-solid exchange. Solid State Ionics 150:207-221

Reviews in Mineralogy & Geochemistry
Vol. 71 pp. 225-251, 2010

Modeling Dislocations and Plasticity of Deep Earth Materials

Philippe Carrez and Patrick Cordier

Université Lille 1
Laboratoire de Structure et Propriétés de l'Etat Solide
UMR CNRS 8008 - *Bat C6*
59655 *Villeneuve d'Ascq Cedex, France*

Patrick.Cordier@univ-lille1.fr

CONTEXT

It is widely recognized that Earth sciences were revolutionized when continental drift proposed by Wegener in 1912 took the more comprehensive form of plate tectonics. Following the early suggestion of Holmes, Hess proposed in the early sixties what missed to Wegener's model: a mechanism and a driving force, i.e., the presence of large flows within the convecting mantle to evacuate the internal heat. The Earth appears now as a dynamic system dominated by flows and deformation processes almost everywhere: convective flow in the liquid core (which is the source of the Earth magnetic field), slow deformation of the hot (viscous?) solid mantle and rigid brittle deformations of the cold crust. Understanding and modeling the deformation mechanisms of Earth materials is thus one of the most important challenges of geophysics. Among the three large topics listed above, we will focus on mantle convection only. Two scientific fields which developed approximately at the same time (i.e., just before and just after the second world war) meet here: geophysics and materials science which aims at explaining how solids can deform (see below). Mantle convection raises however some questions that are specific to this field:

- Which minerals are involved and what are their deformation mechanisms?
- What is the influence of pressure on plastic deformation?
- What is the influence of extremely low strain rates on deformation mechanisms?

Although some very significant advances have been achieved recently on the experimental side (see for instance Karato and Weidner 2008) most of these questions remain largely unsolved and at the limits of our capabilities. It appears thus necessary to complement the usual experimentally-based approach by a numerical, multi-scale modeling approach in order to increase the significance of our models. This multi-scale modeling must be based on the Physics at the relevant scales (see Fig. 1) and will have to be experimentally validated at these scales. The outline of this multi-scale modeling approach of plasticity has already been described in Cordier et al. (2005). The present chapter will focus on the recent developments of numerical modeling of dislocations in minerals.

DISLOCATIONS AND PLASTIC DEFORMATION

Plastic flow is a transport phenomenon carried by the motion of defects: point defects, dislocations, or grain boundaries. Among the various possible deformation mechanisms, dislocation glide is usually the most efficient one. Glide of dislocations can be inhibited (and replaced by other mechanisms) in some cases:

1529-6466/10/0071-0011$05.00 DOI: 10.2138/rmg.2010.71.11

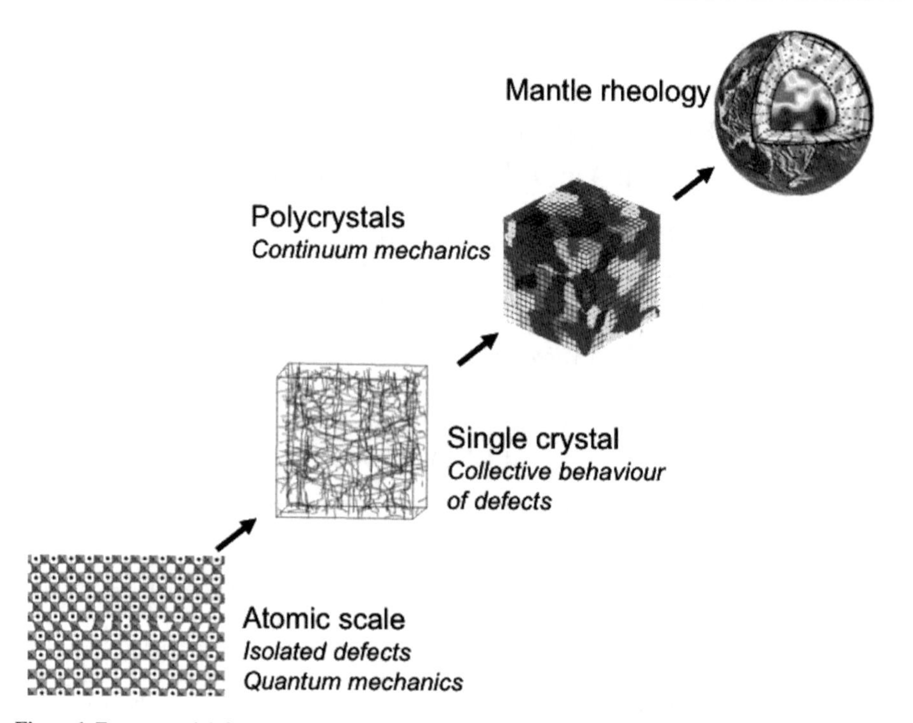

Figure 1. From crystal defects to the Earth's mantle. Understanding mantle rheology requires development of multi-scale numerical tools simulating the behavior of high-pressure silicates in the conditions of the mantle (Cutaway model of the Earth courtesy of the Département de sismologie, IPGP-CNRS).

- For geometrical reasons: for instance by loading along the *c* axis a hexagonal single crystal which deforms only by basal glide.
- When lattice friction is too high: this is the case of ceramics or quasicrystals for instance.

In the previous cases, the materials are brittle unless some other plasticity mechanisms can be activated:

- Nabarro-Herring or Coble creep. In that case, very small grain sizes must be maintained to create short diffusion paths. This represents a situation very far from equilibrium since grain growth is likely to occur if the material is kept for long times at high temperature.
- Grain boundary sliding and grain boundary migration. Same comments as above.
- Mechanical twinning: at low temperature generally and with some minerals only: pyroxenes, sapphire, etc.
- Pure dislocation climb: at high temperature only.

The ability of a material to deform by dislocation glide, and the relative easiness of this mechanism compared to others, are thus key parameters that will determine the mechanical behavior. In particular, dislocation glide is usually associated with non-linear rheologies.

Dislocations glide in response to an applied stress. Whereas forces acting on a dislocation line are well-described within the linear elasticity theory (Peach and Koehler 1950), how fast a dislocation will move under stress is a much more complicated problem since it depends on atomistic core mechanisms (e.g., Cai et al. 2004). In case of high-pressure minerals, the mobility is likely to be controlled by lattice friction and one must wonder whether, under high-pressure and low deviatoric stresses, dislocation glide is possible or whether another deformation mechanism must be activated. The evaluation of lattice friction is a complex problem since it requires modeling of:

- the fine core structure of dislocations
- the associated Peierls potential, and
- the thermally activated mechanisms to overcome this potential that will determine the behavior at finite temperature.

We will describe in the following the numerical models and techniques that are used to develop a modeling approach of the plastic deformation of minerals under high-pressure and high-temperature conditions.

MODELING DISLOCATIONS

The Volterra dislocation

A Volterra dislocation (Volterra 1907) can be introduced in a medium through the following process:

- a cut is made along a surface bounded by a contour (*C*)
- the two parts of the crystal separated by the cut are moved relative to each other by a translation vector of the lattice (further called Burgers vector).
- if necessary, some material can be filled or taken out to preserve the structure of the medium
- the surfaces of the cut are welded together and the applied forces are suppressed

The resulting defect results in a step-like discontinuity across the contour (*C*) (Fig. 2). This simple model leads to simple analytic solutions for the elastic fields surrounding the dislocation line. However, the centre of the dislocations is source of a singularity for the stress and strain fields which must be artificially suppressed by a cut-off radius r_0 which defines the dislocation core.

Two different theoretical approaches can be followed to go beyond the Volterra dislocation and to study dislocation core properties (e.g., Schoeck 1998, 2006): one is the hybrid continuum-atomistic Peierls-Nabarro (PN) model and the other one relies on 'direct' atomistic simulations.

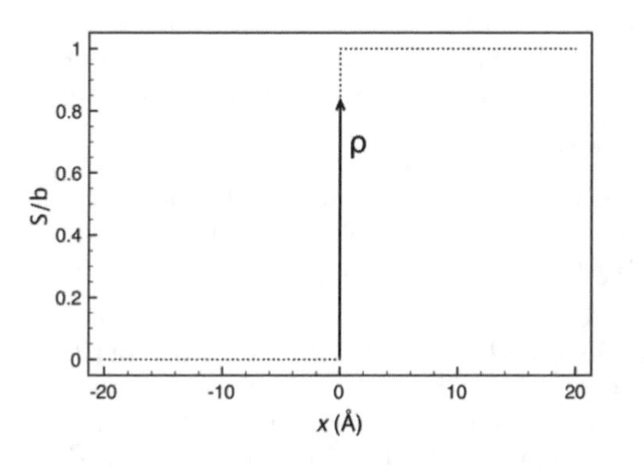

Figure 2. Volterra dislocation of Burgers vector **b**. the shear profile across the dislocation core is a step function.

"Direct" atomistic calculations

Arrangement of atoms near the dislocation core can be obtained by standard static relaxation procedures using atomistic simulations. The main idea of the calculations is to introduce a dislocation by displacing atoms according to the well-known elastic displacement field of the defect. The calculation is performed in two stages. Conventionally, a cylinder of matter centered on the dislocation line is selected (Fig. 3). All atoms belonging to the cylinder are displaced according to the displacement field of the dislocation. Depending on the material, classical isotropic expressions can be used or an anisotropic displacement field may be considered. All the expressions can be found either in Hirth and Lothe (1982) or in Steeds (1973) for full anisotropic treatment. For the sake of simplicity, only isotropic displacement fields for screw and edge dislocations are given in the following.

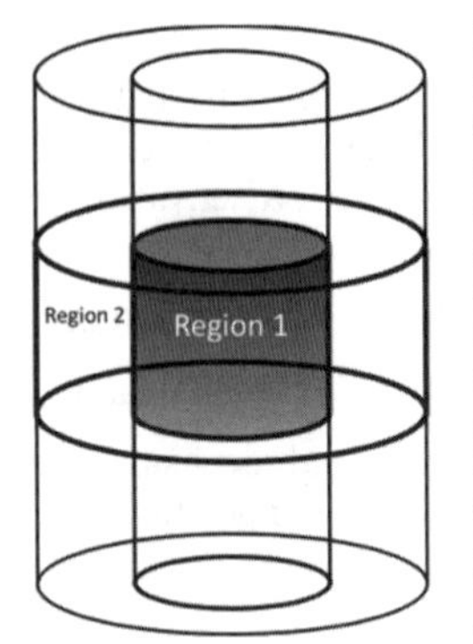

Figure 3. Schematic diagram of the simulation cell used for cluster calculations. The dislocation is introduced at the center of region 1 by displacing all atoms (in both regions 1 and 2) according to the elastic displacement field. Atoms belonging to region 2 are kept fixed during the optimization calculation. Note that a vacuum region is used to terminate the dislocation.

In case of a screw dislocation, in the classical reference frame used since Volterra (i.e., with the dislocation line and Burgers vector **b** along the z axis):

$$u_x = u_y = 0 \qquad u_z = \frac{\mathbf{b}}{2\pi}\tan^{-1}(y/x) \tag{1}$$

whereas for an edge dislocation with a dislocation line along the z axis and the Burgers vector **b** along the x axis, the expressions are more complex and take into account the Poisson ratio ν

$$\begin{aligned} u_x &= \frac{b}{2\pi}\left[\tan^{-1}(y/x) + \frac{xy}{2(1-\nu)(x^2+y^2)}\right] \\ u_y &= -\frac{b}{2\pi}\left[\frac{1-2\nu}{4(1-\nu)}\ln(x^2+y^2) + \frac{x^2-y^2}{4(1-\nu)(x^2+y^2)}\right] \\ u_z &= 0 \end{aligned} \tag{2}$$

Imposing an outer rim of fixed matter (Fig. 3), the central region is allowed to relax to find a low energy configuration. This process involving non-linear elasticity near the core accounts for the arrangement of the atoms and the fine structure of the dislocation core. As these calculations are basically based on elastic theory, they require large cells with thousands of atoms to ensure that elastic theory can be applied. These calculations require accurate description of the interatomic interactions. This can be done by using empirical interatomic potentials. Indeed, the use of empirical potentials is a priori more adapted to the large systems required to model a dislocation core in a complex material. It is however possible that empirical potentials lack accuracy in the highly distorted regions of a dislocation core. On the other hand, first principles calculations can provide a significant improvement of accuracy, but are still limited to a smaller number of atoms due to a high computational cost. Since a few years, this method in which the core structure of an isolated dislocation is calculated is referred as the cluster method (e.g., Woodward 2005). New developments, referred as the dipole approach (e.g., Pizzagalli and Beauchamp 2004, Cai 2005, Clouet 2009), have been proposed to take advantages of three-dimensional boundary conditions. Both techniques lead to similar results (Clouet et al. 2009). The dipole method is compatible with ab intitio calculation, at least for high symmetry materials (Pizzagalli and Beauchamp 2004; Ventelon and Willaime 2007).

All these techniques have been largely used to model core structures in metals (Yamaguchi and Vitek 1973; Duesbery and Vitek 1998; Xu and Moriarty 1998; Woodward and Rao 2002; Wang et al. 2003; Woodward 2005 or Groger et al. 2008), in intermetallic phases (Schroll et al. 1998; Gumbsch and Schroll 1999), in semiconductors such as silicon or diamond-like structure (Koizumi et al. 2000; Miyata and Fujiwara 2001; Pizzagalli et al. 2009) or in ceramics (Woo and Puls 1977; Watson et al. 1999). In minerals the cluster method has been used to model dislocations in zeolite (Walker et al. 2004) and forsterite (Walker et al. 2005a and b) with as main achievement the characterization of 3D cores in [001] screw dislocations (Carrez et al. 2008).

The Peierls-Nabarro (PN) model of dislocations: fundamentals and recent developments

The original PN model (Peierls 1940; Nabarro 1947) represents a useful and efficient approach to calculate the core properties of dislocations (Ren et al. 1995; Joos and Duesbery 1997a; Wang 1996) based on the assumption of a planar core (Schoeck 2005). It has been shown to apply to a wide range of materials (Wang 1996; Bulatov and Kaxiras 1997; Von Sydow et al. 1999; Lu et al. 2000; Lu 2005; Miranda and Scandolo 2005).

The crystal is divided into three regions (Fig. 4). The PN model assumes that the misfit region of inelastic displacement is restricted to the glide plane (between A and B), whereas linear elasticity applies far from it, above and below (regions (a) and (b), Fig. 4). The misfit region of inelastic displacement is characterized by a Generalized Stacking Fault (GSF) energy (also called the γ-surface). A dislocation corresponds to a continuous distribution of shear $S(x)$ along the glide plane (x is the coordinate along the displacement direction of the dislocation in the glide plane) storing a misfit energy $\int\gamma(S(x))dx$. $S(x)$ represents the disregistry across the glide plane and the stress generated by such a displacement can be represented by a continuous distribution of infinitesimal dislocations with density $\rho(x)$ for which the total summation is equal to the Burgers vector **b**. The expression of the total dislocation energy functional taking into account of the misfit energy in region A-B, and elastic energy in regions (a) and (b) is thus given by

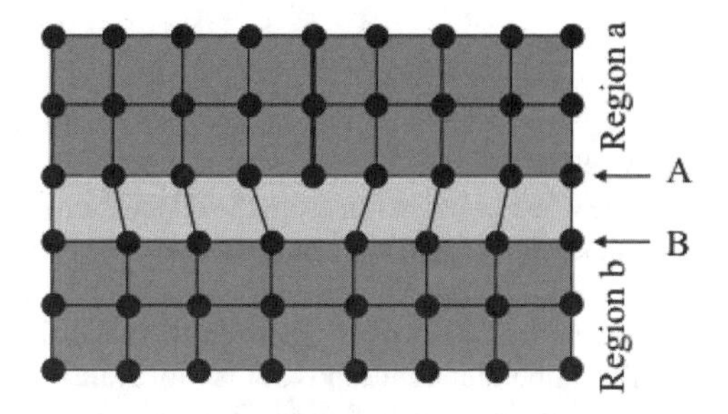

Figure 4. The PN model localizes inelastic displacements in the grey layer between A and B where the core will be allowed to spread. Linear elasticity will apply to regions (a) and (b).

$$U_{tot} = \int\gamma(S(x))dx - \frac{K}{2}\iint\rho(x)\rho(x')\ln|x-x'|dxdx' \tag{3}$$

where $\rho(x) = dS(x)/dx$ and K is a combination of elastic constants, reflecting the energy coefficient of the dislocation character. This coefficient can be calculated within the frame of the Stroh theory (Hirth and Lothe 1982) to take anisotropic elasticity into account.

The equilibrium structure of a dislocation is thus determined by minimizing the above energy functional. A variational derivative of Equation (3) with respect to ρ leads to an integro-differential Equation (4) nowadays referred to as the PN equation.

$$\frac{K}{2\pi}\int_{-\infty}^{+\infty}\frac{1}{x-x'}\left[\frac{dS(x')}{dx'}\right]dx' = \frac{K}{2\pi}\int_{-\infty}^{+\infty}\frac{\rho(x')}{x-x'}dx' = F(S(x)) \tag{4}$$

Originally the PN equation was introduced considering that the restoring force F acting between atoms on either sides of the interface is balanced by the resultant stress of the distribution. The significance of the GSF energy is that for a fault vector $\boldsymbol{S}$, there exists an interfacial restoring stress deriving from γ which has the same formal interpretation as the initial restoring force F introduced originally by Peierls (Vitek 1968).

$$\vec{F}(S) = -\overrightarrow{grad}\gamma(S) \tag{5}$$

The analytical solution. As largely developed since Peierls (1940) and reviewed by Joos and Duesbery (1997a), an analytical solution of the PN equation can be found by introducing a sinusoidal restoring force:

$$F(S(x)) = \tau^{\max} \sin\left(\frac{2\pi S(x)}{b}\right) \tag{6}$$

where τ^{max} is the Ideal Shear Strength (ISS) which is defined as the "maximum resolved shear stress that an ideal, perfect crystal can suffer without plastically deforming" (Paxton et al. 1991). Incorporating this expression in the PN equation leads to a classic solution for the disregistry function:

$$S(x) = \frac{b}{\pi}\tan^{-1}\frac{x}{\zeta} + \frac{b}{2} \tag{7}$$

where $\zeta = Kb/4\pi\tau^{\max}$ represents the half-width of the dislocation distribution $\rho(x)$. One can easily check that the disregistry $S(x)$ across the glide plane verifies the boundary conditions, $S(x) = 0$ when $x \to -\infty$ (far from the dislocation), $S(x) = b$ when $x \to +\infty$ (on the other side) and the summation of $S(x)$ over the whole space is equal to the Burgers vector **b** of the dislocation.

As pointed out by numerous authors (e.g., Lu 2005), one of the main achievements of the PN model is that it provides reasonable estimates of the dislocation core sizes. However, equally important is the possibility to evaluate the Peierls stress, i.e., the stress needed to move the dislocation without help of temperature. However, the original PN equation or the expression of the energy functional (3) is invariant with respect to an arbitrary translation of the dislocation density. Indeed, both hold for an elastic continuous medium which does not take into account of the lattice discreteness whereas $S(x)$ can actually only be defined where an atomic plane is present (e.g., Hirth and Lothe, 1982; Schoeck, 1999). The simplest way to obtain the misfit energy corresponding to the Peierls dislocation and to determine the Peierls stress, is to perform a summation of the local misfit energy at the positions of atoms rows parallel to the dislocation line. The misfit energy (also called the Peierls potential) can be thus considered as the sum of misfit energies between pairs of atomic planes (e.g., Joos et al. 1994; Joos and Duesbery 1997a) and can be written as

$$V_P(u) = \sum_{m=-\infty}^{+\infty} \gamma(S(ma' - u)) \cdot a' \tag{8}$$

where a' is the periodicity of V_P, taken as the shortest distance between two equivalent atomic rows in the direction of the dislocation's displacement. The Peierls stress is then given by:

$$\sigma_P = \max\left\{\frac{1}{b}\frac{dV_P(u)}{du}\right\} \tag{9}$$

In order to solve the misfit energy function analytically, Joos and Duesbery (1997a) introduced a dimensionless parameter $\Gamma = \zeta/a'$. Then, some simple formulas can be derived for the extreme cases of very narrow ($\Gamma << 1$) and widely spread ($\Gamma >> 1$) dislocations:

$$\sigma_p(\Gamma << 1) = \frac{3\sqrt{3}}{8}\tau^{\max}\frac{a'}{\pi\zeta} \tag{10}$$

$$\sigma_p(\Gamma >> 1) = \frac{Kb}{a'}\exp\left(\frac{-2\pi\zeta}{a'}\right) \tag{11}$$

The range of the limits was recently extended to the case $0.2 < \Gamma < 0.5$ where Equation (10) has to be modified to take into account an exponential correction term (Joos and Zhou 2001).

The relevance of this analytical model to predict the easiest slip systems in a given material has been assessed by Wang (1996). Although some trends are observed, this approach has shown its limits and cannot be regarded as fully quantitative.

Numerical solutions based on generalized stacking faults. The reasons for the limitation of the analytical solution is the (strong) hypothesis of a sinusoidal restoring force and the need to know the ISS value. The range of application is thus very limited. However, a larger range of possibilities is offered considering that the restoring force introduced in the PN model is the gradient of the γ-surface.

Nowadays, the GSF energy is relatively easy to calculate using atomistic calculations (either *ab initio* or using semi-empirical potentials). One may note that semi-empirical potentials can be fitted with respect to *ab initio* GSF calculations (e.g., Zimmerman et al. 2000). In order to calculate a γ-surface for a given slip plane, one has to consider a perfect crystal cut across the slip plane into two parts which are subjected to differential displacement through an arbitrary fault vector **S**. The faulted lattice exhibits an extra energy per unit area $\gamma(S)$. The γ-surface is generated by considering all possible **S** vectors in the slip plane. Practically, the calculation is made using supercell methods and minimization of the excess energy by relaxation of atomic positions perpendicularly to the glide plane (Cai et al. 2002; Durinck et al. 2005). Depending on the atomistic calculations techniques used, there are several possibilities to build supercells which have been recently reviewed by Bulatov et al. (2006).

Introducing real restoring forces, calculated at the atomic scale, in the PN equation renders analytical solutions of S or ρ more complicated to evaluate. Among the various ansatz proposed for the expression of $S(x)$ (Kroupa and Lejcek 1972; Hartford et al. 1998; Yan et al. 2004), one may retain the formulation of Joos et al. (1994) based on a set of variational constants. Restricting the treatment to one dimension of the γ-surface along the Burgers vector direction, the disregistry distribution in the dislocation core $S(x)$ can be obtained by searching for a solution in the form:

$$S(x) = \frac{b}{2} + \frac{b}{\pi}\sum_{i=1}^{N}\alpha_i \cdot \arctan\frac{x - x_i}{c_i} \tag{12}$$

where α_i, x_i and c_i are variational constants. The idea is to search for a solution which can be written as a sum of N fractional dislocation densities. Using the previous form of $S(x)$, the infinitesimal dislocation density $\rho(x)$ is given by

$$\rho(x) = \frac{dS(x)}{dx} = \frac{b}{\pi}\sum_{i=1}^{N}\alpha_i \frac{c_i}{(x - x_i)^2 + c_i^2} \tag{13}$$

As the disregistry $S(x)$ and the density $\rho(x)$ must satisfy the normalization condition $\int\rho(x)\, dx = b$, the α_i are constrained by $\sum_{i=1}^{N}\alpha_i = 1$. Substituting Equation (13) into the left-hand side of the PN Equation (4), gives the restoring force

$$F^{PN}(x) = \frac{Kb}{2\pi}\sum_{i=1}^{N}\alpha_i \cdot \frac{x - x_i}{(x - x_i)^2 + c_i^2} \tag{14}$$

The variational constants α_i, x_i and c_i are obtained from a least square minimization of the difference between F^{PN} and the restoring force F derivated from the γ-surface. A typical value of $N = 3$ was used by Joos et al. (1994) for dislocations in silicon. Recent calculations in mineral structures (Carrez et al. 2006, 2007) used $N = 6$, especially for calculations along a GSF characterized by a stable stacking fault. Indeed the existence of a stable stacking fault reflects the tendency of a dislocation to dissociate into two partial dislocations of collinear Burgers vectors.

All above models still present some limitations. Among them, the first regards the strain energy of dislocation that can be unrealistically high especially in case of dislocation with narrow core. The second limitation comes from the constraint that dislocation cores are planar. Indeed the present model can be fully applied to study dissociated collinear dislocations and can be improved by modifying ansatz (Eqns. 12 and 13) to treat the case of non collinear dissociation (complex analytical expression can be derived assuming that the γ-surface is written as a Fourier series, see the work of Schoeck and coworkers, e.g., Schoeck and Krystian 2005) but by essence, the treatment neglects important degrees of freedom as it considers that the dislocation core is restricted to the glide plane. As a consequence, the prediction of the model failed for some screw dislocations known to be spread into several planes such as in most bcc structures. These limitations were addressed by Bulatov and Kaxiras (1997) by developing a semi-discrete variational Peierls-Nabarro (SVPN) framework. In the SVPN approach, Equation (3) is described using a discrete form based on a nodal set.

$$U\left[\{S_i\}\right] = K\sum_{ij}\chi_{ij}\rho_i\rho_j + \Delta x\sum_i \gamma(S_i) \tag{15}$$

where S_i corresponds to the nodal displacement of node i, $\rho_i = (S_i - S_{i-1})/(x_i - x_{i-1})$, $\chi_{ij} = 1.5\phi_{i,i-1}\phi_{j,j-1} + \psi_{i-1}\psi_{j-1} + \psi_{i,j} + \psi_{i,j-1} + \psi_{j,i-1}$ with $\phi_{i,j} = x_i - x_j$ and $\psi_{i,j} = 0.5\phi_{i,j}^2 \ln|\phi_{i,j}|$.

In such an approach, there is no need of analytical expressions and the disregistry function S is simply interpolated linearly between nodal points x_i. It also takes advantages of the nodal description to introduce the effect of an applied stress τ on the dislocation profile. Indeed, to define the Peierls stress, one may imagine two distinct treatments. The first is analogous to Equation (8). Using SVPN, Equation (15) can be estimated for various positions of the dislocation centre that leads to the Peierls barrier. At the same time, an extra effect of stress can be added in Equation (15) and dislocation density can be minimized for any given value of stress. It turns out that an instability is reached around the Peierls stress which leads to the failure of the minimization procedure.

This approach overcomes key deficiencies of the previous PN description. However, despite significant improvements, the SVPN model appears still to be difficult to use for complex dislocation core structures. First, an extended treatment, introducing potential dislocation densities for edge and screw components is necessary to study non-collinear dissociations. Secondly, taking several planes into account requires strong modifications of Equation (15). An example of such modification is presented by Lu (2005) in case of two planes (relevant for cross slip of a screw dislocation, see also Lu et al. 2003). Compared to previously modified version of the PN model (to account for complex core structure, Lejcek and Kroupa 1976; Ngan 1997; Ngan and Zhang 1999), one of the main achievement of the SVPN framework is clearly to highlight the potential effect of the applied stress on the structure of the dislocation core and thus on the value of the expected Peierls stress which may be sensitive to the fact that the core can be modified during the motion of the dislocation.

The Peierls-Nabarro Galerkin method. The concept of nodal description is close to the one of a Finite Element (FE) method. Some coupling between FE methods and discrete representations of dislocation have already been proposed (see for instance the work of Lemarchand et al. (2001)). However, based on a dislocation dynamics approach, the constraint of modeling dislocation with no core structure is a strong limitation to study the Peierls stress. The same kind of arguments applies to the Phase Field methods which are able to reproduce dissociated dislocations core structure but are unable to predict dislocation core width (Shen and Wang 2004). Very recently, Denoual (2004, 2007) rewrote in a very promising way the PN model within the element free Galerkin method (comparable to a FE method). Such treatment should account for complex boundary conditions or material structures and allows for the calculation of core structures as presented in the following. The Peierls-Nabarro Galerkin (PNG) method is a generalization of the PN method to allow for multiple glide planes and complex (possibly tri-dimensional) cores. As in the initial PN model, the dislocation core structure naturally emerges from minimizing an elastic energy (through an approximation of a continuous field representation) and an interplanar potential. For the sake of clarity, let us consider one unique slip plane Σ. In the PNG model, two distinct fields are used: $u(r)$ a three-dimensional displacement field of the volume V and a two dimensional $\mathbf{S}(r)$ field which is expressed in the normal basis of the Σ plane. Thus, $u(r)$ represents a homogenous strain whereas the displacement jump when crossing Σ is measured by $\mathbf{S}$. The problem consists therefore in minimizing the following Hamiltonian H with respect to u and $\mathbf{S}$.

$$H = \int_V \left\{ E^e[u,S] + \frac{1}{2}\Omega\dot{u}^2 \right\} dV + \int_\Sigma E^{isf}[S]d\Sigma \tag{16}$$

In Equation (16), Ω corresponds to the material density. E^e corresponds to the elastic strain energy whereas E^{isf} is the inelastic stacking fault energy which depends on the material and controls the spreading of dislocations. E^{isf} is therefore a function of the GSF energies γ from which all the linear elastic part has been subtracted (Denoual 2007; Pellegrini et al. 2008). In doing so, we strictly comply with the overall linear elastic behavior except for small applied strains. In practice, minimization with respect to S is achieved by means of a time dependent Ginzburg Landau equation whereas the Galerkin method is used to compute the evolution of $u(r)$.

In practice, a simulation can be divided into three phases. A discrete dislocation is first introduced in a finite volume, the boundary of which is given by imposing a displacement equal to the result of a convolution of an elementary elastic solution with the dislocation density (see Pillon and Denoual 2009 for implementations details). The imposed boundary conditions are therefore consistent with a dislocation in an infinite media, without any influence on the core structure. The equilibrium between the displacement jump field S and the density of dislocation ρ, as defined as the derivative of S by the position coordinates, is obtained through a viscous relaxation scheme. During a second step, a homogeneous strain is progressively superimposed at a velocity that allows quasi-static equilibrium, even with noticeable evolution of the core structure. During the last step, a rapid evolution of the dislocation core structure followed by a displacement of one (or more) Burgers vector leads to a relaxation of the measured stress. The ultimate macroscopic stress is the Peierls stress.

Dislocation mobility at finite temperature

In the presence of lattice friction, dislocations move by nucleation and propagation of kink pairs. A kink pair is created when a segment of the dislocation line is brought by thermal activation and stress to a neighboring low-energy position (Fig. 5). Several models have been proposed to describe the kink-pair nucleation along a straight dislocation line (e.g., Caillard and Martin 2003). The first description of the geometry of a kink was proposed by Dorn and Rajnak (1964). The calculation of Dorn and Rajnak was later extended by Seeger (1984) and Suzuki et al. (1991) using a line tension approximation under high stresses (Line Tension model). The low stress regime can be described using a Coulomb interaction model between kinks (Seeger 1984). In 1993, Koizumi and coworkers proposed a new description of the kink-pair nucleation process based on an elastic interaction treatment (Koizumi et al. 1993; Koizumi et al. 1994). All models rely on the knowledge of the Peierls potential V_P as defined before (see for instance Eqn. 8) and the Peierls stress σ_P.

Whatever the level of approximation, the total energy variation ΔH between the initial straight dislocation line and the line carrying a kink-pair is simply written as

$$\Delta H = \Delta E + \Delta P - W \tag{17}$$

where ΔE and ΔP corresponds respectively to the variation of the elastic energy and the variation of the Peierls energy between the initial straight dislocation line and the kinked one, considering that to promote the nucleation of kink, a stress τ has to be applied, the resulting plastic work is W.

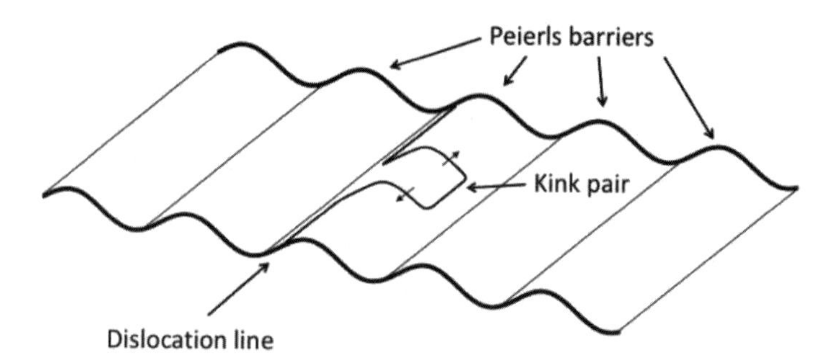

Figure 5. Thermal activation of dislocation glide. At finite temperature and stress below the Peierls stress, the dislocation moves forward by nucleation and propagation sideway of a kink pair.

Whereas numerical treatments of Equation (17) have been proposed by numerous authors (Duesbery and Joos 1996; Joos and Duesbery 1997b; Ngan 1999; Edagawa et al. 2000; Joos and Zhou 2001; Schoeck 2007), we propose here to follow the Elastic Interaction model (Koizumi et al. 1993) applied to a rectangular kink-pair of height h and width w (Fig. 6). In that case, the metastable configuration of the kink-pair is found as a saddle point of the total energy change ΔH with respect to the parameters h and w defining the kink-pair geometry, such that

$$\frac{\partial \Delta H}{\partial h} = \frac{\partial \Delta H}{\partial w} = 0 \tag{18}$$

Technically, differences between the various models of kink nucleation appear in the evaluation of ΔP and mostly ΔE. The choice made here of a rectangular kink pair simplifies the treatment as simple analytical expressions are available (e.g., Carrez et al. 2009b). Under an applied stress τ, a straight dislocation line reaches an equilibrium position x_0 (Fig. 6) from which a neutral kink pair can be nucleated. The formation of a kink-pair of height h induces a variation of the Peierls energy which can be written as

$$\Delta P(h,w) = 2\int_{x_0}^{x_0+h} V_P(x)dx + w\left[V_P(x_0+h) - V_P(x_0)\right] \tag{19}$$

The variation of the elastic energy (isotropic case) can be written as

$$\Delta E(h,w) = \frac{\mu b^2}{2\pi}\left(\begin{array}{l} \sqrt{w^2+h^2} - w - h + w\log\dfrac{2w}{w+\sqrt{w^2+h^2}} \\ -\dfrac{1}{1-\nu}\left(w - \sqrt{w^2+h^2} + h\log\dfrac{h+\sqrt{w^2+h^2}}{w} - h\log\dfrac{h}{er_c}\right) \end{array}\right) \tag{20}$$

for a initial screw dislocation (with kinks of edge character) or

$$\Delta E(h,w) = \frac{\mu b^2}{2\pi}\left(\begin{array}{l} \dfrac{1}{1-\nu}\left(\sqrt{w^2+h^2} - w - h + w\log\dfrac{2w}{w+\sqrt{w^2+h^2}}\right) \\ -w + \sqrt{w^2+h^2} - h\log\dfrac{h+\sqrt{w^2+h^2}}{w} + h\log\dfrac{h}{er_c}\end{array}\right) \tag{21}$$

for an initial edge dislocation (with kinks of screw character).

In Equations (20) and (21), μ corresponds to the shear modulus, ν to the Poisson ratio, **b** to the Burgers vector of the dislocation and r_c corresponds to a cut-off length (usually taken as

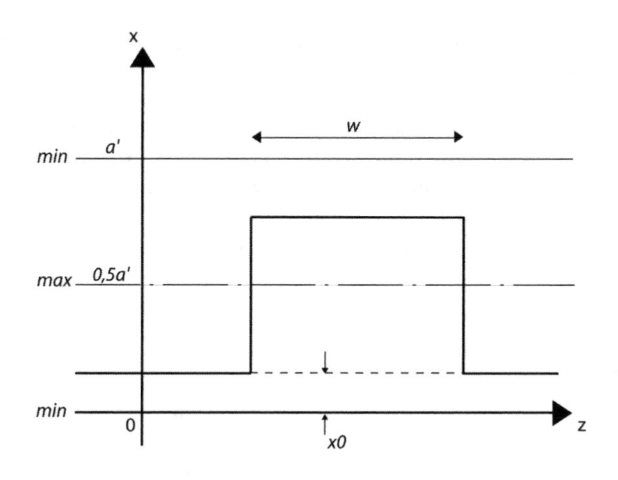

Figure 6. Rectangular kink pair nucleation on the Peierls potential. The period of the Peierls potential a' is indicated with the position of the valleys and hills. x_0 corresponds to the equilibrium position of the initial dislocation under a given applied stress τ.

a fraction of **b**). Finally, the plastic work W is:

$$W(h,w) = \tau bhw \tag{22}$$

Thus, for a given applied stress τ, a simple treatment (as every quantities are parameterized with respect to h and w) of Equation (18) leads to critical values $h^*(\tau)$ and $w^*(\tau)$, from which we can deduce the energy of the critical kink pair configuration $\Delta H^*(\tau)$, i.e., the energy that needs to be supplied by thermal activation.

Finally, one must consider that the critical kink pair energy is one of the key parameter when describing the velocity of a dislocation. Indeed, let us consider the glide of a dislocation between two adjacent Peierls valleys. Without strong restriction, we may assume that the kink pair nucleation controls the mobility. The assumption is supported by the fact that commonly edge component are characterized by a lower Peierls stress compared to screw. One simple reason is the value of the elastic coefficient K larger for edge component (due to the Poisson ratio) which leads to larger edge dislocation core. Therefore, taking into account the nucleation rate of mobile edge kink pairs on the screw line (e.g., Dorn and Rajnak 1964; Guyot and Dorn 1967 or more recently Groger and Vitek 2008), the velocity of a dislocation line can be written as

$$V = \nu_D b \frac{a'}{w^*(\tau)} \exp(-\frac{\Delta H^*(\tau)}{kT}) \tag{23}$$

where ν_D is the Debye frequency and w^* the length of the kink pair in the saddle point configuration. With the velocity of dislocations, mesoscopic scale modeling (using discrete dislocation dynamics, see Cordier et al. 2005) becomes possible.

SELECTED EXAMPLES

$SrTiO_3$ perovskite

Magnesium silicate perovskite (with some Fe and Al) represents approximately half of the Earth's mass and is considered as the most important mineral for the interior of the Earth. Therefore, the rheology of perovskite-structured materials is essential to understand the dynamics of the interior of the Earth. Unfortunately, investigation of the rheology of $MgSiO_3$ perovskite is rendered difficult by the pressure required to reach its stability field and only a few studies have tackled this problem (Karato et al. 1990; Chen et al. 2002; Merkel et al. 2003; Cordier et al. 2004). Alternatively, $SrTiO_3$ represents a good case study to address the issue of perovskite plasticity. The first reason is that $SrTiO_3$ exhibits the ideal (cubic) perovskite structure and hence represents the simplest situation among perovskites. Secondly, having a lot of applications in materials science, $SrTiO_3$ has been the subject of numerous experimental studies. It is thus a valuable material to test and validate the numerical tools.

As generally observed in perovskites (Poirier et al. 1989), the most common slip systems observed in deformed $SrTiO_3$ are $\langle 110\rangle\{1\bar{1}0\}$ and $\langle 100\rangle\{010\}$ (Nishigaki et al. 1991; Matsunaga and Saka 2000; Brunner et al. 2001; Gumbsch et al. 2001). However other dislocations like $\langle 1\bar{1}0\rangle\{001\}$ (Mao and Knowles 1996) or $\langle 001\rangle\{1\bar{1}0\}$ (Jia et al. 2005) have also been reported, following growth processes for instance.

The supercells used for the *ab initio* calculations (using the VASP code, Kresse and Hafner 1993) of the γ-surfaces are shown in Figure 7. Two γ-surfaces have been calculated *ab initio* (Ferré et al. 2008) from which four linear GSF have been extracted that correspond to the potential slip systems in $SrTiO_3$: $\langle 100\rangle\{010\}$, $\langle 100\rangle\{011\}$, $\langle 110\rangle\{001\}$ and $\langle 110\rangle\{1\bar{1}0\}$ (Fig. 8). All the GSF exhibit a single peak shape that can be significantly different from a sinusoidal profile. This is especially true in the case of $\langle 110\rangle\{1\bar{1}0\}$ which is characterized by a plateau around γ^{max}. Whereas low energy paths along the γ-surface correspond to $\langle 100\rangle$ and $\langle 110\rangle$, shapes of γ-surfaces justify going beyond the analytical approach which is clearly poorly adapted.

The shear profiles $S(x)$ and dislocation densities ρ have been thus determined from the PN model (Eqns. 12 to 14). Let us first focus on the case of the $\langle 100\rangle\{011\}$ dislocation. The dislocation density profile is represented on Figure 9 together with an atomistic model directly determined from it. As already pointed out, $SrTiO_3$ is an excellent test material because of the experimental data available. In particular, recent studies have provided high-resolution transmission electron microscopy (HRTEM) micrographs of dislocation cores of very high quality which can be compared to numerical models. Jia et al. (2005) have shown that a combination of hardware correction of the spherical aberration with numerical phase-retrieval techniques could give access to atomic arrangements of cations and anions in the dislocation core of a $\langle 100\rangle$ $\{011\}$ dislocation in $SrTiO_3$. Their study (reproduced in Fig. 10 with permission) provides a direct counterpart to our numerical models. Figure 10b demonstrates that the size of the dislocation (to quote Peierls original paper's title) is perfectly reproduced by the Peierls model.

However, as expected, $\langle 100\rangle\{011\}$ is not one of the easiest slip systems in $SrTiO_3$. Table 1 summarizes the results of the PN modeling of the Peierls stresses. In good agreement with experimental observations, $\langle 110\rangle\{1\bar{1}0\}$ appears to be the easiest slip system with a markedly lower lattice friction (although $\langle 110\rangle$ is not the smallest lattice repeat: 5.571 Å against 3.939 Å for $\langle 100\rangle$ dislocations). The reason for this behavior can be found in the core structure which is largely spread (Fig. 11). The same situation has been found in another cubic perovskite: $CaSiO_3$ (Ferré et al. 2009). This behavior can be anticipated from the shape of the $\langle 110\rangle\{1\bar{1}0\}$ as described above and from the atomic configurations explored by the material during $\langle 110\rangle$ $\{1\bar{1}0\}$ shear. During rigid-body shear along $\langle 110\rangle\{1\bar{1}0\}$, one octahedron is pulled along one corner. Five oxygen atoms out of six remain very close to the silicon atom whereas one of them is pulled away. However, $\langle 110\rangle$ shear brings very rapidly (at ca. 25% shear) another oxygen atom closer. As a result, a standard octahedral site is reconstructed for a 50% shear (Fig. 12). This situation contrasts with what is observed for other slip systems where larger Si-O distances are induced by shear.

Again, the validity of the PN model can be tested by performing 'direct' atomistic calculations of the core structure of the edge $\langle 110\rangle\{1\bar{1}0\}$ dislocation. 'Direct' atomistic calculations were performed using the cluster approach and a set of Buckingham potentials (taken from

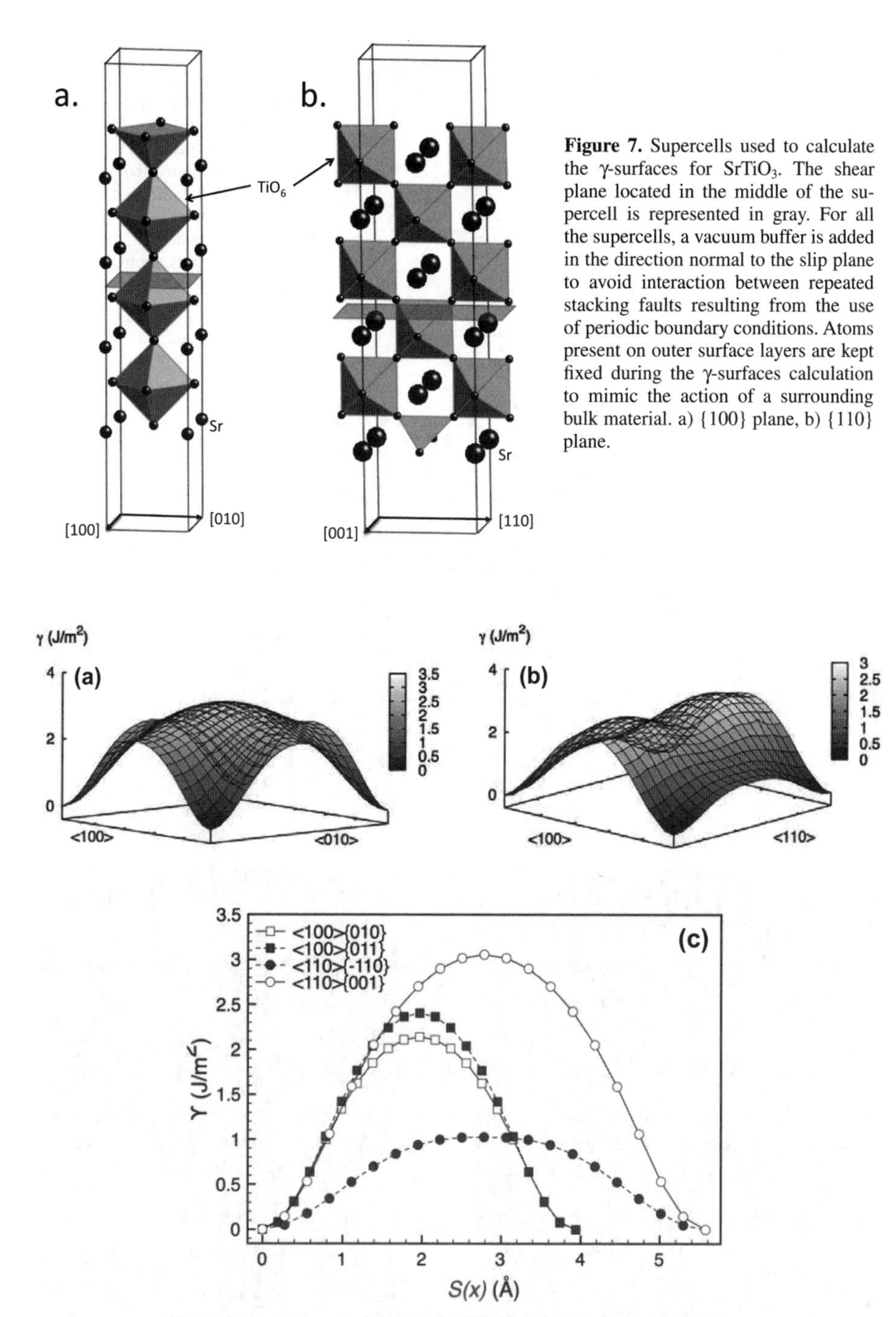

Figure 7. Supercells used to calculate the γ-surfaces for $SrTiO_3$. The shear plane located in the middle of the supercell is represented in gray. For all the supercells, a vacuum buffer is added in the direction normal to the slip plane to avoid interaction between repeated stacking faults resulting from the use of periodic boundary conditions. Atoms present on outer surface layers are kept fixed during the γ-surfaces calculation to mimic the action of a surrounding bulk material. a) {100} plane, b) {110} plane.

Figure 8. γ-surface calculated for $SrTiO_3$. a) {100} plane, b) {110} plane, c) Linear GSF corresponding to the four slip systems of interest.

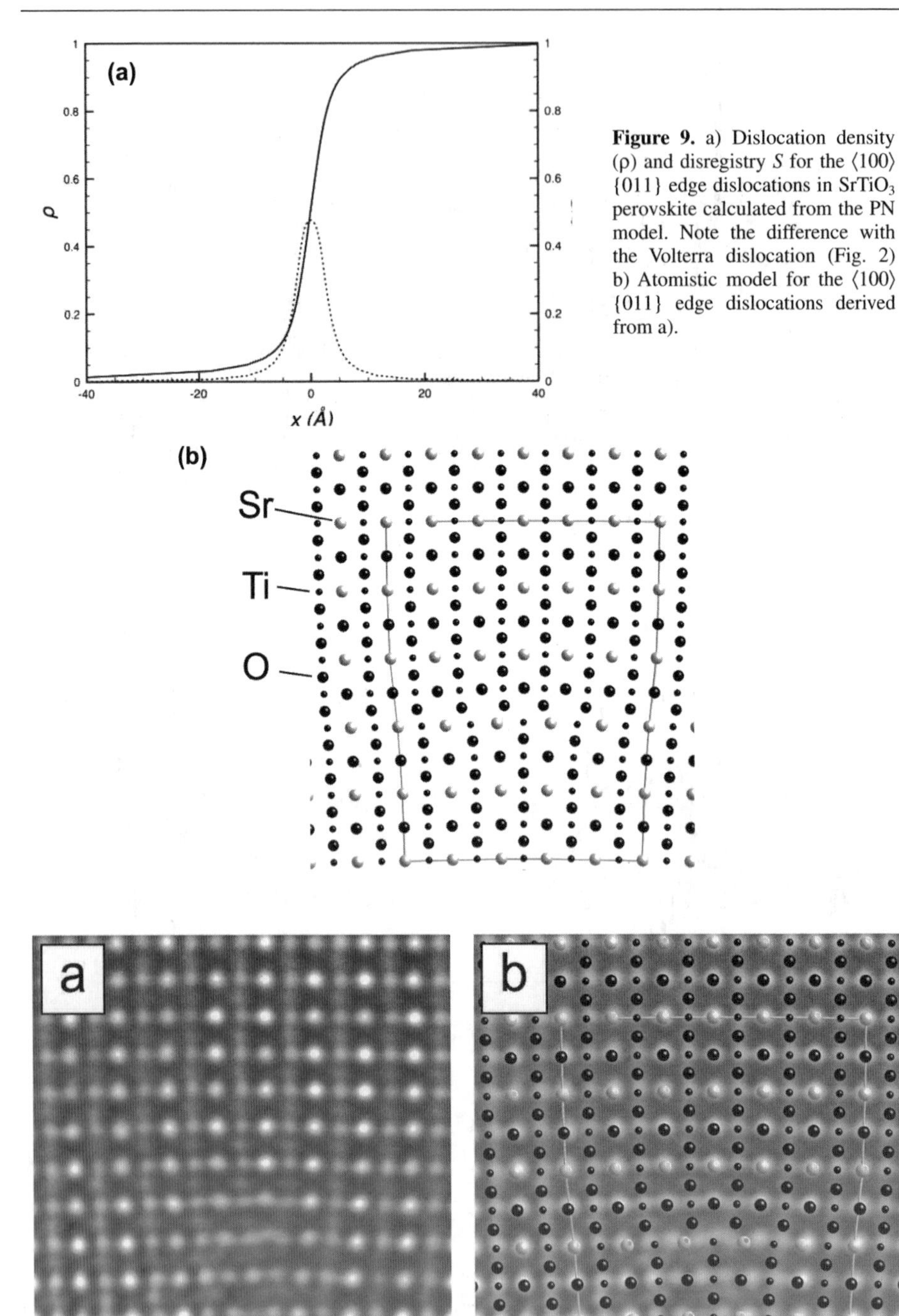

Figure 9. a) Dislocation density (ρ) and disregistry S for the ⟨100⟩ {011} edge dislocations in $SrTiO_3$ perovskite calculated from the PN model. Note the difference with the Volterra dislocation (Fig. 2) b) Atomistic model for the ⟨100⟩ {011} edge dislocations derived from a).

Figure 10. Comparison of the PN model for the ⟨100⟩{011} edge dislocation with experiments. a) Experimental TEM micrograph of Jia et al. (2005) reproduced with permission. b) Calculated structure of Figure 9 superimposed on the experimental image for comparison.

Table 1. Peierls stresses for screw and edge dislocations in $SrTiO_3$.

	σ_p (GPa)	
	Screw	***Edge***
⟨100⟩{010}	0.7	0.6
⟨100⟩{011}	9.9	0.6
⟨110⟩{001}	0.9	1.2
⟨110⟩{110}	0.006	0.004

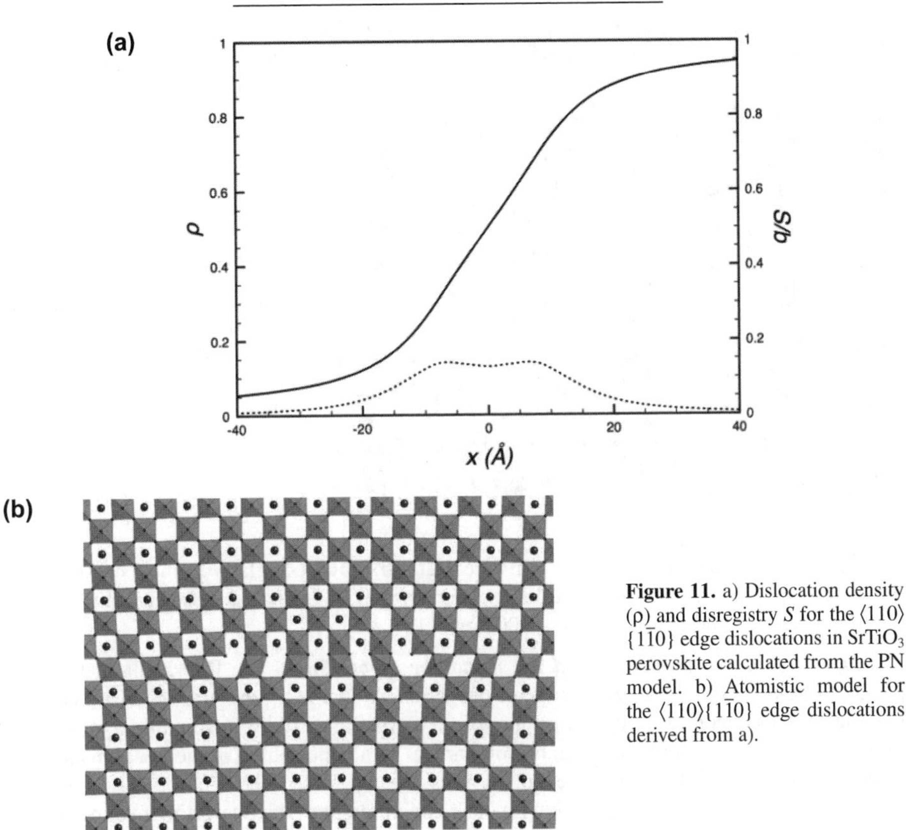

Figure 11. a) Dislocation density (ρ) and disregistry S for the ⟨110⟩ {1$\bar{1}$0} edge dislocations in $SrTiO_3$ perovskite calculated from the PN model. b) Atomistic model for the ⟨110⟩{1$\bar{1}$0} edge dislocations derived from a).

Wohlwend et al. 2009) using the GULP code (Gale and Rohl 2003). The result, presented on Figure 13, is in fair agreement with the PN model with the large spreading of the core due to the low energy of the stacking fault.

Magnesium oxide (MgO)

As an end-member of magnesiowustite (Mg,Fe)O (the second most abundant phase of the lower mantle), MgO is a material of primary importance for Earth sciences. It is also a widely-used ceramic. MgO is an ionic crystal exhibiting a rock salt structure similar to NaCl. The Bravais lattice is face-centred cubic (fcc) and MgO belongs to the space group $Fm\bar{3}m$. At

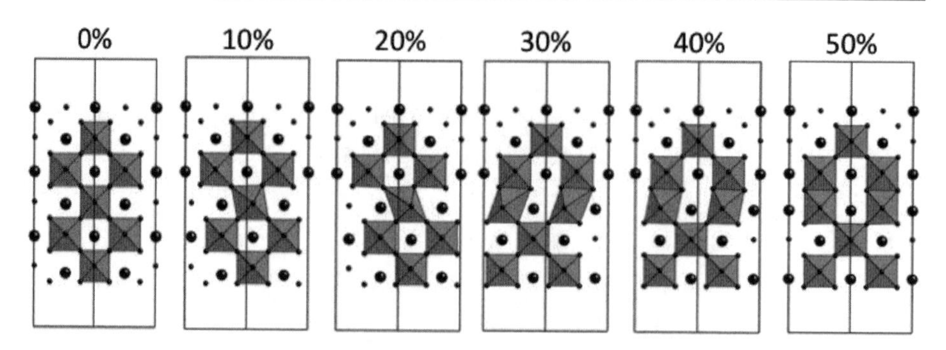

Figure 12. Evolution of the atomic arrangements during shear (0, 10, 20, 30, 40, and 50%) associated with the ⟨110⟩{1$\bar{1}$0} linear GSF. The polyhedral units are represented as a guide for the eye.

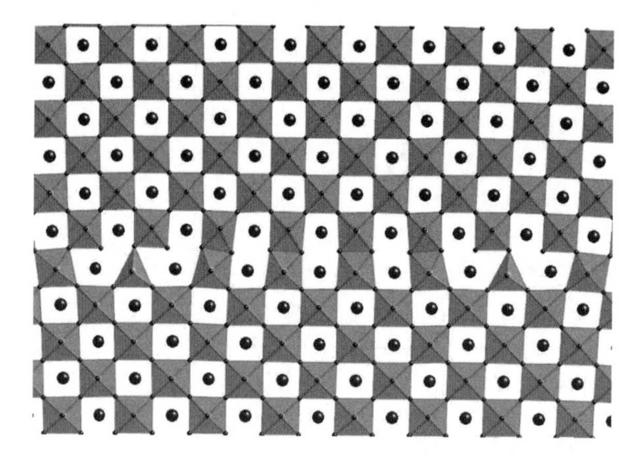

Figure 13. Atomistic calculation of the core structure of ⟨110⟩{110} edge dislocation performed with a cluster of 80 Å of diameter and an outer rim of 16 Å.

ambient pressure, the easiest slip systems are 1/2⟨110⟩{110} (similarly to most crystals with the Rock-Salt structure, Haasen et al. 1985). Slip along 1/2⟨110⟩ is also geometrically possible on other close-packed planes such as {100} or {111}.

Peierls-Nabarro modeling and the influence of pressure. The Rock-Salt structure of MgO is stable under a very wide range of pressures. At ambient pressure, MgO is a ceramic material and has been the subject of numerous experimental studies (providing a large body of data to which our calculations can be compared). As a major phase of the lower mantle, the plastic behavior of MgO is of interest under pressures up to 135 GPa. As for most materials, the elastic constants of MgO increase with increasing pressure. The interesting feature in MgO is the evolution of elastic anisotropy with pressure. The anisotropy ($A = 2C_{44}/(C_{11}\text{-}C_{12}) = 1.57$ at ambient pressure) decreases rapidly with pressure, cancels at ca. 30 GPa ($A = 1$) and is reversed at higher pressure ($A = 0.47$ at 100 GPa). The role of this section is to show how pressure influences the plastic behavior of MgO.

γ-surface calculations for {100}, {110} and {111} have been first calculated using empirical potentials and the GULP code. The results are presented on Figure 14. Whatever the slip plane, the low energy shear paths correspond to 1/2⟨110⟩. As a consequence, an *ab initio* treatment of GSF lines along ⟨110⟩ has been undertaken. In agreement with GULP calculations, 1/2⟨110⟩{110} is the lowest GSF line (Fig. 15) at zero pressure.

Let us first focus on the evolution with pressure of slip on {100} and {110}. Since the γ-surface (Fig. 14) shows that the low energy shear paths in these planes is always collinear

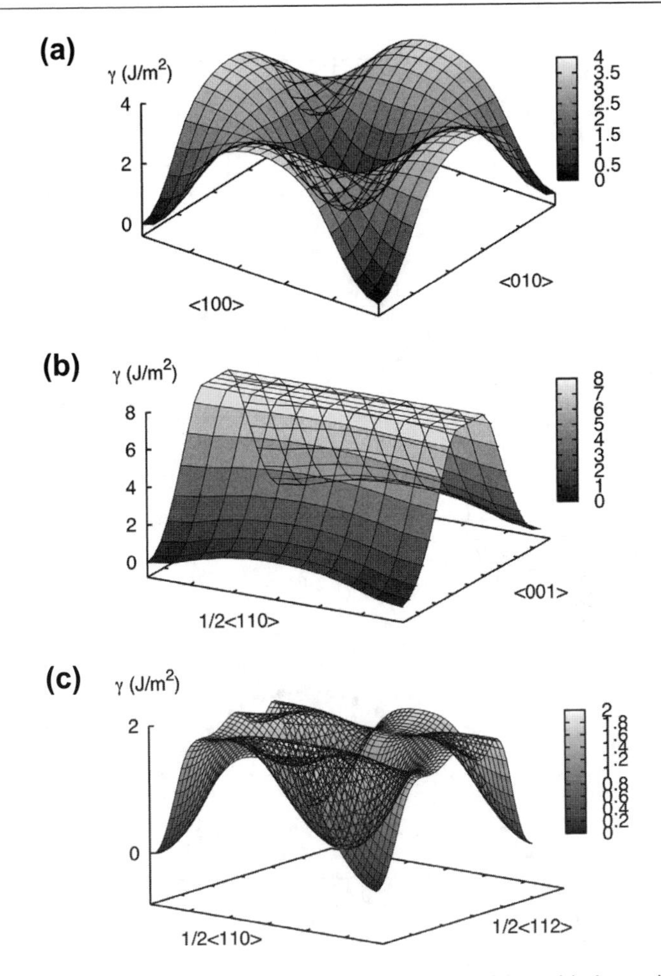

Figure 14. MgO γ-surface calculated at ambient pressure using empirical potentials and the GULP code. a) {100} plane, b) {110} plane, c) {111} plane.

with the Burgers vector direction, it is justified to apply the 1D PN model. The results of these calculations are described in Carrez et al. (2009a) and summarized in Table 2. One can see that the easiest slip system, 1/2⟨110⟩{110}, hardens very much with pressure. At zero pressure, the relatively flat GSF results in a relatively spread dislocation core which bears negligible lattice friction. This is the origin of the ductility of MgO at ambient pressure. When pressure is applied, the core of 1/2⟨110⟩{110} dislocations narrows very rapidly, lowering the mobility of the dislocations. The easiest slip systems 1/2⟨110⟩{110} bears at 100 GPa a lattice friction very comparable to the one of the hard slip system 1/2⟨110⟩{001} at zero pressure. 1/2⟨110⟩{001} also hardens with increasing pressure, but to a less extend. This slip system remains however harder than 1/2⟨110⟩{110} in the whole pressure range.

The case of 1/2⟨110⟩{111} is slightly more complex. As commonly observed in fcc structures, the {111} plane in MgO exhibits a potential staking fault in 1/6⟨112⟩. Whereas it is hardly visible on Figure 14, further *ab initio* calculations performed along 1/6⟨112⟩ as a function of pressure show that the local minimum (indicative of a stable stacking fault) is more and more pronounced as pressure increases (Fig. 15c) to reach a value of 3.05 J/m² under a

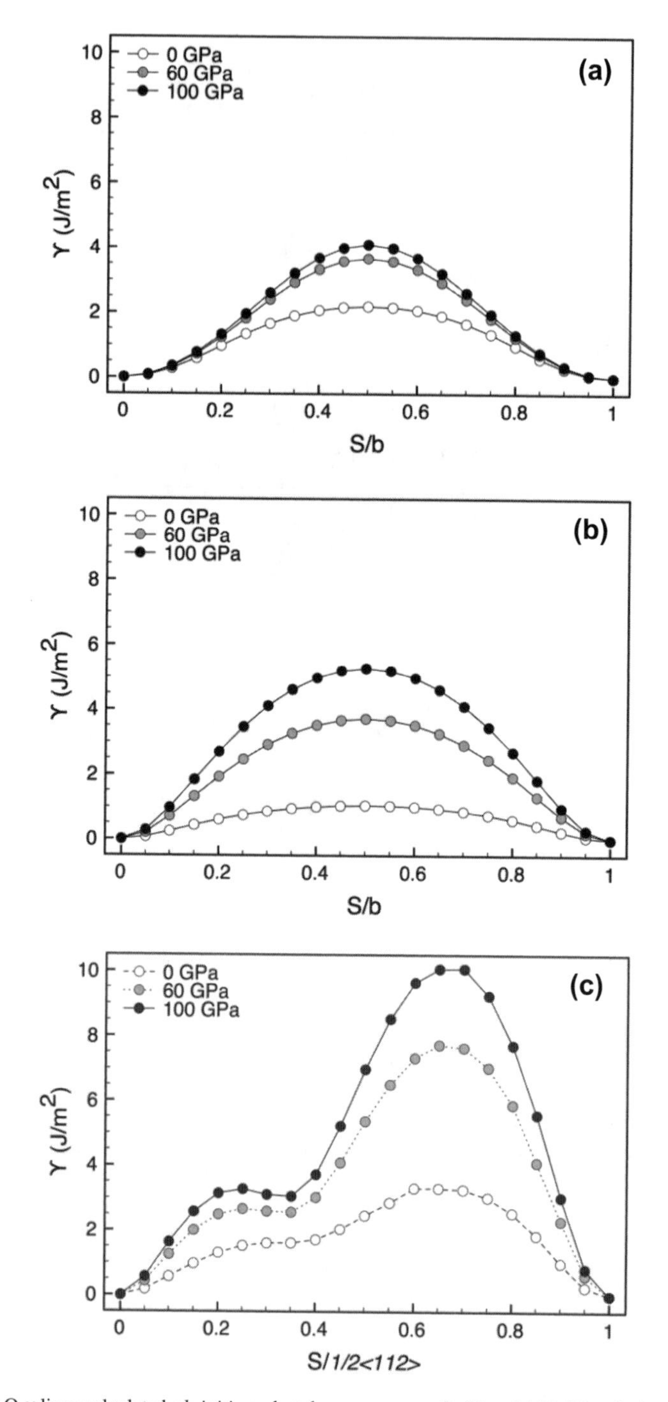

Figure 15. MgO γ-lines calculated *ab initio* and at three pressures: 0, 60 and 100 GPa. a) along *b* (1/2⟨110⟩) in the{100} plane, b) along *b* (1/2⟨110⟩) in the {110} plane, c) along 1/2⟨112⟩ (along dotted line A-B of Fig. 16a) in the {111} plane.

Table 2. Results of the PN model in MgO. ζ corresponds to the half width of dislocation core and σ_p to the Peierls stress.

	1/2⟨110⟩{110}				1/2⟨110⟩{001}			
	Screw		Edge		Screw		Edge	
	ζ/a'	σ_p (GPa)	ζ/a'	σ_p (GPa)	ζ/a'	σ_p (GPa)	ζ/a'	σ_p (GPa)
0 GPa	2.1	0.04	1.7	0.02	1.0	1.53	0.7	1.16
60 GPa	1.0	1.26	0.9	1.44	0.9	1.33	0.5	9.5
100 GPa	0.8	1.37	0.8	1.05	0. 9	2.04	0.5	10.95

pressure of 100 GPa. Splitting of 1/2⟨110⟩{111} into Shockley partials following the reaction $1/2[\bar{1}01] = 1/6[\bar{2}11] = 1/6[\bar{1}\bar{1}2]$ can thus be expected as in fcc metals.

Quasistatic PNG calculations have thus been undertaken using a rectangular volume and a nodal spacing of 12 nodes per Burgers vector 1/2⟨110⟩. The results of the PNG calculation are presented on Figure 16. We see that the dislocation follows the minimum energy path of Figure 16a leading to a splitting of the perfect dislocation into Shockley partials. It is worth noticing that this tendency for core spreading of 1/2⟨110⟩{111} dislocations is only observed under high pressure.

Kink-pair modeling and thermal activation of plastic slip. Calculations of kink-pair energies, for dislocations belonging to 1/2⟨110⟩{110} and 1/2⟨110⟩{001}, have been done at ambient pressure for comparison with available experimental data (Carrez et al. 2009b). These calculations are based on the previous PN modeling (Fig. 17). The elastic parameters appearing in Equations (20) and (21) are derived from the set of elastic constant proposed in Carrez et al. (2009a). The shear modulus μ was taken as 116.5 GPa and Poisson ratios ν were deduced from anisotropic elastic parameter $K(\theta)$ (θ depending on the dislocation character). Assuming $K(0) = \mu$ and $K(90) = \mu/(1 - \nu)$, we used $\nu = 0.18$ and 0.27 for {110} and {100} planes respectively.

The choice of the cut-off length parameter r_c (Eqns. 20 and 21) was parameterized on the core size ζ determined with the PN model. Following Koizumi et al. (1993) we took $r_c = 0.05\zeta$ whatever the slip system.

The calculated variation of the critical geometry of the kink pair h^* and w^* as a function of the applied stress is shown on Figure 18. The critical height h^* evolves monotonically from a' to 0 as τ approaches the Peierls stress. Whatever the slip system, the critical widths w^* is found to decrease rapidly to an almost constant value for intermediate stresses. We note that 1/2⟨110⟩{110} dislocations are characterized by very wide kink-pairs ($w^* \sim 500\ b$ at $0.5\sigma_p$) compared to 1/2⟨110⟩{100} dislocations ($w^* \sim 25\text{-}50\ b$). This feature is strongly related to the value of σ_p/μ. For slip systems with low Peierls stresses, or low Peierls potentials (which is here the case for 1/2⟨110⟩{110}), kink-pairs would be large due to the fact that the line tension governs the kink formation.

The evolution of the critical energy ΔH^* as a function of the applied stress is presented on Figure 19 for the two slip systems of interest. $\Delta H^*(\tau = 0) = \Delta H_0$ corresponds to twice the energy U_k of an isolated kink with a full height a'. Indeed, when τ approaches 0, the separation distance w^* of the kink pair is shown to increase considerably, leading to a configuration where the interaction between kinks can be neglected. The values of ΔH_0, given in Table 3, show that the energy of an isolated kink is nearly three times higher for 1/2⟨110⟩{100} compared to 1/2⟨110⟩{110}.

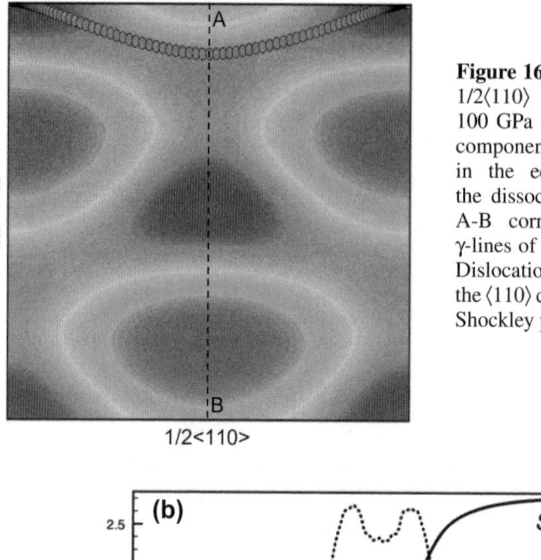

Figure 16. Results of PNG modeling of edge 1/2⟨110⟩ dislocation in {111} plane under 100 GPa of external pressure. a) Disregistry components plotted on the {111} γ-surface in the equilibrium configuration showing the dissociation path into Shockley partials. A-B corresponds to the line where the γ-lines of Figure 15c have been measured. b) Dislocation density (ρ) and disregistry S along the ⟨110⟩ direction (x) and along ⟨112⟩ (y). The Shockley partials separation is nearly 10 Å.

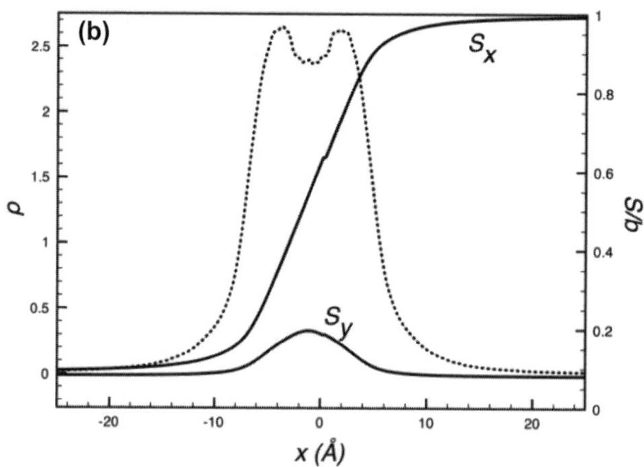

The model can be assessed by comparing the results of our calculations with experimental data of MgO deformation. For that purpose, $\Delta H^*(\tau)$ can be converted into a critical resolved shear stress at constant strain rate and as a function of temperature using the following relationship (Kocks et al. 1975)

$$\Delta H^*(\tau) = CkT \tag{24}$$

This conversion is useful since information for experimental Critical Resolved Shear Stress (CRSS) measurements are available for MgO at ambient conditions (Copley and Pask 1965; Sinha et al. 1973; Sato and Sumino 1980; Barthel 1984). The constant C depends on the strain rate, however, this dependence is small for standard experimental conditions and C is usually in the range 20-30 (e.g., Tang et al. 1998). Figure 20 displays, for each slip systems, the evolution of the predicted CRSS compared to experimental values. Our predictions for 1/2⟨110⟩{110} are slightly below experimental data points, however the evolution of CRSS as a function of T is well reproduced. In particular, the athermal critical temperature $T\mu$ (above which lattice friction is cancelled) is well described. The comparison is even much satisfying for 1/2⟨110⟩{100}. For this slip system, our calculations are in excellent agreement with the athermal temperature as well as with the evolution of CRSS as a function of T.

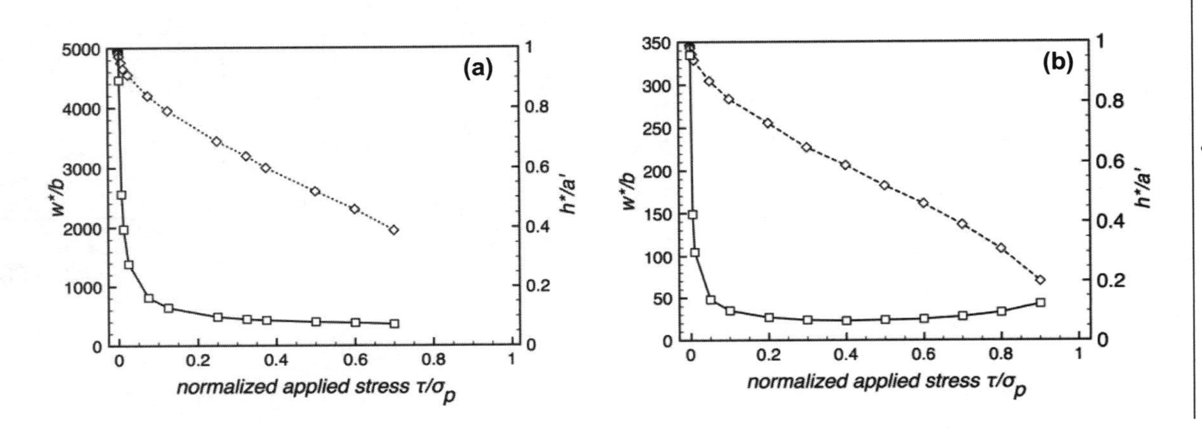

Figure 17. Variation of the Peierls potential V_p as a function of dislocation position u for 1/2⟨110⟩ screw dislocation gliding in {110} plane (a) and in {100} plane (b). V_p potentials, evaluated using Equation (8), is significantly lower for 1/2⟨110⟩{110} compared to 1/2⟨110⟩{100}, as glide is easier in {110}.

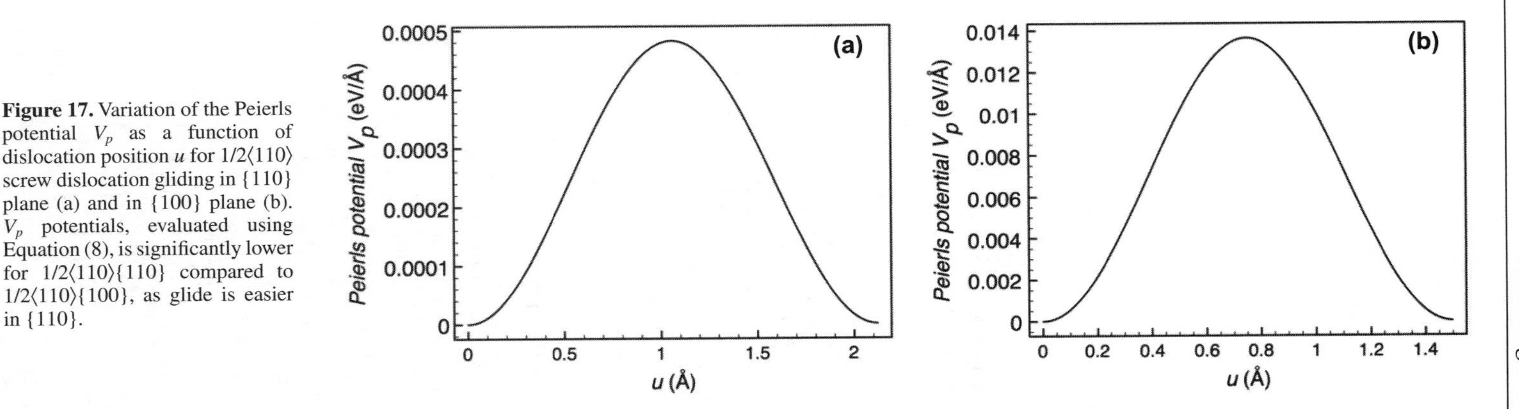

Figure 18. Critical kink geometry (h^*,w^*) as a function of applied stress τ for a screw dislocation gliding in {110} (a) and in{100} (b). The height h^* of kink evolves continuously from a' to 0. Kink pair width w^* diverges under low stress and reaches a plateau around $0.5\sigma_p$.

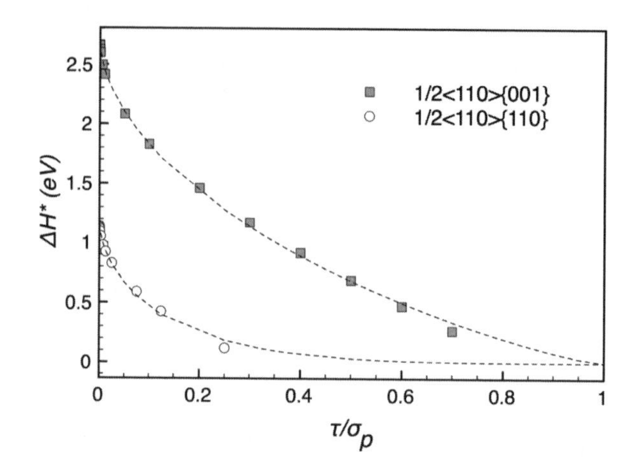

Figure 19. Critical energy ΔH^* for kink pair nucleation geometry as a function of applied stress τ for screw initially straight dislocations.

Table 3. Key features of the kink-pair modeling in MgO. Saddle point configuration (h^*,w^*) and critical energy of a kink-pair nucleation at zero stress ΔH_0. The critical height h^* and width w^* of kink-pair nucleation are given for an applied stress of $0.5\sigma_p$.

	h^*/a'	w^*/b	ΔH_0 (eV)
1/2⟨110⟩{110}	0.52	400	1.16
1/2⟨110⟩{100}	0.52	24	2.69

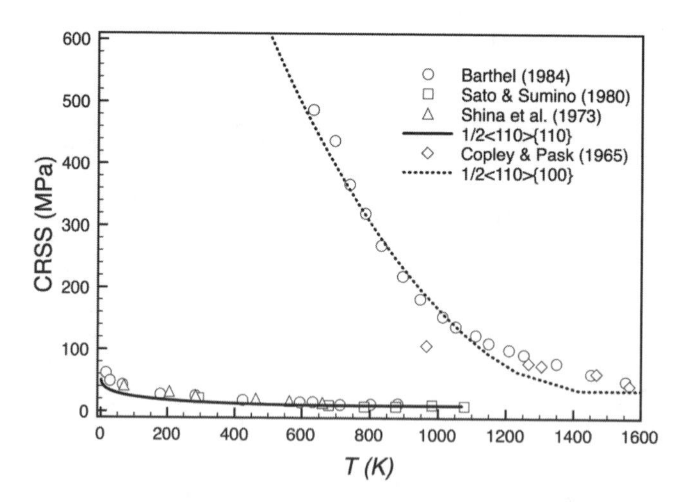

Figure 20. Critical Resolved Shear Stress (CRSS) predicted from the model (lines) plotted against temperature for 1/2⟨110⟩{110} and 1/2⟨110⟩{100} and compared with experimental values (symbols).

Dislocation mobility and influence of strain rate. Having modeled the lattice friction opposed to dislocation glide and how thermal activation and stress combine to overcome this friction, it is possible to determine the velocity of a single dislocation as a function of stress and temperature (and potentially pressure since the input parameters have been calculated at a given pressure) using Equation (23). Figure 21 shows how, at a given temperature (800 K) and a given pressure (0 GPa), the velocity of a dislocation varies as a function of stress. Three regimes can be observed.

In an intermediate stress regime, the dislocation velocity can be fitted to power law of stress (with $n = 4$ at 800 K, see Fig. 21). The stress exponent is strongly dependent on temperature even if one single mechanism is considered: glide. The same conclusion is drawn by Mitchell et al. (2002) with a slightly modified velocity law. This is thus one contribution to the complex stress sensitivity of rheological laws observed in the dislocation creep regime.

At low stress (i.e., low strain rates) and whatever the temperature, the stress sensitivity decreases to give rise to a linear regime. This change is due to the change of the kink regime as observed on Figure 16.

Finally, saturation is observed at high stresses, as predicted by Philibert (1979), as a physical limit that the velocity must reach when stress increases to infinity.

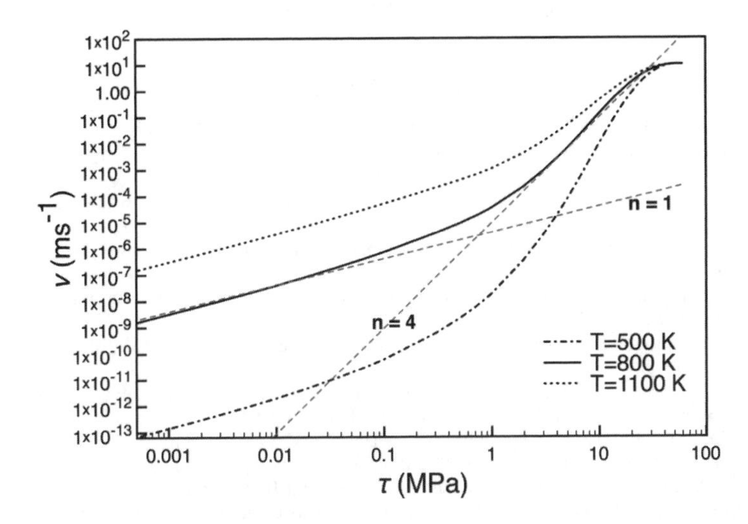

Figure 21. 1/2⟨110⟩{110} dislocation velocity as a function of applied stress. The calculation are based on Equation (23). It is worth noticing that whatever the temperature range, dislocation velocity is linear at low stress whereas at intermediate applied stress, the stress exponent varies with temperature.

CONCLUDING REMARKS

Theoretical and numerical tools suitable for modeling dislocations in complex materials are now available. Dislocations are complex objects and their behavior requires a multiscale and multi-Physics description at the interface between the atomic scale (where quantum mechanics is the most suitable framework to model interactions) and the continuum (which is well described by elastic theories). Several stages will be involved in the modeling:

- the fine geometry of the core must be identified keeping in mind that in some cases, the dislocation core can exhibit several configurations.

- the lattice friction must be modeled since it will determine the geometry of slip (which slip system?) and then the mobility of the dislocations. Pressure can have a very strong influence here.
- the microscopic mechanism for dislocation elementary motion bust be identified and properly modeled to yield a mobility law. This law must apply over a large range of timescales.

The next step (not described in the present chapter) is to transfer this elementary mechanism describing the mobility of a single defect to higher scales. The plasticity of a single crystal (a grain in a rock) is the result of the collective motion of a large number of dislocations which interact at long and short distances. Discrete Dislocation Dynamics is today the most promising technique to address this issue. How individual grains will combine to end up with the rheology of a polycrystal (with possibly several phases) can be described using micromechanical models (using self-consistent or finite-elements methods).

The challenge ahead of us is clearly to adapt and apply these models to the minerals and physical conditions relevant for the interior of the Earth. We have chosen to present in this chapter examples for which comparisons of the models and experimental data are available. This is clearly a major issue and another major challenge ahead is to design a combined approach where experiments performed under extreme conditions and Physics-based modeling combine to yield a rheological model of the interior of the Earth.

REFERENCES

Barthel C (1984) Plastiche Anisotropie von Bleisulfid und Magnesiumoxid. University of Gottingen.

Brunner D, Taeri-Baghbadrani S, Sigle W, Rühle M (2001) Surprising results of a study on the plasticity in strontium titanate. J Am Ceram Soc 84(5):1161-1163

Bulatov VV, Cai W, Baran R, Kang K (2006) Geometric aspects of the ideal shear resistance in simple crystal lattices. Philos Mag 86(25-26):3847-3859

Bulatov VV, Kaxiras E (1997) Semidiscrete variational peierls framework for dislocation core properties. Phys Rev Lett 78(22):4221-4224

Cai J, Lu C, Yap PH, Wang YY (2002) How to affect stacking fault energy and structure by atom relaxation. Appl Phys Lett 81(19):3543-3545

Cai W (2005) Modelling dislocations using a periodic cell. *In :* Handbook of Materials. Modelling. Yip S (ed) Springer, p 813-826

Cai W, Bulatov VV, Chang J, Li J, Yip S (2004) Dislocation core effects on mobility. *In :* Dislocations in Solids. Vol. 12. Nabarro FRN, Hirth JP (eds) North Holland Pub., p 1-80

Caillard D, Martin JL (2003) Thermally Activated Mechanism in Crystal Plasticity. Pergamon

Carrez P, Cordier P, Mainprice D, Tommasi A (2006) Slip systems and plastic shear anisotropy in Mg_2SiO_4 ringwoodite: Insights from numerical modelling. Eur Mineral 18:149-160

Carrez P, Ferré D, Cordier P (2007) Peierls-Nabarro model for dislocations in $MgSiO_3$ post-perovskite calculated at 120 GPa from first principles. Philosl Mag 87(22):3229-3247

Carrez P, Ferré D, Cordier P (2009a) Peierls-Nabarro modelling of dislocations in MgO from ambient pressure to 100 GPa. Modell Simul Mater Sci Eng 17:035010

Carrez P, Ferré D, Cordier P (2009b) Thermal activation of dislocation glide in MgO based on an Elastic-Interaction model of kink-pair nucleation. IOP Conference Series: Mater Sci Eng 3:012011(1-4)

Carrez P, Walker AM, Metsue A, Cordier P (2008) Evidence from numerical modelling for 3D spreading of [001] screw dislocations in Mg_2SiO_4 forsterite. Philos Mag A 88(16):2477-2485

Chen JH, Weidner DJ, Vaughan MT (2002) The strength of $Mg_{0.9}Fe_{0.1}SiO_3$ perovskite at high pressure and temperature. Nature 419(6909):824-826

Clouet E (2009) Elastic energy of a straight dislocation and contribution from core tractions. Philos Mag 89(19):1565-1584

Clouet E, Ventelon L, Willaime F (2009) Dislocation core energies and core fields from first principles. Phys Rev Lett 102:055502

Copley SM, Pask JA (1965) Plastic deformation of MgO single crystals up to 1600°C. J Am Ceram Soc 48:139-146

Cordier P, Barbe F, Durinck J, Tommasi A, Walker AM (2005) Plastic deformation of minerals at high pressure: Multiscale numerical modelling. *In:* Mineral Behaviour at Extreme Conditions. Volume 7. Miletich R (ed) EMU, p 389-415

Cordier P, Ungár T, Zsoldos L, Tichy G (2004) Dislocation creep in $MgSiO_3$ Perovskite at conditions of the Earth's uppermost lower mantle. Nature 428: 37-840

Denoual C (2004) Dynamic dislocation modeling by combining Peierls Nabarro and Galerkin methods. Phys Rev B 70:024106

Denoual C (2007) Modeling dislocation by coupling Peierls-Nabarro and element-free Galerkin methods. Comput Methods Appl Mech Eng 196:1915-1923

Dorn JE, Rajnak S (1964) Nucleation of kink pairs and the peierls mechanism of plastic deformation. Trans Met Soc AIME 230:1052-1064

Duesbery MS, Joos B (1996) Dislocation motion in silicon: the shuffle-glide controversy. Philos Mag Lett 74(4):253-258

Duesbery MS, Vitek V (1998) Plastic anisotropy in BCC transition metals. Acta Mater 46(5):1481-1492

Durinck J, Legris A, Cordier P (2005) Pressure sensitivity of olivine slip systems: first-principle calculations of generalised stacking faults. Phys Chem Miner 32(8-9):646-654

Edagawa K, Koizumi H, Kamimura Y, Suzuki T (2000) Temperature dependence of the flow stress of III-V compounds. Philos Mag A 80(11):2591-2608

Ferré D, Carrez P, Cordier P (2008) Modelling dislocation cores in $SrTiO_3$ with the Peierls-Nabarro model. Phys Rev B 77(1):014106

Ferré D, Cordier P, Carrez P (2009) Dislocation modeling in calcium silicate perovskite based on the Peierls-Nabarro model. Am Mineral 94(1):135-142

Gale JD, Rohl AL (2003) The general utility lattice program. Mol Simulation 29(5):291-341

Groger R, Bayley AG, Vitek V (2008) Plastic deformation of molybdenum and tungsten: I. Atomistic studies of the core structure and glide of $1/2\langle 111 \rangle$ screw dislocations at 0 K. Acta Mater 56:5401

Groger R, Vitek V (2008) Plastic deformation of molybdenum and tungsten: III. Incorporation of the effects of temperature and strain rate. Acta Mater 56:5426

Gumbsch P, Schroll R (1999) Atomistic aspects of the deformation of NiAl. Intermetallics 7:447-454

Gumbsch P, Taeri-Baghbadrani S, Brunner D, Sigle W, Rühle M (2001) Plasticity and an inverse brittle-to-ductile transition in strontium titanate. Phys Rev Lett 87(8):085505(1-4)

Guyot P, Dorn JE (1967) A critical review of the Peierls mechanism. Can J Phys 45:983-1015

Haasen P, Barthel C, Suzuki T (1985) Choice of slip system and Peierls stresses in the NaCl structure. *In:* Dislocations in Solids. Suzuki H, Ninomiya T, Sumino K, Takeuchi S (eds) University of Tokyo Press, p. 455-462

Hartford J, von Sydow B, Wahnström G, Lundqvist BI (1998) Peierls barrier and stresses for edge dislocations in Pd and Al calculated from first principles. Phys Rev B 58(5):2487-2496

Hirth JP, Lothe J (1982) Theory of Dislocations. John Wiley & Sons, Inc, New York

Jia CL, Thust A, Urban K (2005) Atomic-scale analysis of the oxygen configuration at a $SrTiO_3$ dislocation core. Phys Rev Lett 95:225506

Joos B, Duesbery MS (1997a) The Peierls stress of dislocations: an analytic formula. Phys Rev Lett 78(2):266-269

Joos B, Duesbery MS (1997b) Dislocation kink migration energies and the Frenkel-Kontorawa model. Phys Rev B 55(17):11161

Joos B, Ren Q, Duesbery MS (1994) Peierls-Nabarro model of dislocations in silicon with generalized stacking-fault restoring forces. Phys Rev B 50(9):5890-5898

Joos B, Zhou J (2001) The Peierls-Nabarro model and the mobility of the dislocation line. Philos Mag A 81(5):1329-1340

Karato S, Weidner DJ (2008) Laboratory studies of the rheological properties of minerals under deep-mantle conditions. Elements 4(3):191-196

Karato SI (1990) Plasticity of $MgSiO_3$ perovskite: the results of microhardness tests on single crystals. Geophys Res Lett 17(1):13-16

Kocks UF, Argon AS, Ashby MF (1975) Thermodynamics and Kinetics of Slip. Pergamon Press

Koizumi H, Kamimura Y, Suzuki T (2000) Core structure of a screw dislocation in a diamond-like structure. Philos Mag A 80(3):609-620

Koizumi H, Kirchner HOK, Suzuki T (1993) Kink pair nucleation and critical shear stress. Acta Metall Mater 41(12):3483-3493

Koizumi H, Kirchner HOK, Suzuki T (1994) Nucleation of trapezoidal kink pair on a Peierls potential. Philos Mag A 69(4):805-820

Kresse G, Hafner J (1993) *Ab initio* molecular dynamics for liquid metals. Phys Rev B 47:558

Kroupa F, Lejcek L (1972) Splitting of dislocations in the Peierls-Nabarro model. Czech J Phys B 22:813-825

Lejcek L, Kroupa F (1976) Peierls-Nabarro model of non-planar screw dislocation cores. Czech J Phys B 26:528-537

Lemarchand C, Devincre B, Kubin LP (2001) Homogenization method for a discrete-continuum simulation of dislocation dynamics. J Mech Phys Solids 49:1969-1982

Lu G (2005) The Peierls-Nabarro model of dislocations: A venerable theory and its current development. *In* Handbook of Materials Modeling. Volume 1: Methods and Models. S. Yip, (ed) Springer, p 1-19

Lu G, Bulatov VV, Kioussis N (2003) A nonplanar Peierls-Nabarro model and its application to dislocation cross-slip. Philos Mag 83(31-34):3539-3548

Lu G, Kioussis N, Bulatov VV, Kaxiras E (2000) The Peierls-Nabarro model revisited. Philos Mag Lett 80(10):675-682

Mao Z, Knowles KM (1996) Dissociation of lattice dislocations in $SrTiO_3$. Philos Mag A 73(3):699-708

Matsunaga T, Saka H (2000) Transmission electron microscopy of dislocations in $SrTiO_3$. Philos Mag Lett 80(9):597-604

Merkel S, Wenk HR, Badro J, Montagnac G, Gillet P, Mao HK, Hemley RJ (2003) Deformation of $(Mg_{0.9},Fe_{0.1})SiO_3$ Perovskite aggregates up to 32 GPa. Earth Planet Sci Lett 209(3-4):351-360

Miranda CR, Scandolo S (2005) Computational materials science meets geophysics: dislocations and slip planes of MgO. Comp Phys Comm 169:24-27

Mitchell TE, Hirth JP, Misra A (2002) Apparent activation energy and stress exponent in materials with a high Peierls stress. Acta Mater 50:1087-1093

Miyata M, Fujiwa T (2001) *Ab initio* calculation of Peierls stress in silicon. Phys Rev B 63:045206

Nabarro FRN (1947) Dislocations in a simple cubic lattice. Proc Phys Soc London 59:256-272

Ngan AHW (1997) A generalized Peierls-Nabarro model for non-planar screw dislocation cores. J Mech Phys Solids 45(6):903-921

Ngan AHW (1999) A new model for dislocation kink-pair activation at low temperatures based on the Peierls-Nabarro concept. Philos Mag A 79(7):1697-1720

Ngan AHW, Zhang HF (1999) A universal relation for the stress dependence of activation energy for slip in body-centered cubic crystals. J Appl Phys 86(3):1236-1244

Nishigaki J, Kuroda K, Saka H (1991) Electron microscopy of dislocation structures in $SrTiO_3$ deformed at high temperatures. Phys Stat Sol (a) 128:319-336

Paxton AT, Gumbsch P, Methfessel M (1991) A quantum mechanical calculation of the theoretical strength of metals. Philos Mag Lett 63(5):267-274

Peach M, Koehler JS (1950) The forces exerted on dislocations and the stress fields produced by them. Phys Rev 80(3):436-439

Peierls RE (1940) On the size of a dislocation. Proc Phys Soc London 52:34-37

Pellegrini YP, Denoual C, Truskinovsky L (2008) Phase field modeling of non-linear material behavior. *In*: IUTAM Symposium on Variational Concepts with Applications to the Mechanics of Materials. Hackl K (ed) Bochum, in press

Philibert J (1979) Glissement des dislocations et frottement de réseau. *In:* Dislocations et Déformation Plastique. Groh P, Kubin LP, Martin J-L (eds) Ecole d'été d'Yravals, p 101-139

Pillon L, Denoual C (2009) Inertial and retardation effects for dislocation interactions. Philos Mag 89(2): 27-141

Pizzagalli L, Beauchamp P (2004) First principles determination of the Peierls stress of the shuffle screw dislocation in silicon. Philos Mag Lett 84(11):729-736

Pizzagalli L, Demenet J-L, Rabier J (2009) Theoretical study of pressure effect on the dislocation core properties in semiconductors. Phys Rev B 79:045203

Poirier JP, Beauchesne S, Guyot F (1989) Deformation mechanisms of crystals with perovskite structure. *In:* Perovskite: A Structure of Great Interest to Geophysics and Materials Science. Navrotsky A, Weidner D (eds) AGU, Washington DC, p 119-123

Ren Q, Joos B, Duesbery MS (1995) Test of the Peierls-Nabarro model for dislocations in silicon. Phys Rev B 52(18):13223-13228

Sato F, Sumino K (1980) The yield strength and dynamic behaviour of dislocations in MgO crystals at high temperatures. J Mater Sci (15):1625-1634

Schoeck G (1998) Deviations from Volterra dislocations. Philos Mag Lett 77(3):141-146

Schoeck G (1999) The Peierls energy revisited. Philos Mag A 79(11):2629-2636

Schoeck G (2005) The Peierls model: progress and limitations. Mater Sci Eng A 400-401:7-17

Schoeck G (2006) The core structure of dislocations: Peierls model vs. atomic calculations. Acta Mater 54:4865-4870

Schoeck G (2007) The kink chain revisited. Philos Mag 87(11):1631-1647

Schoeck G, Krystian M (2005) The Peierls energy and kink energy in fcc metals. Philos Mag 85(9):949-966

Schroll R, Vitek V, Gumbsch P (1998) Core properties and motion of dislocations in NiAl. Acta Mater 46(3):903-918

Seeger A (1984) Structure and diffusion of kinks in: monoatomic crystals. *In:* Int. Conf. Dislocations 1984. Veyssiere P, Kubin LP, Castaing J (eds) Paris, p 141

Shen C, Wang Y (2004) Incorporation of gamma-surface to phase field model of dislocations: simulating dislocation dissociation in fcc crystals. Acta Mater 52:683-691

Sinha MN, Lloyd DJ, Tangri K (1973) Dislocation dynamics and thermally-activated deformation of MgO single crystals. Philos Mag 28(6):1341-1352

Steeds JW (1973) Introduction to Anisotropic Elasticity of Dislocations. Claredon Press, Oxford

Suzuki T, Takeuchi S, Yoshinaga H (1991) Dislocation Dynamics and Plasticity. Springer-Verlag, Berlin

Tang M, Kubin LP, Canova GR (1998) Dislocation mobility and the mechanical response of BCC single crystals: a mesoscopic approach. Acta Mater 46(9):3221-3235

Ventelon L, Willaime F (2007) Core structure and Peierls potential of screw dislocations in a-Fe from first principles: cluster versus dipole approaches. J Computer-Aided Mater Des 14:85-94

Vítek V (1968) intrinsic stacking faults in body-centered cubic crystals. Philos Mag 18:773

Volterra V (1907) Sur l'équilibre des corps elastiques multiplement connexes. Ann Ecole Norm Super Ser 3 24:401-517

von Sydow B, Hartford J, Wahnström G (1999) Atomistic simulations and Peierls-Nabarro analysis of the Shockley partial dislocations in palladium. Comp Mat Sci 15:367-379

Walker AM, Gale JD, Slater B, Wright K (2005a) Atomic scale modelling of the cores of dislocations in complex materials part 1: methodology. Phys Chem Chem Phys 7:3227-3234

Walker AM, Gale JD, Slater B, Wright K (2005b) Atomic scale modelling of the cores of dislocations in complex materials part 2: applications. Phys Chem Chem Phys 7:3235-3242

Walker AM, Slater B, Gale JD, Wright K (2004) Predicting the structure of screw dislocations in nanoporous materials. Nature Mater 3:715-720

Wang G, Strachan A, Cagin T, Goddard WAI (2003) Atomistic simulations of kinks in $1/2a\langle 111\rangle$ screw dislocations in bcc tantalum. Phys Rev B 68:224101

Wang JN (1996) Prediction of Peierls stresses for different crystals. Mater Sci Eng A 206:259-269

Watson GW, Kelsey ET, Parker SC (1999) Atomistic simulation of screw dislocations in rock salt structured materials. Philos Mag A 79(3):527-536

Wohlwend JL, Behera RK, Jang I, Phillpot SR, Sinnott SB (2009) Morphology and growth modes of metal-oxides deposited on $SrTiO_3$. Surf Sci 603:873-880

Woo CH, Puls MP (1977) The Peierls mechanism in MgO. Philos Mag 35(6):1641-1652

Woodward C (2005) First-principles simulations of dislocation cores. Mater Sci Eng A 400-401:59-67

Woodward C, Rao SI (2002) Flexible *ab initio* boundary conditions: simulating isolated dislocations in bcc Mo and Ta. Phys Rev Lett 88(21):216402

Xu W, Moriarty JA (1998) Accurate atomisitc simulations of the Peierls barrier and kink-pair formation energy for $\langle 111\rangle$ screw dislocations in bcc Mo. Comp Mater Sci 9:348-356

Yamaguchi M, Vitek V (1973) Core structure of nonscrew $1/2\langle 111\rangle$ dislocations on $\{110\}$ planes in bcc crystals I. Core structure in an unstressed crystal. J Phys F: Metal Phys 3:523-536

Yan JA, Wang CY, Wang SY (2004) Generalised-stacking-fault energy and dislocation properties in bcc Fe: a first-principles study. Phys Rev B 70:174105

Zimmerman JA, Gao H, Abraham FF (2000) Generalized stacking fault energies for embedded atom FCC metals. Modelling Simul Mater Sci Eng 8:103-115

Reviews in Mineralogy & Geochemistry
Vol. 71 pp. 253-269, 2010

12

Theoretical Methods for Calculating the Lattice Thermal Conductivity of Minerals

Stephen Stackhouse

Department of Earth and Planetary Science
University of California, Berkeley
307 McCone Hall
Berkeley, California, 94720-4767, U.S.A.

s.stackhouse@berkeley.edu

Lars Stixrude

Department of Earth Sciences
University College London
Gower Street, London, WC1E 6BT, United Kingdom

ABSTRACT

The thermal conductivity of the lower mantle plays a major role in shaping the structure and dynamics of the region. However, the thermal conductivity of lower mantle minerals are, at present, not well constrained, because of difficulties in making measurements at such high pressures. Here we describe the most common theoretical methods available to calculate the thermal conductivity of materials, which provide an invaluable alternative to experimental techniques. In each case the general scheme is given and particular considerations highlighted. The advantages and disadvantages and applicability of the methods are then discussed. We conclude with a short review of theoretical studies of the lattice thermal conductivity of periclase.

INTRODUCTION

The bulk of the lower mantle is composed of ferropericlase and ferromagnesian silicate perovskite (Lee et al. 2004). To understand the lower mantle it is, therefore, essential to constrain the properties of these phases. Of particular importance are their thermal transport properties, which have, from formation, played a major role in shaping the deep Earth. Following segregation, the thermal conductivity of the lower mantle regulated the heat flux from the core and thus had a significant influence on thermal evolution, in particular on the rate of growth of the solid inner core (Lay et al. 2008). In the present Earth, thermal conductivity plays a significant role in determining the structure and dynamics of the lower mantle, controlling the size and stability of thermal upwellings (Dubuffet et al. 1999; Dubuffett and Yuen 2000; Naliboff and Kellogg 2006, 2007). In addition, lateral variations in the thermal structure of the lowermost mantle, which could be related to lateral variations in thermal conductivity, have been shown to influence magnetic field generation (Gubbins et al. 2007; Willis et al. 2007).

The thermal conductivity of the lower mantle can be decomposed into two principle components. The first is the lattice contribution, related to thermal conduction by phonons (lattice vibrations). The second is the radiative contribution related to thermal conduction by photons (electromagnetic radiation). Conduction of heat by electrons is expected to be negligible. The lattice thermal conductivity of ferropericlase and ferromagnesian silicate perovskite has not been measured at lower mantle temperatures and pressures, and geophysical

1529-6466/10/0071-0012$05.00 DOI: 10.2138/rmg.2010.71.12

values are estimated by model extrapolations from low-pressure data (Hofmeister 2007, 2008; Goncharov et al. 2009). Experimental measurements of radiative thermal conductivity remain inconclusive. Though there is general agreement that the high-to-low-spin transition in iron decreases radiative conductivity (Goncharov et al. 2006; Keppler et al. 2007), instead of increasing it as was first thought (Badro et al. 2003, 2004), disagreement remains on values in the low-spin regime. Measurements of the radiative conductivity of silicate perovskite differ by an order of magnitude (Goncharov et al. 2008; Keppler et al. 2008). However, even if the higher values are adopted, it seems probable that the lattice contribution dominates at lower mantle conditions.

Numerous theoretical techniques for determining lattice thermal conductivity have been reported in the literature (e.g., Evans and Morris 1990; Müller-plathe 1997; Daly et al. 2002; Nieto-Draghi and Avalos 2003; de Koker 2009; Tang and Dong 2009) and have been applied to a wide range of materials. However, their application to lower mantle phases is a relatively recent development with only a handful of studies published (Cohen 1998; Shukla et al. 2008; Stackhouse et al. 2008; de Koker 2009; Tang and Dong 2009), all of which focus on periclase, the pure magnesium end-member of ferropericlase. The purpose of the present work is to provide an overview of the most popular theoretical methods for calculating lattice thermal conductivity, discussing their underlying theories, relative merits and shortcomings and highlighting possible pitfalls. In addition, the few theoretical studies of periclase are reviewed. The overall aim is to equip the reader with the knowledge required to select and use the most appropriate method in their own work, after considering the computational resources and time available to them.

FUNDAMENTAL PRINCIPLES

Imagine that a temperature gradient is imposed across a solid, heat flows from the hotter region to the cooler one. Fourier's Law states that the magnitude of the induced heat flux will be proportional to the temperature gradient, and defines thermal conductivity as the property of the solid that relates the two

$$q_i = -k_{ij}\frac{\partial T}{\partial x_j} \tag{1}$$

where q_i is the heat flux, $\partial T/\partial x_j$ the temperature gradient and k_{ij} the second-rank thermal conductivity tensor. For those materials of cubic or isotropic symmetry the thermal conductivity tensor is isotropic (Nye 1957), such that

$$k_{ij} = k\delta_{ij} \tag{2}$$

and thus Fourier's law reduces to

$$q = -k\frac{\partial T}{\partial x} \tag{3}$$

where the temperature gradient and heat flux are in the same direction.

It should be noted that Fourier's law is only applicable to systems in steady-state. For systems out of steady-state we must use the heat conduction equation (Holman 1976; Kreith and Bohn 2001). Focusing on the conductive contribution to heat transfer, for an isotropic system, the heat conduction equation reads

$$\frac{\partial}{\partial x_i}\left(k\frac{\partial T}{\partial x_j}\right)\delta_{ij} = \rho c_P \frac{\partial T}{\partial t} \tag{4}$$

where ρ is the density and C_P the isobaric specific heat. For small homogeneous crystals the spatial gradient in k may be neglected. Taking this into account, and dividing both sides by ρC_P leads to

$$\frac{\partial T}{\partial t} = D \frac{\partial^2 T}{\partial x^2} \tag{5}$$

where D is the thermal diffusivity, related to k via

$$D = \frac{k}{\rho C_P} \tag{6}$$

The phonon contribution to thermal conductivity is related to microscopic dynamics via (Ziman 1960; Srivastava 1990)

$$k = \sum_{\mathbf{q}} \sum_{s=1}^{3N} c_{\mathbf{q},s} v_{\mathbf{q},s}^2 \tau_{\mathbf{q},s} \tag{7}$$

where the sum is over all wave vectors **q** and $3N$ polarization indices s (N is the number of atoms in the primitive cell); $v_{\mathbf{q},s}$ the group velocity and $\tau_{\mathbf{q},s}$ the relaxation time associated with each mode. The mode contribution to the heat capacity is

$$c_{\mathbf{q},s} = \frac{k_B}{V} \frac{x^2 e^x}{(e^x - 1)^2}; \quad x = \frac{h\omega_{\mathbf{q},s}}{k_B T} \tag{8}$$

where $\omega_{\mathbf{q},s}$ is the frequency of the mode; k_B Boltzmann's constant; h Planck's constant; V volume; and T temperature. At lower mantle temperatures, and for typical lower mantle minerals, the mode heat capacities all take on the high temperature limiting value and lattice thermal conductivity may be written as

$$k = \frac{1}{3} C_v \langle v^2 \tau \rangle \tag{9}$$

where C_v is the bulk volumetric isochoric heat capacity and the brackets indicate the average over all modes, or as

$$k = \frac{1}{3} C_v \langle v \Lambda \rangle \tag{10}$$

where $\Lambda = v\tau$ is the phonon mean free path.

THEORETICAL METHODS

In this section we describe the most widely used theoretical methods for calculating lattice thermal conductivity. These are based on either molecular dynamics or lattice dynamics or both, and it is assumed that the reader has a working knowledge of these and concepts associated with them. Information on these topics can be found elsewhere (e.g., Allen and Tildesley 1987; Leach 2001). The descriptions given below are not exhaustive and additional details can be found in some excellent reviews (McGaughey and Kaviany 2006; Chantrenne 2007).

Green-Kubo method

In an equilibrium molecular dynamics simulation the system under investigation has a constant average temperature and an average heat flux of zero. However, at each instant of time a finite heat flux exists due to instantaneous fluctuations in temperature. The popular Green-Kubo method (Green 1954; Kubo 1957), based on the general fluctuation-dissipation theorem

(Kubo 1966), relates the lattice thermal conductivity of the system to the time required for such fluctuations to dissipate

$$k_{ij} = \frac{V}{k_B T^2} \int_0^\infty \langle q_i(0) q_j(t) \rangle dt \tag{11}$$

where k_{ij} is a component of the lattice thermal conductivity tensor (i and $j = x$, y or z), V the volume of the system, k_B Boltzmann's constant, T the temperature of the system, $q_i(0)$ the instantaneous heat flux in the j direction at time zero and $q_j(t)$ the instantaneous heat flux in the i direction at time t. The angular brackets on the right-hand-side indicate an average over time origins. Though the upper limit on the integral is infinite the duration of the simulation must only exceed the relaxation time beyond which the integrand vanishes (Fig. 1). In molecular dynamics simulations time is discretized into time-steps, and thus in practice Equation (11) becomes a summation (Schelling et al. 2002)

$$k_{ij} = \frac{V\Delta t}{k_B T^2} \sum_{m=1}^{M} (N-m) \sum_{n=1}^{N-m} q_i(m+n) q_j(n) \tag{12}$$

where N is the total number of time-steps, each of length Δt, $q_i(m+n)$ the instantaneous heat flux in the i direction at time-step $m+n$ and $q_j(n)$ the instantaneous heat flux in the j direction at time-step n. The instantaneous heat flux, in a given direction, is evaluated from the energy associated with each atom in the simulation

$$\mathbf{q} = \frac{d}{dt} \frac{1}{V} \sum_{i=1}^{N} \mathbf{r}_i \varepsilon_i \tag{13}$$

where $\mathbf{q}$ is the heat flux vector, $\mathbf{r}_i$ the position vector of atom i and ε_i the energy associated with atom i, and the sum is over all N atoms. The energy associated with each atom is the sum of its

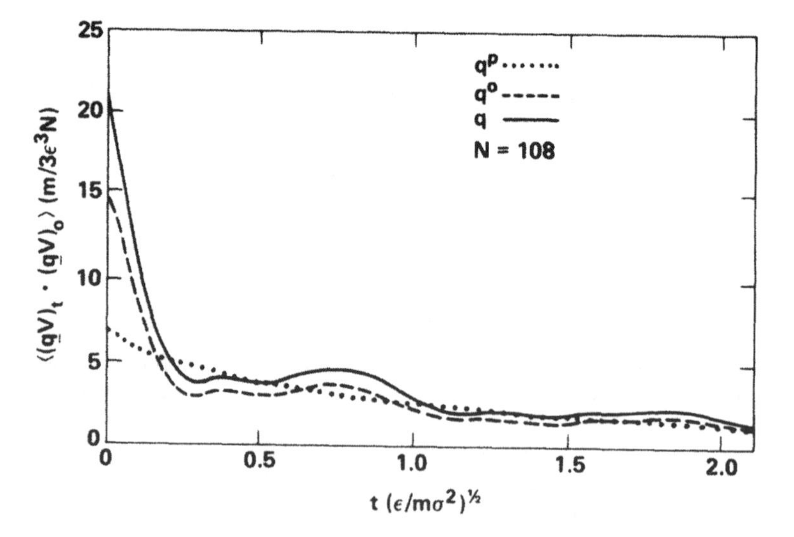

Figure 1. Heat flux autocorrelation function (solid line) of a Lennard-Jones fcc crystal at $T = 0.546\varepsilon/k_B$ or about 25 percent of the melting point, density $N/V = (2)^{1/2}/\sigma^3$, and $N = 108$. Time is non-dimensionalized such that one vibrational period of about $0.25(m\sigma^2/\varepsilon)^{1/2}$. The two dashed lines represent approximations to the auto-correlation function that are valid in the limit of low temperature. m is the atomic mass, σ is the Lennard-Jones length scale and ε is the Lennard-Jones energy scale. Reprinted with permission from Ladd et al. (1986) Physical Review B, Vol. 34, p 5058-5064. Copyright 1986 by the American Physical Society. *http://link.aps.org/doi/10.1103/PhysRevB.34.5058*

kinetic energy and potential energy

$$\varepsilon_i = \frac{1}{2} m_i \mathbf{v}_i^2 + \frac{1}{2} \sum_j^N u_{ij}(r_{ij}) \tag{14}$$

where m_i is the mass of atom i, $\mathbf{v}_i$ the velocity vector of atom i, and $u_{ij}(r_{ij})$ the pair-wise interaction between atoms i and j when separated by a distance r_{ij}. Substituting Equation (14) into Equation (13) we obtain

$$\mathbf{q} = \frac{1}{V}\left[\sum_i^N \mathbf{v}_i \varepsilon_i + \frac{1}{2} \sum_i^N \sum_{j \neq i}^N \mathbf{r}_{ij}(\mathbf{F}_{ij} \cdot \mathbf{v}_i)\right] \tag{15}$$

where $\mathbf{r}_{ij} = \mathbf{r}_i - \mathbf{r}_j$ and $\mathbf{F}_{ij}$ is the force exerted on atom i by atom j. The first term within the square brackets is related to convection and the second to conduction. For more complex potentials additional terms are required (Schelling et al. 2002).

It important to note that in Green-Kubo calculations, just as in many simulation studies, finite-size effect are important and that calculated values converge towards experimental values with increasing system size (Volz and Chen 2000; Sun and Murthy 2006). Such convergence must be checked for in order ensure that calculated values are accurate.

Non-equilibrium molecular dynamics

Non-equilibrium molecular dynamics is the most intuitive theoretical method for determining lattice thermal conductivity, in that it calculates it from the ratio of a heat flux to a temperature gradient, similar to what is done in experimental studies. The simulations can take one of two approaches: either a known heat flux is imposed and the resulting temperature gradient is calculated or a fixed temperature gradient is imposed and the heat flux required to maintain it calculated. However, the former is more usual and we focus on it here.

In general, molecular dynamics simulations are performed in conjunction with a periodic simulation cell, representing the structure of the system under investigation. In such a context, the most common approach to imposing a heat flux is to divide the simulation cell into an even number of equal-size sections, designate one as the hot section and another, half a simulation cell length along, as the cold section, and at regular intervals in time transfer heat from the cold section to the hot section (Fig. 2:A). Since the simulation cell is periodic, heat leaves both sides of the hot section and enters both sides of the cold section, leading to the generation of two heat fluxes in opposing directions and two corresponding temperature gradients (Fig. 2:B).

There are a number of different schemes for transferring heat from the hot section to the cold section. One of the most popular methods involves the regular transfer of heat from the hottest atom in the cold section to the coldest atom in the hot section (Müller-plathe 1997; Nieto-Draghi and Avalos 2003). In particular, at intervals of a fixed number of time-steps, the hottest atom in the cold section is imagined to undergo an elastic collision with the coldest atom in the hot section and the velocities arising from such a collision assigned to the atoms prior to continuation of the simulation. The post-collision velocity of the atom in the cold section is calculated as

$$\mathbf{v}_c' = -\mathbf{v}_c + 2\left[\frac{m_c \mathbf{v}_c + m_h \mathbf{v}_h}{m_c + m_h}\right] \tag{16}$$

and that of the atom in the hot section as

$$\mathbf{v}_h' = -\mathbf{v}_h + 2\left[\frac{m_c \mathbf{v}_c + m_h \mathbf{v}_h}{m_c + m_h}\right] \tag{17}$$

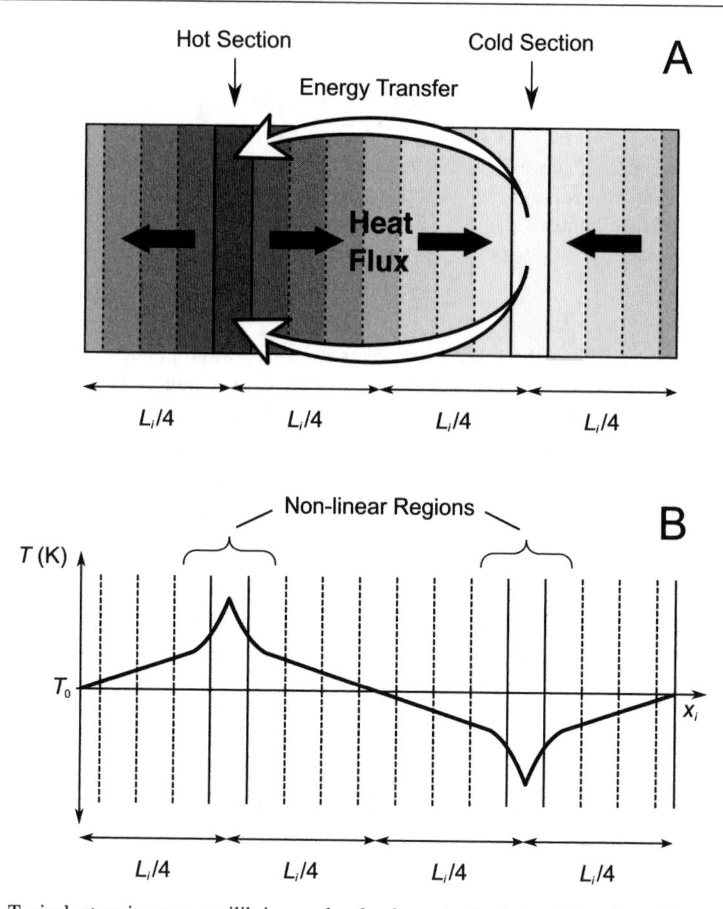

Figure 2. Typical set-up in a non-equilibrium molecular dynamic simulation (A) and resulting temperature profile (B). The simulation cell is divided into sections of equal dimensions, one is designated the 'hot section', another the 'cold section' and at regular intervals heat is transferred from the cold section to the hot section, by modification of the velocities of some or all of the atoms in the two sections. The temperature gradient is non-linear around the hot and cold sections due to the non-Newtonian nature of the heat transfer. Only the linear portion of the temperature gradient is used in the calculation of thermal conductivity.

where m_c and m_h are the respective masses of the atoms in the cold and hot sections, $\mathbf{v}_c$ and $\mathbf{v}_h$ their velocities before the collision and $\mathbf{v}_c'$ and $\mathbf{v}_h'$ their velocities after it. The reasons for using such a construct is that Equations (16) and (17) enable the exchange of heat between the sections, while conserving the total kinetic energy, potential energy and linear momentum of the system. The average heat flux is then determined from

$$q_i = \frac{1}{2AN\Delta t}\sum_{n=1}^{N/\nu_t}\frac{1}{2}m_h(\mathbf{v}_h'(n\nu_t)^2 - \mathbf{v}_h(n\nu_t)^2) \tag{18}$$

where q_i is the average heat flux flowing in the i direction (where $i = x$, y or z), A the cross-sectional area of the simulation cell perpendicular to i, N the total number of time-steps, Δt the time-step, ν_t the frequency of the heat transfers in time-steps, and $\mathbf{v}_h(n\nu_t)$ and $\mathbf{v}_h'(n\nu_t)$ the velocities of the atom in the hot section before and after transfer at time-step $n\nu_t$. The factor half arises because heat flows from both sides of the hot section to both sides of the cold section, with an average of half of the exchanged heat flowing in each of the two directions. Using this method it is impossible to know the resulting heat flux *a priori*. However, by varying the frequency of heat transfers, it is possible to control the magnitude of the heat-flux, and corresponding temperature gradient.

More control on the magnitude of the imposed heat flux is afforded in an alternative approach, which transfers heat from the cold section to the hot section by scaling the velocities of all of the atoms in the two sections (Jund and Jullien 1999). In particular, at regular intervals the velocities of the atoms in the hot and cold sections are adjusted according to

$$\mathbf{v}_i{}' = \mathbf{v}_G + \chi(\mathbf{v}_i - \mathbf{v}_G) \tag{19}$$

where $\mathbf{v}_i$ and $\mathbf{v}_i{}'$ are the velocities of atom i before and after rescaling; $\mathbf{v}_G$ is the velocity of the center of mass of the section before rescaling, computed from

$$\mathbf{v}_G = \frac{\sum_{i=1}^{N} m_i \mathbf{v}_i}{\sum_{i=1}^{N} m_i} \tag{20}$$

where the sum is over all N atoms in the hot or cold section, m_i is the mass of atom i, and χ a scaling coefficient given by

$$\chi = \sqrt{1 \pm \frac{\Delta\varepsilon}{\varepsilon_R}} \tag{21}$$

where $\Delta\varepsilon$ is the amount of heat to be transferred from the cold section to the hot section, and ε_R is the relative energy of the section, defined as

$$\varepsilon_R = \frac{1}{2}\sum_{i=1}^{N} m_i \mathbf{v}_i^2 - \frac{1}{2}\sum_{i=1}^{N} m_i \mathbf{v}_G^2 \tag{22}$$

In Equation (21) the sign is positive for the hot section and negative for the cold section. The heat flux is calculated from

$$q_i = \frac{\Delta\varepsilon}{2A\Delta t} \tag{23}$$

where A is the cross-sectional area of the simulation cell perpendicular to i, and Δt the time-step. The magnitude of $\Delta\varepsilon$ is chosen to give the desired temperature gradient.

For both of the above heat transfer methods, once steady-state is reached, lattice thermal conductivity is calculated from Fourier's Law in one dimension

$$q_i = -k_{ii}\frac{dT}{dx_i} \tag{24}$$

where the temperature gradient, dT/dx_i, is determined from the average instantaneous temperature of each section, T_S, calculated at each time-step from

$$T_S = \frac{1}{3k_B N}\sum_{i=1}^{N} m_i \mathbf{v}_i^2 \tag{25}$$

where the sum is over all N atoms located in the section.

The periodic nature of the simulation cell leads to two temperature gradients, equal in magnitude, but opposite in sign. It is normal to average the temperature of symmetrically equivalent sections to improve statistics. In some studies the difference in the average temperature of symmetrically equivalent sections has been used as an indicator of whether or not steady state has been reached (e.g., Yoon et al. 2004). The process of energy transfer renders the dynamics in the immediate vicinity of the hot and cold sections non-Newtonian and the temperature profile in these regions non-linear (Fig. 2:B). In view of this, the temperature of the sections around the hot and cold sections are discarded during calculation of the temperature gradient. One practical consequence of this phenomenon is that, the simulation cell must be of a sufficient size for the temperature profile to have a linear portion.

The limited simulation cell size tractable with molecular dynamics methods often leads to the issue of finite-size effects. In particular, in calculations of lattice thermal conductivity, unless the length of the simulation cell is many times larger than the phonon mean-free path, one computes a value lower than the true value. This is because of the direct relationship between lattice thermal conductivity and phonon mean free path (Eqn. 10). In a real solid, the mean free path is mainly limited by phonon-phonon scattering, although phonon-defect scattering will also play a role when defects are present, but in non-equilibrium molecular dynamics simulations phonons are also scattered in the hot and cold sections. This leads to a lower mean free path, and thus lower lattice thermal conductivity. It is, however, possible to overcome this problem by performing simulations for cells of different sizes, as has also been done in a number of recent simulation studies (e.g., Schelling et al. 2002; Stackhouse et al. 2008).

The effective phonon mean free Λ_{eff} path for each simulation cell is expressed in terms of that related to phonon-phonon scattering $\Lambda_{ph\text{-}ph}$ and that related to phonon-boundary scattering $\Lambda_{ph\text{-}b}$, which occurs in the hot and cold sections

$$\frac{1}{\Lambda_{eff}} = \frac{1}{\Lambda_{ph-ph}} + \frac{1}{\Lambda_{ph-b}} \tag{26}$$

Since phonons can originate from any point between the hot and cold section and be scattered in the hot or cold sections, the distance that a phonon will travel between scattering events (in the absence of phonon-phonon interactions) will be, on average, one quarter of the length of the simulation cell in the direction of the heat-flux

$$\frac{1}{\Lambda_{eff}} = \frac{1}{\Lambda_{ph-ph}} + \frac{4}{L_i} \tag{27}$$

where L_i is the length of the simulation cell in the i direction. If we substitute Equation (27) into Equation (10) and rearrange, for an isotropic solid, we have

$$\frac{1}{k} = \left[\frac{12}{C_v v}\right]\frac{1}{L_i} + \left[\frac{1}{3} C_v v \Lambda_{ph-ph}\right]^{-1} \tag{28}$$

which means that a plot of k^{-1} against L_i^{-1} should be linear. Therefore by computing k for a range of L_i values, and plotting k^{-1} against L_i^{-1} it is possible to extrapolate to $L_i^{-1} = 0$ to determine the lattice thermal conductivity of a simulation cell of infinite size (Fig. 3), which should be comparable to the infinite-system value.

From a practical point of view, there are several important considerations that must made before performing non-equilibrium molecular dynamics simulations, necessitating some initial experimentation (Müller-Plathe 1997; Schelling et al. 2002; Chantrenne and Barrat 2004; Mountain 2006; Mahajan et al. 2007). The first is that one must choose the number of sections in which to divide the simulation cell. This is not straightforward. Larger sections will contain more atoms, leading to a more accurate estimate of their instantaneous temperature. On the other hand, larger sections will mean fewer sections in total, and thus less data points

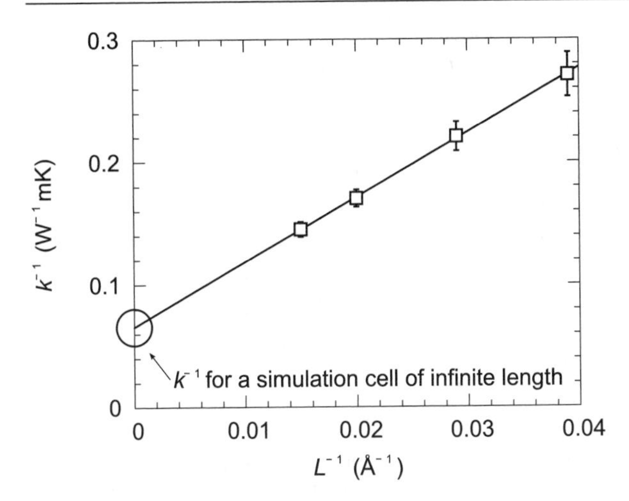

Figure 3. Typical plot of k^{-1} against L^{-1} obtained from a series non-equilibrium molecular dynamics simulation, using simulation cells of different size. Extrapolating back to $L^{-1} = 0$ it is possible to estimate the lattice thermal conductivity of a simulation cell of infinite size, accounting for the artificial phonon-boundary scattering introduced by the heat transfer mechanism.

from which to calculate the temperature gradient. The magnitude of the heat flux is also a concern. Imposing a large heat flux will lead to a large temperature gradient and much faster convergence. However, Fourier's Law is invalid for perturbations outside the linear response regime and care must be taken to ensure that it is still applicable. To assess this one can perform a series of simulations using heat fluxes of decreasing magnitude and check that the calculated lattice thermal conductivity is the same in each case. It is also worth noting that if the temperature gradient is large, then it will be unclear to which temperature the calculated lattice thermal conductivity corresponds. The effect of the size of the cross-sectional area perpendicular to the heat-flux is another factor that is also sometimes considered. This and each of the above factors must be investigated for each system and chosen with care.

Transient non-equilibrium molecular dynamics

In a similar manner to the non-equilibrium molecular dynamics method, the transient non-equilibrium molecular dynamics method (Daly and Maris 2002; Daly et al. 2002) begins by dividing a simulation cell into sections. However, instead of introducing a hot and cold section and imposing a constant one-dimensional heat flux, a sinusoidal temperature perturbation is applied across the sections and lattice thermal conductivity is determined from the rate at which the system re-equilibrates (Fig. 4).

The sinusoidal temperature perturbation is applied to the system according to

$$T(x_i) = T_0 + \Delta T_0 \cos\left[\frac{2\pi x_i}{L_i}\right] \tag{29}$$

where $T(x_i)$ is the temperature a distance x_i across the simulation cell, in the i direction (where $i = x$, y or z), T_0 the equilibrium temperature, ΔT_0 the amplitude of the sinusoidal temperature perturbation and L_i the length of the simulation cell in the i direction. Once applied, the system is allowed to re-equilibrate and the decrease in the amplitude of the sinusoidal temperature perturbation computed as a function of time from

$$\Delta T(t) = \frac{2}{L_i}\int_0^{L_i} T(x_i,t)\cos\left[\frac{2\pi x_i}{L_i}\right]dx_i \tag{30}$$

where $T(x_i,t)$ is the temperature a distance x_i across the simulation cell in the i direction, at a time t.

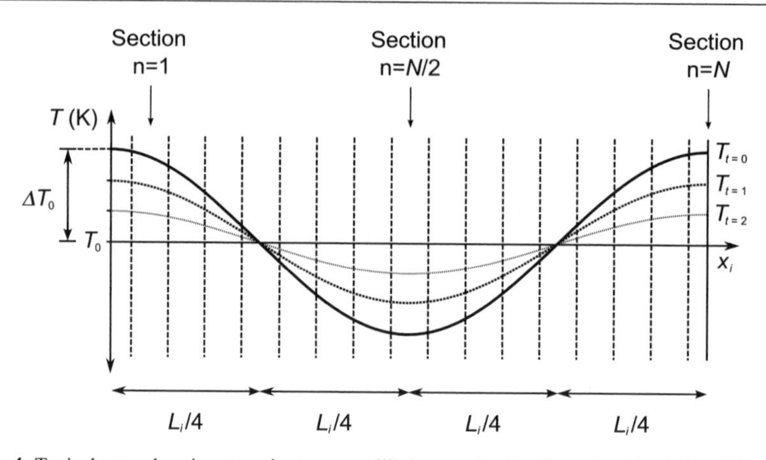

Figure 4. Typical procedure in a transient non-equilibrium molecular dynamics simulation. The system is brought to equilibrium at a temperature of T_0. Then a sinusoidal temperature perturbation of amplitude ΔT_0 is applied, across the simulation cell, and thermal conductivity determined from that rate at which the system re-equilibrates.

The rate at which such a system re-equilibrates can be related to its thermal diffusivity by solving the appropriate heat conduction equation (Holman 1976; Kreith and Bohn 2001)

$$\frac{\Delta T(t)}{\Delta T_0} = e^{-D_{ii}\left(\frac{2\pi}{L_i}\right)^2 t} \tag{31}$$

where D_{ii} is a component of thermal diffusivity tensor. Taking the logarithm gives

$$\ln\left[\frac{\Delta T(t)}{\Delta T_0}\right] = -D_{ii}\left(\frac{2\pi}{L_i}\right)^2 t \tag{32}$$

which allows the determination of D_{ii} from a simple plot of $\ln(\Delta T(t)/\Delta T_0)$ against t. The lattice thermal conductivity of the system can then be calculated from Equation (6), provided that its specific heat capacity and density are known.

In practice, the simulation cell is divided into N sections in the i direction, each of width L_i/N, which we label 1 to N and time is discretized into time-steps, each of length Δt. In this context the initial temperature profile is expressed as

$$T(n) = T_0 + \Delta T_0 \cos\left[\frac{2\pi n}{N}\right] \tag{33}$$

where $T(n)$ is the temperature of section n. The temperature perturbation is applied by rescaling the velocities of all atoms in each section, according to

$$\frac{\mathbf{v}'_n}{\mathbf{v}_n} = \sqrt{\frac{T(n)}{T_0}} \tag{34}$$

where $\mathbf{v}_n$ and $\mathbf{v}_n'$ are the velocities of atoms in section n before and after rescaling. The amplitude of the sinusoidal temperature perturbation is computed as

$$\Delta T(m) = \frac{2}{L_i}\sum_{n=1}^{N} T(n,m)\cos\left[\frac{2\pi n}{N}\right] \tag{35}$$

where $\Delta T(m)$ is the amplitude of the sinusoidal temperature perturbation at time-step m and $\Delta T(n,m)$ the instantaneous temperature of section n at time-step m computed from Equation (25). Thermal diffusivity is then determined from

$$\frac{\Delta T(m)}{\Delta T_0} = e^{-D_{ii}\left(\frac{2\pi}{L_i}\right)^2 m\Delta t} \tag{36}$$

where the time t is replaced by the product of the time-step number m and length Δt.

In addition to similar issues to those encountered in non-equilibrium molecular dynamics simulations, the transient molecular dynamics method suffers from a number of other complications. The sinusoidal temperature perturbation imposes a corresponding sinusoidal variation in thermal pressure across the simulation cell, which upon re-equilibration induces a low frequency vibrational mode, causing atoms to vibrate in the i direction (Daly and Maris 2002; Daly et al. 2002). However, the effect is expected to be small, and can be reduced by using smaller perturbations. More worrisome is an issue regarding the fitting of Equation (36) at short and long times scales, which can only been remedied by applying an *ad hoc* modification to the equation (Daly and Maris 2002; Daly et al. 2002).

Combined Quasiharmonic Lattice Dynamics and Molecular Dynamics Method

In lattice dynamics calculations the potential energy of a system is expressed as a Taylor series expansion of atomic displacements (Maradudin et al. 1963)

$$V = V^0 + \sum_{l,\kappa,i}\frac{\partial V}{\partial u_{l,\kappa,i}} + \frac{1}{2}\sum_{l,\kappa,i;l',\kappa',j}\frac{\partial^2 V}{\partial u_{l,\kappa,i}\partial u_{l',\kappa',j}}u_{l,\kappa,i}u_{l',\kappa',j} + \ldots \tag{37}$$

where V is the potential energy of the system; V^0 the potential energy of the system with all the atoms in their equilibrium positions, $u_{l,\kappa,i}$ the displacement of atom κ in unit cell l in the i direction (where $i = x$, y or z) and $u_{l',\kappa',j}$ the displacement of atom κ' in unit cell l' in the j direction (where $j = x$, y or z). Since the derivatives are evaluated with the atoms at their equilibrium positions, by definition, the first derivative must be zero. It is usual to truncate the expansion at the second derivative as is shown, making what is known as the quasiharmonic approximation.

The second term on the right hand side represents the harmonic inter-atomic interactions and the double derivatives make up the elements of the force constant matrix

$$\Phi_{l,\kappa,i;l',\kappa',j} = \frac{\partial V}{\partial u_{l,\kappa,i}\partial u_{l',\kappa',j}} \tag{38}$$

where $\Phi_{l,\kappa,i;l',\kappa',j}$ is the force exerted in the i direction on atom κ in unit cell l, when atom atom κ' in unit cell l' is displaced a unit distance in the j direction. The elements of the force constant matrix can be determined either by making small displacements of the atoms in one unit cell, with all other atoms held fixed, and determining the forces on all other atoms—the finite displacement method (Kresse et al. 1995; Alfé 2009), or from perturbation theory (Gonze and Lee 1997; Baroni et al. 2001).

The equations of motion of the system can be expressed

$$m_\kappa \frac{\partial^2 u_{l,\kappa,i}}{\partial t^2} = -\sum_{l',\kappa',j}\phi_{l,\kappa,i;l',\kappa',j}u_{l',\kappa',j} \tag{39}$$

where m_κ is the mass of atom κ. If we assume a harmonic solution to Equation (38) we have

$$u_{l,\kappa,i} = \frac{1}{\sqrt{m_\kappa}}u_{\kappa,i}\exp[-i(\omega t + \mathbf{q}\square\mathbf{x}_l)] \tag{40}$$

where ω is the mode frequency, t time, $\mathbf{q}$ mode wave vector and $\mathbf{x}_l$ the coordinates of unit cell l. Substituting this into Equation (39) we obtain

$$\omega^2(\mathbf{q})u_{\kappa,i} = \sum_{\kappa',j} D_{\kappa,i;\kappa',j}(\mathbf{q})u_{\kappa',j} \tag{41}$$

where the dynamical matrix, $D_{\kappa,i;\kappa',j}(\mathbf{q})$, is

$$D_{\kappa,i;\kappa',j}(\mathbf{q}) = \frac{1}{\sqrt{m_\kappa m_{\kappa'}}}\sum_{l'} \Phi_{0,\kappa,i;l',\kappa',j} \exp[-i\mathbf{q}\cdot\mathbf{x}_{l'}] \tag{42}$$

Diagonalization of the dynamical matrix yields the phonon eigenmodes and their frequencies, which allows the determination of both $c_{\mathbf{q},s}$ and $v_{\mathbf{q},s}$ in Equation (7), but in order to calculate lattice thermal conductivity $\tau_{\mathbf{q},s}$ is also needed. In the harmonic approximation the phonon lifetime is infinite and is only limited by anharmonicity, i.e., phonon-phonon interactions, which must be computed using a method other than quasi-harmonic lattice dynamics. In principle anharmonicity will also cause shifts of the mode frequencies, although this effect is typically small for computations of the thermal conductivity (Fig. 5). Estimates of $\tau_{\mathbf{q},s}$ can be obtained from molecular dynamics simulations (de Koker 2009; Turney et al. 2009). One method is to use the Fourier transform of the velocity autocorrelation function to determine the vibrational spectrum at individual wave vectors (Fig. 5) and relate relaxation times to the width of spectral peaks (de Koker 2009). Other techniques include fitting the phonon potential energy autocorrelation function (Turney et al. 2009). Thus by performing both lattice dynamics and molecular dynamics of the same system it is possible to determine its lattice thermal conductivity.

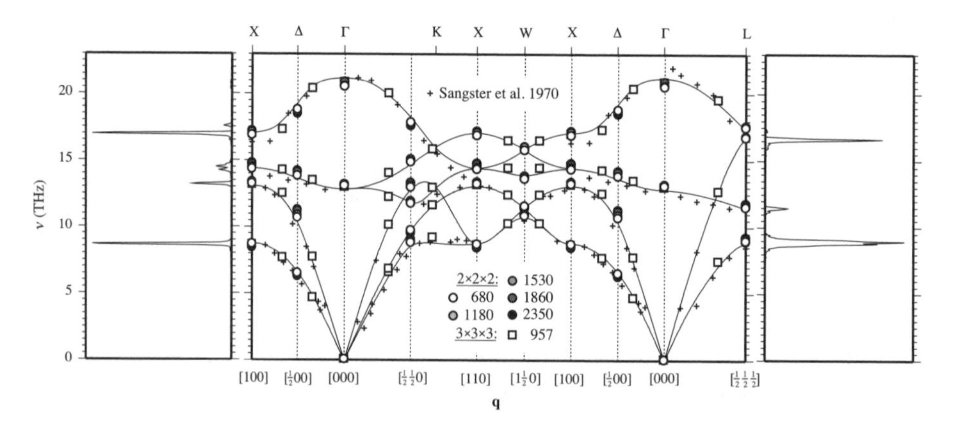

Figure 5. (*center*) Phonon frequencies of MgO periclase calculated from first-principles molecular dynamics (circles; temperature in Kelvin) and first-principles lattice dynamics in the quasi-harmonic approximation (lines). The black crosses are experimental values Sangster et al. (1970). (*left and right panels*) Examples of spectral peaks calculated from first-principles molecular dynamics, at [100] and [½ ½ ½], used to determine phonon relaxation times. Reprinted with permission from de Koker (2009) Physical Review Letters, Vol. 103, Article #125902. Copyright 2009 by the American Physical Society. *http://link.aps.org/doi/10.1103/PhysRevLett.103.125902*

Anharmonic lattice dynamics method

It is also possible to estimate lattice thermal conductivity from lattice dynamics calculations alone, by considering higher order terms in the Taylor expansion (Eqn. 37) (e.g., Broido et al. 2007; Tang and Dong 2009; Turney et al. 2009). In doing so it is possible to determine relaxation times and anharmonic frequencies. Aside from the pertubative treatment of the anharmonic terms, the Taylor series is also typically truncated, neglecting part of the anharmonic contributions, which can lead to approximations in computed thermal conductivity values.

DISCUSSION

Having described the various approaches for calculating lattice thermal conductivity, we now discuss the relative advantages and disadvantages of each, in an attempt to provide some general guidance on selecting the most appropriate method. Of course, as is the case in all theoretical investigations, this will involve making a compromise between speed and precision, and consideration of available computational resources. Since there have been only a handful of investigations that have used the transient non-equilibrium molecular dynamics method and lattice dynamics techniques, we focus on the Green-Kubo and non-equilibrium molecular dynamics methods, which make up the majority of studies in the literature.

If one considers the Green-Kubo method, a clear advantage is that the entire lattice thermal conductivity tensor can be calculated from one simulation. This is in contrast to the non-equilibrium molecular dynamics methods, which necessitates several simulations in each direction, to achieve the same. This could be an important consideration, if the mineral of interest is known to be anisotropic. The Green-Kubo method also requires less experimentation than non-equilibrium molecular dynamics methods, there being no need to investigate the effect of section size or heat flux on results. On the other hand, it can take a long time for the correlation function to decay to zero, thus long simulations are often required. In addition, the Green-Kubo method requires the identification of the self-energy of each atom, which is not straightforward in the context of first-principles calculations, meaning that it is, for the most part, limited to the study of those phases which are well described by a set of empirical pair potentials.

This issue of empirical pair potentials versus first-principles computation of forces is an important one. Due to the large size and length of simulations that must performed in calculations of lattice thermal conductivity, nearly all previous theoretical investigations have utilized empirical pair potentials to describe the forces between atoms. Using such pair potentials large simulation cells can be used, containing thousands of atoms, and it is possible to model polycrystalline systems and the influence of grain boundaries (e.g., Shukla et al. 2008). However, pair potentials are, in general, parameterized using experimental data determined at ambient conditions or values from static first-principles calculations. It is thus uncertain how well they are able to describe the motion of atoms at lower mantle temperature and pressures. On the other hand, forces determined from first-principles, via the Hellmann-Feynman theorem (Hellmann 1937; Feynman 1939) are parameter free and therefore expected to be more reliable.

Non-equilibrium molecular dynamics simulations can be performed using both empirical pair potentials or within the framework of first-principles calculations. Calculating forces from first-principles increases the computational resources required, which means that much smaller simulation cells must be used, leading to increased finite size effects. However, it has been shown that these can be dealt with in a systematic manner, although it should be noted that because of the nature of the reciprocal plot, associated error-bars can be large for systems with high lattice thermal conductivities (Stackhouse et al. 2008). Of the non-equilibrium molecular dynamics methods, the imposed heat-flux method requires smaller simulation cells, as compared to the transient method. This is because the size of the simulation cell limits the number of sections and many more sections are required to define a sinusoidal temperature perturbation than a linear temperature gradient.

It should also be pointed out that all of the molecular dynamics based methods discussed describe the motion of the atoms using classical mechanics, i.e., by solving Newton's equations of motions. This is true even for *ab initio* molecular dynamics simulations, where, although the forces are determined from first-principles, atomic motion is still described by classical mechanics, which generates errors at very low temperatures due to quantum effects on the dynamics (Jund and Jullien 1999). However, this is not expected to be a concern when calculating values for minerals at lower mantle temperatures, far above their estimated Debye temperatures.

Quantum effects are included in lattice dynamics calculations. In addition, like the Green-Kubo method, lattice dynamics should also be able to determine the full lattice thermal conductivity tensor. In general, lattice dynamics studies use forces calculated from first-principles, in order to obtain accurate mode frequencies and lifetimes, but the computational resources required to perform a combined lattice and molecular dynamics investigation is still likely to be less than that required for non-equilibrium molecular dynamics studies, at least for simple structures. In more complex structures, such as silicate perovskite, the task of computing the vibrational spectrum and identifying each mode may prove formidable.

In view of all the above, if computational resources allow, we recommend performing first-principles non-equilibrium molecular dynamics simulations to calculate the lattice thermal conductivity of minerals. This avoids concerns regarding the robustness of empirical pair potentials. However, empirical pair potentials may perform well at lower temperatures and pressures, as will be seen in the next section, and in this case they can be used.

THE LATTICE THERMAL CONDUCTIVITY OF PERICLASE

Periclase is the pure magnesium end-member of ferropericlase, thought to be the second most abundant phase in the lower mantle. The simple rock-salt structure of the phase makes it an ideal mineral on which to test new methods. In all, there have been five studies of the lattice thermal conductivity of periclase: one based on the Green-Kubo method (Cohen 1998); two using non-equilibrium molecular dynamics simulations (Shukla et al. 2008; Stackhouse et al. 2008); one using a combination of lattice dynamics and equilibrium molecular dynamics simulations (de Koker 2009) and one using anharmonic lattice dynamics (Tang and Dong 2009). The phase has thus been studied using almost all of the methods described in the previous section. The results of these investigations are compared with available experimental data in Figures 6. Note that, Tang and Dong (2009) only reported temperature and pressure derivatives. The results of Stackhouse et al. (2008) are still under review and are not shown.

If we first consider the ambient pressure data, one can see that the Green-Kubo calculations of Cohen (1998), based on a non-empirical ionic model, predict values a little lower than experimental values and those from other theoretical studies. The results of the non-equilibrium molecular dynamics simulations that used pair potentials (Shukla et al. 2008) agree well with experiment, in particular, those of Hofmeister and Yuen (2007). This suggests that empirical pair potentials can perform well at low pressures. The first-principles non-equilibrium molecular dynamics calculations of Stackhouse et al. (2008), also agree well with experimental values, in particular, those of Kanamori et al. (1968). Moving to high pressure, we find that the first-principles calculations of de Koker (2009) predicts values in good agreement with the experimental data of Goncharov et al. (2009), which fall between the two theoretical predictions (Fig. 6).

It should be noted that the values determined in the theoretical studies discussed are for perfect periclase single crystals, which do not contain iron impurities or defects. In the lower mantle, ferropericlase exists in polycrystalline form and grain boundaries, and defects, such as iron impurities will increase phonon scattering, decreasing the lattice thermal conductivity of the phase. The lattice thermal conductivity of periclase should therefore be viewed as an upper bound to that of ferropericlase. In their non-equilibrium molecular dynamic simulations, Shukla et al. (2008) investigated the effect of grain boundary scattering on the lattice thermal conductivity of periclase, finding it to be significant at low temperatures, while at higher

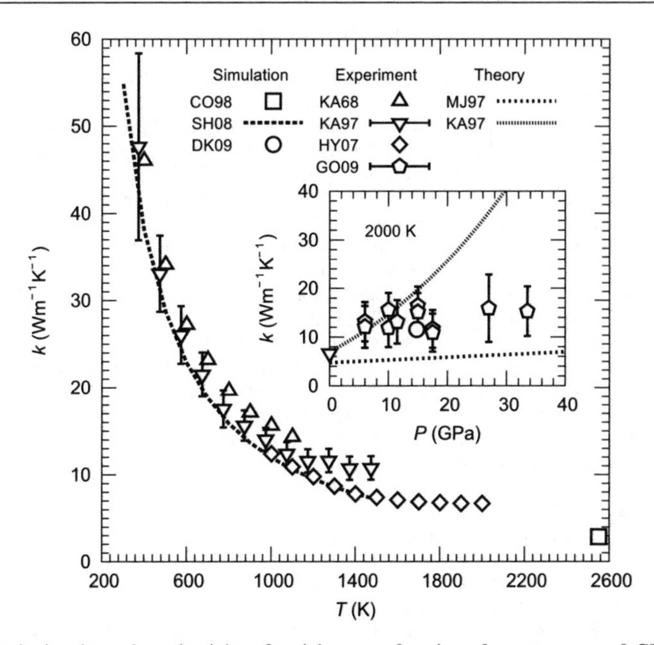

Figure 6. The lattice thermal conductivity of periclase as a function of temperature at 0 GPa (*main*) and as a function of pressure at 2000 K (*inset*). Simulation values: CO90 (Cohen 1998); SH08 (Shukla et al. 2008); DK09 (de Koker 2009). Experimental values: KA68 (Kanamori et al. 1968); KA97 (Katsura 1997); HY07 (Hofmeister and Yuen 2007); GO09 (Goncharov et al. 2009). Theoretical values: MJ97 (Manga and Jeanloz 1997); KA97 (Estimated from results of Katsura 1997). Best agreement with experiment is found for the non-equilibrium molecular dynamics simulations (Shukla et al. 2008).

temperatures, such as those expected in the lower mantle, the effect was small. This is because at high temperatures the phonon mean free path is smaller than the grain size, while at lower temperatures they may be comparable. The influence of grain size will therefore depend on the intrinsic lattice thermal conductivity of the single crystal. The effect of defects, such as iron impurities, remains to be quantified in periclase.

CONCLUSION

Determining the thermal conductivity of lower mantle minerals is important for constraining many important processes in both the past and present deep Earth. In light of the experimental difficulties in measuring thermal conductivity, theoretical methods offer an invaluable alternative. There exist a number of different theoretical techniques, which have been used in the field of materials science, which can also be applied to mantle minerals. Each of these methods have advantages and disadvantages, and the appropriate method must be chosen in a compromise between speed and accuracy. Studies of periclase indicate, at present, that non-equilibrium molecular dynamics simulations are promising and will be readily scaled to more complex crystal structures.

ACKNOWLEDGMENTS

The authors are indebted to Nico de Koker for helpful discussions and an early copy of his manuscript.

REFERENCES

Alfé D (2009) PHON: A program to calculate phonons using the small displacement method. Comput Phys Commun 180:2622-2633

Allen MP, Tildesley DJ (1987) Computer Simulation of Liquids. Clarendon Press, Oxford, United Kingdom

Badro J, Fiquet G, Guyot F, Rueff J-P, Struzhkin VV, Vankó G, Monaco G (2003) Iron partitioning in earth's mantle: towards a deep lower mantle discontinuity. Science 305:383-386

Badro J, Rueff J-P, Vankó G, Monaco G, Fiquet G, Guyot F (2004) electronic transitions in perovskite: possible nonconvecting layers in the lower mantle. Science 300:789-791

Baroni S, de Gironcoli S, Dal Corso A, Giannozzi P (2001) Phonons and related crystal properties from density-functional perturbation theory. Rev Mod Phys 73:515-562

Broido, DA, Malory M, Birner G., Mingo N, Stewart DA (2007) Intrinsic lattice thermal conductivity of semiconductors from first principles. Appl Phys Lett 91:231922

Chantrenne P (2007) Molecular dynamics. Top Appl Phys 107:155-180

Chantrenne P, Barrat J-L (2004) finite size effects in determination of thermal conductivities: comparing molecular dynamics results with simple models. J Heat Trans 126:577-585

Cohen RE (1998) Thermal conductivity of MgO at high pressures. Rev High Pres Sci Tech 7:160-162

Daly BC, Maris HJ (2002) Calculation of the thermal conductivity of superlattice by molecular dynamics simulation. Physica B 316:247-249

Daly BC, Maris HJ, Imamura K, Tamura S (2002) Molecular dynamics calculation of the thermal conductivity of superlattices. Phys Rev B 66:024301

de Koker N (2009) Thermal conductivity of MgO periclase from equilibrium first principles molecular dynamics. Phys Rev Lett 103:125902

Dubuffett F, Yuen DA (2000) A thick pipe-like heat-transfer mechanism in the mantle: nonlinear coupling between 3-D convection and variable thermal conductivity. Geophys Res Lett 27:17-20

Dubuffet F, Yuen DA, Rabinowicz M (1999) Effects of a realistic mantle thermal conductivity on the patterns of 3-D convection. Earth Planet Sci Lett 171:401-409

Evans DJ, Morris GP (1990) Statistical Mechanics of Non-Equilibrium Liquids. Academic Press, London, United Kingdom

Feynman RP (1939) Forces in molecules. Phys Rev 56:340-343

Goncharov AF, Struzhkin VV, Jacobsen SD (2006) reduced radiative conductivity of low-spin (Mg,Fe)O in the lower mantle. Science 312:1205-1208

Goncharov AF, Haugen BD, Struzhkin VV, Beck P, Jacobson SD (2008) Radiative conductivity in the Earth's lower mantle. Nature 456:231-234

Goncharov AF, Beck P, Struzhkin VV, Haugen BD, Jacobsen SD (2009) Thermal conductivity of lower-mantle minerals. Phys Earth Planet Inter 174:24-32

Gonze X, Lee C (1997) Dynamical matrices, born effective charges, dielectric permittivity tensors, and interatomic force constants from density-functional perturbation theory. Phys Rev B 55:10355-10368

Green MS (1954) Markoff random processes and the statistical mechanics of time-dependent phenomena. II Irreversible processes in fluids. J Chem Phys 22:398-413

Gubbins D, Willis AP, Sreenivasan B (2007) Correlation of Earth's magnetic field with lower mantle thermal and seismic structure. Phys Earth Planet Inter 162:256-260

Hellmann H (1937) Einführung in die Quantenchemie. Franz Deuticke, Leipzig, Deutchland

Hofmeister AM (2007) Pressure dependence of thermal transport properties. Proc Nat Acad Sci USA 104:9192-9197

Hofmeister AM (2008) Inference of high thermal transport in the lower mantle from laser-flash experiments and the damped harmonic oscillator model. Phys Earth Planet Inter 170:201-206

Hofmeister AM, Yuen DA (2007) Critical phenomena in thermal conductivity: Implications for lower mantle dynamics. J Geodyn 44:186-199

Holman JP (1976) Heat Transfer. McGraw-Hill, New York, United State of America.

Jund P, Jullien R (1999) Molecular-dynamics calculation of the thermal conductivity of vitreous silica. Phys Rev B 59:13707-13711

Kanamori H, Fujii N, Mizutani H (1968) Thermal diffusivity measurement of rock-forming minerals from 300 degrees to 1100 degrees K. J Geophys Res 73:595-605

Katsura T (1997) Thermal diffusivity of periclase at high temperature and high pressures. Phys Earth Planet Inter 101:73-77

Keppler H, Kantor I, Dubrovinsky LS (2007) Optical absorption spectra of ferropericlase to 84 GPa. Am Mineral 92:433-436

Keppler H, Dubrovinsky LS, Narygina O, Kantor I (2008) Optical absorption and radiative thermal conductivity of silicate perovskite to 125 gigapascals. Science 322:1529-1532

Kreith F, Bohn MS (2001) Principles of Heat Transfer. Brooks-Cole, Pacific Grove, California, United States of America

Kresse G, Furthmuller J, Hafner J (1995) Ab initio force constant approach to phonon dispersion relations of diamond and graphite. Europhys Lett 32:729-734

Kubo R (1957), Statistical mechanical theory of irreversible processes. I. General theory and simple applications to magnetic and conduction problems. J Phys Soc Japan 12:570-586

Kubo R (1966) The fluctuation-dissipation theorem. Rep Prog Phys 29:255-284

Ladd AJC, Moran B., Hoover WG (1986) Lattice thermal conductivity: A comparison of molecular dynamics and anharmonic lattice dynamics. Phys Rev B 34:5058-5064

Lay T, Hernlund J, Buffett BA (2008) Core-mantle boundary heat flow. Nat Geosci 1:25-32

Leach AR (2001) Molecular Modelling: Principles and Applications. Pearson Education Limited, Harlow, United Kingdom

Lee KKM, O'Neill B, Panero WR, Shim SH, Benedetti LR, Jeanloz R (2004) Equations of state of the high-pressure phases of a natural peridotite and implications for the Earth's lower mantle. Earth Planet Sci Lett 223:381-393

Mahajan SS, Subbarayan G, Sammakia BG (2007) Estimating thermal conductivity of amorphous silica nanoparticles and nanowires using molecular dynamics simulations. Phys Rev E 76:056701

Manga M, Jeanloz R (1997) Thermal conductivity of corundum and periclase and implications for the lower mantle. J Geophys Res 102:2999-3008

Maradudin AA, Montroll EW, Weiss GH (1963) Theory of Lattice Dynamics in the Harmonic Approximation. Academic Press, New York and London, United Kingdom

McGaughey AJH, Kaviany M (2006) Phonon transport in molecular dynamics simulations: formulation and thermal conductivity prediction. Adv Heat Trans 39:169-225

Mountain RD (2006) System size and control parameter effects in reverse perturbation nonequilibrium molecular dynamics. J Chem Phys 124:104109

Müller-Plathe F (1997) A simple nonequilibrium molecular dynamics method for calculating the thermal conductivity. J Chem Phys 106:6082-6085

Naliboff JB, Kellogg LH (2006) Dynamic effect of a step-wise increase in thermal conductivity and viscosity in the lowermost mantle. Geophys Res Lett 33:L12S09

Naliboff JB, Kellogg LH (2007) Can large increases in viscosity and thermal conductivity preserve large-scale heterogeneity in the mantle. Phys Earth Planet Inter 161:86-102

Nieto-Draghi C, Avalos JB (2003) Non-equilibrium momentum exchange algorithm for molecular dynamics simulation of heat flow in multicomponent systems. Mol Phys 101:2303-2307

Nye JF (1987) Physical Properties of Crystals: Their Representation by Tensors and Matrices. Oxford University Press, Oxford, United Kingdom

Sangster MJ, Peckham G, Saunderson DH (1970) Lattice dynamics of magnesium oxide. J Phys C 3:1026-1036

Schelling PK, Phillpot SR, Keblinski P (2002) Comparison of atomic-level simulation methods for computing thermal conductivity. Phys Rev B 65:144306

Shukla P, Watanabe T, Nino JC, Tulenko JS, Phillpot SR (2008) Thermal transport properties of MgO and $Nd_2Zr_2O_7$ pyrochlore by molecular dynamics simulation. J Nucl Mater 380:1-7

Srivastava GP (1990) The Physics of Phonons. Taylor & Francis Group, New York, United States of America

Stackhouse S, Stixrude L, Karki BB (2008) The thermal conductivity of periclase (MgO) from first-principles. EOS Trans AGU 89, Fall Meeting Supp. Abs. # MR21C

Sun L, Murthy J (2006), Domain size effects in molecular dynamics simulation of phonon transport in silicon. Appl Phys Lett 89:171919

Tang XL, Dong JJ (2009) Pressure dependence of harmonic and anharmonic lattice dynamics in MgO: A first-principles calculation and implications for lattice thermal conductivity. Phys Earth Planet Inter 174:33-38

Turney JE, Landry ES, McGaughey AJH, Amon CJ (2009) Predicting phonon properties and thermal conductivity from anharmonic lattice dynamics calculations and molecular dynamics simulations. Phys Rev B 79:064301

Volz SG, Chen G (2000) Molecular-dynamics simulation of thermal conductivity of silicon crystals. Phys Rev B 61:2651-2656

Willis AP, Sreenivasan B, Gubbins D (2007) Thermal core-mantle interaction: Exploring regimes for 'locked' dynamo action. Phys Earth Planet Inter 165:83-92

Yoon YG, Car R, Srolovitz DJ, Scandolo S (2004) Thermal conductivity of crystalline quartz from classical simulations. Phys Rev B 70:012302

Ziman JM (1960) Electrons and Phonons. Oxford University Press, Oxford, United Kingdom

Reviews in Mineralogy & Geochemistry
Vol. 71 pp. 271-298, 2010

Evolutionary Crystal Structure Prediction as a Method for the Discovery of Minerals and Materials

Artem R. Oganov

Department of Geosciences, Department of Physics and Astronomy, and New York Center for Computational Sciences
Stony Brook University
Stony Brook, New York, 11794-2100, U.S.A.

artem.oganov@sunysb.edu

Geology Department, Moscow State University
119992 Moscow, Russia

Yanming Ma

National Lab of Superhard Materials
Jilin University
Changchun 130012, P. R. China

Andriy O. Lyakhov

Department of Geosciences
Stony Brook University
Stony Brook, New York, 11794-2100, U.S.A.

Mario Valle

Data Analysis and Visualization Group
Swiss National Supercomputing Centre (CSCS)
Cantonale Galleria 2
6928 Manno, Switzerland

Carlo Gatti

CNR-ISTM, Istituto di Scienze e Tecnologie Molecolari
via Golgi 19
20133 Milano, Italy

ABSTRACT

Prediction of stable crystal structures at given pressure-temperature conditions, based only on the knowledge of the chemical composition, is a central problem of condensed matter physics. This extremely challenging problem is often termed "crystal structure prediction problem," and recently developed evolutionary algorithm USPEX (Universal Structure Predictor: Evolutionary Xtallography) made an important progress in solving it, enabling efficient and reliable prediction of structures with up to ~40 atoms in the unit cell using *ab initio* methods. Here we review this methodology, as well as recent progress in analyzing energy landscape of solids (which also helps to analyze results of USPEX runs). We show several recent applications – (1) prediction of new high-pressure phases of $CaCO_3$, (2) search

1529-6466/10/0071-0013$05.00 DOI: 10.2138/rmg.2010.71.13

for the structure of the polymeric phase of CO_2 ("phase V"), (3) high-pressure phases of oxygen, (4) exploration of possible stable compounds in the Xe-C system at high pressures, (5) exotic high-pressure phases of elements boron and sodium, as well as extension of the method to variable-composition systems.

INTRODUCTION

Crystal structure prediction problem occupies a central place in materials design. Solving this problem would also open new ways for understanding the behavior of materials at extreme conditions, where experiments are difficult (in some cases, prohibitively difficult).

Often the approach has been to compare the free energies of a number of candidate structures (usually taken from analogous materials, or constructed by chemical intuition). Data mining (Curtarolo et al. 2003) is the pinnacle of this approach, as it very efficiently explores databases of known crystal structures and, using correlations between structures adopted by different compounds, indicates a list of likely candidate structures. Problems arise when a totally unexpected and hitherto unknown structure is actually stable (this often happens under pressure, or when the system does not have known good analogs). A number of simpler intuitive empirical schemes (e.g., structure diagrams, polyhedral clusters – see Urusov et al. 1990) have appeared in literature, but their application usually requires a large experimental data set or good understanding of the compound at hand.

Thanks to recent methodological developments, reliable structure prediction can be performed without any prior knowledge or assumptions about the system. Simulated annealing (Deem and Newsam 1989; Pannetier et al. 1990; Boisen et al. 1994; Schön and Jansen 1996), minima hopping (Gödecker 2004) and metadynamics (Martoňák et al. 2003, 2005, 2006) have been used with some success. For small systems, even relaxing randomly produced structures can deliver the stable structure (Pickard and Needs 2006). Here we review the evolutionary algorithm USPEX (Universal Structure Predictor: Evolutionary Xtallography) (Oganov and Glass 2006; Oganov et al. 2006; Glass et al. Hansen 2006) and a small selection of the results it has provided so far. This review is an updated version of the previous account of the methodology (Oganov et al. 2007) and is based on the lectures delivered at the 2009 Erice School on high-pressure crystallography, and at the 2009 MSA Short Course on computational mineral physics. The "*Evolutionary Algorithm USPEX*" section presents basics of the method, the "*Tests of the Algorithm*" section shows several interesting test cases (mostly on systems with a known ground state), while a number of applications to systems where the stable structure is unknown are presented in the "*Some Applications of the Method*" section.

EVOLUTIONARY ALGORITHM USPEX

Several groups attempted the pioneering use of evolutionary algorithms to structure prediction: for crystals (Bush et al. 1995; Woodley et al. 1999; Bazterra et al. 2002; Woodley 2004), colloids (Gottwald et al. 2005) and clusters (Deaven and Ho 1995). The algorithm developed by Deaven and Ho (1995) is perhaps especially interesting as some of its features (real-space representation of structures, local optimization and spatial heredity) are similar to the USPEX method. Their algorithm has successfully reproduced the structure of the C_{60} buckminsterfullerene, but has never been extended to heteroatomic clusters, nor to periodic systems (i.e., crystals). The algorithm of Bush and Woodley (Bush et al. 1995; Woodley et al. 1999; Woodley 2004) was originally developed for crystals and successfully produced a starting model for solving the structure of Li_3RuO_4 (Bush et al. 1995). However, subsequent systematic tests (Woodley 2004; Woodley et al. 1999) showed frequent failures even for rather simple systems containing ~10 atoms/cell. Other drawbacks are that this algorithm requires experimental lattice parameters and simulations are very expensive, unless a cheap and crude heuristic expression is used for fitness. Unlike the Deaven-Ho algorithm and USPEX, in this method structures are represented by binary "0/1" strings, there is no local optimization and no spatial heredity.

In USPEX, structures are represented by fractional coordinates for the atoms and lattice vectors. USPEX operates with populations of structures; from them, parent structures are selected. The fitness of structures is the relevant thermodynamic potential derived from *ab initio* total energy calculations. The worst structures of a population are discarded; for the remaining structures the probability of being selected as parent is a function (e.g., linear) of its fitness rank. A new candidate structure is produced from parent structures using one of three operators: (i) heredity, which combines spatially coherent slabs (in terms of fractional coordinates) of two parent structures, while the lattice vectors matrices are weighted averages of the two parent lattice vectors matrices, (ii) permutation (as in Woodley et al. 1999 and Woodley 2004), which swaps chemical identities in randomly selected pairs of unlike atoms, (iii) lattice mutation, which distorts the cell shape by applying a random symmetric strain matrix. To avoid pathological lattices, all newly produced structures are rescaled to produce a predefined unit cell volume (a reasonable starting value should be supplied in the input, and then allowed to evolve during the run). Heredity enables very broad searches, while preserving already found local fragments of good structures, and introduces ideas of "two-phase" simulations. Permutation facilitates finding the optimal ordering of the atoms; in some situations (for systems with a large range in degree of chemical similarity between different atom types) it may be useful to swap only chemically more similar atoms (e.g., Al-Si in aluminosilicates). Lattice mutation enables better exploration of the neighborhood of parent structures, prevents premature convergence of the lattice, and essentially incorporates the ideas of metadynamics in our search. The action of these variation operators is illustrated in Figures 1 and 2.

Before new candidate structures are relaxed, they are tested against three constraints:

(1) all interatomic distances must be above the specified minimal values;

(2) cell angles must be between 60° and 120°;

(3) all cell lengths must be larger than a specified value (e.g., diameter of the largest atom).

These constraints help to ensure stability of energy calculations and local optimization, and remove only redundant and infeasible regions of configuration space – thus the search is physically unconstrained. If in violation of these constraints, the candidate structure is discarded; otherwise, it is locally optimized (relaxed). Structure relaxations and energy calculations are done by external codes (currently, USPEX is interfaced with VASP (Kresse and Furthmüller 1996), SIESTA (Soler et al. 2002), GULP (Gale 2005)).

The relaxed structures are recorded and used for producing the next generation of structures. A new population of structures is made to contain one or more lowest-enthalpy structures from the previous population and the new structures produced using variation operators. Generation by generation, the above procedure is repeated in a loop.

The first generation usually consists of random structures, but it is possible to include user-specified structures. If lattice parameters are known, runs can be done in the fixed cell, but this is not required and in most cases simulations are done with variable cell shape. We have also improved the algorithm by more exhaustive removal of lattice redundancies (Oganov and Glass 2008). For more details on the USPEX method, see Oganov and Glass (2006) and Glass et al. (2006). A similar evolutionary algorithm was proposed slightly later and independently from us by Abraham and Probert (2006); this method differs from USPEX in the absence of permutation (with potential problems for binary and more complex compounds), different forms of heredity and mutation, and absence of cell rescaling. Recently, we also developed

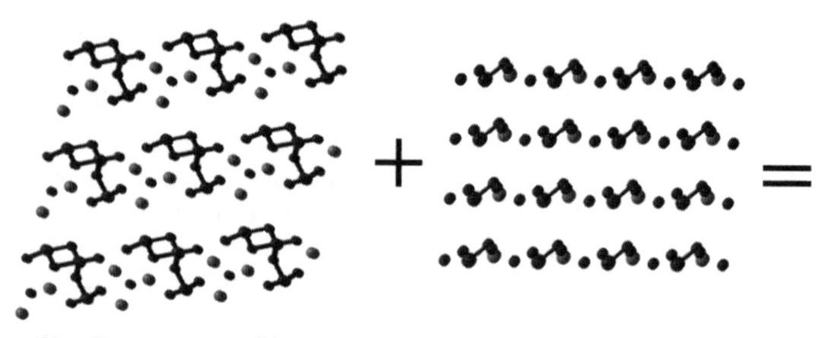

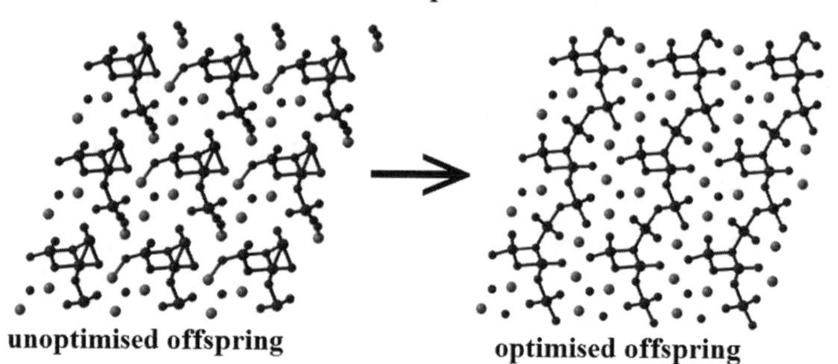

Figure 1. Heredity operator: slices of two parent structures, and the offspring structure before and after local optimization.

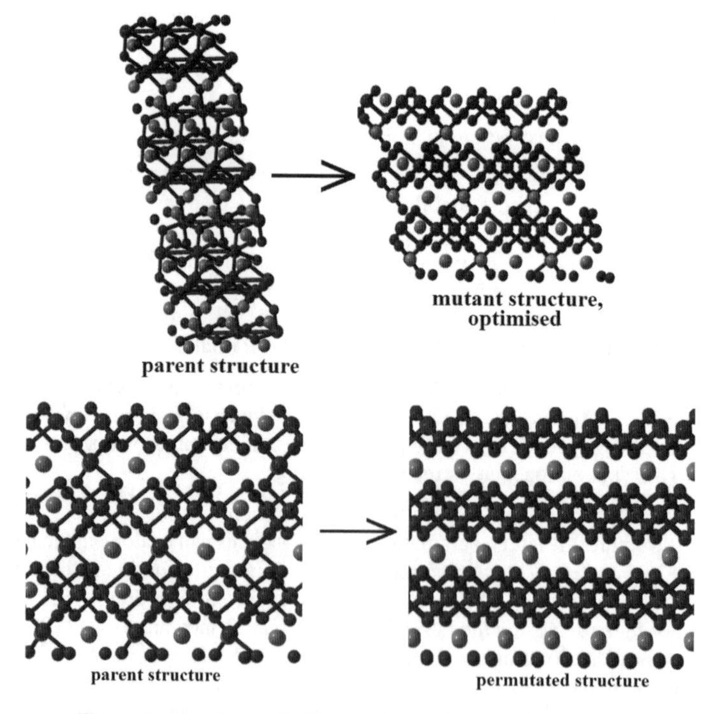

Figure 2. Illustrations of lattice mutation and permutation operators.

an approach, enabling deeper insight into the performance of structure prediction simulations (e.g., see below on similarity matrices) and into the energy landscape that is being sampled during the simulation (Valle and Oganov 2008; Oganov and Valle 2009).

Why is the USPEX methodology successful? One of the reasons is that local optimization creates chemically reasonable local environments for the atoms. Another reason is that evolutionary pressure (through selection) forces the population to improve from generation to generation. Yet another reason is the choice of variational operators. In heredity, local arrangements of atoms (spatially coherent pieces of structures) are partly preserved and combined. This respects the predominant short-ranged interactions in crystals and exploits information from the current population. For large systems it may be advantageous to combine slabs of several structures. On the other hand, for systems with very few atoms (or molecules) in the unit cell heredity becomes obsolete (in the limit of 1 atom/unit cell it is completely useless); these cases, however, are trivial for other variation operators and even for local optimization of random structures. As a general note, a successful evolutionary algorithm needs to maintain a balance between is "learning power" and maintaining diversity of the population. Figure 3 illustrates how, without any prior knowledge, a simulation of boron gradually "learned" about B_{12} icosahedra and arrived at the correct ground-state structure.

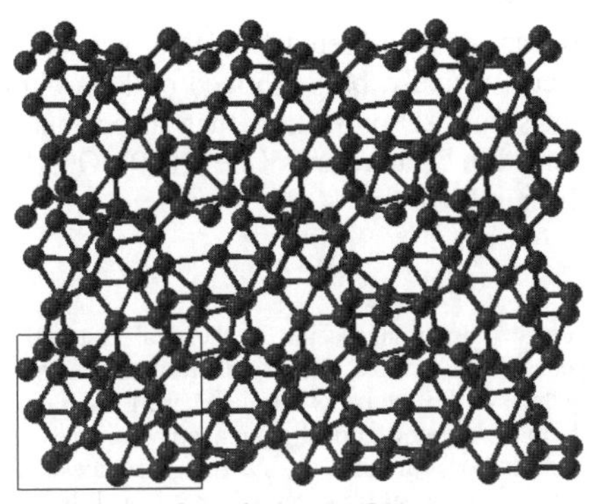

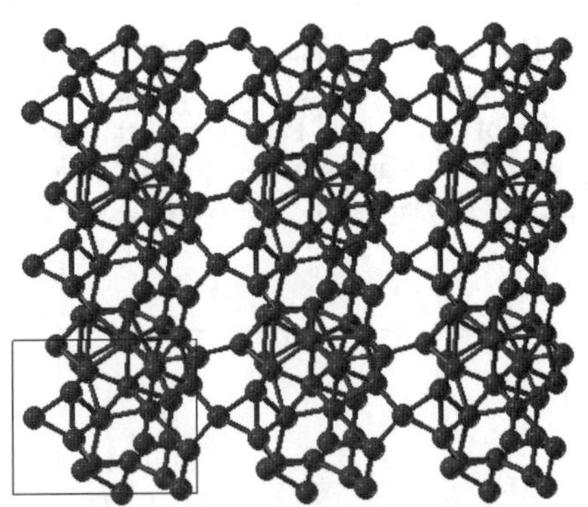

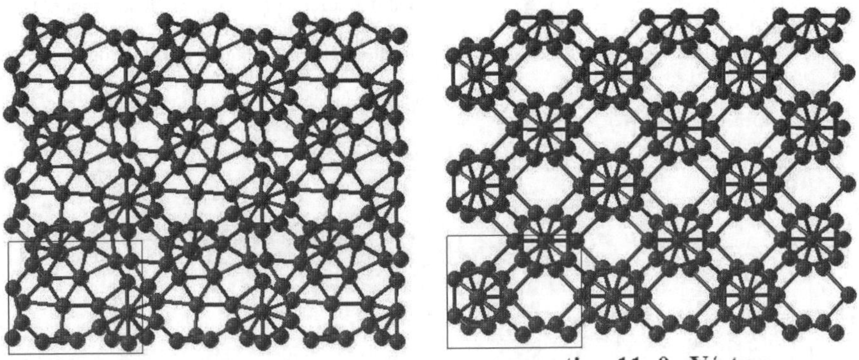

Figure 3. Illustration of an evolutionary search: 24 atoms of boron in a fixed cell. The best structure of the first random generation is 0.22 eV/atom above the ground state and is heavily disordered. In the second generation the best structure already contains an incomplete B_{12} icosahedron, the main building block of the ground-state structure. From Oganov et al. (2009).

Another important reason is that the energy landscapes expected in chemical systems are likely to have an overall "funnel" shape (Fig. 4a), where lowest-energy structures are clustered in the same region of configuration (or order parameter) space. In such cases, evolutionary algorithms are particularly powerful: they "zoom in" on the most promising region of configuration space until the global minimum is found. This "zooming in" is enabled by selection of lower-energy structures as parents for the subsequent generation, and by the form of the variational operators.

Actually, it is possible to test the assumption of an overall benign landscape shape using a recent approach (Oganov and Valle 2009) that enables mapping of energy landscapes. If the landscape has one funnel (like in Fig. 4a), there will be a direct correlation between the "distance" of all structures from the ground-state structure (this abstract "distance" measures the degree of structural dissimilarity) and the energy relative to the ground state – indeed, in many real systems (for example, GaAs with 8 atoms/cell – Fig. 4b) such a correlation is found. Even when more than one funnel is present, the number of funnels is usually small (up to three or four). Such situations arise when very different atomic arrangements are energetically competitive, and such systems are particularly challenging as the algorithm may tend to get stuck in one particular funnel. To avoid this, several tools can be used – including dense random or quasirandom sampling (to cover all funnels), tabu lists or special constraint techniques (to deal with each funnel, or a group of funnels, separately).

The energy-distance correlations (Fig. 4b) can be considered as 1D-projections of multidimensional energy landscapes. Projections can, actually, be performed on an arbitrary number of dimensions. Particular visual insight comes from 2D-projections that can be obtained by interpolating and smoothing the 2D-plots presented in Oganov and Valle (2009). One such depiction of a landscape (for Au_8Pd_4 system) is given in Figure 5.

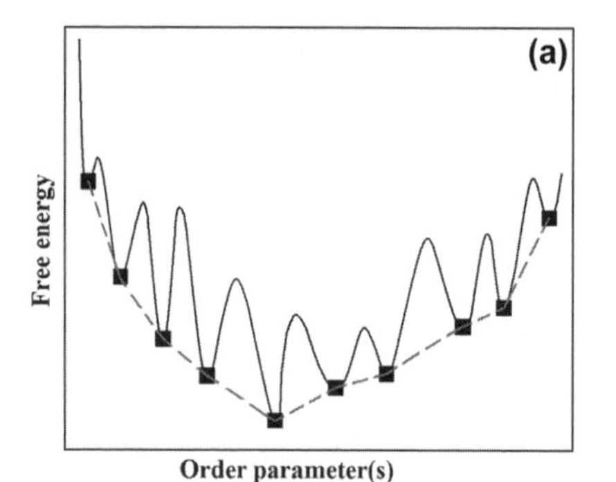

Figure 4. Energy landscapes in chemical systems. (a) A pedagogical cartoon. The original response surface is very "noisy" (i.e., contains very large energy variations, with high barriers). Local optimization reduces this surface to local minima points (black squares). The reduced response surface (dashed line) is well-behaved and has a simple overall shape. This is one of the reasons why the use of local optimization dramatically improves global optimization (Glass et al. 2006). From Oganov et al. (2007). (b) Energy-distance correlation for GaAs (8 atoms/cell). Each point is a locally optimized (i.e., relaxed) structure. The correlation proves that the energy landscape has a simple one-funneled topology. From Oganov and Valle (2009).

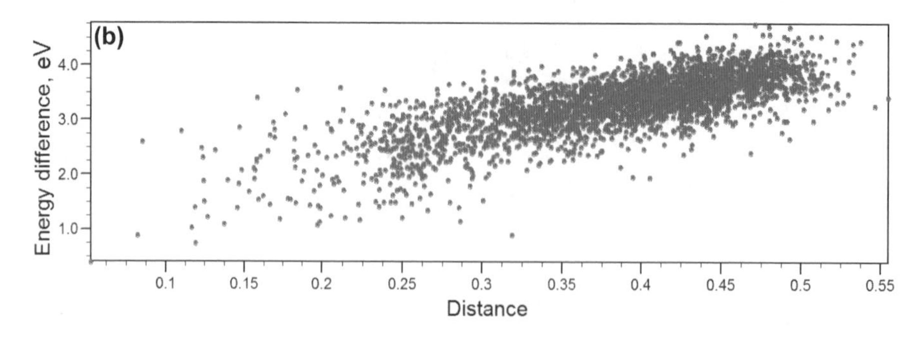

Figure 5. 2D-representation of the energy landscape of Au_8Pd_4 system using method presented in Oganov and Valle (2009). The surface has the same meaning as the dashed line in Figure 4a – it is an interpolation between the points of local minima. Clearly, there is one energy funnel (dark region), which corresponds to different Au-Pd orderings of the underlying fcc-structure.

The overall landscape shape (Figs. 4, 5) implies that, *en route* to the global minimum some of the low-energy metastable minima can be discovered. This is important, as such phases are often interesting as well. Furthermore, metastable structures found during evolutionary simulations provide a deep insight into the structural chemistry of the studied compound. Thus, evolutionary simulations provide three major results – 1) the ground-state structure; 2) a set of low-energy metastable structures; 3) detailed information on the chemical regime of the compound.

OF THE ALGORITHM

To measure the strengths and weaknesses of the algorithm, we consider several issues:

(1) efficiency of finding the global minimum, in particular relative to a simple well-defined search method, the random sampling,

(2) size of systems that can be studied in practice,

(3) how fast the diversity decreases along the evolutionary trajectory.

A number of successful tests have been reported in Oganov and Glass (2006, 2008), Glass et al. (2006), Martoňák et al. (2007), and Oganov et al. (2007). The largest successful test is for a Lennard-Jones crystal with 128 atoms in the (super)cell with variable-cell structure search, which has correctly identified hcp structure as the ground state within 3 generations (each consisting of only 10 structures). For larger Lennard-Jones systems (256 and 512 atoms/cell) we found an energetically very slightly less favorable fcc structure.

The largest test for a chemically complex system is the prediction of the structure of $MgSiO_3$ post-perovskite (Oganov and Ono 2004; Murakami et al. 2004) using a relatively large 80-atom supercell (with fixed supercell parameters) and an empirical potential (Murakami et al. 2004) describing interatomic interactions within a partially ionic model. Local optimization and energy calculations were done using the GULP code (Gale 2005). Previously in Martoňák et al. (2007), we have shown that already in a 40-atom supercell this test is unfeasible using the simple random sampling (with local optimization) (Pickard and Needs 2006): the correct structure was not produced even after 1.2×10^5 random attempts, but was found with 600-950 local optimizations of structures produced by USPEX. With 80 atoms/cell the problem becomes much more complicated (one expects an exponential increase of complexity with system size), but even in this case we correctly produced the post-perovskite structure in a reasonable number (~3200) of local optimizations – see Figure 6.

Figure 7 shows variable-cell *ab initio* results for $MgSiO_3$ at the pressure of 120 GPa. Several runs with somewhat different parameters (but within a reasonable range) have been

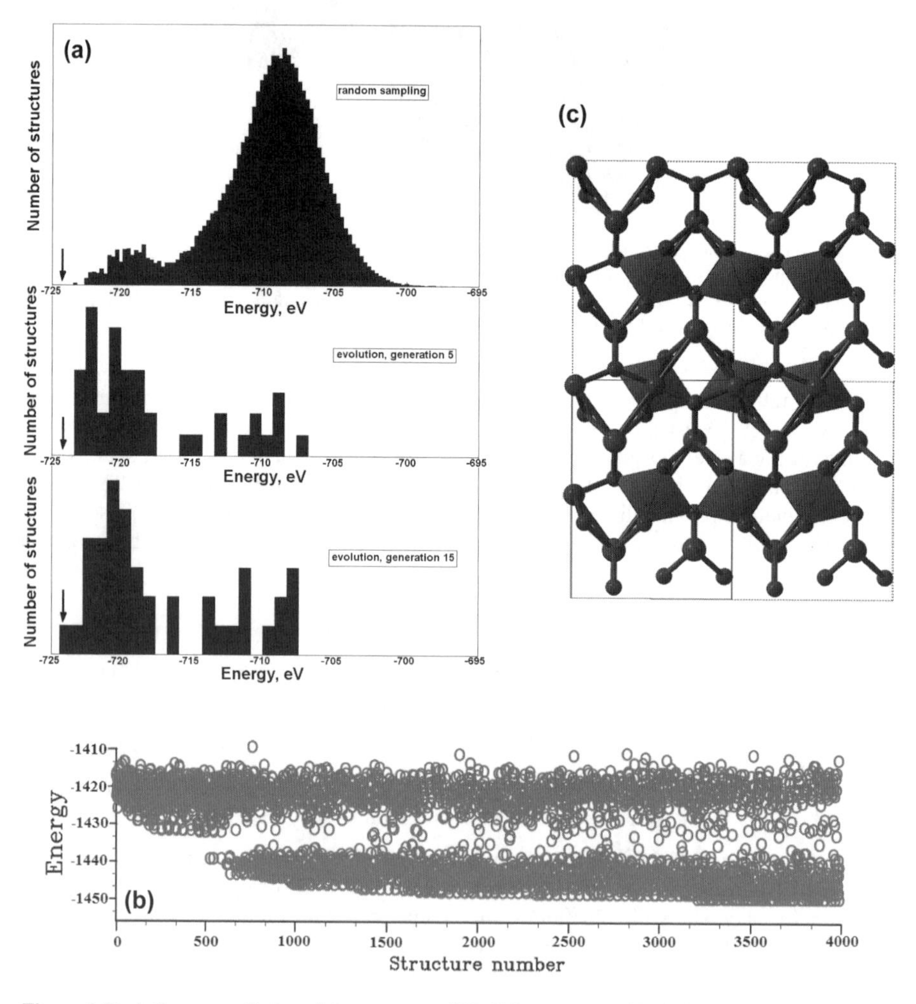

Figure 6. Evolutionary prediction of the structure of $MgSiO_3$ post-perovskite using the experimental cell parameters for a) 40-atom (Martoňák et al. 2007) and b) 80-atom supercells (Oganov and Glass 2008). In both cases, each generation consisted of 41 structures. (a) compares densities of states of optimized structures generated randomly (top) and in the evolutionary run. Random sampling did not find the correct structure within 1.2×10^5 steps, whereas in the evolutionary simulation shown it was found within 15 generations (i.e., 600 local optimizations). Arrows mark the ground-state energy. (b) shows the energies of structures along the evolutionary trajectory for the 80-atom run; (c) shows the structure of post-perovskite was obtained within ~3200 local optimizations. One can see that the density of low-energy structures increases during the simulation.

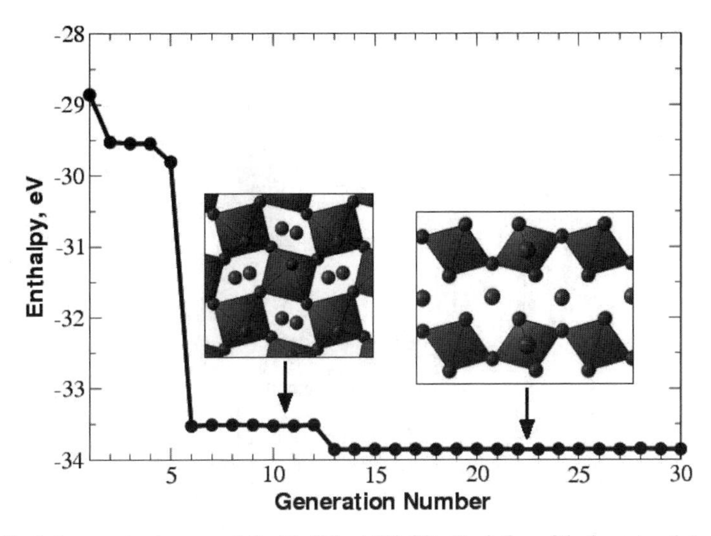

Figure 7. Evolutionary structure search for $MgSiO_3$ at 120 GPa. Evolution of the lowest enthalpy is shown as a function of the generation (insets show the structures of perovskite and post-perovskite phases), from Oganov and Glass (2006).

performed and all produced the correct ground-state structure of post-perovskite. The number of local optimizations performed before this structure was found ranged in different runs between 120 and 390; the longest run is shown in Figure 7.

An example of a very simple test, variable-cell *ab initio* structure search for GaAs with 8 atoms/cell (Oganov et al. 2007), is given in Figure 8. The ground-state structure for systems of such size can be found even by local optimization of a reasonable number of randomly produced structures. The density of states of relaxed random structures (Fig. 8a), obtained from 3000 random structures, has a characteristic multimodal shape, which seems to be a general feature of energy landscapes. The stable zincblende structure has the abundance of ~0.2%, i.e., finding it with random search would on average take ~ 500 local optimizations. In evolutionary simulations (Fig. 8b) it can be found within 3 generations, or just 30 structure relaxations. Similarity matrices for random (Fig. 8c) and evolutionary (Fig. 8d) searches clearly reveal a strong increase of structure similarity (i.e., decrease of diversity, which can be quantified using the approach of Valle and Oganov (2008) and Oganov and Valle (2009)) along the evolutionary run, after finding the global minimum. Even in this extreme case a significant number of dissimilar structures are produced long after the global minimum is found.

Au_8Pd_4 (12 atoms/cell) is an unusual system, where a number of different ordered decorations of the fcc structure have competitive energies. The ground state of this system is unknown, but was investigated in several computational studies (Curtarolo et al. 2005; Sluiter et al. 2006; Barabash et al. 2006; Oganov et al. 2007). Assuming that the ground-state structure should be an ordered variant of the cubic close-packed ("fcc") structure and using the cluster expansion technique with parameters calibrated on a set of *ab initio* energies, Barabash et al. (2006) suggested that there are two energetically nearly degenerate structures (Fig. 9c,d). Our calculations found a new ground-state structure (Fig. 9b) that has been overlooked by the previous cluster-expansion study (Barabash et al. 2006) and turned out to be ~0.1 meV/atom lower in energy than the previously known lowest-energy structures (Fig. 9c,d). Examination of all the produced structures shows that most of them are different ordering schemes of the fcc-structure and the energy differences are in most cases very small (Fig. 9a).

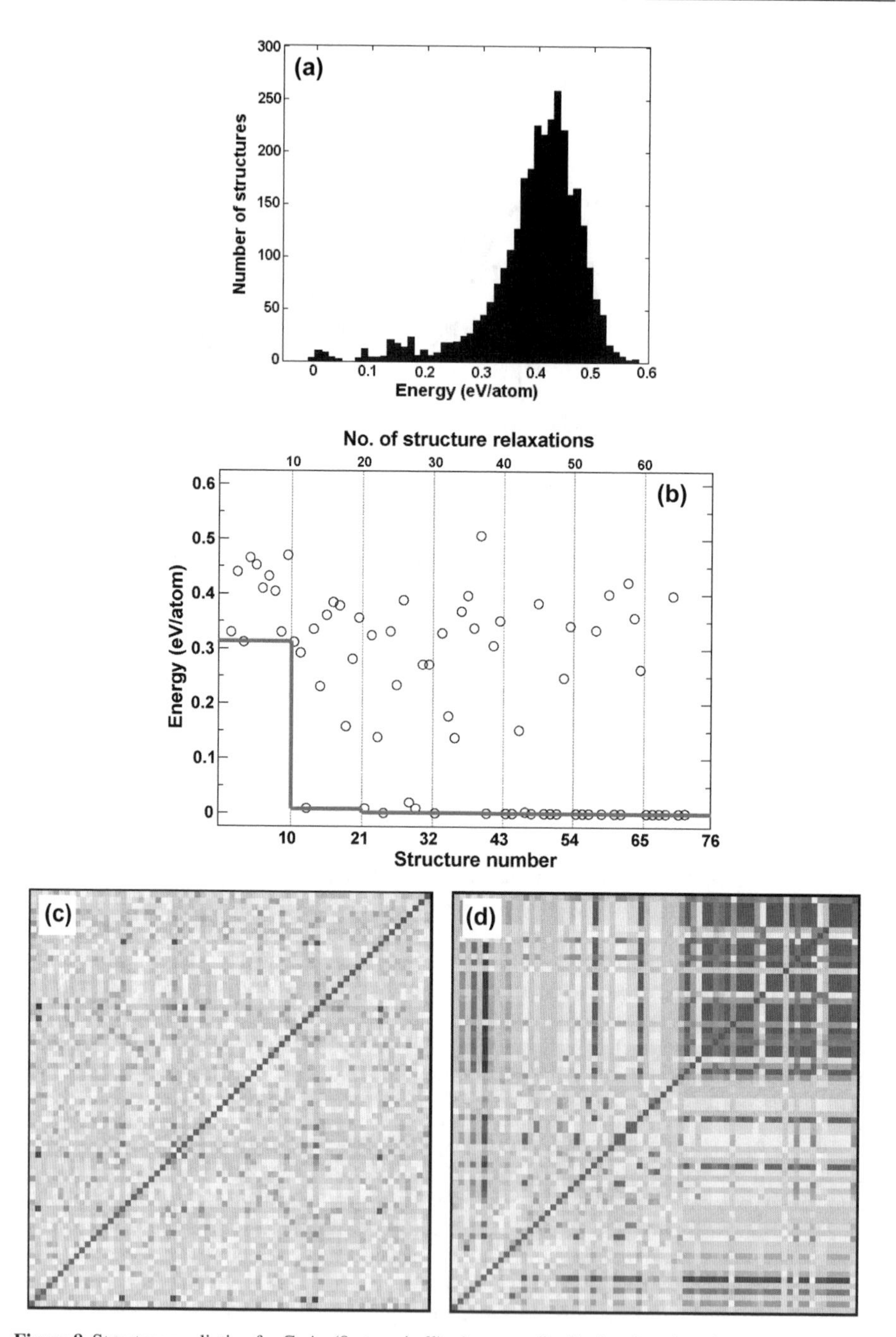

Figure 8. Structure prediction for GaAs (8 atoms/cell): a) energy distribution for relaxed random structures, b) progress of an evolutionary simulation (thin vertical lines show generations of structures, and the grey line shows the lowest energy as a function of generation), c-d) similarity matrices (dimension 70×70) for the random and evolutionary searches, respectively. All energies are relative to the ground-state structure. The evolutionary simulation used a population of 10 structures. Calculations are performed within the GGA (Perdew et al. 1996). From Oganov et al. (2007). Color online.

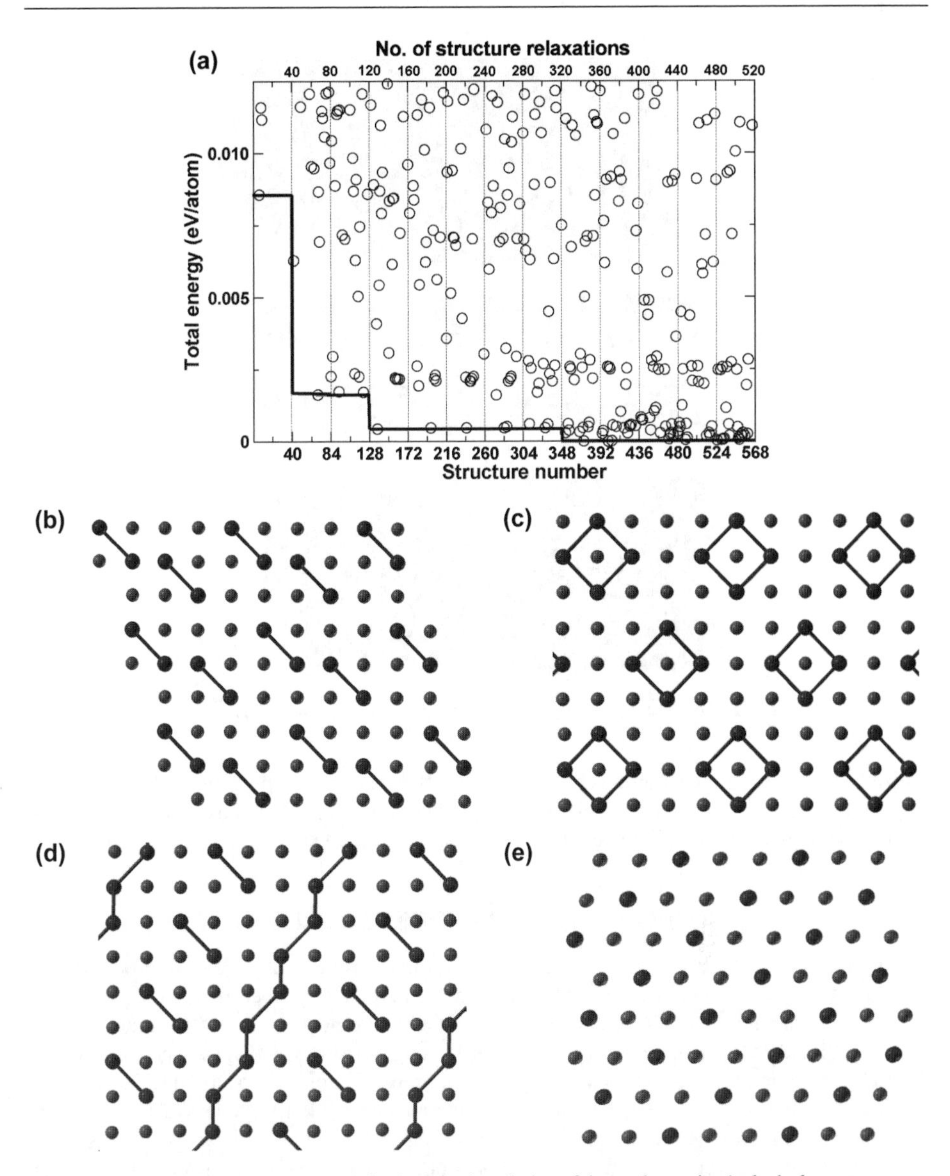

Figure 9. Evolutionary structure search for Au_8Pd_4. a) evolution of the total energies (only the lowest-energy part is shown for clarity), b) the lowest-energy structure found in our evolutionary simulation, c-d) the lowest-energy structures found by cluster expansion in Barabash et al. (2006), e) a suboptimal structure (for discussion, see Oganov et al. 2007, and references therein). Energies are given relative to the ground state.

Periodic boundary conditions suppress decomposition, but when a compound is extremely unstable against decomposition, phase separation can be observed in USPEX simulations. Actually, this happens rather frequently in explorations of hypothetical compositions. A clear example is given by the Cu-C system, which does not have any stable compounds. The tendency to unmixing in this system is very strong and even simulations on small cells show clear separation into layers of fcc-structured Cu and layers of graphite (Fig. 10). When the tendency to unmixing is not so large, simulations on small unit cells may find metastable

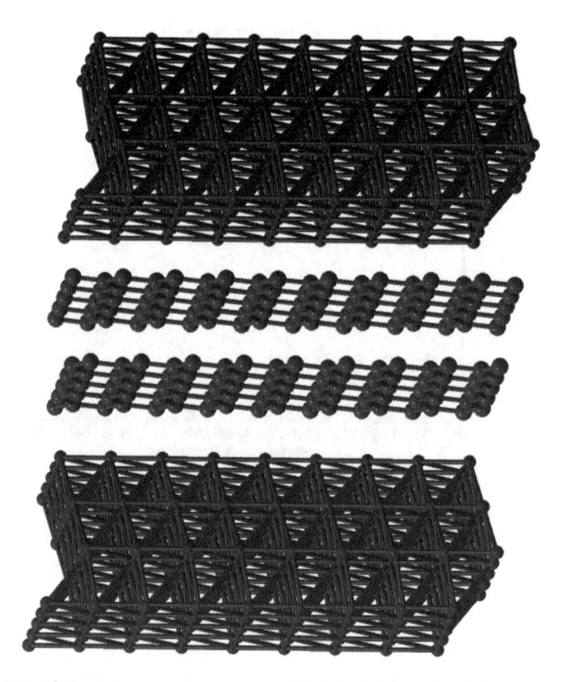

Figure 10. Lowest-energy structure of Cu_2C with 12 atoms/cell at 1 atm.

"mixed" structures. Such structures have the lowest thermodynamic potential only at the given number of atoms in the unit cell; increasing the cell size would lead to phase separation. In the Cu-C system, phase separation is evident already at very small system sizes (Fig. 10).

SOME APPLICATIONS OF THE METHOD

In this section we will review some new insight that has been obtained using our method (see also Oganov and Glass 2006). All structure predictions described here were performed within the generalized gradient approximation (GGA; Perdew et al. 1996) and the PAW method (Blöchl 1994; Kresse and Joubert 1999), using VASP code (Kresse and Furthmüller 1996) for local optimization and total energy calculations. The predicted structures correspond to the global minimum of the approximate free energy surface. For systems where the chosen level of approximation (GGA in cases considered below) is adequate, this corresponds to the experimentally observed structure. Where this is not the case, results of global optimization are invaluable for appraising the accuracy of the approximations.

$CaCO_3$ polymorphs

High-pressure behavior of carbonates is very important for the global geochemical carbon cycle, as high-pressure carbonates of Mg and Ca are expected to contain most of the Earth's oxidized carbon (Shcheka et al. 2006). For $CaCO_3$, there is a well-known transition from calcite to aragonite at ~2 GPa, followed by a transition to a post-aragonite phase at ~40 GPa (Ono et al. 2005b), the structure of which was solved (Oganov et al. 2006) using USPEX, and the predicted structure matched the experimental X-ray diffraction pattern well. Furthermore, Oganov et al. (2006) have predicted that above 137 GPa a new phase, with space group $C222_1$ and containing chains of carbonate tetrahedra, becomes stable. Recently this prediction was verified by experiments (Ono et al. 2007) at pressures above 130 GPa. We note that both post-aragonite and the $C222_1$ structure (Fig. 11) belong to new structure types and could not have been found by analogy with any known structures.

The presence of tetrahedral carbonate-ions at very high pressures invites an analogy with silicates, but the analogy is limited. In silicates, the intertetrahedral angle Si-O-Si is extremely flexible (Lasaga and Gibbs 1987), which is one of the reasons for the enormous diversity of silicate structure types. Figure 12 shows the variation of the energy as a function of the Si-O-Si angle in the model $H_6Si_2O_7$ molecule – method borrowed from Lasaga and Gibbs

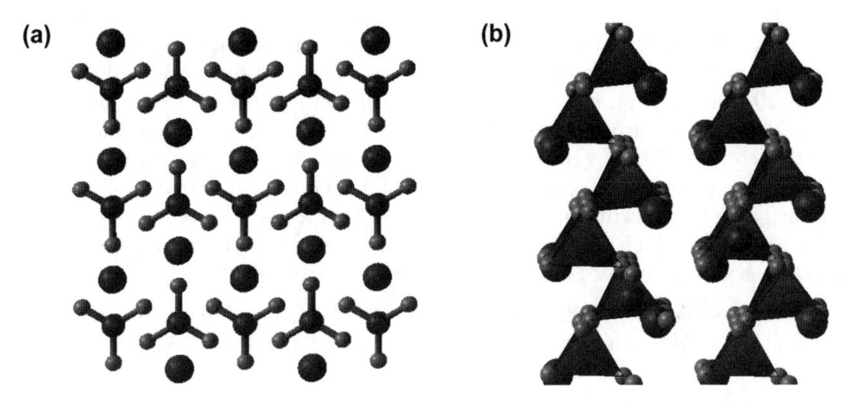

Figure 11. $CaCO_3$ at high pressure. a) structure of post-aragonite phase, b) $C222_1$ phase.

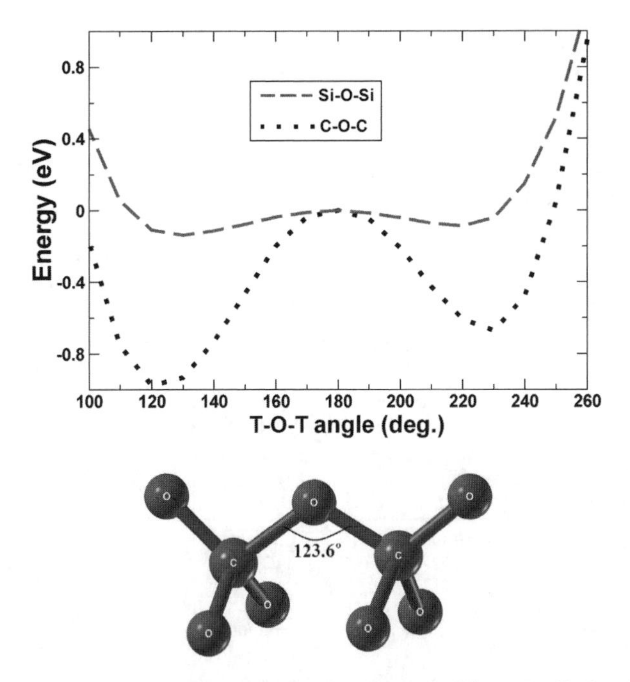

Figure 12. Energy variation as a function of the T-O-T angle (dashed line – T = Si, dotted line – T = C). Calculations were performed on $H_6T_2O_7$ molecules; at each angle all T-O distances and O-T-O valence angles were optimized. Optimum angle C-O-C=124°, Si-O-Si=135°. These calculations were performed with SIESTA code (Soler et al. 2002) using the GGA functional (Perdew et al. 1996), norm-conserving pseudopotentials and a double-ζ basis set with a single polarization function for each atom.

(1987). One can see only a shallow minimum at ∠(Si–O–Si) = 135°, but a deep minimum at ∠(C–O–C) = 124° with steep energy variations for $H_6C_2O_7$ (Fig. 12). This suggests a much more limited structural variety of metacarbonates, compared to silicates. In both $CaCO_3$ and CO_2 the ∠(C–O–C) angles are close to 124° in a wide pressure range.

Polymeric phase of CO_2

High-pressure behavior of CO_2 is still controversial (Bonev et al. 2003). It is known that above ~20 GPa a non-molecular phase (called phase V; Yoo et al. 1999) with tetrahedrally coordinated carbon atoms becomes stable, but its structure is still under debate: in the first experimental study (Yoo et al. 1999) a trydimite structure was proposed, but later theoretical works found it to be unstable (even not metastable) and much less favorable than the β-cristobalite structure (Dong et al. 2000; Holm et al. 2000). At the same time, it was not possible to rule out that there may be even more stable structures. We have performed evolutionary structure searches at 50 GPa, 100 GPa and 150 GPa for systems with 6, 9, 12, 18 and 24 atoms/cell (Oganov et al. 2007, 2008). At all these pressures we confirmed stability of the β-cristobalite structure (Figs. 13 and 14), thus suggesting an experimental re-investigation of phase V of carbon dioxide. CO_2-V is stable against decomposition into diamond and oxygen (the enthalpy of decomposition is very large and increases from 3.3 eV to 3.8 eV between 50 GPa and 200 GPa).

At lower pressures, between 8.9 GPa and 18.9 GPa, the $P4_2/mnm$ phase (see Bonev et al. 2003 for details) is stable, and at even lower pressures (0-8.9 GPa) the *Pa*3 structure is stable

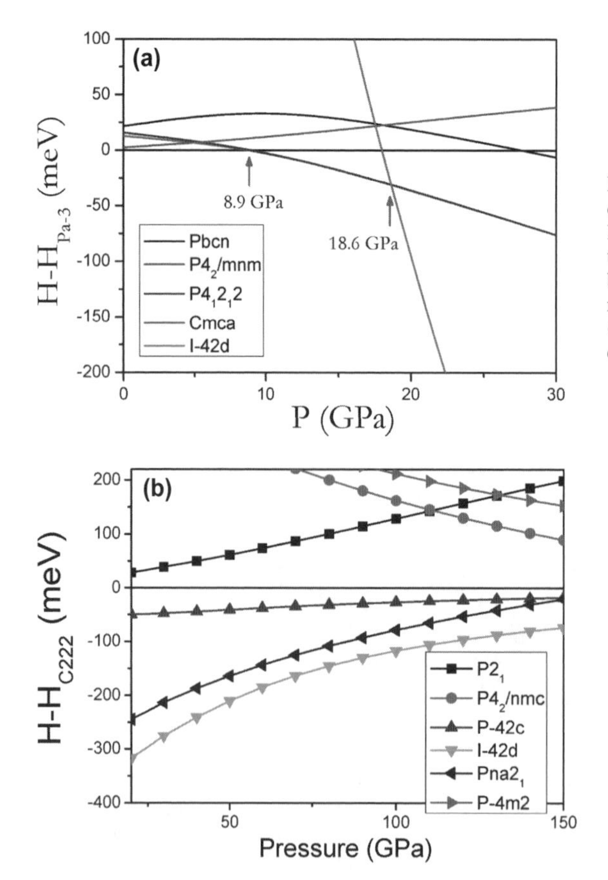

Figure 13. Enthalpies of candidate forms of CO_2: a) in the low-pressure region, relative to the molecular *Pa*3 structure, b) in the high-pressure region, relative to the non-molecular *C*222 structure. From Oganov et al. (2008).

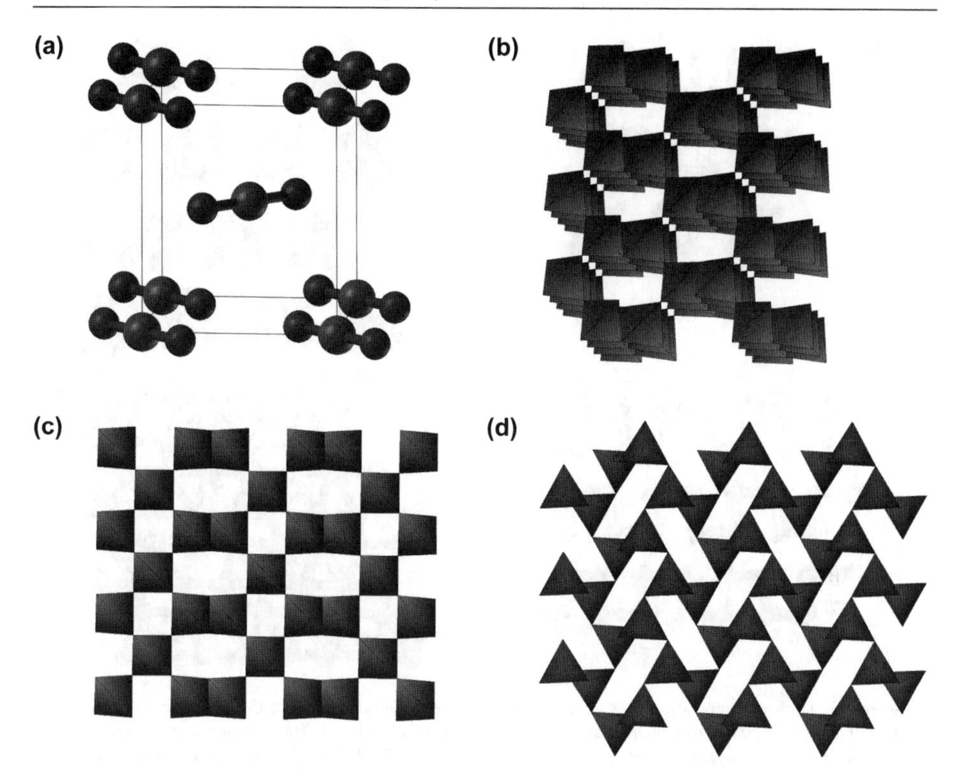

Figure 14. CO_2 structures: a) molecular $P4_2/mnm$ structure, stable at lower pressures than CO_2-V, b) polymeric β-cristobalite-type form of CO_2, suggested to be the structure of phase V and showing carbonate tetrahedra. Structural parameters at 100 GPa: space group $I\bar{4}2d$, $a = b = 3.2906$ Å, $c = 6.0349$ Å, C(0.5; 0; 0.25), O(0.2739; 0.25; 0.125), c) polymeric $C222$ structure, d) metastable polymeric $Pna2_1$ structure. From Oganov et al. (2008).

(Fig. 13). The $Pa3$-$P4_2/mnm$ transition pressure calculated here (8.9 GPa) is consistent with experiment and previous calculation (Bonev et al. 2003).

Semiconducting and metallic phases of solid oxygen: unusual molecular associations

The red ε-phase of oxygen, stable in the pressure range 8-96 GPa, was discovered in 1979 (Nicol et al. 1979), but its structure was solved only in 2006 (Fujihisa et al. 2006; Lundegaard et al. 2006). The metallic (superconducting at very low temperatures; Shimizu et al. 1998) ζ-phase, stable above 96 GPa, was discovered in 1995 (Akahama et al. 1995), and its structure remained controversial for a long time. Neutron diffraction showed that already in the ε-phase (at 8 GPa) there is no long-range magnetic order and likely even no local moments (Goncharenko 2005). The disappearance of magnetism is a consequence of increasing overlap of molecular orbitals with increasing pressure. Ultimately, orbital overlap leads to metallization. To understand high-pressure chemistry of oxygen, we performed extensive structure searches at pressures between 25 GPa and 500 GPa, taking into account only non-magnetic solutions (Oganov and Glass 2006; Ma et al. 2007).

At 25 GPa, we found two particularly interesting structures – one consisting of zigzag chains of O_2 molecules ($Cmcm$ structure of Neaton and Ashcroft 2002 and Oganov and Glass 2006; see Fig. 15b) and one with more complex chains of molecules (see Fig. 15c). These have strong similarities with the experimentally observed structure (Lundegaard et al. 2006; Fujihisa

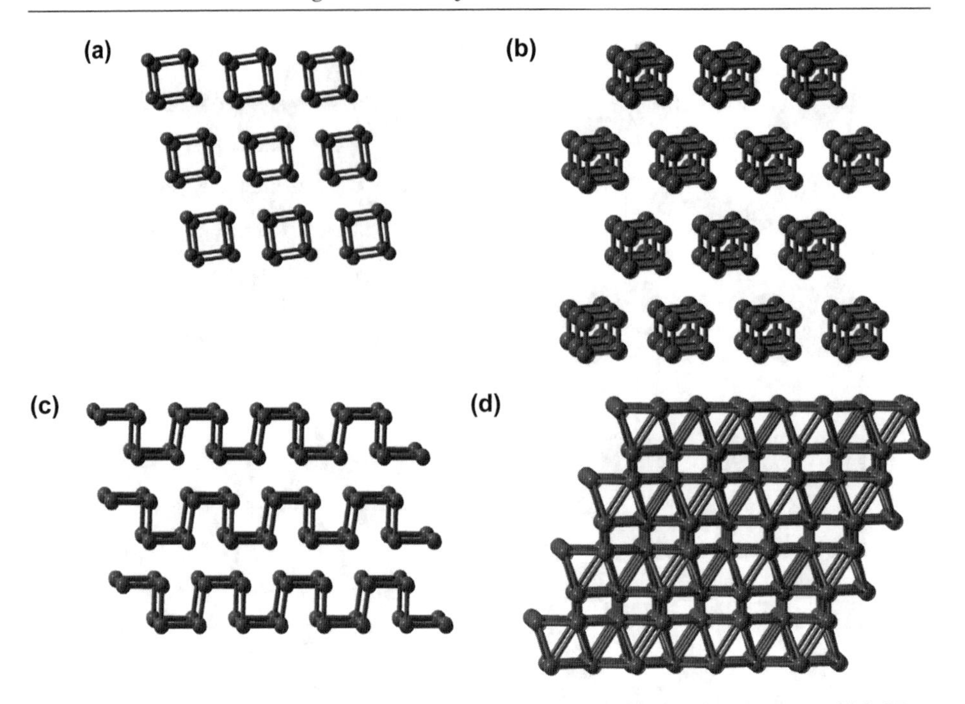

Figure 15. High-pressure structures of oxygen: a) experimentally found ε-O_8 structure at 17.5 GPa (Lundegaard et al. 2006), b) *Cmcm* chain structure (Neaton and Ashcroft 2002; Oganov and Glass 2006), c) metastable $P\bar{1}$ chain structure at 25 GPa (Oganov and Glass 2006), d) *C2/m* structure of the ζ-phase at 130 GPa (Ma et al. 2007). Contacts up to 2.2 Å are shown as bonds. From Oganov et al. (2007).

et al. 2006; see Fig. 15a) consisting of O_8 clusters: all of these structures are molecular, and in all of them each molecule is connected with two other molecules, at distances of ~2.1-2.2 Å (the intermolecular distance is ~1.2 Å). The *Cmcm* structure, first suggested in Neaton and Ashcroft (2002), is the true GGA ground state, but it differs from experiment; as Figure 16a shows, its enthalpy is ~10 meV/atom lower than for the experimentally found structure (Fig. 15a). Metastability of the experimentally studied structure cannot yet be ruled out, but it seems likely that this discrepancy is rather due to deficiencies of the GGA. Molecules in the $(O_2)_4$ clusters interact by weak intermolecular covalent bonds: each O_2 molecule has two unpaired electrons occupying two molecular π*-orbitals, and sharing these electrons with neighboring molecules creates two intermolecular bonds per molecule and a non-magnetic ground state (Ma et al. 2007; Stuedel and Wong 2007). It is well known that DFT-GGA does not perform well for stretched covalent bonds, the root of the problem being in the locality of the exchange-correlation hole in DFT-GGA, whereas the true exchange-correlation hole in such cases is highly delocalized. At high pressure, intermolecular distances decrease, intermolecular bonds become more similar to normal covalent bonds and the true exchange-correlation hole becomes more localized. Therefore, we can apply the GGA with greater confidence for the prediction of the structure of the metallic ζ-phase.

For the ζ-phase, evolutionary simulations at 130 GPa and 250 GPa uncovered two interesting structures with *C2/m* and *C2/c* space groups (Ma et al. 2007). These have very similar enthalpies (Fig. 16a); the *C2/m* structure is slightly lower in enthalpy and matches experimental X-ray diffraction and Raman spectroscopy data very well, better than the *C2/c* structure (Ma et al. 2007). Both structures contain well-defined O_2 molecules; our simulations show that oxygen remains a molecular solid at least up to 500 GPa. Phonon dispersion curves

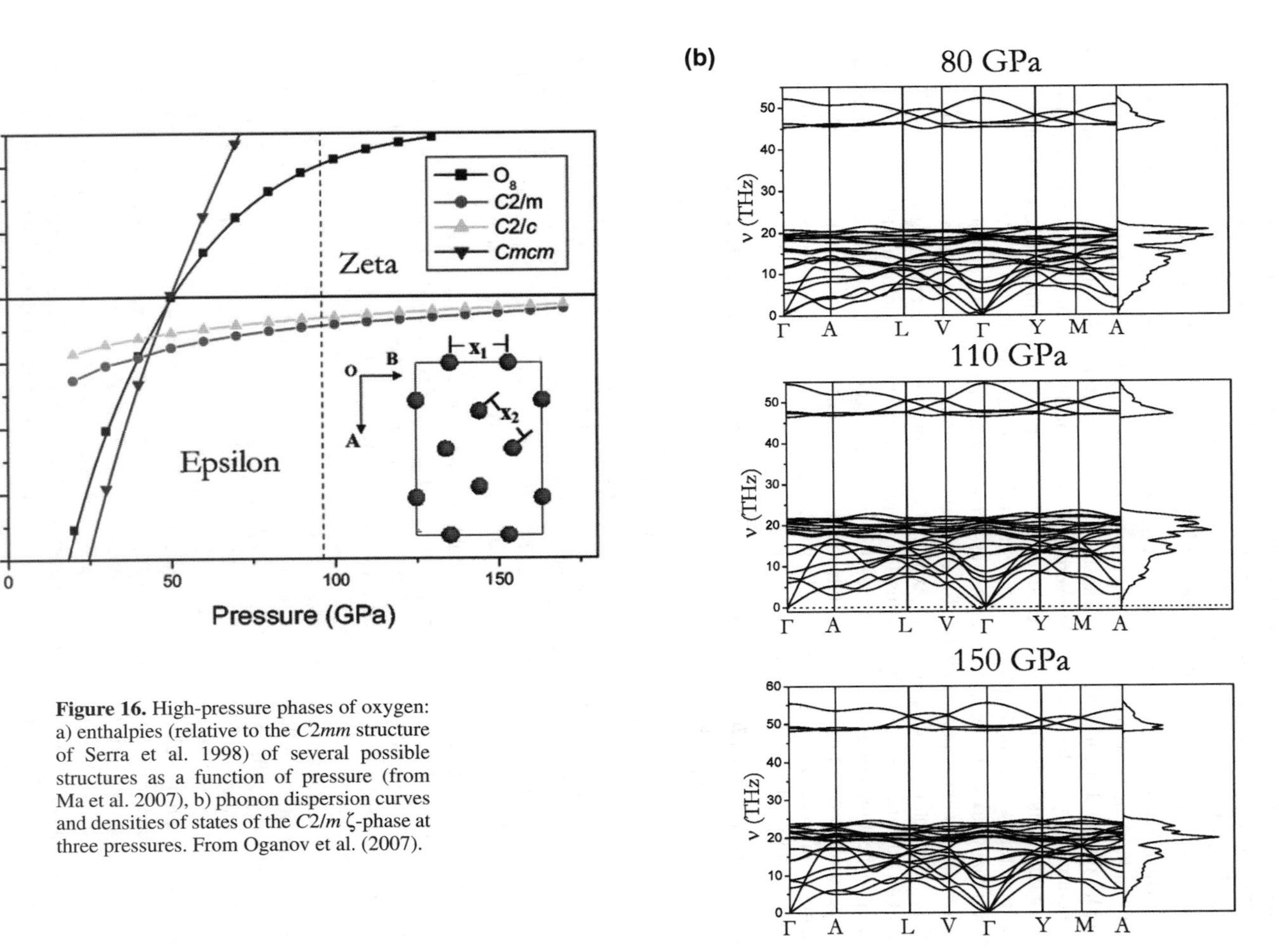

Figure 16. High-pressure phases of oxygen: a) enthalpies (relative to the *C2mm* structure of Serra et al. 1998) of several possible structures as a function of pressure (from Ma et al. 2007), b) phonon dispersion curves and densities of states of the *C2/m* ζ-phase at three pressures. From Oganov et al. (2007).

of the *C2/m* structure (Fig. 16b) contain clearly separated molecular vibrons and show that the structure is dynamically stable, except at 110 GPa, where we see tiny imaginary frequencies in the Γ-V direction, close to the Brillouin zone center. Such soft modes may result in small-amplitude long-wavelength modulations of the structure at very low temperatures.

The ε-ξ transition is isosymmetric, which implies that it is first-order at low temperatures but can become fully continuous above some critical temperature (Christy 1995). Given the small volume discontinuity upon transition and small hysteresis (one can obtain the *C2/m* structure of the ξ-phase by simple overcompression of the ε-O_8 structure, ~5 GPa above the thermodynamic transition pressure), one can expect this critical temperature to be rather low. We note that within the GGA the ε-ξ transition is predicted to occur at 45 GPa (Fig. 16a), much lower than the experimental transition pressure (96 GPa). This has two explanations – (i) as the GGA is expected to perform better for the metallic ζ-phase than for the semiconducting ε-O_8 phase, the enthalpy differences are expected to suffer from non-cancelling errors, (ii) since the ε-ξ transition is not only structural, but also electronic (insulator-metal transition), one might expect metallization at lower pressures than in experiment. Typically, density functional calculations overstabilize metallic states relative to insulating ones, and this is exactly what happens in oxygen. The predicted *C2/m* structure of the ξ-phase was very recently confirmed by single-crystal experiments (Weck et al. 2009).

Reactivity of noble gases: are Xe-C compounds possible at high pressure?

Inducing major changes in the electronic structure of atoms, high pressure may also change their reactivity. For instance, noble (i.e., largely unreactive) metal platinum under pressure easily forms carbide PtC (Oganov and Ono 2004; Ono et al. 2005a) and dinitride PtN_2 (Gregoryanz et al. 2004). One should not confuse chemical reactivity with propensity to phase transitions: so recently it was concluded that gold loses its "nobility" at 240 GPa, when it undergoes an fcc-hcp structural transition (Dubrovinsky et al. 2007). Structural transitions and reactivity are unrelated notions, however: in spite of becoming reactive, Pt does not change its fcc structure, and Cu (not a noble metal by any standards) is only known in one crystalline phase with the fcc structure.

An interesting question is whether noble gases become reactive. Indeed, it was observed that a few percent Xe can be incorporated in quartz (SiO_2) at elevated pressures and high temperatures (Sanloup et al. 2005). A possibility has been suggested by Grochala (2007) that stable Xe-C compounds may be stable at high pressure; indeed, carbon and xenon have similar valence orbital energies (cf. ionization potentials of 12.13 eV and 11.26 eV for Xe and C, respectively) and one expects that pressure would make Xe more reactive (Sanloup et al. 2005). We did simulations at 200 GPa, i.e., above the metallization pressure of Xe (132 GPa; Goettel et al. 1989), when its closed electronic shells are strongly perturbed. These calculations were done within the GGA (Perdew et al. 1996) and on cells containing up to 14 atoms/cell. At this pressure all Xe carbides are extremely unstable (Fig. 17) and their structures (Fig. 18) show clear separation into close-packed Xe layers (i.e., fragments of the elemental Xe structure) and 3,4-connected carbon layers (intermediate between graphite and diamond). The only exception is the 3D-clathrate structure of XeC_8. The observed layering is consistent with the instability to decomposition. Although Xe carbides are unstable at 200 GPa, already at that pressure we observe considerable bonding Xe-C interactions and the effect of Xe on the carbon sublattice is far beyond simple mechanistic size factor – the carbon layers adopt unusual and very interesting configurations that may be prepared in the laboratory under certain conditions.

Boron: novel phase with a partially ionic character

Boron is perhaps the most enigmatic element: at least 16 phases were reported in the literature, but most are believed or suspected to be compounds (rather than forms of the pure element), and until recently the phase diagram was unknown. A number of important results started with experimental findings of J. Chen and V. L. Solozhenko (both arrived independently

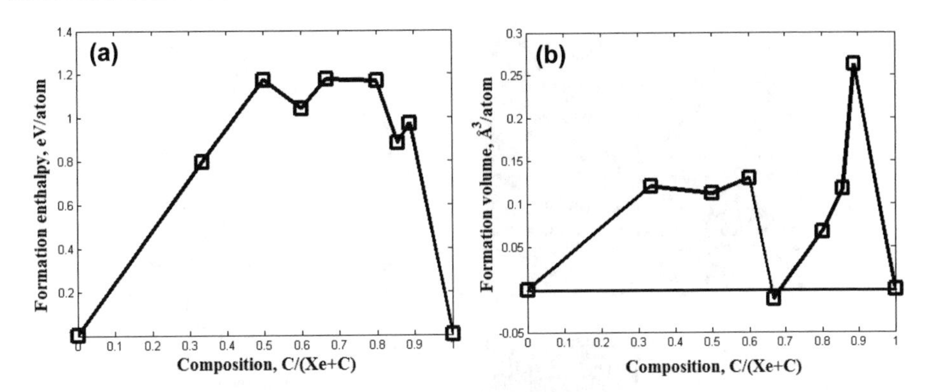

Figure 17. Predicted enthalpy (a) and volume (b) of formation of Xe-C compounds at 200 GPa. The compounds shown are Xe (hcp), Xe_2C, XeC, Xe_2C_3, XeC_2, XeC_4, XeC_6, XeC_8, C(diamond). Note that XeC_2 has a small negative volume of formation and might become stable at much higher pressures. From Oganov et al. (2007).

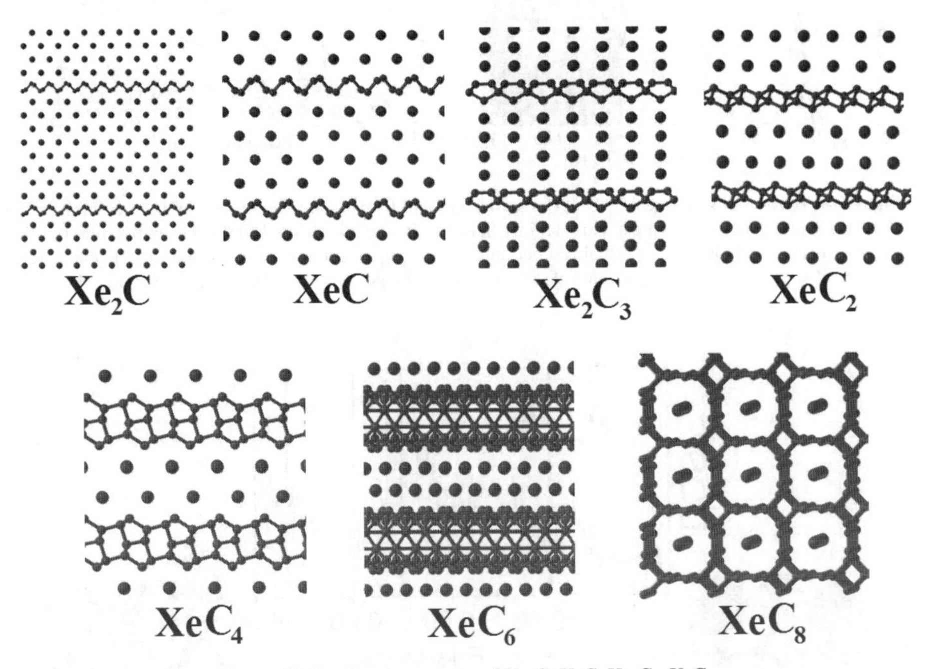

Figure 18. Predicted structures of Xe_2C, XeC, Xe_2C_3, XeC_2, XeC_4, XeC_6, XeC_8 at 200 GPa. From Oganov et al. (2007).

at the same conclusions in 2004) of a new phase at pressures above 10 GPa and temperatures of 1800-2400 K, though at first the structure of this new phase could not be determined from experimental data alone. We found the structure using USPEX and named this phase γ-B_{28} (because it contains 28 atoms/cell). Its structure has space group *Pnnm* and is comprised of icosahedral B_{12} clusters and B_2 pairs in a NaCl-type arrangement. This phase is stable between 19 and 89 GPa, and exhibits sizable charge transfer from B_2 pairs to B_{12} clusters, quite unexpected for a pure element. Details are given in Oganov et al. (2009) and Figures 19 and 20. Figure 21 shows a comparison of theoretical and experimental X-ray powder diffraction profiles.

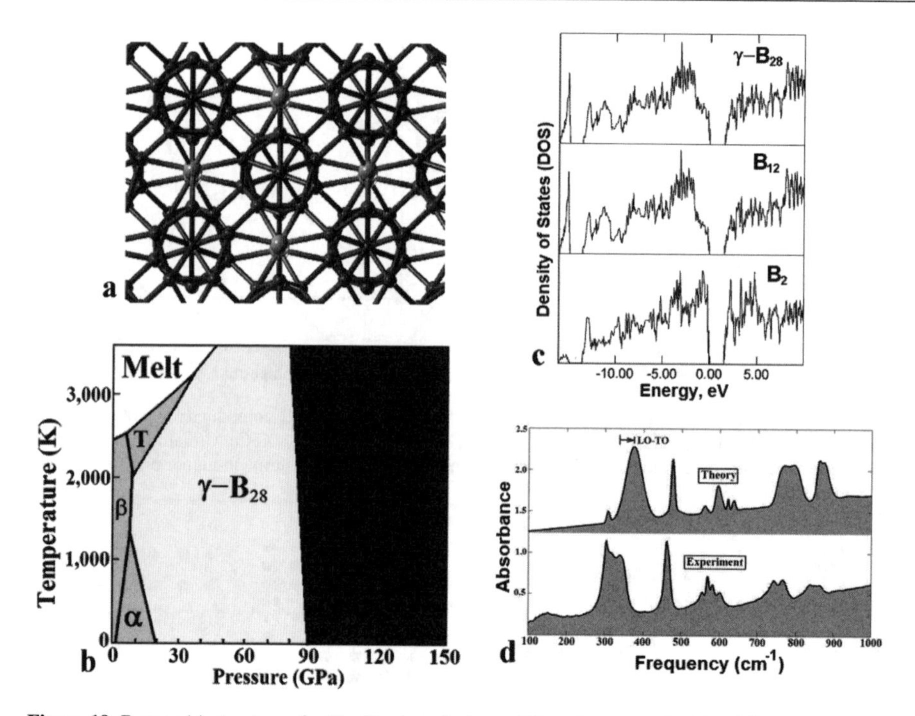

Figure 19. Boron: (a) structure of γ-B_{28} (B_{12} icosahedra and B_2 pairs are marked by different colors). (b) phase diagram of boron, showing a wide stability field of γ-B_{28}. (c) electronic DOS and its projections onto B_{12} and B_2 units (all DOSs are normalized per atom), (d) comparison of theoretical and experimental IR spectra. IR spectra indicate the presence of non-zero Born charges on atoms. From Oganov et al. 2009.

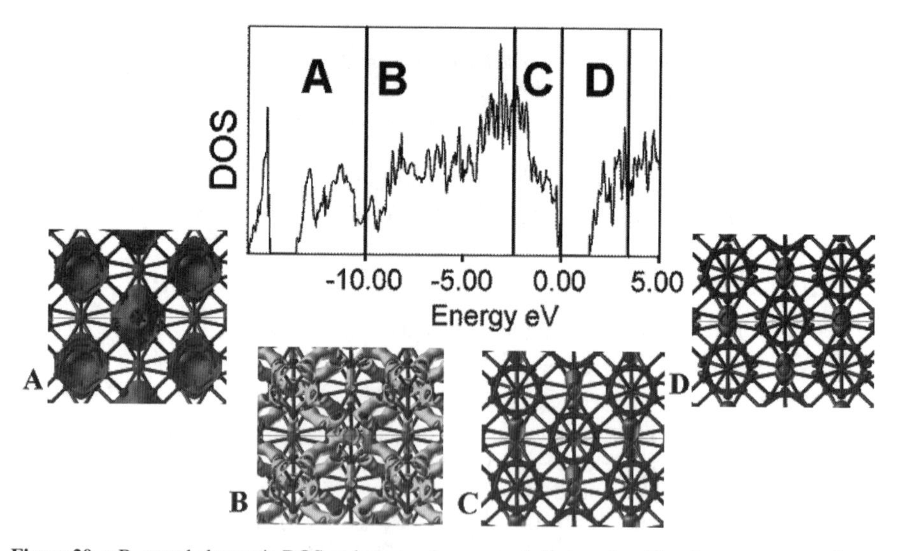

Figure 20. γ-B_{28}: total electronic DOS and energy-decomposed electron densities. Lowest-energy valence electrons are dominated by the B_{12} icosahedra, while top of the valence band and bottom of the conduction band (i.e., holes) are localized on the B_2 pairs. This is consistent with atom-projected DOSs (Fig. 19 c) and the idea of charge transfer $B_2 \rightarrow B_{12}$. Color online.

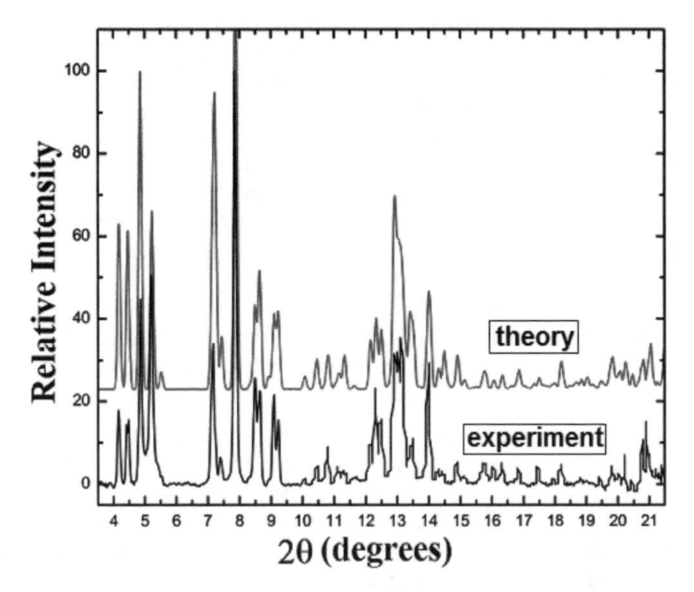

Figure 21. Comparison of theoretical and experimental X-ray powder diffraction profiles of γ-B_{28}. X-ray wavelength $\lambda = 0.31851$ Å. From Oganov et al. 2009.

γ-B_{28} can be represented as a "boron boride" $(B_2)^{\delta+}(B_{12})^{\delta-}$; although the exact value of the charge transfer δ depends on the definition of an atomic charge, for all definitions that we used the qualitative picture is the same. Perhaps the most reliable definition of a charge, due to Bader (1990), gives $\delta \sim 0.5$ (Oganov et al. 2009). Based on the similarity of synthesis conditions and many diffraction peaks, it seems likely that the same high-pressure boron phase may have been observed by Wentorf in 1965. However, material was generally not believed to be pure boron (due to the sensitivity of boron to impurities and lack of chemical analysis or structure determination in Wentorf 1965) and its diffraction pattern was deleted from Powder Diffraction File database. γ-B_{28} is structurally related to several compounds – for instance, B_6P (Amberger and Rauh 1974) or $B_{13}C_2$ (Kwei and Morosin 1996), where the two sublattices are occupied by different chemical species (instead of interstitial B_2 pairs there are P atoms or C-B-C groups, respectively). Significant charge transfer can be found in other elemental solids, and observations of dielectric dispersion (Tsagareishvili et al. 2009), equivalent to LO-TO splitting, suggest it for β-B_{106}. The nature of the effect is possibly similar to γ-B_{28}. Detailed microscopic understanding of charge transfer in β-B_{106} would require detailed knowledge of its structure, and reliable structural models of β-B_{106} finally begin to emerge from computational studies (van Setten et al. 2007; Widom and Mikhalkovic 2008; Ogitsu et al. 2009). It is worth mentioning that γ-B_{28} is a superhard phase, with a measured Vickers hardness of 50 GPa (Solozhenko et al. 2008), which puts it among half a dozen hardest materials known to date.

Sodium: a metal that goes transparent under pressure

A sequence of recent discoveries demonstrated that sodium, a simple s-element at normal conditions, behaves in highly non-trivial ways under pressure. The discovery of an incommensurate host-guest structure (Hanfland et al. 2002), followed by finding of several complex structures (Gregoryanz et al. 2008) in the range of pressures corresponding to the minimum of the melting curve (Gregoryanz et al. 2005), and the very existence of that extremely deep minimum in the melting curve at about 110 GPa – all this evidence points to some unusual changes in the physics of sodium. Later it was shown also that the incommensurate host-guest structure is a quasi-1D-metal (Lazicki et al. 2009), where conductivity is primarily due to chains formed by the guest sublattice. Yet another unusual phenomenon was predicted using USPEX and later (but within the same paper by Ma et al. 2009a) verified experimentally: on further compression sodium becomes a wide-gap insulator! This happens at ~190 GPa, and Figure 22 shows the crystal structure of the insulating "hP4" phase, its enthalpy relative to other structures, and the electronic structure. The structure can be described as a double hexagonal close-packed (dhcp) structure, squeezed more than twice along the *c*-axis, as a result of which sodium atoms have 6-fold coordination. There are 2 inequivalent Na positions: Na1 and Na2, which have the octahedral and trigonal-prismatic coordination, and the hP4 structure can be described as the elemental analog of the NiAs structure type (the same way as diamond is the elemental analog of the zincblende structure type). Calculations suggest that sodium is no longer an s-element; instead, its outermost valence electron has significant s-, p- and d-characters (Fig. 22c). Strongly compressed sodium can be considered as a transition metal, because of its significant d-character.

The band gap is direct, and increases with pressure. At 200 GPa the bandgap calculated with the GW approximation (known to give rather accurate results) is 1.3 eV, and increases to 6.5 eV at 600 GPa. These predictions implied that above 200 GPa sodium will be red and transparent, and at ~300 GPa it will become colorless and transparent (like wide-gap insulators). This has indeed been confirmed in experiments of M. I. Eremets (Ma et al. 2009a) as shown in Figure 23. The insulating behavior is explained by the extreme localization of the valence electrons in the interstices of the structure, i.e., the "empty" space (Fig. 24). These areas of localization are characterized by surprisingly high values of the electron localization function (nearly

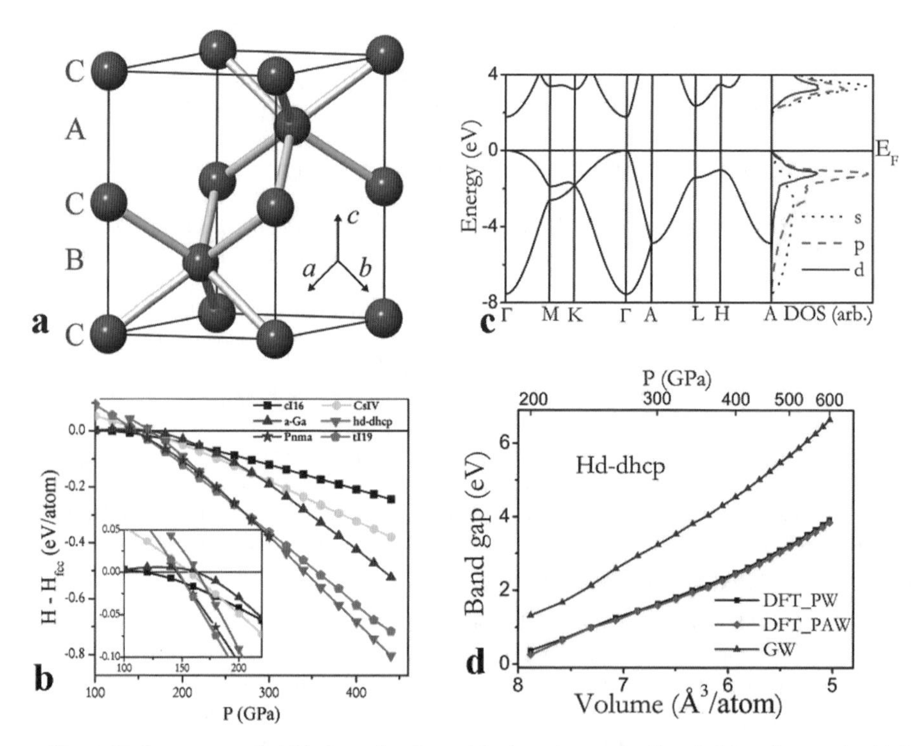

Figure 22. Summary on the hP4 phase of sodium: a) its crystal structure, b) enthalpies of competing high-pressure phases (relative to the fcc structure), c) band structure, d) pressure dependence of the band gap, indicating rapid increase of the band gap on compression. From Ma et al. (2009a).

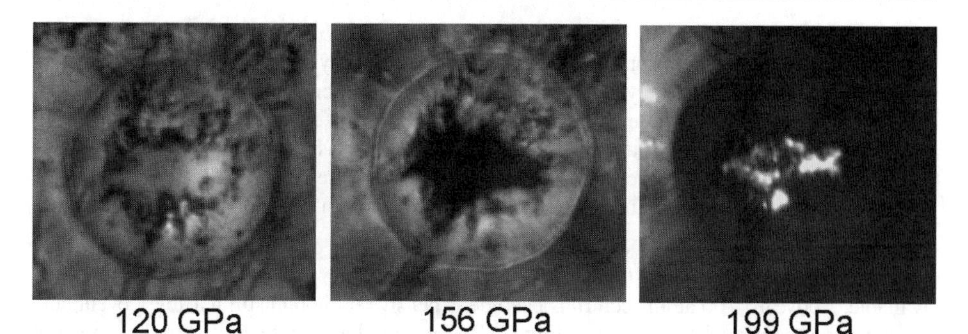

Figure 23. Photographs of sodium samples under pressure. At 120 GPa the sample is metallic and highly reflective, at 156 GPa the reflectivity is very low, and at 199 GPa the sample is transparent. From Ma et al. (2009a).

1.0) and maxima of the total electron density. The number of such maxima is half the number of sodium atoms, and therefore in a simple model we can consider Na atoms as completely ionized (Na^+), and interstitial maxima as containing one entire electron pair. The hP4 structure can also be described as a Ni_2In-type structure, where Na atoms occupy positions of Ni atoms, and interstitial electron pairs in hP4-Na sit on the same positions as In atoms in Ni_2In. At first counter intuitively, the degree of localization of the interstitial electron pairs increases with pressure, explaining the increase of the band gap (Fig. 22d). hP4-Na can be described as an electride, i.e., an ionic "compound" formed by ionic cores and localized interstitial electron pairs. The very fact that sodium, one of the best and simplest metals, under pressure becomes a transparent insulator with localized valence electrons, is remarkable and forces one to reconsider classical ideas of chemistry.

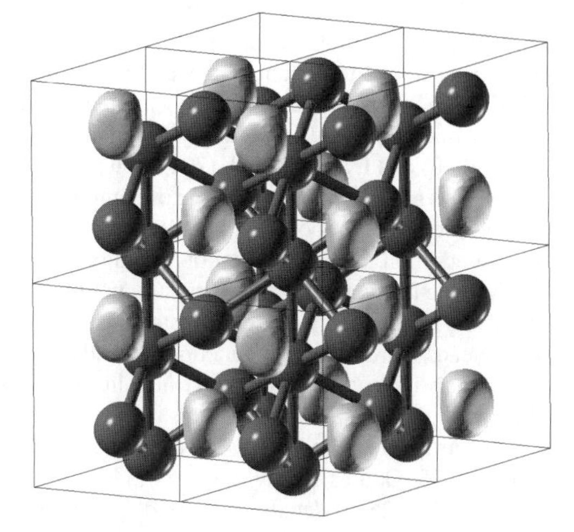

Figure 24. Crystal structure and electron localization function (isosurface contour 0.90) of the hP4 phase of sodium at 400 GPa. Interstitial electron localization is clearly seen.

Interstitial charge localization can be described in terms of (*s*)-*p*-*d* orbital hybridizations, and its origins are in the exclusionary effect of the ionic cores on valence electrons: valence electrons, feeling repulsion from the core electrons, are forced into the interstitial regions at pressures where atomic cores begin to overlap (Neaton and Ashcroft 1999).

CONCLUSIONS

Evolutionary algorithms, based on physically motivated forms of variation operators and local optimization, are a powerful tool enabling reliable and efficient prediction of stable crystal structures. This method has a wide field of applications in computational materials design (where experiments are time-consuming and expensive) and in studies of matter at extreme conditions (where experiments are very difficult or sometimes beyond the limits of feasibility).

One of the current limitations is the accuracy of today's *ab initio* simulations; this is particularly critical for strongly correlated and for van der Waals systems. Note, however, that the method itself does not make any assumptions about the way energies are calculated and can be used in conjunction with any method that is able to provide total energies. Most of practical calculations are done at $T = 0$ K, but temperature can be included as long as the free energy can be calculated efficiently. Difficult cases are aperiodic and disordered systems (for which only the lowest-energy periodic approximants and ordered structures can be predicted at this moment).

We are suggesting USPEX as the method of choice for crystal structure prediction of systems with up to ~50 atoms/cell, where no information (or just the lattice parameters) is available. Above 50-100 atoms/cell runs become expensive (although still feasible), eventually necessitating the use of other ideas within USPEX or another approach, due to the "curse of dimensionality." There is hope of enabling structure prediction for very large (> 200 atoms/cell) systems. The extension of the method to molecular systems (i.e., handling whole molecules, rather than individual atoms) is already available. The first successful step has been made in adapting USPEX to clusters (Schönborn et al. 2009). Similar extensions, relatively straightforward, still need to be done for surfaces and interfaces. One major unsolved problem is the simultaneous prediction of all stable stoichiometries and structures (in a given range of compositions). A pioneering study by Jóhannesson et al. (2002) succeeded in predicting stable stoichiometries of alloys within a given structure type, while an approach for simultaneous prediction of structure and stoichiometry was proposed by Wang and Oganov (2008) and implemented by Trimarchi et al. (2009) and Lyakhov et al. (2010). Here, the basic ideas are: (1) to start with a population (sparsely) sampling the whole range of compositions of interest, (2) allow variation operators to change chemical composition (we lift chemistry-preserving constraints in the heredity operator and, in addition to the permutation operator, introduce a "chemical transmutation" operator), (3) evaluate the quality of each structure not by its (free) energy, but the (free) energy per atom minus the (free) energy of the most stable isochemical mixture of already sampled compounds. This means that this fitness function depends on history of the simulation. Such an approach works surprisingly well. While Trimarchi et al. (2009) introduced a constraint that in each simulation the total number of atoms in the unit cell is fixed, our method has no such constraint, and this proves beneficial and very convenient. An example of a (very difficult) system is given in Figure 25. Odd as it may seem, a binary Lennard-Jones system with 1:2 ratio of radii (see caption to Fig. 25 for details of the model) exhibits a large number of ground states — including the exotic $A_{14}B$ compound and the well-known A_1B_2-type structure, and several marginally unstable compositions (such as A_8B_7, $A_{12}B_{11}$, A_6B_7, A_3B_4, AB_2). The efficiency and reliability of the variable-composition algorithm is illustrated by the fact that a fixed-composition simulation at AB_2 stoichiometry produced results (grey square in the upper graph of Fig. 25) perfectly consistent with the variable-composition runs.

USPEX has been applied to many important problems. Apart from the applications described above, several noteworthy results have been published by us recently. These include the high-pressure post-magnesite phases of $MgCO_3$ (Oganov et al. 2008), polymeric phases of nitrogen (Ma et al. 2009b), superconducting phases of silane (SiH_4) (Martinez-Canales et al. 2009) and germane (GeH_4) (Gao et al. 2008), the latter predicted to have a remarkably high T_C = 64 K (Gao et al. 2008). Its ability to predict not only the ground states, but also low-energy metastable structures has led to the finding of an interesting metastable structure of carbon (Oganov and Glass 2006), which has recently been shown by Li et al. (2009) to match the observed properties of the so-called "superhard graphite," a material scratching on diamond and formed by metastable room-temperature compression of graphite beyond 15 GPa (Mao et al. 2003). One expects many more applications to follow, both in high-pressure research and in materials design.

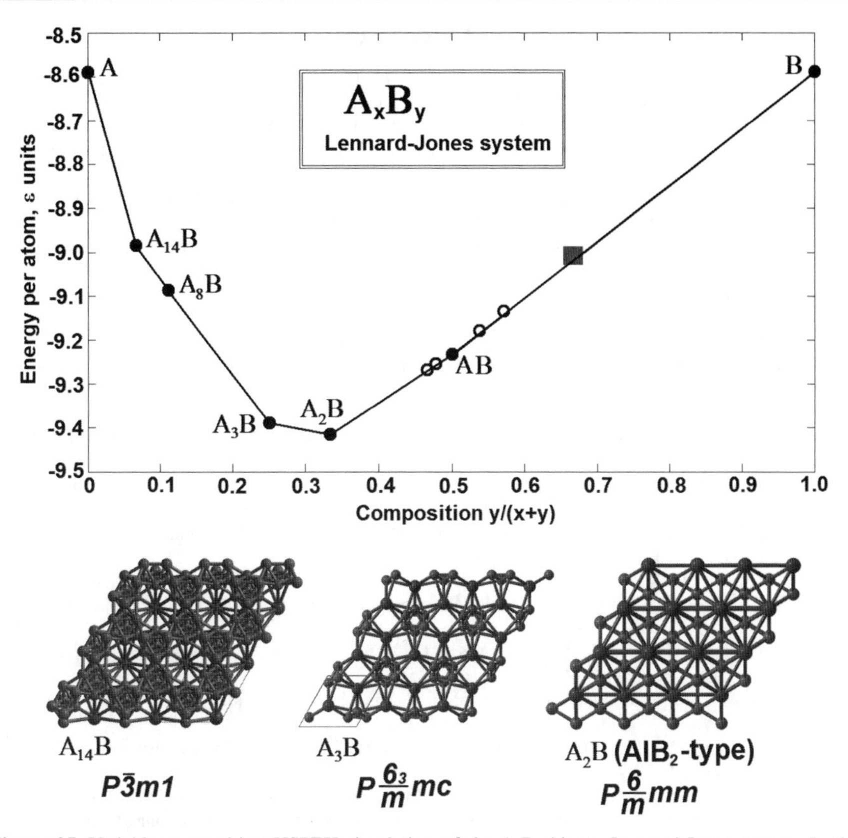

Figure 25. Variable-composition USPEX simulation of the A_xB_y binary Lennard-Jones system. In the upper panel: filled circles – stable compositions, open circles – marginally unstable compositions (A_8B_7, $A_{12}B_{11}$, A_6B_7, A_3B_4). Gray square – fixed-composition result for AB_2 stoichiometry, finding a marginally unstable composition in agreement with the variable-composition results. The lower panel shows some of the stable structures. While the ground state of the one-component Lennard-Jones crystal is hexagonal close packed (hcp) structure, ground states of the binary Lennard-Jones system are rather complex (e.g., $A_{14}B$). Note that the structure found for A_2B belongs to the well-known AlB_2 structure type. The potential is of the Lennard-Jones form for each atomic *ij*-pair: $U_{ij} = \varepsilon_{ij}[(R_{\min,ij}/R)^{12} - 2(R_{\min,ij}/R)^6]$, where $R_{min,ij}$ is the distance at which the potential reaches minimum, and ε is the depth of the minimum). In these simulations we use additive atomic dimensions: $R_{min,BB} = 1.5R_{min,AB} = 2R_{min,AA}$ and non-additive energies (to favor compound formation): $\varepsilon_{AB} = 1.25\varepsilon_{AA} = 1.25\varepsilon_{BB}$.

ACKNOWLEDGMENTS

ARO thanks R. Hoffmann, W. Grochala, R.J. Hemley and R.M. Hazen for exciting discussions. ARO also gratefully acknowledges financial support from the Research Foundation of Stony Brook University and from Intel Corporation. YM's work is supported by the China 973 Program under Grant No. 2005CB724400, the National Natural Science Foundation of China under grant No. 10874054, the NSAF of China under Grant No. 10676011, and the 2007 Cheung Kong Scholars Programme of China. We thank the Joint Supercomputer Center (Russian Academy of Sciences, Moscow) and Swiss National Supercomputing Centre (CSCS) (Manno) for providing supercomputer time. USPEX code is available on request from ARO.

REFERENCES

Abraham NL, Probert MIJ (2006) A periodic genetic algorithm with real-space representation for crystal structures and polymorph prediction. Phys Rev B 73, art. 224104

Akahama Y, Kawamura H, Hausermann D, Hanfland M, Shimomura O (1995) New high-pressure structural transition of oxygen at 96 GPa associated with metallization in a molecular solid. Phys Rev Lett 74:4690-4693

Amberger E, Rauh PA (1974) Structur des borreichen Borphosphids. Acta Crystallogr B 30:2549-2553

Bader (1990) Atoms in Molecules. A Quantum Theory. Oxford University Press: Oxford

Barabash SV, Blum V, Muller S, Zunger A (2006) Prediction of unusual stable ordered structures of Au-Pd alloys via a first-principles cluster expansion. Phys Rev B 74, art. 035108

Bazterra VE, Ferraro MB, Facelli JC (2002) Modified genetic algorithm to model crystal structures. I Benzene, naphthalene and anthracene. J Chem Phys 116:5984-5991

Blöchl PE (1994) Projector augmented-wave method. Phys Rev B 50:17953-17979

Boisen MB, Gibbs GV, Bukowinski MST (1994) Framework silica structures generated using simulated annealing with a potential energy function based on an $H_6Si_2O_7$ molecule. Phys Chem Mineral 21:269-284

Bonev SA, Gygi F, Ogitsu T, Galli G (2003) High-pressure molecular phases of solid carbon dioxide. Phys Rev Lett 91:065501

Bush TS, Catlow CRA, Battle PD (1995) Evolutionary programming techniques for predicting inorganic crystal structures. J Mater Chem 5:1269-1272

Christy AG (1995) Isosymmetric structural phase transitions: phenomenology and examples. Acta Crystallogr B51:753-757

Curtarolo S, Morgan D, Ceder G (2005) Accuracy of ab initio methods in predicting the crystal structures of metals: A review of 80 binary alloys. CALPHAD: Comput Coupling of Phase Diagrams and Thermochem. 29:163-211

Curtarolo S, Morgan D, Persson K, Rodgers J, Ceder G (2003) Predicting crystal structures with data mining of quantum calculations. Phys Rev Lett 91, art. 135503

Deaven DM, Ho KM (1995) Molecular geometry optimization with a genetic algorithm. Phys Rev Lett 75:288-291

Deem MW, Newsam, JM (1989) Determination of 4-connected framework crystal structures by simulated annealing. Nature 342:260-262

Dong JJ, Tomfohr JK, Sankey OF, Leinenweber K, Somayazulu M, McMillan PF (2000) Investigation of hardness in tetrahedrally bonded nonmolecular CO_2 solids by density-functional theory. Phys Rev B62:14685-14689

Dubrovinsky L, Dubrovinskaia N, Crichton WA, Mikhailushkin AS, Simak SI, Abrikosov IA, de Almeida JS, Ahuja R, Luo W, Johansson B (2007) Noblest of all metals is structurally unstable at high pressure. Phys Rev Lett 98:045503

Fujihisa H Akahama Y, Kawamura H, Ohishi Y, Shimomura O, Yamawaki H, Sakashita M, Gotoh Y, Takeya S, Honda K (2006) O_8 cluster structure of the epsilon phase of solid oxygen. Phys Rev Lett 97, art. 085503

Gale JD (2005) GULP: Capabilities and prospects. Z Kristallogr 220:552-554

Gao G, Oganov AR, Bergara A, Martinez-Canalez M, Cui T, Iitaka T, Ma Y, Zou G (2008) Superconducting high pressure phase of germane. Phys Rev Lett 101:107002

Glass CW, Oganov AR, Hansen N (2006) USPEX – evolutionary crystal structure prediction. Comput Phys Commun 175:713-720

Gödecker S (2004) Minima hopping: An efficient search method for the global minimum of the potential energy surface of complex molecular systems. J Chem Phys 120:9911–9917

Goettel KA, Eggert JH, Silvera IF, Moss WC (1989) Optical evidence for the metallization of xenon at 132(5) GPa. Phys Rev Lett 62:665-668

Goncharenko IN (2005) Evidence for a magnetic collapse in the epsilon phase of solid oxygen. Phys Rev Lett 94, art. 205701

Gottwald D, Kahl G, Likos CN (2005) Predicting equilibrium structures in freezing processes. J Chem Phys 122, art. 204503

Gregoryanz E, Degtyareva O, Somayazulu M, Hemley RJ, Mao HK (2005) Melting of dense sodium. Phys Rev Lett 94:185502.

Gregoryanz E, Lundegaard LF, McMahon MI, Guillaume C, Nelmes RJ, Mezouar M (2008) Structural diversity of sodium. Science 320:1054–1057

Gregoryanz E, Sanloup C, Somayazulu M, Badro J, Fiquet G, Mao HK, Hemley RJ (2004) Synthesis and characterization of a binary noble metal nitride. Nat Mater 3:294-297

Grochala W (2007) Atypical compounds of gases, which have been called 'noble'. Chem Soc Rev 36:1632-1655

Hanfland M, Syassen K, Loa I, Christensen NE, Novikov DL (2002) Na at megabar pressures. Poster at 2002 High Pressure Gordon Conference, June 23-28, 2002. Academy Meriden, NH
Holm B, Ahuja R, Belonoshko A, Johansson B (2000) Theoretical investigation of high pressure phases of carbon dioxide. Phys Rev Lett 85:1258-1261
Jóhannesson GH, Bligaard T, Ruban AV, Skriver HL, Jacobsen KW, Nørskov JK (2002) Combined electronic structure and evolutionary search approach to materials design. Phys Rev Lett 88, art. 255506
Kresse G, Furthmüller J (1996) Efficient iterative schemes for ab initio total-energy calculations using a plane wave basis set. Phys Rev B 54:11169-11186
Kresse G, Joubert D (1999) From ultrasoft pseudopotentials to the projector augmented-wave method. Phys Rev B 59:1758-1775
Kwei GH, Morosin B (1996) Structures of the boron-rich boron carbides from neutron powder diffraction: implications for the nature of the inter-icosahedral chains. J Phys Chem 100:8031-8039
Lasaga AC, Gibbs GV (1987) Applications of quantum-mechanical potential surfaces to mineral physics calculations. Phys Chem Minerals 14:107-117
Lazicki A, Goncharov AF, Struzhkin VV, Cohen RE, Liu Z, Gregoryanz E, Guillaume C, Mao HK, Hemley RJ (2009) Anomalous optical and electronic properties of dense sodium. Proc Natl Acad Sci 106:6525-6528
Li Q, Oganov AR, Wang H, Wang H, Xu Y, Cui T, Ma Y, Mao H-K, Zou G (2009) Superhard monoclinic polymorph of carbon. Phys Rev Lett 102:175506
Lundegaard LF, Weck G, McMahon MI, Desgreniers S, Loubeyre P (2006) Observation of an O_8 molecular lattice in the epsilon phase of solid oxygen. Nature 443:201-204
Lyakhov AO, Oganov AR, Wang Y, Ma Y (2010) Crystal structure prediction using evolutionary approach. *In:* Crystal Structure Prediction. Oganov AR (ed) Wiley-VCH, submitted
Ma Y, Eremets MI, Oganov AR, Xie Y, Trojan I, Medvedev S, Lyakhov AO, Valle M, Prakapenka V (2009a) Transparent dense sodium. Nature 458:182-185
Ma Y, Oganov AR, Xie Y, Li Z, Kotakoski J (2009b) Novel high pressure structures of polymeric nitrogen. Phys Rev Lett 102:065501
Ma Y-M, Oganov AR, Glass CW (2007) Structure of the metallic ζ-phase of oxygen and isosymmetric nature of the ε-ζ phase transition: Ab initio simulations. Phys Rev B 76, art. 064101
Mao WL, Mao HK, Eng PJ, Trainor TP, Newville M, Kao CC, Heinz DL, Shu J, Meng Y, Hemley RJ (2003) Bonding changes in compressed superhard graphite. Science 302:425-427
Martinez-Canales M, Oganov AR, Lyakhov A, Ma Y, Bergara A (2009) Novel structures of silane under pressure. Phys Rev Lett 102:087005
Martoňák R, Donadio D, Oganov AR, Parrinello M (2006) Crystal structure transformations in SiO_2 from classical and ab initio metadynamics. Nat Mater 5:623-626
Martoňák R, Laio A, Bernasconi M, Ceriani C, Raiteri P, Zipoli F, Parrinello M (2005) Simulation of structural phase transitions by metadynamics. Z Kristallogr 220:489–498
Martoňák R, Laio A, Parrinello M (2003) Predicting crystal structures: The Parrinello-Rahman method revisited. Phys Rev Lett 90:075503
Martoňák R, Oganov AR, Glass CW (2007) Crystal structure prediction and simulations of structural transformations: metadynamics and evolutionary algorithms. Phase Transitions 80:277-298
Murakami M, Hirose K, Kawamura K, Sata N, Ohishi Y (2004) Post-perovskite phase transition in $MgSiO_3$. Science 307:855-858
Neaton JB, Ashcroft NW (1999) Pairing in dense lithium. Nature 400:141-144
Neaton JB, Ashcroft NW (2002) Low-energy linear structures in dense oxygen: Implications for the epsilon phase. Phys Rev Lett 88:205503
Nicol M, Hirsch KR, Holzapfel WB (1979) Oxygen phase equilibria near 298 K. Chem Phys Lett 68:49-52
Oganov AR, Chen J, Gatti C, Ma Y-Z, Ma Y-M, Glass CW, Liu Z, Yu T, Kurakevych OO, Solozhenko VL (2009) Ionic high-pressure form of elemental boron. Nature 457:863-867
Oganov AR, Glass CW (2006) Crystal structure prediction using ab initio evolutionary techniques: principles and applications. J Chem Phys 124, art. 244704
Oganov AR, Glass CW (2008) Evolutionary crystal structure prediction as a tool in materials design. J Phys Condens Matter 20, art. 064210.
Oganov AR, Glass CW, Ono S (2006) High-pressure phases of $CaCO_3$: crystal structure prediction and experiment. Earth Planet Sci Lett 241:95-103
Oganov AR, Ma Y, Glass CW, Valle M (2007) Evolutionary crystal structure prediction: overview of the USPEX method and some of its applications. Psi-k Newsletter 84:142-171
Oganov AR, Ono S (2004) Theoretical and experimental evidence for a post-perovskite phase of $MgSiO_3$ in Earth's D" layer. Nature 430:445-448
Oganov AR, Ono S, Ma Y, Glass CW, Garcia A (2008) Novel high-pressure structures of $MgCO_3$, $CaCO_3$ and CO_2 and their role in the Earth's lower mantle. Earth Planet Sci Lett 273:38-47

Oganov AR, Valle M (2009) How to quantify energy landscapes of solids. J Chem Phys 130:104504

Ogitsu T, Gygi F, Reed J, Motome Y, Schwegler E, Galli G (2009) Imperfect Crystal and Unusual Semiconductor: Boron, a Frustrated Element. J Am Chem Soc 131:1903-1909

Ono S, Kikegawa T, Ohishi Y (2005a) A high-pressure and high-temperature synthesis of platinum carbide. Solid State Comm 133:55-59

Ono S, Kikegawa T, Ohishi Y (2007) High-pressure phase transition of $CaCO_3$. Am Mineral 92:1246-1249

Ono S, Kikegawa T, Ohishi Y, Tsuchiya J (2005b) Post-aragonite phase transformation in $CaCO_3$ at 40 GPa. Am Mineral 90:667-671

Pannetier J, Bassasalsina J, Rodriguez-Carva jal J, and Caignaert V (1990) Prediction of crystal structures from crystal chemistry rules by simulated annealing. Nature 346:343–345

Perdew JP, Burke K, Ernzerhof M (1996) Generalized gradient approximation made simple. Phys Rev Lett 77:3865-3868

Pickard CJ, Needs RJ (2006) High-pressure phases of silane. Phys Rev Lett 97, art. 045504.

Sanloup C, Schmidt BC, Perez EMC, Jambon A, Gregoryanz E, Mezouar M (2005) Retention of xenon in quartz and Earth's missing xenon. Science 310:1174-1177

Schön JC, Jansen M (1996) First step towards planning of syntheses in solid-state chemistry: Determination of promising structure candidates by global optimization. Angew Chem Int Ed 35:1287–1304

Schönborn S, Goedecker S, Roy S, Oganov AR (2009) The performance of minima hopping and evolutionary algorithms for cluster structure prediction. J Chem Phys 130:144108

Serra S, Chiarotti G, Scandolo S, Tosatti E (1998) Pressure-induced magnetic collapse and metallization of molecular oxygen: The ζ-O_2 phase. Phys Rev Lett 80:5160-5163

Shcheka SS, Wiedenbeck M, Frost DJ, Keppler H (2006) Carbon solubility in mantle minerals. Earth Planet Sci Lett 245:730-742

Shimizu K, Suhara K, Ikumo M, Eremets MI, Amaya K (1998) Superconductivity in oxygen. Nature 393:767-769

Sluiter MHF, Colinet C, Pasturel A (2006) Ab initio calculation of phase stability in Au-Pd and Ag-Pt alloys. Phys Rev B73:174204

Soler JM, Artacho E, Gale JD, Garcia A, Junquera J, Ordejon P, Sanchez-Portal D (2002) The SIESTA method for ab initio order-N materials simulation. J Phys Condens Matter 14:2745-2779

Solozhenko VL, Kurakevych OO, Oganov AR (2008) On the hardness of a new boron phase, orthorhombic γ-B_{28}. J Superhard Mater 30:428-429

Steudel R, Wong MW (2007) Dark-red O_8 molecules in solid oxygen: rhomboid clusters, not S_8-like rings. Angew Chem Int Ed 46:1768-1771

Trimarchi G, Freeman AJ, Zunger A (2009). Predicting stable stoichiometries of compounds via evolutionary global space-group optimization. Phys Rev B 80:092101

Tsagareishvili OA, Chkhartishvili LS, Gabunia DL (2009) Apparent low-frequency charge capacitance of semiconducting boron. Semiconductors 43:14-20

Urusov VS, Dubrovinskaya NA, Dubrovinsky LS (1990) Generation of likely crystal structures of minerals. Moscow State University Press, Moscow

Valle M, Oganov AR (2008) Crystal structure classifier for an evolutionary algorithm structure predictor. IEEE Symposium on Visual Analytics Science and Technology (October 21 - 23, Columbus, Ohio, USA), p 11- 18

van Setten MJ, Uijttewaal MA, de Wijs GA, de Groot RA (2007) Thermodynamic stability of boron: The role of defects and zero point motion. J Am Chem Soc 129:2458-2465

Wang Y, Oganov AR (2008) Research on the evolutionary prediction of very complex crystal structures. IEEE Computational Intelligence Society Walter Karplus. Summer Research Grant 2008 Final Report. *http://www.ieee-cis.org/_files/EAC_Research_2008_Report_WangYanchao.pdf*

Weck G, Desgreniers S, Loubeyre P, Mezouar M (2009) Single-crystal structural characterization of the metallic phase of oxygen. Phys Rev Lett 102:255503

Wentorf RH (1965) Boron: another form. Science 147:49-50

Widom M, Mikhalkovic M (2008) Symmetry-broken crystal structure of elemental boron at low temperature. Phys Rev B77:064113

Woodley SM (2004) Prediction of crystal structures using evolutionary algorithms and related techniques. Struct Bond 110:95-132

Woodley SM, Battle PD, Gale JD, Catlow CRA (1999) The prediction of inorganic crystal structures using a genetic algorithm and energy minimization. Phys Chem Chem Phys 1:2535-2542

Yoo CS, Cynn H, Gygi F, Galli G, Iota V, Nicol M, Carlson S, Hausermann D, Mailhiot C (1999) Crystal structure of carbon dioxide at high pressure: "Superhard" polymeric carbon dioxide. Phys Rev Lett 83:5527-5530

Reviews in Mineralogy & Geochemistry
Vol. 71 pp. 299-314, 2010

Multi-Mbar Phase Transitions in Minerals

Koichiro Umemoto

Department of Geology and Geophysics
University of Minnesota
Minneapolis, Minnesota, 55455, U.S.A.

umemoto@cems.umn.edu

Renata M. Wentzcovitch

Department of Chemical Engineering and Materials Sciences and
Minnesota Supercomputer Institute, University of Minnesota
Minneapolis, Minnesota, 55455, U.S.A.

wentzcov@cems.umn.edu

INTRODUCTION

$MgSiO_3$ perovskite is the most abundant mineral in the Earth's lower mantle. A structural phase transition in this phase to a $CaIrO_3$-type polymorph was discovered in 2004 (Murakami et al. 2004; Oganov and Ono 2004; Tsuchiya et al. 2004). This new polymorph, the so-called post-perovskite (PPV) phase, was produced at pressures and temperatures close to those expected at the core-mantle boundary, 125 GPa and 2,500 K (Murakami et al. 2004). In the Earth, the PPV phase is the final form of $MgSiO_3$. This surprising discovery invited a new question: what is the next polymorph of $MgSiO_3$? $MgSiO_3$ PPV consists of SiO_3 layers intercalated by magnesium (Fig. 1). Therefore, it is natural to expect still other pressure induced transitions to more isotropic close-packed looking structures. This question has acquired further importance since the discovery of terrestrial-type exoplanets: the Super-Earth planet with ~7 Earth masses, GJ876d (Rivera et al. 2005), and the Saturn-like planet with a massive dense core with ~67 Earth masses, D149026b (a dense-Saturn) (Sato et al. 2005). Many others have been found since then. Pressures and temperatures in the mantle of these planets are much higher than in the Earth. There is also a pressing need to understand and model matter in the core of the giants, particularly the solar ones, Jupiter, Saturn, Uranus, and Neptune. In GJ876d, pressure and temperature at its core-mantle boundary was roughly estimated to be ~1 TPa (10 Mbar) and ~4,000 K (Valencia et al. 2006). The gas giants, Jupiter and Saturn, and the icy giants, Uranus and Neptune, have small dense cores surrounded by hydrogen/helium and ice, respectively. Pressures and temperatures at the core-envelope boundaries of these planets have been estimated to be 40 Mbar and 15,000~20,000 K in Jupiter, 10 Mbar and 8,500~10,000 K in Saturn, 8 Mbar and ~8,000 K in Uranus and Neptune (Guillot 2004). To improve modeling of their interiors, there should be a better understanding of possible

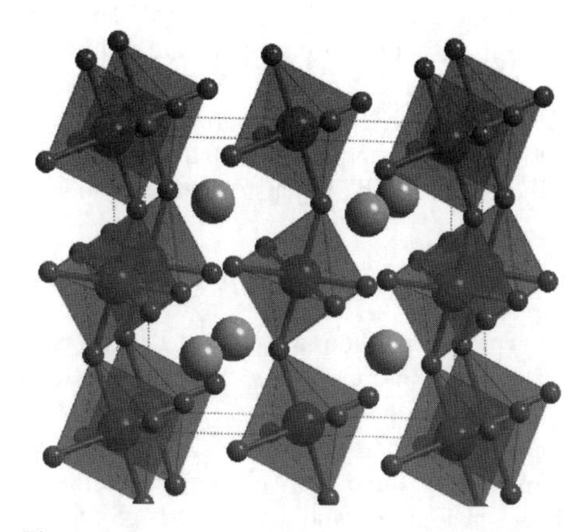

Figure 1. Structure of $MgSiO_3$ post-perovskite (PPV).

1529-6466/10/0071-0014$05.00 DOI: 10.2138/rmg.2010.71.14

phases ($MgSiO_3$ PPV, etc.) and their equations of state under these extreme conditions. Experimentally, it is very challenging to achieve such extremely high pressures and temperatures. The National Ignition Facility (NIF), expected to enter in operation in 2010, offers hope for some data on materials at these conditions. However, first-principles computational methods are very powerful and promising to investigate materials properties at these extreme conditions. They are forging ahead and making predictions to be tested at NIF in the 10^2 Mbar and 10^5 K regime (Umemoto et al. 2006a; Wu et al. 2008; Sun et al. 2008; Umemoto et al. 2008).

In this paper, we discuss first-principles investigations of post-PPV transitions in planet-forming minerals, $MgSiO_3$ and Al_2O_3, at multi-Mbar pressures. We will also discuss low-pressure analogs of these minerals. They can provide experimentally convenient alternatives for exploration of post-PPV transitions. We will see a close relationship between the multi-Mbar chemistry of planet-forming minerals and those of rare-earth sesquisulfides and transition-metal sesquioxides.

COMPUTATIONAL BACKGROUND

This section briefly describes these first principles density functional theory (DFT) (Hohenberg and Kohn 1964; Kohn and Sham 1965) calculations. DFT makes use of approximations for the exchange-correlation (XC) energy. Several XC energy functionals exist, but the most widely used are based on the local-density approximation (LDA) (Ceperley and Alder 1980; Perdew and Zunger 1981) and on the generalized density approximation (GGA), in particular the PBE functional (Perdew et al. 1996). They have been successfully used for many minerals (Wentzcovitch et al. 2010, in this volume). PBE-GGA is also relatively successful for calculations in H_2O-ice. Wave functions are expanded in plane waves in combination with pseudopotentials (Vanderbilt 1990). At multi-Mbar pressures, care must be exercised when generating pseudopotentials. Inter-atomic distances become very small at these extreme pressures, and pseudopotential cut-off radii must be sufficiently small to avoid core overlap. It is often necessary to promote semi-core state to valence. In Umemoto et al. (2006a), 2*s* and 2*p* states of silicon, usually treated as core states, were promoted to valence states. Using the plane wave pseudopotential method, total energy, forces, and stress are efficiently evaluated. Structural search and optimizations are performed at arbitrary pressures using a damped form of variable-cell-shape molecular dynamics (Wentzcovitch 1991; Wentzcovitch et al. 1993). The dynamical stability of optimized structures is assessed by calculation of phonon frequencies. These are calculated by diagonalizing the dynamical matrices obtained using density-functional perturbation theory (Giannozzi et al. 1991; Baroni et al. 2001). The system is dynamically stable if all phonon frequencies are real throughout the Brillouin zone. Phonon frequencies are also used to calculate the Helmholtz free energy within the quasi-harmonic approximation (QHA) (Wallace 1972). All thermodynamic properties, including Gibbs free energies at arbitrary pressures and temperatures, can then be obtained. Knowledge of the Gibbs free energies of two or more phases, phase transformations can be investigated.

DISSOCIATION OF $MgSiO_3$ PPV

The first prediction of a post-PPV transition was the dissociation of $MgSiO_3$ (Umemoto et al. 2006a) into CsCl-type MgO and cotunnite-type SiO_2 at 1.12 TPa (11.2 Mbar) in static enthalpy calculations (Fig. 2). CsCl-type MgO and cotunnite-type SiO_2 have not been synthesized experimentally yet, but they are the most probable candidates of high-pressure forms of MgO and SiO_2 in the dissociation pressure range. The calculated phase boundary shown in Figure 3 indicates that $MgSiO_3$ PPV should not exist any longer in the cores of the gas giants (Jupiter and Saturn) but could survive in cores of the icy giants (Uranus and Neptune). This transition may occur in the mantles of Super-Earths-type exoplanets (Rivera et

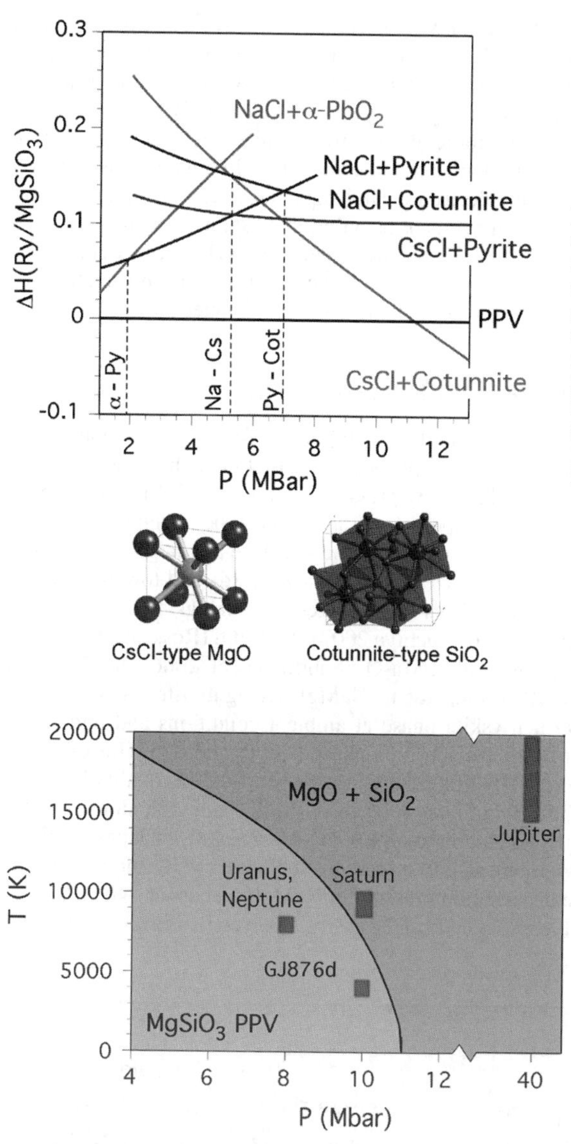

Figure 2. Static enthalpies of aggregation of MgO and SiO_2 in various forms with respect to that of $MgSiO_3$ PPV (Umemoto et al. 2006a). Vertical dashed lines denote static transition pressures of NaCl-type—CsCl-type MgO (5.3 Mbar), a-PbO_2-type—Pyrite-type SiO_2 (1.9 Mbar), and Pyrite-type—cotunnie-type SiO_2 (6.9 Mbar). These values are consistent with those of other calculations (Mehl et al. 1988; Oganov et al. 2003; Oganov et al. 2005).

Figure 3. Dissociation phase boundary in $MgSiO_3$. PPV is predicted to dissociate into CsCl-type MgO and cotunnite-type SiO_2. Rectangles denote estimated pressure-temperature conditions at core-envelope boundaries in the solar giants and in GJ876d.

al. 2005), depending on their masses and temperatures, and in dense larger exoplanets (Sato et al. 2005). Along the dissociation phase boundary the mixture of products was found to be denser than the PPV phase by 1~3%, depending on temperature. The coordination number of silicon increases from 6 to 9 and the averaged bond lengths increase as well. This lowers the vibrational entropy of the dissociation products. Consequently, the Clapeyron slope for this transition is negative. In GJ876d, the dissociation is likely to occur near its CMB (Valencia et al., 2006). The eventual occurrence of this endothermic transition with a large negative Clapeyron slope would be equivalent to the occurrence of the endothermic post-spinel transition near the core of Mars. Geodynamical modeling suggests that this might be the cause of a proposed large martian superplume (Weinstein 1995). Convection in D149026b (Sato et al. 2005), where internal pressures and temperatures should be much higher than in Saturn, could be dramatically affected. A transformation with such large negative Clapeyron slope in the middle of its silicate core-mantle is likely to inhibit convection (Tackley 1995), promote layering, and produce a differentiated core-mantle, with the bottom layer consisting primarily of oxides. Thermal excitation of carriers at the high temperatures relevant for the solar giants and exoplanets affects noticeably important mineral properties. Finite electronic temperature (Mermin 1965; Wentzcovitch et al. 1992) calculations indicated that although the dissociation products are intrinsic semiconductors with electronic band gaps, the carrier concentrations in cotunnite-type SiO_2 become typical of semimetals or heavily-doped semiconductors at 1 TPa and 10,000~20,000 K (Umemoto et al. 2006a). Hence these minerals can be seen essentially as metals with rather high electric and thermal conductivities. This is important information for improving models of the solar giants' interiors and of terrestrial exopolanets.

LOW-PRESSURE ANALOG OF $MgSiO_3$

Although the dissociation of $MgSiO_3$ should be important for the solar giants and terrestrial exoplanets, the predicted dissociation pressure is still too high to be achieved routinely by experiments (~1 TPa). Therefore, low-pressure analogs of $MgSiO_3$ are highly desirable for experimental investigations of properties of the $CaIrO_3$-type structure, including its dissociation. In general, analog compounds with the same structure and larger ions tend to have lower transition pressures. Atomic/ionic sizes increase downwards and towards the left in the periodic table. There are several candidates for low-pressure analogs of $MgSiO_3$, e.g., $CaIrO_3$ (Hirose and Fujita 2005; Tsuchiya and Tsuchiya 2007), $MgGeO_3$ (Ross and Navrotsky 1988; Hirose et al. 2005; Kubo et al. 2006, 2008), Mn_2O_3 (Santillán et al. 2006), and $NaMgF_3$ (Liu et al. 2005; Hustoft et al. 2008). Among them, $NaMgF_3$ (neighborite) is one of the best candidates. It is a stable *Pbnm* perovskite phase at ambient conditions and undergoes a pressure-induced phase transition to the $CaIrO_3$-type phase (Liu et al. 2005; Hustoft et al. 2008). Umemoto et al. (2006b) predicted by first principles that $NaMgF_3$ has qualitatively the same phase diagram as $MgSiO_3$: the PPV transition has a positive Clapeyron slope and the PPV dissociation into CsCl-type NaF and cotunnite-type MgF_2 has a negative Clapeyron slope (Umemoto et al. 2006b). NaF and MgF_2 are also low-pressure analogs of MgO and SiO_2 (Yagi et al. 1983; Haines et al. 2001). The dissociation pressure in $NaMgF_3$ occurs at ~40 GPa (Fig. 4), which is much lower than that of $MgSiO_3$ and can be easily achieved by static compression

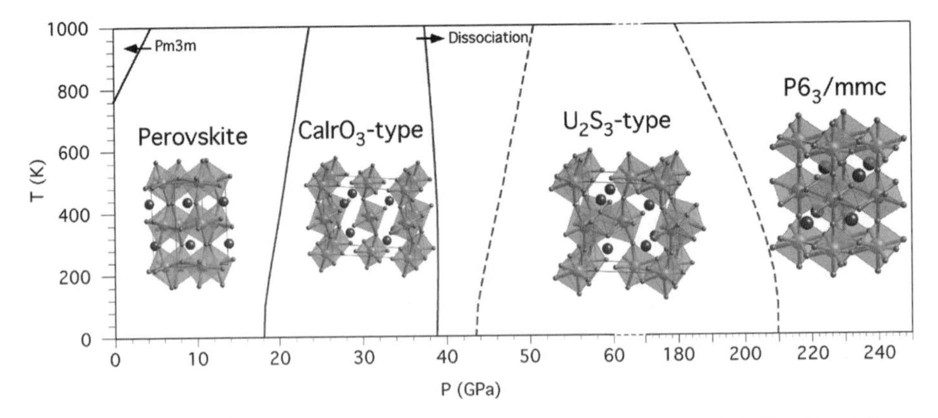

Figure 4. Calculated phase boundaries of $NaMgF_3$ (Umemoto and Wentzcovitch 2006). Two dashed lines denote the metastable phase boundaries between $CaIrO_3$-type and U_2S_3-type, and between U_2S_3-type and $P6_3/mmc$ phases, respectively. The solid line in the upper left corner denotes the experimental phase boundary between orthorhombic perovskite and cubic phases (Zhao et al. 1994).

experiments. Therefore, $NaMgF_3$ should be a good low-pressure analog of $MgSiO_3$ to test the possibility of a dissociation transition. The PPV transition and subsequent dissociation have also been predicted in $CaSnO_3$ (Tsuchiya and Tsuchiya 2006). Despite the latter being an oxide compound, its dissociation pressure was found to be ~70 GPa, which is also quite low compared to that in $MgSiO_3$.

PREDICTION OF POST-PPV CRYSTALLINE PHASES

Although the dissociations of $MgSiO_3$, $NaMgF_3$, and $CaSnO_3$ PPV were predicted, it is not guaranteed that all ABX_3 PPVs should dissociate. Some of them might prefer to undergo a post-PPV transition to another ABX_3 polymorph before dissociating. What could happen if the dissociation is inhibited? Umemoto and Wentzocivtch (2006) investigated the metastable compression of $CaIrO_3$-type $NaMgF_3$ beyond the predicted dissociation pressure (~40 GPa). Up to ~80 GPa, all lattice constants of the $CaIrO_3$-type $NaMgF_3$ decreased as usual. However, there were two anomalies in the behavior of the lattice constants under pressure: increases in lattice constants *a* and *c* at 80 GPa and an abrupt jump of all lattice constants at 150 GPa (Fig. 5). The first anomaly is related to the softening of an acoustic mode at the zone-edge *Y* point (Fig. 6). After superposing the atomic displacements of this particular soft mode to the structure followed by structural re-optimization, new bonds appeared between magnesium and fluorine in adjacent layers. This resulted in a phase transformation from the $CaIrO_3$-type phase to a phase with *Pmcn* symmetry (the *Pnma* symmetry in the standard setting), a sub-group of *Cmcm*. This phase is isostructural with U_2S_3 (Fig. 7b). The magnesium coordination number is 7, larger than that of the $CaIrO_3$-type phase, 6. A possible crystallographic relationship between the $CaIrO_3$-type and the U_2S_3-type structures was discussed by Hyde et al. (1979). It is interesting to note that $UFeS_3$ has the $CaIrO_3$-type structure. Replacement of iron by uranium leads to the U_2S_3-type structure. The second anomaly corresponds to an increase in symmetry of the $CaIrO_3$-type phase and generates a new phase with $P6_3/mmc$ symmetry, a super-group of *Cmcm* (Fig. 7c). The magnesium coordination number in this phase is 8, higher than 6 in the $CaIrO_3$-type and 7 in the U_2S_3-type phases. The structural unit is no longer the MgF_6 octahedron but an MgF_8 parallelepiped. Parallelepipeds share edges to form layers in the *ab* plane. Each layer contacts the adjacent ones at the parallelepipeds' apices along the [0001] direction. Sodium is located in interstitial sites between the MgF_8 parallelepipeds. The sodium coordination number is 11. Sodium and magnesium stack in an ABAC sequence (A: magnesium and B, C: sodium), i.e., the sub-lattice formed by sodium and magnesium has the NiAs structure. The sodium and fluorine sub-lattices have the $IrAl_3$–type structure (Hyde and Andersson 1989). As far as we know, the $P6_3/mmc$ phase has not been identified in any material experimentally so far. Both U_2S_3-type and $P6_3/mmc$ phases are dynamically stable phases of $NaMgF_3$. Therefore, these structures are two potential candidates

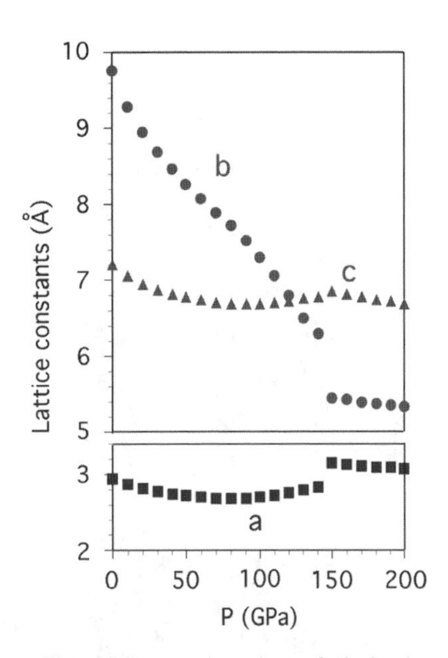

Figure 5. Pressure dependence of calculated lattice constants of $CaIrO_3$-type $NaMgF_3$ (Umemoto and Wentzcovitch 2006).

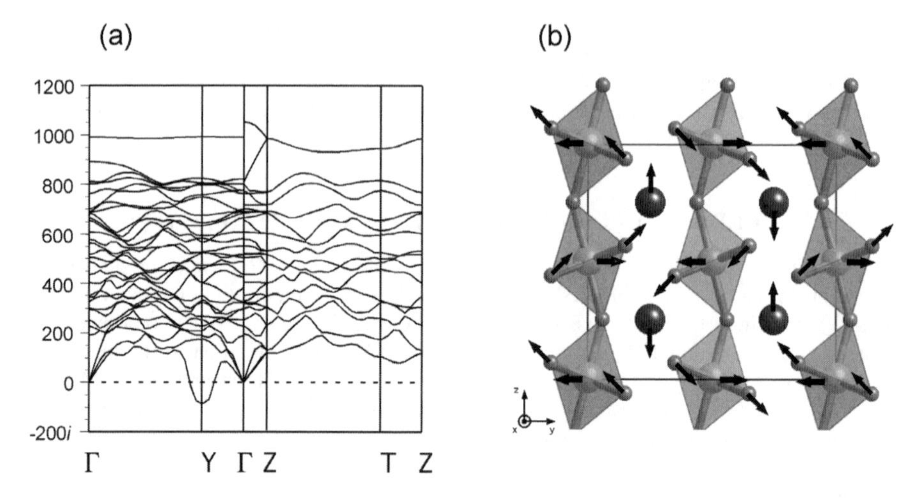

Figure 6. (a) Phonon dispersion of $CaIrO_3$-type $NaMgF_3$ at 100 GPa. (b) Atomic displacements resulting from the unstable phonon at the Y point.

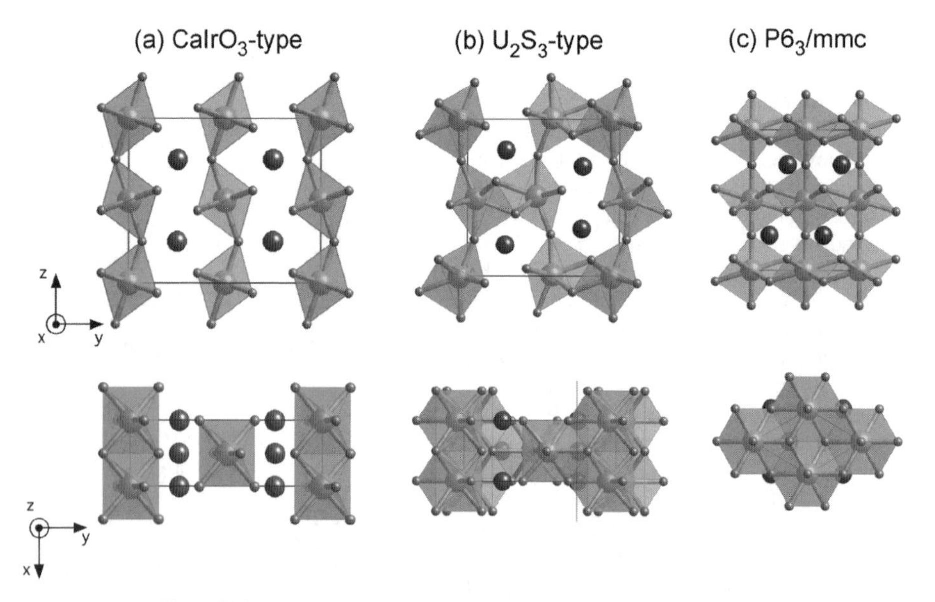

Figure 7. Structures of (a) $CaIrO_3$-type PPV and two possible candidates of post-PPV phases: (b) U_2S_3-type and (c) $P6_3/mmc$ phases.

for post-PPV transitions in $NaMgF_3$ instead of the dissociation (Umemoto and Wentzcovitch 2006). Although they are metastable with respect to the dissociation products, the U_2S_3-type phase could actually be observed experimentally by compressing $CaIrO_3$-type $NaMgF_3$ at sufficiently low temperatures. The energy barrier for the dissociation is expected to be higher than that for the U_2S_3-type transition which is related to a soft mode. In addition, at low temperature, the phase boundaries for the dissociation and for the U_2S_3-type transition are close to each other. Recently Martin et al. (2006) reported diamond anvil cell experiments in *Cmcm* $NaMgF_3$ showing new X-ray diffraction peaks with increasing pressure. However, these peaks could not be attributed only to a mixture of *Cmcm* and its dissociation products, NaF and MgF_2. The superposition of peaks seems to be considerable and the analysis of the published pattern is challenging and inconclusive; but, the new X-ray diffraction pattern may contain peaks related to the U_2S_3-type phase. In $MgSiO_3$, the static transition pressure to the U_2S_3-type phase is calculated to be beyond 1.6 TPa (Umemoto and Wentzcovitch 2006), which is considerably higher than the dissociation pressure of ~1 TPa. In nature these pressures are realized in the interior of the giant planets and exoplanets where temperatures are also expected to be very high, $10^{3.5}$~10^4 K. Since diffusion barriers are expected to be overcome in planetary time scales, dissociation into oxides is still the most likely pressure-induced transition in $CaIrO_3$-type $MgSiO_3$ in planetary interiors.

POST-POST-PEROVSKITE TRANSITION IN Al_2O_3

Two metastable candidate post-PPV structures—U_2S_3-type and $P6_3/mmc$ structures—were proposed for $NaMgF_3$ under pressure, but the most probable transition should be the dissociation into CsCl-type NaF and cotunnite-type MgF_2. However, sometimes dissociation is "forbidden." Al_2O_3 (alumina) is expected to undergo a non-dissociative post-PPV transition. Its highest-pressure phase identified experimentally so far, the $CaIrO_3$-type, has unlikely dissociation products, AlO and AlO_2. For Al_2O_3, two phase transitions have been established so far: corundum—Rh_2O_3(II)-type (Cynn et al. 1990; Marton and Cohen 1994; Thomson et al. 1996; Funamori and Jeanloz 1997; Mashimo et al. 2000; Lin et al. 2004) and Rh_2O_3(II)-type—$CaIrO_3$-type (PPV) (Caracas and Cohen 2005; Oganov and Ono 2005; Tsuchiya et al. 2005; Ono et al. 2006). Al_2O_3 is an important compound in high-pressure technology and geophysics. It is used as window material in shock-wave experiments (McQueen and Isaak 1990). The pressure dependence of the fluorescence line of ruby, Al_2O_3 doped with chromium, serves as a pressure-marker in diamond-anvil-cell experiments (Mao and Bell 1976; Chen and Silvera 1996). In Earth and planetary sciences, Al_2O_3 is a major chemical component in solid solution with $MgSiO_3$ garnet, PV, and PPV. The formation of these solid solutions with Al_2O_3 changes the properties, phase boundaries, electrical conductivity, oxidation state, and spin states of iron in $MgSiO_3$, (e.g., Wood and Rubie 1998; Xu et al. 1998; Zhang and Weidner 1999; Frost et al. 2004; Li et al. 2004; Taneno et al. 2005; Nishio-Hamane et al. 2007).

Static enthalpy calculations by Umemoto and Wentzcovitch (2008) clearly showed a post-PPV transition in Al_2O_3 to a U_2S_3-type phase (Fig. 8). The static transition pressure was predicted to be 373 (380) GPa by LDA (GGA). The $P6_3/mmc$ phase, another candidate for a post-PPV phase, was not found in Al_2O_3 up to 700 GPa. No phonon softening was observed both in the $CaIrO_3$-type and in the U_2S_3-type phases at least up to 700 GPa. These phases should be dynamically stable within this pressure range. Figure 9 shows the calculated phase boundaries of Al_2O_3. The post-PPV transition between the $CaIrO_3$-type and the U_2S_3-type phases has a positive Clapeyron slope, as in $NaMgF_3$.

It is interesting to compare calculated compression curves and those obtained by dynamic compression experiments. In a shock experiment to ~340 GPa, no direct evidence of phase transitions was noticed, although there was an atypical relationship between shock velocity (u_s) and particle velocity (u_p): $u_s = C + Su_p$ in which S (0.957) was unusually small, suggested a sluggish phase transformation (Erskine 1994). However, Figure 10 strongly suggests the presence of all three phases stable below 340 GPa in the raw data (Marsh 1980; Erskine 1994). With shock data alone, it is difficult to resolve phase transitions accompanied by small density changes comparable to the detectability limit.

The transition pressure to the U_2S_3-type polymorph of Al_2O_3 exceeds the pressure at the core-mantle boundary of the Earth (~135 GPa). Therefore the occurrence of U_2S_3-type Al_2O_3 should not alter the current views of the Earth's lower mantle. However, it might affect our

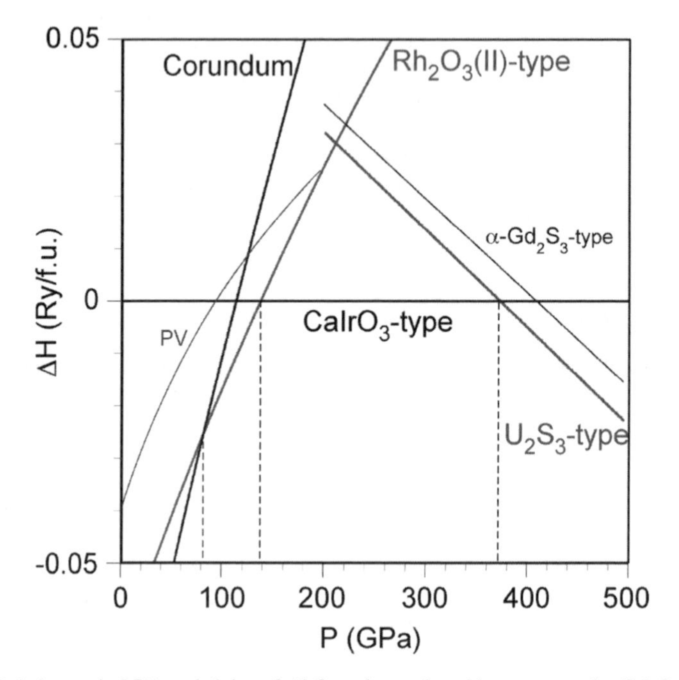

Figure 8. Relative static LDA enthalpies of Al_2O_3 polymorphs with respect to the $CaIrO_3$-type phase (Umemoto and Wentzcovitch 2008). Dashed vertical lines denote corundum—Rh_2O_3(II)-type, Rh_2O_3(II)-type—$CaIrO_3$-type, and $CaIrO_3$-type—U_2S_3-type transition pressures, respectively.

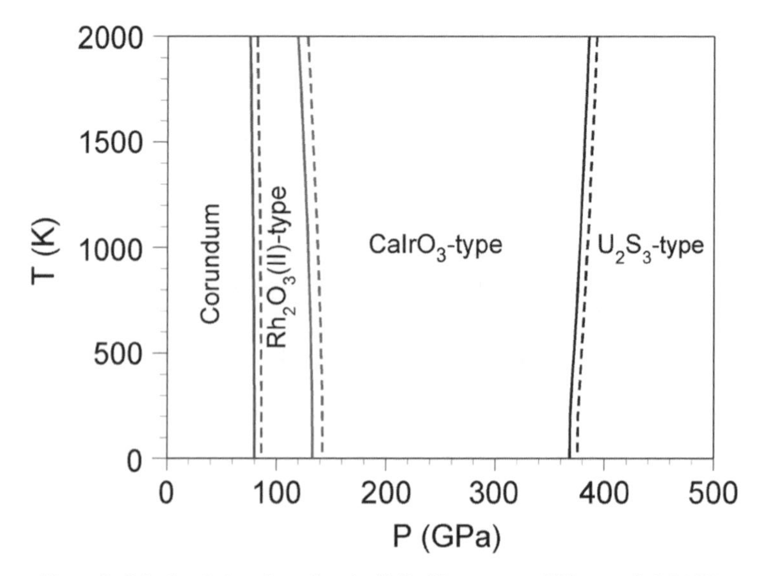

Figure 9. Calculated phase boundary in Al_2O_3 (Umemoto and Wentzcovitch 2008). Solid and dashed lines denote LDA and GGA phase boundaries.

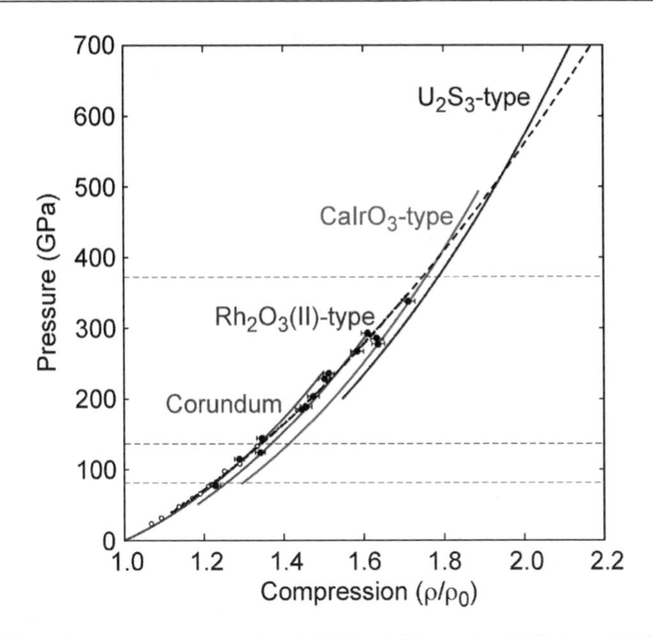

Figure 10. LDA static pressure vs compression (solid lines) (Umemoto and Wentzcovitch 2008). ρ is the calculated density of each phase and ρ_0 is that of corundum at 0 GPa. Horizontal dashed lines represent calculated transition pressures: 82 GPa for corumdum—Rh_2O_3(II)-type, 137 GPa for Rh_2O_3(II)-type—$CaIrO_3$-type, and 373 GPa for $CaIrO_3$-type—U_2S_3-type transitions. Experimental shock data are shown by black (Erskine 1994) and white circles (Marsh 1980). The black dashed line is a compression curve obtained by applying the Rankine-Hugoniot equations to a single linear fit to the Hugoniot data (Erskine 1994).

understanding of the rocky mantles of terrestrial exoplanets since it could change the solubility of Al_2O_3 in $CaIrO_3$-type $MgSiO_3$. The change in coordination number from 6 to 7 in the B site suggests that Al_2O_3 might ex-solve from $MgSiO_3$, in particular, from the smaller 6-fold B site occupied by silicon in the $CaIrO_3$-type structure. The ex-solution of Al_2O_3 could change electrical and thermal conductivities of the $CaIrO_3$-type solid solution, as the incorporation of Al_2O_3 changes the electrical conductivity of $MgSiO_3$ perovskite in presence of iron (Xu et al. 1998).

RARE-EARTH SESQUISULFIDES

Figure 11 shows crystal structures of rare-earth sesquisulfides, $RR'S_3$ (R, R′ = lanthanoid or actinoid), at 1000 °C and ambient pressure. In the $RR'S_3$ family of compounds, we find several structures (corumdum, $GdFeO_3$-type (i.e., perovskite), $CaIrO_3$-type, and U_2S_3-type structures) adopted by $MgSiO_3$ and Al_2O_3. In addition, $LaYbS_3$ adopts the $GdFeO_3$-type structure at high temperatures and the $CaIrO_3$-type structure at low temperatures (Rodier et al. 1983; Mitchell et al. 2004), being reminiscent of the post-perovskite transition in $MgSiO_3$ with a positive Clapeyron slope (Murakami et al. 2004; Oganov and Ono 2004; Tsuchiya et al. 2004). Therefore a close relationship is suggested between the multi-Mbar crystal chemistry of planet-forming minerals and that of rare-earth sesquisulfides. The relationship between $RR'S_3$ crystal structures and cation radii can be summarized as follows: the corundum structure occurs with small R and R′ radii (e.g., Yb_2S_3), the U_2S_3-type structure with large radii, and the $CaIrO_3$-type and the $GdFeO_3$-type structures with large R in the A site and small R′ in the B site (e.g., $NdYbS_3$ and $LaYbS_3$). In R_2S_3, coordination numbers increase with increasing cation radii (Fig. 12). This chemical-pressure effect produces a sequence of structures consistent with that

R

R′	(U)	La	Ce	Pr	Nd	Sm	Gd	Tb	Dy	(Y)	Ho	Er	Tm	Yb	Lu	(Sc)
(U)	η															
La		α	α_c	α_c	α_c	α_c	α_c	α_c	G	G	G	G	G	PV	PV	PV
Ce			α	α_c	α_c	α_c	α_c	α_c	α_1	G	G	G	G	PPV	PPV	PV
Pr				α	α_c	α_c	α_c	α_c	α_c	α_1	α_1	G	G	PPV	PPV	PV
Nd					α	α_c	α_c	α_c	α_c	α_1	α_1	α_1	G	PPV	PPV	PV
Sm						α	α_c	α_c	α_c	α_1	α_1	α_1	α_1	F	F	PV
Gd							α	α_c	α_c	X	X	X	F	F	F	PV
Tb								α	α_c	X	X	X	X	F	F	PV
Dy									α	δ_c	δ_c	δ_c	δ_c	δ_1	δ_1	PV
(Y)										δ	δ_c	δ_c	δ_c	δ_1	δ_1	PV
Ho											δ	δ_c	δ_c	δ_1	δ_1	PV
Er												δ	δ_c	δ_1	δ_1	PV
Tm													δ	δ_1	δ_1	-
Yb														ε	ε_c	-
Lu															ε	-

Figure 11. Polymorph of $RR'S_3$ compounds at 1000 °C (Flahaut 1979). Each symbol denotes the following structural types: η: U_2S_3-type, α: α-Gd_2S_3-type, δ: δ-Ho_2S_3-type, ε: corundum-type, F: $CeYb_3S_6$-type, G: $CeTmS_3$-type, PPV: $NdYbS_3$-type, i.e., $CaIrO_3$-type (see below), I: perovskite, α_c and δ_c: isomorphic solid solutions of the α or δ types, with compositions varying continuous from R_2S_3 to R'_2S_3, α_1 and δ_1: solid solutions of the α or δ types, with a limited solubilities, including the $RR'S_3$ composition, and x: mixture of phases with $RR'S_3$ composition. The $NdYbS_3$-type structure was first reported to have $B22_12$ symmetry which is a subgroup of the *Cmcm* group of the $CaIrO_3$-type structure (Flahaut 1979; Carré and Laruelle 1974). However, the $NdYbS_3$-type structure refined in the $B22_12$ symmetry is nearly indistinguishable from the $CaIrO_3$-type structure. The difference between them is very small. In the $B22_12$ $NdYbS_3$-type structure, the cation at the B site and one kind of sulfur are just slightly shifted from the 4a and 8f Wyckoff positions in the *Cmcm* $CaIrO_3$-type structure. A recent X-ray diffraction study suggested the $NdYbS_3$-type structure should have the *Cmcm* symmetry and be the same as the $CaIrO_3$-type structure (Mitchell et al. 2004).

induced by pressure in Al_2O_3. Therefore, *one might anticipate that rare-earth sesquisulfides are a set of low-pressure analogs of planet-forming silicates and oxides.* $RR'S_3$ structures that do not occur in $MgSiO_3$ and Al_2O_3, may also play an important role. In fact, Al_2O_3 in the Gd_2S_3-type structure, one of the major structures of $RR'S_3$ compounds, and the U_2S_3-type polymorph have very similar enthalpies (see Fig. 8). The Ho_2S_3-type structure is also one of the likely forms between corundum and the Gd_2S_3-type/U_2S_3-type phases, but for Al_2O_3, this phase has very high enthalpy and is energetically unstable.

M_2O_3 SESQUIOXIDES

In addition to rare-earth sesquisulfides, M_2O_3 sesquioxides (M = trivalent cations: group IIIB metals, gallium and indium, and *d*-transition metals) could be another set of low-pressure analogs of planet-forming sesquioxides, especially of Al_2O_3. Ga_2O_3 and Fe_2O_3 exhibit sequences of phase transitions similar to Al_2O_3: corundum—Rh_2O_3(II)-type—$CaIrO_3$-type phases (in Ga_2O_3, corundum phase is preceded by the β phase) (Shim and Duffy 2002; Ono et al. 2005; Ono and Ohishi 2005; Shim et al. 2009, Tsuchiya et al. 2007; Yusa et al. 2008a). The $CaIrO_3$-type structure occurs also in Mn_2O_3 (Santillán et al. 2006).

Caracas and Cohen (2007) performed a series of first-principles calculations of phase transitions in Al_2O_3, Ga_2O_3, Rh_2O_3, and In_2O_3. They showed that In_2O_3 should undergo the

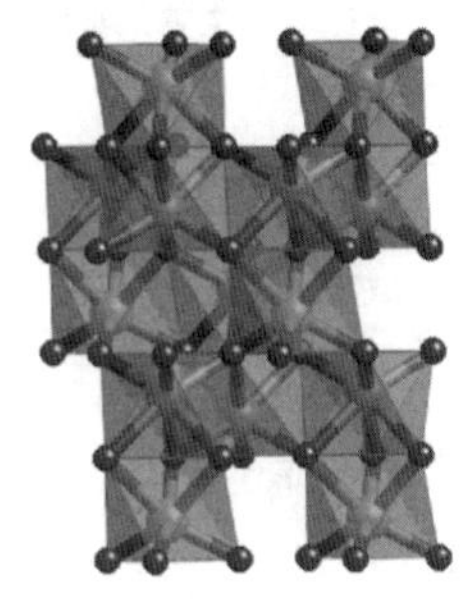

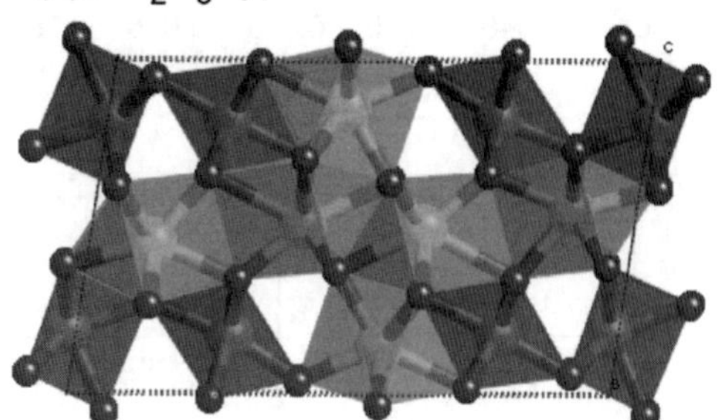

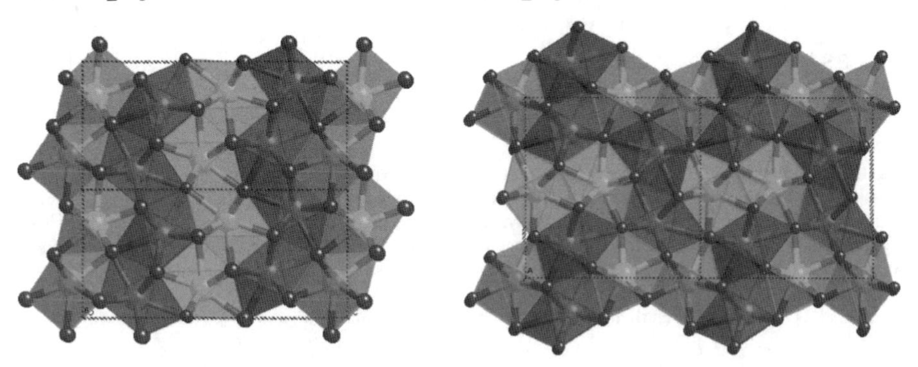

Figure 12. Crystal structures of R_2S_3 (R: lanthanoid or actinoid). The coordination numbers of cations are 6 in corundum, 6 and 7 in Ho_2S_3-type, 7 and 8(=7+1) in Gd_2S_3-type and U_2S_3-type structures.

Rh_2O_3(II)-type—$CaIrO_3$-type phase transition at a pressure as low as 47 GPa. In fact, In_2O_3 does not adopt quite exactly the $CaIrO_3$-type structure experimentally (Yusa et al. 2008b). The PPV transition pressures of Ga_2O_3 and Rh_2O_3 were estimated to be 136 GPa and ~360 GPa, respectively. This is contrary to expectations; The PPV transition pressure in Rh_2O_3 is expected to be lower than in Al_2O_3 and Ga_2O_3, since the ionic radius of rhodium (0.805 Å) is larger than those of aluminum (0.675 Å) and gallium (0.760 Å) (Shannon 1976). Results by Caracas and Cohen (2007) suggest the relationship between crystal structures in M_2O_3 and cation radii is not simple. Indeed, the sequence of phase transitions in M_2O_3 is very complex, as shown in Figure 13. This complexity indicates that *d* electrons may be an important factor for determination of crystal structures. Partially-filled *d* states have preferential orientations in space and should affect the positions of surrounding oxygens. Different *d* states filling factors in different cations may give rise to a variety of crystal structures. For Fe_2O_3, magnetization is also an important factor (Shim et al. 2009). In the case of rare-earth sesquisulfides, the occupation of *f* states varies, depending on the lanthanoid and actinoid atomic species. But the *f* states are quite localized and should hardly affect chemical bonding. Therefore the relationship between crystal structure and ionic radii in rare-earth sesquisulfides is rather simple.

The Gd_2S_3-type structure is a high-pressure form of In_2O_3 and Sc_2O_3 with relatively large ionic radii (Yusa et al. 2008b, 2009). But, in In_2O_3 and Sc_2O_3, the enthalpy differences between the Gd_2S_3-type and the U_2S_3-type phases were found to be very small. In some experimental conditions, both phases might coexist, as in Dy_2S_3 (Meetsma et al. 1991). Ti_2O_3 is very

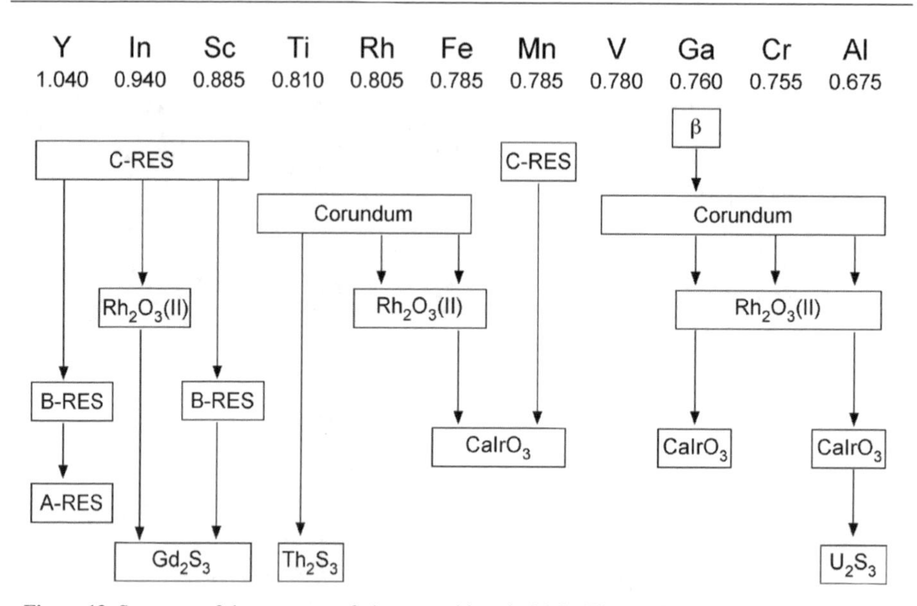

Figure 13. Summary of the sequence of phase transitions in M_2O_3. The numbers represent ionic radii of trivalent cations. Fe and Mn are assumed to be in the high-spin state (Shannon 1976). For simplicity, the monoclinic *I2/a* phase is omitted for V_2O_3 (Dernier and Marezio 1970) and Cr_2O_3 (Shim et al. 2004). This figure is based on the following references for each compound: Y_2O_3 (Atou et al. 1990; Husson et al. 1999; Wang et al. 2009), In_2O_3 (Yusa et al. 2008a,b), Sc_2O_3 (Yusa et al. 2009), Ti_2O_3 (Nishio-Hamane et al. 2009), Rh_2O_3 (Shannon and Prewitt 1970), Fe_2O_3 (Shim and Duffy 2002; Ono et al. 2005; Ono and Ohishi 2005; Shim et al. 2009), Mn_2O_3 (Santillán et al. 2006), Ga_2O_3 (Yusa et al. 2008), and Al_2O_3 (see the text).

interesting. It transforms directly from corundum to the Th_2S_3-type phase (Nishio-Hamane et al. 2009). The Th_2S_3-type structure is isostructural with U_2S_3. The symmetry of both structures is *Pnma*. The difference between them is the type of orthorhombic distortion: lattice constant $a < c$ in the Th_2S_3-type and $a > c$ in the U_2S_3-type structure. So far, no transition from the $CaIrO_3$-type phase has been identified experimentally in M_2O_3. For Ga_2O_3, Mn_2O_3, and Fe_2O_3, whose highest-pressure forms identified experimentally appears to be the $CaIrO_3$-type structure (Shim et al. 2009), we anticipate post-PPV phases with U_2S_3-type or Th_2S_3-type structures.

SUMMARY

Two post-post-perovskite transitions have been identified by first principles calculations: the dissociation of $MgSiO_3$ into CsCl-type MgO and cotunnite-type SiO_2 at ~1 TPa and the transition of Al_2O_3 to the U_2S_3-type phase at ~370 GPa. These transition pressures are currently very challenging for static compression experiments. $NaMgF_3$ may be a good low-pressure analog of $MgSiO_3$. We found a close relationship between the multi-Mbar crystal chemistry of planet-forming minerals and that of rare-earth sesquisulfides. Some of them may be used as low-pressure analogs of planet-forming minerals in the multi-Mbar range. In transition-metal sesquioxides, the Th_2S_3-type phase (very similar to U_2S_3-type) was discovered in Ti_2O_3, suggesting that some transition-metal sesquioxides could serve as low-pressure analog(s) of Al_2O_3. Finally, a new type of dissociation, $FeTiO_3$ perovskite into $(Fe_{1-\delta},Ti_\delta)O$ and $Fe_{1+\delta}Ti_{2-\delta}O_5$, has recently been reported (Wu et al. 2009). This finding suggests that solid solutions (probably with transition-metals) among dissociation products (and in parent phases) could complicate the transition mechanisms we have discussed in this chapter.

ACKNOWLEDGMENTS

Many parts of this article are based on a set of first-principles studies which we performed using the Quantum-ESPRESSO distribution (Giannozzi et al. 2009). These works were supported by NSF grants No. EAR-0135533, EAR-0230319, ITR-0428774 (Vlab), EAR-0757903, EAR-0635990, and ATM-0428774 (Vlab).

REFERENCES

Atou T, Kusaba K, Fukuoka K, Kikumi M (1990) Shock-induced phase transition of M_2O_3 (M=Sc, Y, Sm, Gd, and In)-type compounds. J Solid State Chem 89:378-384

Baroni S, de Gironcoli S, Dal Corso A, Giannozzi P (2001) Phonons and related crystal properties from density-functional perturbation theory. Rev Mod Phys 73:515-562

Caracas R, Cohen RE (2005) Prediction of a new phase transition in Al_2O_3 at high pressures. Geophys Res Lett 32:L06303

Caracas R, Cohen RE (2007) Post-perovskite phase in selected sesquioxides from density-functional calculations. Phys Rev B 76:184101

Carré PD, Laruelle P (1974) Structure Cristalline du Sulfure de Néodyme et d'Ytterbium, $NdYbS_3$. Acta Cryst B30:952-954

Ceperley DM, Alder BJ (1980) Ground state of the electron gas by a stochastic method. Phys Rev Lett 45:566-569

Chen NH, Silvera IF (1996) Excitation of ruby fluorescence at multimegabar pressures. Rev Sci Instrum 67:4275-4278

Cynn H, Isaak DG, Cohen RE, Nicol MF, Anderson OL (1990) A high-pressure phase transition of corundum predicted by the potential induced breathing model. Am Mineral 75:439-442

Dernier PD, Marezio M (1970) Crystal structure of the low-temperature antiferromagnetic phase of V_2O_3. Phys Rev B 2:3771-3776

Erskine D (1994) High pressure Hugoniot of sapphire. *In*: High Pressure Science and Technology—1993. Schidt SC, Shaner JW, Samara GA, Ross M (eds) AIP Press, New York, p 141-143

Flahaut J (1979) Sulfides, selenides, and tellurides. *In*: Handbook on the Physics and Chemistry of Rare-Earths. Gschneidner KA Jr., Eyring LR (eds) North-Holland, Amsterdam, New York, Oxford. 4:1-88

Frost DJ, Liebske C, Langenhorst F, McCammon CA, Trannes RG, Rubie DC (2004) Experimental evidence for the existence of iron-rich metal in the Earth's lower mantle. Nature 428:409-412

Funamori N, Jeanloz R (1997) High-pressure transformation of Al_2O_3. Science 278:1109-1111

Giannozzi P, de Gironcoli S, Pavone P, Baroni S (1991) *Ab initio* calculation of phonon dispersions in semiconductors. Phys Rev B 43:7231-7242

Giannozzi P, Baroni S, Bonini N, Calandra M, Car R, Cavazzoni C, Ceresoli D, Chiarotti GL, Cococcioni M, Dabo I, Dal Corso A, de Gironcoli S, Fabris S, Fratesi G, Gebauer R, Gerstmann U, Gougoussis C, Kokalj A, Lazzeri M, Martin-Samos L, Marzari N, Mauri F, Mazzarello R, Paolini S, Pasquarello A, Paulatto L, Sbraccia C, Scandolo S, Sclauzero G, Seitsonen AP, Smogunov A, Umari P, Wentzcovitch RM (2009) Quantum ESPRESSO: a modular and open-source software project for quantum simulations of materials. J Phys Condens Matter 21:395502

Guillot T (2004) Probing the giant planet. Phys Today 57 (4):63-69

Haines J, Leger JM, Gorelli F, Klug DD, Tse JS, Li ZQ (2001) X-ray diffraction and theoretical studies of the high-pressure structures and phase transitions in magnesium fluoride. Phys Rev B 64:134130

Hirose K, Fujita Y (2005) Clapeyron slope of the post-perovskite phase transition in $CaIrO_3$. Geophys Res Lett 32:L13313-13316

Hirose K, Kawamura K, Ohishi Y, Tateno S, Sata N (2005) Stability and equation of state of $MgGeO_3$ post-perovskite phase. Am Mineral 90:262-265

Hohenberg P, Kohn W (1964) Inhomogeneous electron gas. Phys Rev 136:B864-B871

Husson E, Proust C, Gillet P, Itié JP (1999) Phase transitions in yttrium oxide at high pressure studied by Raman spectroscopy. Mater Res Bull 34:2085-2092

Hustoft J, Catalli K, Shim SH, Kubo A, Prakapenka VB, Kunz M (2008) Equation of state of $NaMgF_3$ postperovskite: Implication for the seismic velocity changes in the D" region. Geophys Res Lett 35:L10309

Hyde BG, Andersson S, Bakker M, Plug CM, O'Keeffe M (1979) The (twin) composition plane as an extended defect and structure-building entity in crystals. Prog Solid St Chem 12: 273-327

Hyde BG, Andersson S (1989) Inorganic Crystal Structures. Wiley, New York

Kohn W, LJ Sham (1965) Self-consistent equations including exchange and correlation effects. Phys Rev 140:A1133-A1138

Kubo A, Kiefer B, Shen G, Prakaoebja VB, Cava RJ, Duffy TS (2006) Stability and equation of state of the post-perovskite phase in $MgGeO_3$ to 2 Mbar. Geophys Res Lett 33:L12S12-12S15

Kubo A, Kiefer B, Shim SH, Shen G, Prakapenka VB, Duffy T (2008) Rietveld structure refinement of $MgGeO_3$ post-perovskite phase to 1 Mbar. Am Mineral 93:965-976

Li J, Struzhkin VV, Mao HK, Shu J, Hemley RJ, Fei Y, Mysen B, Dera P, Prakapenka V, Shen G (2004) Electronic spin state of iron in lower mantle perovskite. Proc Nat Acad Sci USA 101:14027-14030

Lin JF, Degtyareva O, Prewitt CT, Dera P, Sata N, Gregoryanz E, Mao HK, Hemley RJ (2004) Crystal structure of a high-pressure/high-temperature phase of alumina by in situ X-ray diffraction. Nat Mater 3:390-393

Liu HZ, Chen J, Hu J, Martin CD, Weidner DJ, Häusermann D, Mao HK (2005) Octahedral tilting evolution and phase transition in orthorhombic $NaMgF_3$ perovskite under pressure. Geophys Res Lett 32:L04304-04307

Mao HK, Bell PM (1976) High pressure physics: I Megabar in ruby R1 static pressure scale. Science 191:851-852

Marsh S (1980) LASL Shock Hugoniot Data. Univ of California Press, Berkeley, p. 260

Martin CD, Crichton WA, Liu H, Prakapenka V, Chen J, Parise JB (2006) Phase transitions and compressibility of $NaMgF_3$ (neighborite) in perovskite- and post-perovskite-related structures. Geophys Res Lett 33:L11305

Marton FC, Cohen RE (1994) Prediction of a high-pressure phase transition in Al_2O_3. Am Mineral 79:789-792

Mashimo T, Tsumoto K, Nakamura K, Noguchi Y, Fukuoka K, Syono Y (2000) High-pressure phase transformation of corundum (α-Al_2O_3) observed under shock compression. Geophys Res Lett 27:2021-2024

McQueen RG, Isaak DG (1990) Characterizing windows for shock-wave radiation studies. J Geophys Res 95 (B13):21753-21765

Meetsma A, Wiegers GA, Haange RJ, de Boer JL, Boom G (1991) Structure of two modifications of dysprosium sesquisulfide, Dy_2S_3. Acta Crystallogr C47:2287-2291

Mehl MJ, Cohen RE, Krakauer H (1988) Linearized augmented plane wave electronic structure calculations for MgO and CaO. J Geophys Res 93(B7):8009-8022

Mermin ND (1965) Thermal properties of the inhomogeneous electron gas. Phys Rev 137:A1441-A1443

Mitchell K, Somers RC, Huang FQ, Ibers JA (2004) Syntheses, structure, and magnetic properties of several LnYbQ3 chalcogenides, Q=S, Se. J Solid State Chem 177:709-713

Murakami M, Hirose K, Kawamura K, Sata N, Ohishi Y (2004) Post-perovskite transition in $MgSiO_3$. Science 304:855-858

Nishio-Hamane D, Fujino K, Seto Y, Nagai T (2007) Effect of the incorporation of $FeAlO_3$ into $MgSiO_3$ perovskite on the post-perovskite transition. Geophys Res Lett 34:L12307

Nishio-Hamane D, Katagiri M, Niwa K, Sano-Furukawa A, Okada T, Yagi T (2009) A new high-pressure polymorph of Ti_2O_3: implication for high-pressure phase transition in sesquioxides. High Pressure Res 29:379-388

Oganov AR, Gillan MJ, Price GD (2003) *Ab initio* lattice dynamics and structural stability of MgO. J Chem Phys 118:10174-10182

Oganov AR, Ono S (2004) Theoretical and experimental evidence for a post-perovskite phase of $MgSiO_3$ in Earth's D″ layer. Nature 430:445-448

Oganov AR, Gillan MJ, Price GD (2005a) Structural stability of silica at high pressures and temperatures. Phys Rev B 71:064104

Oganov AR, Ono S (2005) The high-pressure phase of alumina and implications for Earth's D″ layer. Proc Nat Acad Sci USA 102:10828-10831

Ono S, Funakoshi K, Ohishi Y, Takahashi E (2005) *In situ* x-ray observation of the phase transformation of Fe_2O_3. J Phys Condens Matt 17:267-276

Ono S, Ohishi Y(2005) *In situ* X-ray observation of phase transformation in Fe_2O_3 at high pressures and high temperatures. J Phys Chem Solids 66:1714-1720

Ono S, Oganov AR, Koyama T, Shimizu H (2006) Stability and compressibility of the high-pressure phases of Al_2O_3 up to 200 GPa: Implications for the electrical conductivity of the base of the lower mantle. Earth Planet Sci Lett 246:326-335

Perdew JP, Zunger A (1981) Self-interaction to density-functional approximations for many-electron systems. Phys Rev B 23:5048-5079

Perdew JP, Burke K, Ernzerhof M (1996) Generalized gradient approximation made simple. Phys Rev Lett 77:3865-3868

Rivera EJ, Lissauer JJ, Butler RP, Marcy GW, Vogt SS, Fischer DA, Brown TM, Laughlin G, Henry GW (2005) A similar to 7.5 $M_\oplus$ plus planet orbiting the nearby star, GJ 876. Astrophys J 634:625-640

Rodier PN, Julien R, Tien V (1983) Polymorphisme de $LaYbS_3$. Affinement des structures des deux varieties. Acta Crystallogr C39:670-673

Ross NL, Navrotsky A (1988) Study of the $MgGeO_3$ polymorphs (orthopyroxene, clinopyroxene, and ilmenite structures) by calorimetry, spectroscopy, and phase equilibria. Am Mineral 73:1355-1365
Santillán J, Shim SH, Shen G, Prakapenka VB (2006) High-pressure phase transition in Mn_2O_3: Application for the crystal structure and preferred orientation of the $CaIrO_3$ type. Geophys Res Lett 33:L15307
Sato B, Fischer DA, Henry GW, Laughlin G, Butler RP, Marcy GW, Vogt SS, Bodenheimer P, Ida S, Toyota E, Wolf A, Valenti JA, Boyd LJ, Johnson JA, Wright JT, Ammons M, Robinson S, Strader J, McCarthy C, Tah KL, Minnti D (2005) The N2K consortium. II. A transiting hot Saturn around HD 149026 with a large dense core. Astrophys J 633:465-473
Shannon RD, Prewitt CT (1970) Synthesis and structure of a new high-pressure form of Rh_2O_3. J Solid State Chem 2:134-136
Shannon RD (1976) Revised effective ionic radii and systematic studies of interatomic distances in halides and chalcogenides. Acta Cryst A32:751-767
Shim SH, Duffy TS (2002) Raman spectroscopy of Fe_2O_3 to 62 GPa. Am Mineral 87:318-326
Shim SH, Duffy TS, Jeanloz R, Yoo CS, Iota V (2004) Raman spectroscopy and x-ray diffraction of phase transitions in Cr_2O_3 to 61 GPa. Phys Rev B 69:144107
Shim SH, Bengtson A, Morgan D, Sturhahn W, Catalli K, Zhao J, Lerche M, Prakapenka V (2009) Electronic and magnetic structures of the postperovskite-type Fe_2O_3 and implications for planetary magnetic records and deep interiors. Proc Nat Acad Sci USA 106:5508-5512
Sun T, Umemoto K, Wu Z, Zhen JC, Wentzcovitch RM (2008) Lattice dynamics and thermal equation of stated of platinum. Phys Rev B 78:024304
Tackley P (1995) On the penetration of an endothermic phase transition by upwellings and downwellings. J Geophys Res 100 (B8):15477-15488
Tateno S, Hirose K, Sata N, Ohishi Y (2005) Phase relations in $Mg_3Al_2Si_3O_{12}$ to 180 GPa: Effect of Al on post-perovskite phase transition. Geophys Res Lett 32:L15306
Thomson KT, Wentzcovitch RM, Bukowinski MST (1996) Polymorphs of alumina predicted by first principles: Putting pressure on the ruby pressure scale. Science 274:1880-1882
Tsuchiya T, Tsuchiya J, Umemoto K, Wentzcovitch RM (2004) Phase transition in $MgSiO_3$ perovskite in the earth's lower mantle. Earth Planet Sci Lett 224:241-248
Tsuchiya J, Tsuchiya T, Wentzcovitch RM (2005) Transition from the Rh_2O_3(II)-to-$CaIrO_3$ structure and the high-pressure-temperature phase diagram of alumina. Phys Rev B 72:020103(R)
Tsuchiya T, Tsuchiya J (2006) New high-pressure phase relations in $CaSnO_3$. Am Mineral 91:1879-1887
Tsuchiya T, Tsuchiya J (2007) Structure and elasticity of *Cmcm* $CaIrO_3$ and their pressure dependences: *Ab initio* calculations. Phys Rev B 76:144119
Tsuchiya T, Yusa H, Tsuchiya J (2007) Post-Rh_2O_3(II) transition and the high pressure-temperature phase diagram of Gallia: A first-principles and x-ray diffraction study. Phys Rev B 76:174108
Umemoto K, Wentzcovitch RM (2006) Potential ultrahigh pressure polymorphs of ABX_3-type compounds. Phys Rev B 74:224105
Umemoto K, Wentzcovitch RM (2008) Prediction of an U_2S_3-type polymorph of Al_2O_3 at 3.7 Mbar. Proc Nat Acad Sci USA 105:6526-6530
Umemoto K, Wetnzcovitch RM, Allen PB (2006a) Dissociation of $MgSiO_3$ in the cores of gas giants and terrestrial exoplanets. Science 311:983-986
Umemoto K, Wentzcovitch RM, Weidner DJ, Parise JB (2006b) $NaMgF_3$: A low-pressure analog of $MgSiO_3$. Geophys Res Lett 33:L15304
Valencia D, O'Connell RJ, Sasselov D (2006) Internal structure of massive terrestrial planets. Icarus 181:545-554
Vanderbilt D (1990) Soft self-consistent pseudopotentials in a generalized eigenvalue formalism. Phys Rev B 41:7892-7895(R)
Wallace D (1972) Thermodynamics of Crystals. John Wiley, Hoboken, NJ
Wang L, Pan Y, Ding Y, Yang Y, Yang W, Mao WL, Sinogeikin SV, Meng Y, Shen G, Mao HK (2009) High-pressure induced phase transition of Y_2O_3 and Y_2O_3:Eu^{3+}. Appl Phys Lett 94:061921
Weinstein SA (1995) The effects of a deep mantle endothermic phase change on the structure of thermal convection in silicate planets. J Geophys Res 100(E6):11719-11728
Wentzcovitch RM (1991) Invariant molecular-dynamics approach to structural phase transitions. Phys Rev B 44:2358-2361
Wentzcovitch RM, Martins JL, Allen PB (1992) Energy versus free energy in first principles calculations. Phys Rev B 45:11372-11374
Wentzcovitch RM, Martins JL, Price GD (1993) *Ab Initio* molecular dynamics with variable cell shape: application to $MgSiO_3$. Phys Rev Lett 70:3947-3950
Wentzcovitch RM, Yu YG, Wu Z (2010) Thermodynamic properties and phase relations in mantle minerals investigated by first principles quasiharmonic theory. Rev Mineral Geochem 71:59-98

Wood BJ, Rubie DC (1998) The effect of alumina on phase transformations at the 660-kilometer discontinuity from Fe-Mg partitioning experiments. Science 273:1522-1524

Wu X, Steinle-Neumann G, Narygina O, Kantor I, McCammon C, Prakapenka V, Swamy V, Dubrovinsky L (2009) High-pressure behavior of perovskite: $FeTiO_3$ dissociation into $(Fe_{1-\delta},Ti_{\delta})O$ and $Fe_{1+\delta}Ti_{2-\delta}O_5$. Phys Rev Lett 103:065503

Wu Z, Umemoto K, Wu A, Zhen JC, Wentzcovitch RM (2008) P-V-T relations in MgO; an ultra-high pressure scale for planetary sciences applications. J Geophys Res 113:B06204

Xu Y, McCammon C, Poe BT (1998) The effect of alumina on the electrical conductivity of silicate perovskite. Science 282:922-924

Yagi T, Suzuki T, Akimoto S (1983) New high-pressure polymorphs ion sodium halides. J Phys Chem Solids 44:135-140

Yusa H, Tsuchiya T, Sata N, Ohishi Y (2008a) Rh_2O_3(II)-type structures in Ga_2O_3 and In_2O_3 under high pressure: Experiment and theory. Phys Rev B 77:064107

Yusa H, Tsuchiya T, Tsuchiya J, Sata N, Ohishi Y (2008b) α-Gd_2S_3-type structure in In_2O_3: Experiments and theoretical confirmation of a high-pressure polymorph in sesquioxide. Phys Rev B 78:092107

Yusa H, Tsuchiya T, Sata N, Ohishi Y (2009) High-pressure phase transition to the Gd_2S_3 structure in Sc_2O_3: a new trend in dense structures in sesquioxides. Inorg Chem 48:7537-7543

Zhang J, Weidner DJ (1999) Thermal Equation of state of aluminum-enriched silicate perovskite. Science 284:782-784

Zhao Y, Weidner DJ, Ko J, Leinenweber K, Liu X, Li B, Meng Y, Pacalo REG, Vaughan MT, Wang Y, Yeganeh-Haeri A (1994) Perovskite at high P-T conditions: An in situ synchrotron X ray diffraction study of $NaMgF_3$ perovskite. J Geophys Res 99(B2):2871-2885

Reviews in Mineralogy & Geochemistry
Vol. 71 pp. 315-335, 2010

Computer Simulations on Phase Transitions in Ice

Koichiro Umemoto

Department of Geology and Geophysics
University of Minnesota
Minneapolis, Minnesota, 55455, U.S.A.

umemoto@cems.umn.edu

INTRODUCTION

Ice has a very rich phase diagram (Petrenko and Whitworth 1999) as shown in Figure 1. Up to now, sixteen crystalline phases have been identified experimentally. All crystalline phases except ice X consist of water molecules connected to four adjacent ones by hydrogen bonds. Broadly speaking, the crystalline phases of ice can be described as follows: the low-pressure forms ($P < \sim 1$ GPa) consist of a unique hydrogen-bond network (1HBN). The high-pressure forms (~ 1 GPa $< P < \sim 80$ GPa) are made up of two interpenetrating networks (2HBN); they are often referred to as self-clathrates. Ice I_h and XI are prototypical forms of low-pressure ice with low density, and ice VII and VIII are those of high-pressure ices with high density. Figure 2 shows atomic structures of ice XI and VIII, hydrogen-ordered phases. Between these prototypical phases, there are many complex phases.

Ices can be seen as hydrogen-bond networks. In Earth science, the hydrogen bond is known to play important roles, when water exists in nominally anhydrous minerals (NAMs) (Keppler and Smyth 2006). Therefore it is crucial to be able to describe properties of ice, i.e., properties of hydrogen bonds computationally, in order to investigate effects of water in NAMs. For planetary science, it is essential to know the sequence of phase transitions and equation of states of the phases up to extremely high pressure for the interior modeling of planets and satellites. In Uranus and Neptune, icy giants, their rocky cores are considered to be surrounded by icy envelopes. Pressure at the core-envelope boundary is estimated to be as high as ~800 GPa (Guillot 2004). Ice may exist also on the surface of many satellites of solar giants (Kouchi et al. 1994; Roush 2001).

In this chapter, we will discuss computational studies on phase transitions in ice: phase boundaries, order-disorder transitions, proton tunneling, pressure-induced amorphization, and predictions of new phases. We will see that first-principles methods are quite effective in theoretical studies of ice. However, despite recent developments of computational power, first-principles molecular-dynamics simulations on very large systems (hydrogen-disordered phases and amorphous) have been still computationally demanding. We will briefly discuss intermolecular model potentials used for molecular-dynamics simulations with the large number of molecules also. There are still other interesting and important properties of phase transitions: melting, superionicity, effects of defects, and so on, but we will not go into these topics.

BACKGROUND OF COMPUTATIONAL METHOD

Within first-principles theoretical framework, a solid is viewed as an interacting many-particle system of nuclei and electrons. Their interactions are calculated using Hartree-Fock (HF)-based quantum-chemistry methods or density-functional theory (DFT) (Hohenberg and Kohn 1964; Kohn and Sham 1965). Sophisticated quantum-chemistry methods, which treat

1529-6466/10/0071-0015$05.00 DOI: 10.2138/rmg.2010.71.15

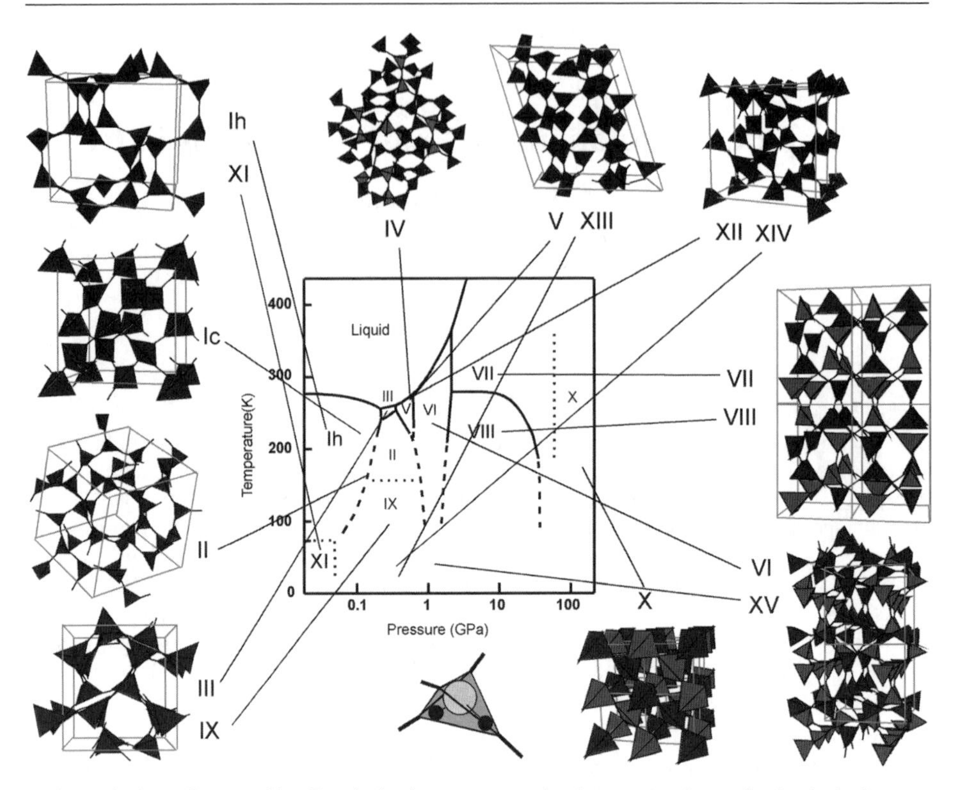

Figure 1. Phase diagram of ice. Tetrahedra denote water molecules connected to each other by hydrogen bonds represented by solid lines. In ice VI, VII, VIII, X, and XV, two interpenetrating lattices are distinguished by different colors of tetrahedra. Ice I_h, I_c, III, IV, V, VI, VII, and XII are hydrogen-disordered. Other phases are hydrogen-ordered. Ice X is a hydrogen-bond-symmetrized form.

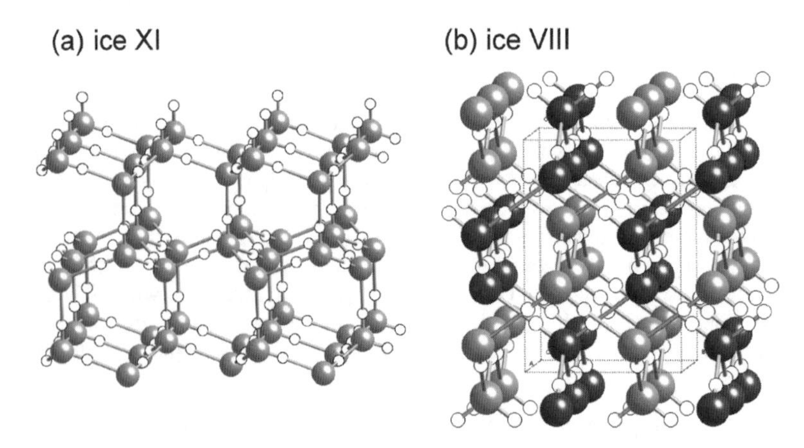

Figure 2. Atomic structures of (a) ice XI and (b) ice VIII. Small white and large gray spheres denote hydrogen and oxygen atoms, respectively. Ice XI is a hydrogen-ordered form of ice I_h, the most usual form of ice. Ice I_h is hexagonal ($P6_3/mmc$) and paraelectric. Ice XI is orthorhombic ($Cmc2_1$) and ferroelectric. For ice VIII, light and dark gray spheres represent two sublattices interpenetrating to each other. Ice VIII is a hydrogen-ordered form of ice VII. Ice VII is cubic ($Pn3m$) and paraelectric. Ice VIII is tetragonal ($I4_1/amd$) and antiferroelectric. In both cases, the hydrogen ordering lowers the symmetry.

the exchange energy explicitly, are computationally quite expensive. They have been used for studies of water molecule and clusters. But the number of their applications to solid ice is not so large. On the other hand, DFT, which is formally exact but requires approximation of the exchange-correlation (XC) functional, allows us to perform calculations with reasonable computational efficiency. First-principles DFT calculations combined with pseudopotentials (Bachelet et al. 1982; Vanderbilt 1990; Troullier and Martins 1991), the plane-wave expansion, and the variable-cell-shape molecular dynamics (Parrinello and Rahman 1980, 1981; Wentzcovitch 1991; Wentzcovitch et al. 1993) have been applied for many minerals, including ice. The essence of a method based on DFT will be described elsewhere (Wentzcovitch et al. 2010, in this volume). To describe the hydrogen bond by DFT, the choice of XC functional is highly important. Functionals based on local-density approximation (LDA) (Ceperley and Alder 1980; Perdew and Zunger 1981) and generalized-gradient approximation (GGA), PW91 (Perdew and Wang 1992), PBE (Perdew et al. 1996) and BLYP (Becke 1988; Lee et al. 1988), have been successfully used for many minerals. As will be discussed later, we need to employ GGA functionals at least to study solid ice.

To perform molecular dynamics simulations of nonperiodic systems of H_2O, e.g., liquid water, melting, and amorphization, it is necessary to prepare large supercells with hundreds or thousands water molecules. Such systems have been still quite difficult to be dealt with by first-principles methods, although applications of first-principles molecular dynamics on hundreds water molecules have started to emerge (e.g., Schwegler et al. 2008). For such large systems, classical molecular dynamics using intermolecular model potentials have been performed. These model potentials assume a rigid-body molecule; intramolecular bond length, the H-O-H angle, and point charges are fixed. Potential energy is represented by the Coulomb interactions between all intermolecular pairs of point charges and the Lennard-Jones term between oxygens. Parameters of model potentials were set to reproduce properties of objective systems, e.g., liquid water at room temperature. Depending on parameters, there are a lot of model potentials. Figure 3 shows geometry and parameters of the TIP4P model potential (Jorgensen et al. 1983), one of model potentials used widely. There are several model potentials derivative from the TIP4P potential for different purposes: TIP4P-Ew (Horn et al. 2004), TIP4P/Ice (Abascal et al. 2005), and TIP4P/2005 (Abascal and Vega 2005). The TIP4P potential is four-site model potential and consists of four interaction sites: two hydrogens, one oxygen, and one dummy site at which negative charge is placed. There are five- and six-site models also (e.g., Mahoney and Jorgensen 2000; Nada and van der Eerden 2003).

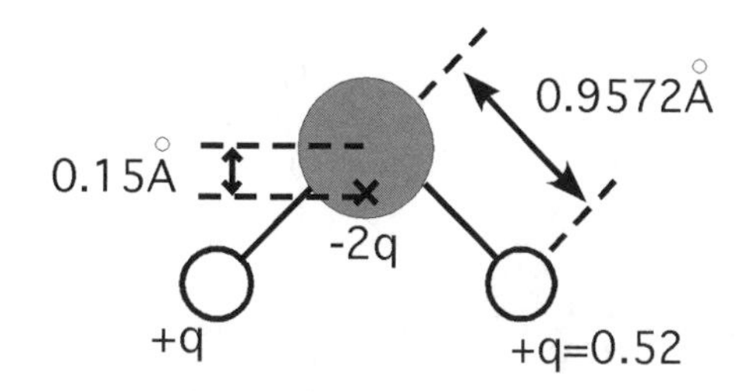

Figure 3. TIP4P potential. White and gray spheres denote hydrogens and oxygen, respectively. Positive charges (+q) are placed at hydrogen atoms, while negative charge (−2q) is moved off oxygen and toward hydrogens at a point (represented by cross) on the bisector of the HOH angle.

EXCHANGE-CORRELATION FUNCTIONAL

H_2O monomer and dimer

Choice of XC functionals in DFT is crucial for calculation of ice. Table 1 shows the dependence of calculated structural parameters of H_2O monomer and dimer on the type of XC functionals and compares DFT results with results of quantum-chemistry calculations and experimental values. Clearly the intermolecular O-O distance of dimer strongly depends on XC functionals, although the intramolecular O-H bond length does not. LDA underestimates the O-O distance, which is primarily determined by the hydrogen bond (O···H), by ~10%. GGAs greatly improve the O-O distance.

Solid ice

Ice I_h is hexagonal and the most usual form of ice. Hamann (1997) calculated volume of Bernal-Fowler ice, which has been frequently used for model of hydrogen-disordered ice I_h (Bernal and Fowler 1933) by several XC functionals (Table 2). We can see the same trend as in the dimer. LDA underestimates volume considerably, while GGAs work much better. Table 3 shows lattice constants of orthorhombic ice XI, a hydrogen-ordered form of ice I_h, by PBE. Also for ice XI, PBE works appropriately, compared with a quantum-chemistry calculation and experimental values.

Ice VIII is a good phase for checking the ability of theoretical methods to describe ice properties, because it is a hydrogen-ordered phase obtained without any impurities, its unit cell consists of the small number (4) of molecules, and its structural parameters have been well constrained experimentally (Kuhs et al. 1984). Table 4 shows that PW91-type, PBE-type, and

Table 1. Structural parameters of H_2O monomer and dimer. "LDA+GC" denotes LDA with gradient corrections (Perdew 1986; Becke 1988).

	Monomer		**Dimer**	**Reference**
	O-H (Å)	**H-O-H (deg)**	**O-O (Å)**	
LDA	0.978	104.4	2.710	Sim et al. (1992)
GGA(PW)	0.980	104.4	2.877	
LDA	0.987	102.1	2.72	Laasonen et al. (1992)
LDA+GC	0.985	101.9	2.93	
LDA	0.978	103.3	2.699	Lee et al. (1993)
LDA+GC	0.987	104.7	2.910	
GGA(PW91)	0.971	104.5		Tse and Klug (1998)
HF+fourth-order perturbation	0.961	103.7		Pisani et al. (1996)
HF+MP2			2.904- 2.913	Frisch et al. (1986)
B3LYP	0.961-0.969	103.7-105.1	2.872- 2.925	Casassa et al. (2005)
Exp	0.95718	104.523	2.980	Benedict et al. (1956); Dyke et al. (1977)

Table 2. Structural parameters of ice I_h at 0 GPa. Volumes calculated by Hamann (1997) are of Bernal-Fowler ice.

		V (Å^3/H_2O)	**Reference**
LDA	Static	26.43	Hamann (1997)
PW91	Static	31.35	
PBE	Static	31.82	
TIP4P	250 K	31.93	Sanz et al. (2004)
SPC/E	250 K	31.69	
Exp.	10 K	32.05	Röttger et al. (1994)
Exp.	250 K	32.51	

BLYP-type GGA can describe structural properties of ice VIII. It should be noted here that a vibrational effect on the hydrogen bond is significant. The hydrogen-bond length with the zero-point motion (ZPM) is longer by ~2.4% than that without the ZPM. On the other hand, the intramolecular bond length hardly changes by the inclusion of the ZPM.

These studies show that not LDA but GGA functionals must be used for DFT studies of ice. PW91, PBE, and BLYP-type functionals have been widely used. Nevertheless, these GGA functionals still have shortcomings. As shown in Tables 1 and 4, compared with experimental values and quantum chemical calculations, the GGA functionals overestimate the intramolecular O-H bond length. To improve this, many functionals have been proposed. According to an extensive study of XC functionals (Dahlke and Truhlar 2005), PBE1W in which one parameter of PBE is tuned, some meta functionals which contain kinetic energy density, and some hybrid GGA functionals which contain exact HF exchange can describe better than the GGA functionals O-H and O···H bonds.

Table 3. Structural parameters of ice XI at 0 GPa.

	a (Å)	*b* (Å)	*c* (Å)	**Reference**
PBE	4.383	7.623	7.163	Umemoto et al. (2004)
HF+fourth-order perturbation	4.50	7.65	7.33	Pisani et al. (1996)
Exp. (5 K)	4.465	7.858	7.292	Line and Whitworth (1996)

Table 4. Structural parameters of ice VIII at 2.4 GPa.

		a (Å)	*c* (Å)	*V* (Å³)	**O-H (Å)**	**O···H (Å)**	**O-O (Å)**	**O···O (Å)**	**Ref.**
GGA(PW91)	static						2.875	2.756	(1)
GGA(BLYP)	static	4.656 (fixed)	6.775 (fixed)	73.44 (fixed)	0.9877		2.8797	2.7724	(2)
GGA(PBE)	static	4.865	6.710	72.80	0.987	1.883	2.870	2.760	(3)
	0K +ZPM	4.727	6.817	76.17	0.986	1.929	2.914	2.797	
	10 K	4.727	6.817	76.16	0.986	1.929	2.914	2.797	
HF	static	4.72	7.11	79.20	0.953	2.002	2.955	2.855	(4)
Exp. (D_2O)	10 K	4.656	6.775	73.44	0.9685	1.911	2.879	2.743	(5)

References: (1) Tse and Klug (1998); (2) Kuo and Klein (2004); (3) Umemoto and Wentzcovitch (2004); (4) Ojamäe et al. (1994); (5) Kuhs et al. (1984)

CRYSTALLINE PHASES

Phase diagram

Determination of complex phase diagram of ice has been challenging for computational studies. In principle, phase diagram is determined by comparing Gibbs free energies of all ice phases. However it is quite difficult to calculate hydrogen-configurational and vibrational entropies for ice phases with the large number of molecules by first principles. The most comprehensive study of the phase diagram of ice (Sanz et al. 2004; Vega et al. 2008) was carried out using empirical model potentials, TIP4P (Jorgensen et al. 1983) and SPC/E (Berendsen et al. 1987) and using the ideal Pauling entropy in fully hydrogen-disordered phases (Pauling 1935) with an additional correction for partially-disordered phases (Howe and Whitworth 1987). Both model potentials have been successfully used for simulations of liquid water. However, the

SPC/E model potential was found to be unsuitable for calculation of the phase diagram of ice; it predicted that ice I_h was stable only at negative pressure and ice III and V were metastable. The TIP4P model potential, on the other hand, was shown to be able to capture the main features of the complex experimental phase diagram at low pressures. Small stability regions of ice III and V were reproduced. Metastability of ice IV and IX was also confirmed. However the TIP4P model potential has two major problems. First, it failed to reproduce the negative Clapeyron slope observed experimentally for ice VII-VIII transition. This indicates that the model potential cannot describe high-pressure properties of ice appropriately. It may be mainly because the model potentials assume the rigid molecule and were designed to reproduce properties of liquid water at low pressures, although the intramolecular O-H bond length increases as pressure increases actually. Secondly, the TIP4P model potential does not predict ferroelectric $Cmc2_1$ structure to be the lowest-energy structure of hydrogen-ordered form of ice I_h, i.e., ice XI. This contradicts neutron-diffraction measurements (Leadbetter et al. 1985; Howe and Whitworth 1989; Line and Whitworth 1996; Jackson et al. 1997; Fukazawa et al. 2006).

Indeed, no model potential has predicted the $Cmc2_1$ structure to have the lowest energy (Buch et al. 1998). In Sanz et al. (2004), ice XI was assumed to be the antiferroelectric $Pna2_1$ phase (Fig. 4), which was calculated to have lower energy than the $Cmc2_1$ phase by the MCY model potential (Davidson and Morokuma 1984). On the other hand, first-principles calculations predicted the $Cmc2_1$ phase to be the lowest-energy configuration for ice XI. A BLYP-DFT calculation showed the $Cmc2_1$ phase has the lowest energy among 16 configurations of an 8-molecule orthorhombic unit cell and 14 configurations of a 12-molecule hexagonal unit cell (Singer et al. 2005; Knight et al. 2006). According to quantum-chemical calculations, the $Cmc2_1$ phase is nearly isoenergetic to $Pna2_1$ (Pisani et al. 1996; Casassa et al. 2005). These first-principles results suggest the energy difference between possible hydrogen configurations of ice XI is very small. Such subtle energy difference may be difficult to be dealt with model potentials. For planetary science, it is an important issue whether ice XI is ferroelectric or antiferroelectric. A neutron diffraction measurement showed that the I_h-XI transition occurred at temperatures at the surfaces of Uranus and its satellites (Fukazawa et al. 2006). Ferroelectric ice sheets with a thickness of several kilometers should become a huge source for an electric field.

Order-disorder transition

Richness of the phase diagram of ice originates in part in hydrogen order-disorder transitions such as I_h-XI (Tajima et al. 1982), III-IX (Whalley et al. 1968; Londono et al. 1993), VII-VIII (Kuhs et al. 1984), V-XIII and XII-XIV (Salzmann et al. 2006), and VI-XV (Salzmann et al. 2009). It is very difficult to determine experimentally such order-disorder

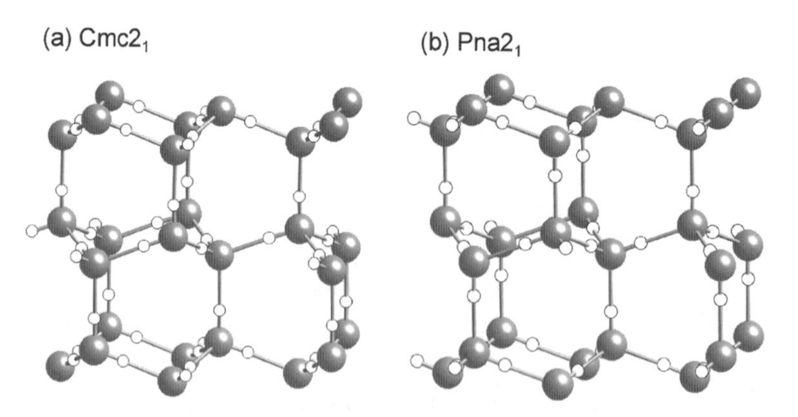

Figure 4. Structure of (a) $Cmc2_1$ and (b) $Pna2_1$ phases.

phase boundaries at low pressures and temperatures, since reorientation of water molecules in pure ice is severely suppressed. To introduce defects and promote hydrogen ordering, impurities were doped: KOH for I_h-XI (Tajima et al. 1982) and DCl for V-XIII, XII-XIV, and VII-VIII (Salzmann et al. 2006, 2009).

Theoretical studies can be very helpful but they also have their intrinsic challenges, i.e., the calculation of entropy and free energy in hydrogen-disordered phases. Metropolis Monte Carlo simulations in large cells have been frequently used. Order-disorder transitions are captured by abrupt jumps of the total energy in simulations for increasing and decreasing temperature. Entropy is calculated by integrating heat capacity ($S = \int dT' C_v/T'$) and then free energy is obtained ($F = E - TS$). Barkema and de Boer (1993) performed a Metropolis Monte Carlo simulation in a 1400-molecule cell for the I_h-XI transition using a model potential; in fact, they studied an order-disorder transition from the antiferroelectric *Pna*2$_1$ phase, because their model potential predicted that ferroelectric *Cmc*2$_1$ phase did not have the lowest energy. Recently a series of Metropolis Monte Carlo simulations using graph invariants were performed for I_h-XI, VII-VIII (Singer et al. 2005; Knight et al. 2006), III-IX (Knight and Singer 2006), V-XIII (Knight and Singer 2008), and VI-XV (Knight and Singer 2005). In these simulations, the energy in supercells consisting of a large number of molecules was expanded using graph invariants with coefficients determined by fitting to energies obtained from first-principles calculations for many hydrogen configurations in smaller cells. For example, for the VII-VIII transition, Metropolis Monte Carlo simulations were performed in a 1024-molecule supercell using parameters determined from BLYP-DFT calculations in a 16-molecule supercell. In the simulations with ascending and descending temperature, the order-disorder transitions were clearly observed by sharp jumps in total energies. This simulation predicted a first-order phase transition with temperature hysteresis, which was also observed experimentally (Pistorius et al. 1968; Song et al. 2003). Transition temperature at ~2.4 GPa was predicted to be ~228 K, lower than experimental value, ~270 K (Pistorius et al. 1968; Pruzan 1994; Pruzan et al. 1997; Song et al. 2003).

Thorough first-principles studies on the order-disorder transitions were performed also. In these studies, first all possible hydrogen configurations were generated. Then, the partition functions were calculated:

$$Z(V,T) = \sum_i w_i \exp\left(-\frac{E_i(V)}{k_B T}\right) \quad (1)$$

where $E_i(V)$ and w_i are the total energies calculated by first-principles and degeneracy of the i-th symmetrically inequivalent configurations. All thermodynamic quantities can be calculated from the partition function. Order-disorder transition temperature can be determined the peak position of heat capacity. This method was applied for several order-disorder transitions in ice and proved to work (Kuo and Klein 2004; Kuo 2005; Kuo et al. 2005; Kuo and Kuhs 2006). The main deficiency of this method is the limitation of the number of molecules in supercells, i.e., the convergence with respect to the number of molecules. For a large supercell of thousands of molecules, it is impossible calculate energy spectrum over all possible configurations with structural optimization; the number of possible configurations in N molecules increases as $\sim(3/2)^N$. And also it has been practically difficult to perform first-principles calculations on the large number of molecules with structural optimization. The small number of molecules leads to a rather sluggish change of the total energy between order and disordered phases and a wide full-width-at-half-maximum of heat capacity.

The two methods described above were applied at only one volume. Therefore they did not determine the phase boundaries over wide pressure. And also the vibrational contribution to the free energy was not considered. Very recently the *phase boundary* of the ice VII-VIII transition was studied by first principles (Umemoto et al. 2010). It was calculated by performing a complete sampling of the ensemble of molecular configurations generated in a 16-molecule supercell.

The partition function of Equation (1) was extended to take the vibrational contribution into account. This approach successfully reproduced important observables: compression curve of ice VII, the negative Clapeyron slope, and the isotope effect.

Ice X: Hydrogen-bond symmetrization

All phases except ice X consist of water molecules. In these phases, between two adjacent oxygens, there are double potential wells. Hydrogen atoms exist at one of two potential wells (ice rule). Therefore hydrogen is away from the midpoint between two oxygens. The intramolecular bonds are distinct from the hydrogen bonds; the former is shorter and the latter is longer. Pressure on ice VII at room temperature does not change the oxygen bcc sublattice up to ~170 GPa, as shown by X-ray measurements (Hemley et al. 1987; Wolanin et al. 1997; Loubeyre et al. 1999; Somayazulu et al. 2008; Sugimura et al. 2008). But the nature of the hydrogen bond changes drastically. Under pressure, double potential wells in ice VII get close to each other and are finally merged into single one. Then there is no difference between intramolecular and hydrogen bonds; both bonds are symmetric. This phase with hydrogen-bond symmetrization is called ice X. The hydrogen-bond symmetrization was predicted by Holzapfel (1972) and Schweizer and Stillinger (1984). When double potential wells are close to each other, the proton tunneling occurs. But the tunneling effect cannot be treated by classical molecular dynamics (here "classical" means that only electrons are treated quantum mechanically and nuclei are not). An *ab initio* path-integral molecular dynamics simulation, which treated *both* electrons and nuclei quantum mechanically (Marx and Parinnelo 1994, 1996; Tuckerman et al. 1996), demonstrated a sequence of changes of the hydrogen-bond character from ice VIII to ice X (Benoit et al. 1998). The hydrogen-bond symmetrization was shown to occur through the four stages: ice VIII, dynamically-disordered ice VII, dynamically-disordered ice X, and ice X. At ~35 GPa where the distance between two oxygens is long, hydrogen existed at one of double potential wells (ice VIII). At ~59 GPa, the spatial distribution of hydrogen became bimodal around the midpoint of two oxygens (dynamically-disordered ice VII). In a parallel classical simulation, hydrogen still existed at one of double potential wells. This indicated the proton tunneling between double potential wells. At ~90 GPa, the spatial distribution of hydrogen became unimodal at the midpoint of two oxygens, although the classical simulation showed that double potential wells were still separated (dynamically-disordered ice X). Finally at ~122 GPa, double potential wells were merged into single one (ice X).

Dynamical stability of ice X was studied by first-principles lattice dynamics (Caracas 2008). At 160 GPa, all phonon frequencies of ice X were positive, and ice X was dynamically stable. Under decompression, a softening of one optic phonon mode was revealed. Around 114 GPa, the branch of this mode became nearly flat and was unstable along the symmetry lines in the Brillouin zone. Displacements of the unstable mode correspond to the bouncing back and forth of hydrogen atoms between two oxygens. The unstable mode is related to the double potential wells. At high pressure, the potential well for hydrogen is single at the midpoint of two oxygens, and ice X is dynamically stable. But at lower pressure (<114 GPa), the single potential is split into the double potential wells. Then hydrogen atoms at the midpoint of two oxygens become unstable and roll down into one of double potential wells. The nearly flat branch of the unstable optic mode suggests the transition from ice X to dynamically-disordered ice VII, even at 0 K. Many incommensurate transitions of hydrogen positions should be simultaneously induced; this may be related to the discussion between pressure-induced amorphization and acoustic phonon collapse (see the section of Amorphization). If the phonon softening had occurred at only one zone-edge point, ice X would have transformed to hydrogen-ordered antiferroelectric ice VIII directly. The splitting of the single potential well was also investigated by other first-principles calculations (Lee et al. 1992, 1993).

Infrared and Raman spectroscopy have been used to investigate transitions from ice VII/VIII to dynamically-disordered ice VII/X experimentally (Hirsch and Holzapfel 1986; Pruzan

1994; Aoki et al. 1996; Goncharov et al. 1996, 1999; Pruzan et al. 1997; Song et al. 2003). The transition from ice VIII to dynamically-disordered VII can be detected by a change of Raman spectrum due to the symmetry change from tetragonal $I4_1/amd$ to cubic $Pn3m$. From ice VII to X, there is no symmetrical change; both have the $Pn3m$ symmetry. In ice VII, under compression, the O-H stretching mode frequencies fall toward zero abruptly. This softening is followed by appearance of new mode characteristic of ice X. Therefore calculation of IR-active and Raman-active phonon frequencies should be powerful theoretical tool to investigate the transition from ice VIII/VII to ice X. However, in the transition zone from ice VIII/VII to ice X, the harmonic approximation is no longer valid, although it works well in low pressure. In this zone, the double potential wells are very close to each other and strongly distorted, and anharmonicity must be taken into account. By first-principles molecular dynamics, infrared and Raman spectrum in the transition zone were calculated from the correlation functions of the total dipole moment (Bernasconi et al. 1998) and the polarizability (Putrino and Parrinello 2002).

Beyond ice X

Experimentally, no higher-pressure phase beyond ice X has been identified so far. A X-ray diffraction measurement showed that an intensity ratio (I_{111}/I_{222}) increased abruptly beyond ~150 GPa (Loubeyre et al. 1999). This abrupt rise of I_{111}/I_{222} was expected to be a sign of a new phase beyond ice X. However, a first-principles molecular dynamics simulation at 300 K up to ~200 GPa reproduced this rise of I_{111}/I_{222} but did not find any transition to a phase beyond ice X (Benoit et al. 2002).

A first-principles molecular dynamics simulation predicted a transition at ~300 GPa from ice X to an orthorhombic phase with the space group *Pbcm* (Benoit et al. 1996). This orthorhombic phase had lower enthalpy than the antifluorite structure, which was previously proposed as post ice X (Demontis et al. 1988). This phase has a distorted hcp oxygen sublattice (Fig. 5). Hydrogen atoms exist at the midpoint of two oxygens. This *Pbcm* phase was predicted also by a dynamical instability of ice X (Caracas 2008). At 429 GPa, in ice X, an acoustic mode became soft at the zone-edge M point. In the supercell which can accommodate displacements induced by the soft mode, structural optimization led to the *Pbcm* phase. According to these two individual first-principles studies, the *Pbcm* phase is expected to be the most promising candidate of a phase beyond ice X. Benoit et al. (1996) named the *Pbcm* phase "ice XI", but roman numerals have been used only for crystalline phases identified experimentally. The phase named "ice XI" is the hydrogen-ordered form of ice I_h. If the *Pbcm* phase is found now, it should be named "ice XVI". In Uranus and Neptune, pressure at the core-envelope boundary is estimated to be ~800 GPa (Guillot 2004). Therefore the transition from ice X to the *Pbcm* phase may play an important role in these icy giants.

Isostructural transition in ice VIII at low pressure?

At low temperature (< ~130 K), ice VIII can be recovered to ambient pressure. Hirsch and Holzapfel (1986) reported that, during decompression of ice VIII, Raman-active frequencies of O-H stretching modes shifted nonlinearly below ~3 GPa, while these frequencies behaved linearly at higher pressures. These nonlinear behaviors could be signs of a phase transition. A neutron diffraction measurement (Besson et al. 1997a) showed a nonlinear variation of the z fractional coordinate of the oxygen atom, z(O), between two plateaus at ~1 GPa and ~7 GPa, suggesting an isostructural transition from ice VIII to ice VIII′. A first-principles calculation also showed an abrupt jump in z(O) between two plateaus (Tse and Klug 1998). However, this isostructural transition has been in controversy. As shown in Figure 6, another first-principles study did not find any plateau in z(O) and did not detect any sign of an isostructural phase transformation, while it well reproduced nonlinear behaviors of Raman-active mode frequencies and volume dependences of structural parameters, which may be precursors of amorphization as will be discussed later (Umemoto and Wentzcovitch 2005a). An abrupt softening of the E_u translational mode frequency, which was considered to be characteristic

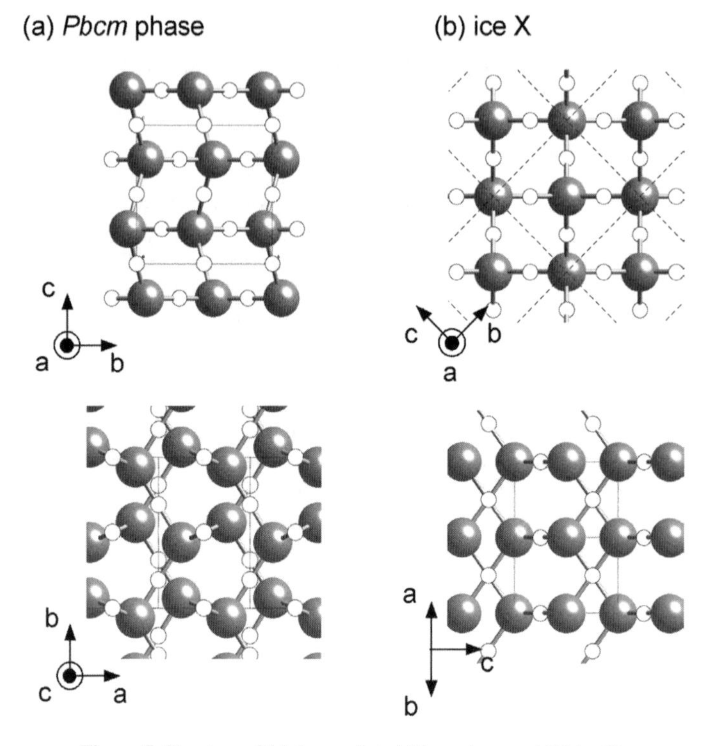

Figure 5. Structure of (a) the predicted *Pbcm* phase and (b) ice X.

of the isostructural transition (Besson et al. 1997a), was not observed by a synchrotron far-infrared spectroscopy (Klug et al. 2004). Moreover, a X-ray diffraction experiment showed the unit-cell axial ratio (c/a) underwent changes at 10~14 GPa, instead of the plateaus of z(O) (Yoshimura et al. 2006). Therefore, further studies are necessary to answer questions of whether or not the isostructural transition in ice VIII at low pressure really exists and what structural parameter would be relevant for this transition.

AMORPHOUS

Amorphization of ice I_h

In 1984, ice I_h was discovered to undergo the pressure-induced amorphization to high-density amorphous (HDA) at 1 GPa and 77 K, at which ice I_h is already metastable (Mishima et al. 1984). After this discovery, pressure-induced amorphization was found in several other materials (Ponyatovsky and Barkalov 1992): α-quartz, SiO_2 (Hemley et al. 1986, 1988), SnI_4 (Fujii et al. 1985), $AlPO_4$ (Kruger and Jeanloz 1990; Polian et al. 1993), and $Co(OH)_2$ (Nguyen et al. 1997). Amorphization of ice I_h is related with the negative Clapeyron slope of the melting line of ice I_h; the observed pressure-temperature condition of amorphization (1 GPa and 77 K) is near the metastable extension of the melting line in the stability field of another high-pressure phase. Later careful calorimetric experiments in ice I_h uncovered a more complex relationship between the amorphization phase boundary and the metastable extension of the melting line (Mishima 1996). They both have negative Clapeyron slopes, but these lines do not coincide; below amorphization temperature, the amorphization phase boundary appears

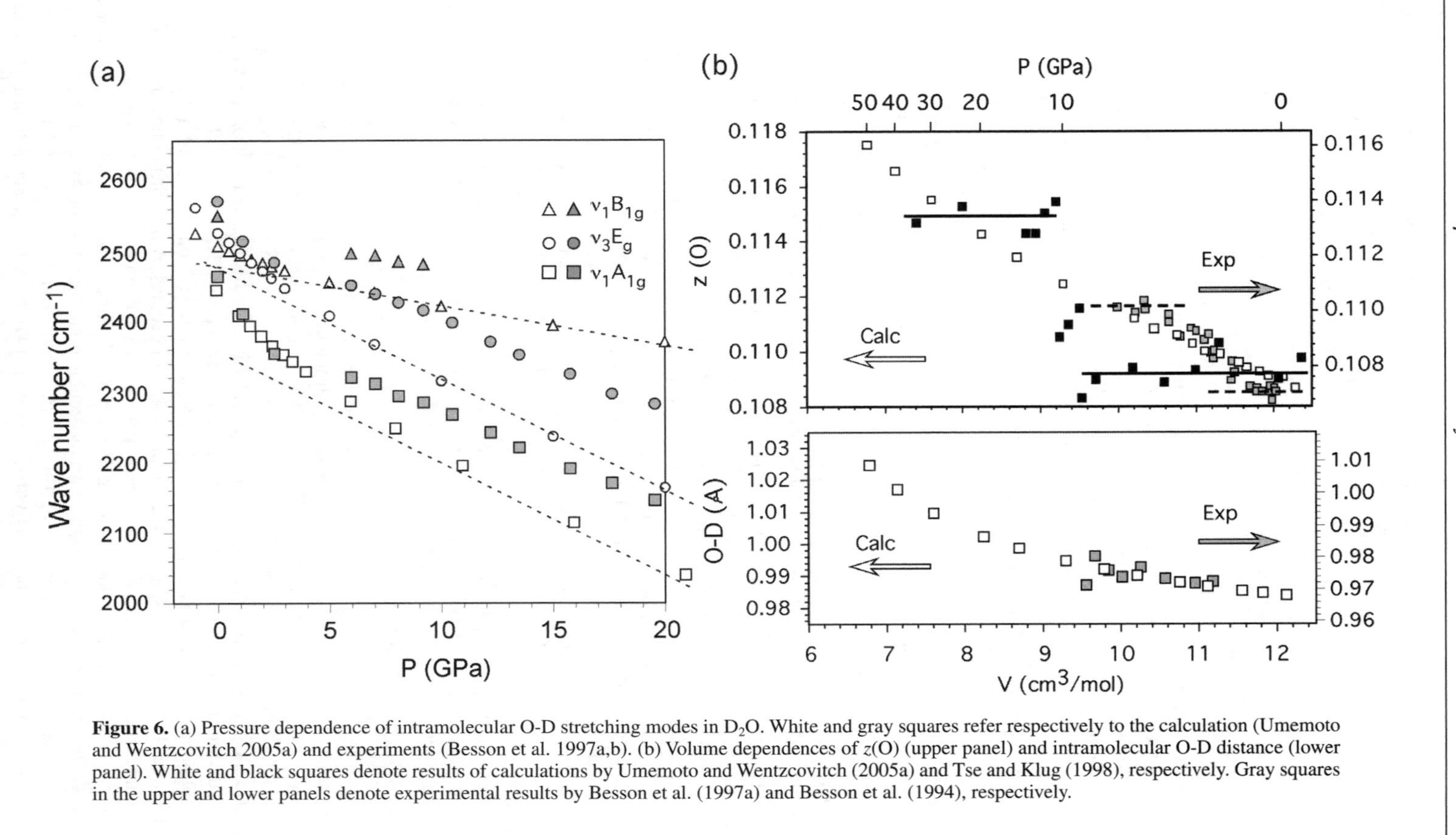

Figure 6. (a) Pressure dependence of intramolecular O-D stretching modes in D_2O. White and gray squares refer respectively to the calculation (Umemoto and Wentzcovitch 2005a) and experiments (Besson et al. 1997a,b). (b) Volume dependences of *z*(O) (upper panel) and intramolecular O-D distance (lower panel). White and black squares denote results of calculations by Umemoto and Wentzcovitch (2005a) and Tse and Klug (1998), respectively. Gray squares in the upper and lower panels denote experimental results by Besson et al. (1997a) and Besson et al. (1994), respectively.

at higher pressure than the extrapolated melting line. Negative Clapeyron slope is a common feature of systems that undergo pressure-induced transitions to higher entropy forms, liquid water and amorphous ice. Pressure decreases volume ($\Delta V < 0$), and amorphization increases entropy ($\Delta S > 0$). Consequently Clapeyron slope, $dP/dT = \Delta S/\Delta V$, is negative.

Analogies exist between pressure-induced amorphization in ice and in α-quartz. In α-quartz, amorphization occurs slightly beyond a hypothetical melting line, i.e., between 21 and 30 GPa in the stability field of stishovite (Hemley et al. 1988). Theoretically, force-field simulations indicated (Chaplot and Sikka 1993) and first-principles calculations (Wentzcovitch et al. 1998) confirmed that pressure in this range softens an acoustic phonon mode at the zone-edge K point. A soft phonon mode induces a phase transition by corresponding unstable phonon mode. Indeed, an intermediate phase with an x-ray diffraction pattern which could result from the soft-phonon phase was found near 21 GPa (Kingma et al. 1993a,b; Wentzcovitch et al. 1998). This zone-edge instability happens to be related to an elastic mechanical instability occurring at zone center (Binggeli et al. 1994). If a crystal is mechanically stable, then the elastic energy and the Born stability criteria are positive (Born and Huang 1954). As a matter of fact, first-principles lattice-dynamical calculations revealed that just before the softening of the K-point mode, the entire lowest acoustic branch along the T line started flattening and collapsed soon after the zone-edge instability at the zone edge. As corresponding acoustic branch was very flat, many incommensurate modes should tend to go soft at the same time. These studies suggest that the process of amorphization is the result of the occurrence of many incommensurate transitions (Baroni and Giannozzi 1998). In short, mechanism of amorphization of α-quartz has been investigated by the Born stability criteria and phonon softening. Amorphization in ice can be also investigated in a similar way.

Pressure-induced amorphization of ice I_h was theoretically investigated by molecular dynamics simulations with the TIP4P model potential (Tse and Klein 1987, 1990; Tse 1992); ice I_h is hydrogen-disordered and quite difficult to be dealt with by first principles, because a supercell with the large number of molecules is necessary. It was demonstrated that at 0.9 GPa and 80 K the long-range order of water molecules largely disappeared. This was accompanied by abrupt decreases of the Born stability criteria, c_{11}-$|c_{12}|$ and $(c_{11}+c_{12})c_{33}-2c_{13}$, to nearly zero and by softening of the transverse acoustic phonons. Therefore molecular dynamics simulations showed that pressure-induced amorphization is induced by mechanical instability. It was also shown computationally that the thermodynamic melting line met the mechanical instability curve at ~160 K (Tse et al. 1999a). A crossover was clarified between two distinct mechanisms of pressure-induced transformations in ice I_h: thermodynamic melting and amorphization, agreeing with experiment by Mishima (1996). These pictures were supported by recent neutron scattering experiments (Strässle et al. 2004, 2007)

Amorphization of ice VIII

Ice VIII also amorphizes. But, opposed to ice I_h, ice VIII transforms to low-density amorphous (LDA) when it is *decompressed* to ambient pressure and then heated to ~130 K (Klug et al. 1989; Balagurov et al. 1991). It is due to an estimated metastable melting line of ice VIII has a positive Clapeyron slope; because $\Delta V > 0$ under decompression, a positive Clapeyron slope gives $\Delta S > 0$. Ice VII, a hydrogen-disordered form of ice VIII, was also found to undergo amorphization under decompression (Klotz et al. 1999). Ice VIII is a desirable phase for a first-principles study of pressure-induced amorphization in ice, since it is hydrogen-ordered and its unit cell is small with 4 molecules. A first-principles study of ice VIII clarified a mechanism of amorphization of ice VIII (Umemoto and Wentzcovitch 2004). Between ~20 GPa and ~4 GPa, intramolecular and hydrogen-bond lengths varied almost linearly. However below ~4 GPa, the intramolecular bond length decreased nonlinearly and the hydrogen bond length behaves oppositely. These nonlinear behaviors can be seen as signs of the imminent collapse of the two hydrogen-bond networks. All Born stability criteria in the tetragonal system (c_{44}, c_{66}, $c_{11}-|c_{12}|$, and $(c_{11}+c_{12})c_{33}-2c_{13}^2$) decreased under decompression (Fig. 7). Among them, the first one to vanish at "−1.2 GPa" (*negative* pressure) was c_{44}. This is consistent with the experimental observation of metastable ice VIII at 0 GPa. Ice VIII at 0 GPa is on the path leading to a mechanical instability. c_{44} determines the restoring force for the long-wavelengths degenerate transverse acoustic (TA) modes propagating in the z direction. The pressure dependence of these TA branches is displayed in Figure 8. At "−1.2 GPa," the entire branches became unstable along the Λ line at once. In ice VIII, *acoustic phonon collapse* is unambiguously linked to the mechanical instability through c_{44}. All TA normal modes

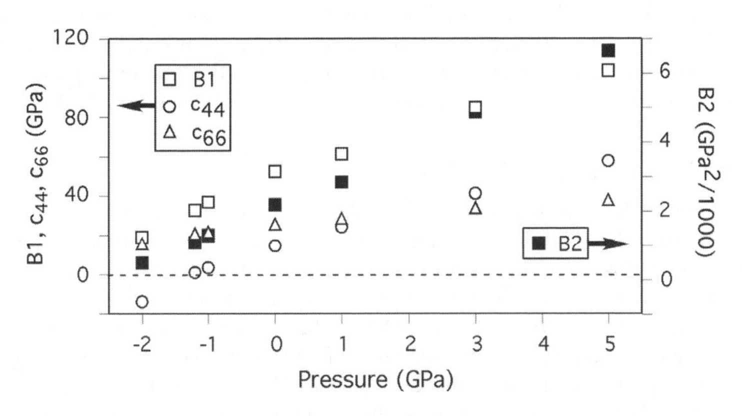

Figure 7. Calculated Born stability criteria of ice VIII (Umemoto and Wentzcovitch 2004).

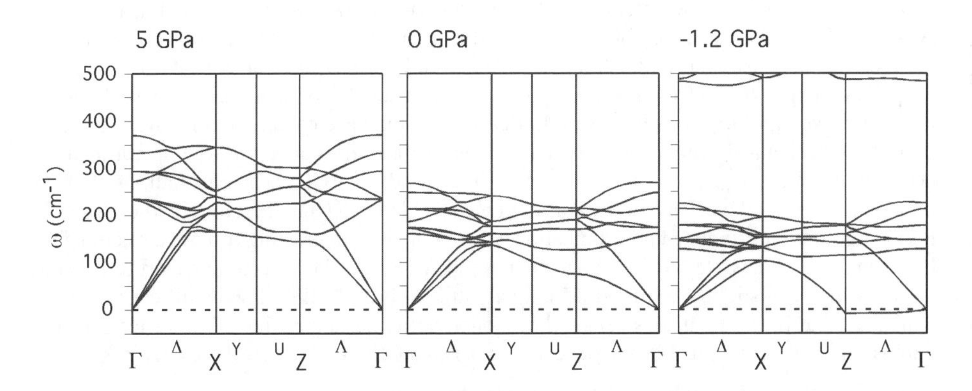

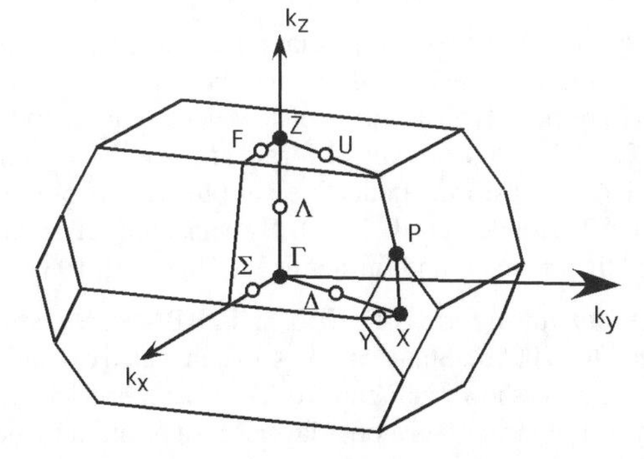

Figure 8. Phonon dispersions of translational modes at 5, 0, and −1.2 GPa (static values). Negative wavenumbers are actually imaginary numbers. On the right, the Brillouin zone of the body-centered tetragonal lattice is shown (Umemoto and Wentzcovitch 2004).

along the Λ line involve hydrogen-bond stretching, some more, some less, depending on the wave number. Therefore, the concurrent instability of all phonons in these branches can be attributed to the critical stretching of the same bond at the critical pressure. Importantly, these modes are not unstable at pressure amorphization occurs experimentally. Ice VIII was found to be mechanically and vibrationally stable at 0 GPa. However, all these phonons should be thermally excited simultaneously when the amorphous is produced by annealing. At 0 GPa, amorphization happens at ~130 K (Klug et al. 1989), while the energies of these phonons lie within ~10 meV, i.e., ~120 K. Hence, this first-principles study gave a reasonable explanation of amorphization of ice VIII by annealing at 0 GPa.

Amorphization of ice VIII was studied using the TIP4P model potential also (Tse et al. 1999b). This model potential study showed that one of the Born stability criteria, $c_{11}-|c_{12}|$, became negative under decompression, being contrary to the first-principles study. This contradiction might be because the intramolecular O-H bond length is fixed in the TIP4P model potential.

Prediction of amorphization of ice XI

Since both ice VII and VIII amorphize under decompression, whether hydrogen atoms are ordered or disordered might not matter to pressure-induced amorphization in ice. So we have one question: Does ice XI, a hydrogen-ordered form of ice I_h, undergo pressure-induced amorphization? To answer this question, ice XI under pressure was studied by first principles (Umemoto et al. 2004). The behavior of ice XI can be summarized as follows: up to 3.3 GPa the lattice parameters decreased monotonically. Adjacent layers parallel to the xy plane shifted laterally with respect to each other along the y axis. The compression mechanism changes beyond 3.5 GPa. At this pressure the lattice parameter a started increasing while the others decreased. This behavior indicates the onset of instability. Inspection of phonon dispersions under pressure revealed the softening of the lowest transverse acoustic branch along the Δ line. Around 1.5 GPa the frequency of the lowest acoustic mode at the Y point started decreasing. As pressure increases the entire branch flattened and softens until at 3.5 GPa the phonon frequency at $\mathbf{q} \approx (0,4/5,0)$ vanished, and the entire acoustic branch went soft (collapsed) (Fig. 9). Beyond 4 GPa, two Born stability criteria of the base-centered lattice went to zero, indicating mechanical instability. Although the phase boundaries at very low temperature have not been constrained, ice XI may be metastable beyond ~1.5 GPa. The calculations showed phonon collapse and mechanical instability in metastable ice XI and therefore predicted ice XI to undergo pressure-induced amorphization, like ice I_h. It is interesting to note a relationship between the phonon softening in ice XI and in ice I_h. Ice XI has the base-centered orthorhombic lattice, which is a slightly distorted hexagonal lattice. So the Brillouin zone of ice XI is very similar to that of ice I_h. In this sense, the phonon softening along the Δ line in ice XI may correspond to that along the Σ line observed experimentally in ice I_h (Strässle et al. 2004).

Amorphization between low ↔ high density transformation

As described in Introduction, the crystalline phases can be classified into two groups: a unique hydrogen-bond network (1HBN) with low density and two interpenetrating networks (2HBN) with high density. Ice I_h and XI are prototypical forms of low-density ices, and ice VII and VIII are those of high-density ices. These prototypical forms are closely related to each other through amorphous phases. Ice I_h transforms under pressure to ice VII preceded by intermediate HDA (Tse and Klein 1987; Hemley et al. 1989). By annealing at ambient pressure, ice VIII transforms to LDA and then to ice I_c and further to I_h (Klug et al. 1989).

Direct transformations between the prototypical forms of 1HBN and 2HBN were clarified by first principles (Umemoto and Wentzcovitch 2005b). Static compression, in which excitations of unstable phonons were suppressed, on ice XI is shown in Figure 10. The $Cmc2_1$ orthorhombic symmetry of ice XI allows displacements of adjacent hexagonal layers in opposite directions

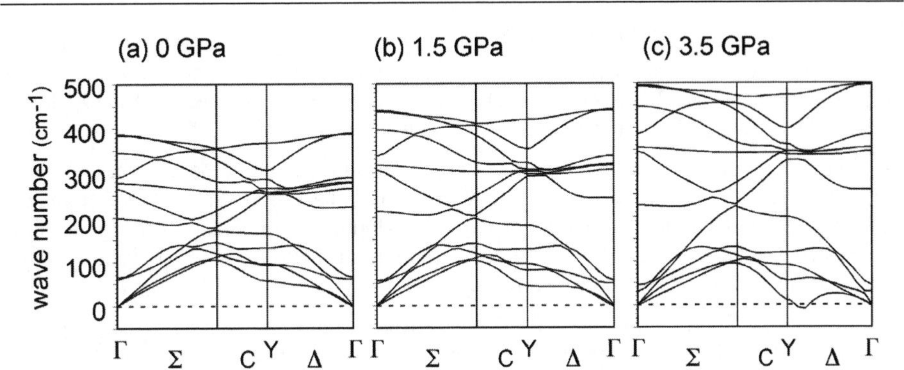

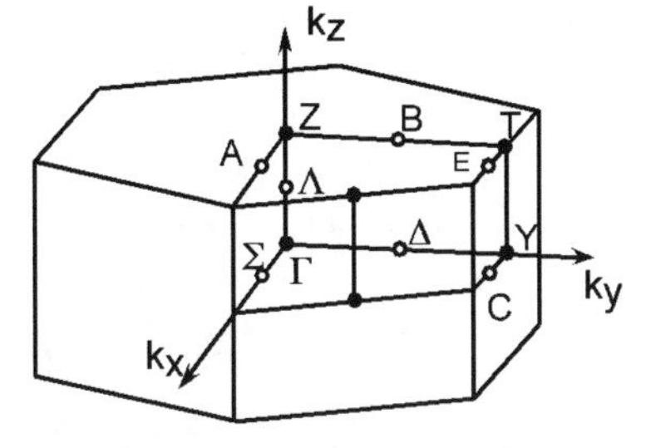

Figure 9. Phonon dispersions of translational modes at 0, 1.5, and 3.5 GPa. Negative wavenumbers are actually imaginary numbers. On the right, the Brillouin zone of the body-centered orthorhombic lattice is shown (Umemoto et al. 2004).

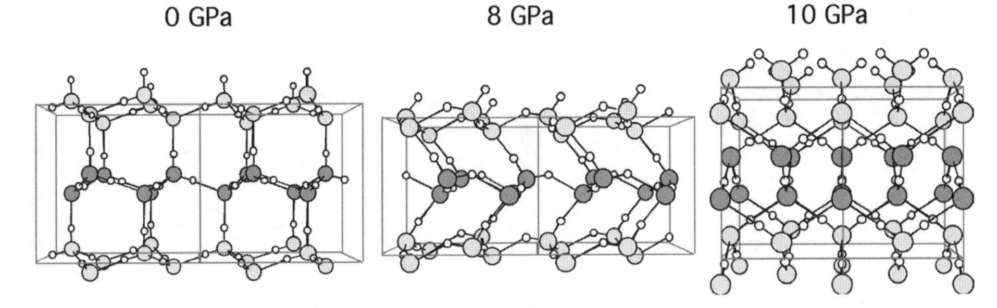

Figure 10. Transition from ice XI to ice VIII-like phase.

(Leadbetter et al. 1985; Jackson et al. 1997); these displacements cannot occur in ice I_h. Under compression, these displacements became larger. Finally at ~9 GPa, some hydrogen bond broke, while new hydrogen bonds reformed. Then, ice XI transformed to a 2HBN structure. The resultant structure is very similar to ice VIII. But it is ferroelectric, while ice VIII is antiferroelectic. We refer to this new phase as ice VIII-like. Ice VIII-like has two molecules per unit cell. Its symmetry is $P4_2nm$ which is a supergroup of $Cmc2_1$, the space group of ice XI. As expected, ice VIII-like is metastable with respect to ice VIII; its enthalpy is higher by 0.01 eV/molecule at 10 GPa. However, it should be stabilized with respect to ice VIII in an external electric field along the [001] direction. Possibly, cooling water or paraelectric VII at high pressure in an electric field might produce ice VIII-like. Quick dynamic compression at low temperature to high pressures might also produce ice VIII-like bypassing intermediate phases and amorphous.

As shown before, ice VIII becomes dynamically unstable at −1.2 GPa because of the softening of the two acoustic branches along the entire Λ line. Unstable ice VIII was computationally annealed at −1.2 GPa in a doubled unit cell, capable of the zone-edge soft mode, and then recompressed back to 0 GPa (Fig. 11). This procedure facilitated hydrogen-bond reconstructions and produced a hexagonal-diamond lattice, a 1HBN phase. This new phase is similar to ice XI, but antiferroelectric. We refer to this phase as ice XI-like. The symmetry of ice XI-like is $P2_12_12_1$. This is the 'M'-type hydrogen-ordering of ice I_h in Howe (1987).

Figure 12 shows the pressure-dependence of the densities at 0 K of the four phases: ice XI, ice XI-like, ice VIII, and ice VIII-like. 1HBN and 2HBN phases are well separated from each other. This diagram has the typical form of a hysteresis loop. Experimental amorphous densities are located inside the loop. They are produced as intermediate steps in the transformation process between low ↔ high density phases.

During transitions from ice XI to ice VIII-like and from ice VIII to ice XI-like, ferroelectricity and antiferroelectricity were kept. This is parallel to the case of hydrogen-disordered and paraelectric ice I_h. Below ~77 K, paraelectric ice I_h transforms to HDA and then to paraelectric ice VII at ~4 GPa in the stability field of antiferroelectric ice VIII; transition to ice VIII requires higher temperature (Hemley et al. 1989). Therefore there is huge energy

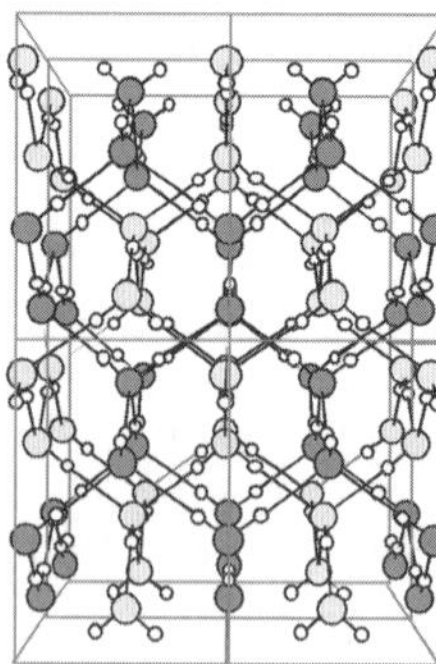

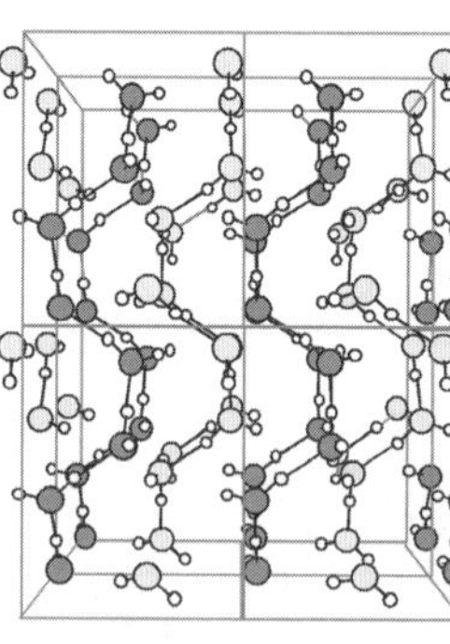

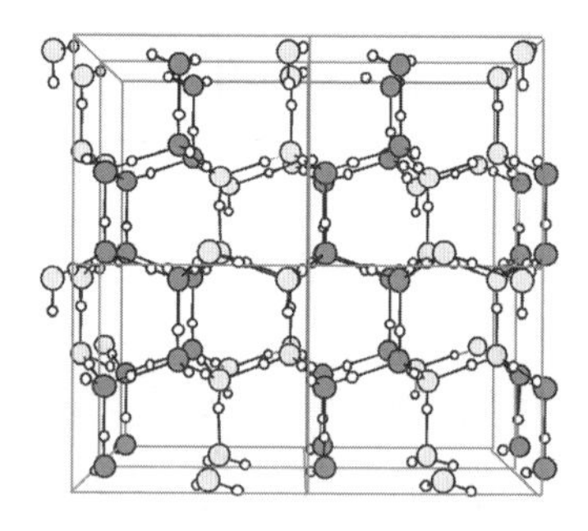

Figure 11. Transition from (a) ice VIII to (c) ice XI-like.

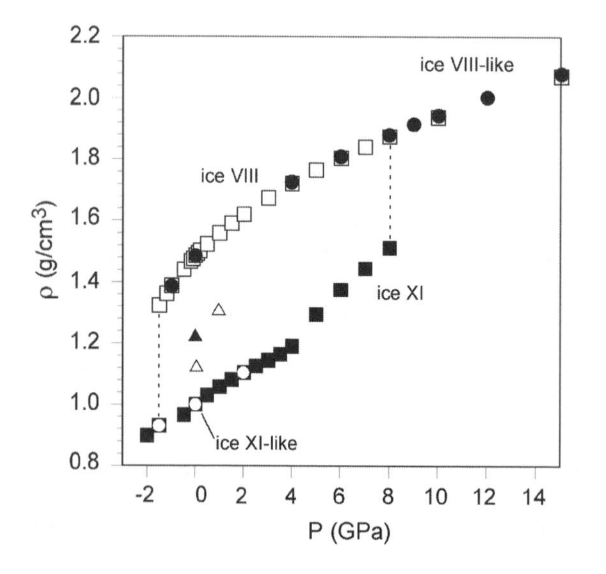

Figure 12. Calculated densities of ice XI (black squares), ice VIII-like (black circles), ice VIII (white squares), and ice XI-like (white circle). White and black triangles denote the experimental densities of high-density amorphous (HDA) (Mishima et al. 1984) and very-high-density amorphous (VHDA) (Loerting et al. 2001), respectively.

barrier in changing dipole reordering. At low temperature, phase transformations even via intermediate amorphous phases may end up in metastable phases, if drastic dipole reordering is required to reach the final stable phase.

ACKNOWLEDGMENTS

Many parts of this article are based on a set of first-principles studies which I performed with collaborators: Renata M. Wentzcovitch, Stefano Baroni, and Stefano de Gironcoli, using Quantum-ESPRESSO package (Giannozzi et al. 2009). These works were supported by NSF grants No. EAR-0135533 (COMPRESS), EAR-0230319, ITR-0428774 (Vlab), EAR-0757903, EAR-0635990, and ATM-0428774 (Vlab).

REFERENCES

Abascal JLF, Sanz E, Garcia Fernandez R, Vega C (2005) A potential mode for the study of ices and amorphous water TIP4P/Ice. J Chem Phys 122:234511

Abascal JLF, Vega C (2005) A general purpose model for the condensed phases of water: TIP4P/2005. J Chem Phys 123:234505

Aoki K, Yamawaki H, Sakashita M (1996) Observation of fano interference in high-pressure ice VII. Phys Rev Lett 76:784-786

Bachelet GB, Hamann DR, Schlüter M (1982) Pseudopotentials that work: from H to Pu. Phys Rev B 26:4199-4228

Balagurov AM, Barkalov OI, Kolesnikov AI, Mironova GM, Ponyatovskii EG, Sinitsyn VV, Fedotov VK (1991) Neutron-diffraction study of phase transitions of high-pressure metastable ice VIII. JETP Lett 53:30-34

Barkema GT, de Boer J (1993) Properties of a statistical model of ice at low temperatures. J Chem Phys 99:2059-2067

Baroni S, Giannozzi P (1998) High pressure lattice instabilities and structural phase transformations in solids from ab-initio lattice dynamics. Mater Res Soc Symp Proc 499:233-241

Becke AD (1988) Density-functional exchange-energy approximation with correct asymptotic behavior. Phys Rev A 38:3098-3100

Benedict WS, Gailar N, Plyler EK (1956) Rotation-vibration spectra of deuterated water vapor. J Chem Phys 24:1139-1165

Benoit M, Bernasconi M, Focher P, Parrinello M (1996) New high-pressure phase of ice. Phys Rev Lett 76:2934-2936

Benoit M, Marx D, Parrinello M (1998) Tunnelling and zero-point motion in high-pressure ice. Nature 392:258-261

Benoit M, Romero AH, Marx D (2002) Reassigning hydrogen-bond centering in dense ice. Phys Rev Lett 89:145501

Berendsen HJC, Grigers JR, Straatsma (1987) Missing term in effective pair potentials. J Phys Chem 91:6269-6271

Bernal JD, Fowler RH (1933) A theory of water and ionic solution, with particular reference to hydrogen and hydroxyl ions. J Chem Phys 1:515-548

Bernasconi M, Silvestrelli PL, Parrinello M (1998) *Ab initio* infrared absorption study of the hydrogen-bond symmetrization in ice. Phys Rev Lett 81:1235-1238

Besson JM, Pruzan Ph, Klotz S, Hamel G, Silvi B, Nelmes RJ, Loveday JS, Wilson RM, Hull S (1994) Variation of interatomic distances in ice VIII to 10 GPa. Phys Rev B 49:12540-12550

Besson JM, Klotz S, Hamel G, Marshall WG, Nelmes RJ, Loveday JS (1997a) Structural Instability in ice VIII under pressure. Phys Rev Lett 78:3141-3144

Besson JM, Kobayashi M, Nakai T, Endo S, Pruzan Ph (1997b) Pressure dependence of Raman linewidths in ices VII and VIII. Phys Rev B 55:11919-11201

Binggeli N, Keskar NR, Chelikowsky JR (1994) Pressure-induced amorphization, elastic instability, and soft modes in α-quartz. Phys Rev B 49:3075-3081

Born M, Huang K (1954) Dynamical Theory of Crystal Lattices. Clarendon, Oxford

Buch V, Sandler P, Sadlej (1998) Simulations of H_2O solid, liquid, and clusters, with an emphasis on ferroelectric ordering transition in hexagonal ice. J Phys Chem B 102:8641-8653

Caracas R (2008) Dynamical Instabilities of ice X. Phys Rev Lett 101:085502

Casassa S, Calatayud M, Doll K, Minot C, Pisani C (2005) Proton ordered cubic and hexagonal periodic models of ordinary ice. Chem Phys Lett 409:110-117

Ceperley DM, Alder BJ (1980) Ground state of the electron gas by a stochastic method. Phys Rev Lett 45:566-569

Chaplot SL, Sikka SK (1993) Comment on "Elastic Instability in α-Quartz under Pressure". Phys Rev Lett 71:2674

Dahlke E, Truhlar DG (2005) Improved density functionals for water. J Phys Chem B 109:15677-15683

Davidson ER, Morokuma K (1984) A proposed antiferroelectric structure for proton ordered ice I_h. J Chem Phys 81:3741-3742

Demontis P, LeSar R, Klein ML (1988) New high-pressure phases of ice. Phys Rev Lett 60:2284-2287

Dyke TR, Mack KM, Muenter JS (1977) The structure of water dimer from molecular beam electric resonance spectroscopy. J Chem Phys 66, 498

Fujii Y, Kowaka M, Onodera A (1985) The pressure-induced metallic amorphous state of SnI,: I. A novel crystal-to-amorphous transition studied by x-ray scattering. J Phys C 18:789-797

Fukazawa H, Hoshikawa A, Ishii Y, Chakoumakos BC, Fernandez-Baca JA (2006) Existence of ferroelectric ice in the universe. Astrophys J 652:L57-L60

Frisch MJ, Del Bene JE, Binkley JS, Schaefer III HF (1986) Extensive theoretical studies of the hydrogen-bonded complexes $(H_2O)_2$, $(H_2O)_2H^+$, $(HF)_2$, $(HF)_2H^+$, F_2H^-, and $(NH_3)_2$. J Chem Phys 84:2279-2289

Giannozzi P et al. (2009) Quantum ESPRESSO: a modular and open-source software project for quantum simulations of materials. J Phys: Condens Matter 21:395502

Goncharov AF, Struzhkin VV, Somayazulu MS, Hemley RJ, Mao HK (1996) Compression of ice to 210 gigapascals: infrared evidence for a symmetric hydrogen-bonded phase. Science 273:218-220

Goncharov AF, Struzhkin VV, Mao HK, Hemley RJ (1999) Raman spectroscopy of dense H_2O and the transition to symmetric hydrogen bonds. Phys Rev Lett 83:1998-2001

Guillot T (2004) Probing the giant planet. Phys Today 57(4):63-69

Hamann DR (1997) H_2O hydrogen bonding in density-functional theory. Phys Rev B 55:R10157-R10160

Hemley RJ, Mao HK, Bell PM, Mysen BO (1986) Raman spectroscopy of SiO_2 glass at high pressure. Phys Rev Lett 57:747-750

Hemley RJ, Jephcoat AP, Mao HK, Zha CS, Finger LW, Cox DE (1987) Static compression of H_2O-ice to 128 GPa (1.28 Mbar). Nature 330:737-740

Hemley RJ, Jephcoat AP, Mao HK, Ming LC, Manghanani MH (1988) Pressure-induced amorphization of crystalline silica. Nature 334:52-54

Hemley RJ, Chen LC, Mao HK (1989) New transformations between crystalline and amorphous ice. Nature 338:638-640

Hirsch KR, Holzapfel WB (1986) Effect of high pressure on the Raman spectra of Ice VIII and evidence for ice X. J Chem Phys 84:2771-2775

Hohenberg P, Kohn W (1964) Inhomogeneous electron gas. Phys Rev 136:B864-B871

Holzapfel WB (1972) On the symmetry of the hydrogen bonds in ice VII. J Chem Phys 56:712-715

Horn HW, Swope WC, Pitera JW, Madura JD, Dick TJ, Hura GL, Head-Gordon T (2004) Development of an improved four-site water model for biomolecular simulations: TIP4P-Ew. J Chem Phys 120:9665-9678

Howe R (1987) The possible ordered structures of ice I_h. J Phys Colloque 48(C1):599-604

Howe R, Whitworth RW (1987) The configurational entropy of partially ordered ice. J Chem Phys 86:6443-6445

Howe R, Whitworth RW (1989) A determination of the crystal structure of ice XI. J Chem Phys 90:4450-4453

Jackson SM Nield V, Whitworth RW, Oguro M, Wilson CC (1997) Single-crystal neutron diffraction studies of the structure of ice XI. J Phys Chem B 101:6142-6145

Jorgensen WL, Chandrasekhar J, Madura JD, Impey RW, Klein M (1983) Comparison of simple potential functions for simulating liquid water. J Chem Phys 79:926-936

Keppler H, Smyth JR (eds) (2006) Water in Nominally Anhydrous Minerals. Rev Mineral Geochem Volume 62. Mineralogical Society of America

Kingma KJ, Meade C, Hemley RJ, Mao HK, Veblen DR (1993a) Microstructural observations of α-quartz amorphization, Science 259:666-669

Kingma KJ, Hemley RJ, Mao HK, Veblen DR (1993b) New high-pressure transformation in α-quartz. Phys Rev Lett 70:3927-3930

Klotz S, Besson JM, Hamel G, Nelmes RJ, Loverday JS, Marshall WG (1999) Metastable ice VII at low temperature and ambient pressure. Nature 398:681-684

Klug DD, Handa YP, Tse JS, Whalley E (1989) Transformation of ice VIII to amorphous ice by "melting" at low temperature. J Chem Phys 90:2390-2392

Klug DD, Tse JS, Liu Z, Gonze X, Hemley RJ (2004) Anomalous transformations if ice VIII. Phys Rev B 70:144113

Kohn W, LJ Sham (1965) Self-consistent equations including exchange and correlation effects. Phys Rev 140:A1133-A1138

Kouchi A, Yamamoto T, Kozasa T, Kuroda T, Greenberg JM (1994) Conditions for condensation and preservation of amorphous ice and crystallinity of astrophysical ices. Astron Astrophys 290:1009-1018

Knight C, Singer SJ (2005) Prediction of a phase transition to hydrogen bond ordered form of ice VI. J Phys Chem B 109:21040-21046

Knight C, Singer SJ, Kuo JL, Hirsch TK, Ojamäe L, Klein ML (2006) Hydrogen bond topology and the ice VII/VIII and I_h /XI proton ordering phase transitions. Phys Rev E 73:056113

Knight C, SJ Singer (2006) A reexamination of the ice III/IX hydrogen bond ordering phase transition. J Chem Phys 125:064506

Knight C, Singer SJ (2008) Hydrogen bond ordering in ice V and the transition to ice XIII. J Chem Phys 129:164513

Kruger MB, Jeanloz R (1990) Memory glass: an amorphous material formed from $AlPO_4$. Science 249:647-649

Kuhs WF, Finney JL, Vettier C, Bliss DV (1984) Structure and hydrogen ordering in ices VI, VII, and VIII by neutron powder diffraction. J Chem Phys 81:3612-3623

Kuo JL, Klein ML (2004) Structure of ice-VII and ice-VIII: a quantum mechanical study. J Phys Chem 108:19634-19639

Kuo JL (2005) The low-temperature proton-ordered phases of ice predicted by *ab initio* methods. Phys Chem Chem Phys 7:3733-3737

Kuo JL, Klein ML, Kuhs WF (2005) The effect of proton disorder on the structure of ice- I_h: A theoretical study. J Chem Phys 123:134505

Kuo JL, Kuhs WF (2006) A first principles study on the structure of ice-vi: static distortion, molecular geometry, and proton ordering. J Phys Chem B 110:3697-3703

Laasonen K, Csajka F, Parrinello M (1992) Water dimmer properties in the gradient-corrected density functional theory. Chem Phys Lett 194:172-174

Leadbetter AJ, Ward RC, Clark JW, Tucker PA, Matsuo T, Suga H (1985) The equilibrium low-temperature structure of ice. J Chem Phys 82:424-428

Lee C, Yang W, Parr RG (1988) Development of the Colle-Salvetti correlation-energy formula into a functional of the electron density. Phys Rev B 37:785-789

Lee C, Vanderbilt D, Laasonen K, Car R, Parrinello M (1992) *Ab Initio* studies on high pressure phases of ice. Phys Rev Lett 69:462-465

Lee C, Vanderbilt D, Laasonen K,Car R, Parrinello M (1993) *Ab initio* studies on the structural and dynamical properties of ice. Phys Rev B 47:4863-4872

Line CMB, Whitworth RW (1996) A high resolution neutron powder diffraction study of D_2O ice XI. J Chem Phys 104:10008-10013

Loerting T, Salzmann C, Kohl I, Mayer E, Hallbrucker A (2001) A second distinct structural "state" of high-density amorphous ice at 77 K and 1 bar. Phys Chem Chem Phys 3:5355-5357

Londono JD, Kuhs WF, Finney JL (1993) Neutron diffraction studies of ices III and IX on under-pressure and recovered samples. J Chem Phys 98:4878-4888

Loubeyre P, LeToullec R, Wolanin E, Hanfland M, Hausermann D (1999) Modulated phases and proton centering in ice observed by X-ray diffraction up to 170 GPa. Nature 397:503-506

Mahoney MW, Jorgensen WL (2000) Five-site model for liquid water and the reproduction of the density anomaly by rigid, nonpolarizable potential functions. J Chem Phys 112:8910-8922

Marx D, Parrinello M (1994) *Ab initio* path-integral molecular dynamics. Z Phys B 95:143-144

Marx D, Parrinello M (1996) *Ab initio* path integral molecular dynamics: Basic ideas. J Chem Phys 104:4077-4082

Mishima O, Calvert LD, Whalley E (1984) 'Melting ice' I at 77 K and 10 kbar: a new method of making amorphous solids. Nature 310:393-395

Mishima O (1996) Relationship between melting and amorphization of ice. Nature 384:546-549

Nada H, van der Eerden JPJM (2003) An intermolecular potential model for the simulation of ice and water near the melting point: A six-site model of H_2O. J Chem Phys 118:7401-7413

Nguyen JH, Kruger MB, Jeanloz R (1997) Evidence for "partial" (sublattice) amorphization in $Co(OH)_2$. Phys Rev Lett 78:1936-1939

Ojamäe L, Hermansson K, Dovesi R, Roetti C, Saunders VR (1994) Mechanical and molecular properties of ice VIII from crystal-orbital *ab-initio* calculations. J Chem Phys 100:2128-2138

Parrinello M, Rahman A (1980) Crystal structure and pair potentials: a molecular-dynamics study. Phys Rev Lett 45:1196-1199

Parrinello M, Rahman A (1981) Polymorphic transitions in single crystals: A new molecular dynamics method. J Appl Phys 52:7182-7190

Pauling L (1935) The structure and entropy of ice and of other crystals with some randomness of atomic arrangement. J Am Chem Soc 57:2680-2684
Perdew JP, Zunger A (1981) Self-interaction to density-functional approximations for many-electron systems. Phys Rev B 23:5048-5079
Perdew JP (1986) Density-functional approximation for the correlation energy of the inhomogeneous electron gas. Phys Rev B R8822-8824
Perdew JP, Wang Y (1992) Accurate and simple analytic representation of the electron-gas correlation energy, Phys Rev B 45:13244-13249
Perdew JP, Burke K, Ernzerhof M (1996) Generalized gradient approximation made simple. Phys Rev Lett 77:3865-3868
Petronko VF, Whitworth RW (1999) Physics of Ice. Oxford Univ. Press, Oxford
Pisani C, Casassa S, Ugliengo P (1996) Proton-ordered ice structures at zero pressure. A quantum-mechanical investigation. Chem Phys Lett 253:201-208
Pistorius CWFT, Rapoport E, Clark JB (1968) Phase diagrams of H_2O and D_2O at high pressures. J Chem Phys 48:5509-5514
Polian A, Grimsditch M, Philippot E (1993) Memory effects in pressure induced amorphous $AlPO_4$. Phys Rev Lett 71:3143-3145
Ponyatovsky EG, Barkalov OI (1992) Pressure-induced amorphous phases. Mater Sci Rep 8:147-191
Pruzan Ph (1994) Pressure effects on the hydrogen bond in ice up to 80 GPa. J Mol Struct 322:279-286
Pruzan Ph, Wolanin E, Gauthier M, Chervin JC, Canny B, Häuesermann D, Hanfland M (1997) Raman scattering and X-ray diffraction of ice in the megabar range. Occurrence of a symmetric disordered solid above 62 GPa. J Phys Chem B 101:6230-6233
Putrino A, Parrinello M (2002) Anharmonic Raman spectra in high-pressure ice from *ab initio* simulations. Phys Rev Lett 88:176401
Roush TL (2001) Physical state of ices in the outer solar system. J Geophys Res 106 (E12):33315-33323
Röttger K, Endriss A, Ihringer J, Doyle S, Kuhs WF (1994) Lattice constants and thermal expansion of H_2O and D_2O ice I_h between 10 and 265 K. Acta Crystallogr B50:644-648
Salzmann CG, Radaelli PG, Hallbrucker A, Mayer E, Finney JL (2006) the preparation and structures of hydrogen ordered phases of ice. Science 311:1758-1761
Salzmann CG, Radaelli PG, Mayer E, Finney JL (2009) Ice XV: a new thermodynamically stable phase of ice. Phys Rev Lett 103:105701
Sanz E, Vega C, Abasca JLF, MacDowell LG (2004) Phase diagram of water from computer simulation. Phys Rev Lett 92:255701
Schwegler E, Sharma M, Gygi F, Galli G (2008) Melting of ice under pressure. Proc Nat Acad Sci 105:14779-14783
Schweizer KS, Stillinger FH (1984) High pressure phase transitions and hydrogen-bond symmetry in ice polymorphs. J Chem Phys 80:1230-1240
Sim F, St-Amant A, Papai I, Salahub DR (1992) Gaussian density functional calculations on hydrogen-bonded systems. J Am Chem Soc 114:4391-4400
Singer SJ, Kuo JL, Hirsch TK, Knight C, Ojamäe L, Klein ML (2005) Hydrogen-bond topology and the ice VII/VIII and ice I_h /XI proton-ordering phase transitions. Phys Rev Lett 94:135701
Somayazulu M, Shu J, Zha CS, Goncharov AF, Tschauner O, Mao HK, Hemley RJ (2008) *In situ* high-pressure x-ray diffraction study of H_2O ice VII. J Chem Phys 128:064510
Song M, Yamawaki H, Fujihisa H, Sakashita M, Aoki K (2003) Infrared investigation on ice VIII and the phase diagram of dense ices. Phys Rev B 68:014106
Strässle Th, Saitta AM, Klotz S, Braden M (2004) Phonon dispersion of ice under pressure. Phys Rev Lett 93:225901
Strässle Th, Klotz S, Hamel G, Koza MM, Schober H (2007) Experimental evidence for a crossover between two distinct mechanisms of amorphization in ice I_h under pressure. Phys Rev Lett 99:175501
Sugimura E, Iitaka T, Hirose K, Kawamura K, Sata N, Ohishi Y (2008) Compression of H_2O ice to 126 GPa and implications for hydrogen-bond symmetrization: Synchrotron x-ray diffraction measurements and density-functional calculations. Phys Rev B 77:214103
Tajima Y, Matsuo T, Suga H (1982) Phase transition in KOH-doped hexagonal ice. Nature 299:810-812
Troullier N, Martins JL (1991) Efficient pseudopotentials for plane-wave calculations. Phys Rev B 43:1993-2006
Tse JS, Klein ML (1987) Pressure-induced phase transformation in ice. Phys Rev Lett 58:1672-1675
Tse JS, Klein ML (1990) Pressure induced amorphization of ice I_h. J Chem Phys 92:3992-3994
Tse JS (1992) Mechanical instability in ice I_h. A mechanism for pressure- induced amorphization. J Chem Phys 96:5482-5486
Tse JS, Klug DD (1998) Anomalous isostructural transformation in ice VIII. Phys Rev Lett 81:2466-2469

Tse JS, Klug DD, Tulk CA, Swainson I, Svensson EC, Loong CK, Shpakov V, Belosludov VR, Belosludov RV, Kawazoe Y (1999a) The mechanisms for pressure-induced amorphization of ice I_h. Nature 400:647-649

Tse JS, Shpakov VP, Belosludov VR (1999b) Vibrational spectrum, elastic moduli and mechanical stability in ice VIII. J Chem Phys 111:11111-11116

Tuckerman ME, Marx D, Klein M, Parrinello M (1996) Efficient and general algorithms for path integral Car-Parrinello molecular dynamics. J Chem Phys 104:5579-5588

Umemoto K, Wentzcovitch RM, Baroni S, de Gironcoli S (2004) Anomalous pressure-induced transition(s) in Ice XI. Phys Rev Lett 92:105502

Umemoto K, Wentzcovitch RM (2004) Amorphization in quenched ice VIII: A first-principles study. Phys Rev B 69:180103(R)

Umemoto K, Wentzcovitch RM (2005a) Theoretical study of the isostructural transformation in ice VIII. Phys Rev B 71:012102

Umemoto K, Wentzcovitch RM (2005b) Low $\leftrightarrow$ high density transformations in ice. Chem Phys Lett 405:53-57

Umemoto K, Wentzcovitch RM, de Gironcoli S, Baroni S (2010) Order-disorder phase boundary between ice VII and VIII obtained by first principles. American Physical Society, APS March Meeting 2010, March 15-19,2010, abstract #X31.009

Vanderbilt D (1990) Soft self-consistent pseudopotentials in a generalized eigenvalue formalism. Phys Rev B 41:7892-7895(R)

Vega C, Sanz E, Abascal JLF, Noya EG (2008) Determination of phase diagrams via computer simulation: methodology and applications to water, electrolytes and proteins. J Phys: Condens Matter 20:153101

Wentzcovitch RM (1991) Invariant molecular-dynamics approach to structural phase transitions. Phys Rev B 44:2358-2361

Wentzcovitch RM, Martins JL, Price GD (1993) *Ab initio* molecular dynamics with variable cell shape: application to $MgSiO_3$. Phys Rev Lett 70:3947-3950

Wentzcovitch RM, da Silva C, Chelikowsky JR, Binggeli N (1998) A new phase and pressure induced amorphization in silica. Phys Rev Lett 80:2149-2152

Wentzcovitch RM, Yu YG, Wu Z (2010) Thermodynamic properties and phase relations in mantle minerals investigated by first principles quasiharmonic theory. Rev Mineral Geochem 71:59-98

Whalley E, Heath JBR, Davidson DW (1968) Ice IX: an antiferroelectric phase related to ice III. J Chem Phys 48:2362-2370

Wolanin E, Pruzan Ph, Chervin JC, Canny B, Gauthier M, Häusermann D, Hanfland M (1997) Equation of state of ice VII up to 106 GPa. Phys Rev B 56:5781-5785

Yoshimura Y, Stewart ST, Somayazulu M, Mao HK, Hemley RJ (2006) High-pressure x-ray diffraction and Raman spectroscopy of ice VIII. J Chem Phys 124:024502

Reviews in Mineralogy & Geochemistry
Vol. 71 pp. 337-354, 2010

16

Iron at Earth's Core Conditions from First Principles Calculations

Dario Alfè

Department of Earth Sciences, and Department of Physics and Astronomy and London Centre for Nanotechnology
University College London
Gower Street, London, WC1E 6BT, United Kingdom

d.alfe@ucl.ac.uk

ABSTRACT

Ab initio techniques, mainly based on the implementation of quantum mechanics known as density functional theory, and more recently quantum Monte Carlo, have now become widely used in the investigation of the high pressure and temperature properties of materials. These techniques have been proven reliable and accurate, and as such can be considered in many cases as complementary to experiments. Here I will describe some applications of *ab initio* techniques to the properties of iron under Earth's core conditions. In particular, I will focus on the description of how to obtain high pressure and high temperature properties, as these are the relevant conditions of interests for the Earth's core. Low temperature properties of solids have often been studied using the quasi-harmonic approximation, which sometimes can retain high accuracy even at temperatures not too far from the melting temperature. However, for solids at high temperature and for liquids the quasi-harmonic approximation fails, and I will describe how using the molecular dynamics technique, coupled with *ab initio* calculations and the thermodynamic integration scheme, it is possible to compute the high temperature thermodynamic properties of both solids and liquids. Examples of the application of these techniques will include the calculation of many thermodynamic properties of iron and its melting curve, which can be used to improve our understanding of the temperature of the Earth's core.

INTRODUCTION

We start the discussion by recalling the structure of the Earth, which can be broadly described in terms of three main shells. The outermost is the crust, with a thickness of only a few tens of kilometers, mainly formed by silicates. Below the crust we find the mantle, which is customarily divided in an upper mantle and a lower mantle, separated by a transition zone. The mantle makes up most of the volume of the Earth, extending to a depth of 2891 km, almost half way towards the center, and like the crust is also mainly formed by silicates, and in particular by $Mg(Fe)SiO_3$ with some significant fraction of $Mg(Fe)O$ and SiO_2. Below the mantle we find the core, which is divided in an outer liquid core extending from 2891 to 5150 km depths and an inner solid core below that, down to the center of the Earth at 6346 km depth. It is widely accepted that the core is mainly formed by iron, possibly with some 5 to 10% of nickel, plus a fraction of unknown light impurities which reduce the density by 2-3% in the solid and 6-7% in the liquid with respect to the density of pure iron under the same pressure-temperature conditions.

Studying the high pressure and temperature properties of core and mantle forming materials is of fundamental importance to the understanding of the formation and evolution of our planet. In particular, knowledge of the thermal structure of the Earth and the thermoelastic

1529-6466/10/0071-0016$05.00 DOI: 10.2138/rmg.2010.71.16

properties of Earth forming minerals will help us to interpret and hopefully predict the behavior of the dynamical processes that occur inside our planet, including the generation of the Earth's magnetic field through the geodynamo, and the convective processes in the mantle which are ultimately responsible for plate tectonics, earthquakes and volcanic eruptions.

Since the core is mainly formed by iron, it is relevant to study its high temperature high pressure properties in details. Here we will address some of these properties using first principles techniques.

The development of theoretical methods based on the very basic laws of nature of quantum mechanics (developed more than 80 years ago), coupled with the recent staggering increase of computer power (~ at least 10,000-fold in the past 15 years), has made it possible to approach these problems from a theoretical-computational point of view. When high level first-principles methods are used, the results are often comparable in quality with experiments, sometimes even providing information in regions of the pressure-temperature space inaccessible to experiments.

With the term *first-principles* we refer here to those calculations based of the very basic laws of nature, in which no empirical adjustable parameter is used. Only fundamental constants of physics are allowed. Specifically, the relevant basic laws of physics are those describing the interactions between nuclei and electrons, i.e., those of quantum mechanics. In practice, approximations to exact quantum mechanics need to be introduced to provide tools that can be actually used; however, as long as these approximations do not involve the introduction of empirical parameters we will still regard those techniques as first-principles.

The exact quantum mechanical treatment of a system containing a large number of atoms is a formidable task. The starting point of nearly every quantum mechanical calculation available is the so called adiabatic approximation, which exploits the large difference of mass between the nuclei and the electrons. Since the electrons are much lighter, they move so much faster that on the time-scale of the nuclei movement the latter can be considered as fixed. Therefore one solves only the electronic problem in which the nuclei are fixed and act as an external potential for the electrons. The energy of the electrons plus the Coulomb repulsion of the nuclei is a function of the position of the nuclei, and can act as a potential energy for the nuclei. This can be mapped in configuration space to create a potential energy surface, which can later be used to study the motion of the nuclei. Alternatively, forces can be calculated as the derivatives of the potential energy with respect to the position of the nuclei, and these can be used to move the atoms around, relax the system, solve the Newton's equations of motion and perform molecular dynamics simulations, or calculate harmonic vibrational properties like phonons. The potential energy can also be differentiated with respect to the simulation cell parameters, which provides information on the stress tensor. The solution of the electronic problem also provides insights into the electronic structure of the system, which can be examined to study physical properties like bonding, charge distributions, magnetic densities, polarizabilities and so forth.

Most first-principles studies of the high pressure and temperature properties of Earth's forming materials are based on the implementation of quantum mechanics known as density functional theory (DFT). This is a technique that was introduced about 45 years ago by Hohenberg and Kohn (HK) (1964), and Kohn and Sham (KS) (1965) in an attempt to simplify the calculation of the ground state properties of materials (in fact, later shown to be useful also for finite temperature properties by Mermin 1965). The basic HK idea was to substitute the cumbersome many-body wave-function of a system containing N particles, which is a function of $3N$ variables, with the particle density, which is only a function of 3 variables. The price to pay for this enormous simplification is a modification of the basic equations of quantum mechanics with the introduction of new terms, one of which, called exchange-correlation (XC) energy, is unfortunately still unknown. However, KS proposed a simple form for the XC functional, known as the local density approximation (LDA) (Kohn and Sham 1965), that would prove later as the insight which has made DFT so successful and so widespread today. More sophisticated XC functionals were developed in the following decades, and are still being developed today, making DFT an evolving technique with increasingly higher accuracy. One additional attractive feature of DFT is the favorable scaling of computational effort with the size of the system. Traditional DFT techniques scale as N^3, where N is the number of electrons in the system, but large effort is being put into so called $o(N)$ techniques, which for some materials already provide a scaling which is only directly proportional to the size of the system (Bowler et al. 2002; Soler et al. 2002).

The limitations in accuracy due to the current state of the art of density functional theory are expected to be progressively removed, either through the formulation of new exchange-correlation functionals, or with the developments of alternative techniques. Among these dynamical mean field theory (Savrasov and Kotliar 2003) and quantum Monte Carlo (Foulkes et al. 2001) (QMC) are probably the most promising on a time scale of 5 to 10 years. In fact, QMC techniques have recently been applied to iron at Earth's core conditions, and I will report results on the zero temperature iron equation of state and the iron melting temperature at 330 GPa.

In the next section I will introduce the main ideas to calculate zero temperature properties of materials, and I will report on some zero temperature properties or iron. In the following section I will move to the description of standard statistical mechanics tools for the calculation of free energies needed at high temperature. This will be done by separating the low temperature regime, where solids can be described within the quasi-harmonic approximation, from the high temperature regime, where the technique of thermodynamics integration is introduced to calculate free energies, both for solids and for liquids. As applications of these techniques, I will report on some high temperature high pressure properties of solid and liquid iron, including its melting curve at Earth's core conditions.

STATIC PROPERTIES

Crystal structures and phase transitions

At zero temperature the main thermodynamic variable is the internal energy of the system E, or more generally at finite pressure p the enthalpy $H = E + pV$, where V is the volume of the system. The simplest possible first principles calculation one can do is the evaluation of the total energy of a system containing a certain number of atoms at fixed lattice sites. To find the most stable configuration of the atoms one simply minimizes the total energy with respect to the atomic positions. This is usually done by evaluating forces, which are then used to move the atoms towards their equilibrium positions. For simple crystal structures this may not be necessary, as the positions may be constrained by symmetry. An example of this is solid iron in either its ambient conditions body-centered-cubic (bcc) crystal structure. As the pressure is increased the crystal structure of the material may change, and in fact in iron we find a transition from bcc to the hexagonal-close-packed (hcp) structure between 10 and 15 GPa (Jephcoat et al. 1986). The pressure at which the phase transformation occurs is defined by the point where the enthalpies of the two crystal structures cross. These enthalpies can be computed using *ab initio* techniques, by computing the energy E and the pressure p of the crystal as function of volume V, and then construct the enthalpy $H = E + pV$. By contrast with bcc, the hcp crystal structure has an additional degree of freedom, coming from the lack of a symmetry relating the hexagonal plane and the direction perpendicular to the plane. This additional degree of freedom, known as the c/a ratio, needs to be optimized for every volume V (we will return below on the issue of the optimal c/a for hcp iron at Earth's core conditions). Once this is done, an enthalpy curve can be constructed and compared with that obtained from the bcc structure. Calculations using DFT with the LDA or various generalized gradient approximations (GGA) have been performed, and it has been shown that in this particular case the LDA gives poor agreement with the experiments, even failing to predict the correct zero pressure crystal structure (Leung et al. 1991; Singh et al. 1991; Körling and Häglund 1992; Zhu et al. 1992; Cho and Scheffler 1996), while the GGA known as PW91 (Wang and Perdew 1991), for example, predicts the transition between 10 and 13 GPa (Alfè et al. 2000), in good agreement with the experimental value which is in the range 10-15 GPa (Jephcoat et al. 1986). In Figure 1, I show a comparison of the DFT-PW91 calculated equation of state with experiments. The calculations of Alfè et al. (2000) have been adapted to contain room temperature thermal expansion (see below), and agree very well with the experimental data (Mao et al. 1990; Dewaele et al. 2006). The figure also reports recent QMC calculations (Sola et al. 2009), which like the DFT-PW91 ones are in good agreement with the experiments.

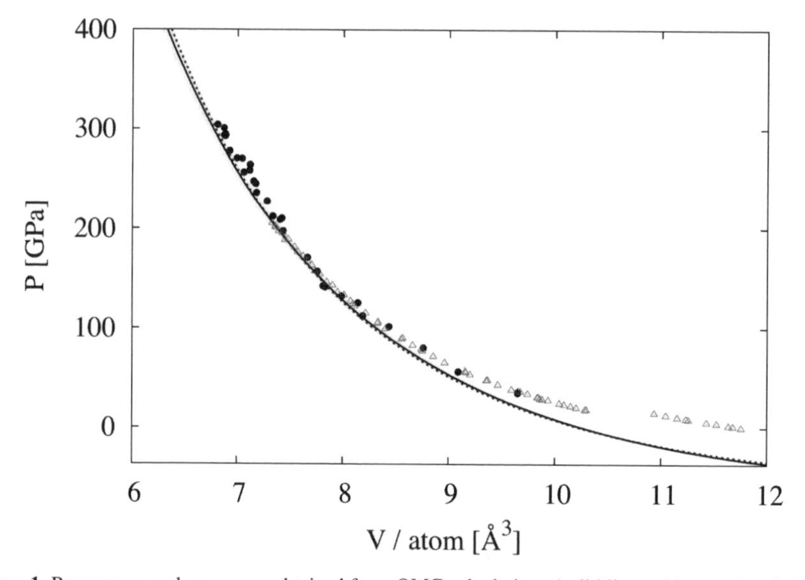

Figure 1. Pressure vs. volume curve obtained from QMC calculations (solid line, with error band). DFT-PW91 results (Alfè et al. 2000) (dotted line) and experimental data (circles (Mao et al. 1990) and open triangles (Dewaele et al. 2006)) are also reported for comparison. Room temperature thermal expansion has been added to the theoretical curves. Notice the discontinuity due to the hcp-bcc phase transition in the values provided by Dewaele et al. (2006). At low pressures, calculations and experiments differ because of the magnetism, which is not taken into account in the theoretical calculations. [Reprinted from Sola et al. (2009) with permission of American Physical Society.]

Elastic constants

Most of what we know about the interior of our planet comes from seismology, and therefore from the elastic behavior of the minerals inside the Earth. The theory of elasticity of crystals can be found in standard books (e.g., Wallace 1998). Briefly, if a crystal is subjected to an infinitesimal stress $d\sigma_{ij}$, with i and j running through the three Cartesian directions in space, then it will deform according to the strain matrix $d\sigma_{ij}$:

$$d\sigma_{ij} = \sum_{k,l} c_{ijkl} d\varepsilon_{kl} \tag{1}$$

The constant of proportionality between stress and strain, c_{ijkl} is a rank-4 tensor of elastic constants. With no loss of generality we can assume $d\sigma_{ij}$ and $d\varepsilon_{ij}$ to be symmetric ($d\sigma_{ij} \neq d\sigma_{ji}$ would imply a non-zero torque on the crystal, which would simply impose an angular acceleration and not a deformation), and therefore the elastic constant tensor is also symmetric.

It is therefore possible to rewrite the rank-2 tensors $d\sigma_{ij}$ and $d\varepsilon_{ij}$ as 6 components vectors, in the Voigt notation, with the index pairs 11, 22, 33, 23, 31, and 12 represented by the six symbols 1, 2, 3, 4, 5 and 6 respectively. In this notation, the stress-strain relation appears as:

$$d\sigma_i = \sum_j C_{ij} d\varepsilon_j \tag{2}$$

with i and j going from 1 to 6. Elastic constants are given as the coefficients C_{ij}. The matrix C_{ij} is symmetric, so that the maximum number of independent elastic constants of a crystal is 21. Because of crystal symmetries, the number of independent constants is usually much smaller. For example, in cubic crystals like bcc Fe there are only three elastic constants, in hcp Fe there are five.

Equation (2) provides the route to the calculation of the elastic properties of materials, and it can be applied both at zero and high temperatures. At zero temperature the components of stress tensor can be calculated as (minus) the partial derivative of the internal energy with respect to the components of the strain:

$$\sigma_{ij} = -\partial E / \partial \varepsilon_{ij} \,|\, \varepsilon \tag{3}$$

Examples of zero temperature calculations of elastic constants include the DFT calculations of Stixrude and Cohen (1995) on the hcp crystal structure of iron at Earth's inner core conditions, who suggested a possible mechanism based on the partial alignment of hcp crystallites to explain the seismic anisotropy of the Earth's inner core.

FINITE TEMPERATURE

The extension to finite temperature properties of materials may simply be obtained by substituting the internal energy E with the Helmholtz free energy F. The pressure p is obtained as (minus) the partial derivative of F with respect to volume, taken at constant temperature:

$$p = -\partial F / \partial V \,|_T \tag{4}$$

If the system of interest is at sufficiently high temperature (above the Debye temperature), the nuclei can be treated as classical particles, and the expression of the Helmholtz free energy F for a system of N identical particles enclosed in a volume V, and in thermal equilibrium at temperature T is (Frenkle and Smit 1996):

$$F = -k_B T \ln \left\{ \frac{1}{N! \Lambda^{3N}} \int_V d\mathbf{R}_1 ... d\mathbf{R}_N \, e^{-\beta U(\mathbf{R}_1,...\mathbf{R}_N;T)} \right\} \tag{5}$$

where $\Lambda = h/(2\pi M k_B T)^{1/2}$ is the thermal wavelength, with M the mass of the particles, h the Plank's constant, $\beta = 1/k_B T$, k_B is the Boltzmann constant, and $U(\mathbf{R}_1,...\mathbf{R}_N;T)$ the potential energy function, which depends on the positions of the N particles in the system, and possibly on temperature, in which case U is the electronic free energy.

The multidimensional integral extends over the total volume of the system V, and it is in general very difficult to calculate. Steinle-Neumann et al. (2001) modeled hcp iron using the *particle in a cell* technique, in which the potential energy experienced by any iron atom in the solid is approximated with an empirical potential fitted to first principle calculations, which coupled with the neglect of correlations between vibrations in different primitive cells makes it possible to integrate analytically Equation (5). They calculated the free energy of the hcp solid as function of various strains applied to the cell at various temperatures, in the same spirit as described in the previous section, and obtained high temperature elastic constants. One of the main conclusions of their work was that at high temperature the elastic behavior of iron is surprisingly different form the zero temperature behavior, and in particular a large increase in the c/a parameter with temperature would result in a different prediction of partial alignment of crystals in the solid core from that inferred using zero temperature elastic constants (Stixrude and Cohen 1995). However, subsequent calculations by Gannarelli et al. (2003, 2005) using the same method did not confirm the large increase of c/a in hcp iron predicted by Steinle-Neumann et al. (2001). The two sets of results are displayed in Figure 2, which also reports some recent experimental data (Ma et al. 2004), confirming the small increase of the c/a ratio with temperature predicted by Gannarelli et al. (2005).

In the general case where the potential energy is not an easy quantity to compute one has to resort to different methods. For example, first principles high temperature elastic constants of iron at core conditions have been calculated by Vočadlo (2007) using molecular dynamics. Here the deformations required in Equation (2) were imposed on the cell much in the same way as in the zero temperature case, but the resulting stresses were calculated as averages over molecular dynamics trajectories.

Using molecular dynamics simulations it is also possible to study the properties of liquids. For example, in a liquid the atoms are free to diffuse throughout the whole volume, and this behavior can be characterized by diffusion coefficients D_α, where α runs over different species in the system. These D_α are straightforwardly related to the mean square displacement of the atoms through the Einstein relation (Allen and Tildesley 1987):

$$\frac{1}{N_\alpha} < \sum_{i=1}^{N_\alpha} |\mathbf{r}_{\alpha i}(t_0 + t) - \mathbf{r}_{\alpha i}(t_0)|^2 > \to 6D_\alpha t, \text{ as } t \to \infty \quad (6)$$

where $\mathbf{r}_{i\alpha}(t)$ is the vector position at time t of the i-th atom of species α, N_α is the number of atoms of species α in the cell, and $< >$ means time average over t_0. The diffusion coefficient

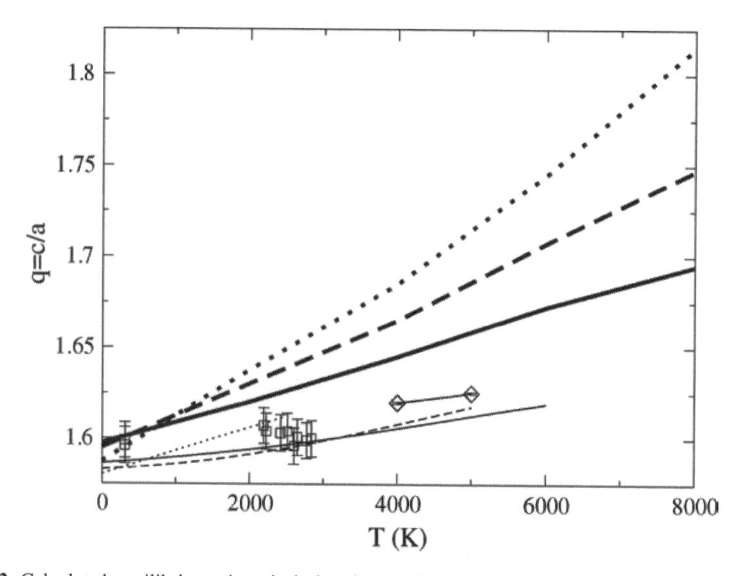

Figure 2. Calculated equilibrium c/a ratio in hcp iron as function of temperature for different volumes. For the work of Gannarelli et al. (2005) (light curves) atomic volumes are 6.97 Å^3 (solid curve), 7.50 Å^3 (dashed curve) and 8.67 Å^3 (dotted curve). For the work of Steinle-Neumann et al. (2001) (heavy curves) volumes are 6.81 Å^3 (solid curve), 7.11 Å^3 (dashed curve) and 7.41 Å^3 (dotted curve). Also shown are the diffraction measurements due to Ma et al. (2004) at 7.73 Å^3/atom (open squares with error bars and the molecular dynamics results of Gannarelli et al. (2005) (open diamonds) at 6.97 Å^3. [Reprinted from Gannarelli et al. (2005) with permission of Elsevier.]

can also be used to obtain a rough estimate of the viscosity η of the liquid, by using the relation between the two stated by the Stokes-Einstein relation:

$$D\eta = \frac{k_B T}{2\pi a} \quad (7)$$

This technique was used by de Wijs et al. (1998) to estimate the viscosity of liquid iron at Earth's core conditions. The Stokes-Einstein relation in Equation (7) is exact for the Brownian motion of a macroscopic particle of diameter a in a liquid of viscosity η. The relation is only approximate when applied to atoms; however, if a is chosen to be the nearest neighbors distance of the atoms in the solid, Equation (7) provides results which agree within 40% for a wide range of liquid metals. To calculate the viscosity rigorously, it is possible to use the Green-Kubo relation:

$$\eta = \frac{V}{k_B T} \int_0^\infty dt < \sigma_{xy}(t)\sigma_{xy}(0) > \quad (8)$$

where σ_{xy} is the off-diagonal component of the stress tensor $\sigma_{\alpha\beta}$ (α and β are Cartesian components). This relation was used in the context of first principles calculations for the first time by Alfè and Gillan (1998b), who first calculated the viscosity of liquid aluminum at ambient pressure and a temperature of 1000 K, showing that the method provided results in good agreement with the experiments, and then applied the method to the calculation of the viscosity of a liquid mixture of iron and sulfur under Earth core conditions. In Figure 3, I show the integral in Equation (8) calculated from 0 to time t for this iron sulfur liquid mixture. In principle, this has to be computed from zero to infinity, as stated in Equation (8), however, in this particular case there is nothing to be gained by extending the integral beyond about 0.2 ps, after which the integrand has decayed to zero and it is dominated by statistical noise. The figure also shows the computed statistical error on the integral, and from this it was possible to infer the value for the viscosity $\eta = 9 \pm 2$ mPa s, in good agreement with that obtained from the diffusion coefficient via the Einstein relation in Equation (7), calculated to be $\eta \sim 13$ mPa s in a previous paper (Alfè and Gillan 1998a).

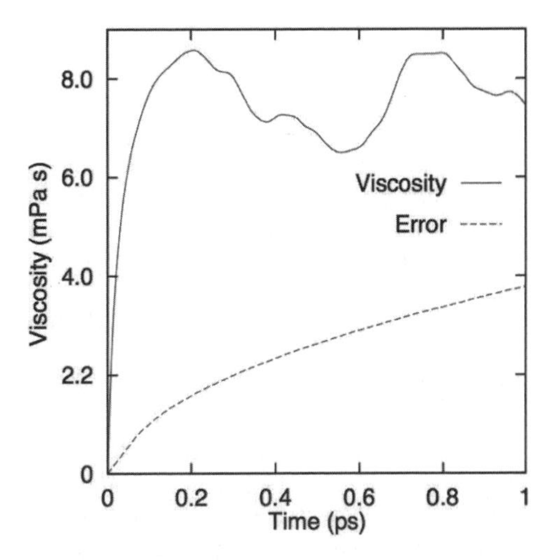

Figure 3. Viscosity integral of the average stress autocorrelation function and its statistical error as a function of time for liquid Fe-S under Earth's core conditions. [Reprinted from Alfè and Gillian (1998b) with permission of American Physical Society.]

In the next two sections I will now introduce the standard techniques to obtain free energies at finite temperature from first principles calculations.

The Helmholtz free energy: low temperature and the quasi-harmonic approximation

For a solid at low temperature, the Helmholtz free energy F can be easily accessed by treating the system in the quasi-harmonic approximation. This is done by expanding the potential (free) energy function U around the equilibrium positions of the nuclei. The first term of the expansion is simply the energy of the system calculated with the ions in their equilibrium positions, $E_{\text{perf}}(V,T)$ (this is a free energy at finite temperature, and therefore depends both on V and T). If the crystal is in its minimum energy configuration the linear term of the expansion is zero, and by neglecting terms of order three and above in the atomic displacements we have that the quasi-harmonic potential is:

$$U_{\text{harm}} = E_{\text{perf}} + \frac{1}{2} \sum_{ls\alpha,l't\beta} \Phi_{ls\alpha,l't\beta} u_{ls\alpha} u_{l't\beta} \tag{9}$$

where $\mathbf{u}_{ls}$ denotes the displacement of atom s in unit cell l, α and β are Cartesian components, and $\Phi_{ls\alpha,l't\beta}$ is the force-constant matrix, given by the double derivative $\partial^2 U/\partial u_{ls\alpha}\partial u_{l't\beta}$ evaluated with all atoms at their equilibrium positions. This force constant matrix gives the relation between the forces $\mathbf{F}_{ls}$ and the displacements $\mathbf{u}_{l't}$, as can be seen by differentiating Equation (9) and ignoring the higher-order an-harmonic terms:

$$F_{ls\alpha} = -\partial U / \partial u_{ls\alpha} = -\sum_{l't\beta} \Phi_{ls\alpha,l't\beta} u_{l't\beta} \tag{10}$$

Within the quasi-harmonic approximation, the potential energy function U_{harm} completely determines the physical properties of the system, and in particular the free energy, which takes the form:

$$F(V,T) = E_{\text{perf}}(V,T) + F_{\text{harm}}(V,T) \tag{11}$$

where the quasi-harmonic component of the free energy is:

$$F_{\text{harm}} = k_B T \sum_{n} \ln(2\sinh(\hbar\omega_n / 2k_B T)) \tag{12}$$

with ω_n the frequency of the nth vibrational mode of the crystal. In a periodic crystal, the vibrational modes can be characterized by a wave-vector $\mathbf{k}$, and for each such wave-vector there are three vibrational modes for every atom in the primitive cell. If the frequency of the sth mode at wave-vector $\mathbf{k}$ is denoted by $\omega_{\mathbf{k}s}$, then the vibrational free energy is:

$$F_{\text{harm}} = k_B T \sum_{\mathbf{k}s} \ln(2\sinh(\hbar\omega_{\mathbf{k}s} / 2k_B T)) \tag{13}$$

The vibrational frequencies $\omega_{\mathbf{k}s}$ can be calculated from first principles, for example using the small displacement method (Alfè 2009a).

As an example of first principles calculations of phonon frequencies using the small displacement method I show in Figure 4 the phonon dispersion relations for bcc iron under ambient conditions, compared with experimental data (Brockhouse et al. 1967). We see that the agreement between theory and experiments is very good almost everywhere in the Brillouin zone, with discrepancies being at worst ~ 3%.

Phonons can also be calculated at high pressure, and as an illustration of this in Figure 5 I show a comparison between DFT-PW91 calculations and nuclear resonant inelastic X-ray scattering (NRIXS) (Seto et al. 1995; Sturhahn et al. 1995) experiments, of phonon density of states of bcc and hcp iron from zero to 153 GPa (Mao et al. 2001). The agreement between theory and experiments is good in the whole pressure region, being slightly better at high pressure.

Once the quasi-harmonic free energy is known, all the thermodynamical properties of the system can be calculated. In particular, the pressure is given by:

$$p = -\partial F / \partial V |_T = -\partial E_{\text{perf}} / \partial V |_T - \partial F_{\text{harm}} / \partial V |_T \tag{14}$$

The last term in the equation above is the ionic component of the thermal pressure, and it is different from zero because the vibrational frequencies $\omega_{\mathbf{k}s}$ depend on the volume of the crystal. In fact, it is easy to see from Equation (13) that even at zero temperature there is a finite

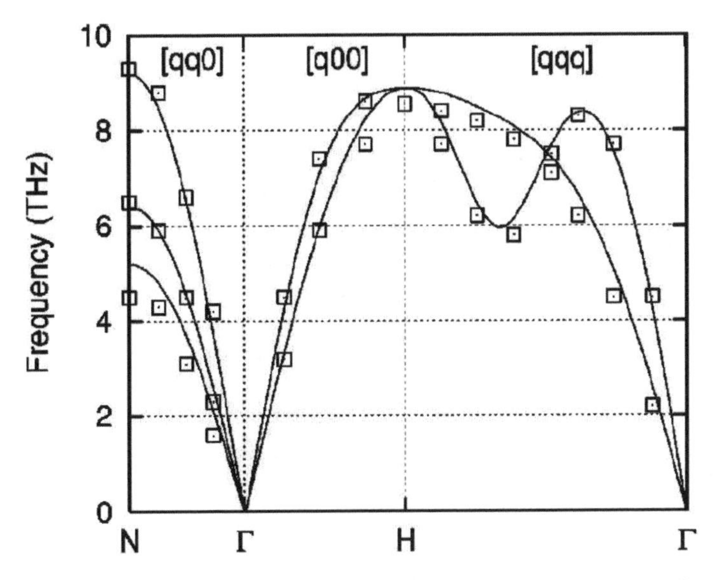

Figure 4. Phonon dispersion relations of ferromagnetic bcc Fe. Lines and open squares show first-principles theory (Alfè et al. 2000) and experiment respectively (Brockhouse et al. 1967). [Reproduced from Alfè et al. (2000) with permission of American Physical Society.]

contribution to the quasi-harmonic free energy, given by:

$$F_{\text{harm}}(V,0)=\sum_{\mathbf{k}s}\frac{\hbar\omega_{\mathbf{k}s}}{2} \tag{15}$$

This zero point energy contribution to the harmonic free energy is also responsible for a contribution to the pressure. Since usually the vibrational frequencies $\omega_{\mathbf{k}s}$ increase with decreasing volume, these contributions are positive, and are responsible for the phenomenon of thermal expansion in solids.

The dependence of $E_{\text{perf}}(V,T)$ on T also means that there is an electronic contribution to the thermal pressure, which is also positive, and in some cases (i.e., iron at Earth's core conditions) can be a significant fraction of the thermal pressure, and a non-negligible fraction of the total pressure (Alfè et al. 2001).

The Helmholtz free energy: high temperature and thermodynamic integration

At high temperature an-harmonic effects in solids may start to play an important role, and the quasi-harmonic approximation may not be accurate enough. Moreover, if the system of interest is a liquid, the quasi-harmonic approximation is of no use. In this section I shall describe a method to calculate the free energy of solids and liquids in the high temperature limit, provided that the temperature is high enough that the quantum nature of the nuclei can be neglected. If this is the case, the Helmholtz free energy F is defined as in Equation (5).

Performing the integral in Equation (5) to calculate F is extremely difficult if the potential energy is not a simple function of the ionic positions. However, it is less difficult to calculate changes in F as some specific variables are changed in the system. For example, we have seen that by taking the derivative of F in Equation (5) with respect to volume at constant T we obtain (minus) the pressure. Therefore, the difference of F between two volumes can be obtained by integrating the pressure p, which can be calculated using a molecular dynamics simulation. Similarly, by integrating the internal energy E one obtains differences in F/T.

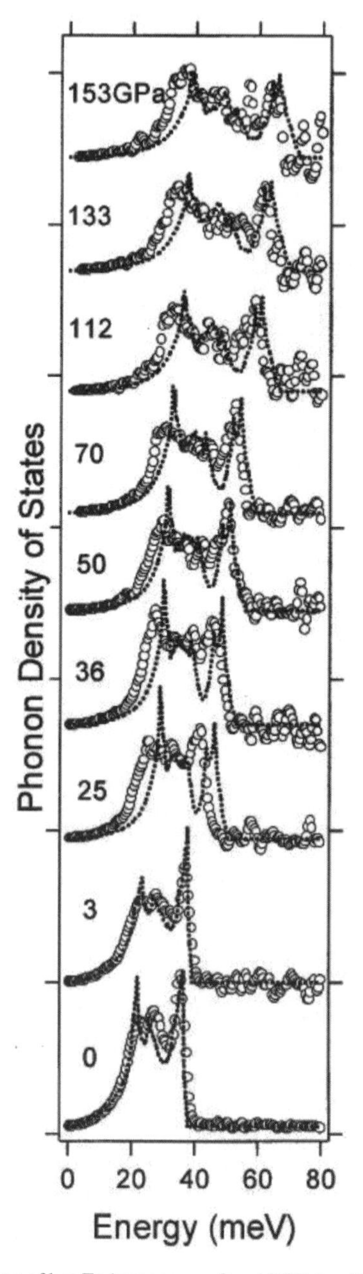

Figure 5. Phonon density of states of bcc Fe (pressure $p = 0$ and 3 GPa) and hcp Fe (p from 25 to 153 GPa). Dotted curves and open circles show first-principles theory and experiment respectively. [Reprinted from Mao et al. (2001) with permission from AAAS.]

It is equally possible to calculate differences in free energy between two systems having the same number of atoms N, the same volume V, but two different potential energy functions U_0 and U_1. This can be done by introducing an intermediate potential energy function U_λ such that for $\lambda = 0$; $U_\lambda = U_0$, and for $\lambda = 1$; $U_\lambda = U_1$, and such that for any value of $0 < \lambda < 1$, U_λ is a continuous and differentiable function of λ. For example, a convenient form is:

$$U_\lambda = (1 - f(\lambda))U_0 + f(\lambda)U \tag{16}$$

where $f(\lambda)$ is an arbitrary continuous and differentiable function of λ in the interval $0 \le \lambda \le 1$, with the property $f(0) = 0$ and $f(1) = 1$. According to Equation (5), the Helmholtz free energy of this intermediate system is:

$$F_\lambda = -k_B T \ln\left\{\frac{1}{N!\Lambda^{3N}} \int_V d\mathbf{R}_1 ... d\mathbf{R}_N \, e^{-\beta U_\lambda(\mathbf{R}_1,...\mathbf{R}_N;T)}\right\} \tag{17}$$

Differentiating this with respect to λ gives:

$$\frac{dF_\lambda}{d\lambda} = \frac{\int_V d\mathbf{R}_1 ... d\mathbf{R}_N e^{-\beta U_\lambda(\mathbf{R}_1,...\mathbf{R}_N;T)} \left(\frac{\partial U_\lambda}{\partial \lambda}\right)}{\int_V d\mathbf{R}_1 ... d\mathbf{R}_N e^{-\beta U_\lambda(\mathbf{R}_1,...\mathbf{R}_N;T)}} = \left\langle \frac{\partial U_\lambda}{\partial \lambda} \right\rangle_\lambda \tag{18}$$

and therefore by integrating dF_λ/d_λ one obtains:

$$\Delta F = F_1 - F_0 = \int_0^1 d\lambda \left\langle \frac{\partial U_\lambda}{\partial \lambda} \right\rangle_\lambda \tag{19}$$

This also represents the reversible work done on the system as the potential energy function is switched from U_0 to U_1. In most cases a suitable choice for the function that mixes U_0 and U_1 is simply $f(\lambda) = \lambda$, and the thermodynamic formula in Equation (19) takes the simple form:

$$\Delta F = F_1 - F_0 = \int_0^1 d\lambda \langle U_1 - U_0 \rangle_\lambda \tag{20}$$

This way to calculate free energy differences between two systems is called thermodynamic integration (Frenkel and Smit 1996). The usefulness of the thermodynamic integration formula expressed in Equation (19) becomes clear when one identifies U_1 with the DFT potential (free) energy function, and with U_0 some classical model potential for which the free energy is easily calculated, to be taken as a reference system. Then Equation (19) can be used to calculate the DFT free energy of the system by evaluating the integrand $\langle U_1 - U_0 \rangle_\lambda$ using FPMD simulations at a sufficiently large number of values of λ and calculating the integral numerically. Alternatively, one can adopt the dynamical method described by Watanabe and Reinhardt (1990). In this approach the parameter λ depends on time, and is slowly (adiabatically) switched from 0 to 1 during a single simulation. The switching rate has to be slow enough so that the system remains in thermodynamic equilibrium, and adiabatically transforms from the reference to the *ab initio* system. The change in free energy is then given by:

$$\Delta F = \int_0^{T_{sim}} dt \frac{d\lambda}{dt} (U_1 - U_0) \tag{21}$$

where T_{sim} is the total simulation time, $\lambda(t)$ is an arbitrary function of t with the property of being continuous and differentiable for $0 \le t \le 1$, $\lambda(0) = 0$ and $\lambda(T_{sim}) = 1$.

Thermodynamic integration can be used to calculate the free energies of both solids and liquids. It is clear from Equation (19) that the choice of the reference system is almost completely irrelevant (of course, the stability of the system cannot change as λ is switched from 0 to 1), provided that ΔF can be calculated in practice. So, if the goal is to obtain *ab initio* free energies, it is essential to minimize the amount of *ab initio* work in order to make the calculations feasible. This is achieved by requiring that: i) the integrand in Equation (19) is a smooth function of λ, ii) the thermal averages $\langle U_1 - U_0\rangle_\lambda$ can be computed within the required accuracy on the time-scales accessible to FPMD and iii) the convergence of ΔF as function of the number of atoms N in the system is again achieved with N accessible to first-principles calculations. Points i), ii), and iii) could obviously be satisfied by a perfect reference system, i.e., a system which differed from the *ab initio* system only by an arbitrary constant. In this trivial case the integrand in Equation (19) would be a constant, and thermal averages could be calculated on just one configuration and with cells containing an arbitrary small number of atoms. The next thing close to a constant is a slowly varying object, and this therefore provides the recipe for the choice of a good reference system, which has to be constructed in such a way that the fluctuations in $U_1 - U_0$ are as small as possible. If this is the case, thermal averages of $U_1 - U_0$ are readily calculated on short simulations. Moreover, $\langle U_1 - U_0\rangle_\lambda$ is a smooth function of λ, so a very limited number of simulations for different values of λ are needed and, finally, convergence of $\langle U_1 - U_0\rangle_\lambda$ with respect to the size of the system is also quick. In fact, if the fluctuations in $U_1 - U_0$ are small enough, one can simply write $F_1 - F_0 \cong \langle U_1 - U_0\rangle_0$, with the average taken in the reference system ensemble. If this is not good enough, the next approximation is readily shown to be:

$$F_1 - F_0 \cong < U_1 - U_0 >_0 - \frac{1}{2k_B T}\left\langle \left[U_1 - U_0 - < U_1 - U_0 >_0\right]^2 \right\rangle_0 \tag{22}$$

This form is particularly convenient since one only needs to sample the phase space with the reference system, and perform a number of *ab initio* calculations on statistically independent configurations extracted from a long classical simulation.

Once the Helmholtz free energy of the system is known, it can be used to derive its thermodynamical properties. For example, it is possible to calculate properties on the so-called Hugoniot line, and compare the results with those obtained in shock-wave experiments. The data that emerge most directly from shock experiments consist of a relation between the pressure p_H and the molar volume V_H on the Hugoniot line, which is the set of thermodynamic states given by the Rankine-Hugoniot formula (Poirier 1991):

$$\frac{1}{2} p_H (V_0 - V_H) = E_H - E_0 \tag{23}$$

where E_H is the molar internal energy behind the shock front, and E_0 and V_0 are the molar internal energy and volume in the zero-pressure state ahead of the front. These experiments are particularly useful in identifying the melting transition. This is done by monitoring the speed of sound, which shows discontinuities at two characteristic pressures p_s and p_l, which are the points where the solid and liquid Hugoniots meet the melting curve. Below p_s, the material behind the shock front is entirely solid, while above p_l it is entirely liquid; between p_s and p_l, the material is a two-phase mixture. To illustrate an example of the quality of the DFT-PW91 predictions of the Hugoniot line I show in Figure 6 the calculations of the $p(V)$ relation on the Hugoniot (Alfè et al. 2002) for solid and liquid iron, compared with the experimental data obtained by Brown and McQueen (1986). We can see that the agreement between the theory and experiments is extremely good. The two theoretical curves come from raw and free energy corrected calculations (see below). In Figure 7, I show a comparison of the calculated speed of sounds of the liquid with those obtained in the shock experiments. Again, the agreement between the two sets of data is extremely good.

Melting

Once the free energies of solid and liquid are known, it is possible to calculate also melting properties, including the whole melting curve of iron under Earth's core conditions (Alfè et al.

1999, 2001, 2002; Bukowinski 1999). It was found that a simple sum of inverse power pair-potentials of the form $U_{\mathrm{IP}}(r) = B/r^{\alpha}$, where r is the distance between two ions and B and α are two adjustable parameters, did an excellent job in describing the energetics of the liquid and the high temperature solid, provided B and α were appropriately adjusted. As mentioned in the previous section, an additional crucial advantage of having a good reference system is that convergence of $F_1 - F_0$ with respect to the size of the system is very rapid, and in fact for both solid and liquid iron it was found that already with 64-atom systems $F_1 - F_0$ was converged to within better than 10 meV/atom, which in turns implied melting temperature converged to better than 100 K with respect to this single technical point. The best estimate for the melting point at the inner-outer core boundary pressure of 330 GPa was $T_m = 6350 \pm 300$ K (Alfè et al. 2002), where the error quoted is the result of the combined statistical errors in the free energies of solid and liquid. Systematic errors due to the approximations of DFT are more difficult to estimate. Some empirical attempts at correcting the DFT-PW91 free energies were based on the differences between the DFT-PW91 zero temperature pressure-volume equation of state and the

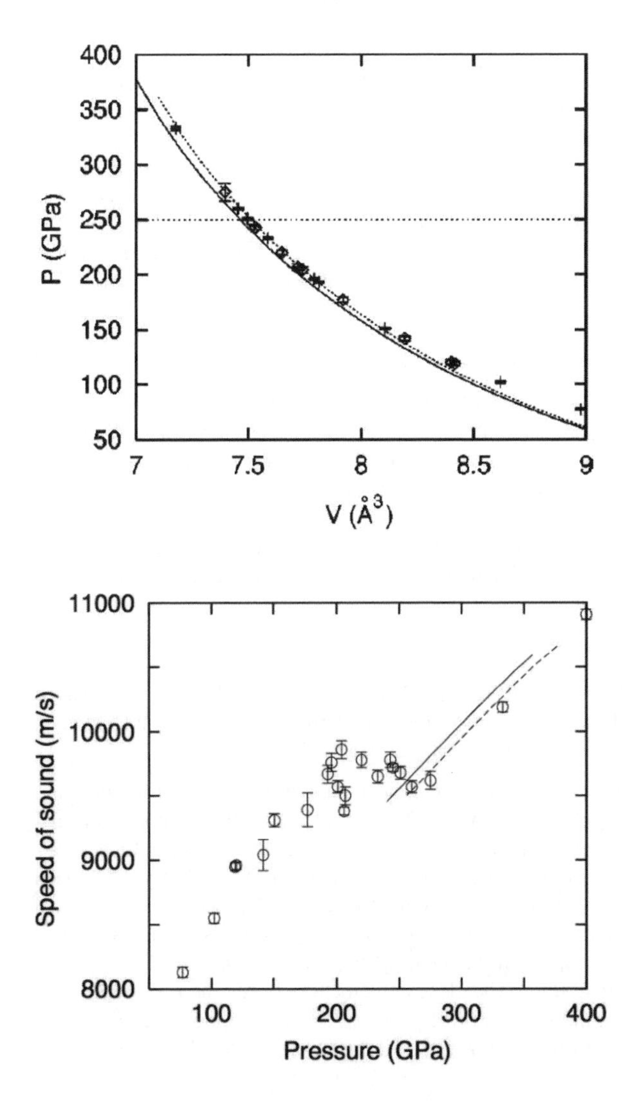

Figure 6. Experimental and first-principles Hugoniot pressure p of Fe as a function of atomic volume V. Symbols show the measurements of Brown and McQueen (1986). Solid curve is first-principles pressure obtained when calculated equilibrium volume of bcc Fe is used in the Hugoniot-Rankine equation; dotted curve is the same, but with experimental equilibrium volume of bcc Fe. The comparison is meaningful only up to a pressure of approximately 250 GPa (horizontal dotted line), at which point the experiments indicate melting. [Reprinted from Alfè et al. (2002) with permission of American Physical Society.]

Figure 7. Longitudinal speed of sound on the Hugoniot. Circles: experimental values from Brown and McQueen (1986); continuous and dashed curves: *ab initio* values from Alfè et al. (2002) without and with free-energy correction (see text). [Reprinted from Alfè et al. (2002) with permission of American Physical Society.]

experimental data, resulting in the free energy corrected results reported in Figures 6 and 7. Very recently, we re-addressed this problem going beyond approximated DFT, using quantum Monte Carlo techniques (see below).

At the present state of knowledge, the experimental understanding of the melting point of Fe under these conditions is still scarce, as experiments based on diamond-anvil cells (DAC) cannot reach these pressures. Moreover, even in the region of the phase space where DAC experiments are possible there is still considerable disagreement between different groups (Boehler 1993; Shen et al. 1998; Ma et al. 2004), and between DAC and shock-wave experiments (Brown and McQueen 1986; Nguyen and Holmes 2004).

On the theoretical side, we mention also the work of Laio et al. (2000) and Belonoshko et al. (2000), who performed DFT based simulations to calculate the melting curve of Fe under Earth's core conditions, although their approach was rather different from ours (Alfè et al. 1999, 2001, 2002, Bukowinski 1999). Instead of calculating free energies, Laio et al. (2000) and Belonoshko et al. (2000) fitted a classical model potential to their first principles calculations, and then used the classical potential to compute the melting curve. To do so, they used the coexistence method, in which solid and liquid are simulated in contact in a box. This method is an alternative route to the calculation of melting curves, and therefore equivalent to the free energy approach. However, Laio et al. (2000) and Belonoshko et al. (2000) found that at the pressure of 330 GPa iron melted at 5400 and 7000 K respectively. The reason of these large differences, and the difference with our value 6350 K, are due to the quality of the classical potentials employed, and in particular to the free energy differences between these classical potentials and the DFT system. This was later investigated by us (Alfè et al. 2002b), and it was shown that it is possible to assess the differences in free energies between the classical potential and the DFT one, and correct for it. In particular, it was shown that at a fixed pressure p, a first approximation of the difference T' in the melting temperature between the classical potential and the *ab initio* system is given by:

$$T' = \Delta G^{ls}\left(T_{\text{mod}}\right) / S^{ls}_{\text{mod}} \tag{24}$$

where S^{ls}_{mod} is the entropy of fusion of the model potential, T_{mod} its melting temperature, and $\Delta G^{ls} = (G^{l}_{\text{ab}} - G^{l}_{\text{mod}}) - (G^{s}_{\text{ab}} - G^{s}_{\text{mod}})$, where G is the Gibbs free energy, the subscripts ab and mod indicate the *ab initio* and the model system respectively, and the superscripts l and s indicate liquid and solid respectively. These differences of Gibbs free energies can be calculated using thermodynamic integration, which if the model potential is not too different from the *ab initio* one can be calculated using the perturbative approach outlined in Equation (22) above. The relation between ΔG, evaluated at constant p, and ΔF, calculated at constant V, is readily shown to be:

$$\Delta G = \Delta F - \frac{1}{2} V \kappa_T \Delta p^2 + o(\Delta p^3) \tag{25}$$

where κ_T is the isothermal compressibility and Δp is the change of pressure when U_{mod} is replaced by U_{ab} at constant V and T. Once these corrections were applied, the results of Belonoshko et al. (2000) came in perfect agreement with ours Alfè et al. (2002a).

The coexistence method is an alternative route to the calculation of melting properties, and as such delivers the same results if applied consistently. For its very nature, the method is intrinsically very expensive, because it requires simulations on systems containing large number of atoms, typically many hundreds or even thousands. For this reason, until very recently it had been only applied to calculations employing classical potentials. However, it has been recently shown that the method can in fact be applied also in the context of first principles calculations, and indeed it has been applied to the calculation of the melting temperature of iron at ICB pressure, with a result of 6390 ± 100 K (Alfè 2009b). The close agreement between the results obtained with the first principles coexistence and the first principles free energies support each other and confirm that the DFT melting temperature of iron at 330 GPa is in the region of 6350 K.

Recently, Sola and Alfè (2009) re-addressed the problem using the quantum Monte Carlo technique. They performed thermodynamic integration between DFT and QMC exploiting Equation (22). This was done by performing long DFT molecular dynamics simulations (using the VASP code; Kresse and Furthmuller 1996), and then calculating QMC (using the CASINO code; Needs et al. 2007) and DFT energies on a selection of statistically independent configurations, which showed that fluctuations in energy differences between QMC and DFT are small and Equation (22) can be used reliably (see Fig. 8). Equation (24) was then used to compute a melting temperature correction of 550 ± 250 K. This correction is small, but maybe not completely negligible in terms of geophysical implications, which I will not discuss here. These results are all displayed in Figure 9, together with a number of experimental data.

I would like to conclude by pointing out that QMC techniques are fairly easy to parallelize, and have already been shown to scale well on tens of thousands of processors. These QMC techniques are 3 or 4 orders of magnitude more expensive than conventional DFT based techniques, however, because of their favorable scaling properties, they are likely to be able to exploit very effectively the computers of the future, and in a decade or two might reach the standards of utilizations that DFT techniques have today.

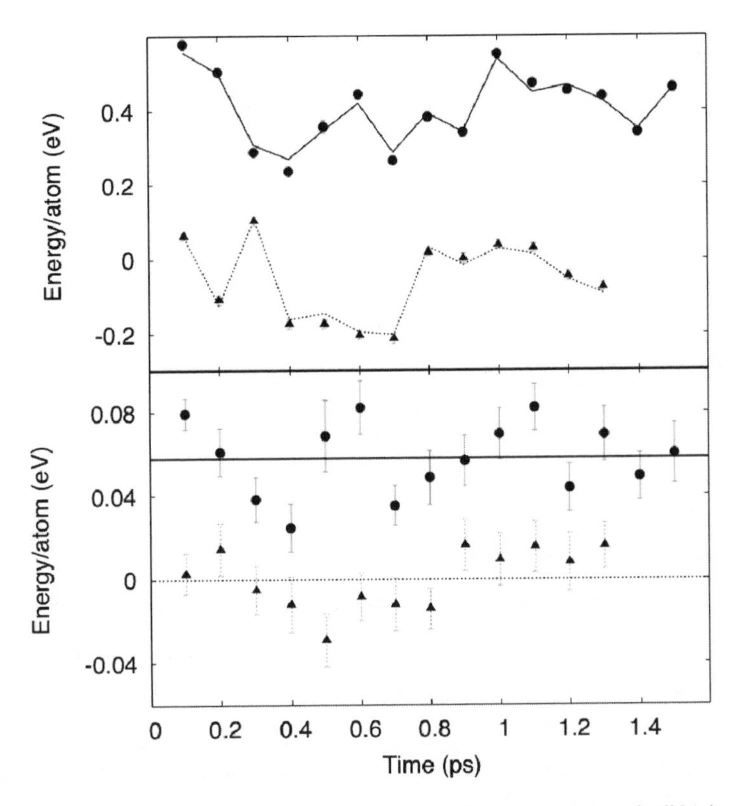

Figure 8. *Top panel:* QMC energies corresponding to configurations representative of solid (triangles) and liquid (dots) iron, generated with DFT molecular dynamics on 64-atom systems. Solid line and dotted line connect DFT energies calculated on the same set of configurations. An offset is added to the energies so that the average value of the QMC and DFT energies is the same, separately in the solid and the liquid. *Bottom panel:* QMC - DFT energy differences on the same configurations. The average QMC - DFT energy difference for the solid is subtracted from all points. Lines represent the average of the energy differences between QMC and DFT in the solid (line at zero energy) and the liquid. [Reprinted from Sola and Alfè (2009) with permission of American Physical Society.]

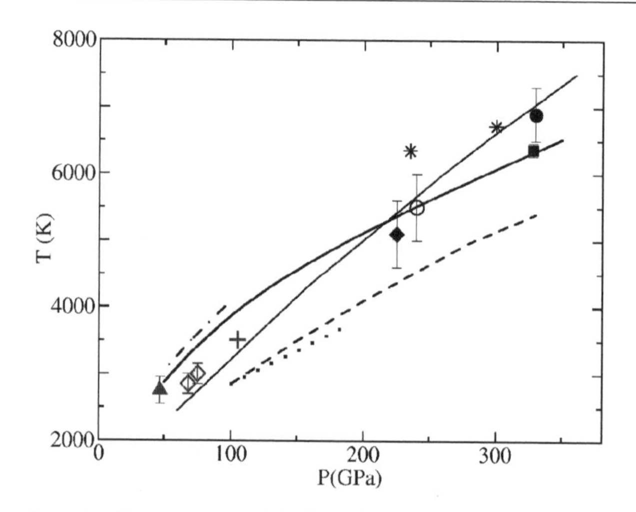

Figure 9. Comparison of melting temperatures (T) of Fe as function of pressure (P) from present calculations with experiments and other *ab initio* results. Filled circle: present DMC results; filled square: melting point from DFT coexistence (Alfè 2009b); solid black line: melting curve from DFT free energies (Alfè et al. 2002a); dashed curve: theoretical results from Laio et al. (2002); light solid curve: theoretical results from Belonoshko et al. (2000); chained and dotted curves: DAC measurements of Williams et al. (1987) and Boehler (1993); open diamonds: DAC measurements of Shen et al. (1998); plus: DAC measurement of Ma et al. (2004); filled triangle: DAC measurement of Jephcoat and Besedin (1996); stars, open circle and filled diamond: shock experiments of Yoo et al. (1993), Brown and McQueen (1986) and Nguyen and Holmes (2004). Error bars are those quoted in original references. [Reprinted from Sola and Alfè (2009) with permission of American Physical Society.]

REFERENCES

Alfè D (2009a) PHON A program to calculate phonons using the small displacement method. Comput Phys Commun 180:2622-2633. Program available at http://chianti.geol.ucl.ac.uk/ dario.

Alfè D (2009b) Temperature of the inner-core boundary of the Earth: melting of iron at high pressure from first-principles coexistence simulations. Phys Rev 79:B060101(R) 1-4

Alfè D, Gillan MJ (1998a) First-principles simulations of liquid Fe-S under Earth. Phys Rev 58:B8248-B8256

Alfè D, Gillan MJ (1998b) First-principles calculation of transport coefficients. Phys Rev 81:L5161-L5164

Alfè D, Gillan MJ, Price GD (1999) The melting curve of iron at the pressures of the Earth's core from ab initio calculations. Nature 401:462-464

Alfè D, Gillan MJ, Price GD (2002a) Iron under Earth. Phys Rev 65:B1651181-11

Alfè D, Kresse G, Gillan MJ (2000) Structure and dynamics of liquid iron under Earth. Phys Rev 61:B132-B14

Alfè D, Price GD, Gillan MJ (2001) Thermodynamics of hexagonal-close-packed iron under Earth. Phys Rev 64:B0451231-16

Alfè D, Price GD, Gillan MJ (2002b) Complementary approaches to the ab initio calculation of melting properties. J Chem Phys 116:6170-6177

Allen MP, Tildesley DJ (1987) Computer simulation of liquids. Oxford Science Publications, Oxford

Belonoshko AB, Ahuja R, Johansson B (2000) Quasi-ab initio molecular dynamic study of Fe melting. Phys Rev 84:L3638-L3641

Boehler R (1993) Temperatures in the Earth's core from melting-point measurements of iron at high static pressures. Nature 363:534-536

Bowler DR, Miyazaki T, Gillan MJ (2002) Recent progress in linear scaling ab initio electronic structure techniques. J Phys Condens Matter 14:2781-2798

Brockhouse BN, Abou-Helal HE, Hallman ED (1967) Lattice vibrations in iron at 296 K. Solid State Commun 5:211-216

Brown JM, McQueen RG (1986) Phase transitions, Grüneisen parameter and elasticity for shocked iron between 77 GPa and 400 GPa. J Geophys Res 91:7485-7494

Bukowinski MST (1999) Earth science: taking the core temperature. Nature 401:432-433
Cho JH, Scheffler M (1996) Ab initio pseudopotential study of Fe, Co, and Ni employing the spin-polarized LAPW approach. Phys Rev 53:B10685-B10689
de Wijs GA, Kresse G, Gillan MJ (1998) First-order phase transitions by first-principles free-energy calculations: the melting of Al. Phys Rev 57:B8223-B8234
Dewaele A, Loubeyre P, Occelli F, Mezouar M, Dorogokupets PI, Torrent M (2006) Quasihydrostatic equation of state of iron above 2 Mbar. Phys Rev 97:L215504 1-4
Foulkes WMC, Mitaš L, Needs RJ, Rajagopal G (2001) Quantum Monte Carlo simulations of solids. Rev Mod Phys 73:33-83
Frenkel D, Smit B (1996) Understanding molecular simulation. Academic Press, San Diego
Gannarelli C, Alfè D, Gillan MJ (2003) The particle-in-cell model for ab initio thermodynamics: implications for the elastic anisotropy of the Earth. Phys Earth Planet Int 139:243-253
Gannarelli CMS, Alfè D, Gillan MJ (2005) The axial ratio of hcp iron at the conditions of the Earth. Phys Earth Planet Inter 152:67-77
Hohenberg P, Kohn W (1964) Inhomogeneous electron gas. Phys Rev 136:B864-B871
Jephcoat AP, Besedin SP (1996) Temperature measurement and melting determination in the laser-heated diamond-anvil cell. Phil Trans Royal Soc London A 354:1333-1360
Jephcoat AP, Mao HK, Bell PM (1986) Static compression of iron to 78 GPa with rare gas solids as pressure-transmitting media. J Geophys Res 91:4677-4684
Kohn W, Sham L (1965) Self-consistent equations including exchange and correlation effects. Phys Rev 140:A1133-A1138
Körling M, Häglund J (1992) Cohesive and electronic properties of transition metals: the generalized gradient approximation. Phys Rev 45:B13293-B13297
Kresse G, Furthmuller J (1996). Efficiency of ab-initio total energy calculations for metals and semiconductors using a plane-wave basis set. Comput Mater Sci 6: 15-50. Efficient iterative schemes for ab initio total-energy calculations using a plane-wave basis set. Phys Rev 54:B11169-B11186
Laio A, Bernard S, Chiarotti GL, Scandolo S, Tosatti E (2000) Physics of Iron at Earth's Core Conditions. Science 287:1027-1030
Leung TC, Chan CT, Harmon BN (1991) Ground-state properties of Fe, Co, Ni, and their monoxides: results of the generalized gradient approximation. Phys Rev 44:B2923-B2927
Ma Y, Somayazulu M, Shen G, Mao HK, Shu J, Hemley RJ (2004) In situ X-ray diffraction studies of iron to Earth-core conditions. Phys Earth Planet Int 143-144:455-467
Mao HK, Xu J, Struzhkin VV, Shu J, Hemley RJ, Sturhahn W, Hu MY, Alp EE, Vočadlo L, Alfè D, Price GD, Gillan MJ, Schwoerer-Böhning M, Häusermann D, Eng P, Shen G, Giefers H, Lübbers R, Wörtmann G (2001) Phonon density of states of iron up to 153 Gigapascals. Science 292: 914-916
Mao K, Wu Y, Chen LC, Shu JF, Jephcoat AP (1990) Static compression of iron to 300 GPa and Fe0.8Ni0.2 alloy to 260 GPa: implications for composition of the core. J Geophys Res 95:21737-21742
Mermin ND (1965) Thermal properties of the inhomogeneous electron gas. Phys Rev 137:A1441-A1443
Needs RJ, Towler MD, Drummond ND, Lopez Rios P (2007) CASINO version 2.1 User Manual. University of Cambridge, Cambridge
Nguyen JH, Holmes NC (2004) Melting of iron at the physical conditions of the Earth's core. Nature 427:339-342
Poirier J-P (1991) Introduction to the physics of the Earth's interior. Cambridge University Press, Cambridge, England.
Savrasov PY, Kotliar G (2003) Linear response calculations of lattice dynamics in strongly correlated systems. Phys Rev 90:L056401 1-4
Seto M, Yoda Y, Kikuta S, Zhang XW, Ando M (1995) Observation of nuclear resonant scattering accompanied by phonon excitation using synchrotron radiation. Phys Rev 74: L3828-L3831
Shen G, Mao HK, Hemley RJ, Duffy TS, Rivers ML (1998) Melting and crystal structure of iron at high pressures and temperatures. Geophys Res 25:L373-L376
Singh DJ, Pickett WE, Krakauer H (1991) Gradient-corrected density functionals: full-potential calculations for iron. Phys Rev 43: B11628-B11634
Sola E, Alfè D (2009) Melting of iron under Earth's core conditions from diffusion Monte Carlo free energy calculations. Phys Rev 103:L078501 1-4
Sola E, Brodholt JP, Alfè D (2009) Equation of state of hexagonal closed packed iron under Earth's core conditions from quantum Monte Carlo calculations. Phys Rev 79:B024107 1-6
Soler JM, Artacho E, Gale JD, Garcia A, Junquera J, Ordejon P, Sanchez-Portal D (2002) The SIESTA method for ab initio order-N materials simulation. J Phys Condens Matter 14:2745-2779
Steinle-Neumann G, Stixrude L, Cohen RE, Gülseren O (2001) Elasticity of iron at the temperature of the Earth's inner core. Nature 413:57-60
Stixrude L, Cohen RE (1995) High-pressure elasticity of iron and anisotropy of Earth's inner core. Science 267:1972-1975

Sturhahn W, Toellner TS, Alp EE, Zhang X, Ando M, Yoda Y, Kikuta S, Seto M, Kimball CW, Dabrowski B (1995) Phonon density of states measured by inelastic nuclear resonant scattering. Phys Rev 74:L3832-L3835

Vočadlo L (2007) Ab initio calculations of the elasticity of iron and iron alloys at inner core conditions: evidence for a partially molten inner core? Earth Plan Sci 254:L227-L232

Wallace DC (1998) Thermodynamics of Crystals. Dover Publications, New York

Wang Y, Perdew JP (1991) Correlation hole of the spin-polarized electron gas, with exact small-wave-vector and high-density scaling. Phys Rev 44:B13298-B13307

Watanabe M, Reinhardt WP (1990) Direct dynamical calculation of entropy and free energy by adiabatic switching. Phys. Rev. 65:L3301-L3304

Williams Q, Jeanloz R, Bass JD, Svendesen B, Ahrens TJ (1987) The melting curve of iron to 250 Gigapascals: a constraint on the temperature at Earth's center. Science 286:181-182

Yoo CS, Holmes NC, Ross M, Webb DJ, Pike C (1993) Shock temperatures and melting of iron at Earth core conditions. Phys Rev 70:L3931-L3934

Zhu J, Wang XW, Louie SG (1992) First-principles pseudopotential calculations of magnetic iron. Phys Rev 45:B8887-B8893

Reviews in Mineralogy & Geochemistry
Vol. 71 pp. 355-389, 2010

First-Principles Molecular Dynamics Simulations of Silicate Melts: Structural and Dynamical Properties

Bijaya B. Karki

Department of Computer Science, Department of Geology and Geophysics
Louisiana State University
Baton Rouge, Louisiana, 70803 U.S.A.

karki@csc.lsu.edu

ABSTRACT

The study of silicate melts from first principles is computationally intensive due to the need of relatively large atomic systems and long simulation runs. Recent advances in hardware and software have made it possible to accurately simulate the liquid phase at pressures and temperatures that are geophysically relevant. This paper reports the details of the methodology used in the context of simulations and subsequent analysis of the output data. The simulations are performed using the parallel first-principles molecular dynamics (FPMD) technique within the framework of density functional theory. Various physical properties including the equation of state, thermodynamics, atomic and electronic structures, self-diffusion and viscosity are obtained from simulations. The position time series are visualized to gain insight into underlying physical mechanisms. We review the recent first-principles studies of three liquids along the MgO-SiO_2 join including $MgSiO_3$ melt to show that the structural and dynamical properties of these liquids are highly sensitive to pressure and temperature.

INTRODUCTION

The liquid state of ceramics and minerals has long been the subject of intensive studies using experiment and theory. From the geophysical viewpoint, the physical properties of mineral liquids, particularly molten silicates, are crucial for our understanding of the cooling and crystallization of Earth's early magma ocean as well as for our understanding of the generation and transport mechanisms of magmas in modern day Earth (e.g., Rigden et al. 1984; Solomatov 2007). For instance, the equation of state of the liquid controls the relative density of partial melts produced by geological processes and co-existing solids, and thus whether these melts will rise or sink. Similarly, the diffusivity and viscosity control the mobility and rate of chemical reactions of liquids with their surroundings. The analysis of melt compositions and xenoliths (e.g., Haggerty and Sauter 1990) suggests that melts can come from very great depths in the mantle. Partial melts are believed to exist in Earth at depths as great as the core-mantle boundary and can be responsible for the ultra low velocity regions in the shallow and deep mantle (Revenaugh and Spikin 1994; Lay et al. 2004). Besides their importance as primary agents of chemical and thermal evolution of the Earth, the mineral liquids are of broader interest reaching beyond geophysics. MgO and SiO_2 – two most important components of silicate melts are considered as prototype oxides and as such they are among the most widely studied materials. Understanding the physics of these liquids is expected to lend considerable insight into the behavior of other oxide liquids including the so-called tetrahedral liquids (e.g., water, BeF_2, $GeSe_2$).

1529-6466/10/0071-0017$05.00 DOI: 10.2138/rmg.2010.71.17

It is important to understand the microscopic (atomic) characteristics of silicate melts and how they control the macroscopic properties. For instance, the degree of polymerization controls the transport properties—the less the structure is polymermized the higher the diffusivity (the lower the viscosity). Experimental information of melt structure is based on the study of mostly glasses and some melts (Hemley et al. 1986; Williams and Jeanloz 1988; Waseda and Toguri 1990; Xue et al. 1991; Farber et al. 1996; Funamori et al. 2004; Myson and Richet 2005; Yamada et al. 2007; Lee et al. 2004, 2008). These studies suggest that the structure exhibits a short- to mid-range order. Investigation of the properties of silicate liquids over large range of pressure, temperature, and composition that are relevant poses tremendous challenges to experiment and theory, which makes it particularly important that multiple complementary methods be developed for their study. Over last one decade, theory has appeared as a complementary approach in the study of Earth materials. From the theoretical point of view, one of the uncertainties in most previous studies (Kubicki and Lasaga 1988, 1991; Rustad et al. 1990; Wasserman et al. 1993; Vocadlo et al. 1996; Winker et al. 2004; Lacks et al. 2007; Adjaouda et al. 2008; Nevins et al. 2009; Martin et al. 2009), which have mostly been based on semi-empirical force fields, is the form of the force-field chosen: various models that have been applied to amorphous and liquid silicates yield significantly different results for the high pressure structure and compression mechanism. Lately, a sophisticated computational approach within quantum mechanical formulation has taken on an increased importance in the study of silicate liquids (e.g., Trave et al. 2002; Pohlmann et al. 2004; Stixrude and Karki 2005; Wan et al. 2007; Mookherjee et al. 2008; de Koker et al. 2008; Karki et al. 2009; Vuilleumier et al. 2009). Such first principles approach is more robust because it makes no assumptions to the nature of bonding or the shape of the charge density and is thus, in principles, equally applicable in the study of a wide variety of materials problems including liquids. These simulations not only calculate the bulk properties of materials but also yield microscopic details revealing underlying physics at the relevant pressure-temperature conditions of Earth's interior. In particular, they can help us to answer several fundamental questions related to molten silicates.

How does the structure of the liquid phase change with increasing pressure and compare with the structure of the solid phase? The equation of state of silicate liquids, their relative density with respect to coexisting solids, and their mobility, which largely depend on the topological characteristics of atomic packing, can be understood on the basis of their structure. Previous theoretical and experimental studies have found that amorphous silicates undergo remarkable pressure-induced changes in structure. These include a gradual pressure-induced increase in the Si-O coordination number, from four-fold at ambient pressure, towards six-fold at higher pressure. Such a structural change mirrors that associated with polymorphic phase transitions in crystalline phase. However, the nature of this coordination change in the amorphous state is still poorly constrained. In particular, the mean pressure at which it occurs, the pressure range over which it takes place, and whether four-fold and six-fold coordinated states are energetically preferred and stable over finite pressure intervals are not known. The structure within the transition interval is also of considerable interest: do four- and six-fold coordination states co-exist over a wide range of pressure, or are intermediate states (i.e., five-fold coordination) also important? The presence of five-fold coordination states has been suggested to enhance mobility (self-diffusion) substantially. Not only short-range structures, of length scale 1 to 3 Å, often expressed in terms of bond lengths and nearest neighbor coordination but also intermediate-range structures of length scales greater than 3 Å are important in liquids.

How does the degree of polymerization vary with pressure? To answer this question requires information about how cations such as Mg, Ca and Na and protons can disrupt the tetrahedral silicate network. For instance, five- and seven-coordinated Si atoms, which do not exist any in crystalline silicates, do exist in molten silicates. What are their abundances and how do they change on compression? It also requires quantitative information on the concentrations of non-bridging oxygen (NBO), bridging oxygen (BO) and oxygen triclusters, which have been experimentally observed in silicate glass (Lee et al. 2008) as a function of pressure. How does NBO transform into bridging oxygen (BO) eventually leading the appearance of oxygen triclusters as the liquid is compressed?

How do the liquid transport properties depend on pressure and temperature, and what is their microscopic origin? The experimental data on diffusion coefficients (Lesher et al. 1996; Poe et al. 1997; Reid et al. 2001; Tinker et al. 2003) and viscosities (Kushiro et al. 1978; Behrens and Schulze 2003; Reid et al. 2003; Dingwell et al. 2004; Tinker et al. 2004) show that the diffusivity and viscosity depend strongly on temperature and pressure in a way which is sensitive to melt structure through the degree of polymerization and the coordination number. For less polymerized structures, the temperature and pressure variations can be well described with the Arrhenius law; as such, the diffusivity (viscosity) decreases (increases) with increasing pressure (e.g., Reid et al. 2001, 2003). However, highly polymerized structures such as liquid silica show the non-Arrhenius dependence on temperature (Giordano and Dingwell 2003) and also show anomalous increase (decrease) in diffusivity (viscosity) with pressure (Bottinga and Richet 1995; Tinker et al. 2003, 2004). What is a theoretical basis to the general trends of the diffusivity and viscosity that have emerged out of the available measured data? Is the same atomic-scale mechanism responsible for both self-diffusion and viscous flow? There is need of evaluating models such as the Stokes-Einstein and Eyring equations for the diffusivity-viscosity relationship (Mungall 2002; Tinker et al. 2004; Nevins et al. 2009), the Vogel-Fulcher-Tammann equation for the non-Arrhenius temperature variation (Giordano and Dingwell 2003), and the Lindemann law of melting based on mean square displacements (Cohen and Gong 1994). Limited success of these models implies a need for a direct determination (experimentally or theoretically) of the transport properties at each pressure-temperature condition of interest.

What is the nature of thermodynamics of mixing and how does it vary with pressure? The calculated equations of state of liquids of several compositions can be compared with the predictions of an ideal solution model to understand the thermodynamics. Experimental data show that thermodynamics of mixing is non-ideal with some degree of liquid immiscibility at low pressures. An important question is whether the non-ideality and immiscibility continue to exist at greater depths. The answers require quantification of the volumes and enthalpies of mixing and understanding of their structural origin and pressure dependencies. The thermodynamics of mixing also affects the physical properties, which show large compositional variations. Studies have suggested that at the ambient pressure, the diffusivities vary by more than 3 orders of magnitude along MgO-SiO_2 join.

A series of geophysically relevant silicate melts (e.g., Stixrude and Karki 2005; Karki et al. 2006a, 2007; Wan et al. 2007; de Koker et al. 2008) along the MgO-SiO_2 join (Lacks et al. 2007) including hydrous phase (Mookherjee et al. 2008; Karki et al. 2009) have recently been investigated from first principles. The major focuses have been on obtaining the equation of state, exploring the structures, understanding the thermodynamics of mixing, and predicting the transport properties at mantle conditions. These simulations have shown that density functional theory provides the over-arching framework to meet these challenges. We have found that the first principles methods provide the ideal complement to the laboratory approach because there are no free parameters in the theory. Moreover, we have shown that theory can provide important insight into the fundamental origins of physical properties in the structure and bonding and promote our understanding of the role of silicate liquids in planetary evolution and magma oceans. In this paper, we illustrate the development/application of the first-principles simulation and visualization methods in the study of the liquid phases of Earth materials. We use MgO, SiO_2 and $MgSiO_3$ liquids as primary examples by presenting relevant results only; other details can be found elsewhere.

The organization of the paper is as follows: First, we introduce the challenges in the liquid simulations. Second, we talk about the methodology in the context of parallel computation and visualization, convergence tests and derivation of physical properties. Then, we present the specific results and discussion on the equation of state, structure and transport properties of melts. Finally, we draw some conclusions and present future directions.

COMPUTATIONAL CHALLENGES

Understanding the behavior of the liquid state poses tremendous challenge for computational scientists. In a solid phase, the constituent atoms interact strongly with each other and often form a crystalline (perfectly ordered) structure. As such, most crystalline properties can be modeled using unit cell containing a relatively few atoms thereby exploiting the crystal symmetry to a great extent. This is no longer true in the case of a liquid phase: The constituent atoms still interact strongly with each other, however, they do not form any long-range correlations. Even the short-range (local) order is temporal and transient fluctuations also occur in the liquid phase.

The temporal (dynamical) nature of the liquid state, which can be only described by statistical measures, requires the construction of accurate and representative ensembles (Allen and Tildesley 1987). This implies two important concerns: First, the ensemble should contain accurate interatomic interactions. Unlike the crystalline phase, it is not straightforward to generate empirical interaction potential that can correctly capture the dynamic nature of the liquid state. In this sense, it is natural to prefer to derive the required interactive forces from a quantum mechanical treatment. The resulting approach can be called the first-principles molecular dynamics (FPMD) method, in contrast to the traditional molecular dynamics method, which uses either empirical or simplified force field model. A widely used FPMD method is based on density functional theory, which explicitly deals with electrons making it computationally very intensive (Hohenberg and Kohn 1964; Kohn and Sham 1965). Second, the representative ensemble must be sufficiently large. Supercells with periodic boundary conditions containing a relatively large number of atoms are commonly used to replicate an ensemble for the liquid state. The system sizes from several tens to a few hundreds of atoms are typical for the first-principle approach. Also, the simulation needs to be run for relatively long time, which can range from a few picoseconds to several picoseconds or even longer to ensure that the time-averaged properties are well converged.

Another important issue in the modeling liquid state is related to the choice of input configuration. Two common approaches are: The first approach is to melt the crystalline phase by thermalizing it at a very high temperature, which is often chosen to be well above the corresponding melting temperature. The system is then quenched to a desired lower temperature. The other approach is to generate random atomic configuration and thermalize at high temperature using pair potentials MD simulation. In this case, the initial configuration does not retain any information of the crystalline phase and the simulations may converge relatively slow. The second approach is particularly important for the simulation of melt compositions that have no homogeneous crystalline analog.

Last important issue is related to analysis of the simulation outputs. FPMD simulations produce massive amounts of three-dimensional and time-dependent data for the atoms and electrons. In essence, these data represent snapshots of the atomic and electronic structures of the liquid state at different instants of time. In recent years, visualization approach has emerged as an attractive approach to uncover important, otherwise hidden information in large-scale scientific data. Several visualization systems (public domain or commercial nature) such as *XCRYSDEN* (Kokaji 1999), *VMD* (Humphrey et al. 1996), *Crystalmaker,* and *Amira* are available, however, using them is always not the best choice. They may not have several specific features that are needed for an efficient visualization of the liquid data due to the complex spatio-temporal behavior of the data. A space-time-multiresolution visualization scheme for atomic data has recently been developed targeting liquid and defective systems (Bhattarai and Karki 2009).

METHODOLOGY

First-principles molecular dynamics simulations

The molecular dynamics simulations compute the phase-space trajectories of atoms by numerically integrating Newton's equations of motion (Allen and Tildesley 1987). In the first-principles method, the forces are computed via density functional theory (Hohenberg and Kohn 1964; Kohn and Sham 1965; Jones and Gunnarson 1989) - an exact theory of the ground state that maps an interacting many-electrons problem to N single-electron problems. It is only recently that FPMD has become widely applicable to complex systems like liquids. The interatomic forces, which are of quantum origin, are computed at each time step of the simulation from a fully self-consistent solution of the electronic structure problem represented by the well-known Kohn-Sham equation:

$$\left[\frac{\hbar^2}{2m}\nabla^2 + V_n(\vec{r}) + V_H(\vec{r},\rho(\vec{r})) + V_{XC}(\vec{r},\rho(\vec{r}))\right]\psi_i(\vec{r}) = \varepsilon_i\psi_i(\vec{r}) \quad (1)$$

Here, the first-term is the kinetic energy, V_n is the ion-electron potential, V_H is the Hartree potential (Coulomb interaction among electrons) and V_{XC} is the exchange-correlation potential (all many-body effects). The electron charge density is given by $\rho(\vec{r}) = \Sigma_i |\psi_i(\vec{r})|^2$.

The self-consistent solutions (wave functions, ψ_i's, and eigenvalues, ε_i's) can be obtained by iteratively solving the KS equation, which requires two approximations. First, the local density approximation (LDA) replaces the exchange-correlation potential at each point r by that of a homogeneous electron gas with a density equal to the local density at point r (Ceperly and Alder 1980). The LDA works remarkably well for silicates and oxides (e.g., Karki et al. 2001). Most previous calculations were based on this approximation, although some calculations have used the generalized gradient approximation (GGA) (Perdew et al. 1996). GGA, which tends to underbind structures: for example, the calculated volume tends to be larger than experimental volume. The LDA and GGA equations of state for crystalline MgO lie systematically below and above the experimental data (Fig. 1). Second, the pseudopotential approximation (Picket 1989) replaces the strong potential due to the nucleus and core electrons by a weaker, more slowly varying potential with the same scattering properties. This allows one to use plane-waves as the basis functions to represent the electronic wave functions and hence to exploit the easiness in obtaining the Hellmann-Feynman forces and quantum mechanical stresses. The results presented here were obtained using ultra-soft pseudopotentials (Kresse and Furthmüller 1994). One can model the interactions between valence electrons and the core using the projector augmented wave method (Kresse and Joubert 1999), which is a frozen-core all-electron approach using the exact wave functions.

One of the widely used parallel FPMD programs is called VASP - Vienna ab-initio simulation package (Kresse and Furthmuller 1996). It has been implemented with a Fortran90 and Message Passing Interface (MPI). It exploits dynamic memory allocation and a single executable, which can be used to simulate a wide range of calculations. Our scalability tests for MgO liquid (Fig. 2) suggest that the use of 64 processors gives an acceptable performance for the simulation systems consisting of order of 100 atoms with about 3 seconds per FPMD step, which does not improve anymore on using more processors. When the system size increases, the time per step increases rapidly, reaching a few minutes for the 512 atoms system on 64 processors. Such a poor scaling is expected for the first-principles calculations. By increasing the number of processors, we can lower the time to about 4 minutes for 128 processors and the

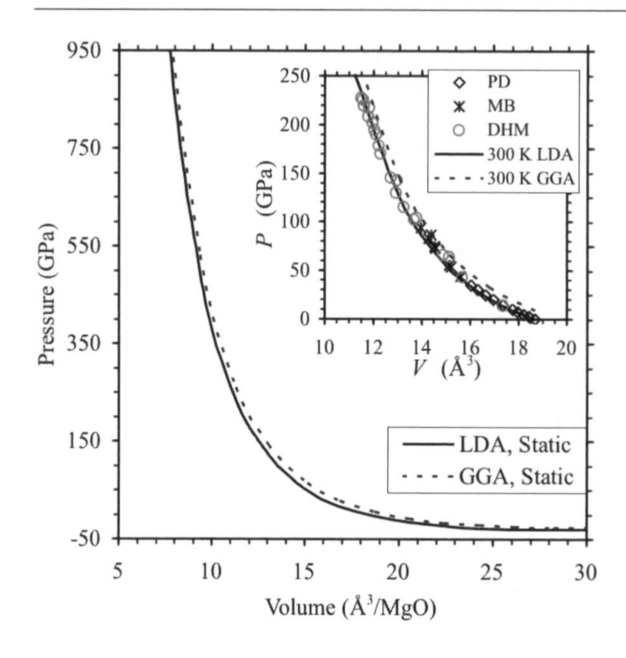

Figure 1. Calculated static equations of state for crystalline MgO using LDA (solid line) and GGA (dashed line) covering a wide compression range of $V/V_0 = 0.3$ to 2.0 (where $V_0 = 18.67$ Å^3 per unit formula). The inset shows comparison with the room temperature experimental data, PD (Perez and Drickmer 1995), MB (Mao and Bell 1979) and DHM (Duffy et al. 1995). Also the 300 K and zero-point corrections are included to the calculations (Karki et al. 2000).

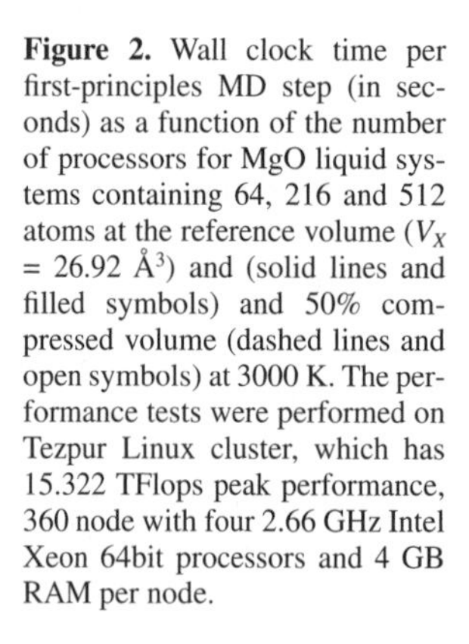

Figure 2. Wall clock time per first-principles MD step (in seconds) as a function of the number of processors for MgO liquid systems containing 64, 216 and 512 atoms at the reference volume (V_X = 26.92 Å^3) and (solid lines and filled symbols) and 50% compressed volume (dashed lines and open symbols) at 3000 K. The performance tests were performed on Tezpur Linux cluster, which has 15.322 TFlops peak performance, 360 node with four 2.66 GHz Intel Xeon 64bit processors and 4 GB RAM per node.

time does not improve much with 256 processors. Note that runs become faster on compression because the plane-wave basis set gets smaller at smaller volumes. Computation speeds up by ~25% on two-fold compression.

The FPMD simulations are particularly suitable for accurate simulations of fluids at high temperatures as quantum effects on the phonon populations are usually only important below the melting point. We use the canonical (*NVT*) ensemble with Nose thermostat (Nose 1984) in which the number of atoms (*N*), volume (*V*) and temperature (*T*) are fixed but the pressure (*P*) and energy (*E*) can vary. Temperature is controlled using an extended Lagrangian formulation in which a degree of freedom is included to represent a reservoir (thermostat). The Lagrangian is

$$L = \frac{1}{2}\sum_i m_i |v_i|^2 - V(\vec{r}_i) + \frac{1}{2}Q\dot{s}^2 - (f+1)k_B T_0 \ln s \quad (2)$$

here s is the new dynamical variable, and Q is the associated mass parameter. T_0 is the externally set temperature and f is the number of degrees of freedom in the system. To speed convergence, Q is chosen so that the period of oscillation of the temperature (or s) is similar to the mean period of atomic vibrations. The nature of the canonical ensemble, for which volume is the independent variable, has considerable analytical advantages. Results show that liquid structure is primarily a function of volume, rather than pressure, and that variations in temperature at constant volume leave mean coordination numbers virtually unchanged. We express the volume as V/V_X, where V_X is the reference volume, which is often taken to be equal, for the number of atoms in the simulation cell, to the experimental volume of the liquid under consideration at the ambient melting point. Alternatively, one can use constant-pressure constant-temperature (*NPT*) ensemble in which the required pressure is maintained by optimizing the cell parameters with variable-cell dynamics. In either ensemble, the pressure can be reported as

$$P(V) = P_{MD}(V) + P_{Pulay}(V) + P_{Emp}(V) \quad (3)$$

Here, P_{MD} is the pressure computed in the simulations. The Pulay correction is

$$P_{Pulay}(V) = P_S(V; E_{cut} = 600\,\text{eV}) - P_S(V; E_{cut} = 400\,\text{eV}) \quad (4)$$

where P_S is the static pressure of corresponding crystalline phase computed with a higher cutoff (such as 600 eV) to yield fully converged results. We find that P_{Pulay} increases monotonically from ~2 to ~6 GPa over the compression regime studied for oxides and silicates. The empirical correction accounts for the well-known over-binding of LDA. Its value was computed as

$$P_{emp} = -P_S(V_{exp}; E_{cut} = 600\,\text{eV}) \quad (5)$$

where V_{exp} is the experimental zero-pressure volume of the solid phase at static conditions computed via the thermodynamic model of Stixrude and Lithgow-Bertelloni (2005). Its value lies between 1 and 2 GPa in most cases. A similar correction (with an opposite sign due to volume overestimation) can be applied to GGA pressure.

Input configuration and simulation schedule

So far only those melt compositions that have homogeneous crystalline analog (i.e., they have mineral-like stoichiometry) have been simulated using the first-principles methods. We generate the configuration by melting a crystalline solid of the desired composition. This requires melting of the solid system at very high temperatures (well above the melting point) followed by cooling to the desired temperature. The simulation cell is always a cube. If the crystalline phase is not a cube, the supercell is strained to a cube. For instance, the simulation of MgO liquid containing 64 atoms uses a 2×2×2 supercell made out of 8-atom cubic cells. Similarly, a 72-atom silica liquid system is a 2×2×3 supercell of 6-atom hexagonal (alpha-quartz structure at low pressure) or tetragonal (stishovite at high pressure) cells, which was strained to a cubic shape. The simulation schedule thus consists of three stages; melting (randomization) at hot temperature, annealing and finally acquiring representative ensemble at the desired temperature. For instance, the 64-atom MgO system of a given volume was first melted and equilibrated at 10,000 K for a period of 5 picoseconds, which means 5,000 FPMD steps, each time step being 1 fs. The system was then quenched to a desired lower temperature, say 7000 K, over a time interval of 3 ps. At this temperature, we first thermalize the system forl ps and collect data over another 4 ps. The temperature was decreased in this way to simulate the liquid at subsequent lower temperatures of 5000, 4000 and 3000 K. The whole process is repeated at every other volume. A series of FPMD runs are usually conducted at several volumes along several isotherms. For each temperature-volume condition, we confirm that the system is in the liquid state by analysis of the radial distribution function and the mean-square displacement as a function of time (Allen and Tildseley 1987). Simulation run durations needed for different liquids vary a lot. For instance, silica liquid requires much longer runs at lower temperatures (see the convergence tests section).

Derivation of physical properties

The FPMD simulations generate results for pressure, volume, energy and temperature, and also microscopic data consisting of atomic-position time series, electronic energies and wave functions. These results can be used to study/derive a wide range of properties of liquids as a function of pressure and temperature, which include the equilibrium thermodynamic properties, atomic and electronic structures, and transport properties.

Thermodynamics. The calculated *P-V-E-T* results can be used to obtain several important thermodynamic properties including bulk modulus, thermal expansivity, specific heat, Gruneisen parameter. We can represent the *P-V-T* equation of state with the Mie-Grüneisen form:

$$P(V,T) = P(V,T_0) + P_{TH}(V,T) \tag{6}$$

Here, $P(V,T_0)$ is the pressure on the reference isotherm ($T = T_0$), which in most cases can be described by the third-order Birch-Murnaghan EoS:

$$P(V,T_0) = 3K_0 f(1+2f)^{5/2}[1+\frac{3}{2}(K_0'-4)] \tag{7}$$

Where $f = ½[(V_0/V)^{2/3} - 1]$ refers to compression (the negative of the strain). $P_{TH}(V,T)$ is the thermal pressure, which generally increases nearly linearly with temperature at all volumes. The increase of thermal pressure on compression is always not uniform: $P_{TH}(V,T) = B(V)(T - T_0)$ with volume dependent coefficient $B(V) = (\partial P/\partial T)_V = [\gamma(V)C_V(V)]/V$, where C_V and γ are isochoric heat capacity and Grüneisen parameter, respectively. As in the case of the thermal pressure, a linear equation is fit to the calculated energy-temperature results: $E_{TH}(V,T) = C_V(V)(T - T_0)$. Using these representations, we can derive other properties including the coefficient of thermal expansivity, $\alpha = (1/V)(\partial V/\partial T)$, isochoric heat capacity, $C_V = (\partial E/\partial T)_V$ and Grüneisen parameter, $\gamma = (V/C_V)(\partial P_{TH}/\partial T)_V$. However, one can use a self-consistent thermodynamic formulation to represent the Helmholtz free energy $F(V,T)$ results given by first principles simulation and use the fundamental relationships to derive various thermodynamic properties that are applicable across wide pressure and temperature ranges (de Koker and Stixrude 2009).

The melting curve (T_M versus P) can be computed using two approaches. The first approach uses the Clausius-Clapeyron relation

$$\frac{\partial T_M}{\partial P} = \frac{\Delta V}{\Delta H / T_M} \tag{8}$$

The volume (ΔV) and enthalpy (ΔH) differences are taken from simulations. The integration of the above equation requires the initial melting temperature often taken from experiment. The second approach, which does not use any experimental input, is based on simulation of a two-phase system (with liquid and solid coexisting) and is computationally much more intensive than the first hybrid approach. The melting curves of MgO (Alfe 2005; de Koker and Stixrude 2009), $MgSiO_3$ (Stixrude and Karki 2005), Mg_2SiO_4 (de Koker et al. 2008) have been calculated using both techniques mentioned earlier. The predicted melting curve and Grueneisn parameter were further used by Stixrude et al. (2009) to obtain the mantle liquidus, solidus and isentrope.

Microscopic structure. The atomic structure of the liquid is usually analyzed in terms of the radial distribution function (RDF) and also by computing time averages of suitable structural quantities of interest such as coordination numbers, bond angles and rings. The RDF can be transformed to the structure factors, which can directly be compared with available

X-rays and neutron scattering data. The radial distribution functions for atomic pair of species α and β is defined as

$$g_{\alpha\beta}(r) = \frac{1}{4\pi\rho_\beta r^2}\left[\frac{dN_\beta(r)}{dr}\right] \tag{9}$$

where ρ_β is the number density of species β and N_β is the number of species β within a sphere of radius r around a selected atom of type α. One can calculate the bond length as the mode of the first peak (r_{peak}) in the partial RDF or a weighted average of all distances covered by the first peak:

$$<r> = \frac{\int_0^{r_{min}} r g_{\alpha\beta}(r)dr}{\int_0^{r_{min}} g_{\alpha\beta}(r)dr} \tag{10}$$

where r_{min} is the position of the minimum after the first peak in the corresponding $g_{\alpha\beta}(r)$. Similarly, the atomic coordination, which is a fundamental quantity used to characterize the local structure, can be calculated for a given species α with respect to another species β using

$$C_{\alpha\beta} = 4\pi\rho_\beta \int_0^{r_{min}} r^2 g_{\alpha\beta}(r)dr \tag{11}$$

The above quantities represent the average atomic structure of liquids so it is necessary to supplement interpretation of these quantities with other structural information by exploring individual snapshots on per atom basis. This requires visualization of the simulation data, which are three-dimensional and time dependent. Similarly, to characterize the electronic structure, we can compute/visualize the electronic density of states, charge distributions, squared wave functions.

Dynamical properties. Dynamical phenomenon for a liquid phase can be characterized by calculating the diffusion coefficient and viscosity. There are two ways of calculating each quantity from FPMD simulations: the Einstein relation, or the fluctuation-dissipation theorem (Green-Kubo relations) (Boon and Yip 1980; Hansen and McDonald 1986; Allen and Tildesley 1989). Here, we discuss the use of the Einstein-formula given by

$$D_\alpha = \lim_{t\to\infty}\frac{\left\langle \Delta r^2(t)\right\rangle_\alpha}{6t} = \lim_{t\to\infty}\frac{\left\langle |\vec{r}(t+t_0)-\vec{r}(t_0)|^2\right\rangle_\alpha}{6t} \tag{12}$$

to calculate the diffusion coefficient (D_α) of a given species (α) from the mean square displacement (MSD). Here, $\vec{r}(t)$ represents particle trajectories that are continuous in Cartesian space, and $\langle\ldots\rangle_\alpha$ denotes an average over all atoms of type α and over different time origins, t_0's. The slope of MSD with respect to t, for range of t within the diffusive regime (where the log-log plot shows a slope of unity) is used to determine the value of D_α. To calculate the shear viscosity (η), we use the Green-Kubo relation

$$\eta = \frac{V}{k_B T}\int_0^\infty \left\langle \sigma_{ij}(t+t_0).\sigma_{ij}(t_0)\right\rangle \tag{13}$$

where σ_{ij}'s (i and $j = x, y, z$) represent the off-diagonal components of stress tensor that is computed directly at every time step of simulation (Nielsen and Martin 1985). We perform the ensemble averaging by calculating the integrand with different time origins, t_0's, and the upper-limit denotes a time at which the autocorrelation function (ACF) decays to zero.

Convergence tests

At each pressure, temperature condition, simulations should be run for long enough to achieve full convergence of the calculated properties, and the finite-size effects should be taken into account appropriately. Here, we present the relevant analysis for MgO and SiO_2, which are two end members of silicate liquids and, more importantly, they represent the class of the structure modifying oxides (such as MgO, CaO) and that of highly polymerized oxides (such as SiO_2, Al_2O_3), respectively. They are expected to provide constraints on the level of convergence from above and below since MgO shows the fastest dynamics (representing the best case of convergence) whereas SiO_2 shows the slowest dynamics (representing the worst case of convergence).

The running time averages of pressure and energy converge well over relatively short run durations for both liquids. Figure 3 shows the energy- and pressure-times series with their running time-averages for MgO liquid at 3000 K. The averages and uncertainties in the energy

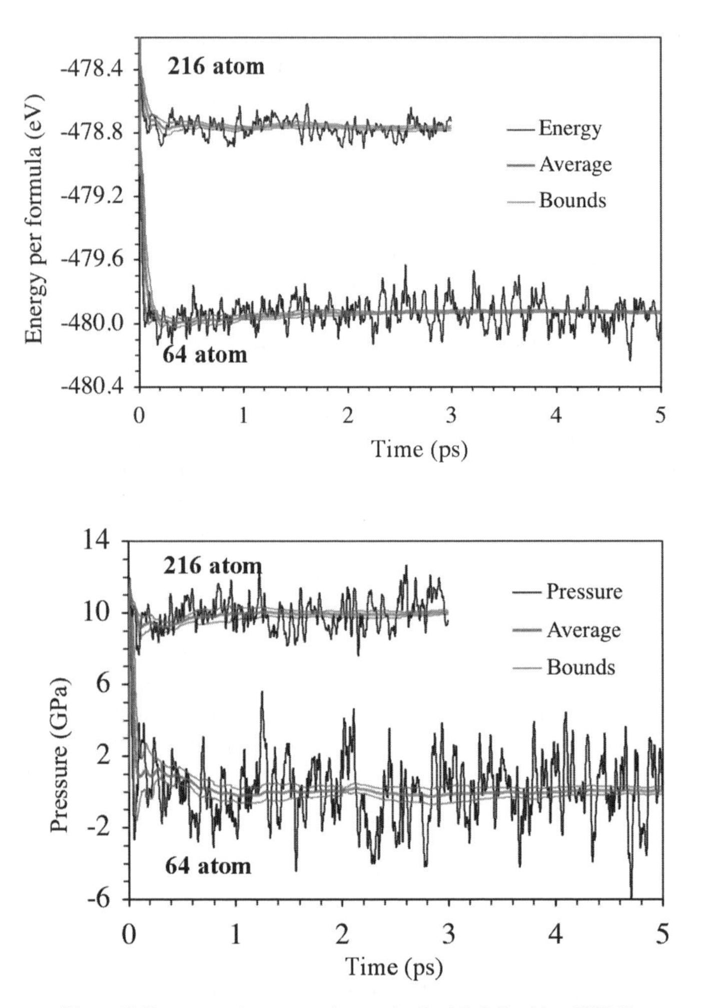

Figure 3. Energy- and pressure-time series for MgO liquid at 3000 K and V_X. The uncertainties (bounds) and averages are also shown.

and pressure were computed by the blocking method (Flyvbjerg and Petersen 1989). The sizes of fluctuations become smaller as the simulation size increases without affecting the averaged quantities significantly. Systems containing atoms in the order of 100 are sufficient to calculate the equilibrium properties including the melt structures.

Calculations of the transport properties use MSD (Eqn. 12) and stress ACF (Eqn. 13). To ensure that the dynamics are in the diffusive regime, we run simulations long enough so that MSD versus time plot clearly shows a linear behavior, and the maximum MSD value exceeds, at least, 10 Å^2. Such a diffusive regime is reached much faster in MgO than in SiO_2 liquid (Fig. 4 top). At the ambient pressure, MgO satisfies the diffusive condition after 2 ps at 3500 K and even earlier at higher temperatures. The situation is more serious in silica; the diffusive condition is satisfied only after 60 ps. Correspondingly, we find that the stress ACF always converges to zero over such durations (Fig. 4 bottom). Note that at 3500 K, ACF of liquid silica decays to zero within 10 ps. As pressure increases, both liquids require much longer runs, e.g.,

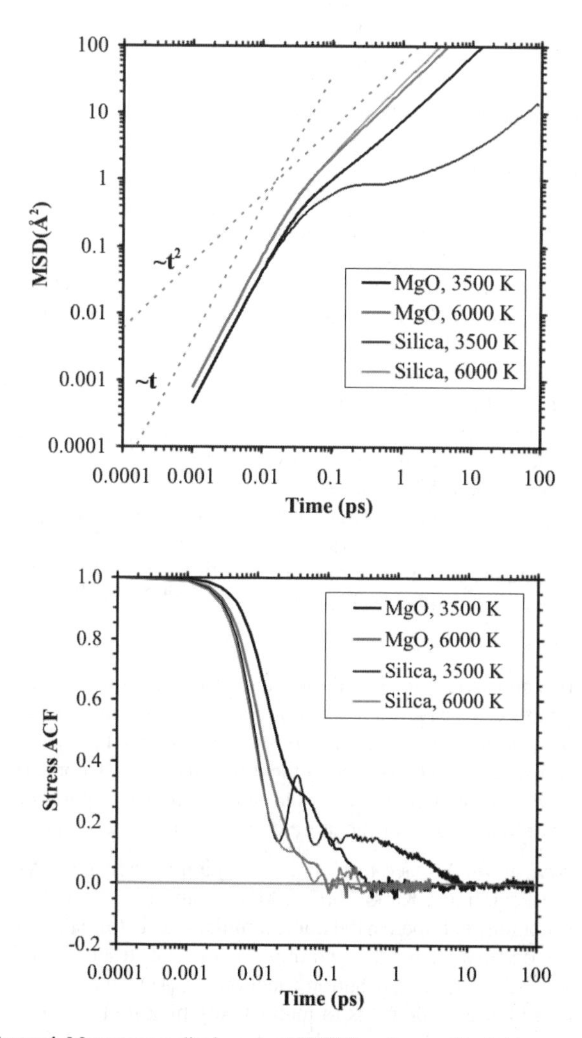

Figure 4. Mean square displacements (MSD) and normalized shear stress autocorrelation functions (ACF) for MgO and SiO_2 liquids at V_X.

exceeding 100 ps at 4000 K and 120 GPa. The MSD and ACF are averaged over different time origins (i.e., different initial configurations). As such, the counting statistics decreases towards the end of the simulation run so the curve tends to be less smooth at large t. In the calculations, it's a good idea to ignore a significant number of steps (e.g., 10% of the total number of steps) both in the beginning (so that that the dynamic regime is reached and shear relaxations occur) and in the end (so that counting statistics is sufficiently large) of the simulation.

Large simulation cell is desirable to decrease the uncertainty in any calculated property. Although they are not significant for equilibrium properties, finite-size effects can be important for transport properties (Rustad et al. 1990; Yeh and Hummer 2004; Zhang et al. 2004). In the case of water and Lennard-Jones system, the correction to the diffusion coefficient scales as the inverse of $N^{1/3}$, where N is the number of atoms, and shear viscosities show no significant size dependence (e.g., Yeh and Hummer 2004). This is true for other liquids, particularly, at high temperatures but silica at temperature below 3500 K requires different forms of scaling functions (Zhang et al. 2004). Moreover, the size effects may be less important at high pressure (e.g., Tsuneyuki and Matsui 1995). By exploring how diffusivity and viscosity scale with the size, we can use the finite-size results to accurately estimate the representative bulk values (i.e., the values in the limit of infinite system). To the date, all first-principles studies of liquids have used system sizes in the order of 100 atoms. In the case of MgO liquid, increasing the size by up to five fold, i.e., up to about 500 atoms shows that the finite-size effects are significant at 3500 K and both diffusivity and viscosity tend to increase with increasing the system size roughly following the inverse $N^{1/3}$ dependence (Fig. 5).

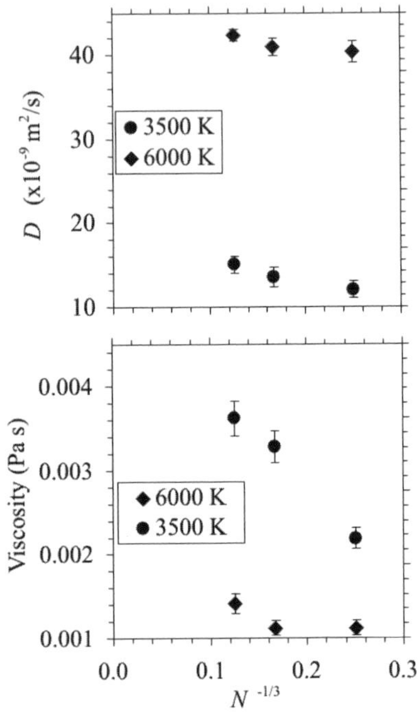

Figure 5. Self diffusion coefficient and viscosity of MgO liquid at 3500 and 6000 K (at V_X) as a function of the number of atoms used in simulations.

Visualization

The simulations consisting of several thousands (or tens or even hundreds of thousands) of steps per run produce massive data for atomic and electronic configurations, which are distributed in a 3D space and vary over time. To fully understand the microscopic structures of liquids requires sophisticated visualization of the direct data (which come from simulations) as well as the additional data (which are generated in the analysis or post-processing stage). We discuss the visualization methods for the atomic and electronic datasets.

Atomistic visualization. A space-time multiresolution visualization system (Bhattarai and Karki 2006, 2009; Bhattarai and Karki 2007, 2009) is particularly suitable for gaining insight into important information contained in the liquid simulations. In this approach, a given position-time series is analyzed at diverse length- and time-scales. The structural analysis starts with the computation of the partial radial distribution functions, $g_{\alpha\beta}(r)$'s using Equation (9). The RDF matrix plot shown in Figure 6 allows us to interactively pick up the critical distances such as the peak distance, r_{peak}, and the distance to the first minimum, r_{min}, and also define a window of arbitrary width for subsequent analyses. For instance, the atomic coordination (Eqn. 11) defined

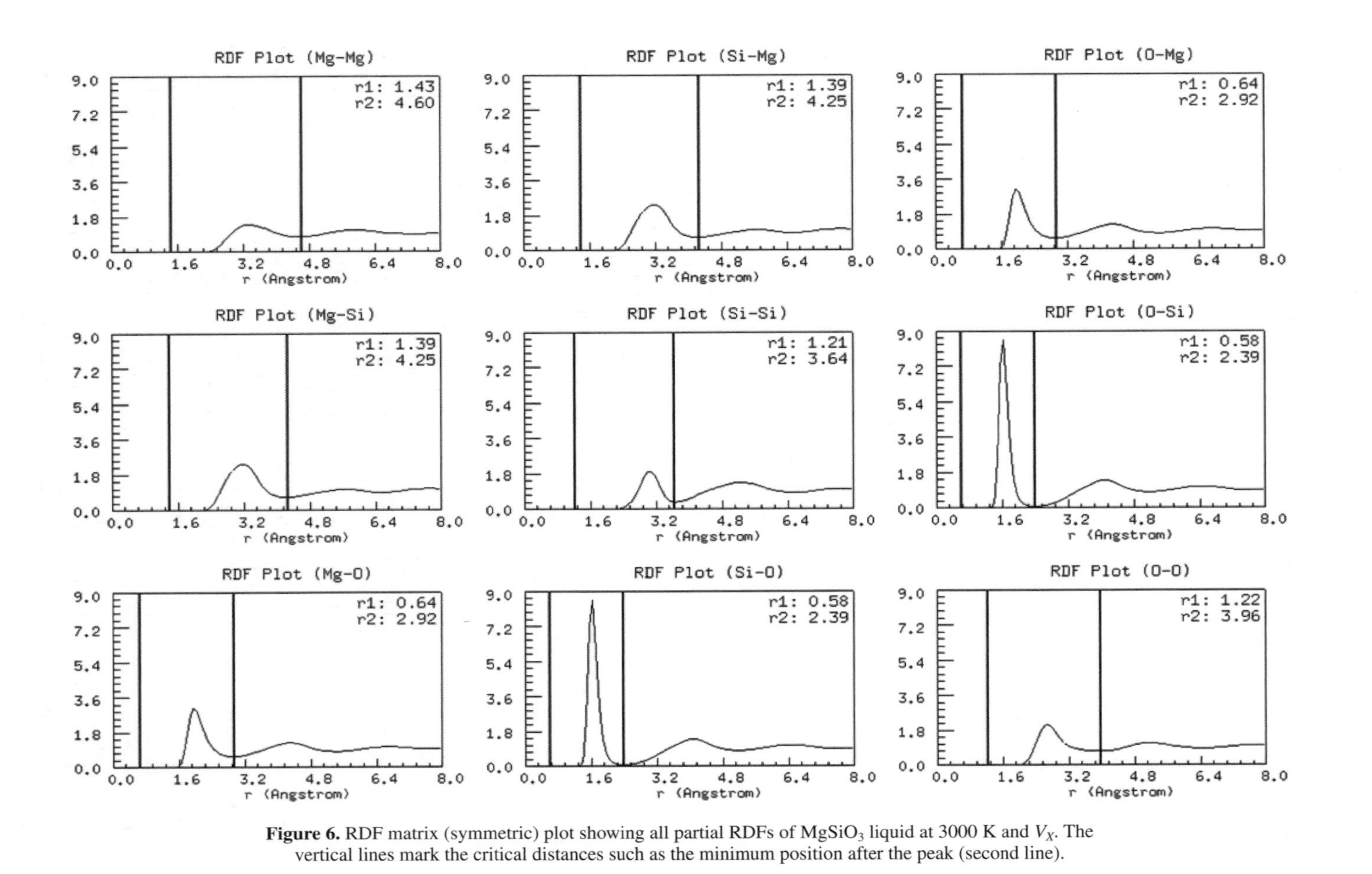

Figure 6. RDF matrix (symmetric) plot showing all partial RDFs of $MgSiO_3$ liquid at 3000 K and V_X. The vertical lines mark the critical distances such as the minimum position after the peak (second line).

for a given species α with respect to another species β uses the corresponding partial RDF information. Here, we talk about the nearest neighbor coordination number so the contributing atoms (of species β) are those that lie within a sphere centered at atom of species α and of radius $r_{\min}$. In a multi-component system containing s species, one can have s^2 coordination types - multivariate information represented by an asymmetric square matrix of order s:

$$\begin{vmatrix} C_{11} & C_{12} & .. & C_{1s} \\ C_{21} & C_{22} & .. & C_{2s} \\ \vdots & \vdots & \vdots & \vdots \\ C_{s1} & C_{s2} & .. & C_{ss} \end{vmatrix} \tag{14}$$

Each mean coordination can be characterized by a number. Both the RDF and mean coordination have limitations as being isotropic. A further analysis decomposes each coordination environment into a variety of coordination species, which depend on space (defined on per atom basis) and time (Bhattarai and Karki 2007, 2009):

$$C^{i}_{\alpha\beta}(t) = \left|\left\{1 \le j \le N : d(i,j) \le r^{\alpha\beta}_{\min} \wedge type(j) = \beta\right\}\right| \quad \text{for} \quad i = 1.....N_{\alpha} \tag{15}$$

We can visualize the coordination data in the form of simple spheres or polyhedra with their values encoded using the color-map (Bhattarai and Karki 2009). Figure 7 encodes the coordination state at the individual atom level for all 9 different coordination environments in $MgSiO_3$ liquid using the number of lines emanating from the centered spheres. The Si-O coordination environment is primarily tetrahedral (13 silicon atoms in four-fold coordinated) with only one three-fold coordinated and two five-fold coordinated silicon atoms. The O-Si coordination plot shows that all oxygen forms bonding with silicon but not necessarily bridging two silicon atoms. The temporal stabilities of the different coordination states can be computed over finite simulation durations and coded in the length of bonds and the size of the spheres (the thicker the bond and larger the sphere, the more stable the corresponding coordination state). In particular, it's very useful to render a mixed environment consisting of two or more coordination types, for instance, a display consisting of Si-O coordination polyhedra, and O-H and H-O coordination spheres (Fig. 8) gives an overall picture of polyhedral framework and water speciation in hydrous silicate melt (Bhattarai and Karki 2009; Karki et al. 2009). In addition to the standard animation and pathline techniques, a variety of displacement data and covariance matrices can be visualized using spheres and ellipsoids to further understand the dynamical behavior.

Electronic visualization. The electronic bands and density of states are usually computed to analyze the electronic structure of a given material system. The DOS can be further decomposed into different angular momentum components and/or projected on individual atomic sites. To explore an overall electron density distribution or contributions from individual states such as valence band maximum or conduction band minimum requires visualization of the volumetric data for electron density or squared wave functions. Such visualization uses the isosurface extraction and clipping methods, which are supported by several visualization systems including XcrysDen (Kokaji 1999). A multiple dataset visualization scheme, which facilitates understanding of important relationships and differences among different data rendered simultaneously, also supports these methods (Khanduja and Karki 2006, 2008). Simulation results suggest that the electronic structure of liquid differs substantially from that of solid. The calculated electronic densities of states of show substantial band broadening as compared with the crystalline state, which is sufficiently large to close the band gap in the liquid giving it a semi-metallic character (Karki et al. 2006a). Also, due to dynamic rearrangement of atoms, the spherical charge density distributions around O sites are significantly perturbed (Fig. 9). One can clearly see that the electron density distribution around O ions forms inter-atomic bridges instead of forming isolated structures (Karki et al. 2006a,b).

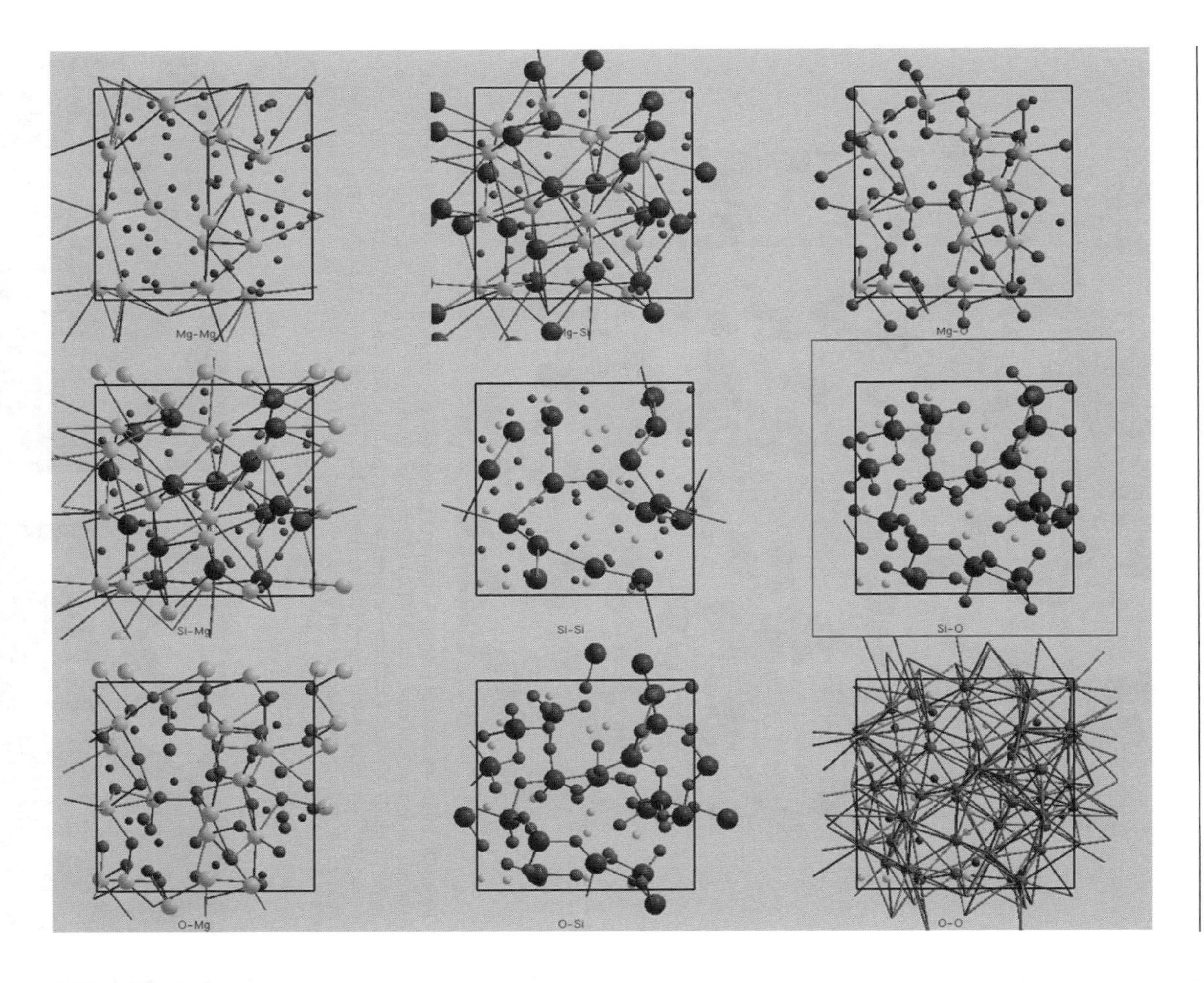

Figure 7. Coordination matrix plot showing nine coordination environments in $MgSiO_3$ liquid at 3000 K and V_X. The atomic species are represented by different sizes and shades: Large (dark), medium (light) and small (medium dark) spheres represent Si, Mg and O atoms, respectively. In each entry, spheres show atomic species pairs involved in that coordination other species is shown by tiny spheres. The number of lines emanating from centered spheres encodes coordination value, $C_{\alpha\beta}{}^{i}(t)$ given by Equation (15). The cutoff distances (r_{min} values) used are from the RDF matrix plot shown in Figure 6.

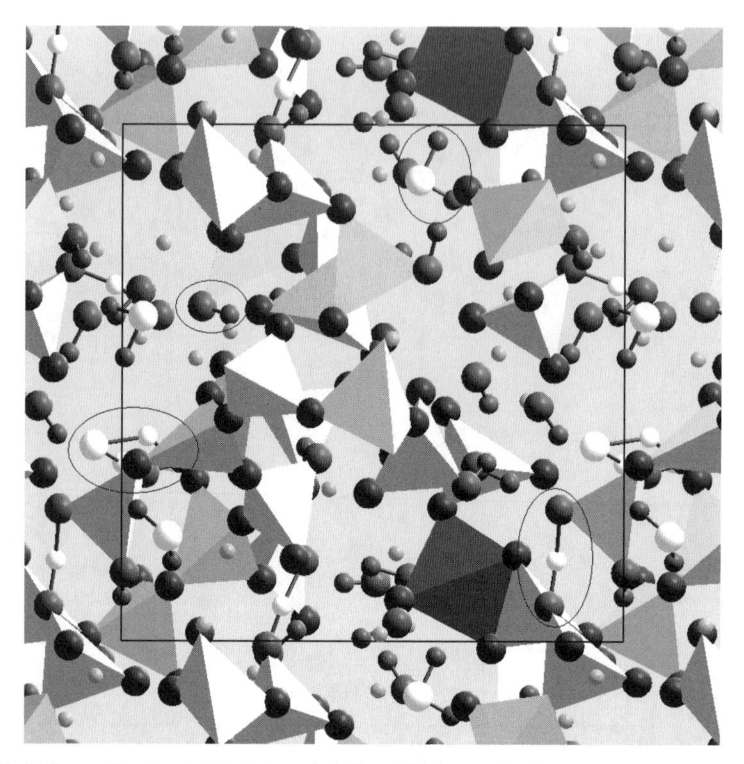

Figure 8. Si-O coordination (polyhedra), and O-H and H-O coordination (spheres) displayed together. The mixed environment represents Si-O coordination polyhedra (4-fold, light; 5-fold, dark) and various forms (marked) of water speciation including hydroxyls (large dark-small dark pairs), water molecules (small dark-large light-small dark triplets), polyhedral bridging (large dark-small light-large dark triplets), and four-atom sequences (such as one in the lower part of the left boundary). The smallest (free) spheres represent Mg atoms. The oxygen atoms (large spheres) bonded to Si atoms are shown at the polyhedral vertices or those to Mg atoms are shown as free spheres.

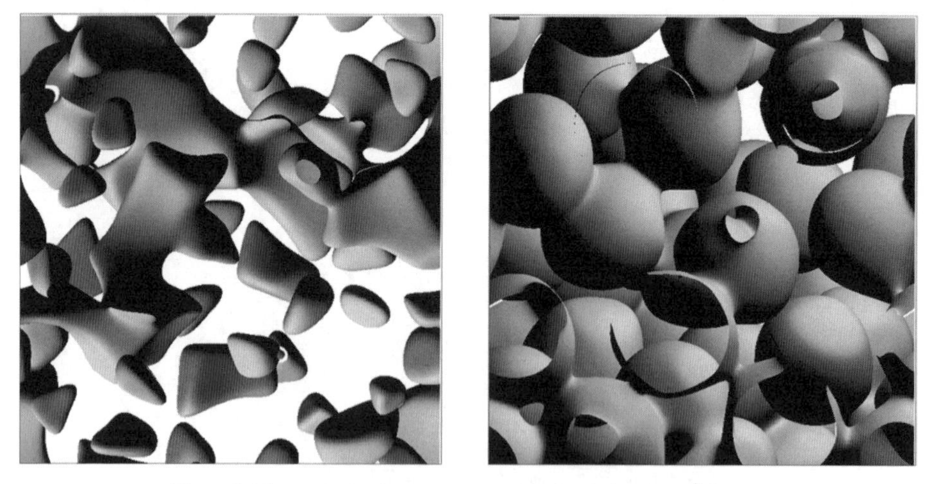

Figure 9. Electronic density isosurfaces of liquid MgO at 0.05 $Å^{-3}$ (left) and 0.15$Å^{-3}$ (right) at zero pressure and 3000 K.

SIMULATION RESULTS AND DISCUSSION

Equations of state and derived properties

The Mie-Grüneisen representation with third order BM equation (Eqn. 6) for the reference isotherm, $P(V,T_0)$, and a simple form of the thermal pressure, $P_{TH}(V,T)$, was shown to accurately describe the liquid thermodynamics in most cases. We find that both the thermal pressure and energy are linear in temperature at each volume. Figure 10 shows the calculated PV isotherms for MgO and SiO_2 liquids. Unlike other liquids, SiO_2 liquid requires a fifth order BM equation and a complex $B(V)$ function since the increase of $P(V,T_0)$ and $P_{TH}(V,T)$ on compression is initially gradual and then becomes more rapid (Karki et al. 2007). The calculated

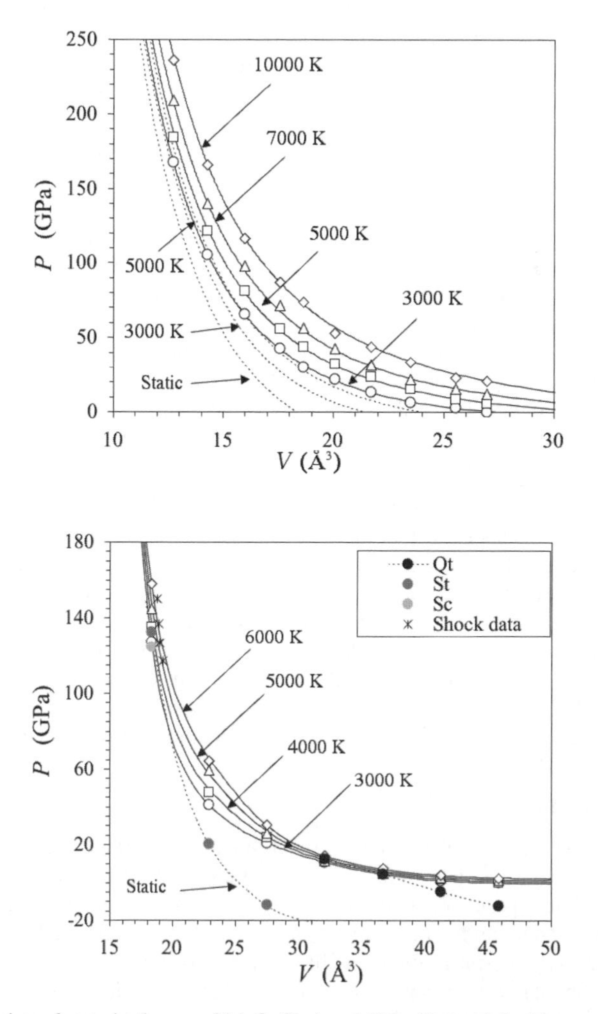

Figure 10. Equation of state isotherms of MgO (Top) and SiO_2 (Bottom) liquids represented by Mie-Gruneisen fits (lines) to the calculated results (symbols). The EoS parameters at T_0 = 3000 K are V_0 = 27.40 (±0.25) Å^3, K_0 = 30.76 (±2.9) GPa and K_0' = 5.03 (±0.33) for liquid MgO whereas the corresponding, are V_0 = 45.8 (±0.2) Å^3/SiO_2, K_0 = 5.2 (±1) GPa, K_0' = 22.5 (±3) and K_0'' = −452.3 (±24) for SiO_2 liquid. Solid phases of MgO and silica (Qt: quartz, St: Stishovite, Sc: scrutinyite) are shown for comparisons. The shock-compression data are shown by asterisks at temperatures (between 4900 and 6500 K) for silica melt (Akins and Ahrens 2002).

P-V isotherms initially remain close and nearly parallel to each other on compression, and then begin to diverge on further compression. This is due to the fact that the thermal pressure increases on compression. The liquid phase is more compressible than the solid phase at the same volume so the volume differences between the solid and liquid phases systematically decrease with compression. For MgO, the volume difference for the 3000 K isotherm is 5.79 Å^3 at zero pressure, and 0.15 Å^3 at 150 GPa. Similarly, for $MgSiO_3$, the volume difference decreases by a factor of 5 over the mantle pressure regime and is 4% at the core-mantle boundary. For silica, the volume of the liquid becomes smaller than that of quartz at about 7 GPa, stishovite at about 90 GPa, and seifertite at about 120 GPa. The liquid-crystal density inversions are possible in MgO and $MgSiO_3$ in the deep mantle due to preferred partitioning of iron in the liquid phase.

First principles results show that the thermodynamic properties of liquids are strongly dependent on compression. For instance, the value of α for MgO liquid is 9.12×10^{-5} K^{-1} at zero pressure, which decreases rapidly on compression reaching 1.68×10^{-5} K^{-1} at 150 GPa along the 5000 K isotherm. Similarly, the isochoric heat capacity and Gruneisen parameter vary significantly on compression, in most cases, linearly (Fig. 11). Again, the exception is the case of silica liquid in which C_V and γ tend to decrease and increase, respectively, with compression in a non-linear way. The variations of C_V and γ of liquids differ significantly from those for solids. Unlike the case of solids, liquid C_V exceeds the Dulong-Petit value ($3Nk_B$). We attribute the larger heat capacity of liquids to the change in the melt structure with increasing temperature. Whereas the mean coordination number remains insensitive to temperature at constant volume, the range of coordination environments increases with increasing temperature (discussed later). The value of γ increases with compression whereas it always decreases during the isostructural compression of solids. The unusual behavior of γ in the liquid can be understood on the basis of the pressure-induced change in liquid structure (Stixrude and Karki 2005). As the coordination number increases on compression, the liquid adopts the values of γ characteristic of the more highly coordinated state.

The calculated equations of state of liquids along the MgO-SiO_2 join can be compared with the predictions of an ideal solution model to understand the thermodynamics of mixing on compression. Our results show that thermodynamics of mixing is strongly non-ideal at low pressures and but approaches ideality at higher pressures (Fig. 12). The volumes of mixing are large negative at zero pressure, decreasing to zero by 25 GPa. This means that the free energies of intermediate compositions become increasingly favorable with compression relative to the end members, so that deep mantle liquid immiscibility is not likely. At the ambient pressure, the enthalpy of mixing shows positive values for silica rich regime ($X_{SiO_2} > 0.7$) and negative values for $X_{SiO_2} < 0.7$. As pressure rises, H_{mix} value decreases strongly initially: above 5 GPa, all H_{mix} values are negative, showing increasingly first order compositional dependence, essentially constant above 25 GPa.

Radial distribution functions

The calculated partial radial distribution functions vary remarkably in their shapes between different atomic pair types as shown in Figure 6. Features common to all functions include the appearance of an initial peak, and a decrease in the magnitude of the fluctuations with increasing distance, approaching unity at the largest distances. These features reflect, respectively, the short-range order and long-range disorder characteristics of the liquid state. The partial RDFs also reflect the dominant role of ionic interactions and charge ordering. The Si-O RDF is best structured exhibiting the tallest and sharpest peaks among all functions and the value at the minimum after the first peak is almost zero (Fig. 13). In contrast, the Mg-O RDF minimum value is significantly higher than zero. Other functions show relatively broader and shorter first peak and, in many cases the minimum and the second peak are poorly defined (Fig. 6). Also, the peaks are located at relatively larger distances. With increasing temperature, the degree of

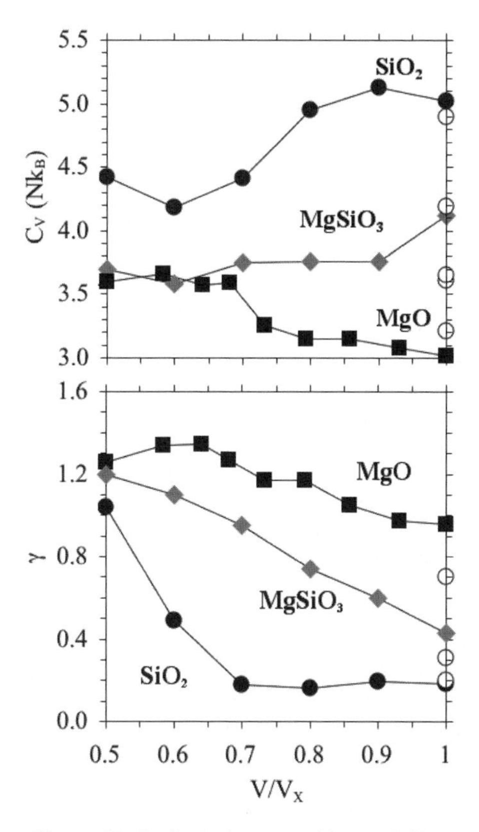

Figure 11. Isochoric heat capacities, and Gruneisen parameters of three liquids (averaged over different temperatures) as a function of compression (Stixrude and Karki 2005; Karki et al. 2006a, 2007). Open symbols represent experimental data (Akins and Ahrens 2002).

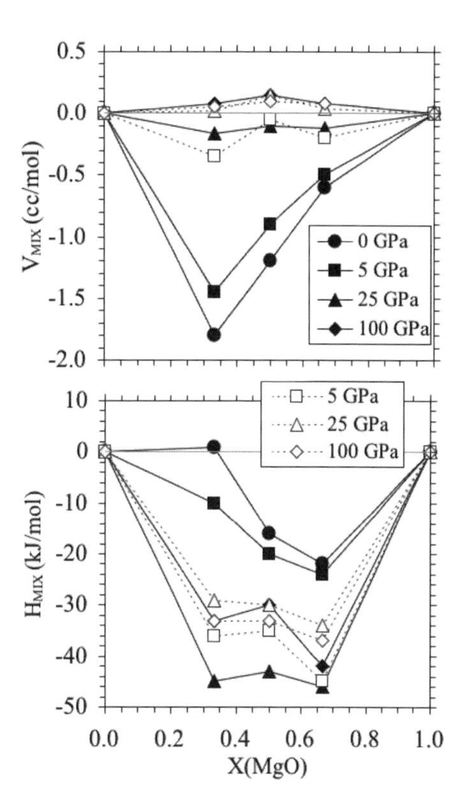

Figure 12. Volume and enthalpy of mixing along MgO-SiO_2 join (covering SiO_2, Mg-Si_2O_5, $MgSiO_3$, Mg_2SiO_4 and MgO phases) at different pressures at 3000 K (solid lines) and 6000 K (dotted lines). From de Koker, Karki and Sixrude, unpub. data.

order decreases, peaks become shorter, wider and less symmetric, and the values at the minima increase. On compression, the first peak in the Si-O RDF gradually decreases in the height; both the peak and minimum positions affected, and a new peak rather appears at much shorter distance than the original second peak (Table 1). In the case of MgO RDF, the heights of the first and second peaks decrease initially and then increase on compression with their positions as well as the minimum position shifting to smaller distances. Following small initial changes, the shape of the O-O RDF changes qualitatively at high compression with a shoulder appearing before the second peak for both SiO_2 and $MgSiO_3$ liquids.

The calculated results compare favorably with experiment (e.g., Waseda and Toguri 1977, 1990; Funamori et al. 2004). The experimental Si-O peak positions at low pressures are comparable to our calculated peak positions and are systematically smaller than our calculated average interatomic distances. Also, the experimental Mg-O peak distance lies between the calculated peak position and calculated average distance. The discrepancy may be due to the limited experimental resolution of the peak asymmetry, which is less serious for the much narrower Si-O peak.

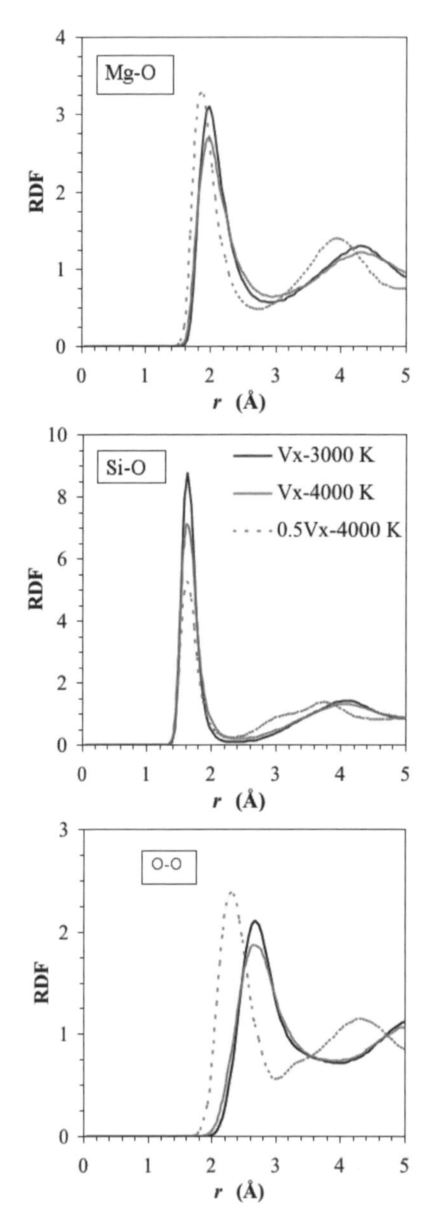

Figure 13. Calculated Mg-O, Si-O and O-O radial distribution functions of $MgSiO_3$ liquid at three different conditions.

Coordination environments

Mean coordination. There are several partial coordination environments, for instance, nine coordination types in the case of $MgSiO_3$ liquid. The calculated mean coordination numbers at all *V-T* conditions studied range from less than one to more than 12. Compression has a profound influence on the mean coordination numbers with their values in most cases monotonically increasing with increasing pressure (Fig. 14). The Si-O and Mg-O coordination numbers increase gradually from ~4 and ~5, respectively, at low compression to ~6 and ~8, respectively, at the highest compression, consistent with the experimental observations of the pressure-induced increase of Si-O coordination in silicate glasses and melts (Williams and Jeanloz 1988). This remarkably simple behavior of coordination in liquids is in contrast to the behavior of solids in which the coordination change occurs in discrete steps, and emphasizes the nature of the liquid as a distinct phase, as opposed to a disordered solid. Note that the effects of temperature on mean coordination at constant volume are small. Whereas the mean Si-O coordination is similar between the enstatite and silica liquids, the O-Si coordination number of enstatite liquid is distinctly smaller than that of silica liquid showing that Mg disrupts the inter-polyhedral linkages (Fig. 15). This is also manifested in the relatively low Si-Si coordination of enstatite liquid, compared to the silica liquid value which varies from ~4 to about ~12 over the compression regime studied (Karki et al. 2007). Also note that Mg-O coordination number of periclase liquid is systematically smaller than that of enstatite liquid.

Si and O coordination species. The Si-O coordination environment represents a mixture of various coordination species (denoted by $_{Z}C_{SiO}$, where Z represents the coordination number) at each volume, temperature condition. At low pressure, four-fold coordination ($_{4}C_{SiO}$) dominates with noticeable contributions from three- and five-fold coordination (Table 2). On isochoric cooling at low compression, the mean Si-O coordination number remains nearly 4 while the proportion of non-tetrahedral defects decreases, consistent with experimental observation that they are essentially absent (less than 0.05%) in silicate glasss (e.g., Stebbins 1991; Stebbins and McMillan 1993). These defects are relatively rare in silica liquid. On compression, contributions from five- and six-fold coordination

Table 1. Calculated positions (in Å) of the first peak (R_{P1}), the minimum after the first peak (R_M) and the second peak (R_{P2}) for Si-O, Mg-O and O-O distribution functions for $MgSiO_3$ liquid. The experimental results at ambient pressure (Waseda and Toguri 1990) are shown.

	Si-O			Mg-O			O-O		
	R_{P1}	R_M	R_{P2}	R_{P1}	R_M	R_{P2}	R_{P1}	R_M	R_{P2}
$V = V_X$									
2500 K	1.625	2.365	4.135	1.965	2.905	4.295	2.675	3.975	5.055
3000 K	1.625	2.375	4.115	1.965	2.920	4.295	2.675	3.975	5.070
4000 K	1.625	2.435	4.035	1.965	2.950	4.305	2.675	3.885	5.075
5000 K	1.625	2.465	4.025	1.965	3.075	4.325	2.675	3.925	5.105
6000 K	1.625	2.475	4.045	1.955	3.075	4.345	2.665	3.895	5.150
Expt	1.620			2.120					
$V = 0.7V_X$									
3000 K	1.635	2.395	3.955	1.965	2.855	4.250	2.525	3.655	4.795
4000 K	1.630	2.395	3.925	1.945	2.950	4.225	2.495	3.595	4.820
6000 K	1.625	2.475	3.915	1.915	2.895	4.175	2.485	3.465	4.795
$V = 0.5V_X$									
3000 K	1.625	2.345	3.755	1.845	2.625	4.005	2.315	3.015	4.275
4000 K	1.620	2.295	3.750	1.845	2.725	3.955	2.295	3.010	4.305
6000 K	1.595	2.375	3.705	1.835	2.750	3.895	2.255	3.060	4.325

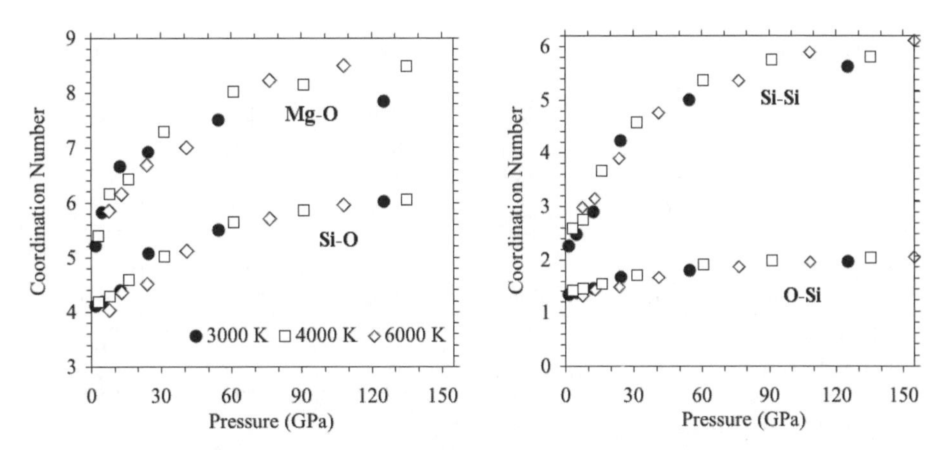

Figure 14. Calculated mean Mg-O, Si-O, Si-Si and O-Si coordination numbers of $MgSiO_3$ liquid along 3000, 4000 and 6000 K isotherms.

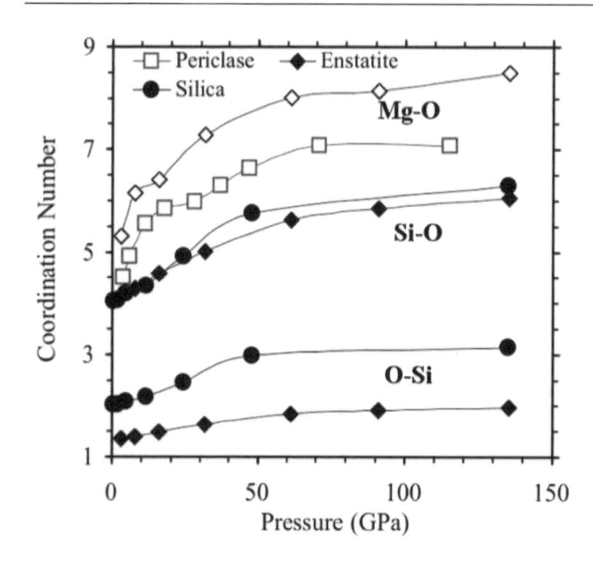

Figure 15. Comparison of Mg-O, Si-O and O-Si coordination between periclase, enstatite and silica liquids at 4000 K.

increase whereas those from three- and four-fold coordination decrease (Fig. 16, left). The preponderance of ${}_5C_{SiO}$ species at mid compression (around 30 GPa) also predicted to occur in silica and other silicate liquids (Angell et al. 1982; Karki et al. 2007) demonstrates that liquid structure is not merely a disordered version of the structure of the crystalline polymorphs, ${}_5C_{SiO}$ being extremely rare in crystalline silicates. The five-fold coordination has been suggested as a transition state that enhances diffusion through the exchange of oxygen. At highest compression, the six-fold coordination dominates with substantial presence of ${}_7C_{SiO}$ and ${}_8C_{SiO}$ species in silica

Table 2. Abundances of various Si-O and O-Si coordination species of enstatite liquid compared to those of silica liquid (*si*). The values in the parentheses in the last column are for 8-fold Si-O coordination.

	${}_3C_{SiO}$	${}_4C_{SiO}$	${}_5C_{SiO}$	${}_6C_{SiO}$	${}_7C_{SiO}$
2500 K, 1.1 GPa	0.48	92.83	6.49	0.20	0
3000 K, 1.8 GPa	1.45	87.77	10.57	0.21	0
0.2 GPa, *si*	0.35	97.64	2.01	0.00	0
4000 K, 3.3 GPa	5.46	72.55	20.59	1.24	0
0.4 GPa, *si*	4.61	87.05	8.16	0.17	0
3000 K, 125.3 GPa	0	0.02	5.66	88.33	5.79 (0.21)
127.1 GPa, *si*	0	0.05	4.00	73.94	19.59 (2.40)

	${}_0C_{OSi}$	${}_1C_{OSi}$	${}_2C_{OSi}$	${}_3C_{OSi}$	${}_4C_{OSi}$
2500 K, 1.1 GPa	2.30	63.01	34.38	0.31	0.31
3000 K, 1.8 GPa	2.49	61.75	35.16	0.59	0.59
0.2 GPa, *si*	0	0.16	98.85	0.99	0.99
4000 K, 3.3 GPa	4.55	56.43	37.23	1.75	0.02
0.4 GPa, *si*	0	2.08	93.94	3.97	0.02
3000 K, 125.3 GPa	3.27	18.46	57.32	20.82	0.13
127.1 GPa, *si*	0	0	5.53	78.86	15.50

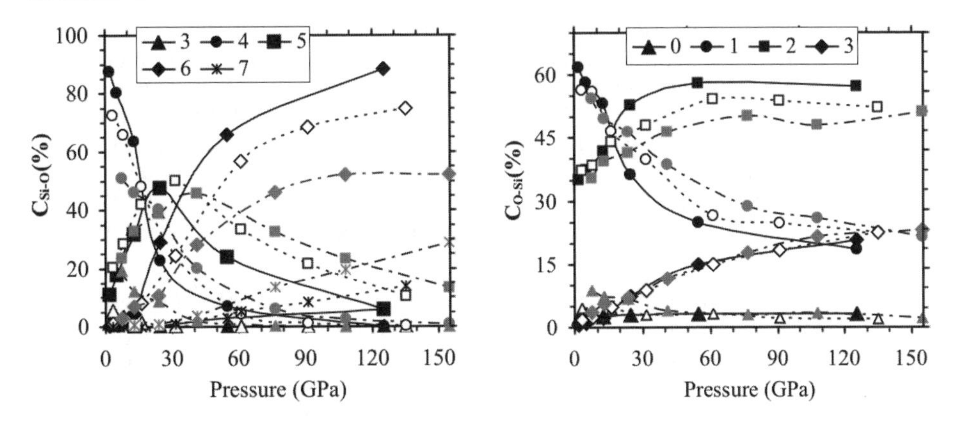

Figure 16. Abundances of different Si-O and O-Si coordination species as a function of compression at 3000 K (solid lines), 4000 K (dotted lines) and 6000 K (dotted-dashed lines) for enstatite liquid as a function of pressure.

liquid thereby raising its mean Si-O coordination above 6 (the value being 6.2 at 127.1 GPa, 3000 K, compared to the corresponding enstatite liquid value of 6 at 125.3 GPa, 3000 K).

Similarly, the relative abundances of different O-Si coordination species change upon compression (Fig. 16, right). At low compression and temperature, enstatite liquid has much less bridging oxygen (i.e., two-fold or over coordinated oxygen) and much more non-bridging oxygen (NBO, i.e., singly coordinated oxygen) compared to silica liquid in which almost all oxygen is involved in bridging (Table 2). A small number of oxygen atoms in enstatite liquid are not bound to any Si atoms but rather to Mg atoms, and as such, they are referred to as non-polyhedral oxygen (NPO) or simply as free oxygen (FO). Whereas both FO and NBO are almost absent in silica liquid, they are prevalent in the silica-poor (e.g., forsterite) and hydrous melts at the cost of bridging oxygen reflecting the tendency of Mg, proton and other cations to depolymerize the melt structure. As enstatite liquid is compressed, the proportions of FO and NBO decrease whereas those of two (BO) and three-fold (O3) oxygen increase. In silica liquid, the proportion of BO also decreases rapidly (to about 5% at the highest compression studied) but the proportion of O3 increases rapidly on compression with the appearance of O4 species (Table 2). Thus, oxygen triclusters (and also quadclusters in silica) become prevalent at high pressure. A recent experimental study has confirmed the formation of oxygen triclusters in $MgSiO_3$ glass at pressures above 12 GPa (Lee et al. 2008).

To explore the relationship between the Si-O and O-Si coordination, we examine the distribution of the quantity $_{Z}Si^{n}$, which is similar to the widely used quantity Q^n defined for four fold Si-O coordination ($Z = 4$). Here, n is the number of bridging oxygen (an oxygen atom which is also bonded to one or more silicon atoms) attached to a silicon atom, which can be in any coordination state in our case. The calculated distribution (Table 3) consists of

Table 3. Abundances (%) of various $_{Z}Si^{n}$ species of enstatite (*en*) and silica (*si*) liquids at 3000 K.

n	1	2	3	4	5	6	7
1.8 GPa, *en*	24.6	34.4	31.2	6.6	1.4	0	0
0.2 GPa, *si*	0	0.1	0.5	97.4	2.0	0	0
125.3 GPa, *en*	0	0	1.1	5.1	46.3	43.9	3.5
127.1 GPa, *si*	0	0	0	0	4.0	73.9	19.6

a mixture of different species with $_Z Si^1$, $_Z Si^2$ and $_Z Si^3$ dominating in enstatitte liquid whereas $_Z Si^4$ species (with other species counting less than 3%) mostly occuring in silica liquid. The $_Z Si^1$ species represents chain-terminating tetrahedra, which are present in entatite liquid at low compression. The absence of $_Z Si^1$ species in silica liquid means that the Si-Si network is complete. As compression increases the distribution becomes wider with increase in the abundances of high-order species; $_Z Si^5$ and $_Z Si^6$ species count for 90% in enstatite liquid at 125.3 GPa whereas $_Z Si^6$ and $_Z Si^7$ species count for 95% in silica liquid at 127.1 GPa. The preponderance of higher order species in compressed liquids is consistent with the increased $_2C_{\mathrm{OSi}}$, $_3C_{\mathrm{OSi}}$ and $_4C_{\mathrm{OSi}}$ abundances.

How the high coordination species are formed from the low coordination species (for example, the appearance of five-fold defects in an otherwise tetrahedral environment) is expected to give useful insight into compression and diffusion mechanisms. Visualization of position-time series reveals a few relevant reactions. First, $MgSiO_3$ liquid has a large number of non-bridging oxygen atoms which can contribute to coordination increase as previously suggested based on experimental studies of sodium silicates (Wolf et al. 1990; Xue et al. 1991; Farber and Williams 1996): an NBO bonded to a neighboring silicon atom becomes bonded to the silicon atom under consideration (Fig. 17). This increases the Si-O coordination number (Z) as well as the number of bridging oxygen (n):

$$\mathrm{NBO} + {}_Z\mathrm{Si}^n \rightarrow \mathrm{BO} + {}_{Z+1}\mathrm{Si}^{n+1} \tag{16}$$

The two polyhedra share a corner but not necessarily an edge or a face unless they are already sharing another corner or edge. As pressure increases, the number of NBO (one-fold coordination) decreases whereas the number of BO (two-fold coordination) increases (Fig. 16, right) so the reaction (Eqn. 16) can be a primary mechanism for the pressure-induced coordination increase in enstatite liquid in the low compression regime.

Also increased on compression is the three-fold O-Si coordination (Fig. 16, right), i.e., the number of oxygen triclusters (O3). This requires a mechanism in which a BO is added to the coordination shell of the Si atom under consideration as shown in Figure 18 (Wolf et al. 1990):

$$\mathrm{BO} + {}_Z\mathrm{Si}^n \rightarrow \mathrm{O3} + {}_{Z+1}\mathrm{Si}^{n+1} \tag{17}$$

The Si atom gains an O atom, which already bridges other two Si atoms so both Z and n increase. Thus, it involves a closing of the Si-O-Si angle between two tetrahedra, which brings a fifth oxygen within the coordination shell of one of the Si atoms. The product is a shared edge between a pentrahedron and a tetrahedron, and a three-coordinated oxygen (O3). The Si-Si coordination number is unchanged by this process as is the network connectivity: no new connections are formed between coordination environments. This mechanism is primarily responsible for the formation of 5-fold coordination states in the silica liquid in which the number of NBO is negligible at low compression. At very high compression, not only two-fold oxygen but also three-fold oxygen form bonds to Si to form higher order Si-O coordination species.

Finally, silica poor liquids such as Mg_2SiO_4 (de Koker et al. 2008) or hydrous $MgSiO_3$ (Karki et al. 2009) and even hydrous silica (Pohlmann et al. 2004; Anderson et al. 2008) contain a significant number of free oxygen (not coordinated with any silicon atom), which can contribute to coordination increase: an NPO enters into the coordination shell of the silicon atom thereby increasing the coordination number but without affecting the number of bridging oxygen (Fig. 19):

$$\mathrm{NPO} + {}_Z\mathrm{Si}^n \rightarrow \mathrm{NBO} + {}_{Z+1}\mathrm{Si}^{n} \tag{18}$$

Both the mean Si-O and O-Si coordination increase.

Compression enhances (suppresses) the abundances of high (low) coordination species, which are manifested through enhancement of corner, edge and even face sharing polyhedra.

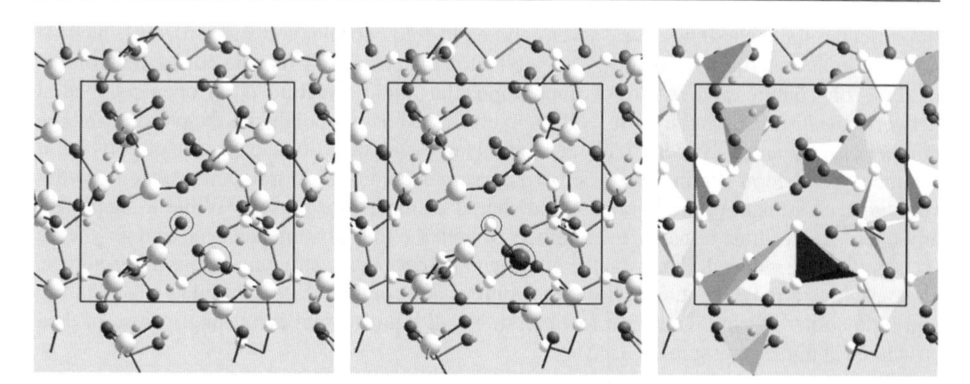

Figure 17. Coordination increase through transformation of non-bridging oxygen (marked dark medium sphere) to bridging oxygen (marked light medium sphere). As a result, Si atom changes from a four-fold state (marked light large sphere) to a five-fold one (marked dark large sphere). Mg atoms are shown as small spheres. The right figure shows the newly formed dark pentahedron. The visualization snapshots are for $MgSiO_3$ liquid at V_X and 3000 K.

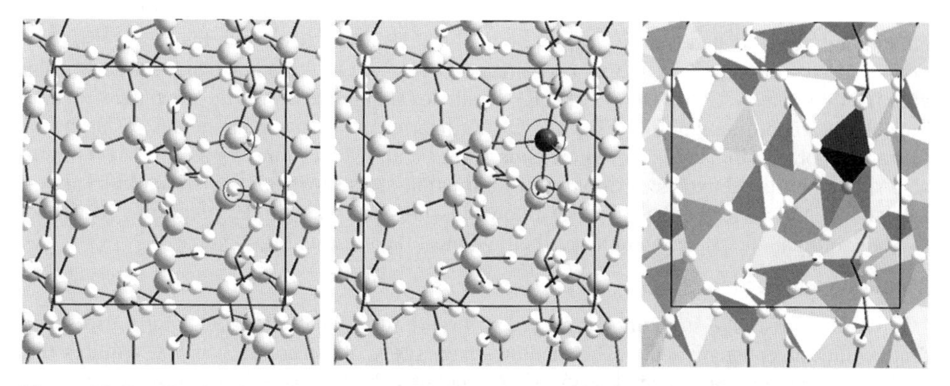

Figure 18. Coordination increase through bridging oxygen. A four-fold coordinated Si atom (marked light large sphere) turns into a five-fold coordinated Si atom (marked dark large sphere) through the transformation of BO (marked light small sphere) to O3 (marked darker small sphere). The right figure shows the newly formed dark pentahedron. The visualization snapshots are for SiO_2 liquid at V_X and 3000 K.

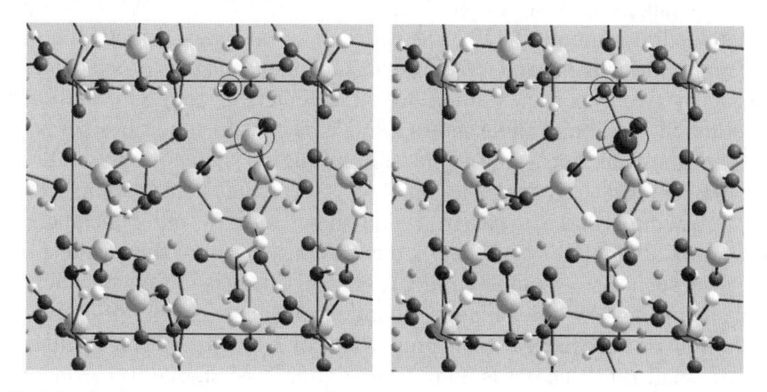

Figure 19. Coordination increase from four-fold (marked light large sphere) to five-fold (marked dark large sphere) through transformation of free oxygen (marked black medium sphere) to non-bridging oxygen (marked dark medium sphere) as marked. The free small spheres represent Mg atoms whereas the bonded small spheres represent H atoms. The visualization snapshots are for hydrous $MgSiO_3$ liquid at V_X and 3000 K (Karki et al. 2009).

In particular, the melt shows five-, seven- and eight-fold coordination states for Si, which do not exist in any crystalline silicates. The polyhedral distortion and also the polyhedral volume increase on compression. At low compression, the liquid structure is an open polyhedral network. In silica poor and hydrous silicate melts, free-floating polyhedra are present, and the number of dead-end polyhedra in the network is large with occasional existence of polyhedral dimmers. Polyhedral sharing of one or two corners is common. At mid compression a wider variety of coordination environments and shared elements are present. While a significant amount of free volume remains, the numbers of both the free-floating and dead-end polyhedra are significantly reduced. Finally at the highest compression, the structure begins to resemble a close-packing arrangement, with little free volume, and common face- as well as edge-sharing and little corner sharing. Unlike at low pressures, all liquids adopt increasingly close-packed structures at higher pressures.

Mg and O coordination species. Mg and O coordination environments consist of various coordination species whose abundances change on compression. Five and six-fold coordination species have been detected experimentally in silicate glasses (Li et al. 1999; Kroeker and Stebins 2000; Shimoda et al. 2007). In $MgSiO_3$ liquid, the five-fold Mg-O coordination state (41.4% at V_X, 3000 K) dominates at low pressure with significant presence of four-fold (20.9%) and six-fold (27.6%) coordinated Mg. Correspondingly, a significant fraction of O atoms are either singly coordinated (28.5%) or not bonded (10.3%) with Mg while the rest consists of doubly (43.8%) and triply (16.3%) coordinated O atoms. On the other hand, the eight-fold coordination state (42.2% at $0.5V_X$, 3000 K) dominates at high pressure with seven (32.4%) and nine (17.4%) coordination. Increased Mg-O coordination arises due to an increased sharing of O atoms; more O atoms (87%) are now coordinated with two or more Mg atoms. MgO liquid also shows similar abundances of Mg-O coordination species (36.3% four-fold, 43.5% five-fold and 14.4% six-fold at V_X, 3000 K) at low pressures and with similar pressure effects; the O-Mg coordination species abundances being almost the same as Mg-O values.

Medium range order

While short-range structures discussed so far are relevant at 1 to 3 Å, liquids also show features at longer length scale. The medium-range order occurring at 3 to 6 Å requires the analysis of interpolyhedral angles and ring structures. In essence, such intermediate-range structures represent how the short-range units (SiO_n polyhdera) are linked. Here, we discuss such structures in silica—a tetrahedral liquid because tetrahedra are excellent ring formers (Karki et al. 2007). The distributions of the O-Si-O and Si-O-Si bond angles in liquid silica show interesting changes on compression. The average O-Si-O angle at low pressures is close to the ideal tetrahedral angle (109.47°). The distribution becomes bi-modal on compression; the first peak representing the characteristic octahedral angle and the second peak representing the angle made by two opposite O atoms with the center Si atom. On the other hand, the distribution of the Si-O-Si bond angle shows a single peak at all conditions but becomes more asymmetric with the peak and small-angle tail shifting to smaller angles. The distribution extending below 120° down to angles as small as 70° contains contributions from edge- and face-sharing polyhedra. Unlike α-quartz, which shows only 6- and 8-membered rings, the liquid phase shows a variety of Si-Si ring structures varying from the smallest 3-membered ring to large rings such as 10-membered ring (Fig. 20). Changes in ring statistics can account for the initial compression of silica liquid (Stixrude and Bukowinski 1989; Karki et al. 2007). Our results show an initial decrease in the proportions of 3-membered rings with compression. The number of small rings increases upon further compression, due to the increase in Si-O coordination number. These are primarily associated with the formation of 5-fold coordinated Si, which does not influence the network connectivity. Moreover, the characteristic ring size increases on compression.

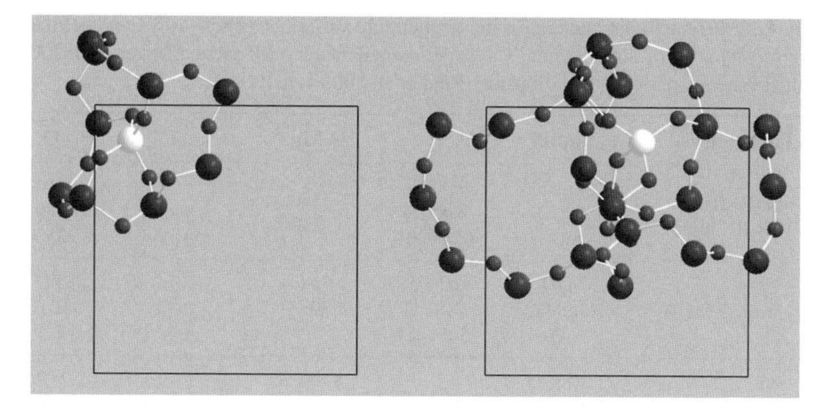

Figure 20. Different Si-Si rings for a given silicon atom (white large sphere) in silica liquid. Left shows the smaller rings (3-, 4- and 5-member) and right shows the larger rings (8, 9- and 10-member). The silicon atoms (large spheres) are bonded via oxygen atoms (small spheres).

Diffusion coefficients and viscosities

Unlike the thermodynamics and structures, the dynamical properties, particularly, viscosity of liquids, have not been widely studied from first-principles. Diffusion can be characterized by calculating the mean square displacements and diffusion coefficients for different atomic species (Eqn. 12). The calculations suggest that the dynamical behavior differs substantially between MgO and SiO_2—two end members of silicate liquids. The MSD, in general, shows two temporal regimes (Fig. 4, top). The first is the ballistic regime (for relatively short times) in which the atoms move without interacting strongly with their neighbors and MSD is proportional to t^2. The second one is the diffusion regime (for long times) in which MSD is proportional to t. However at lower temperatures and higher compression, particularly, in the case of silica liquid an intermediate regime appears where MSD increases rather slowly due to the so-called cage effect in which the atoms are temporarily trapped within the cages made by their neighbors (Fig. 4, top). A diffusive regime is thus reached much faster in MgO than in SiO_2.

A complete set of temperature-pressure results for the diffusivity of MgO can be described accurately with the Arrhenius relation:

$$D(P,T) = D_0 \exp[-(E_D + PV_D) / RT \qquad (19)$$

where E_D and V_D are the activation energy and volume, respectively, whose values are given in Table 4. However, this is not true in the case of silica as its diffusion coefficients show anomalous pressure dependence. At 4000 K or lower temperature their diffusivities increase with pressure and then decrease upon further compression (Karki et al. 2007). This means that the activation volume is negative at low pressures and becomes positive at high pressures. An Arrhenius representation of the temperature variations at V_X gives the activation energies of 3.43 and 3.27 eV for silicon and oxygen diffusion, respectively. Including in the Arrhenius analysis only the results in the pressure regime where the diffusivity decreases on compression, we find much smaller activation energy (about 2 eV) and the positive activation volume (about 1.2 Å^3) for the total diffusivity. The large differences in the activation energies emphasize the difference in the diffusional behavior between low and high pressures.

The calculated partial MSD-time curves for $MgSiO_3$ liquid behave normally and exceed 10 Å^2 during simulation runs performed at each volume-temperature condition (Fig. 21). An intermediate slow regime appears and the beginning of the diffusive regime is delayed at lower temperatures and higher compression. The partial diffusion coefficients follow mostly

Table 4. Arrhenius fit parameters for the temperature variations of self-diffusion coefficients and viscosity at V_X of three liquids. The experimental data are for dacite (Tinker et al. 2004[a]), basalt (Lesher et al.1996[b]) and diopside (Reid et al. 2003[c]) melts.

Parameters	Liquid	η	Mg	Si	O
η_0 (Pa s) or D_0 ($\times 10^{-9}$ m^2/s)	MgO	0.00035	170		210
	$MgSiO_3$	0.00025	240	370	380
	SiO_2	0.000001		24400	23500
E_η or E_D (kJ/mol)	MgO	62 (7)	76 (4)		84 (5)
	$MgSiO_3$	81 (11)	89 (5)	130 (8)	118 (6)
	SiO_2	305 (48)		326	311
V_D (Å^3)	MgO		1.4 (0.2)		1.2 (0.2)
	$MgSiO_3$		1.7 (0.3)	1.6 (0.2)	1.5 (0.3)
	SiO_2			1.2 (0.5)	1.2 (0.5)
E_η or E_D (kJ/mol) Expt		275[a]		179[b], 220[c]	174[b], 220[c]

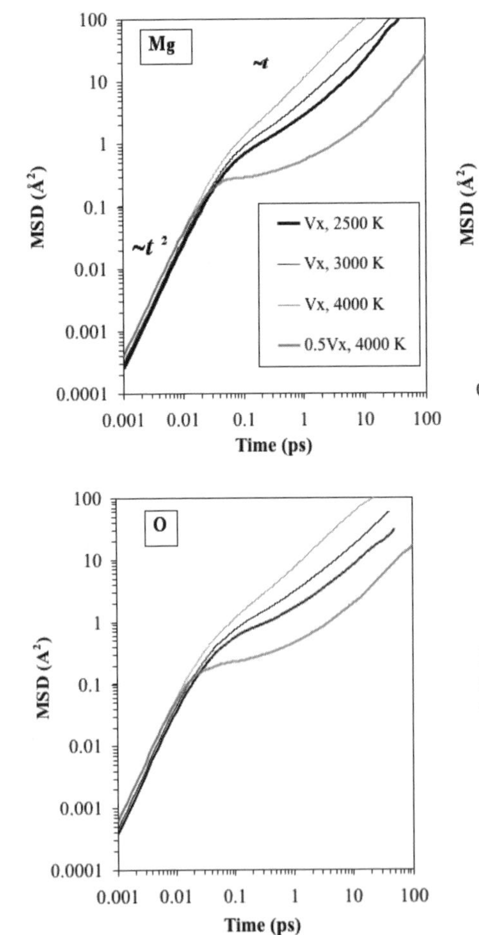

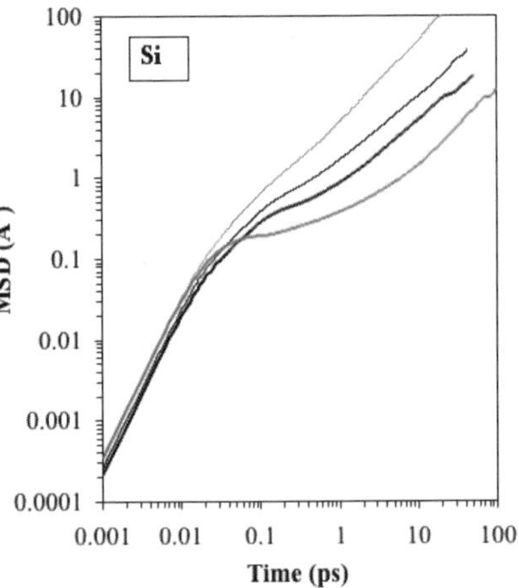

Figure 21. MSD versus time plots for three species of the $MgSiO_3$ liquid at different conditions.

the general trend that they increase with temperature and decrease with pressure (Fig. 22). However, both framework ions, Si and O, show weak anomalous diffusion at 3000 K. Such pressure-induced anomaly was previously predicted in anhydrous $MgSiO_3$ liquid (Kubicki and Lasaga 1991) and found to be more pronounced in silica-rich (e.g., Lacks et al. 2007) liquids. Since the anomaly is relatively weak, the Arrhenius relation can be fit to the entire results (Table 4). Magnesium is the fastest moving species with the smallest activation energy and largest activation volume whereas silicon is the slowest moving species. The diffusion coefficients of all species become increasingly similar at higher pressures. The competing effects of pressure and temperature on the diffusivity make the diffusion coefficients nearly constant (within a factor of two) along the melting curve covering the entire mantle pressure regime.

We discuss the viscosity results obtained from first principles only at low pressures (Fig. 3). With decreasing temperature, the viscosities of three liquids diverge with silica liquid becoming much more viscous than the periclase and enstatite liquids. This is consistent with the prediction that the silica liquid have much smaller diffusion coefficients than other two liquids. Arrhenius representation $\eta(T) = \eta_0\exp(E_\eta/RT)$ of viscosity-temperature results at V_X gives similar activation energies for the periclase and enstatite liquids but a much larger value for the silica liquid.

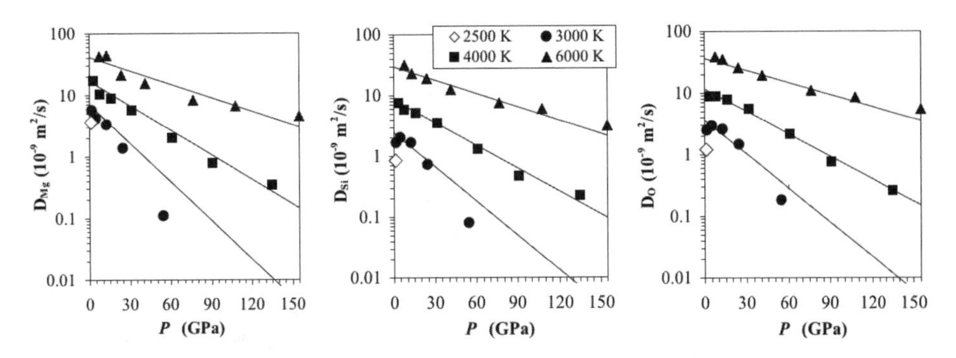

Figure 22. Diffusivities for Mg, Si and O as a function of pressure at 3000 K (circles), 4000 K (squares) and 6000 K (triangles) for enstatite liquid. The lines are Arrhenius fits. Also shown are the results at 2500 K (diamonds).

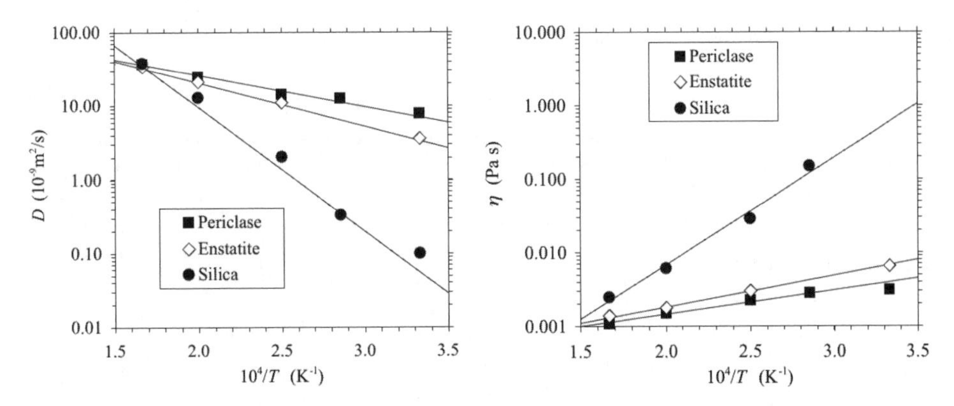

Figure 23. Total diffusivities and viscosity of periclase, enstatite and silica liquids at the reference volume V_X. The lines are Arrhenius fits.

In a dynamical system, a given atom may form bond with different atom(s) at different time instants. The related inherent process of bond formation and rupture in the liquid is thus expected to result in a wide range of bond lifetimes. We calculate the lifetime distributions of Mg-O and Si-O bonds as shown in Figure 24. All distributions show the maximum abundance of the bonds with the lifetimes lying between 0 and 100 fs, and the abundance of the bonds with longer lifetimes decreases rapidly according to some power law. The effects of compression on bond lifetimes are shown to be substantial. Both compression and temperature enhance the bond-breaking rates (Table 5). However, not all events contribute to ionic diffusion; only oxygen transfer in which one Si-O bond is broken and a different Si-O bond is formed. As a result, the oxygen atom gets transferred from one silicon atom to another silicon atom. One can estimate the proportion of oxygen transfers by counting the number of distinct Si-O bonds that exist during a finite time period. Let N_B be such number for a period of 10 ps. We can estimate the ratio of the transfers to the bond-breaking events as $(N_B - n_B) / (10\alpha_B)$, where n_B is the average number of Si-O bonds per step (= 16 × mean Si-O coordination number) as shown in Table 5. The transfer rate is higher at higher temperature. With increasing compression at 3000 K, the ratio tends to remain the same or increase slightly initially and then decreases on further compression. This is consistent with the higher Si and O diffusivities at higher temperatures and on initial compression. Also relevant to dynamical behavior is the polyhedral (coordination) stability since some coordination states are considered to act as activated complex for atomic diffusion. At V_X, the four-fold Si-O coordination is more stable and the five-fold coordination state is less stable in silica liquid than in enstatite liquid (Fig. 25).

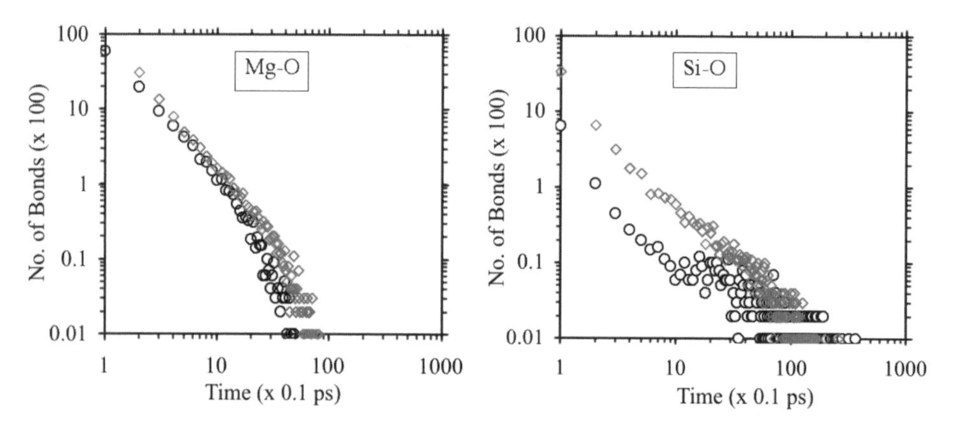

Figure 24. Distributions of Mg-O and Si-O bond lifetimes for $MgSiO_3$ liquid at V_X, 3000 K (circles) and $0.7V_X$, 3000 K (diamonds).

Table 5. Various Si-O bond related parameters (α_B = rate of bond-breaking events, n_B = average number of Si-O bonds per step, N_B = number of distinct Si-O bonds) for $MgSiO_3$ liquid at different conditions.

V, *T* (K), *P*(GPa)	α_B (events/ps)	n_B	N_B (during 10 ps)	$(N_B - n_B)$ / $(10\alpha_B)$
V_X, 3000, 1.8	42.6	65.4	207	0.33
V_X, 4000, 5.0	83.1	66.6	383	0.38
$0.9V_X$, 3000, 3.3	54.2	67.0	252	0.34
$0.7V_X$, 3000, 24.7	118.5	81.1	275	0.16

The $MgO–SiO_2$ system shows large compositional variation of dynamical properties. At the ambient pressure, the diffusivities and viscosities vary by more than 3 orders of magnitude along this join, with the silica melt characterized by lowest (highest) diffusivity (viscosity) (Fig. 23). As pressure rises, the compositional differences are strongly suppressed, for instance, the calculated diffusivities of melts in this system systematically converge and vary by less than an order of magnitude across the entire range of compositions at lower mantle pressures (Fig. 26). In particular, the diffusion coefficients of silica liquid unlike others show two distinct regimes, a low-compression anomalous regime showing a temperature-dependent diffusivity maximum and a high-compression regime in which the diffusion coefficient decreases with increasing pressure. We find that the diffusivity maximum is closely associated with the presence of five-fold coordinated Si (Angell et al. 1982; Karki et al. 2007). The pentahedral coordination

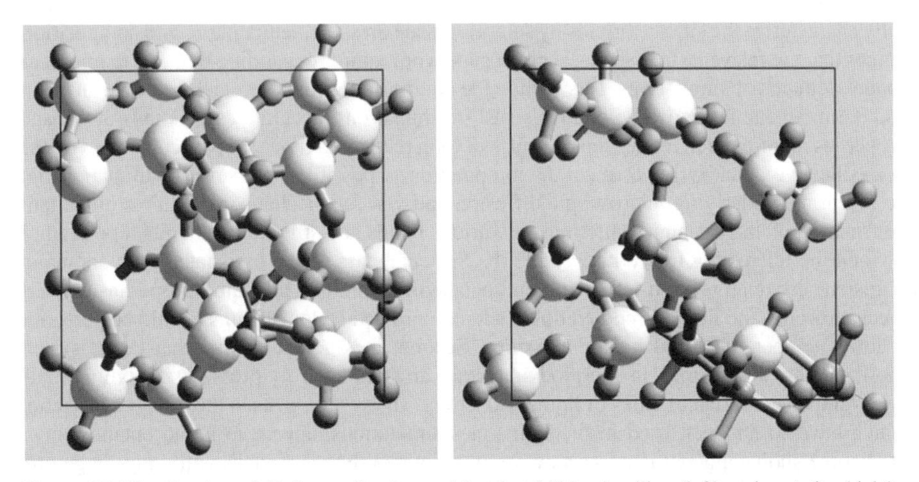

Figure 25. Visualization of Si-O coordination and bond stabilities in silica (left) and enstatite (right) liquids. The size of the light (four-fold) and dark (five-fold) centered spheres encodes the stability of the Si-O coordination states and the thickness of the bonds encodes the stability of the Si-O bonds. Note that the coordination stability is defined as the fraction of the total simulation time during which a silicon atom is in that particular state whereas the bond stability is defined as the time fraction during which a silicon atom is bonded to a particular oxygen atom (small spheres).

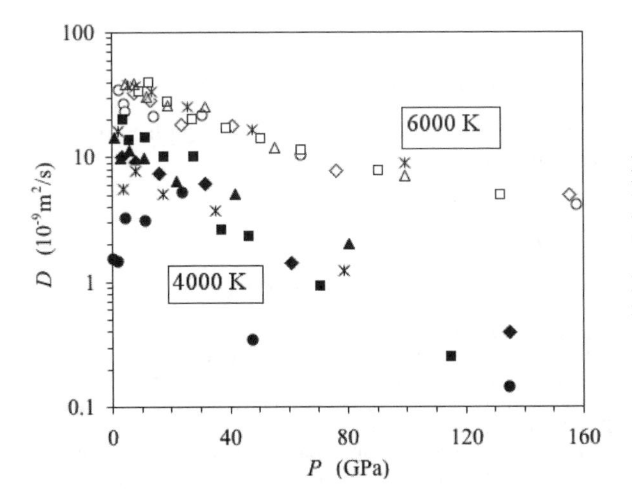

Figure 26. Total diffusivities of five liquids along the $MgO-SiO_2$ join as a function of pressure at 4000 K (filled symbols) and 6000 K (open symbols). SiO_2, $MgSi_2O_5$, $MgSiO_3$, Mg_2SiO_4 and MgO are, respectively, represented by circles, asterisks, squares, triangles and squares.

environments can be considered as transition states for atomic diffusion. In particular, the pressure at which five-fold coordinated Si is most abundant coincides with the pressure of the diffusivity maximum at 4000 K. Our results are thus consistent with a picture in which silica liquid behaves like a strong liquid at low pressure and low temperature, and a fragile liquid at high pressure and high temperature.

CONCLUSION AND FUTURE DIRECTIONS

We illustrate the applications of first-principle molecular dynamics simulation method in the study of geophysically important liquid phases at deep Earth conditions. We present various challenges associated with liquid simulations and show how they can be overcome. First-principles investigations of silicate liquids have just begun. A lot is yet to be done in order to fully characterize the silicate liquids by sampling wide pressure, temperature and compositional ranges that are relevant. In essence, such an endeavor requires expanding both the temporal and spatial domains of simulations. First, while the equilibrium properties such as liquid structures and thermodynamics converge relatively fast in simulations, accurate calculations of transport properties require much longer runs because of long relaxation times. In particular, the viscosity is a collective quantity so its accurate computation is possible with several simulation runs. To reveal the underlying microscopic diffusion and viscous mechanisms requires quantitative information on the dynamical lifetimes of various bonds and structural subunits. For example, it would be interesting to see how stable the Si-pentahedra (often associated with anomalous diffusion) are relative to Si-tetrahedra and octahedra. Note that all liquid properties converge much slower as the liquid is cooled down and compressed so longer runs should be performed at lower temperatures and higher pressures. Second, it is generally true that larger systems show smaller fluctuations in thermodynamical variables such as pressure implying smaller uncertainties in the calculations of liquid properties. The system sizes in the order of 100 atoms, which have so far been used in first-principles simulations, appear to be acceptable only to predict the short-range structures. However, the liquid although it lacks with long-range order has an intermediate-range order with length scale 3 to 6 Å or larger, requiring systems with a dimension of larger than 10 Å. Unlike the equilibrium properties, the dynamic properties are more likely to be sensitive to the size thus requiring not only longer runs but also larger systems. Bigger systems also allow us to study realistic melt compositions, which contain many components at small concentrations For instance, simulation cell for model basalt composition (67% diopside and 33% anorthite) should contain cations Mg, Ca, Al and Si in sufficient amounts to correctly capture compositional effects on melt structure.

ACKNOWLEDGMENTS

This work was supported by the NSF grant (EAR 0809489). Computing facilities were provided by CCT at Louisiana State University.

REFERENCES

Adjaouda O, Steinle-Neumanna G, Jahnb S (2008) Mg_2SiO_4 liquid under high pressure from molecular dynamics. Chem Geol 256:185-192

Akins JA, Ahrens TJ (2002) Dynamic compression of SiO_2: a new interpretation. Geophys Res Lett 29:1394

Alfe D (2005) Melting curve of MgO from first-principles simulations. Phys Rev Lett 94:235701

Allen MJ, Tildesley DJ (1987) Computer Simulation of Liquids. Oxford University Press, Oxford.

Anderson KE, Grauvilardell LC, Hirschmann MM, Siepmann JI (2008) Structure and speciation in hydrous silica melts. 2. Pressure effects. J Phys Chem 112:13015-13021

Angell CA, Cheeseman PA, Tamaddon S (1982) Pressure enhancement of ion mobilities in liquid silicates from computer-simulation studies to 800-kilobars. Science 218: 885-887

Behrens H, Schulze F (2003) Pressure dependence of melt viscosity in the system $NaAlSi_3O_8CaMgSi_2O_6$. Am Mineral 88:1351-1363

Bhattarai D, Karki BB, Stixrude L (2006) Space-time multiresolution atomistic visualization of MgO and $MgSiO_3$ liquid data. Vis Geosci DOI: 10.1007/s10069-006-0003-y

Bhattarai D, Karki BB (2007) Atomistic visualization: on the-fly data extraction and rendering. Proc 45th ACM Southeast Regional Conf., p 437-442

Bhattarai D, Karki BB (2009) Atomistic visualization: Space-time multiresolution integration of data analysis and rendering. J Mol Graphics Modell 27:951-968

Boon JP, Yip S (1980) Molecular Hydrodynamics. Dover, New York

Bottinga Y, Richet P (1995) Silicate melts: The anomalous pressure dependence of the viscosity. Geochim Cosmochim Acta 59:2725-2731

Ceperley DM, Alder BJ (1980) Ground state of the electron gas by a stochastic method. Phys Rev Lett 45:566-569

Cohen RE, Gong Z (1994) Melting and melt structure of MgO at high pressures. Phys Rev B 50:12301-12311

de Koker N, Stixrude L, Karki BB (2008) Thermodynamics, structure, dynamics, and melting of Mg_2SiO_4 liquid at high pressure. Geochim Cosmochim Acta 72:1427-1441

de Koker N, Stixrude L (2009) Self-consistent thermodynamic description of silicate liquids, with application to shock melting of MgO periclase and $MgSiO_3$ perovskite. Geophys J Int 178:162-179

Duffy TH, Hemely RJ, Mao HK (1995) Equation of state and shear strength at multimegabar pressures: magnesium oxide to 227 GPa. Phys Rev Lett 74:1371-1374.

Farber DL, Williams Q (1996) An *in situ* Raman spectroscopic study of $Na_2Si_2O_5$ at high pressures and temperatures: Structure of compressed liquids and glasses. Am Mineral 81:273-283.

Flyvbjerg H, Petersen HG (1989) Error estimates on averages of correlated data. J Chem Phys 91:461-466

Funamori N, Yamamoto S, Yagi T, Kikegawa t (2004) Exploratory studies of silicate melt structure at high pressure and temperature by *in situ* X-ray diffraction. J Geophys Res 109:B03203

Giordano G, Dingwell DB (2003) Non-Arrhenian multicomponent melt viscosity: a model. Earth Planet Sci Lett 208:337-349

Haggerty SE, Sautter V (1990) Ultradeep (greater than 300 kilometers), ultramafic upper mantle xenoliths. Science 248: 993-996

Hansen JP, McDonald IR (1986) Theory of Simple Liquids. Academic, London.

Hemley RJ, Mao HK, Bell PM, Mysen BO (1986) Raman spectroscopy of SiO_2 glass at high pressure. Phys Rev Lett 57:747-750

Hohenberg P, Kohn W (1964) Inhomogeneous electron gas. Phys Rev 136:B864-B871

Humphrey W, Dalke A, Schulten K (1996): VMD - Visual Molecular Dynamics. J Mol Graphics 14:3-38

Jones R, Gunnarson O (1989) The density functional formalism, its application and properties. Rev Mod Phys 61:689-746

Karki BB, Bhattarai D, Stixrude L (2006a) First principles calculations of the structural, dynamical and electronic properties of liquid MgO. Phys Rev B 73:174208

Karki BB, Bhattarai D, Stixrude L (2006b) A first-principles computational framework for liquid mineral systems. CMC: Com Mat Cont 3:107-118

Karki BB, Bhattarai D, Stixrude L (2007) First principles simulations of liquid silica: structural and dynamical behavior at high pressure. Phys Rev B 76:104205

Karki BB, Bhattarai D, Mookherjee M, Stixrude L (2009) Visualization-based analysis of structural and dynamical properties of simulated hydrous silicate melt. Phys Chem Miner doi:10.1007/s00269-009-0315-1

Karki BB, Wentzcovitch RM, Gironcoli de S, Baroni S (2000) High pressure lattice dynamics and thermoelasticity of MgO. Phys Rev B 61:8793-8800

Karki BB, Stixrude L, Wentzcovitch RM (2001) Elastic properties of major materials of earth's mantle from first principles. Rev Geophys 39:507-547

Khanduja G, Karki BB (2006) A systematic approach to multiple datasets visualization of scalar volume data. Proc Int Conf Computer Graphics Theory and Applications (GRAPP'06), p 59-66

Khanduja G, Karki BB (2008) Exploiting data coherency in multiple dataset visualization. Proc 10th IASTED Int Conf Computer Graphics and Imaging (CGIM'08) p 261-266

Kohn W, Sham L (1965) Self-consistent equations including exchange and correlation effects. Phys Rev 140:A1133-A1138

Kokaji A (1999) XcrySDen – a new program for displaying crystalline structures and electron densities. J Mol Graphics Modell 17:176-179

Kresse G, Hafner J (1994) Norm-conserving and ultrasoft pseudopotentials for first-row and transition-elements. J Phys C 6:8245-8257

Kresse G and Furthmuller J (1996) Efficiency of ab-initio total energy calculations for metals and semiconductors using a plane-wave basis set. Comp Mater Sci 6:15-50

Kresse G, Joubert D (1999) From ultrasoft pseudopotentials to the projector augmented-wave method. Phys Rev B 59:1758-1775

Kroeker S, Stebbins JF (2000) Magnesium coordination environments in glasses and minerals: New insight from high field magnesium-25 MAS NMR. Am Mineral 85:1459-1464

Kubicki JD, Lasaga AC (1988) Molecular dynamics simulations of SiO_2 melt and glass – ionic and covalent models. Am Mineral 73:941-955

Kubicki JD, Lasaga AC (1991) Molecular dynamics simulations of pressure and temperature effects on $MgSiO_3$ and Mg_2SiO_4 melts and glasses. Phys Chem Miner 17:661-673

Kushiro I (1978) Viscosity and structural-changes of albit ($NaAlSi_3O_8$) melt at high pressures. Earth Planet Sci Lett 41:87-90

Lacks DJ, Rear DB, Van Orman JA (2007) Molecular dynamics investigation of viscosity, chemical diffusivities and partial molar volumes of liquids along the MgO-SiO_2 joins as functions of pressure. Geochim Cosmochim Acta 71:1312-1323

Lay T, Garnero EJ, Williams Q (2004) Partial melting in a thermo-chemical boundary at the base of the mantle. Phys Earth Planet Inter 146:441-467

Li D, Peng M, Murata T (1999) Coordination and local structure of magnesium in silicate minerals and glasses: Mg K-edge XANES study. Can Mineral 37:199-206

Lee SK, Cody GD, Fei Y, Mysen BO (2004) Nature of polymerization and properties of silicate melts and glasses at high pressure. Geochim Cosmochim Acta 68:4189-4200

Lee SK, Li JF, Cai YQ, Hiraoka N, Eng PJ, Okuchi T, Mao HK, Meng Y, Hu MY, Chow P, Shu J, Li B, Fukui H, Lee BH, Kim HN, Yoo CS (2008) X-ray Raman scattering study of $MgSiO_3$ glass at high pressure: Implication for triclustered $MgSiO_3$ melt in Earth's mantle. Proc Nat Acad Sci USA 105:7925-7929

Lesher CE, Hervig RL and Tinker D (1996) Self-diffusion of network formers (silicon and oxygen) in naturally occurring basaltic liquid. Geochim Cosmochim Acta 60:405-413

Mao HK, Bell PM (1979) Equation of state of MgO and ε-Fe under static pressure conditions. J Geophys Res 84:4533-4536

Martin B, Spera FJ, Ghiorso M, Nevins D (2009) Structure, thermodynamic and transport properties of molten Mg_2SiO_4: molecular dynamics simulations and model EOS. Am Mineral 94:693-703

Mookherjee M, Stixrude L, Karki BB (2008) Hydrous silicate melt at high pressure. Nature 452:983-986

Mungall J (2002) Empirical models relating viscosity and tracer diffusion in magmatic silicate melts. Geochim et Cosmochim Acta 66:125-143

Mysen BO, Richet P (2005) Silicate Glasses and Melts, Properties and Structure. Elsevier, Amsterdam

Nevins D, Spera FJ, Ghiorso MS (2009) Shear viscosity and diffusion in liquid $MgSiO_3$: Transport properties and implications for terrestrial planet magma ocean. Am Mineral 94:975

Nielsen OH, Martin RM (1985) Stresses in semiconductors: *Ab initio* calculations on Si, Ge, and GaAs. Phys Rev B 32:3792-3805

Nose S (1984) A unified formulation of the constant temperature molecular dynamics method. J Chem Phys 81:511-519

Perez-Albuerne EA, Drickamer HG (1965) Effect of high pressures on the compressibilities of seven crystals having NaCl or CsCl structure. J Chem Phys 43:1381-1387

Perdew JP, Burke K, Ernzerhof M (1996) Generalized gradient approximation made simple. Phys Rev Lett 77:3865-3868

Pickett WE (1989) Pseudopotential methods in condensed matter applications. Comp Phys Rep 9:115-197

Poe BT, McMillan PF, Rubie DC, Chakraborty S, Yarger J, Diefenbacher J (1997) Silicon and oxygen self diffusivities in silicate liquids measured to 15 gigapascals and 2800 Kelvin. Science 276:1245-1248

Pohlmann M, Benoit M, Kob W (2004) First-principles molecular dynamics simulations of a hydrous silica melt: structural properties and hydrogen diffusion mechanism. Phys Rev B 70:184209

Reid JE, Poe BT, Rubie DC, Zotov N, Wiedenbeck M (2001) The self-diffusion of silicon and oxygen in diopside ($CaMgSi_2O_6$) liquid up to 15 GPa. Chem Geol 174:77-86

Reid JE, Suzuki A, Funakoshi KI, Terasaki T, Poe BT, Rubie DC, Ohtani E (2003) The viscosity of $CaMgSi_2O_6$ liquid at pressures up to 13 GPa. Phys Earth Planet Inter 139:45-54

Revenaugh J, Sipkin SA (1994) Seismic evidence for silicate melt atop the 410-km mantle discontinuity. Nature 369:474-476

Rigden SM, Ahrens TJ, Stolper EM (1984) Densities of liquid silicates at high-pressures. Science 226:1071-1074

Rustad JR, Yuen DA, Spera FJ (1990) Molecular-dynamics of liquid SiO_2 under high-pressure. Phys Rev A 42:2081-2089

Shimoda K, Tobu Y, Hatakeyama M, Nemoto T, Saito K (2007) Structural investigations of Mg local environments in silicate glasses by ultra-high field ^{25}Mg 3Q MAS NMR spectroscopy. Am Mineral 92:695-698

Solomatov VS (2007) Magma oceans and primordial mantle differentiation. *In:* Treatise on Geophysics, Vol. 9. Schudbert G (ed) Elsevier, p 91-120

Stebbins JF (1991) NMR evidences for five-coordinated silicon in a silicate glass at atmospheric pressure. Nature 351:638-639

Stebbins JF, McMillan P (1993) Compositional and temperature effects on five-coordinated silicon in ambient pressure silicate glasses. J Noncrystal Sol 160:116-125

Stixrude L, Bukowinski MST (1989) Compression of tetrahedrally bonded SiO_2 liquid and silicate liquid-crystal density inversion. Geophys Res Lett 16:1403-1406

Stixrude L, Karki BB (2005) Structure and freezing of $MgSiO_3$ liquid in Earth's lower mantle. Science 310:297-299

Stixrude L, Koker N, Sun N, Mookherjee M, Karki BB (2009) Thermodynamics of silicate liquids in deep Earth. Earth Planet Sci Lett 278:226-232

Stixrude L, Lithgow-Bertelloni C (2005) Thermodynamics of mantle minerals. I. Physical Properties. Geophys J Int 162:610-632

Tinker D, Lesher CE, Hutcheon (2003) Self-diffusion of Si and O in diopside-anorthite melt at high pressure. Geochim Cosmochim Acta 67:133-142

Tinker D, Lesher CE, Baxter GM, Uchida T, Wang Y (2004) High-pressure viscometry of polymerized silicate melts and limitations of the Eyring equation. Am Mineral 89:1701-1708

Trave A, Tangney P, Scandolo S, Pasquarello A, Car R (2002) Pressure-induced structural changes in liquid SiO_2 from *ab initio* simulations. Phys Rev Lett 89:245504

Tsuneyuki S, Matsui Y (1995) Molecular dynamics study of pressure enhancement of ion mobilities in liquid silica. Phys Rev Lett 74:3197-3200

Vocadlo L, Price GD (1996) The melting of MgO - computer simulation via molecular dynamics. Phys Chem Miner 23:42-49

Vuilleumier R, Sator N, Guillot B (2009) Computer modeling of natural silicate melts: What can be learned from *ab initio* simulations. Geochim Cosmochim Acta 73:6313-6339

Wan TK, Duffy TS, Scandolo S, Car R (2007) First principles study of density, viscosity and diffusion coefficients of liquid $MgSiO_3$ at conditions of the Earth's deep mantle. J Geophys Res 112:B03208

Waseda Y, Toguri JM (1977) The structure of molten binary silicate systems $CaO\text{-}SiO_2$ and $MgO\text{-}SiO_2$. Metal Trans B 8B:563-56

Waseda Y, Toguri JM (1990) Structure of silicate melts determined by X-ray diffraction. *In*: Dynamics Process of Material Transport and Transformation in Earth's Interior. Marumo F (ed) Terra Sci, Tokyo, p 37-51

Wasserman E.A, Yuen DA, Rustad JR (1993) Molecular dynamics study of the transport properties of perovskite melts under high temperature and pressure conditions. Earth Planet Sci Lett 114:373-384

Williams Q, Jeanloz R (1988) Spectroscopic evidence for pressure-induced coordination changes in silicate-glasses and melts. Science 239: 902-905

Winkler A, Horbach J, Kob W, Binder K (2004) Structure and diffusion in amorphous aluminum silicate: A molecular dynamics study J Chem Phys 120:384-393

Wolf GH, Durben DJ, McMillan PF (1990) High-pressure Raman spectroscopic study of sodium tetrasilicate ($Na_2Si_4O_9$) galls. J Chem Phys 93:2280-2288

Xue X, Stebbins JF, Kanzaki M, McMillan PF, Poe B (1991) Pressure-induced silicon coordination and tetrahedral structural changes in alkali silicate melts up to 12 GPa: NMR, Raman, and infrared spectroscopy. Am Mineral 76:8-26

Yamada A, Inoue T, Urakawa A, Funakoshi K, Funamori N, Kikegawa T, Ohfuji H, Irifune T (2007) *In situ* X-ray experiment on the structure of hydrous Mg-Silicate melt under high pressure and high temperature. Geophys Res Lett 34: L10303

Yeh IC, Hummer G (2004) System-size dependence of diffusion coefficients and viscosities from molecular dynamics simulations with periodic boundary conditions. J Phys Chem 108:15873-15879

Zhang G, Guo G, Refson K, Zhao Y (2004) Finite-size effect at both high and low temperatures in molecular dynamics calculations of the self-diffusion coefficient and viscosity of liquid silica. J Phys C 16: 9127-9135

Reviews in Mineralogy & Geochemistry
Vol. 71 pp. 391-411, 2010

Lattice Dynamics from Force-Fields as a Technique for Mineral Physics

Julian D. Gale and Kate Wright

Department of Chemistry/Nanochemistry Research Institute
Curtin University of Technology
P.O. Box U1987
Perth, WA 6845, Australia

INTRODUCTION

Theory and computation play an increasingly important role in the field of mineral physics by allowing the scientist to probe environments, such as the deep Earth, that are challenging or impossible to access extensively by experiment. Quantum mechanical methods are often the technique of choice, usually based on Kohn-Sham density functional theory as the computationally most practical approach for solids. Although calculations at this level can already be performed on thousands of atoms (Soler et al. 2002; Cankurtaran et al. 2008), the ability to sample nuclear configuration space is often restricted. While density functional theory is typically considered the defacto standard, it is important to remember that with current functionals the results will typically be quantitatively in error with respect to experiment, with occasional qualitative errors (Bilic and Gale 2009). The strength of the method is that the errors are generally systematic and can be anticipated *a priori*.

Despite the ever-increasing scope of electronic structure theory for condensed phases, there are still many problems that will lie beyond their reach for the foreseeable future. Consequently, there remains a need for more approximate, but efficient, techniques to complement quantum mechanical studies. Semi-empirical Hamiltonians and tight binding represent one possibility, but if even greater speed is required then force-field methods are a valuable option. As will be discussed later, the boundaries between the aforementioned approaches are continually becoming blurred as the sophistication of force-fields increases. Beside the greater speed, force-field methods have the advantage of a clear conceptual connection between the functional form and the underlying physics. Hence much can be learnt about what physical interactions are important to describe the properties of a system.

The use of force-field methods is widespread from biological modelling to mineralogy, aided by the availability of many different programs for interatomic potential simulation. By far the majority of this work involves the use of molecular dynamics (MD), which is a powerful technique for sampling complex potential energy surfaces (PES). MD is appropriate for studying an ensemble of configurations connected by barriers of the order of magnitude of thermal energy (k_BT) at a given temperature. The simplicity of requiring only that the energy and forces can be calculated as a function of nuclear configuration on a smooth and continuous PES has led to this widespread availability of software. Of course the ability to produce fancy color movies to enliven talks is an added bonus! The temptation is to become hypnotized by the moving particles into believing that every force-field calculation has to be done with molecular dynamics. One of the foci of the present article is to espouse the point of view that there is an alternative to molecular dynamics, namely lattice dynamics, and that it is important to contemplate which method is most appropriate for a given system under particular conditions of temperature and pressure. Of course, it should also be remembered

1529-6466/10/0071-0018$05.00 DOI: 10.2138/rmg.2010.71.18

that there are further alternatives, such as Monte Carlo, that can also be very effective for certain classes of problem.

In this article, we aim to briefly describe the current status of the static lattice/lattice dynamical approach to mineral physics, including the derivation of force-fields. Although this will be done in the context of one particular piece of software, namely the General Utility Lattice Program (GULP) (Gale 1997; Gale and Rohl 2003), we recognize that there is a long history of lattice dynamical programs spanning more than 50 years. Many of the possibilities to be described are also applicable to other current lattice dynamics codes, such as PARAPOCS (Wall et al. 1993) and SHELL (Taylor et al. 1998) from the groups of Professors Steve Parker and Neil Allan, respectively. Having described the underlying methods, we will present a selection of examples to illustration the application of the techniques to minerals.

METHODOLOGY

Interatomic potentials

Fundamentally it is known that the energy of a material is a complex many-body function of both the nuclear and electronic positions, as well as their momenta. However, this seemingly insurmountable complexity can be reduced through a number of approximations, usually starting with that of Born-Oppenheimer to separate the nuclear and electronic degrees of freedom. Starting from this point, force-field methods seek to describe the potential energy surface by treating the electrons implicitly and expressing the energy in terms of only the nuclear coordinates (supplemented in some cases by further degrees of freedom that mimic aspects of the electronic behavior). Formally we can consider the internal potential energy, U, as being expanded in terms of contributions that depend on increasing numbers of nuclei:

$$U_{total} = \sum_{i=1}^{N} U_i^1(\mathbf{r}_i) + \frac{1}{2}\sum_{i=1}^{N}\sum_{j=1}^{N} U_{ij}^2(\mathbf{r}_i,\mathbf{r}_j) + \frac{1}{6}\sum_{i=1}^{N}\sum_{j=1}^{N}\sum_{k=1}^{N} U_{ijk}^3(\mathbf{r}_i,\mathbf{r}_j,\mathbf{r}_k) + \ldots$$

While this expression is exact, the expansion is only useful if it can be truncated. Fortunately for many systems the contributions do decrease as one goes from the two-body energy (U^2) to the three-body energy (U^3) to the four-body energy (U^4), etc, while the self-energy term (U^1) can often be neglected since usually only relative energies are required rather than absolute values. Where the above series can be truncated depends on the type of system of interest. For simple ionic systems the two-body contribution alone can be sufficient, while more covalent materials require up to four- or six-body terms to capture all the conformational preferences. In contrast, to describe metals accurately all the way from gas phase clusters to the bulk material requires expressions that depend simultaneous on two to three nearest neighbor coordination shells. For a solid represented by periodic boundary conditions, thus making it quasi-infinite, the contribution of any given term must also be restricted to a subset of the total number of atoms. Interactions decay with the distance between the atoms and so cut-offs are applied to limit the distance range of a given term.

By appealing to physical concepts we can determine reasonable functional forms for each of the important terms in the interaction energy. For the majority of minerals, where it is convenient to think of them as ionic solids, the leading term that typically contributes more than 90% of the binding energy (based on formal charges) is the electrostatic interaction;

$$U_{electrostatic}^2(r_{ij}) = \frac{q_i q_j e^2}{4\pi\varepsilon_0 r_{ij}}$$

where q_i and q_j are the charges in atomic units on atoms i and j, respectively, and r_{ij} is the distance between the atoms. Because the rate of decay of the Coulomb interaction with distance is slower than the rate of increase of the number density of interactions (which is proportional to r^2), the summation of the electrostatic interactions is a conditionally convergent series and the energy depends on the macroscopic charge and dipole moment of the crystal (Deleeuw et al. 1980). In practice, the Ewald summation technique (Ewald 1921), in which the Coulomb interaction is partially transformed into reciprocal space, makes it feasible to calculate this contribution for the most important case of a charge neutral system in 3-D with zero dipole moment. Similar considerations apply for 2-D boundary conditions, which are important in surface simulation, and so an analogous summation method, due to Parry (1975), can be utilized. For 1-D boundary conditions, that are appropriate to polymers or screw dislocations, for example, the electrostatic energy is now absolutely convergent, but can still be accelerated through the use of improved algorithms (Saunders et al. 1994).

An alternative to the above lattice sum techniques for evaluating the electrostatic energy is the summation method of Wolf et al. (1999). This method attempts to converge the energy purely in real space by evaluating the real space component of the Ewald sum (the standard Coulomb term multiplied by a complementary error function), while imposing charge neutrality on the cut-off sphere via a boundary image term. Although there is little advantage to this method for small unit cells, it can be computationally very effective for large systems of lower periodicity when combined with a domain decomposition algorithm. This has been used to study systems containing up to a million particles within GULP (Walker et al. 2004).

Aside from the electrostatic interactions, there are at least two other physical contributions that should be included. Firstly, at short-range the repulsion that arises from the overlap of electron densities must be described to prevent ions of opposite charge from coalescing. Secondly, there is always an attractive force between atoms due to the coupling of spontaneous fluctuations in the charge density, which gives rise to the van der Waals contribution. These two interactions are collectively described most commonly by either a Lennard-Jones or Buckingham potential, depending on the favoured form of the repulsive term:

$$U^2_{Lennard-Jones}\left(r_{ij}\right) = \frac{A}{r_{ij}^m} - \frac{C}{r_{ij}^n} \qquad m > n$$

$$U^2_{Buckingham}\left(r_{ij}\right) = A\exp\left(-\frac{r_{ij}}{\rho}\right) - \frac{C}{r_{ij}^6}$$

While the Buckingham potential has the physically correct exponential dependence of the repulsion on distance for weak overlap, it suffers from the problem that the potential becomes asymptotic to $-\infty$ as r tends to zero, if $C > 0$. This can be corrected by damping of the dispersion term at short-range, as proposed by Tang and Toennies (1984).

Where covalency is important, for example in molecular ions such as hydroxide and carbonate, then inclusion of explicitly bonded two-body potentials, as well as angle-bending three-body contributions may be necessary. There is an extensive literature on the development of molecular mechanics style force-fields for covalent materials where interactions are determined by bonding connectivity. This approach is supported by GULP, including the ability to differentiate interactions based on bond order (single vs. double vs. aromatic, etc). However, since this approach is relatively rare for minerals it will not be considered further in the present article.

The combination of fixed charges with some form of short-range repulsive potential and van der Waals attraction is generally known as a rigid ion model. One of the most important neglected contributions at this level of description is the polarizability of ions. For oxides and sulphides the second electron affinity is endothermic and so the formal oxidation state is only achievable due to the electrostatic potential of the surrounding cations. Thus the polarizability of these anions is highly significant. Neglect of this polarizability leads to stabilization of high symmetry structures in the phase diagram over distorted forms. For example, the experimentally observed monoclinic distortion of silicalite at low temperatures is not reproduced by a rigid ion model (Bell et al. 1990).

There are two classes of approach to the inclusion of ion polarizability. Arguably the most natural approach is to assign point-ion polarizabilities to each atomic species, since this can be readily generalized to any order of polarizability tensor. This methodology has been extensively applied by Wilson, Madden and co-workers to a range of ionic materials (Madden and Wilson 1996). The alternative approach is the so-called shell model of Dick and Overhauser (1958). Here a simple mechanical model for dipole polarizability is employed in which an atom is described by two particles, a core and a massless shell, where the former conceptually represents the core electrons and nucleus, while the latter mimics the valence electrons. The two particles are coupled by a short-range interaction (usually a harmonic spring though other forms are possible) and they are Coulombically screened from each other. The effective in-crystal polarizability, α, of this shell model species is given by;

$$\alpha = \frac{q_{shell}^2}{k_{shell} + k_{environment}}$$

where q_{shell} is the charge on the shell, k_{shell} is the second derivative of the interaction energy between the core and shell (the spring constant), and $k_{environment}$ is the second derivative acting on the shell due to the embedding medium. The advantage of the shell model over point-ion polarizabilities is that the polarizability can be seen to be a function of the environment and is naturally damped as the material is compressed. On the downside, there is no natural method for extending the scheme to quadrupolar and higher order contributions, though the breathing shell model can capture some effects of ion compressibility.

Regardless of the method chosen for inclusion of polarization, this increases the cost of a simulation since the dipole (and higher if appropriate) moments must be allowed to interact self-consistently. Thus an iterative solution is necessary with respect to determining the ground state energy for a given nuclear configuration. The extra cost during a geometry optimization is less though since the polarizability variables can be minimized concurrently with the nuclear coordinates such that the Born-Oppenheimer surface is only achieved at the minimum. Special treatment is also needed during molecular dynamics; here the polarizability variables can either be minimized for every configuration to perform Born-Oppenheimer dynamics, or, Car-Parrinello dynamics (Car and Parrinello 1985) can be used in which a ficitious mass is assigned to these variables so that they oscillate about the ground state surface (Lindan and Gillan 1993). Both approaches require care since systematic error propagation can lead to deviations in the conserved quantity.

At present in GULP only the shell model approach is available, though it is hoped to add point-ion polarizability as option in the future. Point dipoles can currently be included, but without self-consistent coupling of the induced dipole - induced dipole interactions. There are also other sophisticated interatomic potential functional forms available in GULP and more information about these can be found elsewhere (Gale and Rohl 2003). Some examples will be highlighted later when we discuss the most recent developments in mineral force-field modeling.

Lattice dynamics versus molecular dynamics

Once the energy can be defined as a function of nuclear coordinates, the challenge is then to explore the potential energy surface. In general the PES will consist of a series of local minima separated by a sequence of transition states. For an ordered solid, one of these minima will be the global minimum while the rest represent metastable structures. For disordered systems, such as liquids or amorphous solids, the concept of a global minimum is no longer relevant. To determine the overall thermodynamic properties of a system requires the partition function that contains the information regarding the probability of the system occupying each state. The exploration of this configuration space can be conducted either via statistical or systematic methods. Lattice dynamics involves the systematic characterization of a PES by computing the properties of each individual minima and transition state, if kinetic information is required. Statistical methods, in contrast, start from one or more points in phase space and execute a series of moves to explore the accessible configurations. Two popular approaches are common here; Monte Carlo and molecular dynamics. In Monte Carlo an importance sampling of the PES is made to achieve the Boltzmann weighted averages by random steps. Molecular dynamics instead starts by assigning a velocity distribution to the atoms and then evolves via the numerical solution of Newton's equations of motion. Of these two statistical approaches, molecular dynamics potentially yields far greater information since time correlated properties are available. Both suffer from uncertainty though since there is no guarantee that the nuclear configurations have been adequately sampled.

Lattice dynamics involves performing a harmonic analysis of the potential energy surface about stationary points. By computing the second derivative matrix and weighting by the inverse square roots of the masses, m, of the cores, i and j, involved in each element the dynamical matrix is obtained:

$$D_{i\alpha,j\beta} = \frac{1}{\left(m_i m_j\right)^{1/2}} \left\{ \left(\frac{\partial^2 U}{\partial \alpha_i \partial \beta_j} \right) + \frac{4\pi}{V} \delta_{k\Gamma} \frac{\left(\mathbf{k}^{\Gamma} \mathbf{q}_i^{born}\right)_{\alpha} \left(\mathbf{k}^{\Gamma} \mathbf{q}_j^{born}\right)_{\beta}}{\left(\mathbf{k}^{\Gamma} \varepsilon^{\infty} \mathbf{k}^{\Gamma}\right)_{\alpha\beta}} \right\}$$

Here the second term in the expression only applies to the gamma point and represents the non-analytic correction for the LO-TO splitting based on the direction of approach to gamma, k^{Γ}, the cell volume, V, the Born effective charges, and the high frequency dielectric constant tensor (Baroni et al. 2001). For models containing non-nuclear degrees of freedom, such as shell model coordinates, dipoles or variable charges, there is a correction to the first second derivative term above that must also be evaluated (Catlow and Mackrodt 1982).

Diagonalization of the dynamical matrix yields the square of the vibrational frequencies as eigenvalues, while the normal modes are defined by the eigenvectors. Through summing over the vibrational energy levels, the vibrational partition function can be determined leading to all the key thermodynamic properties via the standard expressions of statistical mechanics. An important feature of lattice dynamics that is absent from standard statistical approaches is that the nuclear energy levels are quantized. Classical behavior of the nuclear dynamics is only recovered as the conditions approach the Debye temperature, which for ionic minerals can be in excess of 1000 K. Of course at high temperatures harmonic lattice dynamics breaks down due to the neglect of anharmonic effects, though at elevated pressures the regime of validity may be extended. Anharmonicity can be accounted for in lattice dynamics via more sophisticated techniques, such as vibrational self-consistent field (VSCF) (Bowman 1986), that allow for phonon-phonon coupling. However, the cost is greatly increased and so molecular dynamics becomes more readily tractable.

Whether it is appropriate to use lattice dynamics or a statistical approach obviously depends on the system and conditions. Molecular dynamics is an appropriate technique where the activation energies to pass between minima are not more than a few times $k_B T$, where k_B is Boltzmann's constant and T is the temperature. Here the energy surface is strongly anharmonic and the system must be described by averaging over all these states. However, if a system is strongly localized in a single state at low to moderate temperatures then molecular dynamics may represent an expensive way to convolute a well-defined answer with some statistical noise. Here a lattice dynamical calculation can yield a more accurate answer in a fraction of the computer time. If in doubt, it is prudent to perform an initial static exploration of the energy surface to determine which technique is more appropriate. Alternatively, if molecular dynamics is to be used, first explore the energy surface using an accelerated dynamics scheme, such as metadynamics (Laio and Parrinello 2002). Approximately mapping out the free energy landscape with respect to any key variables of interest can ensure that you are starting from the right place.

Finding stationary points

Of pivotal importance to lattice dynamics is the ability to efficiently find and characterize stationary points on the potential energy surface. Optimization methods start from the premise that the energy at a configuration **x** can be expanded in terms of a displacement, $\delta\mathbf{x}$, about this point:

$$U(x+\delta x)=U(x)+\frac{\partial U}{\partial x}\delta x+\frac{1}{2}\frac{\partial^2 U}{\partial x^2}(\delta x)^2+\ldots$$

For a truly harmonic energy surface (an approximation that becomes valid close to a stationary point), the expansion can be truncated at the second order term. Based on this the displacement required to take the system to the nearest stationary point is:

$$\delta x=-\left(\frac{\partial^2 U}{\partial x^2}\right)^{-1}\left(\frac{\partial U}{\partial x}\right)$$

Once a system is in the harmonic region, a single step is sufficient to converge an optimization if the exact inverse of the second derivative (Hessian) matrix is known. Away from the harmonic region, the displacement vector is approximate and therefore it is necessary to apply the process iteratively, in combination with a line search to find the best scale factor for the displacement vector, until the forces drop below the convergence criteria.

Calculating the inverse of the Hessian can be the most expensive step in the quest for a stationary point and therefore more efficient, but approximate, techniques can be used. A common algorithm to use is the updating of the Hessian according to either the BFGS (Shanno 1970) or DFP methods. Here either the exact Hessian matrix or an approximation to it is calculated for one configuration and then the matrix is updated based on the change in first derivatives during a step, rather than recalculating from scratch, thus avoiding the cost of matrix inversion. Provided the angle between the search vector and forces remains small enough, this approximate Hessian is likely to be adequate. Because GULP is focused towards the calculation of high order analytical derivatives of the energy, the above Newton-Raphson scheme based on the use of Hessian is the default leading to highly converged optimizations. For large system sizes the cost of the Hessian, both in computer time and memory, may become excessive and so other algorithms based only on the first derivatives are also available, such as conjugate gradients and limited memory BFGS (Nocedal 1980).

A particular advantage of routinely being able to calculate the analytic Hessian matrix is that the location of higher order stationary points becomes facile. In particular, the use of the Rational Function Optimizer (RFO) (Banerjee et al. 1985) is especially powerful. This algorithm applies a level shift to the Hessian matrix to ensure that the state being sought on the PES has a predetermined number of negative eigenvalues. By specifying zero this will lead to checking that the converged result is a genuine minimum, subject to any symmetry constraints, while requesting a value of one negative eigenvalue will lead to a transition state. The more widely used methods for determining activation barriers, such as Nudged Elastic Band (Henkleman et al. 2000), are based on first derivatives only. RFO can locate a transition state without *a priori* specification of a target product, as well as achieving high precision for the final state through being quadratically convergent. In the case of GULP, transition state searching methods based on connecting two specified minima are also available to identify the region of a particular transition state of interest, which can be subsequently refined using RFO.

Here we generally find that the synchronous transit approach (Dewar et al. 1984) is slightly more effective than nudged elastic band for the present implementation.

Mineral properties

Having located a minimum energy configuration, the availability of analytic second derivatives makes it possible to routinely determine a myriad of mineral properties. A list of the properties that are currently available in GULP is presented in Table 1. Further illustration of the calculation of properties will be given in the applications section. It should be noted that some of the properties, particularly the dielectric ones, can only be meaningfully calculated for models that contain a mimic for the electronic charges.

One property that merits particular mention is that of thermal expansion. While this quantity is often considered to lie within the realm of molecular dynamics, since simulation within the NPT ensemble naturally leads to the cell parameters at a specified temperature, this actually represents an unnecessarily painful and formally incorrect means of determining this property in the low temperature limit. As already highlighted, molecular dynamics is only valid once the system

Table 1. Mineral properties that can be obtained with GULP for an optimized structure. Properties that can be derived from the result of calculations on several configurations are not included.

Property	Notes
Elastic constant tensor	
Elastic compliance tensor	
Bulk modulus	Voight, Reuss and Hill definitions
Wave velocities	S-wave or P-wave
Compressibility	
Young's moduli	
Poisson's ratios	
Piezoelectric constants	Strain or stress
Dielectric constant tensor	Static, high frequency, or frequency dependent
Refractive indices	Static, high frequency, or frequency dependent
Phonon frequencies	Any point in the Brillouin zone
Phonon densities of states	Includes spherical averaging of non-analytic correction about gamma point.
Phonon dispersion curves	Includes non-analytic correction for dispersion vector
Oscillator strengths	
Reflectivity	
Born effective charge tensors	
Infra-red intensities	
Raman intensities	Approximate method for homogeneous materials
Mean kinetic energy of phonons	
Heat capacity	Constant volume – constant pressure possible, but involves more calculation.
Entropy	From vibrational partition function – not disorder.
Electric field gradient tensor	Leading to asymmetry parameter for NMR.
PDFs	Available version 3.5.
$S(Q,\omega)$	Under development.

approaches the Debye temperature. Furthermore, thermal expansion coefficients for ionic materials can be of the order of 10^{-5}/K. Therefore, considerable care is required to extract thermal expansion coefficients from molecular dynamics given the fluctuations within the method.

Lattice dynamics allows the free energy to be calculated based on the vibrational partition function at a given temperature. Through the use of the quasiharmonic approximation, the expansion of the unit cell can be found through the use of derivatives of the free energy with respect to strain, as demonstrated in the pioneering work of Parker and co-workers (Price and Parker 1989). Initially the free energy derivatives were determined numerically, but following the derivation of analytic formulae by Kantorovich (1995) this more efficient approach was implemented into both Shell (Taylor et al. 1998) and GULP (Gale 1998). Once the sampling of the phonons is sufficiently converged with respect to the free energy, the thermal expansion can be readily determined by a series of minimizations at different temperatures with the same level of precision as for a conventional optimization.

In principle the availability of analytic free energy first derivatives makes it feasible to optimize both internal coordinates and cell parameters as a function of temperature. However, this creates a fundamental inconsistency in that the phonons are being determined based on a harmonic analysis of the second derivatives of the internal energy about an assumed stationary point in this quantity, while the coordinates are being driven to a minimum in the free energy instead. These demands are mutually incompatible wherever there is a substantial vibrational contribution to the force. Consequently, it is better to make what is known as the Zero Static Internal Stress Approximation (ZSISA) in which the atomic coordinates are minimized with respect to the internal energy while only the cell responds to the free energy contribution. This second approach demonstrates stability to far greater temperatures before breaking down in the face of anharmonic contributions.

Derivation of force-fields

The single largest barrier to commencing research using non-*ab initio* methods is the determination of the approximate model and its parameters. Unfortunately, although there exists a "universal force field" in name, one doesn't exist in practice. Rule-based approaches to the development of interatomic potential parameters for every element in the periodic table are appealing, but involve significant compromise due to the wide variety of environments that an atom can find itself in. Therefore, despite the fact that GULP comes complete with a library for the Universal Force Field (UFF) (Rappe et al. 1992), this is usually not the best choice – especially for minerals. If one is lucky then there will already be a well-tested literature model for your mineral of interest. Otherwise you will be faced with the prospect of deriving a new force-field, or at least adapting or augmenting an existing one.

In the present section we shall aim to offer a few pointers as to how to tackle the process of deriving a force-field model, based on the methods available in GULP. Clearly the starting point is to contemplate the physical interactions that are likely to be important for your mineral and use this insight to determine the functional form of the model. This can often be an iterative process, since it is best to start off with the simplest possible choice that might work and then to increase complexity as the deficiencies become apparent.

For the majority of minerals the starting point will be to choose the charges to be assigned to the species since the electrostatic interactions tend to dominate. Formal oxidation states can often be a good practical choice, especially when combined with the shell model for anions and other polarizable species. This will ensure the maximum degree of compatibility between compounds of different stoichiometry, as well as making it feasible to study the defect chemistry for non-charge neutral cases. If there are molecular fragments present, for example carbonate, then the formal charge on the whole species can be preserved, while allowing non-integer charges for the individual atoms within the molecule.

In the absence of a shell model, the formal charges may be too strongly cohesive and so the use of partial charges becomes necessary. This is usually only the case for cations where the charge state is +3 or higher. If multiple compounds are to be studied then the use of a consistent scaling factor (<1) for the charge is efficacious in maintaining compatibility. This factor can be employed as a fitting parameter or fixed by other arguments. One seemingly obvious way to determine partial charges is through using the results of quantum mechanical calculations. However, the partitioning of electron density into atomic parts is a famously arbitrary task. Definitions such as Mulliken charges are notoriously sensitive to the choice of basis set and so represent perhaps the worst choice. For molecular systems the use of electrostatic potential derived charges is a logical approach since the aim of the force is to reproduce this potential. In the case of solids this option is lost and data from molecular fragments doesn't transfer well to the condensed phase for ionic materials. Our contention is that the optimal choice for force-field development is to fit the Born effective charge tensor;

$$q_{i,\alpha\beta}^{born} = \frac{\partial \mu^{\alpha}}{\partial \beta_i}$$

where α and β are Cartesian directions and μ represents the dipole moment vector. This tensor is a quantity that is available from first principles methods that does not depend on partitioning and relates to the response of the charge distribution to atomic displacements. Note, that even though the dipole moment may be ill-defined for a periodic solid, its derivative is straight forward to calculate and unambiguous.

The use of fixed charges, be they partial or formal, has one significant downside in that the long-range dissociation of bonds cannot be accurately described. This process will usually lead to the formation of neutral fragments that are inconsistent with the bulk charges. For distortions of the solid material, including the diffusion of ions within the condensed phase, this is not a problem but represents a limitation when the density of the system is decreased. This can occur for the surfaces of the material or even more acutely for molecular fragments *in vacuo*. A natural solution to this is to use geometry dependent charges and this is a possibility allowed for by GULP. The "on the fly" determination of charges clearly increases the computational cost significantly, but with the advance of computing power this represents a diminishing barrier. Inclusion of the charge transfer process has the benefit of capturing high order polarization contributions that go beyond the shell model.

The simplest method of determining charges "on the fly" is via bond increments, as used in the well-known Gasteiger technique (Gasteiger and Marsili 1980). Here the charge only depends on the connectivity, rather than the specific bond lengths, and so the charges are usually only determined once at the initial configuration to avoid discontinuities in the energy surface. A more sophisticated, and truly geometry dependent, approach is to use electronegativity equalization (Sanderson 1951) to determine the charges where the electrostatic energy of a given atom is specified by:

$$U_{i,electrostatic} = \chi_i q_i + \frac{1}{2}\mu_i q_i^2 + \frac{1}{2}\sum_{j \neq i}^{N} V_{ij}$$

Here χ_i and μ_i are the electronegativity and hardness of the atom, respectively, and are usually determined based on a quadratic fit to the first ionization energy and electron affinity of the species. The final term represents the sum over all the other atoms in the system of the electrostatic potential at site *i*. While the potential may be calculated as the bare Coulomb interaction of the point charges, more realistic screened interactions based on Slater orbitals or Gaussian charge distributions have also been utilized, as is the case in the QEq model (Rappe and Goddard 1991).

Once the above expression for the electrostatic energy is defined, the charges at any given geometry can be determined by finding the set of values that yield the same effective electronegativity (i.e., the first derivative of the electrostatic energy at the site with respect to charge) at all atoms, subject to the constraint of fixed total charge. Originally this was formulated as a set of linear equations to be solved by matrix factorization, although for large systems functional minimization using the first derivatives with respect to the charges is well known to be more efficient. Once the charges are determined that minimize the energy then the forces can be found straightforwardly due to the Hellmann-Feynman theorem. However, the determination of the second derivatives requires the simplified equivalent of linear-response theory.

Special consideration is required for the QEq method (Rappe and Goddard 1991) when hydrogen is present. Here the hardness is defined as a function of the charge for this element, in recognition of the fact that the change in hydrogen from proton to hydride is particularly extreme. In the case of this model the expressions to be solved to determine the charges are non-linear in form and therefore must be solved iteratively even when using matrix algebra. The equations for QEq originally presented do not satisfy the Hellmann-Feynman theorem and therefore the forces must be corrected for the charge derivative contributions. In GULP a slightly modified set of expressions is used to avoid this extra computational effort, while the charges obtained differ only slightly from those obtained with the original formulation.

Having determined a suitable set of charges, or charge determination scheme, for the system of interest the remaining contributions to the potential model can be established. If the system is of low symmetry or the dielectric properties are important (as is the case for charged defects for example) then inclusion of a shell model is usually necessitated. Short-range repulsion must be included between all species of opposite charge and van der Waals attraction can be added between polarizable anionic species, such as oxide and sulfide.

Once the functional form of the force-field is chosen, the remaining challenge is to determine the unknown parameters and coefficients within these expressions through some form of fitting. However, some values are best determined by physical insight, rather than being allowed to fit to data directly. The most common example of this is the C_6 coefficient of a Lennard-Jones or Buckingham potential. If allowed to vary as a fitting parameter physically unreasonable magnitudes can be obtained. This is because dispersion represents a relatively uniform background cohesive interaction that can become highly correlated with the short-range repulsion. Given that estimates are available for in crystal polarizabilities of many of the key anions, it is better to fix the dispersion contribution based on the resulting estimates of physically reasonable C_6 terms. For example, oxide-oxide C_6 coefficients will usually lie in the region of 30 eVÅ^6.

Two broad categories of approach to potential fitting exist, namely empirical and *ab initio*. In empirical fitting, as the name suggests, the parameters are fitted to reproduce a training set of experimental data using a least squares procedure:

$$F_{fit} = \sum_{observables} w\left(f_{observed} - f_{calculated}\right)^2$$

Each observable can be weighted with a coefficient, w, according to the reliability of the data and the relative importance in reproducing that particular value. Of course, these are fuzzy criteria and so there are an infinite number of possible fitted potential sets depending on the weighting criteria. Furthermore, empirical fitting is limited by the quality, consistency and range of experimental data. For example, the structure may have been measure at a different temperature to the mechanical properties and yet they are usually fitted while neglecting this discrepancy. Similarly, the experimental data may only contain information regarding the potential energy surface close to the minimum and so the ability to describe other regions depends on the physicality of the underlying model. The alternative to using empirical data is to use information from *ab initio* or first principles techniques instead. Here consistency of observables can be ensured, while any region of the potential energy surface can be explored during training provided the underlying quantum mechanical technique is capable of describing it. The downside of this approach is that most quantum mechanical methods that can currently be applied to condensed phases contain systematic errors and so the force-field generated will reproduce the world according to density functional theory, which may be a different place from reality!

Regardless of whether the data is empirical or theoretical, the important thing in the creation of any force-field model is to use as extensive a training set as possible, preferably spanning the structures of multiple phases, their energy differences, curvature related properties and information relating to strong variations in density.

Several different strategies for performing least squares fitting are available. If using quantum mechanical data then the energy surface can be directly fitted, usually with weights that emphasize the importance of correctly describing the regions close to the minimum on the curve. For empirical fitting, the choice is more complex. The conventional approach to fitting an experimental structure is to minimize the forces on the atoms and stresses on the cell at this configuration. In the case of a shell model special care is needed to ensure that the shells are minimized during the fit too so that the system remains close to its polarization ground state. In GULP this is achieved using so-called "simultaneous" fitting, to reflect the fact that the shell displacements are concurrently used as variables in the procedure (Gale 1996). Although this conventional fitting technique can be useful for simple high symmetry systems it is found to be sub-optimal in general. This is because while the forces may be reduced, the error in the structure (which is usually what is considered to be important) depends on the product of the forces and the inverse Hessian matrix. Hence, if the errors in the Hessian become larger then lower forces need not lead to better reproduction of the structure. Consequently, an alternative approach known as "relaxed" fitting is preferable. Here the structure is optimized at each step of the least squares procedure and the error in the displacements explicitly determined. In addition, the properties can be calculated for a true stationary point, rather than at the experimental structure where the forces may be non-zero. Clearly this relaxed fitting approach is an order of magnitude more expensive than the conventional strategy, but the results are often considerably better. Furthermore, the method becomes faster as the fit converges since fewer optimization steps are needed. Of course before it is possible to use relaxed fitting it is a prerequisite to have a sufficiently good force-field to be able to optimize the structure in the first place. Hence, it can be consider as much a force-field refinement procedure rather than an initial fitting method.

Evolution of a force-field: Silica potentials through the ages

To illustrate the development of force-field models it is instructive to take a simple example and examine how things have changed over the last three decades. Silica represents an ideal case in many respects since it exists in a wide range of polymorphs from dense phases through to nanoporous zeolitic frameworks, as well as being of mineralogical relevance. To make things especially interesting the stable polymorph under ambient conditions, α-quartz, possesses a relatively low symmetry structure that transforms from a tetrahedral silicon coordination environment to octahedral at higher pressure on forming the rutile-structured stishovite. Furthermore, some of the zeolitic polymorphs exhibit the interesting phenomenon of negative thermal expansion (Couves et al. 1993), which provides a more demanding test of any model.

At the outset it is important to point out that there have been many different silica force-fields since the dawn of lattice dynamics and simulations. Our aim is not to achieve a comprehensive review of the area, but to give a personal reflection on some of the developments made and to seek forgiveness for any omissions. The lattice dynamics of silicate minerals arguably began with the pioneering studies of Parker, Catlow and co-workers (Catlow et al. 1982; Parker 1983; Parker et al. 1983) who showed that the simulation of a range of silicate minerals could be successfully achieved using a two-body model with shells for oxygens and formal charges. Although the use of formal charges for silica was initially very controversial, since it contradicts the view that silicates are significantly covalent, they can be justified based on the agreement with the Born effective charge tensors obtained from first principles methods when the shell contribution is included. While this early work proved the usefulness of the ionic model as a technique for studying silicates, there were issues in generating a truly universal Si-O potential. A key step forward was made by Sanders et al. (1984) when they explored the addition of a three-body angle-bending term to describe partial covalency about silicon. A further modification was to include a small, but significant, C_6 coefficient for the Si-O two-body potential, which leads to a notable improvement. With this model they were able to reproduce a range of properties for α-quartz and proposed that this potential would be transferable to zeolitic phases of silica, as it indeed turned out to be.

Over the years there have been a number of tweaks to the form and parameters of the Sanders et al. (1984) model, but in essence the successful ingredients have remained and the results have proved hard to surpass. That said, there are a couple of problems faced by the potential and both relate to the use of the three-body potential. Firstly, the equilibrium angle is set to be tetrahedral and thus the potential does not transfer well to stishovite and six-fold coordinated phases. Secondly, when studying the formation of silicon or oxygen vacancies (and certain other defects) there is an uncertainty regarding the energy. This arises from the fact that there is no distance decay of the three-body term and so the dissociation limit is ill-defined.

In parallel to the development of silica models where the focus was primarily on the lattice dynamics of ordered phases, there was a similar effort to create a model suitable for molecular dynamics studies of glassy materials. Although the shell model has been used for this, usually in a modified form in which the discontinuous three-body term is replaced by one that decays with the Si-O distance, there was a strong emphasis on avoiding the use of shells due to the increase in computational resources required. Arguably the first work to dynamically simulate silica was that of Woodcock et al. (1976) who used a formally charged rigid ion model. The model developed was found to be only partially successful, and so the quest for a simple dynamical force-field was later revisited by Tsuneyuki et al. (1988). This led to a rigid ion potential, known as TTAM, based on the use of a partial charge for silicon of +2.4. Here the parameters were determined based on Hartree-Fock cluster calculations for SiO_4^{4-}, which was stabilized by placing it in the embedding field of four point charges. Despite this cluster based approach to parameterization, the model was found to yield reasonable structures and bulk moduli for the polymorphs α-quartz, α-cristobalite, coesite and stishovite. Unfortunately coesite was found to be more stable than α-quartz in the low temperature limit and the transition to stishovite occurred at a significantly lower pressure than experiment.

Not long after the appearance of the TTAM model, came a similar model from van Beest et al. (1990) now commonly referred to as the BKS potential. A similar strategy was employed in that quantum mechanical cluster calculations were performed (though based on neutral species here) to initially define the parameters. However, in this case further refinement was conducted against bulk data. The functional form consisted of a simple Buckingham two-body potential plus Coulomb interactions, again with a partial charge of +2.4 for silicon.

A major feature of the BKS model is that the C_6 coefficients are unphysically large; for example the O-O dispersion coefficient is 175 eVÅ^6, approximately 5-6 times greater than the expected magnitude. Even Si-O possesses a C_6 value of 133.5381 eVÅ^6, which is surprising considering the low polarizability of a positively charge silicon ion. All of these indicators suggest that the partial charge used is too low, and therefore the electrostatic underbinding has been compensated by the van der Waals terms. This is confirmed by first principles calculations where the diagonal elements of the Born effective charge tensor are between +3 and +3.6 for silicon (Gonze et al. 1992). To further complicate matters the original potential used cut-offs of 10 Å for the Buckingham contributions, but because of the large magnitude of the attractive term this leads to noticeable discontinuities in the energy surface. This can be corrected through the use of a taper function, cut-and-shifting of the potential, or an Ewald-style lattice sum for the C_6 contribution.

Being similar to the TTAM model it is not surprising that the BKS model also suffers from similar failures. Again coesite is found to be more stable than α-quartz, while six-coordinate polymorphs are now even more stable than this, probably due to the dispersion term favoring high density forms. With the advent of the metadynamics algorithm (Laio and Parrinello 2002) for accelerating the dynamical exploration of phase diagrams it has now been possible to extensively scan the configuration space for materials looking for new structures and this has been applied to silica (Martonak et al. 2006). This recently led to the revelation that the BKS model has the anatase structure as the lowest ground state discovered so far (Martonak et al. 2007). This structure is not known to have a stability field for silica, either experimentally or within GGA-based first principles studies.

After the above early efforts to develop silica potentials, there was an explosion in interest leading to many different force-fields and so this overview naturally becomes more selective. Following the failures of the simple rigid ion models, Vashishta et al. (1990) proposed a more sophisticated treatment for silica suitable for molecular dynamics. Here they combined many ingredients, such as damped point-ion polarizabilites and distance-dependent three-body terms, including for the Si-O-Si angle. Despite the low silicon charge of +1.6 the model was found to work well within the limited set of tetrahedral phases tested.

During the early 1990s there was a particular shift toward exploring potential models that include geometry dependent charges, with that of Alavi et al. (1992) being one of the first. Here something more akin to the Gasteiger approach to charge determination is used where the shift of charge between near neighbors as a function of bond length is mimicked, rather than the global flow of charge, as is the case for electronegativity equalization. Soon after, Jiang and Brown (1995) studied the diffusion of oxygen in silicon. This raised what remains one of the greatest challenges in force-field models, namely the simulation of the interface of two different types of materials. In order to describe oxygen in silicon a model was required that could describe both silica and silicon, as well as intermediate states between these two extremes. Jiang and Brown decided to adopt two models that at the time were believed to work well for the two end members; the Stillinger-Weber potential for silicon and the BKS model for silica. To connect these two potentials, the authors used a charge-dependent bond-softening expression so that the Si-Si interaction from silicon was turned off in silica. Similarly the charge on a silicon atom became a function of the number of oxygen neighbors according to the sum of cosine tapering functions that switch between charge transfer of 0.6 and 0 with varying bond length. This elegant, yet simple, approach would in principle allow the coupling of any two models, though the mixing of shell and non-shell models could prove complex.

Following the work of Jiang and Brown there were various further developments for the simulation of silicon/silica interfaces, including the model of Yasukawa (1996) that uses the Tersoff bond order potential for silicon combined with a full charge equalization. This has been subsequently refined through the addition of non-quadratic charge self-energies, as well as explicit coupling of the short-range and Coulomb terms, to form the COMB potential (Yu et al. 2007). Comparison of the phase stabilities for a number of polymorphs shows that this model corrects the failings of some of the previous approaches.

A clear trend in the community that naturally leads on from the desire to include quantum mechanical effects into the underlying models is towards the creation of so-called "reactive force-fields". Here the functional form of the potential is chosen so as to describe bond dissociation in a smooth fashion such that the connectivity of a material is not pre-defined. In the hydrocarbon field this led to the widely used REBO model of Brenner and co-workers (Brenner et al. 2002) that has proved a useful tool in simulation of nanostructured carbon.

Although it has many commendable features, a weakness of the REBO model has been in the treatment of long-range interactions, which have had to be grafted on in an *a posteri* manner, if used at all. A more sophisticated method has been developed by van Duin and co-workers, known as ReaxFF (van Duin et al. 2001). Here van der Waals forces and Coulomb interactions with variable charges are included from the start. Furthermore, an elegant feature is the inclusion of three bond order terms to represent single, double and triple bond contributions as the distance decreases. The latest form of ReaxFF (Chenoweth et al. 2008) has now been incorporated into GULP and includes the options to use domain decomposition and iterative charge solution such that large systems can be handled efficiently.

ReaxFF has been parameterized for silicon/silicon oxide systems (van Duin et al. 2003) and the original paper included evaluation of the model for a number of silica polymorphs. In Table 2 we show a more extensive evaluation of the force-field for a range of known and hypothetical forms of silica based on structures with the stoichiometry of MX_2. It should be noted that the detailed equations associated with ReaxFF have evolved with time and therefore there may be a number of small differences between the current form and that at the time of the original work. Hence, discrepancies are possible between the results presented here and those of the original model, though for many of the quantities given in the published manuscript there is reasonable agreement.

As observed in the previous study (van Duin et al. 2003), ReaxFF also suffers problems when it comes to the correct prediction of the global minimum for silica. We find that both α-cristobalite and tridymite are more stable than α-quartz, while van Duin et al. (2003) had the former material as almost isoenergetic with the experimental ground state, but slightly higher. Given the similar distorted tetrahedral arrangements in all three phases it is quite a challenge to discriminate between these materials. Where ReaxFF does represent a considerable advance over some of the other force-fields discussed is that none of the high coordination number

Table 2. Calculated structure, energies relative to α-quartz, and properties for silica when adopting different MX_2 crystal structures using the ReaxFF force field.

Polymorph	Relative energy (kJ/mol)	Density (g cm^{-3})	Bulk modulus: Reuss (GPa)	Bulk modulus: Voight (GPa)	Phonon stable at Γ / mode if unstable (cm^{-1})	Charge on Si (a.u.)
α-quartz	0.0	2.547	87	95	Yes	1.383
Coesite	+3.0	2.952	99	100	Yes	1.382
α-cristobalite	−1.6	2.175	58	58	Yes	1.403
β-cristobalite	+0.1	2.137	120	120	Yes	1.406
Tridymite	-0.6	2.147	62	66	Yes	1.394
Stishovite	+82.4	4.361	766	2709	Yes	1.282
α-PbO_2	+76.3	4.345	741	783	Yes	1.283
Anatase	+400.5	2.591	27	99	No 77i	0.937
Baddeleyite	+49.5	3.750	198	225	Yes	1.295
Brookite	+74.6	3.867	205	210	Yes	1.258
Pyrite	+121.8	4.739	1796	1796	Yes	1.277
Marcasite	+82.4	4.361	766	2709	Yes	1.282
Fluorite	+282.4	4.434	54201	54201	Yes	1.179

polymorphs are lower in energy than α-quartz and the energy difference for stishovite is close to experiment. Surprisingly, we note that stishovite is less ionic than α-quartz despite the higher coordination number of silicon to oxygen. All structures were found to be phonon stable with the exception of anatase where a single imaginary mode was identified at the Γ point, indicating that a lowering of symmetry is required. The highest frequency modes for the polymorphs were significantly overestimated with respect to experimental values for an Si-O stretch, with α-quartz having a maximum value of 1836 cm^{-1} at Γ.

For all of the polymorphs studied we have also evaluated the mechanical properties by using central finite differencing of the analytic first derivatives with respect to the strains and internal coordinates. All quantities were checked for sensitivity to the size of the finite difference step and were found to be sufficiently converged. Although the full elastic constant tensor is available we just quote the bulk modulus according to two definitions, as computed from this matrix, for brevity. Here we do find significant differences with respect to the original paper; for example α-quartz is found to be three times harder than experiment in the present work whereas the original publication exhibited good agreement with known values. Furthermore, we find that the hardness of the materials is uniformly overestimated. Particularly striking is the bulk modulus of the fluorite-structured phase, which would be two orders of magnitude harder than diamond if it could be made. This noteworthy value was cross-checked by computing the bulk modulus from a pressure-volume curve to ensure that it wasn't an artifact of the finite difference method. While some of the error in the bulk moduli may be a consequence of changes in the functional form of ReaxFF, it appears that the lack of explicit inclusion of curvature information during the parameterization (energy-distance curves are fitted but this is not weighted towards the second derivative at the minimum) is limiting the quality of the mechanical properties. To stress the importance of the inclusion of second derivative checks during fitting, the recent reactive force-field for the perovskite material $BaZrO_3$ (van Duin et al. 2008) was fitted with the aim of stabilizing the experimental cubic form, but on examination of the phonons was found to give a distorted ground state structure.

Before leaving the topic of silica and reactive force-fields, it is also worth noting the work from the Garofalini group (Mahadevan and Garofalini 2007). Here they have developed a fixed charge model for water that allows for dissociation of the O-H bond, built on earlier water models that have a Gaussian charge distribution for the particles. Although the formal charge states for proton transfer in the gas phase are incorrect, this is not an issue for proton transfer in condensed phases where the charge remains distributed. Hence, by using a temperature and pressure dependent short-range potential for one interaction it has been possible to obtain good results for water over a range of conditions. Of particular relevance here is that they have been able to simulate the reactivity of water with hydroxylated surfaces of silica (Mahadevan and Garofalini 2008a,b), showing that the force-field calculations accord well with the results of first principles simulations. This may represent a promising future direction for the study of water incorporation in minerals without the complexity of some other models.

APPLICATIONS

In this section we will briefly highlight some of the applications involving lattice dynamics in the study of minerals, with particular emphasis on high pressure/temperature environments.

Some of the earliest uses of force-field calculations in mineral physics were aimed at elucidating the properties and defect behavior of mantle minerals, in particular the α, β, and γ polymorphs of Mg_2SiO_4. Price et al. (1987b) developed the so-called THB1 forcefield, combining parameters derived for Mg- and Si-oxides with the 3-body O-Si-O term of Sanders et al. (1984) and a shell model for oxygen polarizibility. This forcefield has been hugely successful and widely applied to a range of Mg-Si-O minerals. Initially, THB1 was used to predict the lattice dynamics of the three polymorphs. The calculated thermodynamic properties were in good agreement with experiment although the 0 K phase transition pressures were too high and too low for the α-β and β-γ transitions, respectively. Price et al. (1987a) attributed this failure to the inability of the model to accurately simulate the Si_2O_7 dimer in the β-phase. Despite the limitations of this model, it has been used extensively to investigate the properties of mantle silicates.

One area of particular importance where force-field calculations have played a major role is that of understanding defects and defect processes in minerals. Studies range from investigating the controls on uptake of impurities by mantle silicates (Wright and Catlow 1994; Purton et al. 1996, 1997; Gatzemeier and Wright 2006), though determination of diffusion migration mechanisms (Walker et al. 2003; Bejina et al. 2009) to modelling the cores of dislocations (Walker et al. 2004, 2005).

Simulation of hydrogen defects in the nominally anhydrous minerals (NAMs) of the Earth's mantle has been ongoing for almost 20 years. Most NAMs contain some water, present as hydrogen defects bound to lattice oxygen. These defects can profoundly influence the physical and mechanical properties of NAMs, even when present in only trace amounts. The identity of the charge compensating defects in NAMs had been the subject of considerable debate in the literature, as has the extent to which these defects bind with the hydrogen. Forsterite is the most extensively studied of all the NAMs, and has been investigated using, DFT, force-field, and hybrid QM/MM approaches (see Wright 2006 for a review). The results are surprisingly consistent between the different methods. All show that in the case of pure forsterite, hydrogen incorporation via a hydrogarnet type defect, as illustrated in Figure 1, is energetically favoured over mechanisms involving magnesium defects. However, more recent DFT calculations on Ti-bearing forsterite (Walker et al. 2007) indicate that impurities could facilitate uptake of hydrogen.

The partitioning of impurities between different solid phases and between solid and melts is relevant to many processes occurring within the Earth's crust and mantle. Purton et al. (1996,

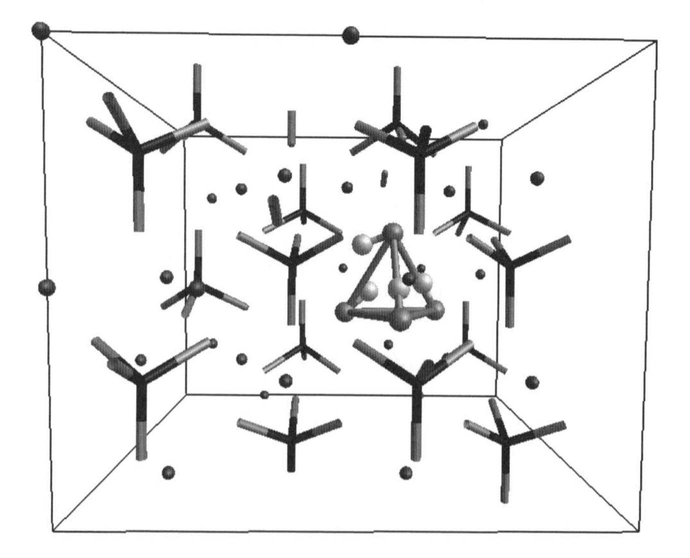

Figure 1. Calculated structure for the hydrogarnet defect in forsterite at 10 GPa. The atoms shown in ball and stick mode represent the hydroxyl group nest about the vacant silicon site, while other atoms are shown purely in stick mode for clarity.

1997) performed calculations on both isovalent and heterovalent substitutions in forsterite, diopside and enstatite to obtain solution energies and provide insights into the controls on impurity incorporation. Their work showed that solution energies show a parabolic dependence on ionic radii and in the case of heterovalent substitutions, the solution energy of a trivalent cation was dependent on the nature of the charge compensating species. Watson et al. (2000) investigated the defect chemistry of $MgSiO_3$ perovskite at high pressure and its ability to form solid solutions with $CaSiO_3$. At the time, experimental observations gave conflicting results. The simulations indicated that, under lower mantle conditions, solution of Ca into $MgSiO_3$ would be extremely unfavorable, and any mixture would exsolve into two distinct phases.

More recently, the study of defects in minerals using computational methods has been extended to cover the structure of dislocations, where force-field calculations (e.g., Walker et al. 2004, 2005) are complemented by studies using DFT (e.g., Durinck et al. 2005). Modelling of dislocation cores is an application that requires the use of very large simulations cells containing hundreds of thousands of atoms where it is necessary to use the Wolf sum and domain decomposition algorithms. In addition, short-range cut-offs are reduced to keep computational costs to a minimum. The output from these calculations can give insights into the atom scale displacements around the dislocation core that would not always be apparent from experimental observation. For example, introduction of a screw dislocation into the structure of a simple zeolite material (Fig. 2) changes the channel structure such that molecular transport will be enhanced in the direction of the dislocation line, but retarded normal to the line (Walker et al.

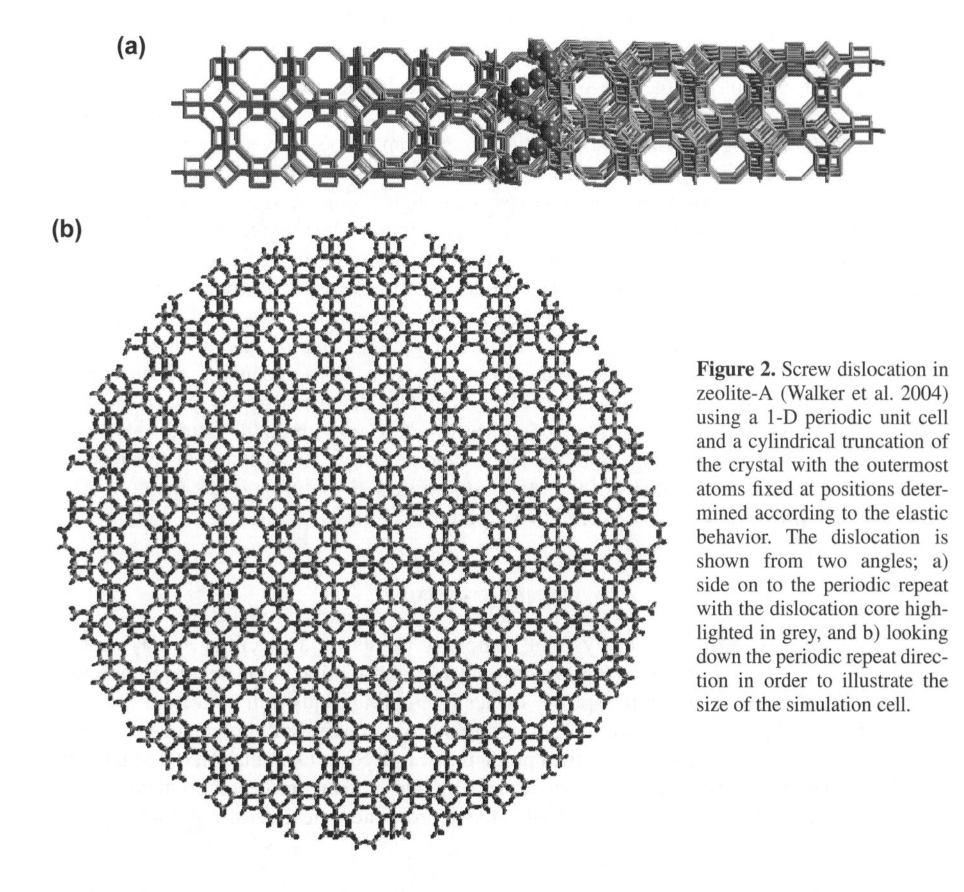

Figure 2. Screw dislocation in zeolite-A (Walker et al. 2004) using a 1-D periodic unit cell and a cylindrical truncation of the crystal with the outermost atoms fixed at positions determined according to the elastic behavior. The dislocation is shown from two angles; a) side on to the periodic repeat with the dislocation core highlighted in grey, and b) looking down the periodic repeat direction in order to illustrate the size of the simulation cell.

2004). This has important implications for the use of zeolitic materials as molecular sieves.

So far in this section we have only looked at silicate minerals. Although these, along with carbonates, are by far the most commonly simulated minerals, there are many studies on other important Earth materials including oxides, phosphates, sulfates and sulfides using force-field methods.

Studies of sulfide minerals using force-fields are limited as they commonly exhibit a high degree of covalent or metallic bonding with complex magnetic states. There are, however, parameter sets for ZnS (Hamad et al. 2002; Wright and Gale 2004) and for FeS_2 (Sithole et al. 2003). Both systems exhibit polymorphism, with two phases existing that are very close in energy. In the ZnS system, cubic sphalerite is more stable than wurtzite up to temperatures of around 1200 °C and although the correct order can be obtained at low temperature by the use of a four-body parameter (Wright and Gale 2004), the order is not reversed at high temperature. DFT methods can also have difficulty with correctly determining the energy of such polymorphs as demonstrated for the pyrite and marcasite phases of FeS_2 (Spagnoli et al. in prep). The potentials of Sithole et al. (2003) predict that pyrite will be more stable than marcasite by 3.4 $kJmol^{-1}$, in excellent agreement with the experimental value of 3.9 $kJmol^{-1}$ (Lennie and Vaughan 1992). The high pressure behavior of pyrite is of interest due to the presence of Fe-S in the cores of planetary interiors. Experimental data on pyrite compression is variable and apparently dependent on experimental stress conditions although experiments by Merkel et al. (2002) have overcome these problems. The predicted elasticity of pyrite up to 44GPa from simulation (Sithole et al. 2003) is in good agreement with the data of Merkel et al. (2002) Furthermore, the simulations predict that marcasite behaves in a similar fashion to pyrite.

As a final example of the use of lattice energy calculations, we consider the case of cation disorder and substitutions within minerals. Here the rate of diffusion is typically slow and therefore sampling of order-disorder phenomena cannot be efficiently conducted using molecular dynamics. The approach taken to such problems highlights the power of combining several different computational techniques to tackle a single problem. As described by Vinograd et al. (2004, 2007), the foundation is to generate a force-field model that accurately describes the structure and properties of the ordered end-member minerals, as validated against experiment and first principles methods. This force-field model can then be used to compute the energy, or even free energy, of a range of cation distributions within a supercell description. Despite the speed of interatomic potentials it remains a challenging task to explore the enormous number of configurations possible over a wide range of composition. Consequently, the use of cluster expansion methods, by training against the force-field data, can dramatically extend the scope of the calculations, making it feasible to perform Monte Carlo simulations to determine the free energy of mixing over a range of temperatures and pressures. As an example, this has been used to study the thermodynamics of the pyrope-majorite solid solution in the range of 1073-3673 K and at pressures of up to 20 GPa (Vinograd et al. 2006).

CONCLUSION

Calculations based on the analysis of stationary points on a potential energy surface, coupled to lattice dynamics, have long made a valuable contribution in the field of mineralogy. Although the increase in computing power makes it tempting to just run molecular dynamics for everything, lattice dynamics still has an important role to play, both in the development of new force-fields and in making theoretical predictions. Indeed an examination of the potential energy surface and key barrier heights should be a precursor to any molecular dynamics study for a solid phase in order to determine what is the appropriate timescale of relevance.

Into the future the complexity of force-field models will undoubtedly increase as interatomic potentials seek to invade the traditional preserves of quantum mechanics, such as reactivity. The price that can often be paid for this is the loss of the intuitive connection between functional form and physical motivation, while parameters become increasingly coupled. Arguably a more promising route is that of increased exploitation of empirical valence bond theory. Here the underlying simplicity of the description of molecular fragments can be retained whilst allowing more correctly for the crossing of potential energy curves associated with different states. However things develop, the use of force-fields, informed by quantum mechanical information, will remain a powerful tool for the study of minerals.

ACKNOWLEDGMENTS

The authors gratefully acknowledge the support of the Australian Research Council through the Discovery program. We would also like to thank Andrew Walker for providing the figures of the screw dislocation in zeolite-A, as well as Adri van Duin for valuable discussions regarding ReaxFF.

REFERENCES

Alavi A, Alvarez LJ, Elliott SR, McDonald IR (1992) Charge-transfer molecular-dynamics. Philos Mag B 65:489-500

Banerjee A, Adams N, Simons J, Shepard R (1985) Search for stationary-points on surface. J Phys Chem 89:52-57

Baroni S, de Gironcoli S, Dal Corso A, Giannozzi P (2001) Phonons and related crystal properties from density-functional perturbation theory. Rev Mod Phys 73:515-562

van Beest BWH, Kramer GJ, van Santen RA (1990) Force-fields for silicas and aluminophosphates based on ab initio calculations. Phys Rev Lett 64:1955-1958

Bejina F, Blanchard M, Wright K, Price GD (2009) A computer simulation study of the effect of pressure on Mg diffusion in forsterite. Phys Earth Planet Int 172:13-19

Bell RG, Jackson RA, Catlow CRA (1990) Computer-simulation of the monoclinic distortion in silicalite. J Chem Soc-Chem Comm 782-783

Bilic A, Gale JD (2009) Ground state structure of $BaZrO_3$: A comparative first-principles study. Phys Rev B 79:174107

Bowman JM (1986) The self-consistent-field approach to polyatomic vibrations. Acc Chem Res 19:202-208

Brenner DW, Shenderova OA, Harrison JA, Stuart SJ, Ni B, Sinnott SB (2002) A second-generation reactive empirical bond order (REBO) potential energy expression for hydrocarbons. J Phys-Cond Matter 14:783-802

Cankurtaran BO, Gale JD, Ford MJ (2008) First principles calculations using density matrix divide-and-conquer within the SIESTA methodology. J Phys-Cond Matter 20:294208

Car R., Parrinello M (1985) Unified approach for molecular-dynamics and density-functional theory. Phys Rev Lett 55:2471-2474

Catlow CRA, Mackrodt WC (1982) Theory of simulation methods for lattice and defect energy calculations in crystals. Lecture Notes in Physics 166:3-20

Catlow CRA, Thomas JM, Parker SC, Jefferson DA (1982) Simulating silicate structures and the structural chemistry of pyroxenoids. Nature 295:658-662

Chenoweth K, van Duin ACT, Persson P, Cheng MJ, Oxgaard J, Goddard WA (2008) Development and application of a ReaxFF reactive force field for oxidative dehydrogenation on vanadium oxide catalysts. J Phys Chem C 112:14645-14654

Couves JW, Jones RH, Parker SC, Tschaufeser P, Catlow CRA (1993) Experimental-verification of a predicted negative thermal expansivity of crystalline zeolites. J Phys-Cond Matter 5:L329-L332

Deleeuw SW, Perram JW, Smith ER (1980) Simulation of electrostatic systems in periodic boundary-conditions. 1. Lattice Sums and dielectric-constants. Proc Roy Soc London A 373:27-56

Dewar MJS, Healy EF, Stewart JJP (1984) Location of transition-states in reaction-mechanisms. J Chem Soc-Faraday Trans II 80:227-233

Dick BG, Overhauser AW (1958) Theory of the dielectric constants of alkali halide crystals. Phys Rev 112:90-103

Durinck J, Legris A, Cordier P (2005) Pressure sensitivity of olivine slip systems: first-principle calculations of generalised stacking faults. Phys Chem Miner 32:646-654
Ewald PP (1921) Die berechnung optischer und elektrostatisher gitterpotentiale. Ann Phys 64:253-287
Gale JD (1996) Empirical potential derivation for ionic materials. Philos Mag B 73:3-19
Gale JD (1997) GULP: A computer program for the symmetry-adapted simulation of solids. J Chem Soc-Faraday Trans 93:629-637
Gale JD (1998) Analytical free energy minimization of silica polymorphs. J Phys Chem B 102:5423-5431
Gale JD, Rohl AL (2003) The General Utility Lattice Program (GULP). Mol Simul 29:291-341
Gasteiger J, Marsili M (1980) Iterative partial equalization of orbital electronegativity - a rapid access to atomic charges. Tetrahedron 36:3219-3228
Gatzemeier A, Wright K (2006) Computer modelling of hydrogen defects in the clinopyroxenes diopside and jadeite. Phys Chem Miner 33:115-125
Gonze X, Allan DC, Teter MP (1992) Dielectric tensor, effective charges, and phonons in alpha-quartz by variational density-functional perturbation-theory. Phys Review Lett 68:3603-3606
Hamad S, Cristol S, Catlow CRA (2002) Surface structures and crystal morphology of ZnS: Computational study. J Phys Chem B 106:11002-11008
Henkelman G, Uberuaga BP, Jonsson H (2000) A climbing image nudged elastic band method for finding saddle points and minimum energy paths. J Chem Phys 113:9901-9904
Jiang Z, Brown RA (1995) Atomistic calculation of oxygen diffusivity in crystalline silicon. Phys Rev Lett 74:2046-2049
Kantorovich LN (1995) Thermoelastic properties of perfect crystals with nonprimitive lattices .1. General-theory. Phys Rev B 51:3520-3534
Laio A, Parrinello M (2002) Escaping free-energy minima. Proc Natl Acad Sci 99:12562-12566
Lennie AR, Vaughan DJ (1992) Kinetics of the marcasite-pyrite transformation - an infrared spectroscopic study. Am Mineral 77:1166-1171
Lindan PJD, Gillan MJ (1993) Shell-model molecular-dynamics simulation of superionic conduction in CaF_2. J Phys-Cond Matter 5:1019-1030
Madden PA, Wilson M (1996) 'Covalent' effects in 'ionic' systems. Chem Soc Rev 25:339-350
Mahadevan TS, Garofalini SH (2007) Dissociative water potential for molecular dynamics simulations. J Phys Chem B 111:8919-8927
Mahadevan TS, Garofalini SH (2008a) Dissociative chemisorption of water onto silica surfaces and formation of hydronium ions. J Phys Chem C 112:1507-1515
Mahadevan TS, Garofalini SH (2008b) Dissociative chemisorption of water onto silica surfaces and formation of hydronium ions (vol 112, 1514,2008). J Phys Chem C 112:5694-5694
Martonak R, Donadio D, Oganov AR, Parrinello M (2006) Crystal structure transformations in SiO_2 from classical and ab initio metadynamics. Nature Mater 5:623-626
Martonak R, Donadio D, Oganov AR, Parrinello M (2007) From four- to six-coordinated silica: Transformation pathways from metadynamics. Phys Rev B 76:014120
Merkel S, Jephcoat AP, Shu J, Mao HK, Gillet P, Hemley RJ (2002) Equation of state, elasticity, and shear strength of pyrite under high pressure. Phys Chem Miner 29:1-9
Nocedal J (1980) Updating quasi-Newton matrices with limited storage. Mathematics of Computation 35:773-782
Parker SC (1983) Prediction of mineral crystal-structures. Solid State Ionics 8:179-186
Parker SC, Catlow CRA, Cormack AN (1983) Prediction of mineral structure by energy minimization techniques. J Chem Soc-Chem Commun 936-938
Parry DE (1975) Electrostatic potential in surface region of an ionic-crystal. Surf Sci 49:433-440
Price GD, Parker SC (1989) Computer modelling of phase transitions in minerals. Adv Sol State Chem 1:295-327
Price GD, Parker SC, Leslie M (1987a) The lattice-dynamics and thermodynamics of the Mg_2SiO_4 polymorphs. Phys Chem Miner 15:181-190
Price GD, Parker SC, Leslie M (1987b) The lattice-dynamics of forsterite. Mineral Mag 51:157-170
Purton JA, Allan NL, Blundy JD (1997) Calculated solution energies of heterovalent cations in forsterite and diopside: Implications for trace element partitioning. Geochim Cosmochim Acta 61:3927-3936
Purton JA, Allan NL, Blundy JD, Wasserman EA (1996) Isovalent trace element partitioning between minerals and melts: A computer simulation study. Geochim Cosmochim Acta 60:4977-4987
Rappe AK, Casewit CJ, Colwell KS, Goddard III WA, Skiff WM (1992) UFF, a full periodic table force field for molecular mechanics and molecular dynamics simulations. J Am Chem Soc 114:10024-10035
Rappe AK, Goddard WA (1991) Charge equilibration for molecular-dynamics simulations. J Phys Chem 95:3358-3363
Sanders MJ, Leslie M, Catlow CRA (1984) Interatomic potentials for SiO_2. J Chem Soc-Chem Commun 1271-1273
Sanderson RT (1951) An interpretation of bond lengths and a classification of bonds. Science 114:670-672

Saunders VR, Freyriafava C, Dovesi R, Roetti C (1994) On the electrostatic potential in linear periodic polymers. Comp Phys Commun 84:156-172

Shanno DF (1970) Parameter selection for modified Newton methods for function minimization. Siam J Numerical Anal 7:366-372

Sithole HM, Ngoepe PE, Wright K (2003) Atomistic simulation of the structure and elastic properties of pyrite (FeS_2) as a function of pressure. Phys Chem Miner 30:615-619

Soler JM, Artacho E, Gale JD, Garcia A, Junquera J, Ordejon P, Sanchez-Portal D (2002) The SIESTA method for ab initio order-N materials simulation. J Phys-Cond Matter 14:2745-2779

Tang KT, Toennies JP (1984) An improved simple-model for the van der Waals potential based on universal damping functions for the dispersion coefficients. J Chem Phys 80:3726-3741

Taylor MB, Barrera GD, Allan N, Barron THK, Mackrodt WC (1998) Shell: A code for lattice dynamics and structure optimisation of ionic crystals. Comp Phys Commun 109:135-143

Tsuneyuki S, Tsukada M, Aoki H, Matsui Y (1988) 1st-principles interatomic potential of silica applied to molecular-dynamics. Phys Rev Lett 61:869-872

van Duin ACT, Dasgupta S, Lorant F, Goddard WA (2001) ReaxFF: A reactive force field for hydrocarbons. J Phys Chem A 105:9396-9409

van Duin ACT, Merinov BV, Han SS, Dorso CO, Goddard WA (2008) ReaxFF reactive force field for the Y-doped $BaZrO_3$ proton conductor with applications to diffusion rates for multigranular systems. J Phys Chem A 112:11414-11422

van Duin ACT, Strachan A, Stewman S, Zhang QS, Xu X, Goddard WA (2003) ReaxFF(SiO) reactive force field for silicon and silicon oxide systems. J Phys Chem A 107:3803-3811

Vashishta P, Kalia RK, Rino JP, Ebbsjo I (1990) Interaction potential for SiO_2 - a molecular-dynamics study of structural correlations. Phys Rev B 41:12197-12209

Vinograd VL, Gale JD, Winkler B (2007) Thermodynamics of mixing in diopside-jadeite, $CaMgSi_2O_6$-$NaAlSi_2O_6$, solid solution from static lattice energy calculations. Phys Chem Miner 34:713-725

Vinograd VL, Sluiter MHF, Winkler B, Putnis A, Halenius U, Gale JD, Becker U (2004) Thermodynamics of mixing and ordering in pyrope-grossular solid solution. Mineral Mag 68:101-121

Vinograd VL, Winkler B, Putnis A, Kroll H, Milman V, Gale JD, Fabrichnaya OB (2006) Thermodynamics of pyrope-majorite, $Mg_3Al_2Si_3O_{12}$-$Mg_4Si_4O_{12}$, solid solution from atomistic model calculations. Mol Simul 32:85-99

Walker AM, Gale JD, Slater B, Wright K (2005) Atomic scale modelling of the cores of dislocations in complex materials part 2: applications. Phys Chem Chem Phys 7:3235-3242

Walker AM, Hermann J, Berry AJ, O'Neill HS (2007) Three water sites in upper mantle olivine and the role of titanium in the water weakening mechanism. J Geophys Res 112, doi:10.1029/2006JB004620

Walker AM, Slater B, Gale JD, Wright K (2004) Predicting the structure of screw dislocations in nanoporous materials. Nat Mater 3:715-720

Walker AM, Wright K, Slater B (2003) A computational study of oxygen diffusion in olivine. Phys Chem Miner 30:536-545

Wall A, Parker SC, Watson GW (1993) The extrapolation of elastic-moduli to high-pressure and temperature. Phys Chem Miner 20:69-75

Watson GW, Wall A, Parker SC (2000) Atomistic simulation of the effect of temperature and pressure on point defect formation in $MgSiO_3$ perovskite and the stability of $CaSiO_3$ perovskite. J Phys-Cond Matter 12:8427-8438

Wolf D, Keblinski P, Phillpot SR, Eggebrecht J (1999) Exact method for the simulation of Coulombic systems by spherically truncated, pairwise r^{-1} summation. J Chem Phys 110:8254-8282

Woodcock LV, Angell CA, Cheeseman P (1976) Molecular-Dynamics Studies of Vitreous State - Simple Ionic Systems and Silica. J Chem Phys 65:1565-1577

Wright K (2006) Atomistic models of OH defects in nominally anhydrous minerals. Rev Mineral Geochem 62:67-83

Wright K, Catlow CRA (1994) A computer-simulation study of (Oh) defects in olivine. Phys Chem Miner 20:515-518

Wright K, Gale JD (2004) Interatomic potentials for the simulation of the zinc-blende and wurtzite forms of ZnS and CdS: Bulk structure, properties, and phase stability. Phys Rev B 70:035211

Yasukawa A (1996) Using an extended Tersoff interatomic potential to analyze the static-fatigue strength of SiO_2 under atmospheric influence. JSME Int J Series A-Mechanics and Material Engineering 39:313-w320

Yu JG, Sinnott SB, Phillpot SR (2007) Charge optimized many-body potential for the Si/SiO_2 system. Phys Rev B 75:085311

Reviews in Mineralogy & Geochemistry
Vol. 71 pp. 413-436, 2010

An Efficient Cluster Expansion Method for Binary Solid Solutions: Application to the Halite-Silvite, NaCl-KCl, System

Victor Vinograd and Björn Winkler

Institute of Geosciences
University of Frankfurt
Frankfurt a.M., Germany

v.vinograd@kristall.uni-frankfurt.de b.winkler@kristall.uni-frankfurt.de

ABSTRACT

The system NaCl-KCl is used as an example to introduce the reader to a variety of concepts, which form the modern solid solution theory. The chapter starts with the discussion of the chemical stress both end members experience in the process of mixing. This stress is treated as a homogeneous deformation of the end members within the concept of the virtual crystal approximation, VCA. It is shown that the VCA predicts the existence of a positive excess enthalpy in the NaCl-KCl solid solution, but drastically overestimates its magnitude. The inaccuracy of the VCA is attributed to its failure to describe local relaxations. Further, the local relaxations are linked to the effects of intracrystalline reactions of NaNa + KK = 2NaK type, and it is shown that the excess enthalpy can be written as a sum of such interactions acting at various distances within the supercell. The decomposition of the enthalpy of mixing into pairwise interactions is then discussed in the relation to a more general cluster expansion technique, which considers effective interactions from clusters of various sizes and shapes including many-body interactions. Then we argue that a wide range of materials interesting for geosciences can be described within the concept of pairwise interactions only and introduce the supercell expansion method dealing with the effective pairwise interactions. It is then shown that the calculation of the effective pairwise interactions can be performed with different methods. The traditional approach consisting in the calculation of the static energies of a large number of randomly varied structures and solving for the pairwise interactions with the least squares method is compared to a deterministic algorithm, which is based on the consideration of the excess energies of the structures with double defects inserted at all possible pairwise distances within the supercell. This Double Defect Method, DDM, is then applied to the system NaCl-KCl. The effective interactions obtained are used to construct an Ising-type Hamiltonian, from which the temperature dependent mixing enthalpy is evaluated with the Monte Carlo method. Then the temperature-dependent enthalpies are thermodynamically integrated to obtain the free energies of mixing and to construct the temperature-composition diagram. A comparison of the calculated diagram with the experimentally known phase relations shows that a certain thermodynamic effect is missed in the simulations. This effect is identified as the excess vibrational entropy and the DDM calculations are repeated within the quasi-harmonic approximation, which evaluates both the static and the vibrational components of the free energy. These calculations are used to derive the temperature-dependent pairwise interactions, which via the Monte Carlo simulations lead to a much better agreement with the experiment. Finally we discuss applications of the DDM approach to solid solutions with coupled substitution and to multicomponent solutions.

1529-6466/10/0071-0019$05.00 DOI: 10.2138/rmg.2010.71.19

INTRODUCTION

This chapter is concerned with the thermodynamics of mixing in solid solutions. Current research in this field is undergoing a rapid transition from traditional descriptive models, tightly linked to experimental data, to predictive models based on atomistic simulations, which do not require direct experimental input. Our aim is to outline the progress in the field of atomistic simulations and to show how the available simulation tools can be applied to study phase separation in the system NaCl-KCl. Starting from this simple example, we will discuss approaches to modeling of solid solutions with more complex substitution schemes.

The aim of any atomistic simulation study of mixing is to interpolate between the properties of the chemically different end members. The simplest approach is the direct proportional averaging of the properties of the end members, which leads to the "virtual crystal approximation" (VCA). To be able to mix the end members, e.g., KCl and NaCl, with large and small volumes, respectively, one needs to deform the structures of both end members such that their volumes are equal at a given composition. Alternatively, one can understand this process such that the cations K^+ and Na^+ with different ionic radii are hybridized producing a cation of intermediate size. In force-field models the hybridization is achieved by averaging parameters of the interatomic potentials (Winkler et al. 1991), while in quantum mechanical calculations this is done by introducing a special "averaging" operator (Bellaiche and Vanderbilt 2000). This approach is very convenient because after the properties of the hybrid atom are defined, the solid solution phase can be treated as a pure phase. The VCA is often able to predict reasonable average properties of a solid solution system with given composition (cell parameters, bond distances, bulk moduli) (e.g., Winkler et al. 2002, 2004; Vinograd et al. 2007; Wilson et al. 2008). However, it is inadequate, when the aim is the prediction of the thermodynamic mixing functions, such as the enthalpy, entropy and free energy.

To see why it does not work, let us consider the enthalpy of mixing of a binary NaCl-KCl mixture. In the virtual crystal approximation the enthalpy of mixing arises due to the homogeneous deformation of the lattices of the end members. The volume of KCl is reduced by $|V_x - V_{KCl}|$, while the volume of NaCl is expanded by $|V_x - V_{NaCl}|$. Both these deformations require an increase in the lattice energy, as the end members are deformed away from their equilibrium volumes. Assuming that at small deformations around the equilibrium volume the energy is a parabolic function of volume change, the energy increase due to the deformation of the end member i is

$$\Delta E_i = \tfrac{1}{2}\frac{d^2E_i}{dV^2}(V_x - V_i)^2 = \tfrac{1}{2}K_iV_i\left(\frac{V_x - V_i}{V_i}\right)^2 \quad (1)$$

where K_i and V_i are the bulk modulus and the standard volume of the end member i and V_x is the volume of the solid solution with the composition x. When the solid solution obeys Vegard's law (the lattice parameter linearly depends on the composition, $a = a_i + \Delta a(1 - x_i)$), Equation (1) can be written as follows

$$\Delta E_i \approx \tfrac{1}{2}K_iV_i\left(\frac{3\Delta a(1 - x_i)}{a_i}\right)^2 \quad (2)$$

where a_i is the lattice parameter of the end member i and Δa is the difference between the lattice parameters of the end members. By summing the energy effects proportionally to the end member mole fractions, the contribution of the volume deformation energy to the enthalpy of mixing of the solid solution at zero pressure can be written as

$$\Delta H_{VD} = x_A x_B^2 W_A + x_B x_A^2 W_B \quad (3)$$

where

$$W_i = \tfrac{9}{2} K_i V_i \left(\frac{\Delta a}{a_i} \right)^2 \tag{4}$$

Using K_{KCl} = 18.1 GPa, K_{NaCl} = 24.9 GPa (Hearmon 1979), a_{NaCl} = 5.64 Å, a_{KCl} = 6.295 Å (Wyckoff 1963) and $x_{\text{K}} = x_{\text{Na}} = 0.5$, one obtains ΔH_{VD} = 9.3 kJ. Thus the enthalpy of the virtual crystal is larger by this amount than that of the mechanical mixture of NaCl and KCl. The problem is that the calculated effect exceeds the experimental value of the enthalpy of mixing of the disordered solution (Barrett and Wallace 1954) by a factor of 2, meaning that the VCA greatly overestimates the amount of the deformation energy accumulated within the solid solution.

The reason is that the VCA does not consider local relaxation. For example, in the NaCl-structure, second nearest neighbor pairs of Na-Na or K-K cations interact through the Cl^- anions. The structure is particularly rigid along this direction, and thus at intermediate compositions (and intermediate cation-cation distances) the energy of Na-Na and K-K pairs increases. This effect is included in the value of 9.3 kJ/mol calculated using the VCA. However, the VCA does not take into account a different behavior of pairs of Na-K type, which inevitably form in the solid solution. Indeed, the joining of large and small cations creates the possibility for Cl^- anions to move closer to Na^+ and thus to keep Na-Cl and K-Cl distances closer to their relaxed values in pure NaCl and KCl as shown in Figure 1. This allows to minimize the stress locally and thus to decrease the excess enthalpy. Figure 1 also illustrates the point that the ability of a certain configuration to relax is proportional to the number of dissimilar AB-type pairs (in this example Na-K). The "right" configuration in Figure 1 is more "relaxed" than the "left" configuration as it contains more pairs of AB type.

This example is important in two aspects. First, it shows that the excess enthalpy in an (A,B)*R* solid solution can be associated with the stress the end members A*R* and B*R* experience due to the need of geometrical adjustment to each other. Second, it suggests that this stress can be reduced by local relaxations due to the formation of pairs of dissimilar atoms, which more easily adjust to the average lattice at an intermediate composition. The formation of AB-type pairs from pairs of AA- and BB-types occurs via the intracrystalline reactions

$$\text{AA} + \text{BB} = 2\text{AB} \tag{5}$$

Naturally, these reactions play an important role in the solid solution theory.

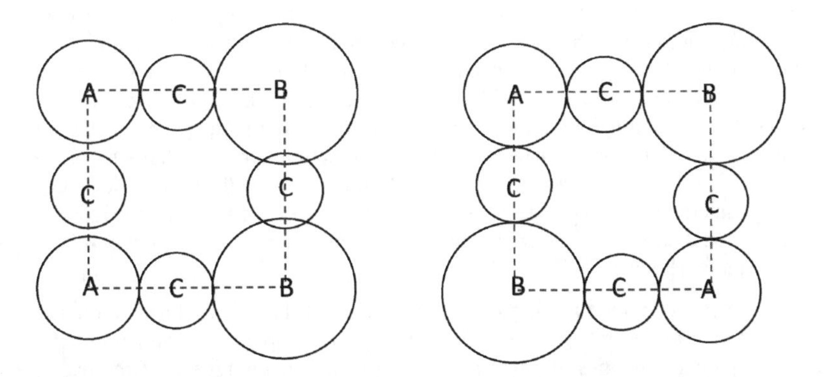

Figure 1. The unrelaxed (*left*) and the relaxed (*right*) configurations with size mismatch. The relaxation effect can be attributed to the reaction AA + BB = 2 AB. A, B and C emulate Na, K and Cl, respectively, in the (Na,K)Cl structure. Dashed lines show the second neighbor pairs.

The importance of such reactions has been recognized in the studies of ferromagnetism, where it was shown that the spontaneous magnetization can be modeled by considering a system of regularly arranged atoms with up (U) and down (D) spins, where the pairwise interactions of UU + DD = 2 UD type favor their parallel alignment (Ising 1925; Bragg and Williams 1934; Bethe 1935; Onsager 1944; Cowley 1950). It was recognized that the pairwise interactions cause a shift in the probabilities of pairs of AA,BB and AB types, P_{ij}, from the values, which can be expected for a statistically random distribution of the exchangeable particles, the phenomenon known as the short-range ordering (SRO). For example, the relation $P_{AB} = P_A P_B$ is not valid when the enthalpy of the reaction (5) deviates from zero. The SRO can greatly affect both the configurational enthalpy and the entropy of a system with disorder. While the effect of SRO on the configurational enthalpy can be easily evaluated in terms of the pairwise probabilities and the pairwise interactions (see below), it is not so for the configurational entropy. A general algorithm for the derivation of analytical equations for the entropy of systems with SRO for lattices with different space group symmetry (the Cluster Variation Method) was developed by Kikuchi (1951). It has also become clear that the reactions (5) can formally explain the sign of the excess mixing effects in solid solutions: depending on the sign of the reaction enthalpy, the enthalpy of mixing can be positive or negative (Guggenheim 1935). Indeed, it is easy to show that in the symmetric case, $W_A = W_B$, the volume deformation energy can be written in the form $\Delta H_{VD} = x_A x_B W$, and thus shown to be proportional to the fraction of AB-type pairs in the disordered solution. When one further assumes that only one type of pairwise interactions is important (e.g., the nearest neighbor interactions), the sign of the excess enthalpy will be determined by the sign of W. It has been recognized, however, that many solid solutions that show positive enthalpy of mixing at high temperatures can show negative enthalpies at certain intermediate compositions at lower temperatures. The latter effects indicate long-range ordering, i.e. the formation of supercell structures. It was shown that the combination of phase separation and ordering tendencies can be modeled by attributing effects of different signs to the effective interactions acting at different distances (e.g., Van Baal 1973; Sanchez and de Fontaine 1980; Burton and Kikuchi 1984; Mohri et al. 1985; Inden and Pitsch 1991). The success of such studies has motivated the development of a general algorithm which allowed to split the total excess effect into the effective cluster interactions in solid solutions (not necessarily limited to pairwise interactions) called the cluster expansion method (Connolly and Williams 1983; Sanchez et al. 1984). The applications of this method were mainly focused on alloys. In the method of Connolly and Williams (1983) the effective cluster interactions (ECI) were derived from volume dependent energies of a few ordered structures with intermediate compositions. The resulting ECIs were also volume dependent. Later Ferreira et al. (1988) and Wei et al. (1990) showed that when the equilibrium volume of an alloy is approximately independent of the state of order, the volume dependence of the effective interactions can be separated out as a uniform composition dependent function, G, which gives a positive contribution to the mixing enthalpy, while the remaining chemical energy, ε, can be expanded in terms of volume independent effective interactions of different signs associated with clusters of various shape. The composition dependent function G can be identified with the volume deformation energy discussed above. It has been shown that the "G–ε" theory often permits a better prediction of phase relations in systems with size mismatch (Ferreira et al. 1988, Wei et al. 1990) relative to the theories, which consider only the chemical interactions. Later, the "G–ε" theory was succeeded by a more rigorous concept of the "constituent strain energy," which took into account anisotropy of the elastic properties of a solid solution (Laks et al. 1992).

Although the latter approach is considered the state-of-the-art in solid solution theory (e.g., Wolverton et al. 2000; Barabash et al. 2008), its practical implementation involving prior calculation of epitaxial energies and the expansion of the remaining part of the excess energy in pairwise interactions formulated in the reciprocal space is rather difficult. Moreover, the constituent-strain formalism (Laks et al. 1992) is developed for cubic system only. Therefore, the search for efficient cluster expansion algorithms is continuing. The present study exploits a recently proposed easy-to-do expansion named the Double Defect Method (Vinograd et al. 2009a), which is discussed below.

DECOMPOSITION OF THE EXCESS ENTHALPY INTO PAIRWISE INTERACTIONS

Let us consider a supercell created from the unit cell of an end member AR, by increasing the cell dimensions by integer factors, e.g., 2×2×2. Let us also require that the size of this supercell is sufficient to include the most important ordering schemes. This is equivalent to requiring that the supercell includes pairs of the exchangeable atoms up to the distance $L_{n_{\max}}$, while the strength of the effective interactions between the exchangeable atoms is decreased to effectively zero values when n approaches $n_{\max}$. The last condition ensures that the effective interactions determined using the supercell of the given size can be used to predict the excess enthalpies in an infinitely large system. The supercell is thus the minimal part of the lattice that samples the most important ordering interactions. Under the assumption that the interactions are pairwise additive (many-body interactions are ignored), the excess energy of any configuration in a supercell of an (A,B)R solid solution can be associated with interactions between the exchangeable (A,B) atoms only. The interactions of A and B with the remainder of the structure R will cancel out when the energy of the mechanical mixture of AR and BR end-members is subtracted from the energy of the configuration. The excess enthalpy of a configuration k with the composition $x_A = P_A$ can be written as:

$$\Delta H_k = \tfrac{1}{2} N \sum_n Z_n \Big(\sum_{i,j} P_{ij(n,k)} H_{ij(n)} - \sum_i P_i H_{ii(n)} \Big) \tag{6}$$

where N is the number of the exchangeable atoms in the supercell, Z_n is the coordination number (the number of neighbors at the n-th distance, $P_{ij(n,k)}$ is the probability of finding an ij pair at the n-th distance in the supercell, and $H_{ij(n)}$ is the effective (average) interaction energy between atoms i and j at the same distance. The energy of the mechanical mixture (the last sum in Eqn. 6) is represented with AA and BB pairs only, which occur with the probabilities P_A and P_B, respectively. Noting that the single site probabilities, P_i, can be written as sums of the pair probabilities, $P_A = P_{AB(n)} + P_{AA(n)}$ and $P_B = P_{BA(n)} + P_{BB(n)}$, Equation (6) can be rewritten as follows

$$\Delta H_k = \sum_n f_{AB(n,k)} J_n \tag{7}$$

where $f_{AB(n,k)} = \tfrac{1}{2} N Z_n P_{AB(n,k)}$ is the number of AB pairs at the n-th distance in this structure and

$$J_n = H_{AB(n)} + H_{BA(n)} - H_{AA(n)} - H_{BB(n)} \tag{8}$$

is the effective (averaged over all configurations) pair interaction at the distance n.

Equations (6)-(8) are valid for a solid solution of any fixed composition, x_A. However, it is important to keep in mind that the J_n parameters may depend on x_A. When the Js are assumed to be composition independent, their values can be determined by solving the system of k equations of type (7) for n unknowns ($k > n$) with a least squares method. Each equation corresponds to a supercell structure which energy is calculated explicitly. The sets of the AB-type frequencies, $f_{AB(n,k)}$, corresponding to these structures should be different. For example, the three structures shown in Figure 2 are characterized with different sets of $f_{AB(1,k)}$, $f_{AB(2,k)}$ and $f_{AB(3,k)}$ values. This permits expanding their excess enthalpies in terms of J_1, J_2 and J_3. Equation (7) can be extended to include many-body interactions what is done in the generalized cluster expansion method (Connolly and Williams 1983; Sanchez et al. 1984). In this chapter, due to reasons given below, many-body interactions will not be considered. Consequently, here we will avoid discussion of a large field of research related to the thermodynamics of alloys,

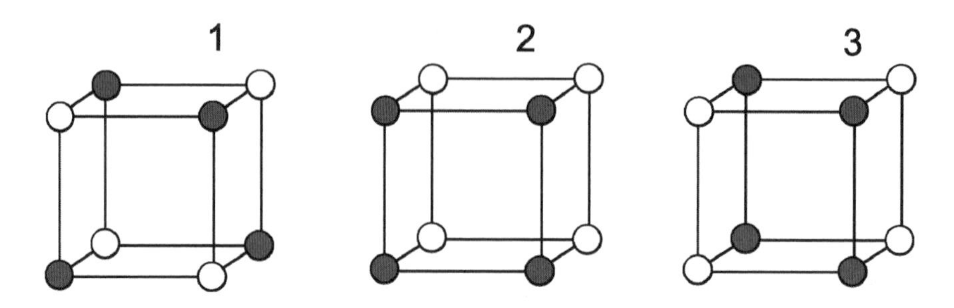

Figure 2. Three supercells structures of an $A_{0.5}B_{0.5}$ solid solution used to expand the excess enthalpy in terms of the pairwise interactions. Note that each structure is characterized by a unique set of numbers of AB pairs at the first, second and third near-neighbor distances.

where many-body interactions cannot be ignored. The readers interested in this field are referred to the review articles (Zunger 1994; van de Walle 2002; Ruban and Abrikosov 2008). This study is focused on minerals. Although one could argue that minerals are just a certain class of "oxidized" alloys, the physical picture of the mixing process in oxides is different. In contrast to alloys, the chemical interactions in minerals are mediated by "inert" anionic groups such as O^{2-}, SiO_4^{2-}, AlO_6^{9-}, CO_3^{2-}, Cl^-, etc. Due to this mediation the interactions between the exchangeable particles become indirect. This makes the ordering interactions less sensitive to many-body effects. In contrast to the situation in an alloy, in an oxide material a pair of interacting cations is not likely to be directly affected by a third cation, as this cation would not be able to come sufficiently close to the pair due to the repulsive force from the mediating anion. Therefore, one expects that the ordering interactions in oxides closely obey pairwise additivity.

This simplification does not imply, however, that the thermodynamic description of minerals is easier than that of the alloys. In alloys the arrangements of the exchangeable atoms can often be described with BCC, FCC or HCP lattices, while minerals typically have lower symmetries. Alloys, due to the high symmetry, develop a variety of ordered superstructures, which can be conveniently used as input configurations for the cluster expansion procedure. The ordered structures due to their symmetry can be converted to smaller supercells. Therefore, it is easier to calculate their total energies with *ab initio* methods. The cluster expansions for BCC, FCC and HCP alloys can now be performed in a fully automatic way using the ATAT package (van de Walle et al. 2002; van de Walle 2009).

This approach does not always work for minerals. Ordering patterns in minerals typically show less variability, while the low symmetry permits only few possibilities for the development of a superstructure, which requires a further decrease in symmetry. The number of such superstructures is usually too small for them alone to be used as the basis set for the cluster expansion. Therefore, in mineralogical studies the supercell structures whose energies are to be calculated, are often chosen at random and the energy optimizations are performed with space group symmetry *P*1 irrespective to whether ordering is present or not. In contrast to alloys, in minerals one is often concerned with ordering of charged species, typically cations, which also have different sizes. The elastic strain perturbations caused by difficulties of fitting cations of different sizes into the average lattice are coupled with electrostatic field gradients and this makes the ordering interactions long ranged. This long-ranged nature of the ordering process dictates the use of large supercells. Consequently, a type of the cluster expansion, which excludes many-body interactions, but includes all possible pairwise interactions within relatively large supercells has become popular in geosciences (e. g. Becker et al. 2000; Bosenick et al. 2000, 2001; Warren et al. 2001; Vinograd et al. 2006, 2007a; Vinograd and Sluiter 2006). It has been shown that the effective interactions determined with this method can be used to predict reasonable excess mixing in many petrologically and geochemically important solid solutions. The disadvantage is that to determine the effective interactions with confidence one has to use hundreds of randomly generated structures. The large number of these structures prohibits the use of *ab initio* methods, and the force field approach is then the only alternative. Another disadvantage is that as soon as the *J*s are assumed to be configuration and composition independent, the predicted mixing functions in a binary (A,B)*R* solid solution are necessary symmetric with respect to $x_A = 0.5$. One possibility for correcting this drawback is to evaluate the volume dependence of the energies of the input configurations, which can be subsequently recast into the volume and composition dependence of *J*s (Connolly and Williams 1983). This approach is difficult, however, due to the large number of structures to be considered. The other alternative is to introduce asymmetry in the models by adding a configuration independent term to Equation (7) (e.g., Vinograd and Sluiter 2006; Vinograd et al. 2007a,b). This term can be either calculated explicitly with Equations (3) and (4) or fitted together with the *J*s to the excess energy. The problem is that the direct evaluation of this term for non-cubic minerals is difficult, while in the fitting procedure the W_i parameters correlate severely with the *J*s. The method introduced in the next paragraph avoids both these drawbacks.

THE DOUBLE DEFECT METHOD (DDM)

The DDM (Hoshino et al. 1993; Vinograd et al. 2009a) is based on the notion that in the limit of infinite dilution the effective pair interactions in a binary solid solution can be directly obtained with the help of a simple computational experiment. Let us consider AB, BA, AA and BB pairs placed at the *n*-th distance within the matrix of the pure "A" end-member. In this case the differences in the enthalpies of the supercells of the same size containing AB, BA, AA and BB pairs will be the same as the differences between $H_{AB(n)}$, $H_{BA(n)}$, $H_{AA(n)}$, and $H_{BB(n)}$ terms of Equation (8). One can also note that the A atoms, which belong to the pairs, will be indistinguishable from the A atoms of the matrix. Hence, the AA pair will disappear, while the AB and BA pairs will reduce to single B-type defects within the A-matrix. This means that in the diluted limit of the end-member A*R* the *Js* can be computed as the differences in total energies of sufficiently large supercells with single, double and no B defects.

$$J_{A(n)} = 2H_{B_1A_{m-1}} - H_{A_m} - H_{B_2A_{m-2}(n)} \tag{9}$$

By adding and subtracting twice the mechanical mixture energy

$$0 = -2(\frac{m-1}{m}H_{A_m} + \frac{1}{m}H_{B_m}) + (\frac{2m-2}{m}H_{A_m} + \frac{2}{m}H_{B_m}) \tag{10}$$

to the right hand side of Equation (9), one obtains

$$J_{A(n)} = 2\Delta H_B - \Delta H_{BB(n)} \tag{11}$$

where ΔH_B and $\Delta H_{BB(n)}$ are the excess enthalpies of the supercells with single B- and double BB-defects, respectively. Similarly, in the limit of the B*R* end member

$$J_{B(n)} = 2\Delta H_A - \Delta H_{AA(n)} \tag{12}$$

In the case of periodic boundary conditions the insertion of a double defect generates D_n AA or BB pairs, where D_n is the degeneracy factor. Equations (11) and (12) thus transform as follows

$$J_{A(n)} = (2\Delta H_B - \Delta H_{BB(n)}) / D_n \tag{13}$$

$$J_{B(n)} = (2\Delta H_A - \Delta H_{AA(n)}) / D_n \tag{14}$$

Further, one observes that due to Equation (6) the *J*s derived with Equations (13) and (14) should obey the self-consistency relations

$$\Delta H_{A} = \tfrac{1}{2} N \sum_{n} Z_n P_A J_{A(n)} \tag{15}$$

and

$$\Delta H_{B} = \tfrac{1}{2} N \sum_{n} Z_n P_B J_{B(n)} \tag{16}$$

Equations (15)-(16) show that the enthalpies of the single defect structures do not depend on the configuration since for the supercell of the end member B_m with a single A-defect $P_{AB(n,k)} = P_A$ and for the supercell of the end member A_m with a single B-defect $P_{AB(n,k)} = P_B$. The last two equations put severe constraints on the values of the *J*s, while they require that a certain combination of the *J*s should give a configuration-independent property, namely the excess enthalpy of a single defect structure. Practical calculations show that the self-consistency constraints implied by Equations (15) and (16) can be easily destroyed due to errors inherent in the determination of the energies of the single and double defect structures. Therefore, a correction procedure is needed.

This procedure is based on noting that the meaning of the *J*s as the ordering energies implies that J_n should vanish when the distance between the two defects is infinitely long. Therefore,

$$2\Delta H_B = \Delta H_{BB(\infty)} \tag{17}$$

and

$$2\Delta H_A = \Delta H_{AA(\infty)} \tag{18}$$

where $\Delta H_{BB(\infty)}$ and $\Delta H_{AA(\infty)}$ are the energies of the double defect structures with the defects at infinitely large separation. It is clear that in the case of a supercell of limited size with periodic boundary conditions an interaction between a single defect and its periodic images cannot be avoided, therefore, twice ΔH_A is not exactly equal to $\Delta H_{AA(\infty)}$ and thus Equations (17) and (18) are slightly violated.

To achieve the self-consistency, Equations (13) and (14) can be rewritten as follows

$$J_{A(n)} = (\Delta H_{BB(\infty)} - \Delta H_{BB(n)}) / D_n \tag{19}$$

and

$$J_{B(n)} = (\Delta H_{AA(\infty)} - \Delta H_{AA(n)}) / D_n \tag{20}$$

where $\Delta H_{BB(\infty)}$ and $\Delta H_{AA(\infty)}$ are considered as adjustable parameters. Our experience shows that a very fine tuning of the values of $\Delta H_{BB(\infty)}$ and $\Delta H_{AA(\infty)}$ in Equations (19) and (20) relative to $2\Delta H_B$ and $2\Delta H_A$ is usually required for Equations (15) and (16) were fulfilled.

The two sets of the *J*s characterize the pair ECIs at two extremes along the composition axis. The variation of the *J*s at intermediate compositions is assumed to be a linear combination of the *J*s calculated in the A- and B-limits:

$$J_{(n)} = x_A J_{A(n)} + x_B J_{B(n)} \tag{21}$$

This equation can be shown to be consistent with the subregular behavior of the enthalpy of mixing in the limit of complete disorder.

The method has been recently applied to the calcite-magnesite system (Vinograd et al. 2009a). It has been shown that the existence of the intermediate dolomite phase and the correct shape of the temperature composition phase diagram can be predicted from the *J*s computed in the diluted limits. Here the same approach is applied to NaCl-KCl system.

THE SYSTEM NaCl-KCl

Quantum mechanical calculations

A 2×2×2 supercell (32 Na/K and 32 Cl atoms) allows the double defects to be located at 5 different distances in the range of 4-10 Å. The total energies of the single and double defect structures were calculated with density functional theory in the GGA approximation using the ultrasoft pseudopotentials integrated in the CASTEP distribution (Clark et al. 2005). The plane wave expansion cutoff was 410 eV. The k-point sampling was according to a 2×2×2 Monkhorst-Pack grid. The results of the calculations are given in Table 1 and in Figure 3. The excess energies obtained could be directly used to estimate the *J*s. However, due to reasons explained below, such calculations were not attempted. The present results are used only as a test for the force-field model, which is introduced in the next paragraph.

Calculation of the excess energies based on the empirical force-field model

The present force-field model for NaCl is only slightly modified from that described by Dove (1993). The model includes electrostatic interaction of formal charges of 1 and −1 for

Table 1. The excess enthalpies of the single and double defect structures in the NaCl-KCl solid solution in 2×2×2 supercell computed with density functional theory (DFT) and with the force-field model (FF).

Single defects			
Type	**DFT 0 GPa**	**FF 0 GPa**	**FF 2 GPa**
K in NaCl	0.20920	0.24755	0.27194
Na in KCl	0.14810	0.18854	0.20851

Double defects					
n	L_n, (Å)	D_n	**DFT 0 GPa**	**FF 0 GPa**	**FF 2 GPa**
KK defects in NaCl					
1	3.974	2	0.38970	0.46134	0.50582
2	5.620	2	0.49540	0.57962	0.65901
3	6.883	2	0.38210	0.45024	0.49178
4	7.948	4	0.41500	0.50070	0.54640
5	9.734	8	0.39730	0.46970	0.51277
NaNa defects in KCl					
1	3.974	2	0.30220	0.36426	0.40365
2	5.620	2	0.33210	0.39248	0.44810
3	6.883	2	0.29710	0.35933	0.39392
4	7.948	4	0.31580	0.38310	0.42173
5	9.734	8	0.30250	0.36726	0.40342

Note: The energy values are in eV per 2×2×2 supercell.

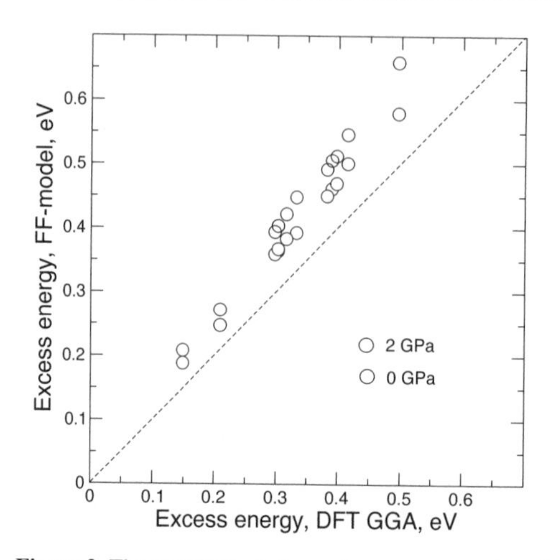

Figure 3. The correlation between the excess energies of single and double defect structures in NaCl-KCl calculated with DFT and with the force-field model. All calculations are performed in the static limit. The values are in eV per 2×2×2 supercell.

Na and Cl ions and the short-range repulsive interaction which is modeled with the Born-Mayer potential, $\varphi(r) = A\exp(-r/\rho)$ (Born and Mayer 1932). The ρ parameter (ρ = 0.3164 Å) was derived from the minimum of the lattice energy by substituting the equilibrium lattice constant of NaCl of 5.64 Å (Wyckoff 1963) and the experimental value of the lattice energy of 764 kJ/mol (Kittel 1966). The A parameter (A = 1239.53 eV) was derived by fitting to the bulk modulus of NaCl (24.9 GPa; Hearmon 1979). Similar calculations were performed for KCl using the lattice constant 6.291 (Wyckoff 1963), the bulk modulus (18.1 GPa), and the lattice energy of 693.7 kJ/mol (Kittel 1966). The fitted values were ρ = 0.3155 Å and A = 2796.99 eV. The predicted and modeled properties of NaCl and KCl are compared in Table 2. The derived force-field model was used to calculate the excess enthalpies of the single and double defect structures in the 2×2×2 supercell at 0 and 2 GPa. The results are compared to the values obtained from quantum mechanical calculations in Table 1 and in Figure 3. The comparison shows that the excess enthalpies of the single- and double-defect structures predicted with the force-field model correlate linearly with the results of the DFT calculation. The force-field model predicts values, which are about 15% too high. The force-field calculations at 2 GPa show that the defect energies increase further and are about 10% higher than at 0 GPa (Fig. 3). This shows that the excess energies are sensitive to compression. Since DFT in the general gradient approximation predicts slightly larger volumes than those observed experimentally (Table 1), one could argue that the excess energies predicted with DFT are slightly underestimated. This supports the accuracy of the values predicted with the force-field model. Due to this reason the pairwise interactions in the static limit were calculated using the excess energies obtained with the force-field model. The results of these calculations are shown in Figure 4.

The enthalpy of mixing in the limit of the complete disorder

The Js can straightforwardly be used to calculate the enthalpy of mixing in the disordered limit with Equation (6). The probabilities of AB pairs in Equation (2) can be substituted with the product $P_A P_B$). Since the Js vary with composition, the enthalpy of the disordered solution is asymmetric with respect to x = 0.5. This asymmetric function can be visualized as a linear interpolation between two symmetric functions of different amplitudes which result from using $J_{A(n)}$ or $J_{B(n)}$ (Fig. 5). One observes that the enthalpy of mixing passes nearly exactly through the excess enthalpies of the two structures with single defects. This is not a coincidence. The probability of finding an AB pair in a structure with a single A-defect is $P_A P_{B/A} = P_A$. This holds because the conditional probability (the probability to find a B-atom at a given distance from an A-atom) is equal to 1. Since in the diluted composition limit

$$P_A(1 - P_A) \approx P_A \tag{22}$$

the high-temperature enthalpy is bound to pass almost exactly through the excess enthalpy of the single defect structure. Thus, as Sluiter and Kawazoe (2002) have shown, the high-temperature

Table 2. The simulated structural and physical properties of NaCl and KCl in comparison to the experimental data: a is the lattice parameter, K is the bulk modulus and E is the cohesive energy.

	NaCl			KCl		
	EXP	**FF**	**DFT**	**EXP**	**FF**	**DFT**
a (Å)	5.640	5.643	5.695	6.295	6.275	6.349
K (GPa)	24.90	24.53		18.10	18.39	
E (kJ/mol)	764.0	764.2		693.7	696.1	

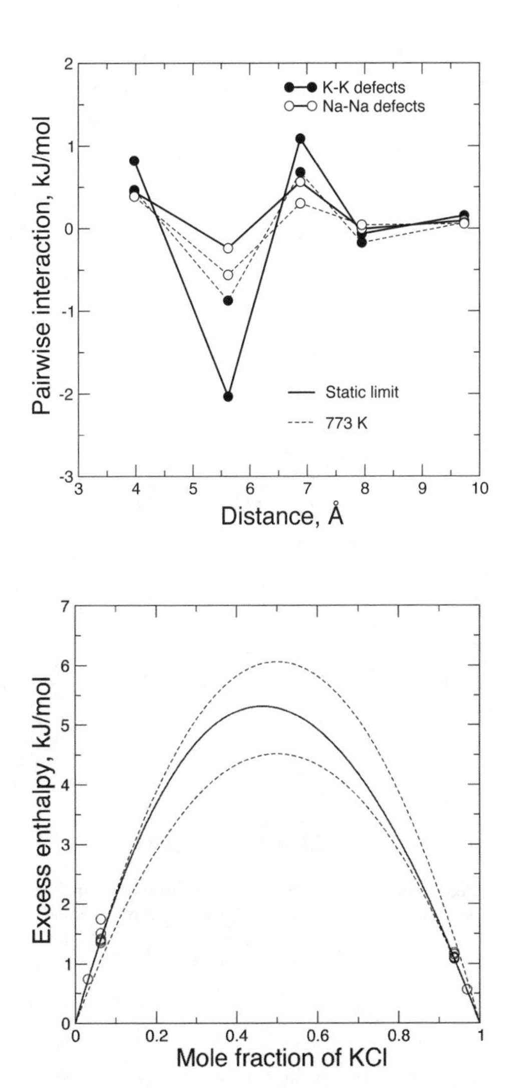

Figure 4. The effective pairwise interactions in NaCl-KCl system in the static limit (0 K) and at 773 K.

Figure 5. The excess enthalpy in NaCl-KCl system in the limit of complete disorder. Circles are the excess enthalpies of supercell structures with single and double defects. The values are in kJ/(mole of Cl).

enthalpy can be predicted from the excess enthalpies of just two single defects. The predicted type of asymmetry with the maximum shifted to NaCl-rich compositions is consistent with the shape of the enthalpy of mixing in the system (Barrett and Wallace 1954).

Monte Carlo simulations of the effect of the temperature on the enthalpy of mixing

Fast convergence of the ECIs as a function of interatomic separation (Fig. 4) suggests that Equation (4) is applicable for calculation of excess enthalpies in a much larger supercell. Here we consider a 12×12×12 supercell containing 2592 exchangeable atoms. The average enthalpies calculated with supercells of this size are practically indistinguishable from the averages, which would correspond to an infinitely large supercell. The average enthalpy at a given temperature and composition can be calculated with canonical Monte Carlo simulation employing the Metropolis algorithm (Metropolis et al. 1953). At each simulation step a pair of dissimilar exchangeable atoms is chosen randomly and then swapped. The swap can be either accepted or rejected (atoms are returned to their original positions) depending on the enthalpy change, ΔH, due to the swap and the current value of the stochastic variable ξ, $0 \leq \xi \leq 1$. When $\Delta H \leq 0$, the new configuration is accepted with certainty, when $\Delta H > 0$, the new configuration is accepted, if $\xi \leq \exp(-\Delta H/(kT))$, or rejected otherwise. The sequence of configurations produced with this algorithm is known to converge after a large number of steps to the Boltzmann probability distribution

$$p_i = \frac{e^{-\Delta H_i/(kT)}}{\sum_i e^{-\Delta H_i/(kT)}} \tag{23}$$

where p_i is the probability to find the configuration with the excess enthalpy ΔH_i and k is the Boltzmann constant. The temperature-dependent excess enthalpy can be calculated by direct averaging of the energies of a converged sequence of the configurations. Here the simulations were performed with the composition-dependent Js on a grid of 32 compositions between NaCl and KCl and 21 temperatures between 273 and 1273 K. Each run consisted of 20 billion Monte Carlo steps where the last 10 billion steps were used to calculate the averages. Figure 6 shows the enthalpy isotherms calculated with this method. The configurational Gibbs free energies of mixing were calculated from Monte Carlo averaged excess free energies via a λ–integration (Myers et al. 1998; Dove 2001; Warren 2001):

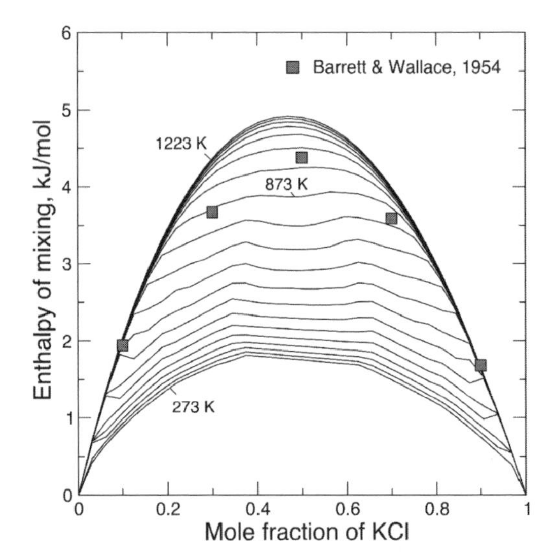

Figure 6. The enthalpy of mixing isotherms in NaCl-KCl system calculated with the Monte Carlo method.

$$\Delta G = \Delta G_0 + \int_0^1 \Delta H_\lambda d\lambda \tag{24}$$

The integration constant ΔG_0 is the Gibbs free energy of mixing of an (A,B)R solid solution with completely random distribution of A and B atoms at a given temperature. This constant can be calculated theoretically:

$$\Delta G_0 = \Delta H_0 + RT(x_A \ln x_A + x_B \ln x_B) \tag{25}$$

Where ΔH_0 is the excess free energy of mixing in the limit of complete disorder and ΔH_λ is the average enthalpy of the system in a state with a non-equilibrium intermediate degree of chemical disorder, λ; $0 < \lambda < 1$. The state $\lambda = 1$ corresponds to an equilibrated system at a given temperature, while the states with $\lambda < 1$ correspond to an artificial disorder that is introduced on top of the equilibrium disorder at the same temperature. This artificial disorder is simulated by scaling the Js according to the equation $J_n^\lambda = J_n\lambda$. In our simulations, λ was gradually increased from 0 to 1 with a step size of 0.04. The integral describes the change in the Gibbs free energy of mixing of a system at a fixed temperature from the state with zero ordering energy ($\lambda = 0$) to its equilibrium state determined with the nominal values of the Js ($\lambda = 1$). The free energy isotherms are plotted in Figure 7.

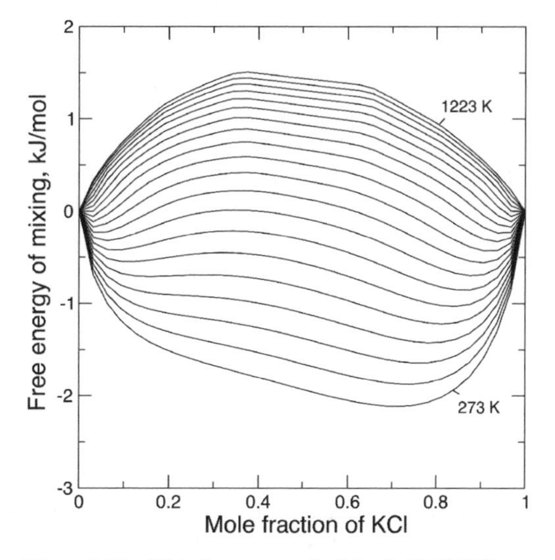

Figure 7. The Gibbs free energy of mixing in NaCl-KCl system calculated by thermodynamic integration of the Monte Carlo results.

The phase diagram

The Gibbs free energies of mixing were converted to a phase diagram by comparing the free energy at each composition x_i along an isotherm to the free energy of a mechanical mixture of two phases with the same total composition. If there is a pair of compositions x_k and x_j, that has lower free energy, the solution with composition x_i is unstable or metastable. The miscibility gap was thus outlined (Fig. 8). The predicted phase boundary lies significantly above the experimental solvus (Barrett and Wallace 1954).

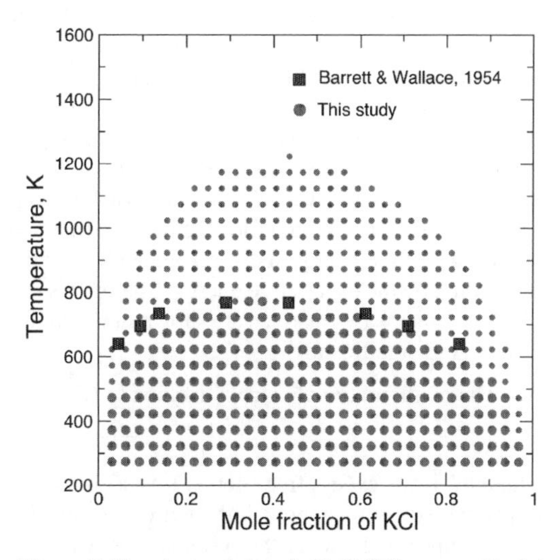

Figure 8. The phase relations in NaCl-KCl system. Shaded circles are the unstable compositions selected based on the common tangent analysis of the free energy isotherms. The small circles correspond to the simulations with the temperature independent Js. The large circles correspond to the simulation with the temperature dependent Js (see text). The squares are the experimental data of Barrett and Wallace (1954).

The excess vibrational free energy

The model based on the energies of the double defect structures calculated in the static limit is in poor agreement with the experimental data. This implies the existence of non-negligeable excess entropy of mixing. Similar conclusion was obtained in the recent modelling studies of the NaCl-KCl system (Burton and van de Walle 2006; Urusov et al. 2007). Burton and van de Walle (2006) have shown that when the phase diagram is calculated based on the DFT GGA results obtained in the static limit, the solvus closes at about 1600 K, which is in nearly two times higher than the experimental value. Including the excess vibrational entropy in the calculations permitted to shift the boundary to a much lower temperature. The force-field modelling study of Urusov et al. (2007) has shown that the expected vibrational entropy in the random NaCl-KCl solid solution at intermediate composition is about 2.5 J/K/mol. To improve consistency with the experimental data, we have repeated the calculation of the excess energies of the double and single defect structures using quasi-harmonic lattice dynamics calculations within the zero static internal stress approximation (ZSISA) (Allan et al. 1996) as implemented in the program GULP (Gale 1997; Gale and Rohl 2003). In these calculations the minimized function is similar to the Helmholtz free energy, which, besides the static lattice energy, includes the thermally averaged contribution from vibrational states. The calculations were performed at 273 and 1273 K. The excess energies of the single and double defect structures were then derived from the free energies. The results are shown in Table 3. The *J*s in the limit of A*R* end member were obtained via a linear fit to the *J*s derived from 273 and 1273 K sets via the equation

$$J_{\mathrm{A}(n)} = J^{H}_{\mathrm{A}(n)} - TJ^{S}_{\mathrm{A}(n)} \tag{26}$$

and with an analogous equation in the limit of B*R* end member. Additional calculation at 773 K confirmed the nearly linear variation of the *J*s in this temperature interval. The fitted values of the *J*s are given in Table 4. The model with the temperature dependent *J*s predicts a significantly smaller tendency to phase separation. The excess free energy of mixing at 773 K in the completely disordered solution is plotted in Figure 9. The magnitude of this function is nearly twice smaller than the excess enthalpy calculated with the temperature independent *J*s. Monte Carlo simulations with the temperature dependent *J*s were performed in a completely analogous way to the simulations with the temperature independent *J*s. At each given temperature the *J*s are constants. The difference is that the property, which is calculated after the averaging over the converged Monte Carlo runs, is not the excess enthalpy. The calculated property is the excess free energy, which includes both the static and vibrational contributions. This function is defined here not with respect to the free energy of mixing of an ideal solid solution, as is usually done for regular solid solutions, but with respect to the configurational entropy at a given temperature times the temperature. Below we will show that the configurational entropy depends on the temperature. The Gibbs free energy of mixing, which includes the configurational entropy term is calculated by the thermodynamic integration of the excess free energy. This is done with Equations (23)-(25) where the excess enthalpy is substituted by the excess free energy. The results of these calculations are shown in Figure 10. The predicted phase relations are shown in Figure 8. The agreement with the experimental data is improved significantly.

The configurational entropy isotherms (Fig. 11) can be calculated with the equation:

$$S = (\Delta F - \Delta G) / T \tag{27}$$

where ΔF is the excess free energy. It is seen that at very high-temperatures and in the diluted limits the isotherms approach the entropy of ideal mixing. At intermediate compositions and lower temperatures, when the system is above the solvus, the configurational entropy is smaller than that of the ideal mixing due to the short-range ordering. These features of the entropy function confirm the correctness of the thermodynamic integration procedure.

The configurational entropy is only a part of the excess mixing entropy. The vibrational entropy is plotted in Figure 12. To calculate this function we have assumed that the vibrational entropy is represented by the derivatives of the *J*s with respect to the temperature, while the *J*s

Table 3. The excess free energies of the double defect structures at 273, 773 and 1273 K. The values which correspond to the infinitely large distance are obtained from the self-consistent fitting procedure.

	n	L_n, (Å)	ΔF_{KK}	ΔF_{NaNa}
273 K				
	1	3.974	0.39475	0.29824
	2	5.620	0.48915	0.31789
	3	6.883	0.38451	0.29645
	4	7.948	0.43178	0.31469
	5	9.734	0.40301	0.30253
		Inf.	0.40246	0.30541
773 K				
	1	3.974	0.28646	0.18998
	2	5.620	0.34186	0.22922
	3	6.883	0.27760	0.19341
	4	7.948	0.31990	0.20250
	5	9.734	0.29469	0.19648
		Inf.	0.29299	0.20163
1273 K				
	1	3.974	0.17744	0.08118
	2	5.620	0.19363	0.15456
	3	6.883	0.16998	0.08989
	4	7.948	0.20726	0.08976
	5	9.734	0.18560	0.08989
		Inf.	0.18275	0.09824

Note: The enthalpy values are in eV per 2×2×2 supercell.

Table 4. The *J*s used in The Monte Carlo simulations. The temperature-dependent *J*s were obtained by a linear fit to the Js calculated from the excess free energies of the double defect structures at 273 K and 1273 K.

The Js calculated in the static limit			
n	L_n, (Å)	J_{Na}	J_K
1	3.974	0.8207	0.4434
2	5.620	−2.0322	−0.2374
3	6.883	1.0888	0.5620
4	7.948	−0.0642	−0.0058
5	9.734	0.1549	0.0927
The temperature-dependent Js			
n	L_n, (Å)	J_{Na}^H	J_K^H
1	3.974	0.8174	0.3674
2	5.620	−2.0225	0.2795
3	6.883	1.0845	0.4859
4	7.948	−0.0636	−0.0429
5	9.734	0.1543	0.0738
n	L_n, (Å)	J_{Na}^S *10³	J_K^S *10³
1	3.974	-0.444	0.052
2	5.620	1.443	−1.243
3	6.883	−0.511	−0.200
4	7.948	−0.135	0.121
5	9.734	−0.110	−0.013

Note: The values are in eV.

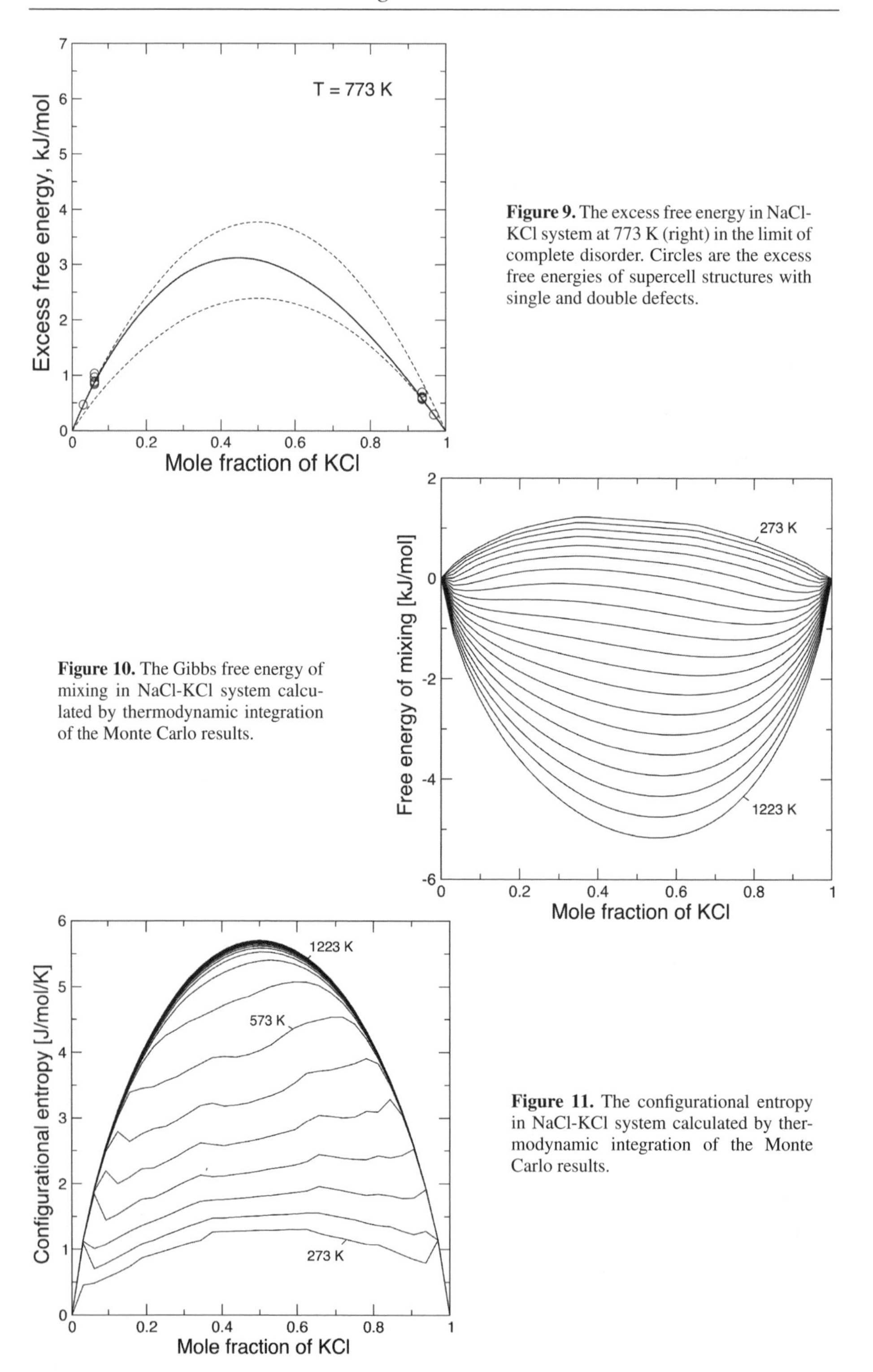

Figure 9. The excess free energy in NaCl-KCl system at 773 K (right) in the limit of complete disorder. Circles are the excess free energies of supercell structures with single and double defects.

Figure 10. The Gibbs free energy of mixing in NaCl-KCl system calculated by thermodynamic integration of the Monte Carlo results.

Figure 11. The configurational entropy in NaCl-KCl system calculated by thermodynamic integration of the Monte Carlo results.

linearly depend on the temperature. This implies that the vibrational entropy does not directly depend on the temperature. There is however an indirect dependence of the vibrational entropy on the temperature through the dependence of the cation distribution on the temperature. Evidently, the configurations with the lowest excess free energies are sampled more often, which gives an advantage to the configurations with higher excess and configurational entropies at higher temperatures. Therefore, the excess entropy increases at the intermediate compositions and particularly so at higher temperatures. The excess volume shows a similar behavior due to the correlation of the vibrational entropy with the volume (Vinograd and Sluiter 2006). The excess volume will not be investigated here in detail.

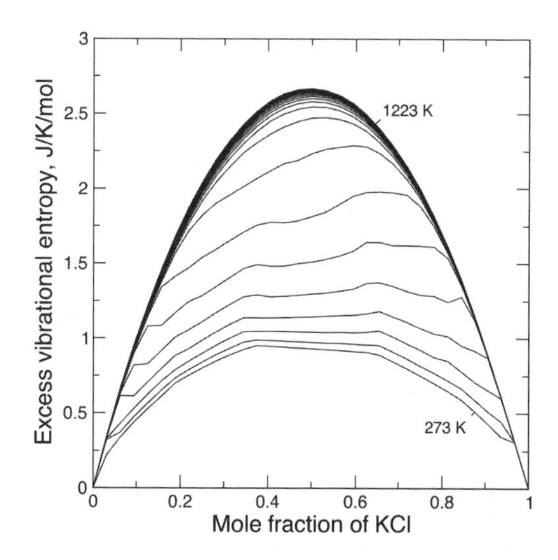

Figure 12. The thermal average excess vibrational entropy in NaCl-KCl system calculated with the Monte Carlo method.

The obtained model can be further tested against the calorimetric data of Barrett and Wallace (1954). Since the calorimetric studies measure the enthalpy, but not the free energy, the relevant simulated property is the enthalpy in the static limit. However, one has to keep in mind that the samples used in the calorimetric measurements were synthesized at 903 K. This means that the cation distribution in the samples was affected by the excess vibrational entropy and therefore was more disordered than the static model implies. Thus, the statistical distribution in the samples should be simulated with the temperature dependent *J*s, while the excess enthalpy should be calculated with the *J*s, which correspond to the static limit. In the other words, to make the comparison with the experiment, one needs first to reproduce the correct cation distribution by performing calculations with the temperature dependent *J*s and then perform the configurational averaging of the enthalpy with the *J*s, which correspond to the static limit. The enthalpy of mixing corrected for the effect of additional disorder is plotted in Figure 13. The experimental data lie close to the 623 K isotherm. This result is not surprising recalling that Barrett and Wallace (1954) have annealed their samples at certain temperatures below the synthesis conditions. The exact annealing temperatures were not given.

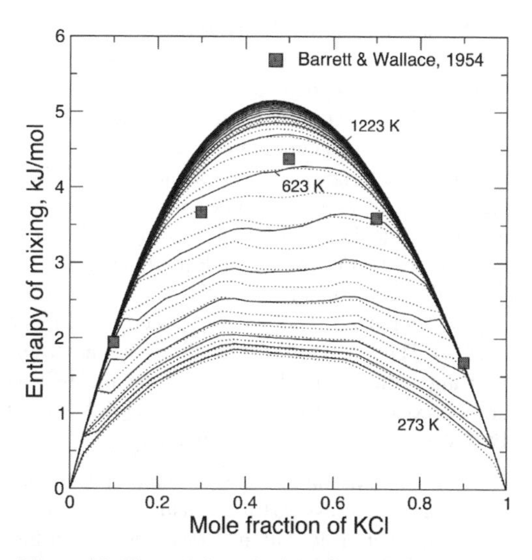

Figure 13. The enthalpy of mixing in NaCl-KCl system. Dotted lines show the enthalpy isotherms in the static limit. Solid lines show the enthalpy isotherms corrected for the additional disorder due to the excess vibrational entropy.

DISCUSSION AND CONCLUSIONS

In this chapter we have outlined the state-of-the-art in solid solution theory and introduced the Double Defect Method (DDM). The DDM has many features in common with the currently popular approach based on randomly varied supercell structures. Similarly the DDM aims at the determination of the effective pairwise interactions. However, instead of permuting a large number of possible configurations with intermediate compositions and then solving a least squares problem, the DDM samples only few configurations in the diluted composition ranges, which are selected based on a deterministic principle. The main advantage is the minimization of the number of the sampled structures. The whole simulation approach adopted here consisted of the following steps:

- construction of a supercell,
- distance analysis permitting identification of crystallographically different pairs,
- calculation of the static energies of supercell structures with paired defects inserted at the specified distances,
- calculation of the free energies of the structures at several temperatures,
- evaluation of the *J*s and Monte Carlo simulation of the distribution of the exchangeable species over a dense grid of compositions, temperatures and λ values,
- thermodynamic integration and calculation of the phase diagram with the common tangent method.

We have seen that this algorithm gave an almost quantitative description of the phase separation phenomenon in the NaCl-KCl system. A recent study of the system calcite-magnesite (Vinograd et al. 2009a) has shown that the DDM is able to predict the formation of the dolomite phase and thus is applicable to systems with ordering. The nice feature of the DDM is its ability to evaluate the compositional dependence of the *J*s and thus to correctly predict an asymmetry of the functions of mixing.

Most importantly, the DDM permits a deeper understanding of the process of mixing. In the Introduction we have stated that the solid solution theory has always tried to reconcile configuration-dependent and configurational-independent contributions to the enthalpy of mixing. The DDM does not require these contributions to be treated separately. The non-configurational behavior appears as a combined effect of all the *J*s. For example, the quantity $\sum_n Z_n J_{A(n)}$ is configuration independent. This follows from noting that it is proportional to the excess enthalpy of a single defect structure, ΔH_A, while the latter value does not depend on the position of the defect within the supercell. Furthermore, the DDM reveals a dual meaning of the *J*s. Indeed, as follows from Equation (6), the *J*s are the contributions to the excess enthalpy due to the formation of AB pairs. This first meaning of the *J*s as partial excess energies is widely recognized. The second, less obvious, meaning of the *J*s follows from Equations (11) and (12). These equations show the *J*s are the differences between the chemical stresses due to inserting of the double defects at the specific distances, L_n, relative to the "non-configurational" stress produced by two non-interacting defects. The non-trivial observation is that a sum of these stresses gives the excess enthalpy.

This observation has several important implications. The sum of the *J*s reflects the non-configurational stress due to a misfit from a single defect. Therefore, *J*s with a positive sign play a dominant role within the expansion. On the other hand, the signs of the *J*s reflect the relative rigidity of a structure along specific directions. A negative sign of a particular J_n does not mean that the insertion of the corresponding double defect causes a negative excess enthalpy. It only means that the stress due to the two defects inserted at the particular distance L_n is smaller than the stress due to two single (non-interacting) defects. The contribution of this particular J_n to the excess enthalpy is negative, however, the total excess enthalpy of the corresponding double defect is positive, while it is comprised of all the *J*s, where positive interactions dominate. Chemical ordering can be understood as a change in the probability distribution of the exchangeable particles leading to the preferential formation of AA- and BB-type pairs along the directions in the structure, which correspond to a low chemical stress. This, in turn, means that AB-type pairs tend to be formed along the directions, which correspond to a high structural rigidity. Naturally, these directions in the structure either correspond to the shortest distances between the neighbors or to directions in which the interactions between the neighbors are mediated by extra species that reduce space between the neighbors and thus increase the rigidity of the structure in the particular direction. In NaCl structure this is certainly the second neighbor distance, where the cations interact via the Cl^- ions. Figure 4 shows that the J_2 is by far the strongest interaction. The structure reacts to this local stress by increasing the proportion of Na-Cl-K pairs along this direction, thus permitting the stress relaxation via the shift of Cl^- towards Na. A similar phenomenon has been observed in grossular-pyrope solid solution, where the strongest interaction is the J_3 that is mediated by a rigid SiO_4 grouping (Bosenick et al. 2000; Vinograd and Sluiter 2006). The increase in the number of AB-type pairs is particularly effective at the intermediate compositions. This explains why ordering often develops at the 50:50 ratio, as, for example, in the calcite-magnesite system. These observations imply that the double defect structures can be used as the sensors permitting to identify the most important ordering interactions. Thus with the DDM it becomes possible to predict the type of ordering in the concentrated range of a solid solution by investigating the diluted limits only. The other consequence is that the average magnitude of the *J*s depends mostly on the distances between the exchangeable atoms. It follows that the pressure and the temperature affect the magnitude of the *J*s in the opposite ways. The increase in the pressure decreases the interatomic distances and increases the average rigidity of the structure. Consequently, the magnitude of the *J*s increases. The temperature has the opposite effect to the pressure.

The NaCl-KCl system represents the simplest type of a solid solution. The gratifying conclusion of the present study is that such a system can be simulated with the available tools with a quantitative accuracy. On the other hand, we have seen that achieving this accuracy required rather sophisticated calculations, which in other more complex cases might become too demanding. Obviously, the results obtained in the static limit (which is the style of calculations commonly adopted in earth sciences) could be significantly altered when the effects of the excess vibrational entropy come into play. For example, significant excess vibrational entropy effect has been measured in pyrope-grossular garnets (Haselton and Westrum 1980; Dachs and Geiger 2006) and in Na-K feldspars (Haselton et al. 1983). In this study we had an advantage of using an accurate force-field model what made the quasi-harmonic lattice dynamics calculations inexpensive. In other cases one might be forced to use *ab initio* methods, which in non-static conditions could be extremely tedious. Therefore, there is little hope that *ab initio* methods will soon completely substitute force-field approaches. A combination of *ab initio* and force-field methods seems to be currently the most efficient approach.

Many solid solutions, which are of interest to geosciences, are based on more complex substitution schemes and often include more than two components. The description of such solid solutions naturally requires much computational effort. In this respect the DDM is particularly attractive, as it offers a nice combination of accuracy and simplicity. In the recent *ab initio* study of Jung et al. (2010) the DDM was successfully applied to modeling of solid solutions between $MgSiO_3$ and Al_2O_3 with perovskite and ilmenite structures, where the mixing of Al/Si and Al/Mg occurs on two separate sublattices. The study predicted that the ilmenite and perovskite solid solutions have strikingly different mixing behavior: the ilmenite system phase separates, what is consistent with the experimentally known phase relations in Mg-Al-Si-O system (Kubo and Akaogi, 2000), while perovskite forms an ordered structure of intermediate composition with a strongly negative excess enthalpy. The subsequent *ab initio* calculations confirmed the predicted value of the excess enthalpy. The study of Jung et al. (2010) thus showed that the DDM permits prediction of ordered states even in cases of coupled substitutions. Finding such structures by random permutation of the arrangements of Al, Mg and Si cations is virtually impossible. Recently, Vinograd et al. (2009b) were able to apply the DDM to the ternary solid solution between calcite, magnesite and rhodochrosite and to predict the miscibility gaps in ternary space in semi-quantitative agreement with the experiment (Goldsmith and Graf 1957).

This chapter would not be complete if we did not comment on the problems in solid solution theory, which remain to be solved. One of such problems is the non-negligible effect of the interfacial energy on the results of Monte Carlo simulations of systems with phase separation. Since Monte Carlo simulations are performed in supercells containing only few thousand of exchangeable atoms, a large fraction of these atoms is involved in building the interface. This problem appears in the present study too and its consequences can be seen, for example, in the features of the excess enthalpy function (Fig. 6). The low temperature isotherms do not vary linearly with the composition as it is expected for a system with phase separation. On the contrary, these isotherms reveal a positive contribution to the excess enthalpy, which can be associated with the strain energy at the interface. In the intermediate composition range the interfacial energy flattens out revealing the regime of a single interface in the preferred crystallographic orientation (Fig. 14a). At less concentrated compositions ($0.2 < x_K < 0.35$) the system phase separates by forming a cylinder-shaped domain build by the admixed component (Fig. 14b). At more diluted compositions a nearly spherical domain is formed (Fig. 14c). The size of this domain decreases as the total composition of the system approaches the diluted limit. Consequently, the interface energy decreases. At low temperatures, due to the existence of a high interfacial energy, the phase separation occurs via a first order transition. This implies that the homogeneous disordered system that is in equilibrium with the phase separated system should have an additional excess enthalpy that compensates for the interfacial energy. This dictates the homogeneous system to have a very high enthalpy and, consequently, a high configurational entropy. Our simulations show that at low temperatures the configurational entropy of the disordered phase that is in equilibrium with the phase separated system achieves the maximum. The maximum entropy regime is indeed observed in Figure 11 as the "collisions" between different isotherms. The points of collisions lie on the curve of the ideal mixing. The same phenomenon is reflected in the rapid change in the Monte Carlo patterns simulated below and above the phase boundary (Fig. 14c,d). The configuration of Figure 14d is simulated at a temperature that is marginally higher than that of the configuration of Figure 14c. A drastic change in the degree of disorder is obvious. The first order nature of the phase separation process is an artifact of the simulation setup that does not avoid the formation of the interface. Due to the presence of the interfacial energy, the free energy has a positive component, which would not have existed in the thermodynamic limit. Therefore, at low temperatures, the predicted miscibility gap is wider than it should be in reality. Fortunately, at higher temperatures, where the compositions of the exsolved phases become less dissimilar, the interfacial energy decreases. Therefore, the gap closing temperature can be predicted correctly. We hope that in future studies it will be possible to avoid the problems associated with the interface by performing simulations with two supercells whose compositions can change due to the swaps across the supercells. In such a setup the interface disappears, as the both phases can be simulated with periodic boundary conditions.

On the other hand, there are problems where the interfacial energy must be explicitly included in the simulation. The studies of solid solution formation at the solid-aqueous interface (e.g., Astilleros et al. 2006; Perez-Garrido et al. 2009) invariably point out to the important role of the surface energy, which can hinder the solid solution formation even at a very high supersaturation. The thermodynamics of mixing of such systems is yet to be developed.

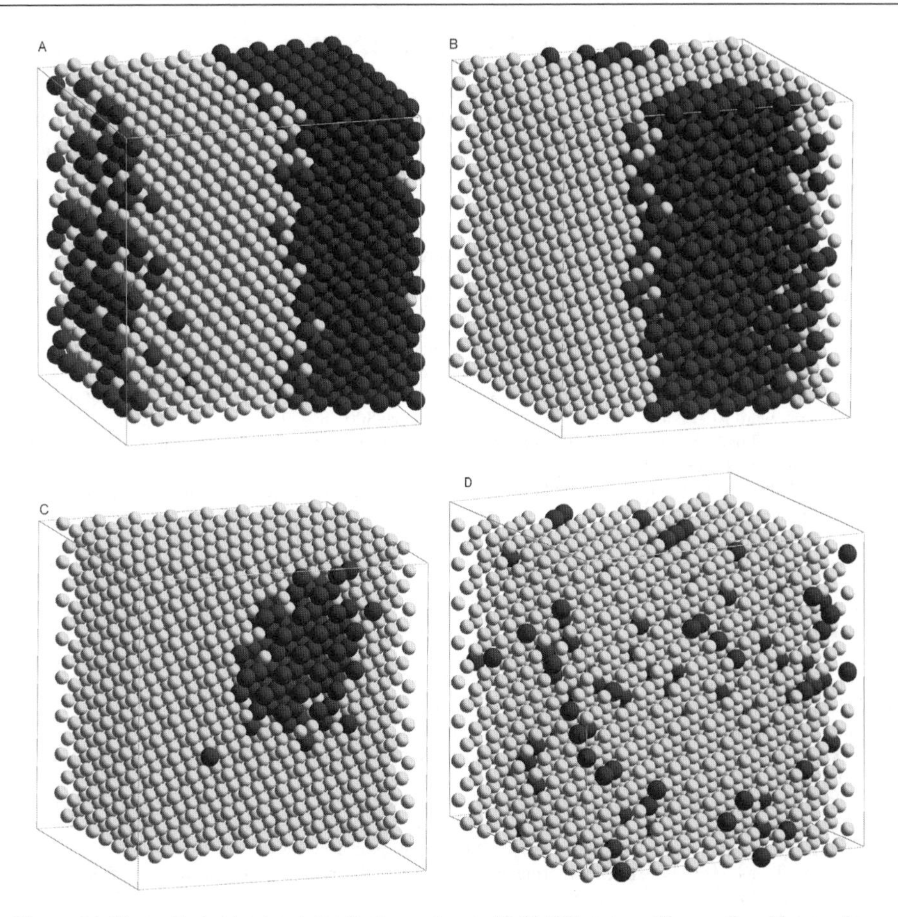

Figure 14. Monte Carlo simulated distribution patters in NaCl-KCl system. The small and large spheres represent Na and K atoms, respectively. Cl atoms are not shown. A: $x_K = 0.4375$, $T = 373$ K; B: $x_K = 0.25$, $T = 373$ K; C: $x_K = 0.0625$, $T = 373$ K; D: $x_K = 0.0625$, $T = 423$ K.

Much effort is needed in the understanding of the thermodynamics of systems where chemical ordering is coupled to magnetic ordering. The recent study of Harrison and Becker (2001) has shown that the superexchange interactions between Fe^{2+} and Fe^{3+} significantly affect the phase separation in the hematite-ilmenite, Fe_2O_3-$FeTiO_3$, system, while, on the other hand, the phase separation causes important changes in the magnetic properties of these minerals. Thus it is necessary to develop simulation tools, which would permit an accurate calculation of the exchange interaction constants. DDM calculations with non-relaxed spin configuration of the admixed atoms could probably be useful. It is also necessary to develop simulation approaches for systems where the chemical ordering is combined with spin state or electronic state transitions (e.g., Badro et al. 2003; Umemoto et al. 2008; Narygina et al. 2009). The new phenomenon, which currently available tools cannot easily treat, is that an atom with a chemically distinct property is introduced into the system not merely as a result of changing the composition, but as a result of a change in an intensive parameter, e.g., the pressure. Finding solutions to these problems will greatly advance the solid solution theory and will lead to a better understanding of phase relations in many geochemically and industrially important systems.

ACKNOWLEDGMENTS

VLV gratefully acknowledges the research fellowship from the Helmholtz Society (The Virtual Institute for Advanced Solid-Aqueous Radio-Geochemistry). We also wish to thank J.D. Gale, V. Milman and N. Paulsen for the help with calculations. The helpful comments by M.T. Dove are greatly appreciated.

REFERENCES

Allan NL, Barron THK, Bruno JAO (1996) The zero static internal stress approximation in lattice dynamics, and the calculation of isotope effects on molar volumes. J Chem Phys 105:8300–8303

Astilleros JM, Pina CM, Fernandez-Diaz L, Prieto M, Putnis A (2006) Nanoscale phenomena during the growth of solid solutions on calcite {104} surfaces. Chem Geol 225:322–335

Badro J, Fiquet G, Guyot F, Rueff JP, Struzhkin VV, Vanko G, Monaco G (2003) Iron partitioning in Earth's mantle: Toward a deep lower mantle discontinuity. Science 300:789-791

Barabash SV, Ozolins V, Wolverton C (2008) First-principles theory of the coherency strain, defect energetics, and solvus boundaries in the PbTe-$AgSbTe_2$ system. Phys Rev B 78:214109(7)

Barrett WT, Wallace WE. (1954) Studies of NaCl-KCl solid solutions. I. heats of formation, lattice spacings, densities, shottky defects and mutual solubilities. J Am Chem Soc 76:366–369

Becker U, Fernandez-Gonzalez A, Prieto M, Harrison R, Putnis A (2000) Direct calculation of thermodynamic properties of the barite/celestite solid solution from molecular principles. Phys Chem Mineral 27:291–300

Becker U, Pollock K (2002) Molecular simulations of interfacial and thermodynamic mixing properties of the grossular - andradite garnets. Phys Chem Mineral 29:52–64

Bellaiche L, Vanderbilt D (2000) Virtual crystal approximation revisited: application to dielectric an piezoelectric properties of perovskites. Phys Rev B 61:7877–7882

Bethe HA (1935) Statistical theory of superlattices. Proc R Soc London A150:552–575

Born M, Mayer JE (1932) Zur Gittertheorie der Ionenkristalle. Z Phys 75:1-18

Bosenick A, Dove MT, Geiger CA (2000) Simulation studies on the pyrope-grossular solid solution. Phys Chem Mineral 27:398–418

Bosenick A, Dove MT, Myers ER, Palin EJ, Sainz-Diaz CI, Guiton BS, Warren MC, Craig MS, Redfern SAT (2001) Computational methods for the study of energies of cation distributions: applications to cation-ordering phase transitions and solid solutions. Mineral Mag 65:193-219

Bragg WL, Williams EJ (1934) Effect of thermal agitation on atomic arrangement in alloys. Proc R Soc London A145:699–730

Burton BP, Kikuchi R (1984) Thermodynamic analysis of the system $CaCO_3$-$MgCO_3$ in the tetrahedron approximation of the Cluster Variation Method. Am Mineral 69:165–175

Burton BP, van de Walle A (2006) First-principles phase diagram calculations for the system NaCl-KCl: The role of excess vibrational entropy. Chem Geol 225:222–229

Clark SJ, Segall MD, Pickard CJ, Hasnip PJ, Probert MJ, Refson K, Payne MC (2005) First principles methods using CASTEP. Z Kristallogr 220:567–570

Connolly JWD, Williams AR (1983) Density-functional theory applied to phase transformations in transition-metal alloys. Phys Rev B 27:5169–5172

Cowley JM (1950) An approximate theory of order in alloys. Phys Rev 77:669–675

Dachs E, Geiger CA (2006) Heat capacities and entropies of mixing of pyrope-grossular ($Mg_3Al_2Si_3O_{12}$-$Ca_3Al_2Si_3O_{12}$) garnet solid solutions: A low-temperature calorimetric and a thermodynamic investigation. Am Mineral 91:894–906

Dove MT (2001) Computer simulation of solid solutions. *In*: Solid Solution in Silicate and Oxide Systems. EMU Notes in Mineralogy. Volume 3. Geiger CA (ed) Eötvös University Press, Budapest, p 225–249

Dove MT (1993) Introduction to Lattice Dynamics. Cambridge University Press, Cambridge

Ferreira LG, Mbaye AA, Zunger A (1988) Chemical and elastic effects on isostructural phase diagrams: The e-G approach. Phys Rev B 37:10547–10570

Gale JD (1997) Gulp - a computer program for the symmetry-adapted simulation of solids. J Chem Soc Faraday Trans 93:629–637

Gale JD, Rohl AL (2003) The General Utility Lattice Program (GULP). Mol Simulations 29:291–341

Goldsmith JR, Graf DL (1957) The system CaO-MnO-CO_2: solid solution and decomposition relations. Geochim Cosmochim Acta 11:310–334

Guggenheim EA (1935) Statistical mechanics of regular solutions. Proc R Soc London A148:304–312

Harrison RJ, Becker U (2001) Margnetic ordering in solid solutions. *In*: Solid Solution in Silicate and Oxide Systems. EMU Notes in Mineralogy, Volume 3. Geiger CA (ed) Eötvös University Press, Budapest, p 349–383

Haselton HT Jr, Hovis GL, Hemingway BS, Robie RA (1983) Calorimetric investigation of the excess entropy of mixing in analbite-sanidine solid solutions: lack of evidence for Na,K short-range order and implications for two-feldspar thermometry. Am Mineral 86:398–413

Haselton HT Jr, Westrum FF Jr (1980) Low-temperature heat capacities of synthetic pyrope, grossular and pyropegrossular. Geochim Cosmochim Acta 44:701–709

Hearmon RFS (1979) The elastic constants of crystals and other anisotropic materials. *In*: Landolt-Börnstein Tables, III, Volume 11. Hellwege KH, Hellwege AM (eds) Springer Verlag, Berlin, p 854

Hoshino T, Schweika W, Zeller R, Dederichs PH (1993) Impurity-impurity interactions in Cu, Ni, Ag, and Pd. Phys Rev B 47:5106–5117

Inden G, Pitsch W (1991) Atomic ordering. *In*: Materials Science and Technology. Volume 5. Haasen P (ed) VCH Press, Weinheim, p 497–545

Ising E (1925) Beitrag zur Theorie des Ferromagnetismus. Z Physik 31:253–258

Jung DY, Vinograd VL, Fabrichnaya OB, Oganov AR, Schmidt MW, Winkler B (2010) Thermodynamics of mixing in $MgSiO_3$ – Al_2O_3 perovskite and ilmenite from ab initio calculations. Earth Planet Sci Lett, submitted

Kikuchi R (1951) Theory of cooperative phenomena. Phys Rev 81:988–1003

Kittel C (1966) Introduction to Solid State Physics. John Wiley & Sons, New York

Kubo A, Akaogi M (2000) Post-garnet transitions in the system $Mg_4Si_4O_{12}$-$Mg_3Al_2Si_3O_{12}$ up to 28 GPa: phase relations of garnet, ilmenite and perovskite Phys Earth Planet Inter 121:85-102

Laks DB, Ferreira LG, Froyen S, Zunger A (1992) Efficient cluster expansion for substitutional systems. Phys Rev B 46:12587–12605

Metropolis NI, Rosenbluth AW, Rosenbluth MN, Teller AN, Teller E (1953) Equation of state calculations by fast computing machines. J Chem Phys 21:1087–1092

Mohri T, Sanchez JM, de Fontaine D (1985) Binary ordering prototype phase diagrams in the cluster variation approximation. Acta Metall Mater 33:1171–1185

Myers ER, Heine V, Dove MT (1998) Some consequences of Al/Al avoidance in the ordering of Al/Si tetrahedral framework. Phys Chem Mineral 25:457–464

Narygina O, Mattesini M, Kantor I, Pascarelli S, Wu X, Aquilanti G, McCammon C, Dubrovinsky L (2009) High-pressure experimental and computational XANES studies of (Mg,Fe)(Si,Al)O_3 perovskite and (Mg,Fe)O ferropericlase as in the Earth's lower mantle. Phys Rev B 79:174115

Onsager L (1944) Crystal statistics. I. A two-dimensional model with an order-disorder transition. Phys Rev 65:117–149

Palin EJ, Harrison RJ (2007) A Monte Carlo investigation of the thermodynamics of cation ordering in 2-3 spinels. Am Mineral 92:1334–1345

Pina CM, Garcia A (2009) Nanoscale anglesite growth on the celestite (001) face. Surface Science 603(17):2708-2713

Purton JA, Allan NL, Lavrentiev MY, Todorov IT, Freeman CL (2006) Computer simulation of mineral solid solutions. Chem Geol 225:176–188

Perez-Garrido C, Fernandez-Diaz L, Pina CM, Prieto M (2007) In situ AFM observations of the interaction between calcite (10-10) surfaces and Cd-bearing aqueous solutions. Surf Sci 602:5499–5509

Ruban AV, Abrikosov IA (2008) Configurational thermodynamics of alloys from first principles: effective cluster interactions. Rep Prog Phys 71:046501-1-30

Sanchez JM, de Fontaine D (1980) Ordering in fcc lattices with first- and second-neighbor interactions. Phys Rev B 21:216–228

Sanchez JM, Ducastelle F, Gratias D (1984) Generalized cluster description of multicomponent systems. Physica 128A:334–350

Sluiter MHF, Kawazoe Y (2002) Prediction of mixing enthalpy of alloys. Europhys Lett 57:526–532

Umemoto K, Wentzcovitch RM, Yu YG, Requist R (2008) Spin transition in (Mg,Fe)SiO_3 perovskite under pressure 276:198-206

Urusov VS, Petrova TG, Leonenko EV, Eremin NN (2007) A computer simulation of halite-sylvite (NaCl-KCl) solid solution local structure, properties, and stability. Moscow University Geology Bulletin 62:117–122

van Baal CM (1973) Order disorder transformations in a generalized Ising alloy. Physica 64:571–586

van de Walle A (2009) Multicomponent multisublattice alloys, nonconfigurational entropy and other additions to the Alloy Theoretic Automated Toolkit. CALPHAD 33:266–278

van de Walle A, Asta M, Ceder G. (2002) The Alloy Theoretic Automated Toolkit: A user guide. CALPHAD 26:539–553

Vinograd VL, Sluiter MHF, Winkler B (2009a) Subsolidus phase relations in the $CaCO_3$-$MgCO_3$ system predicted from the excess enthalpies of supercell structures with single and double defects. Phys Rev B 79:104201–104209

Vinograd VL, Burton BP, Gale JD, Allan NL, Winkler B (2007) Activity-composition relations in the system $CaCO_3$-$MgCO_3$ predicted from static structure energy calculations and Monte Carlo simulations. Geochim Cosmochim Acta 225:304–313

Vinograd VL, Paulsen N, Winkler B (2009b) Thermodynamics of mixing in the ternary rhombohedral carbonate solid solution $(CaMg,Mn)CO_3$ from atomistic simulations. CALPHAD (in press). doi: 10.1016/j.calphad.2010.01.002

Vinograd VL, Sluiter MHF (2006) Thermodynamics of mixing in pyrope - grossular, $Mg_3Al_2Si_3O_3$-$Ca_3Al_2Si_3O_{12}$, solid solution from lattice dynamics calculations and Monte Carlo simulations. Am Mineral 91:1815–1830

Vinograd VL, Sluiter MHF, Winkler B, Putnis A, Hålenius U, Gale JD, Becker U (2004) Thermodynamics of mixing and ordering in the pyrope-grossular solid solution. Mineral Mag 68:101–121

Vinograd VL, Winkler B, Gale JD (2007) Thermodynamics of mixing in diopside-jadeite, $CaMgSi_2O_6$-$NaAlSi_2O_6$, solid solution from static lattice energy calculations. Phys Chem Minerals 34:713–725

Vinograd VL, Winkler B, Putnis A, Kroll H, Milman V, Gale JD, Fabrichnaya OB (2006) Thermodynamics of pyrope - majorite, $Mg_3Al_2Si_3O_{12}$-$Mg_4Si_4O_{12}$, solid solution from atomistic model calculations. Mol Simulations 32:85–99

Vinograd VL, Juares-Arellano EA, Lieb A, Knorr K, Schnick W, Gale JD, Winkler B (2007) Coupled Al/Si and O/N order/disorder in $BaYb[Si_{4-x}Al_xO_xMg_4N_{7-x}]$ sialon: neutron powder diffraction and Monte Carlo simulations. Z Kristallogr 222:402–415

Warren MC, Dove MT, Myers ER, Bosenick A, Palin EJ, Sainz-Diaz CI, Guiton BS (2001) Monte carlo methods for the study of cation ordering in minerals. Mineral Mag 65:221–248

Wei SH, Ferreira LG, Zunger A (1990) First-principles calculation of temperature-composition phase-diagrams of semiconductor alloys. Phys Rev B 41:8240–8269

Wilson DJ, Winkler B, Juares-Arellano EA, Friedrich A, Knorr K, Pickard CJ, Milman V (2008) Virtual crystal approximation study of nitridosilicates and oxonitridoaluminosilicates. J. Phys Chem Solids 69:1861–1868

Winkler B, Dove MT, Leslie M (1991) Static lattice energy minimization and lattice dynamics calculations on aluminosilicate minerals. Am Mineral 76:313–331

Winkler B, Pickard C, Milman V (2002) Applicability of a quantum mechanical 'virtual crystal approximation' to study Al/Si disorder. Chem Phys Lett 362:266–270

Winkler B, Milman V, Pickard C (2004) Quantum mechanical study of Al/Si disorder in leucite and bicchulite. Mineral Mag 65:819–824

Wolverton C, Ozolins V, Zunger A (2000) Short-range-order types in binary alloys: a reflection of coherent phase stability. J Phys Condens Matter 12: 2749-2768

Wyckoff RWG (1963) Crystal Structures. John Wiley and Sons, New York

Zunger A (1994) First principles statistical mechanics of semiconductor alloys and intermetallic compounds. *In*: NATO ASI on Statics and Dynamics of Alloy Phase Transformations. Turchi P, Gonis A (eds) Plenum Press, New York, p 361–419

Reviews in Mineralogy & Geochemistry
Vol. 71 pp. 437-463, 2010

Large Scale Simulations

Mark S. Ghiorso

OFM Research Inc.
7336 24th Ave NE
Seattle, Washington, 98115, U.S.A.

ghiorso@ofm-research.org

Frank J. Spera

Department of Earth Science & Institute for Crustal Studies
University of California
Santa Barbara, California, 93106, U.S.A.

spera@geol.ucsb.edu

INTRODUCTION

In this chapter we will focus on molecular dynamics (MD) simulations with large numbers of atoms ($N > 1000$). Typically, *ab initio* or **F**irst **P**rinciples **M**olecular **D**ynamics (FPMD) calculations are performed with smaller clusters of say 100 or fewer atoms. This choice is largely due to limitations in computing resources. The simulation procedure we will explore in this chapter is known as **E**mpirical **P**air-**P**otential **M**olecular **D**ynamics (EPPMD). EPPMD does not rely upon estimating the forces between atoms with **D**ensity **F**unctional **T**heory (DFT) as an approximation to the solution of the quantum mechanical problem, which is the computationally costly aspect of FPMD. Rather, EPPMD estimates forces between atoms from an empirical parameterization of a classical description of the potential energy, whose spatial derivative captures the pairwise-additive attractive and repulsive forces between all atoms. Because the interatomic force calculation is classical, its computational cost is minimal, and this encourages the application of EPPMD to simulations involving large numbers of atoms. Additionally, whereas FPMD simulations are typically run for durations on the order of a few picoseconds, EPPMD simulations may be extended to much longer durations (2-10 nanoseconds), which permits investigation of transport properties such as shear viscosity and phonon thermal conductivity (Fig. 1) using Green-Kubo theory (Kubo 1966). The longer durations of EPPMD runs (with up to order 10^7 time steps) are again a consequence of the computational efficiency derived from the classical force field approximation. In this chapter we are going to refer to long duration molecular dynamics simulations involving large numbers of atoms as **L**arge **S**cale **S**imulations (LSS).

The main advantage of LSS is to improve the counting statistics associated with estimating equilibrium properties, including of course the equilibrium structure, as well as the ability to monitor dynamical fluctuations over a long enough time interval to estimate transport properties with better precision (Nevins and Spera 2007). In any MD simulation there is an inherent uncertainty in every ensemble-averaged property, for example, fluctuations in temperature (σ_T) and pressure (σ_P) scale as $N^{-1/2}$ (McQuarrie 2000) in calculations carried out in the NEV or microcanonical ensemble[1]. This scaling implies that the uncertainty in temperature (T) or pressure

1 A calculation where the total number of atoms, the total energy of those atoms, and the volume occupied by those atoms is fixed by initial conditions of the simulation.

1529-6466/10/0071-0020$05.00 DOI: 10.2138/rmg.2010.71.20

(P) — that is the location of the state point in T-P space — is ~9 times smaller in a MD simulation of, say, 8000 atoms compared to one with 100 atoms; a significant difference when one is faced with the task of constructing an **E**quation **o**f **S**tate (EOS) or an Arrhenian description of a transport property such as the liquid shear viscosity over a broad region of T-P space. Similarly, equilibrium structures are determined by estimating a **R**adial **D**istribution **F**unction (RDF) for atom-pairs (e.g., O about Si, O about O, etc.), and these RDFs are more accurately determined when there are more atoms available for counting. An example is illustrated in Figure 2, for an important geoliquid, $MgSiO_3$. Both cubes in Figure 2 contain the same number density of atoms and are dawn to scale; the positions of the atoms are projected onto the front face of the cube. It is obvious from examination of the figure that an RDF developed for O about Mg, for example will be better estimated by averaging 1600 Mg positions in the right box compared to 20 in the left. The consequences become especially acute if long range RDF features (second and third nearest neighbor configurations) are the subject of interest.

An advantage of LSS when computed using EPPMD is that a large number of state points may be simulated using a well tested potential to explore subtleties in the thermodynamic relations, without the expenditure of an excessive amount of computing resources. An excellent illustration of this is the work of Poole et al. (2005) on the density minimum in liquid water and its energetic relation to the liquid-liquid critical point found in the supercooled amorphous state. These authors performed 2,718 state point simulations (with N ~ 5000) to explore the energy surface at sufficient detail to resolve the effect.

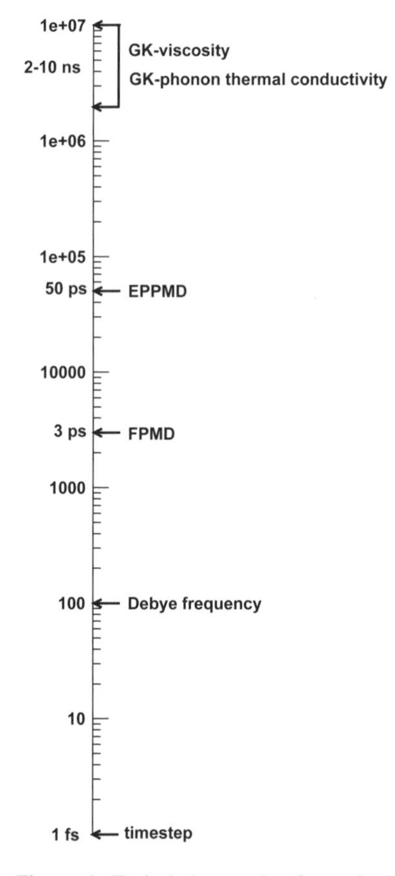

Figure 1. Typical time scales for various molecular dynamics calculations. The value labeled as "Debye frequency" refers to the reciprocal of $(3N/4\pi V)^{1/3}v_s$, where N is the number of atoms in the simulation, V is the volume and v_s is the sound speed of the material of interest; a typical value is shown.

The important caveat to appreciate when evaluating the advantages of LSS computed using EPPMD is the necessity of utilizing a well-tested and verified pair-potential. There is, of course, no sense in doing a large number of fast computations if the results of the effort are nonsensical. But, if the empirical pair-potential of interest has been parameterized from first principles studies and/or can be shown to yield results that are supported by experimental observations, then its use can be justified, especially if the computational regimen is supplemented and cross-checked against *ab initio* calculations. In some cases, where it is necessary to account for electron transfer (as in atoms with variable valence states) or where the spin state configuration of the outer electron orbitals is at issue, EPPMD cannot be easily applied, and a full quantum mechanical description of the forces must be used.

In considering the application of EPPMD in LSS it is important to appreciate that all MD simulations — even DFT based FPMD — are approximations (Cohen et al. 2008). It is critical to understand the limitations of these approximations and to choose a technique that is suitable

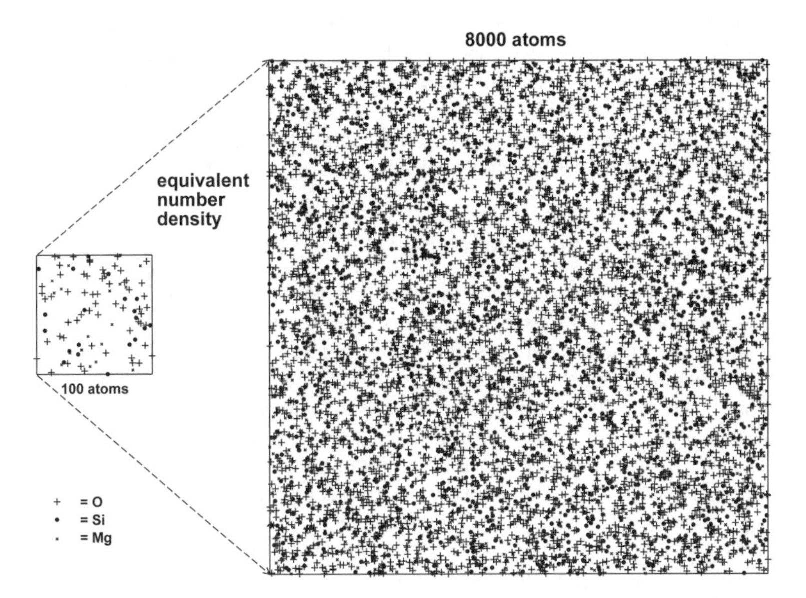

Figure 2. Visual comparison of the number density of atoms an MD simulation with 100 atoms (left) and 8000 atoms (right). The atoms are randomly placed in the cube and do not overlap; all atoms are projected to the front face. Atoms are coded to indicate O, Si and Mg, with the overall stoichiometry: $MgSiO_3$.

for the problem at hand. In this regard an analogy may be drawn between EPPMD and FPMD techniques and the geophysical interpretation of Earth structure from seismologic observations. Although no geophysicist would seriously contemplate that the Earth's interior is radially symmetric, huge advances in understanding the structure and mineralogical constitution of the Earth were made in the last century by making precisely this assumption. More recently, the assumption of radial symmetry has been relaxed and seismic tomography has revealed a more richly detailed and refined picture. But, to a decent approximation, many problems in geochemistry and geophysics can be productively addressed assuming radial symmetry. The EPPMD method is like the seismological assumption of radial symmetry. It is not as accurate as FPMD based calculations, but its careful application in computational chemistry can provide an essential data set of materials and transport properties upon which more rigorous and costly calculations may be contextualized.

Our interests in LSS are directed towards the thermodynamic and transport properties of silicate melts and glasses, especially at elevated pressure. LSS is ideally suited to investigating these substances because of their chemical complexity and their propensity to develop marked changes in short-range structure as temperature and (especially) pressure varies. The ultimate goal of our work is to formulate approximate but useful equations of state and transport properties for multicomponent liquids in the system CaO-MgO-FeO-SiO_2-TiO_2-Al_2O_3-K_2O-Na_2O-H_2O with a focus on melts of natural composition. These results can then be used with analogous properties of crystalline phases to compute phase equilibria within terrestrial planets, and address mass and heat transport phenomena central to magma dynamics.

We are going to focus this chapter on what to do with LSS results obtained on amorphous materials, using as an example EPPMD simulations recently completed on molten $MgSiO_3$. We think that it is better to illustrate the analysis of MD data using a concrete example rather than talk about that analysis in the abstract. We hope that our approach will provide the reader with a template for performing a rigorous evaluation of LSS results that can readily be applied to other amorphous materials. The chapter will begin with a formalism for thermodynamic analysis of MD state point arrays. Having developed the thermodynamic treatment, we will then touch on issues of polyamorphism and the calculation of the shock Hugoniot. Evaluation of melt structure will follow and we will explore the relation between structure and density and other thermodynamic properties. Following that we will examine transport properties. Finally we will comment on the feasibility of LSS studies that are based upon FPMD.

THERMODYNAMIC ANALYSIS

Traditionally, thermodynamic models are constructed from physical experimental data because thermodynamics provides a theoretical framework for interpolating and extrapolating these measurements without the necessity of performing a new experiment for every situation of interest. Additionally, thermodynamic analysis allows an experiment of one kind, say the measurement of the density of a liquid, to be used in estimating another property that would be difficult or impossible to measure, for instance, the variation of the Gibbs free energy with pressure. When the experimental basis for determining the properties of a system is founded on MD simulations however, the utility of constructing thermodynamic models from the results is less obvious. Why not simply compute the properties of the system at every point of interest? This reasoning is especially acute if the basis for the simulation is EPPMD running on multiprocessor machines in optimized mode for which the computational cost is modest. The answer to this argument is that thermodynamics provides more than a convenient interpolative scheme. The principles of thermodynamics permit conclusions to be drawn about the topology of the energy surface and a model facilitates the representation and analysis of that surface as a smooth continuous function. It is essential both that a model be constructed from MD simulation data and that this model be fully constrained by a sufficient number of state points to adequately embody the topological features of the energy surface. Once thermodynamic models for multicomponent liquids are available for instance, solid-liquid equilibria and the phenomenon of melting can be better explored using all the standard tools of equilibrium thermodynamics.

Perhaps the best way to illustrate the arguments of the previous paragraph is to develop an example of thermodynamic analysis of EPPMD data in LSS that we have recently calculated for $MgSiO_3$ liquid (Spera et al. 2010; see also Nevins et al. 2009). The data are obtained using NEV simulation and the pair-potential from Oganov et al. (2001), derived specifically for $MgSiO_3$ composition based on DFT calculations and semi-empirical methods utilizing ionization potential, electron affinity and other chemical properties for O, Mg and Si. The Oganov pair-potential energy between two atoms i and j has the Buckingham (1938) form:

$$V(r_{ij}) = \frac{q_i q_j e^2}{4\pi\varepsilon_0 r_{ij}} + A_{ij} e^{-r_{ij}/B_{ij}} - \frac{C_{ij}}{r_{ij}^6} \tag{1}$$

where q_i, q_j are the effective charges on species i and j, r_{ij} is the distance between the pair i-j, e is the charge of the electron, ε_0, is the vacuum permittivity, A_{ij} and C_{ij} are energy parameters describing repulsive and van der Waals attractive forces, respectively and B_{ij} is an e-folding length characterizing the radially-symmetric decay of electron repulsion energy between atom pair i-j.

In Figure 3 we illustrate results of 72 EPPMD LSSs of $MgSiO_3$ liquid ($N = 8000$) based on the Oganov pair-potential. These results are typical of state point grids computed by NEV MD in that computations are performed along chosen isochors with internal energies selected to map out isotherms. The state points in Figure 3 have uncertainties in pressure smaller than the size of the symbols, a direct consequence of using 8000 atoms in the simulations. The associated temperature fluctuations are on the order of 25 K. These results constitute a data set from which a thermodynamic model may be constructed.

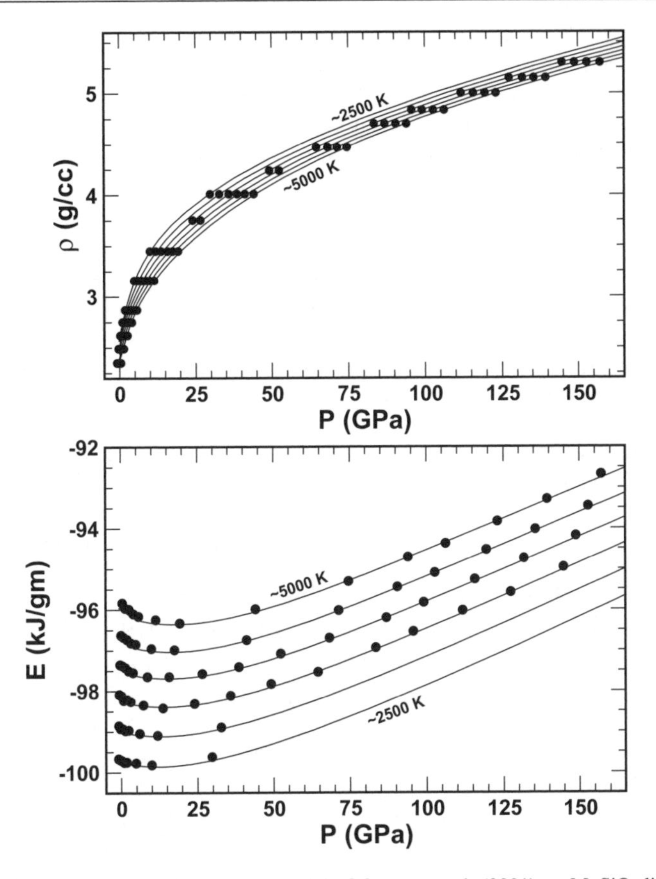

Figure 3. EPPMD using LSS and the pair-potential of Oganov et al. (2001) on $MgSiO_3$ liquid. Results were computed using the NEV ensemble along 16 isochors at evenly spaced temperatures over the range 2500-5000 K. *Upper panel*: density (ρ) plotted as a function of pressure (*P*). *Lower panel*: Internal energy (*E*) plotted as a function of pressure. Pressure uncertainty (i.e., MD pressure fluctuations) are smaller than the size of the symbols.

How to go about making this thermodynamic model generates some important considerations. The rule is: always formulate first an expression for the energy (either the internal energy, enthalpy, Helmholtz energy, or Gibbs free energy) and formally manipulate this expression using thermodynamic identities to retrieve internally consistent equations for the derivative properties. Importantly, when formulating a thermodynamic model for any substance, one should not start by fitting an EOS — that is a relationship between pressure, temperature and density or volume — to the data set, unless it can be established that the EOS is a proper derivative of an energy function; in this context a necessary (but not sufficient condition) is that the thermodynamic identify:

$$V\alpha K + K\frac{\partial\alpha}{\partial P} = V + \frac{1}{K}\frac{\partial K}{\partial T} \tag{2}$$

(Prigogene and Defay 1954; V is the volume, α is the isothermal coefficient of thermal expansion, K is the isothermal bulk modulus, T is the temperature, and P is the pressure) must hold for any adopted EOS.

A thermodynamic model will be created for these MD data by developing an equation for the Helmholtz energy. As a starting point we will use a temperature-potential energy scaling law derived by Rosenfeld and Tarazona (1998) from a fundamental-measure energy functional for hard spheres and thermodynamic perturbation theory. Their postulated expression relates the potential energy (E_P) of a substance to temperature according to

$$E_P = a + bT^{3/5} \tag{3}$$

where a and b are unspecified functions of volume but not temperature (note that at absolute zero, a is identical to the internal energy of the material). One does not need to assume that Equation (3) holds; MD simulation results can be used to test the applicability of Rosenfeld-Tarazona (RT) scaling to dense high temperature silicate liquids. Clearly, when MD generated values of E_P are plotted versus $T^{3/5}$, linear correlations are expected if RT scaling is valid. Potential energy-temperature relations are plotted in Figure 4 for the state point arrays shown in Figure 3. Indeed, RT scaling holds with model errors on the order of 0.01% in E_P. We have found previously (Nevins et al. 2009; Martin et. al. 2009; Ghiorso et al. 2009; Spera et al. 2009) that the Rosenfeld-Tarazona scaling law applies equally well to a wide variety of silicate melts — polymerized and unpolymerized, strong and fragile — over a broad range of compressions and temperatures. Although originally proposed to be applicable to "cold, dense" liquids, RT scaling appears to extend to "hot, dense" geoliquids as well. It is important to emphasize that application of RT scaling to thermodynamic modeling of silicate liquids is an assumption that is easily tested directly using simulation results. To date, this test has succeeded in every case that we have examined.

From Equation (3), the internal energy (E) is given by

$$E = E_P + \frac{3n}{2}RT \tag{4}$$

where n is the number of atoms in the formula unit of the simulated substance (i.e., for $MgSiO_3$, $n = 5$) and R is the universal gas constant in units of energy/g-K. The right-hand term in Equation

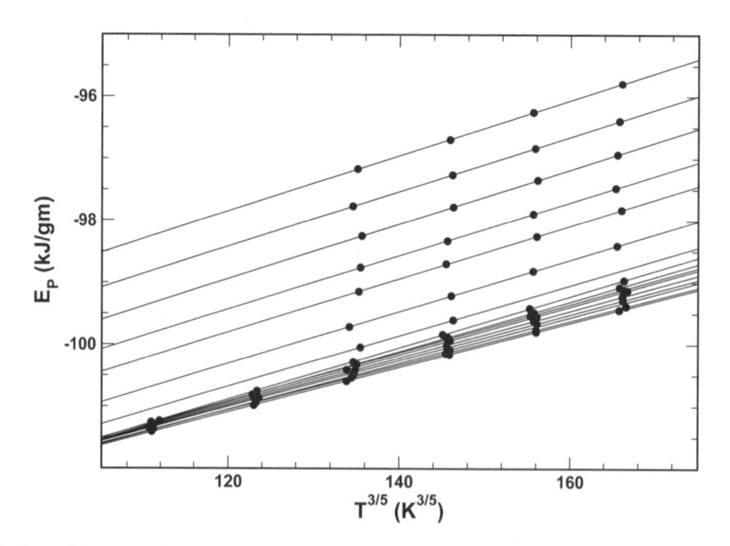

Figure 4. Test of the Rosenfeld-Tarazona (1998) scaling law for the data set of Figure 3. Potential energy (E_p) is plotted as a function of temperature (T). Deviations from the linear scaling relation, $E_p = a(V) + b(V)\,T^{3/5}$, are less than 0.01%.

(3) is the classical high-temperature limit of the kinetic energy, which can be shown to hold for these MD simulation results (to within 0.01%). Substitution of Equation (3) into Equation (4) gives a Rosenfeld-Tarazona compatible model expression for the internal energy

$$E = a + bT^{3/5} + \frac{3n}{2}RT \tag{5}$$

A derivation of the Helmholtz energy (A) from Equation (5) was first developed by Saika-Voivod et al. (2000). Following their algorithm, we start with the thermodynamic identity,

$$A = E - TS \tag{6}$$

which illustrates that the Helmholtz energy depends on the entropy of the system (S). Fortunately, an expression for entropy is easily constructed. Start by writing the total derivative of S as

$$dS = \left(\frac{\partial S}{\partial T}\right)_V dT + \left(\frac{\partial S}{\partial V}\right)_T dV \tag{7}$$

The two partial derivatives in Equation (7) may be rewritten using the first law of thermodynamics ($dE = TdS - PdV$), under the assumption of reversibility) as

$$dS = \frac{1}{T}\left(\frac{\partial E}{\partial T}\right)_V dT + \left[\frac{1}{T}\left(\frac{\partial E}{\partial V}\right)_T + \frac{P}{T}\right]dV \tag{8}$$

and this expression may be integrated along a path from some reference volume (V_0) and reference temperature (T_0) to V,T_0 and then from V,T_0 to V,T:

$$S - S_{V_0,T_0} = \frac{1}{T_0}\int_{V_0}^{V}\left[\left(\frac{\partial E}{\partial V}\right)_T + P_{T_0}\right]dV + \int_{T_0}^{T}\frac{1}{T}\left(\frac{\partial E}{\partial T}\right)_V dT \tag{9}$$

Substitution of Equation (5) into Equation (9) and carrying out the integrations results in a Rosenfeld-Tarazona compatible model expression for the entropy

$$S = S_{V_0,T_0} + \frac{1}{T_0}\left[a - a_{V_0} + T_0^{3/5}\left(b - b_{V_0}\right) + \int_{V_0}^{V} P_{T_0}dV\right] - \frac{3}{2}\left(\frac{1}{T^{2/5}} - \frac{1}{T_0^{2/5}}\right)b + \frac{3n}{2}R\ln\left(\frac{T}{T_0}\right) \tag{10}$$

The subscripts V_0 and T_0 denote evaluation of the quantity under the specified conditions. It is worth noting that the entropy can only be known to within an arbitrary constant (S_{V_0,T_0}). From Equations (6) and (10) the Helmholtz energy may be written:

$$A = a + T^{3/5}b + \frac{3n}{2}RT - TS_{V_0,T_0} - \frac{T}{T_0}\left[a - a_{V_0} + T_0^{3/5}\left(b - b_{V_0}\right) + \int_{V_0}^{V} P_{T_0}dV\right] \tag{11}$$

$$+T\frac{3}{2}\left(\frac{1}{T^{2/5}} - \frac{1}{T_0^{2/5}}\right)b - \frac{3n}{2}RT\ln\left(\frac{T}{T_0}\right)$$

and by standard thermodynamic transformations both the Gibbs free energy ($G = A + PV$) and the enthalpy ($H = E + PV$) may be derived from Equations (5) and (11).

A Rosenfeld-Tarazona compatible EOS is obtained from Equation (11) by differentiation,

$$P = -\left(\frac{\partial A}{\partial V}\right)_T = \left(\frac{T}{T_0} - 1\right)\frac{da}{dV} + \frac{5}{2}T^{3/5}\left[\left(\frac{T}{T_0}\right)^{2/5} - 1\right]\frac{db}{dV} + \frac{T}{T_0}P_{T_0} \tag{12}$$

There are several interesting features of this expression that are worth taking a moment to ponder. Let us first focus on the last term in the expression involving P_{T_0}. This term is the pressure variation along the T_0 isotherm. To evaluate it, either the ρ-P relations of state points along some reference isotherm must be interpolated or alternately, these points may be fitted to some appropriate function. Of course, this function would be nothing more than an *isothermal* equation of state, such as the 3rd order Birch-Murnaghan equation (Birch 1939, 1952):

$$P_{T_0} = \frac{3}{2}K\left(x^7 - x^5\right)\left[1 - \frac{3}{4}(4 - K')\left(x^2 - 1\right)\right] \qquad x = \left(\frac{V_{T_0,P_0}}{V_{T_0}}\right)^{1/3} \tag{13}$$

which is derived from strain theory, or the Universal EOS (Vinet et al. 1986):

$$P_{T_0} = \frac{3K(1-x)e^{\eta(1-x)}}{x^2} \qquad \eta = \frac{3}{2}(K' - 1) \qquad x = \left(\frac{V_{T_0}}{V_{T_0,P_0}}\right)^{1/3} \tag{14}$$

which is derived from a generalized bonding potential function, or alternately by a purely empirical expression (e.g., Saika-Voivod et al. 2000):

$$P_{T_0} = \sum_i \frac{c_i}{V_{T_0}} \tag{15}$$

In all cases, the parameters of the representations (V_{T_0,P_0}, K, K', c_i) are constants. The temperature dependence of the pressure in the EOS embodied in Equation (12) is determined in part from scaling P_{T_0} by T/T_0 and in part from the volume dependence of the intercepts and slopes of the E_P versus $T^{3/5}$ relations. The functional form of this expression is not necessarily compatible with alternate temperature dependent extensions of the more commonly used equations of state, for example the thermal pressure correction to the Universal EOS (Vinet et al. 1987) or the Mie-Grüneisen extension of the Birch-Murnaghan equation. The reverse is also true: These alternate temperature-dependent EOS formalisms will not recover the E_P vs. $T^{3/5}$ scaling of Rosenfeld-Tarazona theory. In other words, the practice of applying a "pressure correction" to the Universal or Birch-Murnaghan EOS is generally incompatible with RT scaling.

In order to use Equation (12) in practice for the dataset displayed in Figure 3 we must obtain $a(V)$ and $b(V)$ as differentiable functions of volume and we must select a way to represent P_{T_0}. A polynomial expansion in volume works well for the Rosenfeld-Tarazona parameters, i.e.

$$a = \sum_{i=0} a_i V^i \qquad b = \sum_{i=0} b_i V^i \tag{16}$$

(Fig. 5) and the Universal EOS (Eqn. 14) recovers the ρ-P relations along the 4000 K isotherm (with V_0 = 0.408031 cc/g, K = 13.6262 GPa, and K' = 7.66573). The smooth curves plotted in Figure 3 are computed from this parameterization and Equations (5) and (12).

Having derived a thermodynamic model from the MD state points, one may explore a few derivative thermodynamic properties as an illustration. The temperature derivative of the pressure (Eqn. 12) along an isochore is the "thermal pressure" coefficient, which for the Rosenfeld-Tarazona form is given by

$$\alpha K = \left(\frac{\partial P}{\partial T}\right)_V = \frac{1}{T_0}\frac{da}{dV} + \frac{3}{2}T^{-2/5}\left[\frac{5}{3}\left(\frac{T}{T_0}\right)^{2/5} - 1\right]\frac{db}{dV} + \frac{P_{T_0}}{T_0} \tag{17}$$

and is equal to the product of the isothermal coefficient of thermal expansion (α) and the isothermal bulk modulus (K). Plots of the thermal pressure coefficient for our example liquid are shown in Figure 6. For many solids, αK is approximately a constant, and this fact is utilized

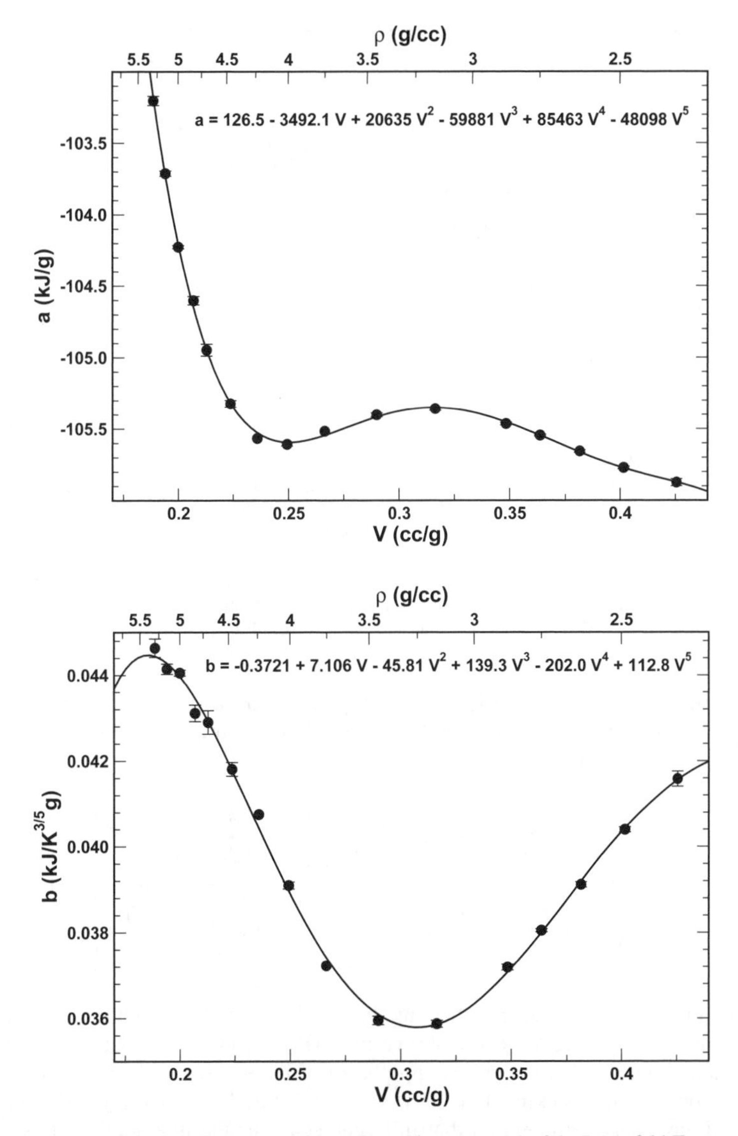

Figure 5. Polynomial functions fitted to the slopes (b) and intercepts (a) of the Rosenfeld-Tarazona scaling relations (Fig. 4), parameterized as a function of specific volume (V). Uncertainties on a and b are plotted as one sigma and are derived from fitting straight lines to the scaling relations.

in many temperature-dependent EOS formulations, including that of Vinet et al. (1987). The constancy of αK is certainly not the case for simulated $MgSiO_3$ liquid and one reason for this may be rooted in the observation that the structure (cation-oxygen coordination) of this liquid changes dramatically as a function of pressure below ~50 GPa (this issue is investigated in greater detail below). The pressure variation of the bulk modulus,

$$K = -\left(\frac{T}{T_0} - 1\right) V \frac{d^2 a}{dV^2} - \frac{5}{2} T^{3/5} \left[\left(\frac{T}{T_0}\right)^{2/5} - 1\right] V \frac{d^2 b}{dV^2} - \frac{T}{T_0} V \frac{dP_{T_0}}{dV} \tag{18}$$

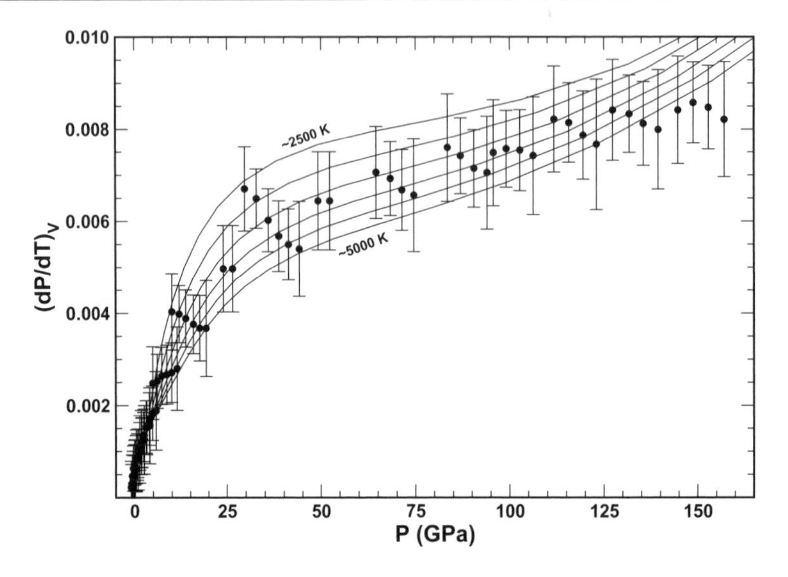

Figure 6. Thermal pressure (= αK) for the data set of Figure 3. The points are calculated by finite difference of simulation results along computed isochors. Uncertainties are plotted as one sigma and estimated by error propagation. The curves are model estimates calculated from Equation (17).

by contrast behaves as might be expected of a typical solid and to first order (and perhaps surprisingly) is linear and not influenced by changes in melt structure. It follows then that the isothermal coefficient of thermal expansion,

$$\alpha = -\frac{\frac{1}{T_0}\frac{da}{dV} + \frac{3}{2}T^{-2/5}\left[\frac{5}{3}\left(\frac{T}{T_0}\right)^{3/5} - 1\right]\frac{db}{dV} + \frac{P_{T_0}}{T_0}}{V\left\{\left(\frac{T}{T_0} - 1\right)\frac{d^2a}{dV^2} + \frac{5}{2}T^{3/5}\left[\left(\frac{T}{T_0}\right)^{2/5} - 1\right]\frac{d^2b}{dV^2} + \frac{T}{T_0}\frac{dP_{T_0}}{dV}\right\}} \tag{19}$$

will exhibit a pressure dependence reflecting the non-linearity in αK and that is borne out by the MD simulation data (Fig. 7). These observations (Figs. 6 and 7) demonstrate an interesting generalization about the thermodynamic properties of liquids that undergo a strong pressure dependent cation-oxygen packing. If a matrix of second partial derivatives (the Hessian matrix) is formed of any energetic potential that fully characterizes the thermodynamic state of the liquid, say the Helmholtz free energy, then the on-diagonal elements of that matrix, e.g.,

$$\frac{\partial^2 A}{\partial V^2} = -\frac{\partial P}{\partial V} = \rho K \tag{20}$$

and

$$\frac{\partial^2 A}{\partial T^2} = -\frac{\partial S}{\partial T} = -\frac{C_V}{T} = \frac{3}{5}\frac{b(V)}{T^{7/5}} - \frac{3n}{2}\frac{R}{T} \tag{21}$$

will not be strong functions of the changes in cation-oxygen packing. This conclusion is illustrated in Figure 8 by noting that the pressure dependence of the plotted quantities is not dramatically different below 50 GPa, where $SiO^{[4]}$ and $SiO^{[5]}$ dominate (see below), when

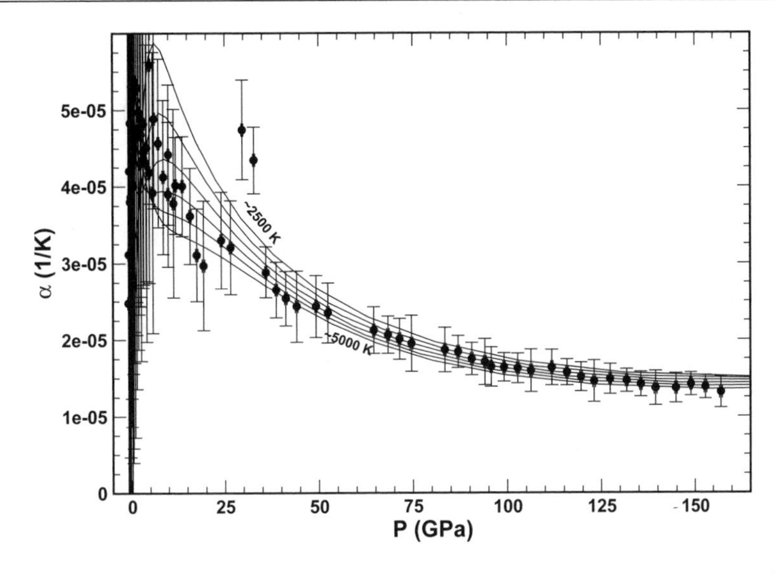

Figure 7. The isothermal coefficient of thermal expansion (α) for the data set of Figure 3. The points are calculated by finite difference of simulation results and uncertainties are plotted as one sigma and estimated by error propagation. The curves are model estimates calculated from Equation (19).

compared to higher pressures, where SiO[6] dominates[2]. By contrast, the off-diagonal Hessian elements, e.g.,

$$\frac{\partial^2 A}{\partial T \partial V} = -\frac{\partial P}{\partial T} = -\alpha K \tag{22}$$

will reveal a sensitivity to melt structure, as illustrated in Figure 6, where the pressure-dependence of the thermal pressure coefficient is seen to be dramatically different below and above 50 GPa. The on-diagonal Hessian elements are responding to the bulk properties of the liquid, what in the thermodynamics of solid solutions is normally referred to as the vibrational contribution to the energy. Energetic contributions arising from changes in the basic configurational units or building blocks that comprise the liquid are typically second order and while present in the on-diagonal elements, they are overwhelmed by the vibrational contribution and are only readily detectable in the off-diagonal terms of the Hessian. These considerations explain why the thermal Grüneisen parameter,

$$\gamma = \frac{\alpha K V}{C_V} \tag{23}$$

which is essentially a scaling of the thermal pressure coefficient, displays such a strong dependence on pressure below 50 GPa and essentially no pressure dependence above that (Fig. 9).

Polyamorphism

A thermodynamic model that has been constructed from LSS MD data can be used to investigate the topology of the energy surface and to probe that topology for evidence of phase instability. This is an especially useful technique when it is applied to simulation data on

2 Notation: XY[n] implies n atoms of type "Y" surround, in nearest neighbor coordination, the central atom "X."

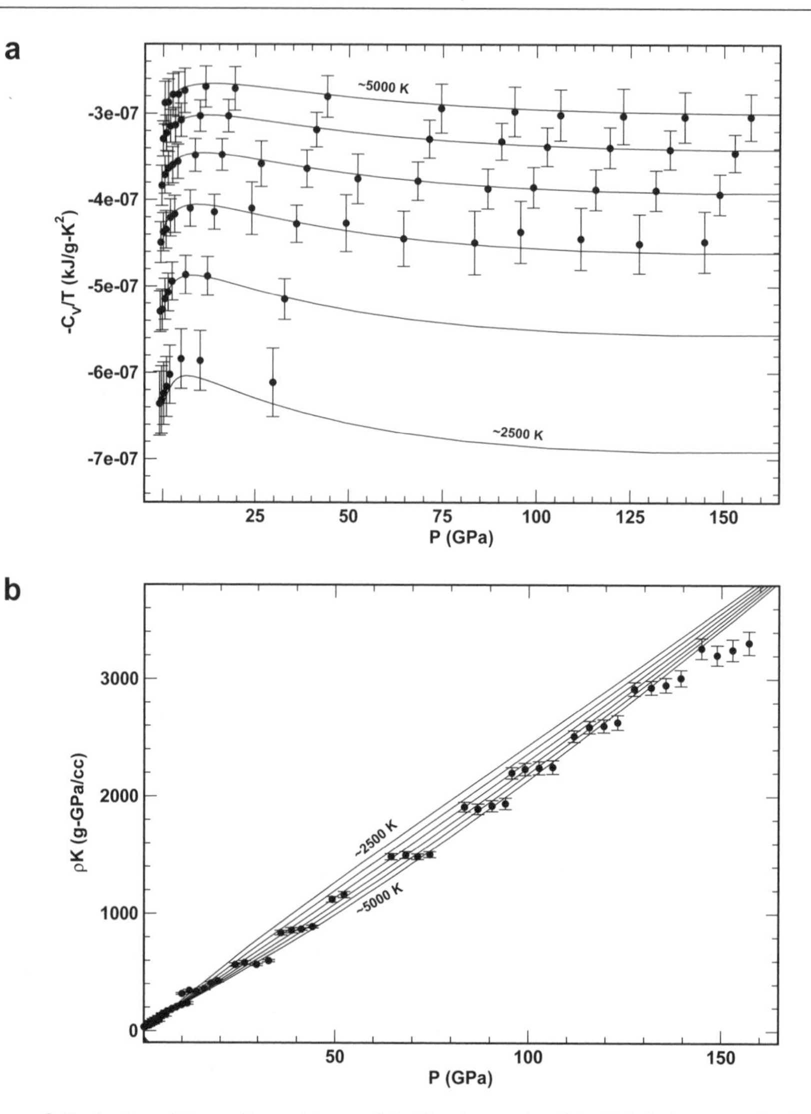

Figure 8. Evaluation of the on-diagonal terms of the Hessian matrix of the Helmholtz energy for the data set of Figure 3. The points are calculated by finite difference of simulation results and uncertainties are plotted as one sigma and estimated by error propagation. The curves are model estimates calculated from Equation (21) (Fig. 8a) and Equation (20) (Fig. 8b).

liquids, where an analysis of the topology of the energy surface affords an understanding of the phenomena of polyamorphism. Polyamorphism refers to the equilibrium coexistence of two or more amorphous phases that have the same composition, temperature and pressure but differ in structure and hence density. It is analogous to polymorphism which occurs when a crystalline phase undergoes a phase transition from one structural form to another (e.g., quartz to coesite) and is similar to compositional unmixing, which occurs in isostructural solid and liquid solutions when phases of differing composition coexist in thermodynamic equilibrium.

Polyamorphism has been observed experimentally in liquid phosphorous (Katayama et al. 2000, 2004) and other compounds and has been identified in molecular dynamics studies

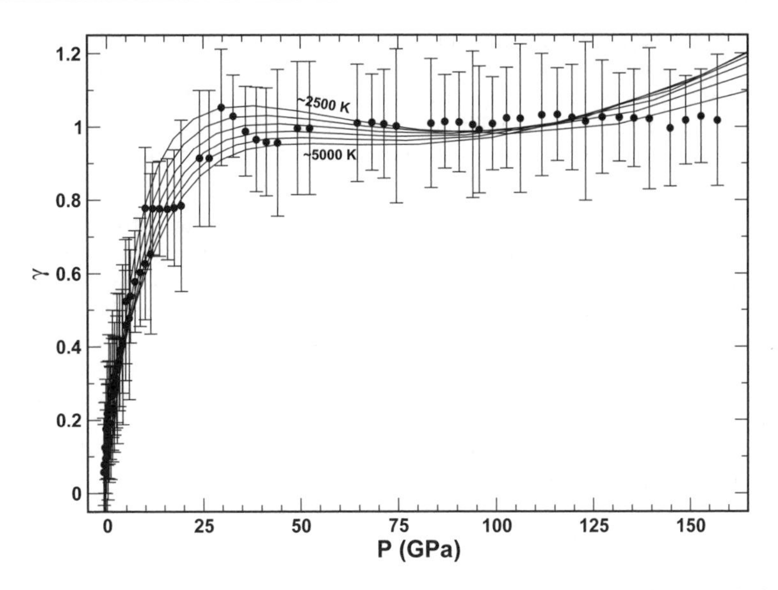

Figure 9. The thermal Grüneisen parameter (ϒ) for the data set of Figure 3. The points are calculated by finite difference of simulation results and uncertainties are plotted as one sigma and estimated by error propagation. The curves are model estimates calculated from Equation (23).

of liquid H_2O (Poole et al. 2005), SiO_2 (Saika-Voivod et al. 2000), and liquid $CaAl_2Si_2O_8$ (Ghiorso et al. 2009). Tanaka (2000) discusses the thermodynamics of polyamorphism and presents methods for developing models that describe its energetic consequences.

A sufficient criteria for the development of polyamorphism is violation of the so-called condition of mechanical stability (Prigogene and Defay 1954) which states that the derivative, $\partial^2 A/\partial V^2$ must be positive for a homogeneous phase. This condition is equivalent to $K > 0$ (Eqn. 20). Using the Rosenfeld Tarazona EOS (Eqn. 12), the locus of points that define a zero in K is:

$$0 = -\left(\frac{T}{T_0} - 1\right)\frac{d^2 a(V)}{dV^2} - \frac{5}{2}T^{3/5}\left[\left(\frac{T}{T_0}\right)^{2/5} - 1\right]\frac{d^2 b(V)}{dV^2} - \frac{T}{T_0}\frac{dP(T_0,V)}{dV} \tag{24}$$

and a solution of this expression corresponds to an implicit surface in V-T-P space. This surface is known as the spinodal. Note, that although pressure is not explicit in Equation (24), its value is implied by any (V,T) pair *via* the EOS (Eqn. 12, or equivalent). Equation (24) or its equivalent is usually solved by selecting a T or V (= 1/ρ) and calculating the remaining variable by iteration. This process will always result in either two real roots, one real root, or two complex roots. Solutions of the first kind correspond to the limbs of the spinodal, while a solution of the second kind gives the critical point for the onset of unmixing.

Utilizing the Rosenfeld-Tarazona EOS calibrated above for $MgSiO_3$ liquid, we can illustrate calculation of the spinodal and its critical end point. Our first consideration in performing this examination is to consider qualitatively the topology of the energy surface and address the question of whether we expect polyamorphism to occur at all. The answer is "yes" and we know this by examining the Helmholtz energy in the low-temperature limit. From Equation (11), $\lim_{T\to 0} A = a$, which tells us that in a thermodynamic model of the liquid that follows a Rosenfeld-Tarazona scaling law, the variation of Helmholtz free energy at zero T is simply given by $a(V)$. This function is plotted in Figure 5a, and we note that the curve has two points of inflection, which means that within the volume interval interior to these inflection points, $\partial^2 A/\partial V^2$ is negative and the liquid is mechanically unstable. The inflection points are the limbs of the spinodal, in this case at zero Kelvin. It is also worth noting from Figure 5a that the limbs of the spinodal *do not* correspond to the densities of hypothetical equilibrium coexisting liquid phases. These equilibrium compositions are exterior to the spinodal, that is at both lower and higher densities than the spinodal points, and are defined by the intersection of a line of mutual tangency to the $a(V)$ curve at both locations.

Now that we know that a spinodal exists for this liquid, we can solve Equation (24) by choosing suitable temperatures, finding volume-roots until a temperature is reached — the critical temperature (T_C) — where the points of inflection coalesce to a single critical volume (V_C) or density (ρ_C). The results of these calculations are plotted in Figure 10 for our example LSS $MgSiO_3$ data set. We find that in this case the critical point for liquid-liquid unmixing is at a temperature well below the stable liquid field, that is within the thermodynamic stability field of solid $MgSiO_3$ (enstatite), so it is very unlikely that polyamorphism will ever be observed in this liquid. However, other silicate liquid compositions, such as $CaAl_2Si_2O_8$ liquid, appear to have critical end points for liquid-liquid unmixing within the stable liquid field (Ghiorso et al. 2009), a prediction that awaits verification by physical experimentation. The general expectation is that polyamorphism is an expected phenomena in molten silicates because they tend to form strongly bonded covalent localized structural units (Kurita and Tanaka 2005).

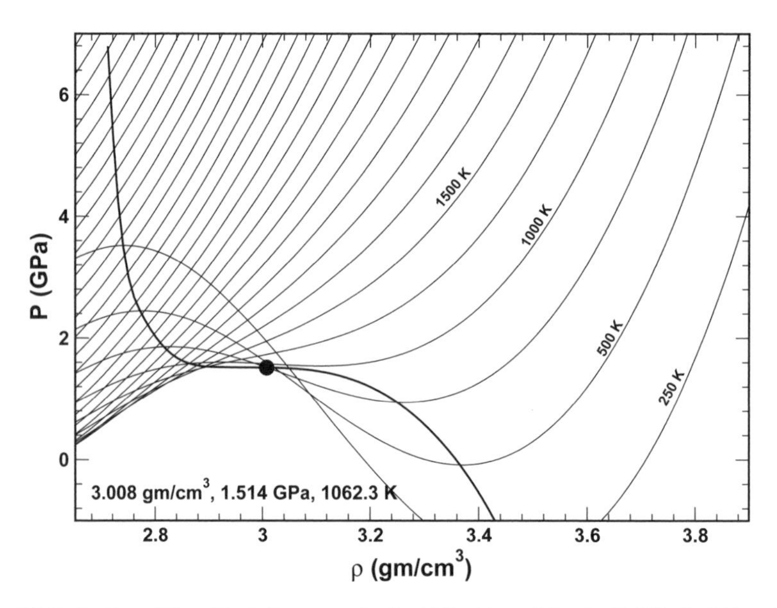

Figure 10. Evaluation of the critical phenomena for liquid-liquid unmixing in $MgSiO_3$ liquid based upon the data set of Figure 3. The critical point for the onset of unmixing is shown by the *large dot*; its critical density, pressure and temperature are indicated. The *heavy curve* traces the spinode, which encloses a region of mechanical instability. The remaining curves are isotherms calculated from the model EOS (Eqn. 12).

Hugoniot

One final example of what can be done with a thermodynamic model derived from LSS MD studies involves calculation of the shock-wave Hugoniot. The Hugoniot is a univariant curve in *V-T-P*-space that characterizes the final states of a series of shock experiments of varying intensity that originate from the same set of initial conditions (Anderson 1989). The Hugoniot curve is modeled by finding *V-T-P* points that simultaneously satisfy an EOS and a constraint

$$E - E_i = \frac{1}{2}(P - P_i)(V_i - V) \tag{25}$$

(Anderson 1989) that results from a combination of the conservation of mass, momentum and energy equations. The subscript i refers to the initial shock conditions.

For a substance that obeys the Rosenfeld-Tarazona EOS (Eqn. 12), Equation (25) reduces to

$$C_1 T^{3/5} + C_2 T = C_3 \tag{26}$$

where

$$C_1 = b + \frac{5}{4}\frac{db}{dV}(V_i - V) \tag{27}$$

$$C_2 = \frac{3n}{2}R - \frac{V_i - V}{2T_0}\left(\frac{da}{dV} + \frac{5}{2}T_0^{3/5}\frac{db}{dV} + P_{T_0}\right) \tag{28}$$

and

$$C_3 = a_{V_i} - a + \frac{1}{2}\left(\left.\frac{da}{dV}\right|_{V_i} - \frac{da}{dV}\right)(V_i - V) + T_i^{3/5}\left(b_{V_i} + \frac{5}{4}\left.\frac{db}{dV}\right|_{V_i}(V_i - V)\right) + T_i\left[\frac{3n}{2} - \frac{V_i - V}{2T_0}\left(\left.\frac{da}{dV}\right|_{V_i} + \frac{5}{2}T_0^{3/5}\left.\frac{db}{dV}\right|_{V_i} + P_{T_0}\right)\right] \tag{29}$$

In Figure 11 we plot a calculated Hugoniot for our example $MgSiO_3$ liquid for initial conditions of 1832 K and zero pressure (the metastable melting temperature for enstatite). The inset of this figure shows the Hugoniot curve translated into alternate coordinates

$$U_s = V_i\left(\frac{P - P_i}{V_i - V}\right)^{\frac{1}{2}} \qquad U_p = \left[(P - P_i)(V_i - V)\right]^{\frac{1}{2}} \tag{30}$$

where U_s and U_p are the shock and particle velocities, respectively. The calculated Hugoniot shows very high-shock temperatures at elevated pressures (~10000 K at 125 GPa). For comparison, we have also plotted an isentrope for this liquid calculated by solving Equations (10) and (12) simultaneously while holding the system entropy at a value defined by the initial conditions of the shock. Note that the isentrope also climbs in T with increasing P, but far less dramatically than the Hugoniot. The data points are from Akins et al. (2004) with properties inferred from the experimental shock wave studies of solids and glasses that underwent melting during compression.

STRUCTURAL FEATURES

LSS MD studies are ideally suited for investigation of structural changes that take place in a substance as a function of T and P. This is especially true for non-crystalline materials, where the absence of both long-range order and lattice symmetry constraints requires the statistical evaluation of nearest neighbor atomic coordinations to fully characterize the structural state. The difficulty of this task is further exacerbated if the material is chemically complex, because under these circumstances more atoms are required to sample all possible structural configurations. As an example, consider the LSS MD study on $MgSiO_3$ used in the previous

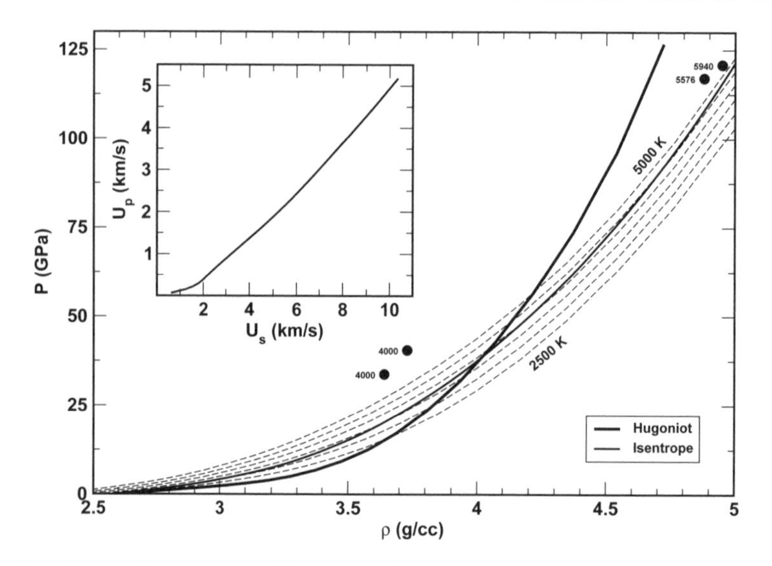

Figure 11. Calculated Hugoniot curve (Eqns. 16-28; *heavy curve*) and isentrope (Eqn. 10; *normal curve*) for $MgSiO_3$ liquid based upon the data set of Figure 3. *Dashed curves* are isotherms calculated from the model EOS (Eqn. 12). The dots indicate coordinates obtained from shock waves studies (Akins et al. 2004) and are labeled according to estimated shock temperature. The inset shows the Hugoniot curve mapped into the alternate coordinate space defined in terms of particle (U_p) and shock (U_s) velocities.

section. For amorphous substances, short-range nearest neighbor structure is determined by statistical analysis of atom locations using partial pair correlation functions (radial distribution functions, RDF) expressed as:

$$g_{ij}(r) = \frac{V}{N^2}\left\langle \sum_{i=1}^{N} \sum_{j=1, i \neq j}^{N} \delta(r - r_{ij}) \right\rangle \tag{31}$$

For two atoms i and j, Equation (31) gives the normalized averaged distribution of atom i around a central j atom within a defined cut off distance. V is the volume of the MD primary cell and N the number of particles. The brackets denote averaging. To illustrate this concept, the RDF for oxygen around a central oxygen (oxygen about oxygen, hereafter) is plotted in Figure 12 for the 3500 K isotherm and for pressures between zero and 150 GPa. The function values give the average number of O-atoms around a given O-atom as a function of distance about that atom. By convention, the average coordination number of oxygen about oxygen is defined by numerical integration of $g_{\text{O-O}}(r)$ up to a distance from the central atom corresponding to the first minimum in the RDF that follows the first maximum. This cut off distance is uniquely determined by the position of the minimum for each isobar shown in Figure 12. We plot these average O-O coordination numbers in Figure 13, along with values calculated in an identical way from RDFs for oxygen about silicon and oxygen about magnesium. The results in this example are fascinating for a variety of reasons. The most striking observation is the abrupt change in average coordination number of O-O between about 65 and 85 GPa. What is happening is that the minimum in the O-O RDF abruptly shifts location over this pressure interval, as can be seen in Figure 12 (with details illustrated in the inset). The same abrupt shift in minima location is *not* displayed by the RDFs for Si-O or Mg-O, and one can see from Figure 13 that the bulk liquid density, which is essentially a linear function of average Si-O coordination number, does not respond to this fundamental rearrangement of the oxygen packing. We will see later that the transport properties are affected by this rearrangement.

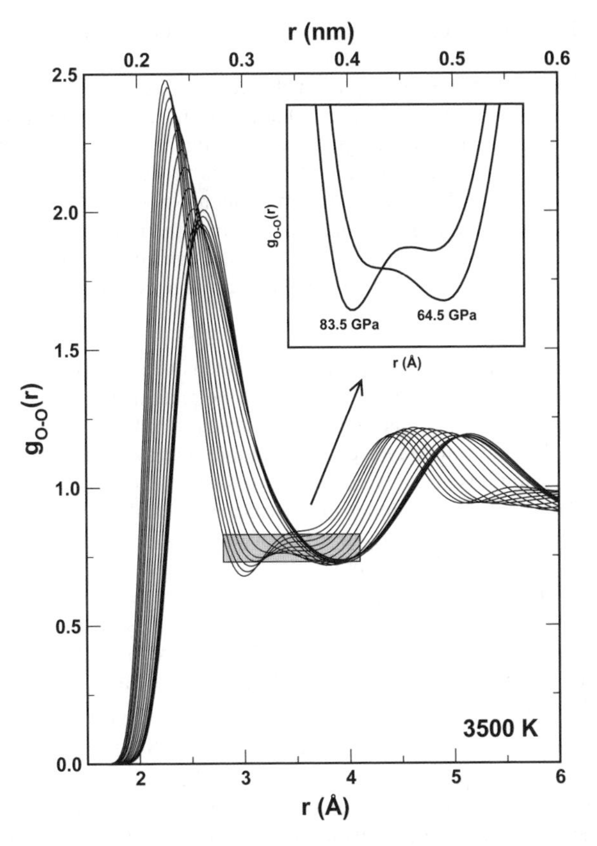

Figure 12. Radial distribution functions (Eqn. 31) for oxygen-oxygen pairs calculated from simulation results for the data set of Figure 3 plotted as a function of distance from the central oxygen atom. Inset shows a blow up of the shift in the position of the minimum over the pressure interval ~65 to ~85 GPa.

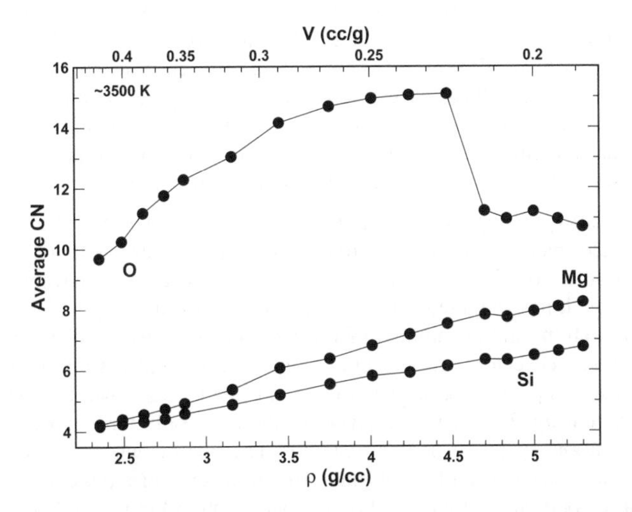

Figure 13. Average coordination number (CN) for oxygen about oxygen, oxygen about Mg and oxygen about silicon calculated from simulation results for the data set of Figure 3.

Further details of the structural variation can be scrutinized by counting the number of configurations of a given kind up to the cut-off radius of the RDF and forming a distribution of occurrences of each unit. For example, in the case of oxygen about oxygen, we would count the number of $OO^{[1]}$, $OO^{[2]}$, $OO^{[3]}$, etc., that are found within the cut off radius for each oxygen atom in the simulation and average these quantities to obtain a probability distribution or number abundance of particular configurations. This procedure allows us to examine the "speciation" of the liquid as a function of changing intensive variables. Figure 14 shows these results along a ~3500 K isotherm for the liquid, and from them one can visualize the changes in Si-O coordination as a function of pressure and understand better the oxygen about oxygen rearrangements that accompany the abrupt change in average coordination number.

Microscopic to macroscopic

One of the great advantages of calculating an extensive array of state points using MD simulations of materials is the opportunity to explore the relationship between the microscopic and the macroscopic, that is, the relation between structural change and the variation in bulk physical properties. We have already seen for example that qualitatively the bulk density of our example $MgSiO_3$ liquid is strongly correlated to the average oxygen coordination number of silicon (Fig. 13). There are a number of approaches to making these relationships more quantitative and to explore more deeply the kinds of structural transformations that accompany changes in T and P. In a liquid like $MgSiO_3$ for instance, the variation in oxygen coordination number of Si is probably not independent of that of Mg, since the same oxygen atoms are likely involved in both cation coordination polyhedra. The degree of correlation amongst changes in structural features that take place as a function of T and P, reduces the number of degrees of freedom and directly impacts the *configurational* entropy associated with the effects.

A simple way to visualize these correlations is to construct a cluster dendrogram from a correlation coefficient matrix that is in turn constructed from the abundances of various species or configurational units along a specified isobar or isotherm. For example, Figure 15 illustrates a cluster dendrogram obtained from examination of the abundances of cation-oxygen and oxygen-oxygen nearest neighbor coordination polyhedra at 12 state points along the ~3500 K isotherm for our $MgSiO_3$ example liquid. For those unfamiliar with cluster analysis, the degree of correlation is represented by vertical linkages which are positioned to indicate high to low correlation by their location right to left on the diagram. Thus, $SiO^{[6]}$ and $MgO^{[7]}$ are very highly correlated as are $SiO^{[7]}$ and $MgO^{[9]}$, but the first group is more weakly correlated to the second. All four configurations however, are uncorrelated to the abundance of $SiO^{[4]}$. The dendrogram enables a visual assessment of the extent of correlation between structural units in the liquid and from it we can conclude that there are essentially three principal groupings or combinations of structural units that vary independently as a function of pressure along the isotherm. This result is important because it emphases that the number of "independent" modes of structural variation is three, despite the myriad ways of counting the atomic configurations.

To further quantify this assessment of structural correlation, a principal component analysis (PCA) can be performed on the same matrix of correlation coefficients (R) used in making the cluster dendrogram of Figure 15. PCA performs an eigenvector/eigenvalue decomposition of the R-matrix, yielding weighted linear combinations (eigenvectors) of the abundances of structural configurations and the amount of variance (eigenvalue) associated with each linear combination. The eigenvectors are, of course, uncorrelated and may be ranked in order of decreasing variance. As would be expected from our dendrogram, we find that the first three eigenvectors account for 99% of the variance, so there are only three independent measures of structural variation operative in this example. The coefficients of these three principal eigenvectors demonstrate which of the original structural configurations contribute and in what proportion. These coefficients are plotted on the bar graph in Figure 16. Focusing just on the first eigenvector, we can see that it embodies the transformation from low- to high-coordination

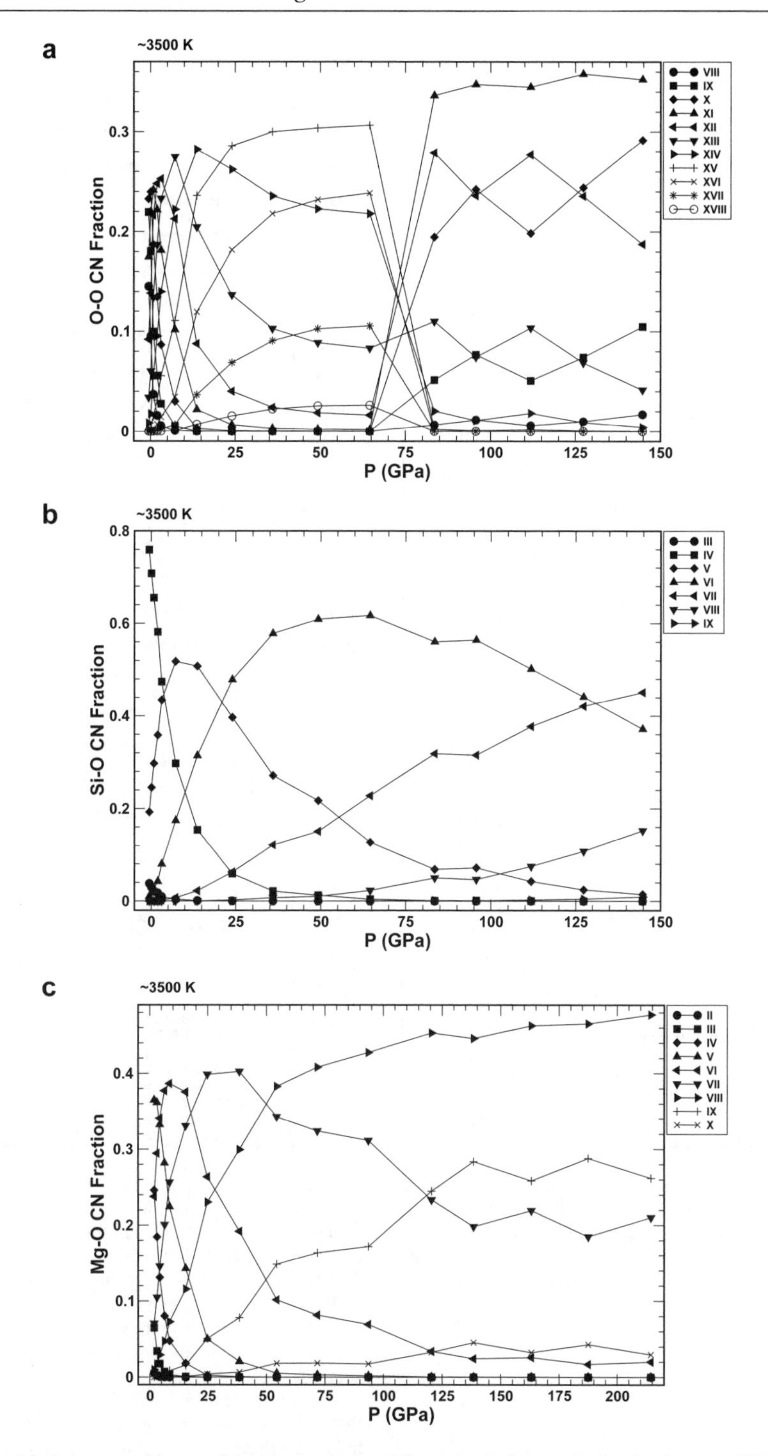

Figure 14. Oxygen packing configurations calculated from simulation results for the data set of Figure 3. Concentrations are plotted as fractional abundances. (a) Oxygen about oxygen configurations. (b) Oxygen about silicon configurations. (c) Oxygen about magnesium configurations.

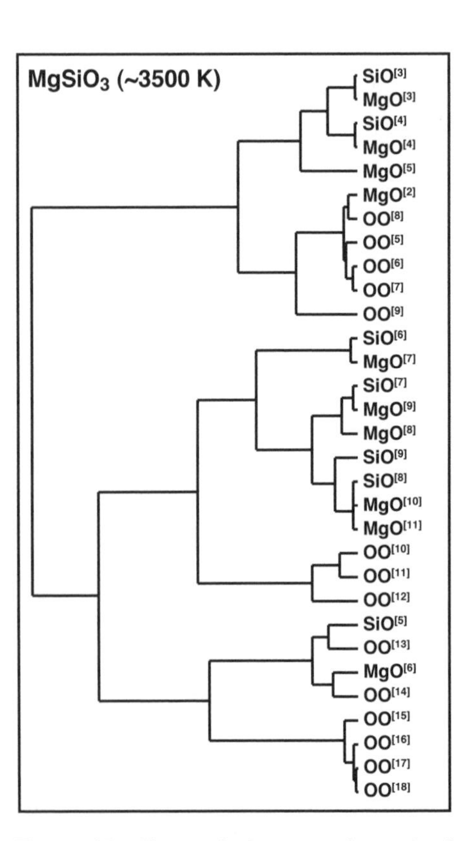

Figure 15. Cluster dendrogram of correlated abundances of O-O, Si-O, and Mg-O configurations (i.e., Fig. 14). Correlations are denoted by the vertical linkages which are placed from right to left to indicate strong to weak correlation. Groupings are computed by averaging to reduce the correlation matrix.

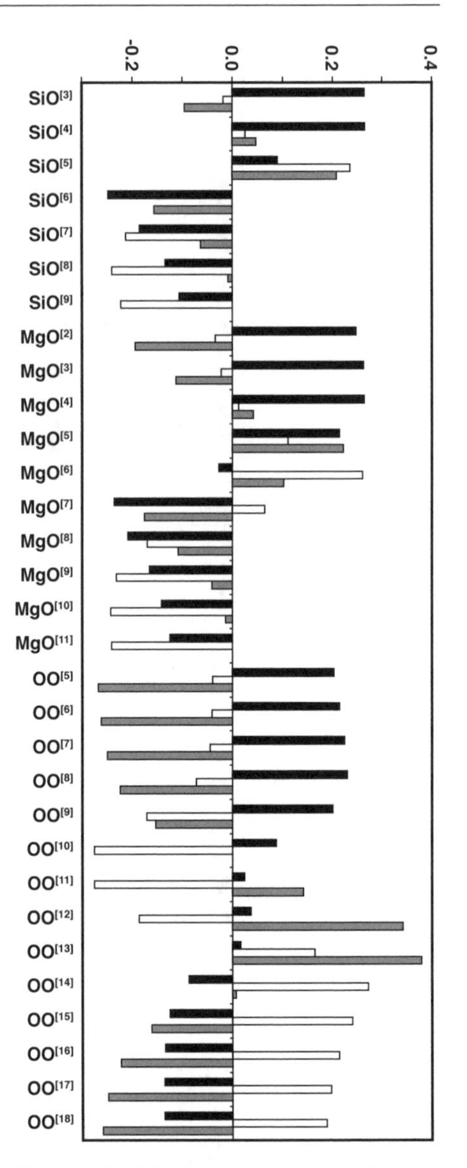

Figure 16. Principal component analysis of the matrix of correlation coefficients relating abundances of O-O, Si-O, and Mg-O configurations (i.e., Fig. 14) in $MgSiO_3$ liquid (data set of Fig. 3). The coefficients of the first three eigenvectors are plotted, with coefficients for each configuration indicated by the length of the vertical bar in the graph. Note, some configurations contribute positively and some negatively to the eigenvectors and that each eigenvector is denoted by a different fill pattern. The first three eigenvectors have associated eigenvalues indicating that they account for 99% of the observed structural variance in the liquid over the entire pressure range of this isotherm (~3500 K).

number of O about Si,Mg. The second eigenvector largely accounts for intermediate Si-O, Mg-O and O-O configurations, while the third captures the O-O coordination number shift at pressures of ~ 70 GPa. With the PCA, we can make the quantitative connection between structure and some macroscopic property.

We will illustrate this connection by constructing an equation for the volume of $MgSiO_3$ liquid on the assumption of ideal mixing of Si-O, Mg-O, and O-O configuration units, i.e.,

$$V = \sum_i v_{\mathrm{Si},i} \left\{ \mathrm{SiO}^{[i]} \right\} + \sum_i v_{\mathrm{Mg},i} \left\{ \mathrm{MgO}^{[i]} \right\} + \sum_i v_{\mathrm{O},i} \left\{ \mathrm{OO}^{[i]} \right\} \tag{32}$$

where the $v_{x,i}$ are partial molar volumes of the "species" $XO^{[i]}$ and the curly brackets denote the fractional abundance of each which respect to the total Si, Mg and O in a mole of $MgSiO_3$. The analysis of the previous paragraphs shows that not all $XO^{[i]}$ species vary in abundance independently. To fit the model embodied in Equation (32), regression techniques that deal properly with this correlation must be employed. One such technique is Singular Value Analysis (SVA, Press et al. 1999) which uses PCA to estimate the number of "independent" variables in the least squares problem and performs the regression analysis in this "reduced" variable space. The output of a SVA procedure is a set of regression coefficients on the original correlated variables that are not overdetermined because they incorporate the intrinsic correlation determined from the PCA analysis. We use SVA to fit Equation (32) to the volume-structure data for $MgSiO_3$ liquid along the 3500 K isotherm; the three principal eigenvectors of structure variation (Fig. 16) are used to perform the regression. The model coefficients so obtained are plotted in Figure 17. The model recovers the volume of the liquid along the entire isotherm with errors on recovery never exceeding 1%.

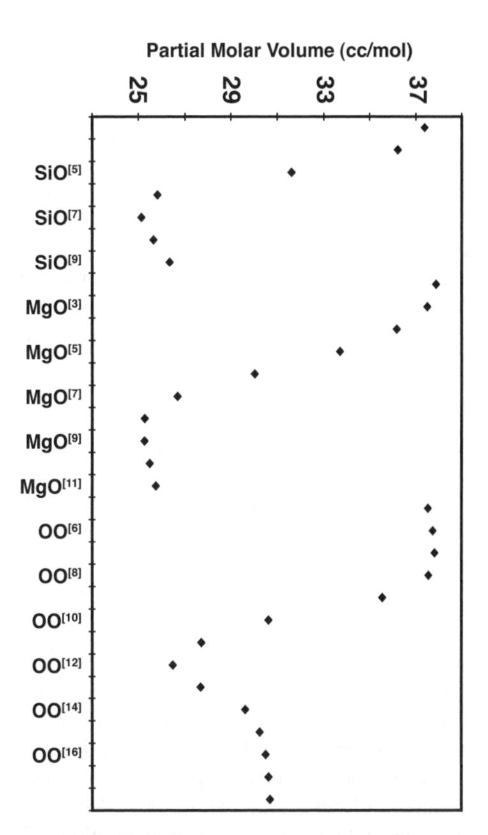

Figure 17. Partial molar volumes of O-O, Si-O, and Mg-O configurational units (i.e., Fig. 14) in $MgSiO_3$ liquid (data set of Fig. 3) along the ~3500 K isotherm over the pressure range zero to ~150 GPa. Values are estimated using singular value decomposition fitting techniques utilizing results from the principal component analysis displayed in Figure 16. These estimates are strongly correlated as they are derived from only three independent measures of structural variation. See text for further explanation.

The first thing that should be noticed about the model "partial molar volumes" plotted in Figure 17 is that the trends for each element form a logical pattern. There is a general decease in the intrinsic volume of $SiO^{[i]}$ and $MgO^{[i]}$ units as i increases; the reversal of this trend for $i > 7$ (Si) and > 9 (Mg) may be real or may be an artifact of low abundances of these configurations in the liquid (Fig. 14). The minimum in volume of $OO^{[12]}$ is especially interesting. An icosahedron or polyhedron with 20 faces and 12 vertices is one of the classical Platonic solids and represents an optimal way to pack equal volume spheres. The volume of a single regular oxygen icosahedron is about $2.18d_O^3$ where d_O is the diameter of an oxygen anion. Taking the diameter of oxygen to be 0.275 nm gives a volume of 27.3 cc/mol for a mole of oxygen icosahedra, surprisingly close to the volume minimum for the $OO^{[12]}$ species shown on Figure 17. Perhaps the most important consequence of this analysis is the realization that once volume is successfully parameterized in terms of abundances of structural units in the liquid, then the internal energy and all dependent thermodynamic properties are similarly parameterized because E is consistent with the Rosenfeld-Tarazona scaling law (Eqn. 5). The $a(V)$ and $b(V)$ functions of volume (Fig. 5) now become functions of melt structure. Obviously, this analysis could be pursued farther by investigating multiple isotherms and developing a more comprehensive formalism, but we will not develop the example further here. We hope that we have illustrated, however, the potential for LSS MD studies to elucidate the relationship between microscopic and macroscopic properties and to demonstrate the importance of carrying out a comprehensive post-simulation analysis of MD results so that variation of physical properties can be understood in the context of fundamental measures of material structure.

TRANSPORT PROPERTIES

We turn now from discussing static equilibrium properties determined from LSS MD studies to the estimation of transport properties. By transport properties we mean diffusivity, viscosity, thermal conduction, and related irreversible phenomena that can be studied by examining dynamical fluctuations within the simulation volume. Even more so than with equilibrium properties, the estimation of transport properties by MD benefits enormously from LSS and especially from long-duration simulations (Fig. 1). Here we focus upon the shear viscosity of liquid $MgSiO_3$ and related self-diffusion of O, Si and Mg. Once again, we will emphasize the strong correlation that exists between the macroscopic and microscopic worlds.

Viscosity

Shear viscosity can be computed from LSS MD results using linear response theory embodied as the Green-Kubo (GK) relations. Shear viscosity is determined by studying the temporal decay of appropriate stress components (both off and on-diagonal; Rapaport 1995). The GK expression for the shear viscosity is given by integration of the stress (pressure) autocorrelation function,

$$\eta = \frac{V}{3kT}\int_0^\infty \left\langle \sum_{x<y} \sigma_{xy}(t)\sigma_{xy}(0) \right\rangle dt \tag{33}$$

where η is the shear viscosity, V is the system volume, T is the temperature, k is Boltzmann's constant, and σ_{ij} refers to the ij component of the stress tensor. In addition to σ_{xy}, used in Equation (33), the decay of temporal autocorrelations: σ_{xz}, σ_{yz}, $(\sigma_{xx} - \sigma_{yy})$ and $(\sigma_{yy} - \sigma_{zz})$, also provide independent GK values for the shear viscosity. The reported MD-derived viscosity at a given state point is the average of the five independently computed values following the method detailed in Nevins and Spera (2007). Viscosity estimates are made from long simulation runs of ~2 ns, with a sampling window width of ~2, ~5, or ~10 ps, and a time interval between the start of successive sampling windows of ~10 fs. As an approach is made to the computer glass transition, longer and longer simulation durations are required in order to compute a precise shear viscosity. Details may be found in the study of Morgan and Spera (2001).

Continuing with our example involving $MgSiO_3$ liquid, we plot in Figure 18 shear viscosity computed using the GK method at 17 state points along a nominal 3500 K isotherm. The viscosity increases by a factor of ~ 32 from low pressure to 100 GPa and is extrapolated to increase by a factor of 110 across the entire terrestrial mantle (to the core-mantle boundary at 135 GPa). This increase has important dynamical implications for convective mixing,

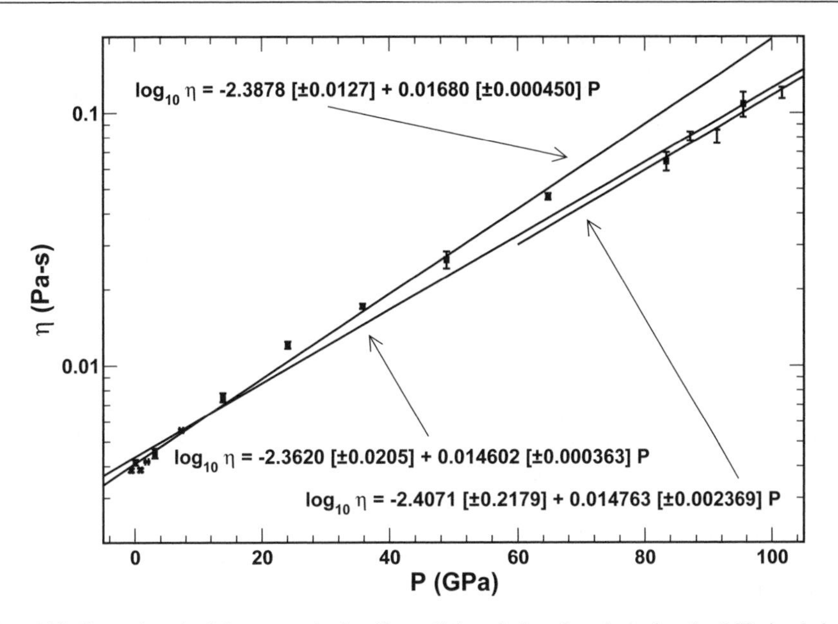

Figure 18. Shear viscosity (η) computed using Green-Kubo relations from long-duration MD simulations using the pair-potential of Oganov et al. (2001) on $MgSiO_3$ liquid. Error bars are one sigma estimates. Calculations were performed along a nominal ~3500 K isotherm. Note the break in slope of the viscosity-pressure relations over the pressure interval ~ 65-85 GPa.

chemical stratification and crystal fractionation in the terrestrial planet magma oceans. The activation energy (Q) for viscous flow, defined from the Boltzmann relation, $\eta \sim e^{Q/RT}$, has a pressure dependence of ~1 cc/mol, but Figure 18 demonstrates that there is an abrupt change in slope in $\log_{10} \eta$ vs. P at ~70 GPa. This decrease in Q occurs over the same pressure interval where the oxygen about oxygen coordination polyhedra display a dramatic reconfiguration (Fig. 14a). Recall, that over the same pressure interval there are no significant rearrangements of Si-O or Mg-O configurations (Figs. 14b,c) nor are there significant perturbations in pressure dependences of the thermodynamic properties, demonstrating that in this liquid the anion-anion oxygen packing strongly influences relative atomic motion and by inference other transport phenomena. The average oxygen about oxygen coordination number in the high-pressure region is ~11, quite close to the maximum packing efficiency associated with icosahedra. The difference from the ideal limit of CN = 12 may be related to the irregularity of packing presumably associated with the agitated thermal state or intrinsic mobility of oxygen in the high-temperature (~ 3500 K) liquid.

Self-diffusivity

Another transport property that can be readily examined with LSS MD is self-diffusivity. In a continuous system, the diffusion coefficient D is defined by combining Fick's first and second laws to give,

$$\frac{\partial C}{\partial t} = D\nabla^2 C \tag{34}$$

where $C(r,t)$ is the local density or concentration of some atom as a function of position (r) and time (t). Equation (34) applies both to diffusion of one species through another in response to a spatial gradient in chemical potential and to self-diffusion, which is the relative motion of one species generated by dynamical fluctuations in an otherwise equilibrium state. MD results can be used to compute diffusivities associated with chemical potential gradients, but this requires more effort because the effects of all atoms must be taken into account; self-diffusivities are simpler to compute and are addressed here. At the discrete particle level concentration, C may be written using the Dirac delta function, $\delta(r)$, as

$$C(\vec{r},t)=\sum_{j=1}^{N_a}\delta\left(\vec{r}-\vec{r}_j(t)\right) \tag{35}$$

For large t, the Einstein expression (McQuarrie 2000) for the Mean Square Displacement (MSD) is related to the diffusion coefficient by

$$D=\frac{1}{6N_a t}\left\langle\sum_{j=1}^{N_a}\left[\vec{r}_j(t)-\vec{r}_j(0)\right]^2\right\rangle \tag{36}$$

where N_a refers to the number of atoms of species "a" and the quantity in brackets represents the MSD of the ath atom type. Unfolded atomic trajectories are used in the calculation of the self-diffusivity (Rapaport 1995). At each state point, the MSD for a particular species is accumulated from the unfolded atom trajectories and a plot of MSD *versus* time is made. Following a brief (<1 ps) ballistic transport regime not of immediate interest, the MSD average over all atoms of the same type becomes linear, and the slope of this average is directly proportional to the self-diffusivity.

Self-diffusivity of O, Mg and Si are plotted in Figure 19 for our example $MgSiO_3$ liquid along the ~3500 K isotherm (see Nevins et al. 2009 for additional isotherms). At very low P (~ 0 GPa) there is a small region of "anomalous" diffusion for Si and O (and perhaps Mg) in which increasing pressure increases the diffusivity; there is also a cross over in Mg and O self-diffusivity at ~ 2 GPa. Of special significance are the breaks in slope ~70 GPa, where the rate of decrease of all diffusivities diminishes over a 10 GPa interval. This observation implies that the activation volume (V_i°) for tracer diffusion, given by $(\partial \ln D_i/\partial P)_T = -V_i^{\circ}/RT$, is smaller at this pressure, where the average O-O mean coordination number decreases abruptly, than at pressures above or below the transition interval. Just below this pressure the mean oxygen around oxygen CN is ~15 as noted on Figure 13. These packing polyhedra become

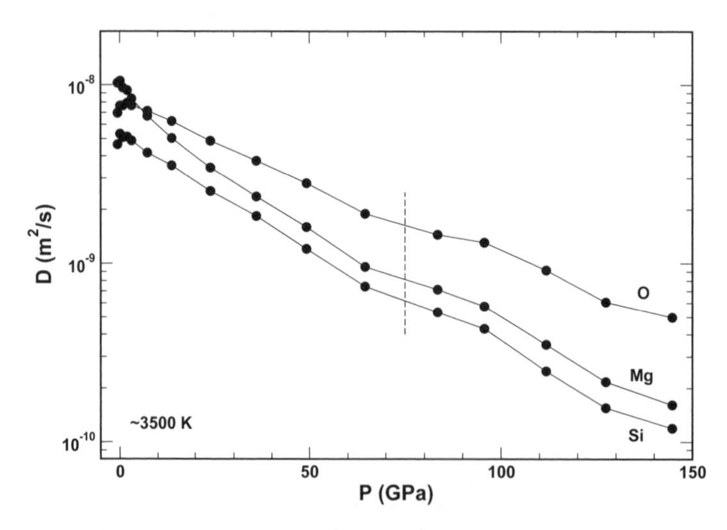

Figure 19. Self-diffusivity (D) of O, Mg, and Si calculated from MD simulation results for the data set of Figure 3. Note the displacement over the pressure interval ~ 65-85 GPa.

unstable with respect to the more efficiently packed irregular icosahedra of mean CN ~11. Clearly, the oxygen packing exerts a first-order effect on the mobility of oxygen, especially its pressure-dependence. The recurring theme is that microscopic structure does indeed map into macroscopic properties and that LSS allows one to investigate the connections rather easily.

Eyring relation

Because independent values of self-diffusivity and shear viscosity are available from the LSS MD simulations, the applicability of the Eyring et al. (1982) relation can be addressed. Typically the Eyring or closely-related Stokes-Einstein relation is used to estimate shear viscosity from self-diffusion data (or vice versa) by assuming a length scale ("size") of the atom or cluster of atoms involved in the activated process of atom mobility. Ambiguity generally arises, however, in selecting the "size" of the activated cluster *a priori*.

The Eyring formulation is based on a phenomenological picture involving a jump of a particular atom from one coordination environment to another. These environments can be characterized by a size related to the atom cluster involved in the diffusive event. The Eyring relationship between self-diffusivity and viscosity is

$$\xi\left(\frac{V}{nN_A}\right)^{1/3} = \frac{kT}{D\eta} \tag{37}$$

where n is the number of atoms per formula unit, V is the molar volume, N_A is Avogadro's number and $\xi+1$ is the number of atoms in the activated complex. The ratio $kT/\eta D$ defines a scale-length, λ, associated with the size of the "activated cluster" involved in atomic mobility and viscous shear flow. In Figure 20, both $kT/\eta D$ and ξ are plotted for oxygen along the nominal 3500 K isotherm for our $MgSiO_3$ example data set. As would be expected, λ decreases rapidly with pressure from around 1.8 nm at 1 bar to 0.2 nm for P ~150 GPa. The number of atoms ($\xi+1$) in the activated cluster changes from a ~7-9 at low pressure to a smaller activated unit of 2-3 atoms according to the Eyring relation. The plateau in λ and ξ for $65 < P < 85$ GPa is consistent

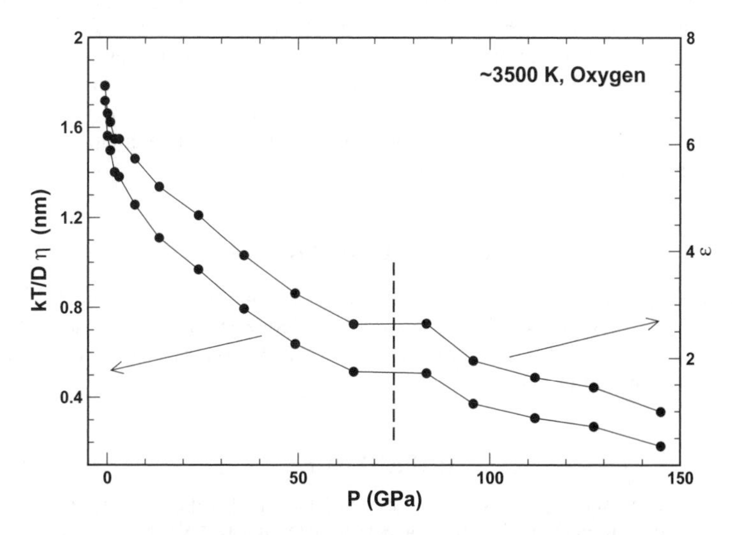

Figure 20. Evaluation of the Eyring relation (left vertical axis) computed from shear viscosity (Fig. 18) and oxygen self-diffusion data (Fig. 19) for the data set of Figure 3. On the right vertical axis the Eyring characteristic length is translated into a quantity reflecting the number of atoms (ξ+1) in the activated complex for transport.

with the unusual behavior in the shear viscosity and self-diffusion discussed previously. The rapid reorganization of the oxygen packing in the melt over the interval 65-85 GPa abates but does not negate the trend in λ and ξ.

THE FUTURE — LSS IN THE CONTEXT OF FPMD

We close this chapter with a few remarks concerning the future of LSS MD studies. Historically, EPPMD is the method generally chosen for LSS. The reason for this is that large numbers of atoms can be used in the simulation without the prohibitive consumption of computing resources. The LAMMPS code (Plimpton 1995) for example, a modified version (Nevins 2009) of which was utilized for all of the example results provided in this chapter, has computational costs that scale as $N \log N$ for long-range forces, permitting calculations with large N to be handled routinely. By contrast, FPMD uses DFT for estimation of forces between atoms, and the majority of commonly used computational packages that implement this technique have computational costs that scale at N^3 (or the cube of the number of electrons, Ordejón et al. 2000). Of course, the advantage of a DFT description of atomic forces is that it is closer approximation to a complete quantum mechanical solution, and therefore presumably more accurate. While it may be argued that EPPMD on well characterized systems is entirely adequate as long as results are spot checked against FPMD, there is considerable more uncertainty involved in using EPPMD as an exploratory tool in temperature, pressure and composition regimes where the underlying pair-potential has not been adequately tested. And, as mentioned above, there are material calculations involving free electrons for example that are simply beyond the capabilities of the EPPMD method. On the other hand, transport property estimates of shear viscosity and thermal conductivity are generally thought to be beyond the reach of current FPMD methodology.

The future for LSS based upon DFT is not gloomy, however. Computational advances (e.g., SIESTA: Soler et al. 2002) demonstrate methods for the calculation of total energies and forces that scales linearly with the number of atoms in the system. Codes based on these algorithmic advances enable electronic structure and MD computations on larger systems (N~10^3) with modest computational costs. In the near future, we can envision large-scale FPMD simulations of long duration, and the capability of creating a dense grid of state points that enclose the *P-T* range of interest. LSS using FPMD will potentially provide us with the best data source for thermodynamic modeling and exploration of the causal relationship between microscopic structure and macroscopic properties of geomaterials. This kind of calculation may very well be within reach in the not too distant future.

ACKNOWLEDGMENTS

Material support was provided by the National Science Foundation *via* EAR-0838182, ATM-0425059 and EAR-0440057. We also acknowledge support from the Mineralogical Association of America. This is OFM Research contribution #15 and UCSB Institute of Crustal Studies contribution 09-231.

REFERENCES

Akins JA, Luo SN, Asimow PD, Ahrens TJ (2004) Shock-induced melting of $MgSiO_3$ perovskite and implications for melts in Earth's lowermost mantle. Geophys Res Lett 31:14

Anderson DL (1989) Theory of the Earth. Blackwell Scientific Publications

Birch F (1939) The variation of seismic velocities within a simplified Earth model in accordance with the theory of finite strain. Bull Seismol Soc Am 29:463-479

Birch F (1952) Elasticity and constitution of the Earth's interior. J Geophys Res 57:227-286

Buckingham RA (1938) The classical equation of state of gaseous Helium, Neon and Argon. Proc R Soc London A 168:264-283
Cohen AJ, Mori-Sanchez P, Yang W (2008) Insights into current limitations of density functional theory. Science 321:792-794
Eyring H, Henderson D, Stover BJ, Eyring EM (1982) Statistical Mechanics and Dynamics. John and Wiley and Sons
Ghiorso MS, Nevins D, Cutler I, Spera FJ (2009) Molecular dynamics studies of $CaAl_2Si_2O_8$ liquid: II. equation of state and a thermodynamic model. Geochim Cosmochim Acta 73:6937-6951, DOI: 10.1016/j.gca.2009.08.012
Katayama Y, Inamura Y, Mizutani T, Yamakata M, Utsumi W, Shimomura O (2004) Macroscopic separation of dense fluid phase and liquid phase of phosphorus. Science 306:848-851
Katayama Y, Mizutani T, Utsumi W, Shimomura O, Yamakata M, Funakoshi K (2000) A first-order liquid–liquid phase transition in phosphorus. Nature 403:170-173
Kubo R (1966) The fluctuation-dissipation theorem. Rep Prog Phys 29:255-284
Kurita R, Tanaka H (2005) On the abundance and general nature of the liquid–liquid phase transition in molecular systems. J Phys Cond Matter 17:L293-L302
Martin B, Spera FJ, Ghiorso M, Nevins D (2009) Structure, thermodynamic and transport properties of molten Mg_2SiO_4: Molecular dynamics simulations and model EOS. Am Mineral 94:693-703
McQuarrie DA (2000) Statistical Mechanics, 2nd edition. University Science Books
Morgan NA, Spera FJ (2001) The Glass Transition, Structural Relaxation and Theories of Viscosity: A Molecular Dynamics Study of Amorphous $CaAl_2Si_2O_8$. Geochim Cosmochim Acta 65:4019-4041
Nevins D (2009) Understanding Silicate Geoliquids at High Temperatures and Pressures Through Molecular Dynamics Simulations. PhD dissertation, University of California, Santa Barbara, Santa Barbara
Nevins D, Spera FJ (2007) Accurate computation of shear viscosity from equilibrium molecular dynamics simulations. Molec Simul 33:1261-1266
Nevins D, Spera FJ, Ghiorso MS (2009) Shear viscosity and diffusion in liquid $MgSiO_3$: Transport properties and implications for magma ocean stratification. Am Mineral 94:975-980
Oganov AR, Brodholt JP, Price GD (2001) The elastic constants of $MgSiO_3$ perovskite at pressures and temperatures of the Earth's mantle. Nature 411:934-937
Ordejón P, Sanchez-Portal D, Garcia A, Artacho E, Junquera J, Soler JM (2000) Large scale DFT calculations with SIESTA. RIKEN Rev 29:42-44
Plimpton S (1995) Fast Parallel Algorithms for Short-range Molecular Dynamics. J Comput Phys 117:1-19
Poole P, Saika-Voivod I, Sciortino F (2005) Density minimum and liquid-liquid phase transition. J Phys Cond Matter 17:L431-L437
Press WH, Teukolsky SA, Vetterling WT, Flannery BP (1999) Numerical Recipes in C, 2nd edition. Cambridge University Press
Prigogene I, Defay R (1954) Chemical Thermodynamics. Longmans Green and Co.
Rapaport DC (1995) The art of molecular dynamics simulation. Cambridge University Press
Rosenfeld Y, Tarazona P (1998) Density functional theory and the asymptotic high-density expansion of the free energy of classical solids and fluids. Molec Phys 95:141-150
Saika-Voivod I, Sciortino F, Poole PH (2000) Computer simulations of liquid silica: Equation of state and liquid-liquid phase transition. Phys Rev E 63:011202
Soler JM, Artacho E, Gale JD, Garcia AI, Junquera J, Ordejon P (2002) The SIESTA method for ab initio order-N materials simulation. J Phys Cond Matter 14:2745-2780
Spera FJ, Nevins D, Cutler I, Ghiorso MS (2009) Structure, thermodynamic and transport properties of $CaAl_2Si_2O_8$ liquid: Part I. molecular dynamics simulations. Geochim Cosmochim Acta 73:6918-6936, 10.1016/j.gca.2009.08.011
Spera FJ, Nevins D, Ghiorso MS (2010) Structure, equation of state and transport properties of liquid $MgSiO_3$: comparison of pair-potential molecular dynamics, first principles molecular dynamics and experimental data. Geochim Cosmochim Acta, submitted
Tanaka H (2000) General view of a liquid-liquid phase transition. Phys Rev E 62:6968-6976
Vinet P, Ferrante J, Smith JR, Rose JH (1986) A universal equation of state for solids. J Phys C: Solid State Phys 19:L467-L473
Vinet P, Smith JR, Ferrante J, Rose JH (1987) Temperature effects on the universal equation of state of solids. Phys Rev B 35:1945-1953

Reviews in Mineralogy & Geochemistry
Vol. 71 pp. 465-484, 2010

Thermodynamics of the Earth's Mantle

Lars Stixrude and Carolina Lithgow-Bertelloni

Department of Earth Sciences
University College London
Gower Street
London WC1E 6BT, United Kingdom

l.stixrude@ucl.ac.uk

INTRODUCTION

Central to the goals of mineral physics is an elucidation of how material behavior governs planetary processes. Planetary accretion, differentiation into crust, mantle, and core, ongoing processing by magmatism, dynamics and thermal evolution, and the generation of magnetic fields, are all processes controlled by the physical properties and phase equilibria of planetary materials. Knowledge of these processes has advanced in large measure by progress in the study of material behavior at extreme conditions of pressure and temperature.

Thinking of a planet as an experimental sample emphasizes the relationship between planetary processes and material behavior via properties that control its response to natural perturbations. We may explore the response of the planet to a sudden increase in energy, provided, for example, by a giant impact of the type that is thought to have formed Earth's moon (Canup 2004). Responses include an increase in temperature, controlled by the heat capacity, a decrease in density controlled by the thermal expansivity, and phase transformations controlled by the free energy. Phase transformations also contribute to changes in temperature and density via the heats and volumes of transformation. A giant impact also perturbs the stress state, to which the planet responds via compression (controlled by the bulk modulus), adiabatic heating (Grüneisen parameter) and phase transformations (free energy).

A planet is an unusual experimental sample because its pressure is self-generated via gravitational self-compression, and its temperature via adiabatic compression and radioactive decay. Indeed one of the major challenges in understanding planetary-scale processes is that the characteristic pressure (1 Mbar) and temperature (several thousand K) are so large. Consider an Earth-like planet in which the mantle makes up 2/3 of the mass and the core makes up the remainder. The pressure P_M at the base of the mantle depends approximately linearly on planetary mass M: P_M(Mbar)~1.4(M/M_E), as does the temperature at the base of the mantle due to adiabatic compression T_M(K) ~1600+1000(M/M_E), assuming the potential temperature to be set by silicate melting, as it is on Earth (Valencia et al. 2006).

First principles theory of the kind discussed elsewhere in this volume has played a major role in the elucidation of planetary processes in large part because it is able to access with ease the extreme conditions of planetary pressure and temperature. Density functional theory in particular has had a tremendous impact because it is equally applicable to all conditions of planetary pressure and temperature and in principle to all elements of the periodic table (Kohn 1999), although some major elements are more problematic than others, such as Fe in oxides and silicates (Gramsch et al. 2003; Cococcioni 2010; Wentzcovitch et al. 2010). Methods with in-principle even greater scope and accuracy, including improved functionals and Quantum Monte Carlo, are being developed (Mitas and Kolorenč 2010; Perdew and Ruzsinszky 2010). By comparison, static experiments have only recently been able to access pressure-temperature

1529-6466/10/0071-0021$05.00 DOI: 10.2138/rmg.2010.71.21

conditions of Earths' core-mantle boundary routinely and to perform a variety of precise measurements *in situ* (Murakami et al. 2004, 2007; Mao et al. 2006). Continued advances in theory and experiment are both important because density functional theory is not exact and produces small, systematic errors that are well understood, but persistent, while experiments at extreme conditions can produce contradictory results because they are often pushing the envelope of what is technologically feasible.

As experimental samples, planetary mantles are unusual in at least one other respect that presents challenges to first principles theory and experiment: they are chemically complex and heterogeneous. In Earth's mantle there are at least 6 essential chemical components (SiO_2, MgO, FeO, CaO, Al_2O_3 Na_2O), that stabilize a large number of impure phases over the entire mantle pressure-temperature regime, and a host of minor elements that may also play an important role in planetary processes, such as H_2O and CO_2 (Hirschmann 2006; Ohtani and Sakai 2008). The length scale of heterogeneity is known to range from the grain scale represented by adjoining phases of contrasting physical properties, such as olivine and spinel in a typical xenolith, to that of lithologic banding, represented by pyroxenite veins in obducted peridotites, and may include much larger scales as well (Allegre and Turcotte 1986). Mantle convection appears not to be as efficient at homogenizing major element composition as had once been thought and may be compatible with planet-scale variations in bulk composition between upper and lower mantle, perhaps driven by lithologic density contrasts (Xie and Tackley 2004).

Understanding planetary processes then demands a method complementary to first principles calculations and experiment that is able to interpolate among and extrapolate from necessarily limited results to the full chemical richness that is typical of silicate mantles. Here we review a method, based on new developments in thermodynamic theory and a careful analysis of the information provided by experiment and first principles theory that permits the construction of realistic models of planetary interiors. This method allows one to fully capture the heterogeneity inherent in the relevant multi-component, multi-phase equilibria and the physical properties of the multi-phase assemblages. The method is complementary to first principles theory and experiment and indeed uses results on simple systems to build up predictive power for more complex and relevant assemblages. There have been many previous attempts to construct thermodynamic models of Earth's mantle. We show below that all of these have important limitations. We show how our method can be used to address mantle heterogeneity on all length scales ranging from that of the subducting slab to the possibility of mantle-wide radial variations in bulk composition.

OUR APPROACH AND PREVIOUS WORK

The key difference between our approach and previous models is our adherence to thermodynamic self-consistency (Stixrude and Lithgow-Bertelloni 2005b). This adherence has the important advantage that all thermodynamic relations including the Maxwell relations and the Clapeyron equation are uniquely satisfied. There are practical advantages as well: because so many thermodynamic quantities are intimately linked, diverse experimental and first principles results can be used independently to constrain more robustly the thermodynamic model. For example, phase equilibria and equation of state data both constrain the volume of a phase, placing redundant constraints on the equation of state. Another important difference is the scope of our model. Whereas many models of the mantle have been primarily focused on either phase equilibria, or physical properties, our approach is to self-consistently describe both, differing fundamentally from so-called hybrid models in which some sub-set of physical properties are computed self-consistently and phase equilibria are determined independently and non-self-consistently (Ita and Stixrude 1992; Cammarano et al. 2003; Hacker et al. 2003). The important point being that phase transformations influence physical properties as much as the effects of pressure and temperature on single phases over the mantle range, a feature that our approach naturally encompasses.

Our approach makes use of four sets of thermodynamic principles that, except for the last, have featured in other contexts, but have not before been combined in application to Earth's mantle:

Fundamental thermodynamic relations

A fundamental thermodynamic relation is a single functional relationship that contains complete information of all equilbrium properties of all equilibrium states of a particular species (Callen 1960; Stixrude and Bukowinski 1990). All thermodynamic properties are computed as derivatives of the fundamental relation with respect to its natural variables, guaranteeing self-consistency. For example

$$\frac{\partial^3 G}{\partial P \partial P \partial T} = \frac{\partial^3 G}{\partial T \partial P \partial P} = \frac{\partial}{\partial P}(V\alpha)_T = \frac{\partial}{\partial T}\left(\frac{V}{K}\right)_P \tag{1}$$

where G is the Gibbs free energy, P is pressure, T is temperature, V is volume, α is thermal expansivity, and K is the isothermal bulk modulus. We write the fundamental relation in analytic form with analytically computable derivatives so that relations of this type are satisfied exactly. Some previous models of mantle thermodynamics violate the Maxwell relations (Sobolev and Babeyko 1994).

Euler form

In our model the fundamental thermodynamic relation is written in Euler form, which is always possible because of the first order homogeneity of thermodynamic functions. The importance of the Euler form, in which the thermodynamic potential appears as an absolute quantity, e.g., $G = G(P,T)$, is important for applications to planetary mantles and differs from the more common use of the differential form

$$dG = -SdT + VdP \tag{2}$$

which must be integrated to determine the value of G at the pressure and temperature of interest. The integration is typically performed along a path that proceeds in two legs: first upwards in temperature at ambient pressure and second upwards in pressure at elevated temperature (Berman 1988; Fei and Saxena 1990; Ghiorso and Sack 1995; Holland and Powell 1998). The advantage of the differential approach is that it maintains close contact with experimental measurements at ambient pressure of the volume and the entropy S (via integration of measurements of the heat capacity). The differential approach is adequate as long as the temperature and pressure are not too high, but has at least three difficulties in application to Earth's mantle:

1) The temperature in the middle of Earth's mantle is approximately 2000 K and may be 4000 K at its base. This temperature far exceeds melting points at ambient pressure. This means that the initial leg of the integration path must extend into the super-solidus regime where no thermodynamic data on the crystalline phase of interest exist and where the crystalline phase of interest may be mechanically unstable.

2) The term VdP dominates over most of Earth's mantle and must be treated with greater care than is possible using the differential form. For example, the change in Gibbs free energy along a mantle isotherm far exceeds the difference in Gibbs free energy between 300 K and 2000 K. Moreover, the increase in G along the isotherm quickly surpasses in magnitude even the Gibbs free energy of formation from the elements at depths as shallow as the transition zone.

3) Some mantle phases (e.g. post-perovskite) are unquenchable to ambient pressure, rendering the first leg of the integration path ill-defined.

Legendre transformations

While the Gibbs free energy is the natural thermodynamic potential for treating phase equilibria, the Helmholtz free energy is much more convenient for describing the equation of state. Legendre transformations provide a simple way of moving from one thermodynamic potential to the other (Alberty 2001). So for example, the Helmholtz free energy F is related to G by

$$G(P,T) = F(V,T) + PV(P,T) \tag{3}$$

We may then describe the variation of the Gibbs free energy with pressure in terms of the most successful account of the equation of state, which is formulated as the volume derivative of Helmholtz free energy (Birch 1978). This overcomes the poor convergence of equations of state based on power series expansions of G in P, or the Murnaghan equation, as used in many thermodynamic models popular in the Earth sciences (Ghiorso and Sack 1995; Holland and Powell 1998).

Anisotropic generalization

A final key feature of our model and a new theoretical development is the generalization of the thermodynamic machinery to the consideration of deviatoric stress and strain in such a way that fully self-consistent computation of phase equilibria and the elastic constant tensor is possible. This final step is essential for making contact with seismological observations and we are not aware of another thermodynamic model that has this capability. Previous mantle models have either not specified the shear modulus or other components of the elastic moduli (Mattern et al. 2005), or have done so non-self-consistently (Kuskov 1995; Hama and Suito 1998). Our derivation is given fully elsewhere (Stixrude and Lithgow-Bertelloni 2005b) and builds on previous work in relating thermodynamic potentials to elasticity (Wallace 1972; Davies 1974). The analysis is based on a polynomial expansion of the thermodynamic potential in the Eulerian finite strain with special attention paid to the distinction between the large finite strain associated with the Earth's internal pressure and the much smaller strain of general symmetry applied by a passing seismic wave.

THERMODYNAMIC THEORY

Our thermodynamic method has been derived elsewhere and our purpose here is to summarize the theory, sketch its derivation, and highlight some of its key features (Stixrude and Lithgow-Bertelloni 2005a,b).

The Gibbs free energy of the multi-phase assemblage

$$G(\sigma_{ij},T,n_\beta) = \sum_\beta n_\beta \mu_\beta(\sigma_{ij},T,n_\beta) = \sum_\beta n_\beta \left[G_\beta(\sigma_{ij},T) + RT \ln a_\beta \right] \tag{4}$$

where the sum is over all species (end-members) in the model n_β is the number of moles of species β, μ_β is the chemical potential, G_β is the Gibbs free energy of pure species β, R is the universal gas constant a_β is the activity, and the stress is related to pressure by

$$\sigma_{ij} = -P\delta_{ij} + \tau_{ij} \tag{5}$$

where τ_{ij} is the deviatoric stress. We assume that the quantity $RT\ln f_\beta$ is independent of pressure and temperature, where f_β is the activity coefficient that relates activity to concentration (Ita and Stixrude 1992). This assumption permits non-ideal enthalpy of solution, but neglects the contribution of non-ideailty to other physical properties, such as volume or entropy, because such contributions are small compared with uncertainties in these properties at mantle pressure and temperature. We also neglect surface energy, which may be significant for very small grains (~1 nm).

The Gibbs free energy of the pure species G_β is specified by the Legendre transformation of the Helmholtz free energy

$$F(E_{ij},T) = F_0(0,T_0) + F_c(E_{ij},T_0) + \Delta F_q(E_{ij},T) \tag{6}$$

where we have suppressed the index β, the first three terms on the right hand side are respectively the reference value at the natural configuration, and the contributions from compression at ambient temperature (the so-called "cold" part), and lattice vibrations in the quasi-harmonic approximation. These are expected to be the most important for the physical properties of mantle phases. Additional contributions to order-disorder transitions such as the α-β quartz transition (Landau term), and a magnetic term to account for spin disorder are discussed in a forthcoming publication and will not be further addressed here. The E_{ij} is the Eulerian finite strain relating the "natural" state at ambient conditions with material points located at coordinates a_i, to the "final" state with coordinates x_i. Thermodynamic quantities are related to the strain S_{ij} that relates the final state to the initial or pre-stressed state with coordinates X_i. We assume that the initial state is one of hydrostatic stress, appropriate to the earth's interior, and that the final state differs slightly from the initial state, corresponding to the small amplitude of seismic waves. We view the natural configuration as that at ambient pressure and temperature, corresponding to the "master" configuration of Davies (1974).

We have argued that the following form of the Helmholtz free energy is appropriate

$$\rho_0 F_c(E_{ij},T) = \frac{1}{2} b^{(1)}_{ijkl} E_{ij} E_{kl} - \frac{1}{6} b^{(2)}_{ijklmn} E_{ij} E_{kl} E_{mn} + \frac{1}{24} b^{(3)}_{ijklmnop} E_{ij} E_{kl} E_{mn} E_{op} + \ldots \tag{7}$$

$$\rho_0 F_q(E_{ij},T) = \rho_0 nRT \left[\frac{\theta(E_{ij})}{T} \right]^{-3} \int_0^{\theta(E_{ij})/T} \ln\left(1 - e^{-t}\right) t^2 dt \tag{8}$$

where ρ is the density, n is the number of atoms in the formula unit, θ is the Debye temperature and subscript 0 indicates values at ambient conditions. We find the equation of state and isothermal elastic constants by taking the appropriate strain derivatives and evaluating in the initial state; correct to fourth order in the strain

$$\rho_0 F = \rho_0 F_0 + \frac{1}{2} b^{(1)}_{iikk} f^2 + \frac{1}{6} b^{(2)}_{iikkmm} f^3 + \frac{1}{24} b^{(3)}_{iikkmmoo} f^4 + \rho_0 \Delta F_q \tag{9}$$

$$P = \frac{1}{3}(1+2f)^{5/2} \left[b^{(1)}_{iikk} f + \frac{1}{2} b^{(2)}_{iikkmm} f^2 + \frac{1}{6} b^{(3)}_{iikkmmoo} f^3 \right] + \gamma \rho \Delta U_q \tag{10}$$

$$c_{ijkl} = (1+2f)^{5/2} \left\{ \begin{aligned} & c_{ijkl0} + \left(3K_0 c'_{ijkl0} - 5c_{ijkl0}\right) f \\ & + \left[6K_0 c'_{ijkl0} - 14 c_{ijkl0} - \frac{3}{2} K_0 \delta^{ij}_{kl} (3K'_0 - 16) \right] f^2 \end{aligned} \right\} \tag{11}$$

$$+ \left[\gamma_{ij}\gamma_{kl} + \frac{1}{2}\left(\gamma_{ij}\delta_{kl} + \gamma_{kl}\delta_{ij}\right) - \eta_{ijkl} \right] \rho \Delta U_q - \gamma_{ij}\gamma_{kl} \rho \Delta (C_V T)$$

where U is the internal energy, C_V is the isochoric heat capacity, c_{ijkl} is the elastic constant tensor, $\delta^{kl}_{ij} = -\delta_{ij}\delta_{kl} - \delta_{ik}\delta_{jl} - \delta_{il}\delta_{jk}$, the prime indicates pressure derivative and the parenthetical superscripts provide a convenient way of distinguishing among the coefficients when alternating between standard and Voigt notation (Davies 1974). We have assumed an isotropic state of initial finite strain

$$E_{ij} = -f \delta_{ij} \tag{12}$$

$$f = \frac{1}{2}\left[\left(\frac{\rho}{\rho_0}\right)^{2/3} - 1\right] \tag{13}$$

The Grüneisen parameter $\gamma = V(\partial P/\partial U)_V$, the Grüneisen tensor

$$\gamma_{ij} = -\frac{\partial \ln \theta}{\partial S_{ij}} \tag{14}$$

and $\gamma_{ii} = \gamma\delta_{ii}$ for an isotropic material. The adiabatic elastic moduli follow from $c_{ijkl}{}^S = c_{ijkl} - \gamma_{ij}\gamma_{kl}\rho C_V T$.

The isothermal bulk modulus and the shear modulus of an isotropic material follow from $3K = c_{ijkl}\delta_{ij}\delta_{kl}/3$ and $G = c_{44} = (c_{11} - c_{12})/2$

$$K = (1+2f)^{5/2}\left[K_0 + (3K_0K_0' - 5K_0)f + \frac{27}{2}(K_0K_0' - 4K_0)f^2\right] \tag{15}$$
$$+(\gamma+1-q)\gamma\rho\Delta U_q - \gamma^2\rho\Delta(C_V T)$$

$$G = (1+2f)^{5/2}\left[G_0 + (3K_0G_0' - 5G_0)f + \left(6K_0G_0' - 24K_0 - 14G_0 + \frac{9}{2}K_0K_0'\right)f^2\right] \tag{16}$$
$$-\eta_S\rho\Delta U_q$$

where $q = (\partial\ln\gamma/\partial\ln V)$, and η_S is the shear part of the tensor

$$\eta_{ijkl} = \frac{\partial \gamma_{ij}}{\partial S_{kl}} \tag{17}$$

which for an isotropic material is

$$\eta_{ijkl} = \gamma q\delta_{ij}\delta_{kl} + \eta_S\left(\delta_{ik}\delta_{jl} + \delta_{il}\delta_{jk} - \frac{2}{3}\delta_{ij}\delta_{kl}\right) \tag{18}$$

We assume that the strain dependence of the Debye temperature is expressed as a power series in the finite strain

$$\theta^2 = \theta_0^2\left[1 + 6\gamma_0 f + \frac{1}{2}(-12\gamma_0 + 36\gamma_0^2 - 18q_0\gamma_0)f^2 + \ldots\right] \tag{19}$$

which has been evaluated at the initial state.

We assume that the activity has two contributions, an ideal contribution, given by ideal mixing of unlike cations on multiple sites (Thompson 1969) and a non-ideal contribution given by the symmetric regular solution formulation. We assume that the excess Gibbs free energy is independent of pressure and temperature. This seems justified by the fact that excess volume and entropy are either unmeasured or small for most mantle phases. Moreover, when they can be measured precisely, as in the case of the excess volume, the measurements exist only at 1 bar and may not be representative of the excess volume at the pressure of interest, where it may even be of opposite sign.

Some of the scope and goals of the model are illustrated by an example (Fig. 1). We have computed the phase equilibria and physical properties of a typical mantle bulk composition (Workman and Hart 2005) along a self-consistently computed isentrope with a potential temperature of 1600 K, approximately that required by the generation of mid-ocean ridge basalt (McKenzie and Bickle 1988). The phase equilibria are consistent with experimental results

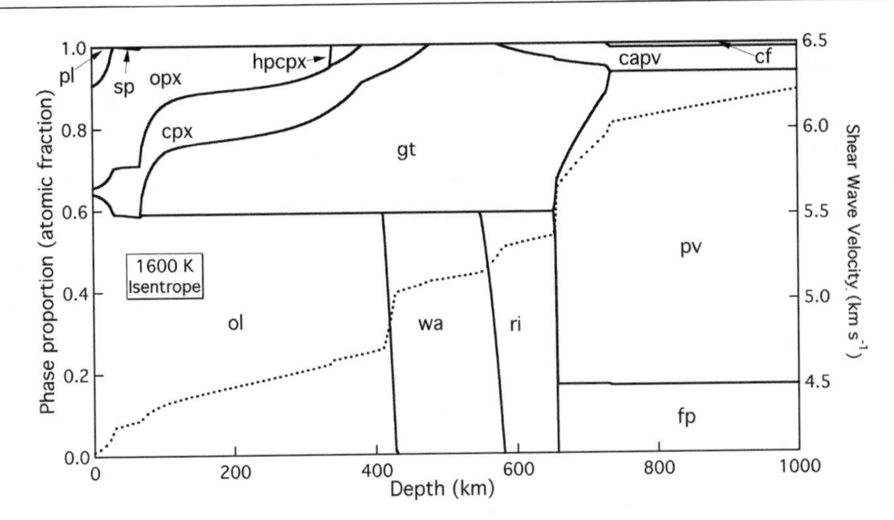

Figure 1. Computed phase equilibria (solid, left axis) and self-consistent shear wave velocity (dashed, right axis) along a self-consistently computed isentropic temperature profile assuming a potential temperature of 1600 K, according to our thermodynamic theory and our latest set of published parameters (Xu et al. 2008). Phases are: plagiolcase (plg), spinel (sp), olivine (ol), orthopyroxene (opx), clinopyroxene (cpx), high-pressure Mg-rich clinopyroxene (hpcpx), garnet (gt), wadsleyite (wa), ringwoodite (ri), akimotoite (ak), Calcium silicate perovskite (capv), Magnesium-rich silicate perovsite (pv), ferropericlase (fp), and Calcium-Ferrite structured phase (cf).

and with constraints imposed by xenoliths, including the transition from spinel-bearing to garnet-bearing assemblages, and the gradual dissolution of pyroxenes into garnet (Boyd 1989; Collerson et al. 2000). The computed shear wave velocity is also consistent with experimental results and bears many similarities to that seismologically observed in the mantle, in particular the locations of major "discontinuities" near 410 and 660 km depth, and the steeper velocity gradient in between these two boundaries (Dziewonski and Anderson 1981). The way in which such computations form a foundation from which the thermal and chemical state of Earth can be determined is illustrated further in the Applications section below.

CONSTRAINING AND TESTING THE MODEL

The thermodynamic theory contains a total of ten free parameters for each of the mantle species. These are F_0, V_0, K_0, K_0', G_0, G_0', θ_0, γ_0, q_0, η_{S0}, respectively the values at ambient conditions of the Helmholtz free energy, volume, bulk modulus and its pressure derivative, the shear modulus and its pressure derivative, the Debye temperature, its logarithmic volume derivative the Grüneisen parameter, and the logarithmic volume derivative of γ, and the second derivative of the Debye temperature with respect to shear strain. In addition the activities of mantle species in solution are specified by symmetric regular solution parameters $W_{\alpha\beta}$. These are defined only for those phases for which thermochemical data or phase equilibria data demand non-ideal contributions to the Gibbs free energy.

The values of the parameters are determined by comparing primarily with experimental data (Stixrude and Lithgow-Bertelloni 2005b; Xu et al. 2008). Sufficient experimental information now exists to constrain all the relevant parameters for at least the most abundant species of all relevant phases. In cases where no experimental data are available, we estimate the values of the parameters from systematic relationships, for example Birch's law (Anderson et al. 1968), or from the results of first principles theory.

First principles calculations are most powerful as a means of identifying functional relationships among thermodynamic quantities and testing the forms assumed in the construction of our thermodynamic theory. First principles calculations are able to sample much greater ranges of pressure and temperature and with much greater precision than experiment, making functional relationships clear. Density functional theory also yields predictions of key material properties that have not yet been measured, but because it is not exact, and because experimental progress is rapid, experimental values are preferred where these are available.

Three examples show the ability of first principles calculations to reveal and test functional relationships:

Elastic constants

The ability of the Eulerian finite strain expansion to capture the physics of high pressure elasticity is illustrated by a comparison of our formulation with first principles predictions of the elastic constants of $MgSiO_3$ perovskite (Fig. 2) (Karki et al. 1997). The Eulerian finite strain formulation (Eqns. 15,16) is able to represent first principles results accurately with only two parameters (moduli and their first pressure derivatives). In contrast, an alternative formulation, using the Lagrangian finite strain fails completely, as does a simple linear function of pressure.

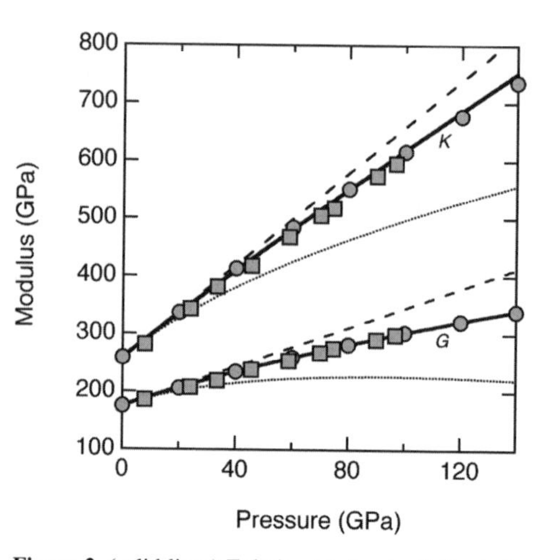

Figure 2. (solid lines) Eulerian third-order finite strain versus (short-dashed lines) Lagrangian third-order finite strain expressions and (long-dashed lines) a linear extrapolation compared with (circles) first principles results (Karki et al. 1997) and (squares) experimental results (Murakami et al. 2007) for the room temperature isothermal bulk modulus (K) and the shear modulus (G) of $MgSiO_3$ perovskite. For the purposes of this comparison, all curves were computed using values of $K_0 = 259$, $G_0 = 175$, $K_0' = 4.0$, and $G_0' = 1.7$ taken from the first principles study. (Modified from Stixrude and Lithgow-Bertelloni 2005b).

Grüneisen parameter and q

First principles predictions show that the volume dependence of the Grüneisen parameter differs significantly from the usual assumption of constant q. Instead, q itself is found to decrease substantially on compression (Karki et al. 2000a,b; Oganov et al. 2001a; Oganov and Dorogokupets 2003). These patterns are captured by our thermodynamic theory and indeed served as inspiration for choosing to describe the volume dependence of the vibrational frequency as a power series in finite strain (Eqn. 19) (Fig. 3). In fact previous results perhaps show a somewhat stronger volume dependence of q than predicted by third order Eulerian finite strain theory, although the scatter is substantial. The value of q at high pressure is important because it controls lateral variations with temperature of the bulk sound velocity (Isaak et al. 1992).

Temperature dependence of the shear modulus

First principles calculations show that the temperature derivative of the shear modulus depends strongly on compression (Oganov et al. 2001b; Wentzcovitch et al. 2004), while many non-self-consistent mineralogical models in the past have assumed that dG/dT is independent of pressure (Cammarano et al. 2003). The agreement between the first principles studies is remarkable because they are based on different approximations to the exchange-correlation functional, and different methods for computing high temperature properties (molecular dynamics vs. lattice dynamics). Our thermodynamic theory, in which the volume dependence of η_S is determined by Equations (14) and (17) and the primitive form of Equation (19), appears

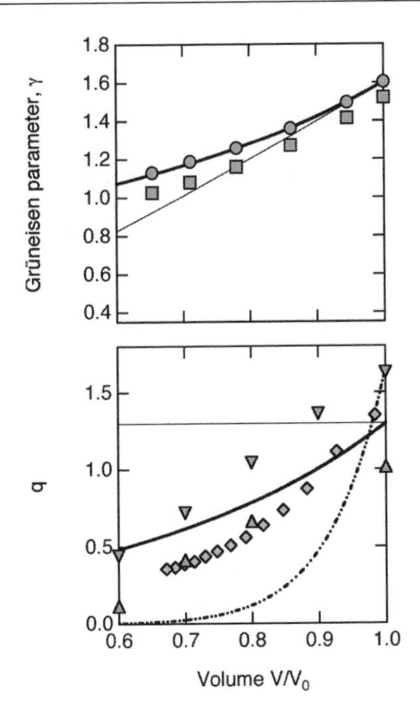

Figure 3. (top) Grüneisen parameter, γ and (bottom) q of MgO periclase from (bold solid lines) third order Eulerian finite strain theory (Eqns. 14,17-19), first principles calculations at 1000 K (circles) (Karki et al. 2000b) and (squares) (Oganov and Dorogokupets 2003), *ab initio* ionic models (diamonds) (Agnon and Bukowinski 1990), (up-triangles) (Inbar and Cohen 1995), an experimental analysis of the Hugoniot (double-dotted line) (Speziale et al. 2001), a combination of theory and experiment (down-triangles) (Anderson et al. 1993) and (thin solid lines) the approximation that $q = q_0$ is constant. For periclase we used $V_0 = 77.24$ Å^3 per unit cell to generate these plots (Karki et al. 2000b). The Eulerian finite strain results are calculated assuming the following values taken from the first principles calculations for the purposes of this comparison: $\gamma_0 = 1.60$, $q_0 = 1.3$ for periclase (Karki et al. 2000b). The two first principles calculations disagree in the value of γ_0 but show the same functional form. (Modified from Stixrude and Lithgow-Bertelloni 2005b).

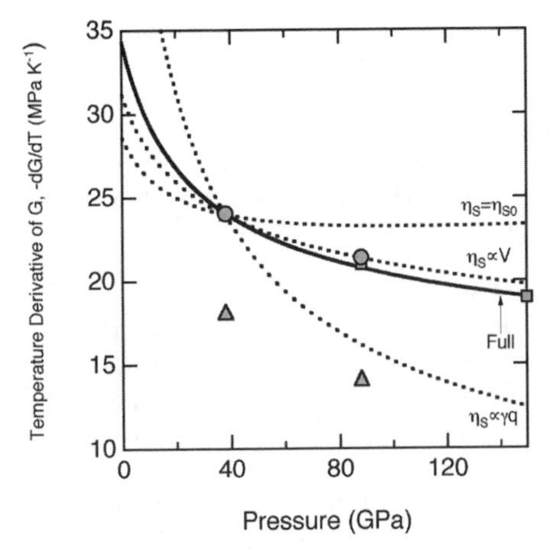

Figure 4. Temperature derivative of the shear modulus of $MgSiO_3$ perovskite from third order Eulerian finite strain theory (bold solid line) compared with density functional theory (squares, 2000-3000 K) (Wentzcovitch et al. 2004), (circles, 1500-3500 K) (Oganov et al. 2001b), an *ab initio* ionic model (PIB-Potential Induced Breathing; triangles) (Marton and Cohen, 2002) and (dashed lines) different approximations to the volume dependence of η_S: η_S arbitrarily assumed to be constant ($\eta_S = \eta_{S0}$), proportional to the volume, V, or proportional to $\eta_V = \gamma q$. The value of η_{S0} for each curve is set so that it passes through the first principles point at 38 GPa. (Modified from Stixrude and Lithgow-Bertelloni 2005b).

to capture the volume dependence predicted by density functional theory (Fig. 4). Accurate values of this parameter are critical in making comparisons between mineralogical models and seismological models as the shear wave velocity and its lateral variations in the mantle are well constrained and because the shear wave velocity is more sensitive than the longitudinal wave velocity to variations in temperature and bulk composition.

SCALING

How can we relate the physical properties of the heterogeneous, multi-phase mantle to those of laboratory scale samples, or to those of notional perfect crystals as probed with first principles theory? There are two important issues when considering physical properties of the mantle as revealed by seismology: scaling in space and in time.

For perfect, homogeneous crystals, scaling from laboratory length scales to geophysical length scales is essentially exact (Stixrude and Jeanloz 2007). Typical grain sizes in the laboratory and in the mantle far exceed the interatomic spacing, so that dispersion of the acoustic modes away from the Brillouin zone center is negligible, even for grains as small as one micron. In first principles calculations, it is straightforward to compute the elastic wave velocities in the geophysically relevant limit, i.e., in the limit $k \rightarrow 0$, where k is a reciprocal space vector. This unity of length scales is modified by heterogeneity: e.g., the contrast in elastic properties between neighboring grains, and anelasticity (dissipation) in real rocks.

A seismic wave passing through any part of the mantle senses the elastic response of a heterogeneous composite consisting of many grains of differing size, shape, orientation, and elasticity. Therefore, while knowledge of the elastic properties of individual phases is necessary for determining the seismic wave velocity, it is not sufficient. In principle, we must also specify the geometry and elasticity of each grain; of order 10^{21} grains sampled by a typical seismic wavelength of 50 km. It is not possible, nor useful to specify the grain geometry in such detail. Instead, one may place rigorous bounds on the elastic properties of the composite from knowledge of the elastic properties of the constituents.

The approaches of Hashin-Shtrikam (Hashin and Shtrikman 1963) and Voigt-Reuss-Hill (Hill 1963) place bounds on the effective elastic moduli of a heterogeneous aggregate (Fig. 5). The bounds depend only on properties that are determined by equilibrium thermodynamics: the elastic moduli and volume fractions of the constituent phases. The Hashin-Shtrikman bounds are rigorous under the assumption that the constituent phases are distributed randomly. The Voigt-Reuss bounds correspond, respectively to conditions of uniform strain and uniform stress across the constituent grains. Neither of these two limits is physically realizable, except in special geometries, since uniform strain violates mechanical equilibrium, and uniform stress entails grain-grain gaps and overlap. An advantage of the Voigt-Reuss bounds are that they encompass special arrangements of phases, such as layering (shape-preferred orientation), that are often encountered in the Earth (Backus 1962)

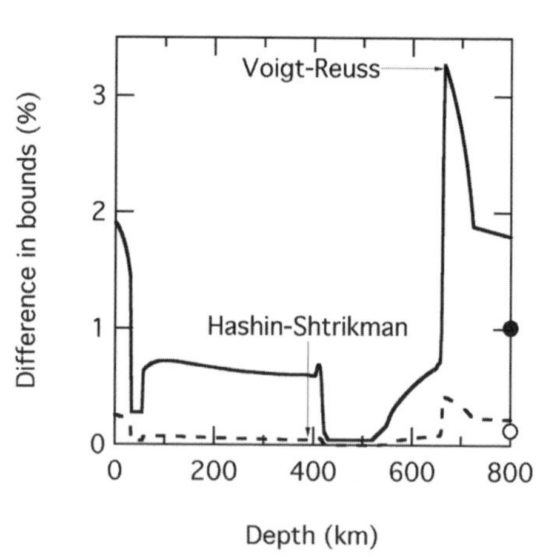

Figure 5. Computed differences between the Voigt and Reuss (solid) and Hashin-Shtrikman (dashed) bounds for S-wave velocity in a model mantle composition (Workman and Hart 2005) along a self-consistent 1600 K isentrope. Symbols represent the values at the core-mantle boundary (2891 km depth). (Modified from Stixrude and Jeanloz 2007).

$$M_R^* = \left(\sum_{\alpha} \frac{\phi^{\alpha}}{M^{\alpha}} \right)^{-1} < M^* < \sum_{\alpha} \phi^{\alpha} M^{\alpha} = M_V^* \tag{20}$$

where ϕ^{α} and M^{α} are the volume fraction and elastic modulus of phase α, and asterisks indicate effective moduli, and subscripts R and V indicate, respectively Reuss and Voigt bounds. The so-called Voigt-Reuss-Hill average is the simple average of the Voigt and Reuss bounds and is not theoretically justified as a best estimate of the effective elastic moduli. Empirically, one finds for many aggregates, including those typical of the mantle that the Voigt-Reuss-Hill average falls within the narrower Hashin-Shtrikman bounds (Watt et al. 1976). In Earth's mantle, the bounds are in any case not wide, and seldom exceed experimental uncertainty.

Seismic wave velocities depend on time scale because elastic wave propagation is dissipative in the seismic band (Anderson and Given 1982). The importance of dissipative processes such as defect migration and grain-boundary sliding are measured by the quality factor, Q defined in terms of the energy loss per cycle

$$Q^{-1} = \frac{dE}{E} \tag{21}$$

Dissipation has two important consequences, both of which can be measured seismologically: attenuation and dispersion, which refers to the dependence of the elastic wave velocity on frequency or wavelength. Experimental and seismological observations are consistent with a simple model in which attenuation is assumed to depend slightly on frequency over a broad range of frequencies (the absorption band), reflecting the wide range of dissipative mechanisms present in real materials (Anderson and Given 1982). In the limit of small dissipation, the velocity is (Jackson et al. 2002)

$$V(P,T,\omega) = V(P,T,\infty)\left[1 - \frac{1}{2}\cot\left(\frac{\alpha\pi}{2}\right)Q^{-1}(P,T,\omega)\right] \tag{22}$$

where V is the velocity, ω is the frequency, and α is an empirically determined parameter that describes the frequency dependence of Q with $\alpha = 0.26$ a typical experimental value. Because first principles theory and experiment (MHz-GHz) determine elastic properties in the infinite frequency limit, a correction for dispersion must be applied before they may be compared with seismologically determined velocities. For values of Q similar to the lowest found in one-dimensional seismological models (~80) (Romanowicz 1995), the velocity at seismic frequencies is ~1% lower than that in the infinite frequency limit.

APPLICATIONS

Origin of the low velocity zone

The origin of the low velocity zone remains enigmatic, even though this region is seismologically well established and geodynamically important (Gutenberg 1959). The low velocity zone is a region of the upper mantle in which velocity decreases with increasing depth, and is associated with a zone of reduced viscosity (asthenosphere) that may be an essential enabler of plate tectonics (Richards et al. 2001). The presence of low velocity zones in Earth-like planets is an inevitable consequence of a thermal boundary layer: the geothermal gradient is so steep that the influence of temperature on the velocity exceeds that of pressure (Stixrude 2007). The key question regarding the low velocity zone of Earth's mantle is quantitative: are the velocities so low as to require partial melt to explain them? The answer can only come from comparison to the predictions of mineralogical models of known thermodynamic state.

We have found that partial melt is not required to explain the low velocity zone, and that variations of temperature along plausible geotherms in sub-solidus mantle assemblages are sufficient to explain the seismological observations within uncertainty (Stixrude and Lithgow-Bertelloni 2005a) (Fig. 6). Our mineralogical models predict variations in the depth extent and slowness of the low velocity zone with lithospheric age that are seen seismologically. Moreover, the lowest velocities predicted are consistent with a range of seismological models of the same lithospheric age. The only part of the low velocity zone that probably requires partial melt is in the immediate vicinity of the ridge (age <5 Ma). Our conclusions differ from previous analyses based on non-self-consistent thermodynamic models in which the seismic wave velocity was assumed to vary linearly with temperature from room temperature to mantle

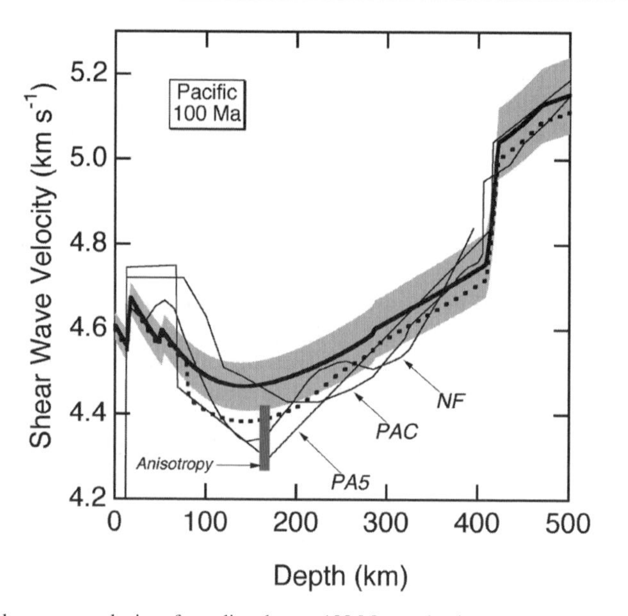

Figure 6. The shear wave velocity of pyrolite along a 100 Ma conductive cooling geotherm in the elastic limit (bold line), and including the effects of dispersion according to the seismological attenuation model QR19 (Romanowicz 1998) (bold dashed). The shading represents the uncertainty in the calculated velocity. The mineralogical model is compared with seismological models (light lines) PAC (Graves and Helmberger 1988), NF110+ (Nishimura and Forsyth 1989), and PA5 (Gaherty et al. 1999). The approximate magnitude of SH-SV anisotropy in the mantle is indicated by the vertical bar. [Used by permission of American Geophysical Union from Stixrude and Lithgow-Bertelloni (2005).]

temperatures (Birch 1969; Schubert et al. 1976). In fact, the velocity varies non-linearly with temperature particularly at temperature near room temperature and below as required by the third law, and a realistic account of the physics predicts much lower solid-phase velocities at mantle conditions.

Our study highlighted two important questions (Stixrude and Lithgow-Bertelloni 2005a). First, while our models are consistent with seismology to within mutual uncertainty, there appears to be a systematic tendency for seismological models to predict slower minimum velocities. We found that the origin of this small remaining discrepancy is unclear, but that it might be explained by small super-adiabatic variations in temperature, or inhomogeneities in composition. Second, the variation of velocity with depth deeper than the low velocity zone is much greater in seismological models. We speculated that this may be due to sub-adiabatic temperature gradients, or an increase with depth of a silica-rich component. Indeed, one recent study, which also used our thermodynamic model, showed that an increase with depth of a basaltic component could explain the high gradient zone (Cammarano et al. 2009). However, the amount of basalt enrichment required over a relatively narrow depth interval (from 5% at 250 km to 35% at 400 km) may be unrealistic, and the question of the origin of the high gradient zone remains open.

Origins of lateral heterogeneity

Seismic tomography reveals lateral variations in the elastic structure of the mantle that can often be correlated with known tectonic features at the surface (Romanowicz 2008). Many features are plausibly linked to lateral variations in temperature such as those thought to be associated with subducting slabs. Tomography holds out the promise of mapping mantle convection in the present-day Earth. Yet quantifying the origins of lateral variations in velocity remains difficult. One of the reasons is that lateral variations in temperature alone cannot account for all of the observed structure. Lateral variations in the shear wave velocity V_S that far exceed those in the longitudinal wave velocity V_P (Bolton and Masters 2001), regions in which lateral variations in V_S and the bulk sound velocity V_B are anti-correlated (Ishii and Tromp 1999), and anomalies with sharp boundaries (Ni et al. 2002), all point to a non-thermal origin of a significant part of the mantle's three-dimensional structure.

Understanding the three-dimensional structure of the mantle requires the unraveling of three possible origins of lateral heterogeneity in Earth's mantle: lateral variations in temperature, bulk composition, and phase. Each of these contributions may have similar magnitudes. Lateral variations in bulk composition have also received considerable attention, although the origin of such lateral variations remains obscure (Trampert et al. 2004).

We have shown that lateral variations in phase assemblage have an influence on three-dimensional mantle structure comparable to lateral variations in temperature over the upper 800 km of the mantle (Stixrude and Lithgow-Bertelloni 2007). This part of Earth is replete with phase transformations, all of which have finite Clapeyron slopes, and most of which occur over a depth interval that is a substantial fraction of the wavelength of seismic probes. Phase transitions are likely to be important in explaining lateral heterogeneity near the core-mantle boundary as well (Hirose 2006).

At any depth, the mantle is made of several different phases, with distinct elastic properties. As temperature varies, the relative proportions, and compositions of these phases change, contributing to laterally varying structure. In a multi-phase assemblage, the variation with temperature of any property X, such as velocity or density, at constant pressure has two contributions

$$\left(\frac{\partial \ln X}{\partial T}\right)_P = \left(\frac{\partial \ln X}{\partial T}\right)_{P,\vec{n}} + \left(\frac{\partial \ln X}{\partial \vec{n}}\right)_{P,T}\left(\frac{\partial \vec{n}}{\partial T}\right)_P \tag{21}$$

where $\vec{n}$ is the vector specifying the amounts of all end-member species of all phases. The first term on the right hand side may be called the isomorphic part: the derivative is taken at constant amounts and compositions of all coexisting phases. The second term may be called the metamorphic part and accounts for the variations in phase proportions (and compositions) with temperature. The magnitude of the metamorphic contribution

$$\left(\frac{\partial \ln X}{\partial \vec{n}}\right)_{P,T}\left(\frac{\partial \vec{n}}{\partial T}\right)_P \approx f\Gamma\frac{\Delta \ln X}{\Delta P} \tag{22}$$

where f is the volume fraction of the mantle composed of the transforming phases, $\Delta \ln X$ is the relative difference in the property X between the two transforming phases, ΔP is the pressure range over which the transition occurs, and Γ is the effective Clapeyron slope. The effect is largest for sharp transitions, such as the olivine to wadsleyite transition, and in this case may also be described in terms of the topography of the transition. It is sensible to describe lateral variations in velocity due to phase equilibria in terms of topography when the topography exceeds the width of the phase transformation

$$\left(\frac{\partial P}{\partial T}\right)_{eq} \delta T > \Delta P \tag{23}$$

where δT is the anticipated magnitude of lateral temperature variations. The influence of phase transformations is also important for broad transitions such as pyroxene to garnet. In the upper mantle, most phase transitions have positive Clapeyron slopes. This means that, except in the vicinity of the 660 km discontinuity, phase transitions systematically increase the temperature dependence of seismic wave velocities throughout the upper 800 km of the mantle.

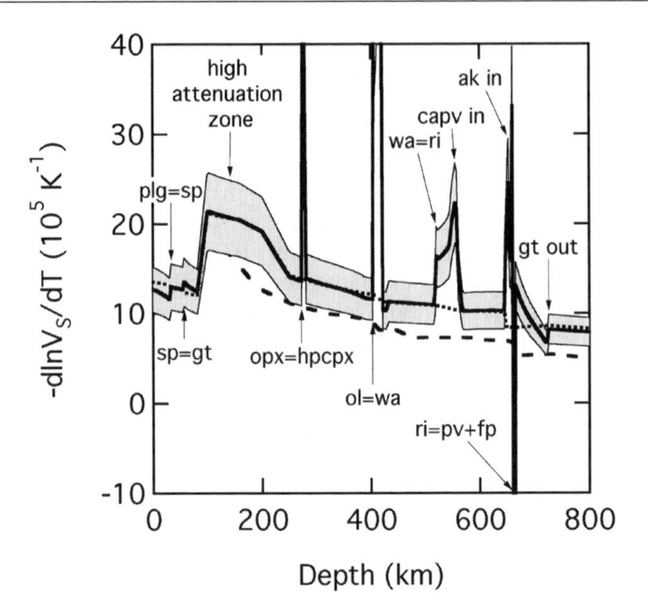

Figure 7. Temperature derivative of the shear wave velocity along the 1600 K isentrope (bold solid) and the isomorphic contribution (short dashed). The shading indicates nominal uncertainties. We compare with the results of (Cammarano et al. 2003) (long dashed). The locations of the high attenuation zone and various phase transitions are indicated with abbreviations defined in Figure 1. [Used by permission of Elsevier from Stixrude and Lithgow-Bertelloni (2007).]

The influence of phase transformations is illustrated by the temperature derivative of the shear wave velocity (Fig. 7). The derivative is positive throughout the upper 660 km of the mantle and of greater magnitude than many previous studies that neglected the influence of phase transformations. The value of the scaling varies rapidly with depth and undergoes large excursions in the vicinity of phase transformations. Several phase transformations, in addition to those responsible for the 410 and 660 km discontinuities exhibit a substantial signal. At the latter transformation, the metamorphic contribution is large and negative, reflecting the negative sign of the Clapeyron slope (Eqn. 22).

The metamorphic contribution to the thermal expansivity has an important influence on mantle dynamics that has been recognized for some time (Christensen 1995). In models of mantle convection, this contribution is often approximated only by the ringwoodite=per ovskite+ferropericlase transition. In contrast, we find that there are important metamorphic contributions over much of the upper 800 km of the mantle (Fig. 8). We illustrate by computing the complete thermal expansivity and also the mean value of the thermal expansivity over a finite temperature interval (δT). The latter quantity shows the anticipated features: the magnitude of the peaks associated with phase transformations is reduced in amplitude ($\Delta\ln\rho/\delta T$) as compared with the local temperature derivative, and spread over a broader depth interval ($\sim\Gamma\delta T$). The metamorphic term may influence dynamics near the core-mantle boundary as well (Nakagawa and Tackley 2006). A more realistic account of the thermal expansivity of multi-phase assemblages should be included in mantle flow codes (Nakagawa et al. 2009).

Our analysis highlights one of the challenges in interpreting seismic tomographic models in terms of temperature anomalies. The magnitude and even sign of the metamorphic contribution varies rapidly with depth, and much more rapidly than the typical depth-resolution of seismic tomography. This means that seismological observations (travel times and normal mode frequencies) are not sensitive to all the depth variations in $d\ln V_S/dT$ that we predict

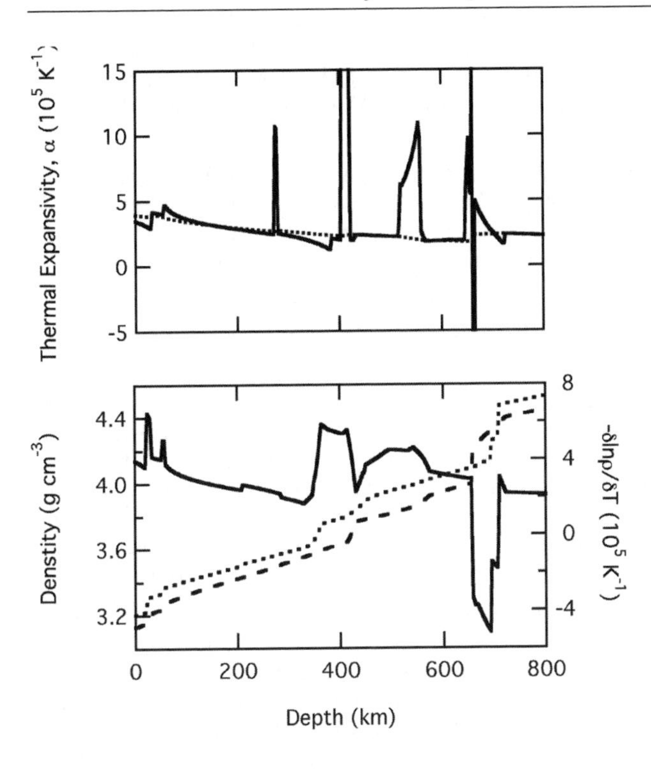

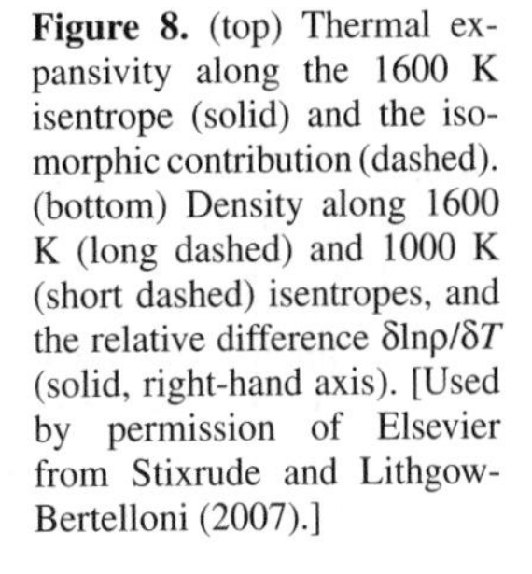
Figure 8. (top) Thermal expansivity along the 1600 K isentrope (solid) and the isomorphic contribution (dashed). (bottom) Density along 1600 K (long dashed) and 1000 K (short dashed) isentropes, and the relative difference $\delta\ln\rho/\delta T$ (solid, right-hand axis). [Used by permission of Elsevier from Stixrude and Lithgow-Bertelloni (2007).]

should exist in the mantle. Indeed, tomographic models generally find a smooth variation of the magnitude of heterogeneity with depth (Romanowicz 2003). A naïve interpretation of tomographic models, by simply scaling the observed heterogeneity to temperature using the full thermodynamically computed $d\ln V_S/dT$, produces unphysical variations with depth of the temperature heterogeneity. This problem can be overcome by inverting seismological observables directly for three-dimensional variations in temperature using our mineralogical model (Cammarano and Romanowicz 2007).

Influence of lithologic heterogeneity

Partial melting is a powerful agent of chemical differentiation that has existed throughout much of Earth's history. Most melting today occurs at mid-ocean ridges where the mantle is differentiated into a silica-rich, basaltic crust and a silica-depleted, harzburgitic layer. This differentiated package is returned to the mantle at subduction zones.

An analysis of the mass balance and diffusion rates suggests that the mantle may be composed entirely of differentiated material (Xu et al. 2008). Geochemical tracers indicate that most or all of the mantle has been differentiated at least once at a mid-ocean ridge. Moreover, the subducted differentiated material is unlikely to re-equilibrate over geologic time as the rates of chemical diffusion are so slow.

Based on this analysis, we have proposed an alternative model of the composition of the mantle (Xu et al. 2008). We view the mantle as a mechanical mixture of basalt and harzburgite. A basalt fraction of 18% produces an overall bulk composition equal to that of the MORB source and is consistent with our knowledge of subduction rates and the thickness of the oceanic crust. This picture differs considerably from the usual notion that the mantle is to first order chemically homogeneous and pyrolitic. In reality some re-equilibration between the subducted basalt and harzburgite must occur, possibly aided by fluids. In the limit of complete re-equilibration, the mechanical mixture would revert to pyrolite. We may therefore view lithologic heterogeneity in the mantle in terms of two end-members: the mechanically mixed (MM) mantle in which no re-equilibration occurs, and the equilibrium assemblage (EA) in which lithologic heterogeneity is erased immediately upon subduction.

We have found that the mechanically mixed mantle has significantly different elastic properties from the equilibrium assemblage (Fig. 9). The reason is that the phase equilibria in the mechanical mixture are different. For example, while free silica is stable in basaltic compositions, it is not stable in pyrolite. Schematically, the phase relations may be viewed as

$$2MgSiO_3\ (EA) = Mg_2SiO_4\ (MM;\ harzburgite) + SiO_2\ (MM;\ basalt) \qquad (24)$$

so that the pyroxene component in the equilibrium assemblage dissociates into olivine in the harzburgite and free silica in the basalt.

The seismic wave velocities in the mechanical mixture agree better with seismological models as compared with the equilibrium assemblage (Fig. 9). In MM, the velocity is faster and the variation of velocity with depth is greater in the transition zone. In MM, the velocity gradient is steeper also in the high gradient zone between the low velocity zone and the 410, also leading to better agreement with seismology. Both MM and EA are substantially slower along a 1600 K isentrope than the lower mantle. The origin of this discrepancy is unclear, but may be related to deviations from isentropy in the lower mantle or radial variations in bulk composition.

The mechanically mixed model of the mantle gives one a natural way of thinking about large-scale chemical heterogeneity and a plausible means of producing it (Fig. 10). In principle, describing chemical heterogeneity in the mantle is a daunting task because there are at least 6 important oxide components. But not all variations in chemical composition are geologically plausible. The mechanical mixture focuses on a type of heterogeneity that is produced by a

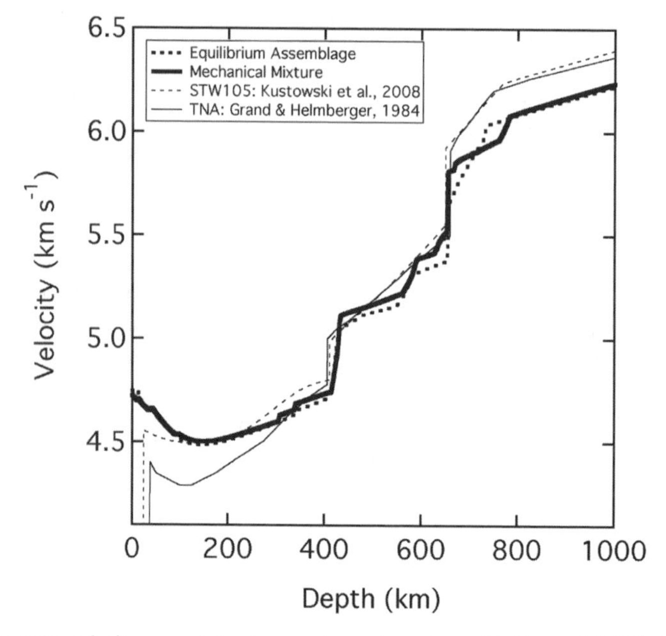

Figure 9. Shear wave velocity computed using our thermodynamic model (Xu et al. 2008) along a 1600 K isentrope modified by a half-space cooling upper thermal boundary layer (100 Ma) for (bold solid) the mechanical mixture (MM) and (bold dashed) the equilibrium assemblage (EA) compared with (thin dashed) global (Kustowski et al. 2008) and (thin solid) regional (Grand and Helmberger 1984) seismological models.

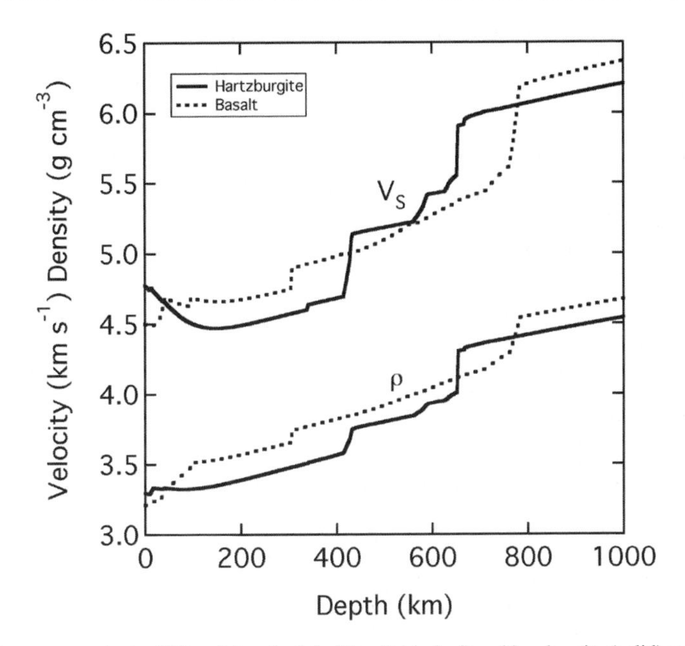

Figure 10. Shear wave velocity (V_S) and density (ρ) of basalt (dashed) and harzburgite (solid) computed using our thermodynamic model (Xu et al. 2008) along a 1600 K isentrope modified by a half-space cooling upper thermal boundary layer (100 Ma).

well-understood process (mid-ocean ridge melting) and that must be returned to the mantle. Moreover, the physical properties of the two components of the mechanical mixture are sufficiently different that they may separate dynamically and that variations in the basalt fraction should be visible seismologically throughout much of the mantle.

CONCLUSIONS AND OUTLOOK

The thermodynamic theory of Earth's mantle has developed sufficiently so that it is now possible to account self-consistently for all aspects of equilibrium that contribute to the generation of Earth structure and dynamics. The theory appears to be applicable to the entire range of sub-solidus conditions encountered in Earth's mantle. These developments have been informed by first principles theory, especially in the design and testing of functional relationships. Density functional theory has also provided predictions of key material properties that have not yet been measured experimentally.

Future challenges include expanding the scope of the thermodynamic theory to include other important classes of phases such as silicate melts (de Koker and Stixrude 2009) and metallic systems relevant to the core, and physical behavior (magnetic collapse, and electronic transitions), which will permit exploration of other regions of the mantle including the lower mantle, and other geological processes, including melt generation and differentiation, that have played a major role in Earth's history.

REFERENCES

Agnon A, Bukowinski MST (1990) Thermodynamic and elastic properties of a many-body model for simple oxides. Phys Rev B 41(11):7755-7766

Alberty RA (2001) Use of Legendre transforms in chemical thermodynamics - (IUPAC Technical Report). Pure Appl Chem 73(8):1349-1380

Allegre CJ, Turcotte DL (1986) Implications of a 2-component marble-cake mantle. Nature 323(6084):123-127

Anderson DL, Given JW (1982) Absorption-band Q model for the earth. J Geophys Res 87(NB5):3893-3904

Anderson OL, Oda H, Chopelas A, Isaak DG (1993) A thermodynamic theory of the Grüneisen ratio at extreme conditions: MgO as an example. Phys Chem Miner 19:369-380

Anderson OL, Schreiber E, Lieberman RC (1968) Some elastic constant data on minerals relevant to geophysics. Rev Geophys 6(4):491-524

Backus GE (1962) Long-wave elastic anisotropy produced by horizontal layering. J Geophys Res 67(11):4427-4400

Berman RG (1988) Internally-consistent thermodynamic data for minerals in the system Na_2O-K_2O-CaO-MgO-FeO-Fe_2O_3-Al_2O_3-SiO_2-TiO_2-H_2O-CO_2. J Petrol 29(2):445-522

Birch F (1969) Density and compostion of the upper mantle: first approximation as an olivine layer. *In*: The Earth's Crust and Upper Mantle. Hart PJ (ed) American Geophysical Union, Washington, DC, p. 18-36

Birch F (1978) Finite strain isotherm and velocities for single-crystal and polycrystalline NaCl at high-pressure and 300-degree-K. J Geophys Res 83:1257-1268

Bolton H, Masters G (2001) Travel times of P and S from the global digital seismic networks: Implications for the relative variation of P and S velocity in the mantle. J Geophys Res-Solid Earth 106(B7):13527-13540

Boyd FR (1989) Compositional distinction between oceanic and cratonic lithosphere. Earth Planet Sci Lett 96(1-2):15-26

Callen HB (1960) Thermodynamics. John Wiley and Sons, New York

Cammarano F, Goes S, Vacher P, Giardini D (2003) Inferring upper-mantle temperatures from seismic velocities. Phys Earth Planet Inter 138(3-4):197-222

Cammarano F, Romanowicz B (2007) Insights into the nature of the transition zone from physically constrained inversion of long-period seismic data. Proc Nat Acad Sci USA 104(22):9139-9144

Cammarano F, Romanowicz B, Stixrude L, Lithgow-Bertelloni C, Xu WB (2009) Inferring the thermochemical structure of the upper mantle from seismic data. Geophys J Int 179(2):1169-1185

Canup RM (2004) Dynamics of lunar formation. Ann Rev Astron Astrophys 42:441-475

Christensen U (1995) Effects of phase transitions on mantle convection. Ann Rev Earth Planet Sci 23:65-87

Cococcioni M (2010) Accurate and efficient calculations on strongly correlated minerals with the LDA+U method: review and perspectives. Rev Mineral Geochem 71:147-167

Collerson KD, Hapugoda S, Kamber BS, Williams Q (2000) Rocks from the mantle transition zone: Majorite-bearing xenoliths from Malaita, southwest Pacific. Science 288(5469):1215-1223

Davies GF (1974) Effective elastic-moduli under hydrostatic stress .1. Quasi-harmonic theory. J Phys Chem Solids 35(11):1513-1520

de Koker NP, Stixrude L (2009) Self-consistent thermodynamic description of silicate liquids, with application to shock melting of MgO periclase and $MgSiO_3$ perovskite. Geophys J Int 178:162-179

Dziewonski AM, Anderson DL (1981) Preliminary reference earth model. Phys Earth Planet Inter 25:297-356

Fei YW, Saxena SK (1990) Internally consistent thermodynamic data and equilibrium phase-relations for compounds in the system MgO-SiO_2 at high-pressure and high-pressure and high-temperature. J Geophys Res-Solid Earth Planets 95(B5):6915-6928

Gaherty JB, Kato M, Jordan TH (1999) Seismological structure of the upper mantle: a regional comparison of seismic layering. Phys Earth Planet Inter 110(1-2):21-41

Ghiorso MS, Sack RO (1995) Chemical mass-transfer in magmatic processes. 4. A revised and internally consistent thermodynamic model for the interpolation and extrapolation of liquid-solid equilibria in magmatic systems at elevated-temperatures and pressures. Contrib Mineral Petrol 119(2-3):197-212

Gramsch SA, Cohen RE, Savrasov SY (2003) Structure, metal-insulator transitions, and magnetic properties of FeO at high pressures. Am Mineral 88(2-3):257-261

Grand SP, Helmberger DV (1984) Upper mantle shear structure of North-America. Geophys J Royal Astron Soc 76(2):399-438

Graves RW, Helmberger DV (1988) Upper mantle cross-section from Tonga to Newfoundland. J Geophys Res-Solid Earth and Planets 93(B5):4701-4711

Gutenberg B (1959) Physics of the Earth's Interior. Academic Press, New York

Hacker BR, Abers GA, Peacock SM (2003) Subduction factory - 1. Theoretical mineralogy, densities, seismic wave speeds, and H_2O contents. J Geophys Res-Solid Earth 108(B1):2029, doi: 10.1029/2001JB001127

Hama J, Suito K (1998) High-temperature equation of state of $CaSiO_3$ perovskite and its implications for the lower mantle. Phys Earth Planet Inter 105(1-2):33-46

Hashin Z, Shtrikman S (1963) A variational approach to the theory of the elastic behaviour of multiphase materials. J Mech Phys Solids 11(2):127-140

Hill R (1963) Elastic properties of reinforced solids - some theoretical principles. J Mech Phys Solids 11(5):357-372

Hirose K (2006) Postperovskite phase transition and its geophysical implications. Rev Geophys 44(3):Art. No. RG3001, doi: 10.1029/2005RG000186

Hirschmann MM (2006) Water, melting, and the deep Earth H_2O cycle. Ann Rev Earth Planet Sci 34:629-653

Holland TJB, Powell R (1998) An internally consistent thermodynamic data set for phases of petrological interest. J Metamorph Geol 16(3):309-343

Inbar I, Cohen RE (1995) High pressure effects on thermal properties of MgO. Geophys Res Lett 22:1533-1536

Isaak DG, Anderson OL, Cohen RE (1992) The relationship between shear and compressional velocities at high pressures: reconciliation of seismic tomography and mineral physics. Geophys Res Lett 19:741-744

Ishii M, Tromp J (1999) Normal-mode and free-air gravity constraints on lateral variations in velocity and density of Earth's mantle. Science 285(5431):1231-1236

Ita J, Stixrude L (1992) Petrology, elasticity, and composition of the mantle transition zone. J Geophys Res-Solid Earth 97(B5):6849-6866

Jackson I, Gerald JDF, Faul UH, Tan BH (2002) Grain-size-sensitive seismic wave attenuation in polycrystalline olivine. J Geophys Res-Solid Earth 107(B12):2360

Karki BB, Stixrude L, Clark SJ, Warren MC, Ackland GJ, Crain J (1997) Elastic properties of orthorhombic $MgSiO_3$ perovskite at lower mantle pressures. Am Mineral 82:635-638

Karki BB, Wentzcovitch RM, de Gironcoli S, Baroni S (2000a) Ab initio lattice dynamics of $MgSiO_3$ perovskite at high pressure. Phys Rev B 62(22):14750-14756

Karki BB, Wentzcovitch RM, de Gironcoli S, Baroni S (2000b) High-pressure lattice dynamics and thermoelasticity of MgO. Phys Rev B 61(13):8793-8800

Kohn W (1999) Nobel lecture: electronic structure of matter-wave functions and density functionals. Rev Mod Phys 71(5):1253-1266

Kuskov OL (1995) Constitution of the Moon. 3. Composition of middle mantle from seismic data. Phys Earth Planet Inter 90(1-2):55-74

Kustowski B, Ekstrom G, Dziewonski AM (2008) Anisotropic shear-wave velocity structure of the Earth's mantle: A global model. J Geophys Res-Solid Earth 113(B6):B06306, doi: 10.1029/2007JB005169

Mao WL, Mao HK, Sturhahn W, Zhao JY, Prakapenka VB, Meng Y, Shu JF, Fei YW, Hemley RJ (2006) Iron-rich post-perovskite and the origin of ultralow-velocity zones. Science 312(5773):564-565

Marton FC, Cohen RE (2002) Constraints on lower mantle composition from molecular dynamics simulations of $MgSiO_3$ perovskite. Phys Earth Planet Inter 134(3-4):239-252

Mattern E, Matas J, Ricard Y, Bass J (2005) Lower mantle composition and temperature from mineral physics and thermodynamic modelling. Geophys J Int 160(3):973-990

McKenzie D, Bickle MJ (1988) The volume and composition of melt generated by extension of the lithosphere. J Petrol 29(3):625-679

Mitas L, Kolorenč J (2010) Quantum Monte Carlo studies of transition metal oxides. Rev Mineral Geochem 71:137-145

Murakami M, Hirose K, Kawamura K, Sata N, Ohishi Y (2004) Post-perovskite phase transition in $MgSiO_3$. Science 304(5672):855-858

Murakami M, Sinogeikin SV, Hellwig H, Bass JD, Li J (2007) Sound velocity of $MgSiO_3$ perovskite to Mbar pressure. Earth Planet Sci Lett 256(1-2):47-54

Nakagawa T, Tackley PJ (2006) Three-dimensional structures and dynamics in the deep mantle: Effects of post-perovskite phase change and deep mantle layering. Geophys Res Lett 33(12):L12S11, doi: 10.1029/2006GL025719

Nakagawa T, Tackley PJ, Deschamps F, Connolly JAD (2009) Influence of MORB bulk composition on 3-D spherical models of thermo-chemical mantle convection with self-consistently calculated mineral physics. Geochim Cosmochim Acta 73(13):A929-A929

Ni SD, Tan E, Gurnis M, Helmberger D (2002) Sharp sides to the African superplume. Science 296(5574):1850-1852

Nishimura CE, Forsyth DW (1989) The anisotropic structure of the upper mantle in the Pacific. Geophys J Oxford 96(2):203-229

Oganov AR, Brodholt JP, Price GD (2001a) Ab initio elasticity and thermal equation of state of $MgSiO_3$ perovskite. Earth Planet Sci Lett 184:555-560

Oganov AR, Brodholt JP, Price GD (2001b) The elastic constants of $MgSiO_3$ perovskite at pressures and temperatures of the Earth's mantle. Nature 411(6840):934-937

Oganov AR, Dorogokupets PI (2003) All-electron and pseudopotential study of MgO: Equation of state, anharmonicity, and stability. Phys Rev B 67(22):224110, doi: 10.1103/PhysRevB.67.224110

Ohtani E, Sakai T (2008) Recent advances in the study of mantle phase transitions. Phys Earth Planet Inter 170(3-4):240-247

Perdew JP, Ruzsinszky A (2010) Density functional theory of electronic structure: a short course for mineralogists and geophysicists. Rev Mineral Geochem 71:1-18

Richards MA, Yang WS, Baumgardner JR, Bunge HP (2001) Role of a low-viscosity zone in stabilizing plate tectonics: Implications for comparative terrestrial planetology. Geochem Geophys Geosys 2: 2000GC000115

Romanowicz B (1995) A global tomographic model of shear attenuation in the upper-mantle. J Geophys Res-Solid Earth 100(B7):12375-12394

Romanowicz B (1998) Attenuation tomography of the earth's mantle: A review of current status. Pure Appl Geophys 153(2-4):257-272

Romanowicz B (2003) Global mantle tomography: progress status in the past 10 years. Ann Rev Earth Planet Sci 31:303-328

Romanowicz B (2008) Using seismic waves to image Earth's internal structure. Nature 451(7176):266-268

Schubert G, Froidevaux C, Yuen DA (1976) Oceanic lithosphere and asthenosphere - thermal and mechanical structure. J Geophys Res 81(20):3525-3540

Sobolev SV, Babeyko AY (1994) Modeling of mineralogical composition, density and elastic-wave velocities in anhydrous magmatic rocks. Surveys in Geophysics 15(5):515-544

Speziale S, Zha CS, Duffy TS, Hemley RJ, Mao HK (2001) Quasi-hydrostatic compression of magnesium oxide to 52 GPa: Implications for the pressure-volume-temperature equation of state. J Geophys Res-Solid Earth 106(B1):515-528

Stixrude L (2007) Properties of rocks and minerals - Seismic properties of rocks and minerals, and structure of the Earth. *In:* Mineral Physics, Vol. 2. Price GD (ed) Elsevier, Amsterdam, p 7-32

Stixrude L, Bukowinski MST (1990) Fundamental thermodynamic relations and silicate melting with implications for the constitution of D''. J Geophys Res 95:19311-19325

Stixrude L, Jeanloz R (2007) Constraints on seismic models from other disciplines - Constraints from mineral physics on seismological models. *In*: Seismology and Structure of the Earth, Vol. 1. Dziewonski AM, Romanowicz B (eds) Elsevier, Amsterdam, p. 775-803

Stixrude L, Lithgow-Bertelloni C (2005a) Mineralogy and elasticity of the oceanic upper mantle: Origin of the low-velocity zone. J Geophys Res-Solid Earth 110(B3):B03204, doi: 10.1029/2004JB002965

Stixrude L, Lithgow-Bertelloni C (2005b) Thermodynamics of mantle minerals - I. Physical properties. Geophys J Int 162(2):610-632

Stixrude L, Lithgow-Bertelloni C (2007) Influence of phase transformations on lateral heterogeneity and dynamics in Earth's mantle. Earth Planet Sci Lett 263(1-2):45-55

Thompson JB (1969) Chemical reactions in crystals. Am Mineral 54(3-4):341-375

Trampert J, Deschamps F, Resovsky J, Yuen D (2004) Probabilistic tomography maps chemical heterogeneities throughout the lower mantle. Science 306(5697):853-856

Valencia D, O'Connell RJ, Sasselov D (2006) Internal structure of massive terrestrial planets. Icarus 181(2):545-554

Wallace DC (1972) Thermodynamics of Crystals. John Wiley and Sons, New York

Watt JP, Davies GF, Connell RJO (1976) The elastic properties of composite materials. Rev Geophys Space Phys 14:541-563

Wentzcovitch RM, Karki BB, Cococcioni M, de Gironcoli S (2004) Thermoelastic properties of $MgSiO_3$-perovskite: Insights on the nature of the Earth's lower mantle. Phys Rev Lett 92(1):018501, doi: 10.1103/PhysRevLett.92.018501

Wentzcovitch RM, Yu YG, Wu Z (2010) Thermodynamic properties and phase relations in mantle minerals investigated by first principles quasiharmonic theory. Rev Mineral Geochem 71:59-98

Workman RK, Hart SR (2005) Major and trace element composition of the depleted MORB mantle (DMM). Earth Planet Sci Lett 231(1-2):53-72

Xie SX, Tackley PJ (2004) Evolution of helium and argon isotopes in a convecting mantle. Phys Earth Planet Inter 146(3-4):417-439

Xu WB, Lithgow-Bertelloni C, Stixrude L, Ritsema J (2008) The effect of bulk composition and temperature on mantle seismic structure. Earth Planet Sci Lett 275(1-2):70-79